Birkhäuser

More information about this series at http://www.springer.com/series/4927

Twenty Five Years of Modern Tsunami Science Following the 1992 Nicaragua and Flores Island Tsunamis. Volume I

Edited by
Utku Kânoğlu · Yuichiro Tanioka · Emile A. Okal ·
Maria Ana Baptista · Alexander B. Rabinovich

Previously published in *Pure and Applied Geophysics* (PAGEOPH),
Volume 176, No. 7, 2019

Birkhäuser

Editors
Utku Kânoğlu
Department of Aerospace Engineering
Middle East Technical University
Ankara, Turkey

Emile A. Okal
Department of Earth and Planetary Sciences
Northwestern University
Evanston, IL, USA

Alexander B. Rabinovich
Department of Fisheries and Oceans
Institute of Ocean Sciences
Sidney, BC, Canada

P.P. Shirshov Institute of Oceanology
Russian Academy of Sciences
Moscow, Russia

Yuichiro Tanioka
Institute of Seismology and Volcanology
Hokkaido University
Sapporo, Japan

Maria Ana Baptista
Instituto Superior de Engenharia de Lisboa
Instituto Politécnico de Lisboa
Lisbon, Portugal

Instituto Dom Luiz
Universidade de Lisboa
Lisbon, Portugal

ISSN 2504-3625
Pageoph Topical Volumes
ISBN 978-3-030-43750-3

Cover illustration: A chaotic scenery of destroyed houses, restaurants and fisher boats observed in the lowest parts of the Masachapa beaches in the early morning of September 2, 1992, when W.S. as a researcher of Nicaragua Geosciences Institute (INETER) was allowed to enter the disaster area together with troops of the Nicaraguan military, which—to prevent looting—had closed completely the access during the night after the tsunami had affected all the Pacific coast of Nicaragua. (Courtesy Wilfried Strauch)

This book is published under the imprint Birkhäuser, www.birkhauser-science.com by the registered company Springer Nature Switzerland AG
The registered company address is: Gewerbestrasse 11, 6330 Cham, Switzerland

Contents

Pure Appl. Geophys. 176 (2019), 2757–2769

https://doi.org/10.1007/s00024-019-02266-5

Pure and Applied Geophysics

Introduction to "Twenty Five Years of Modern Tsunami Science Following the 1992 Nicaragua and Flores Island Tsunamis, Volume I"

Utku Kânoğlu,[1] Yuichiro Tanioka,[2] Emile A. Okal,[3] Maria Ana Baptista,[4,5] and Alexander B. Rabinovich[6,7]

Abstract—Twenty-two papers on tsunamis are included in the Pure and Applied Geophysics topical issue "Twenty five years of modern tsunami science following the 1992 Nicaragua and Flores Island tsunamis: Volume I," reporting on the frontiers of tsunami science and research. The first three papers overview significant tsunamis of 1992–2018 and discuss the problems of tsunami cataloguing. The main focus of the next four papers is on specific details of historical tsunami events and field surveys. First, three papers are related to thorough analyses of several historical events based on macroseismic, seismological, and tsunami observations, tide gauge data, and modelling results: the 1907 Sumatra "tsunami earthquake," the 1941 Andaman Islands earthquake, and five great tsunamis in the Pacific Ocean (1946, 1952, 1957, 1960 and 1964). The last paper of the section concerns results of the field survey of the 2017 Bodrum-Kos tsunami. The reconstruction of the tsunami sources is the main target of the four following papers, with four events examined in detail: the historical 1810 Baja California, 1992 Flores Island, 2012 Haida Gwaii and 2015 Chilean (Illapel) tsunamis. A set of three papers address problems associated with landslide-generated tsunamis in three different regions: a modelling of the 2017 landslide and tsunami at Karrat Fjord, Greenland; a probabilistic analysis of the hazard from the Indus Canyon in the NW Indian Ocean; and a study of the landslide-induced tsunami hazard along the US East Coast. The next section, including three papers, reports on comparisons between different types of tsunami models, on numerical modelling of tsunami waves in the Caspian Sea, and on the modelling of magnetic signals at Easter Island, following the 2010 and 2015 Chilean tsunamis. The last group of five papers discusses tsunami hazard assessment and warning for various regions of the world oceans, including Alaska, the eastern and western Mediterranean, Australia, the Northeast Atlantic and the entire Pacific Ocean; one specific aspect of these studies is the compilation and efficient application of observed data, in particular, from DARTs.

Key words: Tsunami observations and detection, DARTs, tsunami modelling, tsunami earthquake, Pacific Ocean, Mediterranean Sea, Caspian Sea, tsunami warning and hazard assessment, landslide generated tsunami, tsunami statistics and probability.

1. Introduction

On 2 September 1992, an M_w 7.7 earthquake occurred near the coast of Nicaragua (Satake et al. 1993). This relatively moderate earthquake produced a disproportionately destructive tsunami with a maximum runup height up to 10 m that killed approximately 170 people. The earthquake was identified as a "tsunami earthquake," a term introduced by Kanamori (1972) to characterize an event whose tsunami is significantly larger than expected from its magnitudes, especially conventional ones.

Three months later, on 12 December 1992, an M_w 7.8 earthquake hit the northern coast of Flores Island, Indonesia, and generated another catastrophic tsunami with maximum runup of 26 m caused by a submarine landslide triggered by the earthquake. The earthquake and tsunami killed about 2600 people and injured 500 more. The total damage exceeded US$100 million (Yeh et al. 1994).

These two events, which occurred following 28 relatively quiet years after the 1964 Alaska M_w 9.2

[1] Department of Aerospace Engineering, Middle East Technical University, 06800 Ankara, Turkey. E-mail: kanoglu@ metu.edu.tr
[2] Institute of Seismology and Volcanology, Hokkaido University, Sapporo, Japan. E-mail: tanioka@mail.sci.hokudai.ac.jp
[3] Department of Earth and Planetary Sciences, Northwestern University, Evanston, IL 60208, USA. E-mail: e-okal@northwestern.edu
[4] Instituto Superior de Engenharia de Lisboa, Instituto Politécnico de Lisboa, Lisbon, Portugal. E-mail: mavbaptista@gmail.com
[5] Instituto Dom Luiz, Universidade de Lisboa, Lisbon, Portugal.
[6] Department of Fisheries and Oceans, Institute of Ocean Sciences, 9860 West Saanich Road, Sidney, BC V8L 4B2, Canada. E-mail: a.b.rabinovich@gmail.com
[7] P.P. Shirshov Institute of Oceanology, Russian Academy of Sciences, 36 Nakhimovsky Pr., Moscow 117997, Russia.

earthquake and tsunami, opened a 25-year period of numerous devastating earthquakes and tsunamis, including two of the most destructive natural disasters in recent human history: the 26 December 2004 Sumatra tsunami that killed about 230,000 people and impacted at least 16 nations directly in the Indian Ocean (Synolakis and Kong 2006), and the 11 March 2011 Tohoku (East Japan) tsunami that killed almost 20,000 people and destroyed the Fukushima Dai-ichi nuclear power plant (Satake et al. 2013a; Synolakis and Kânoğlu 2015). At the same time, these events marked the beginning of "modern tsunami science era" (Okal 2019a).

The 1992 Nicaragua and Flores Island tsunamis were the first tsunamis subjected to comprehensive International Post-Tsunami Surveys, which become systematic following later significant tsunamis (Synolakis and Okal 2005; Okal 2019a). Precise estimations of the tsunami runup along the affected coasts became the main source of key scientific information about the events, and was essential for the long-term forecast of future events and for the estimation of tsunami risk for a specific region. However, as the field surveys followed scientific progress, much more complete information was collected for the events, see Arcos et al. (2019). In addition, the information collected was found to be very valuable for the development and verification of numerical models (Synolakis et al. 2008).

Further, the 1992 Nicaragua and Flores Island tsunamis initiated the establishment of the Tsunami Bulletin Board (TBB) network (tsunami_bb@infolist.nws.noaa.gov), proposed by Dr. Frank González of the Pacific Marine Environmental Laboratory (PMEL), National Oceanic and Atmospheric Administration (NOAA), Seattle, Washington, USA. For several years the TBB was supported by PMEL/NOAA but then, when the number of participants exceeded several hundred, it was transferred to the International Tsunami Information Center (ITIC), UNESCO/NOAA, Honolulu, Hawaii, USA. The TBB development was extremely effective, as it built a link between practically all tsunami scientists throughout the entire world. It became possible in a trice to exchange information, data, and numerical modelling results, and to consult among colleagues. The greatest value of the TBB became evident during major catastrophic events, like the 2004 Sumatra and 2011 Tohoku tsunamis. At present, the ITIC TBB turned out to be a modern tsunami communication system. Further, tsunami science became a "science without borders"; the level of the scientific international cooperation in tsunami research appears to be higher than in any other geophysical discipline and it is common for tsunami papers to have joint international teams of authors, the present issue being a spectacular example.

The crucial factor promoting enormous progress in tsunami science is an incredible enhancement in tsunami measurements. Most of coastal tide gauges working before 1992 were built to record tides, storm surges, seasonal oscillations and other low-frequency processes (cf. Pugh and Woodworth 2014). These were either analogue "pen-and-paper" gauges, or digital instruments with poor time resolution (with a 15–20 min time step). Such instruments were quite adequate to measure tides and other long-period processes but were inappropriate for tsunami monitoring. Specifically, the 1992 Nicaragua and Flores Island tsunamis, as well as other tsunamis of the early 1990s, advanced the reconstruction of the tide gauge network in the Pacific Ocean (e.g. Rabinovich and Stephenson 2004). However, most of the NOAA tide gauges still had insufficient time resolution (6 min), while in the Indian and Atlantic oceans, coastal tide gauges were either still analogue or had 6–15 min sampling. At that point in time, the common opinion was that there are no tsunamis in these oceans and there was no reason to have higher resolution. Only the catastrophic 2004 Sumatra tsunami in the Indian Ocean, which was globally observed, including in far remote regions of the North Pacific and North Atlantic (e.g. Rabinovich et al. 2006; Titov et al. 2005a), initiated the total upgrade of the worldwide sea level network. Modern digital tide gauges are high-resolution instruments with 15–60 s sampling, which is perfect for tsunami measurements. As a result, each new strong tsunami brings enormous amount of high-quality data from several hundred instruments. These new data enabled to investigate tsunamis with a much higher accuracy than was possible before, and to significantly improve the general tsunami understanding.

However, coastal measurements only indirectly characterize open-ocean tsunami characteristics.

These observations are strongly distorted by various coastal processes, including refraction, reflection, nonlinearity, friction, scattering by coastal and bottom irregularities, high noise level caused by wind waves and swell, wave trapping, shelf resonance, and especially the harbour resonance of local water basins (bays, inlets, and harbours where tide gauges are normally situated). Hence, deep-ocean tsunami measurements are crucial to understand the generation mechanism and physical properties of tsunami waves and to provide effective and timely tsunami warning. The first record of a tsunami in the open ocean was obtained by Filloux (1982), but it was made by an autonomous bottom pressure sensor and the tsunami was detected only after the instrument was retrieved from the bottom, i.e. about 1.5 years after the tsunami actually occurred (14 March 1979). The first real-time measurements of tsunami waves in the open ocean, and the continuous monitoring of these waves for operational tsunami warning, began in the 1990s and resulted in the creation of the Deep-ocean Assessment and Reporting of Tsunamis (DART)[1] system, which yielded precise real-time tsunami measurements (with an accuracy of about 0.1 cm), made possible by data transmission via modem and satellite communication technology (Mofjeld 2009; Rabinovich and Eblé 2015).

Before 2005, there were seven DARTs working in the Pacific Ocean (six from NOAA, USA, and one Chilean), and two of them provided 15-s data during the 2004 Sumatra tsunami in the Indian Ocean. High-quality records of this tsunami were extracted from both DARTs, even though the instruments were ~ 18,000 to 22,000 km from the source area, and the amplitudes of the recorded waves were only ~ 1 cm (Rabinovich et al. 2011, 2017a). As a result of the catastrophic 2004 Sumatra tsunami, a worldwide network of DART stations was deployed along the most seismically active zones of the World Ocean, which presently includes 45 instruments in the Pacific Ocean, 10 in the Indian Ocean and 7 in the Atlantic Ocean. These instruments became the main backbone of the modern International Tsunami Warning System (e.g. Mofjeld 2009; Titov 2009) and one of the most important sources of invaluable scientific information. Moreover, after the destructive 2011 Tohoku tsunami, Japan established the DONET and S-net arrays of high-resolution precise cable stations off the Pacific coast of Japan that include more than 200 bottom gauges (e.g. Rabinovich and Eblé 2015). Also, in 2009 the Ocean Network Canada (ONC), installed on the shelf—continental slope of Vancouver Island a seafloor network of geophysical observatories,[2] which was found to be very effective to measure tsunami waves and has already recorded great many Pacific tsunamis, beginning with the 2009 Samoa event (Thomson et al. 2011). As a result, any new tsunami brings enormous amount of comprehensive information about tsunami properties in the open ocean that can be used for the verification of numerical models (Synolakis et al. 2008), and for real-time reconstruction of the source during operational tsunami warning. While before 1992, the main problem in tsunami studies was the absence of reliable observational data, we now face the opposite situation, when we are overwhelmed by the amount of available coastal and open-ocean tsunami data.

Another important problem, where an incredible progress was achieved in the last 25 years, is the numerical modelling of tsunamis. Before 1992, numerical simulations were used mostly for specific modelling experiments with artificial sources, rather than to numerically reconstruct and examine real events. At present time, numerical modelling has become a main tool of tsunami research: it enables scientists to examine particular features of historical and recent tsunami events, in order to provide real-time numerical modelling and effective forecast of actual tsunamis (e.g. Titov et al. 2005b), and to reconstruct with high accuracy initial sources of tsunamis. Also, numerical modelling is now the main instrument for local and regional tsunami zoning, i.e. for the estimation of maximum expected tsunami wave heights and possible inundation along specific coastal areas of interest.

[1] DART is an effective network of deep-ocean stations, elaborated by NOAA for monitoring tsunami waves and early tsunami warning.

[2] This network was previously known as NEPTUNE-Canada (Canadian North-East Pacific Underwater Networked Experiments); it was established by the University of Victoria, British Columbia, Canada. (http://www.oceannetworks.ca/).

Several essential factors determined this progress, including advanced numerical models, enormous improvement in modern computers, the availability of high-resolution bathymetry/topographic data, and improved knowledge of the characteristics of seismic sources. Observational data and, first of all, open-ocean tsunami measurements, now play a determinative role in the model "tuning" and its adaptation to a specific problem or event.

While there are many modern numerical models to simulate propagation of tsunami waves, two of them became most widespread: (1) The Method Of Splitting Tsunami (MOST) model (Titov and González 1997; Titov and Synolakis 1998; Titov et al. 2016) and the Tohoku University's Numerical Analysis Model for Investigation (TUNAMI) (Imamura 1996). Both models are widely used both for operational and scientific purposes. Most importantly, tsunami modelling and tsunami measurements/detection, which are two independent components of tsunami studies, now constitute joint parts of united tsunami analyses: Tsunami modelling helps to identify tsunami waves (even for low signal-to-noise cases) and to explain the observed properties of the propagating waves, while the observations help to verify and fit the model. Specifically, the combination of modelling and observations allow a better reconstruction of the tsunami source (e.g. Fujii and Satake 2007; Wei et al. 2008; Titov et al. 2005b), and have been crucial in explaining such subtle effects as the influence of the Earth elasticity and seawater stratification on open-ocean tsunami propagation (Watada et al. 2014; Allgeyer and Cummins 2014).

A subject of great importance, which began to be actively developed only in the last 25 years, is that of *paleotsunami* studies (e.g., Clague et al. 2000; Tappin 2007). These have allowed scientists to go deeply into tsunami history and find reliable information about events that occurred hundreds, or even thousands, of years ago. In this way, paleotsunami findings fundamentally improve tsunami statistics and supply scientists with invaluable information to estimate tsunami risk for specific coastal areas. Even for recent events, like for example, the 2011 Tohoku tsunami, paleotsunami information proved to be very useful, significantly supplementing our knowledge about that particular event (e.g. Goto et al. 2012). Probably the greatest paleotsunami finding was the discovery of the Cascadia Subduction Zone earthquake and major tsunami of 26 January 1700, based on tsunami signatures revealed in sediment deposits in Oregon, Washington and British Columbia, supplemented by Japanese historical chronicles (Satake et al. 1996; Atwater et al. 2005). In addition to studies of onshore sediments, researchers have also studied the turbidite event history of the Cascadia Basin and identified 19 turbidite events within the last 10,000 years, i.e. approximately one event in 500–530 years (Goldfinger et al. 2012).

The catastrophic events of 1992–2018, their intensive investigation, the tremendous progress in tsunami science, hazard mitigation and warning, all resulted in thousands of papers, hundreds of books and professional reports, and in many topical issues and book article collections. Two of them—Bernard and Robinson (2009) and Kânoğlu et al. (2015)—tried to summarize modern tsunami knowledge and the latest achievements in tsunami science.

Among the numerous special issues on tsunamis published by various journals, *Pure and Applied Geophysics* (PAGEOPH) topical issues are probably the most significant. These issues are published under the umbrella of the Joint Tsunami Commission, a part of the International Union of Geodesy and Geophysics (IUGG). This commission, established following the great 1960 Chile Tsunami, has held biannual International Tsunami Symposia (ITS), and published special volumes of selected papers, initially under the leadership of Professor Kenji Satake (University of Tokyo). Altogether, 13 such volumes were published in PAGEOPH during 1992–2019 (Satake and Imamura 1995; Satake et al. 2007, 2011a, b, 2013a, b, Cummins et al. 2008, 2009; Rabinovich et al. 2015a, b, 2017b, 2018; Geist et al. 2016). In addition, the two catastrophic tsunamis of 2010 (Chile) and 2011 (Tohoku), as well as other strong events, generated so much attention and substantial new information, that they motivated the publication of an extra, intersession volume (Rabinovich et al. 2014). Moreover, high interest regarding the Illapel (Chile) earthquake and tsunami of 16 September 2015 resulted in a topical collection of regular PAGEOPH papers "Chile-2015" that were later published as a book (Braitenberg and

Rabinovich 2017). In addition, another PAGEOPH topical volume, specifically related to landslide-generated tsunamis, was prepared by Bardet et al. (2003), in the wake of the 1998 Aitape, Papua New Guinea destructive landslide generated tsunami. All these volumes document research at the frontiers of the tsunami science and reflect progress made continuously in tsunami warning and hazard mitigation. Many papers published in these volumes have become classics and been highly cited.

The Commission proposed preparation of the present volume during the 28th International Tsunami Symposium, which was held in Bali on 21–23 August 2017, followed by a field trip to Babi Island, Flores on 24–25 August. The field trip brought participants back to the scene of an event that had occurred 25 years earlier, when a large earthquake on 12 December 1992 in northern Flores generated a tsunami that claimed 2600 lives, and destroyed 18,000 houses (Yeh et al. 1994, Kânoğlu and Synolakis 1998). From this standpoint, the topical volume "Twenty Five Years of Modern Tsunami Science Following the 1992 Nicaragua and Flores Island Tsunamis, Volume I" should be considered as a highly valuable continuation of the previous topical Pure and Applied Geophysics (PAGEOPH) tsunami issues and, at the same time, as an illustration of current development and progress in tsunami science. In this context, it includes overview papers by Okal (2019a), Gusiakov et al. (2019) and Arcos et al. (2019), specially prepared for this issue. Altogether, this issue comprises 22 papers, with additional papers to be published in the next volume.

2. *Twenty-Five Years of Progress in Tsunami Science*

Okal (2019a) critically evaluates forty-seven tsunamis of geological origin starting with the 1992 Nicaragua event, and discusses how these events helped shape field, experimental, theoretical, and/or societal aspects of tsunami science. Okal (2019a) listed principal developments as follows: from the geophysical standpoint, the author earmarks the development of the W-phase inversion for the low-frequency moment tensor of the parent earthquake and the abandonment of the concept of a maximum earthquake magnitude for a given subduction zone, controlled by simple plate properties (Kagan and Jackson 2013; Stein and Okal 2007). Further, from the standpoint of hydrodynamics, he lists the development of tsunami numerical models to simulate tsunami runup over a dry land, and recent investigations of current velocities, during interaction of tsunamis with harbours. Further, Okal (2019a) emphasises the need for research on the real time identification of "tsunami earthquakes", whose tsunamis are larger than expected from their seismic magnitudes, and the continuous need for further education of the populations at risk, in coastal areas.

Gusiakov et al. (2019) present an analysis of global tsunami occurrences from 1992 to 2016. This 25-year period represents less than 1% of the total catalogue duration (ca. 4000 years), but contains almost 12% of the total number of historical events and nearly 54% of the total runup and inundation depth measurements, illustrating the data obtained in field surveys, notably for the two trans-oceanic mega-tsunamis in this period—the 26 December 2004 Indian Ocean and the 11 March 2011 Tohoku tsunamis, with 1625 and 6271 total measurements, respectively. Two hundred ninety recorded tsunamis and tsunami-like events are listed in the catalogue; ninety five of them featured maximum runup or inundation height greater than 1 m causing damaging or potentially damaging effects, and thirty-two destructive tsunamis resulted in human fatalities.

Arcos et al. (2019) provide the evolution of the global tsunami field survey database colocated at the NOAA National Centers for Environmental Information (NCEI) and World Data Service (WDS) for Geophysics. Even though post-tsunami field surveys had taken place prior to the 2 September 1992 Nicaragua tsunami (such as for the 1964 Southern Alaska, 1976 Mindanao, and 1983 Sea of Japan tsunamis), the first formal International Tsunami Survey Team (ITST) was formed in the aftermath of the Nicaraguan tsunami (Satake et al. 1993). Since then, ITSTs, organized after each event, not only resulted in international and interdisciplinary collaboration, but also in the standardisation of survey methodologies and guidance (Kong 2011), which eventually led to the publication of the post-tsunami field survey guide (UNESCO-IOC 2014). The standardisation of measurements of maximum runup, maximum water level and inundation extent have allowed their comparison across continents, events and time. Moreover, much more complete data are now captured during recent post-tsunami surveys, i.e. not just only maximum runup or inundation depth measurements, but also information regarding wave sequencing and current velocities. Tsunami field survey databases also provide reliable information for historical events, which could be used for verification of tsunami numerical models as outlined by Synolakis et al. (2008), or for identifying the nature of the source. As an example, ITST survey data proved crucial in identifying the source of the 1998 Papua New Guinea tsunami as a landslide, as opposed to a seismic dislocation (Okal and Synolakis 2004).

3. *Studies of Historical Tsunami Events and Field Surveys*

Martin et al. (2019) reassess the 4 January 1907 Sumatra, Indonesia tsunami (earthquake) with a multidisciplinary approach identifying two large earthquakes occurring within an hour of each other but with clearly different seismological character. Across the Indian Ocean, they compiled 88 observations, obtained from individual accounts as well as from reports published in local newspapers, of the tsunami generated by the mainshock. Relative to its instrumental magnitude, the mainshock produced an anomalously large tsunami, which qualifies the event as a "tsunami earthquake". Yet, the shaking on Nias was severe, which is unusual for a tsunami earthquake. Martin et al. (2019) identified a large, second event occurring 53 min after the mainshock, which resolves the paradox of the surprisingly high intensities following the slow 1907 mainshock. The local S'mong tradition, a probable result of the 1907 events, may have prevented heavy loss of life on Simeulue in 2004 and 2005, since both were preceded by severe shaking (McAdoo et al. 2006).

Okal (2019b) presents a modern seismological study of the large earthquake of 26 June 1941 in the Andaman Islands along the India-Burma plate boundary. The author shows that the 1941 earthquake took place under the Andaman accretionary prism and consisted of a composite event, whose nucleating phase, with a strike-slip mechanism, is incompatible with spectral amplitudes of mantle Rayleigh and Love waves. This event resulted in a small tsunami due to the combination of the generation mechanism, and of the location of the source under the islands and in shallow water, leading to a reduction in amplitude when transitioning to a deeper basin, under Green's law.

Rabinovich et al. (2019) examine analogue tide gauge records on the coast of British Columbia (BC) for the five major trans-Pacific tsunamis that occurred between 1946 and 1964. A comprehensive frequency analysis of each event is presented, suggesting a strong influence of local and regional topography on the incoming waves, since the dominant frequencies of the waves and their evolution in time were similar at each station for different events, but differed considerably among sites for a particular tsunami. A common feature of all these events is the fact that they penetrated deep inside the narrow channels, fjords and inlets typical of the BC coast, which suggests that these morphological features do not hinder the incoming waves but rather can strongly enhance them through resonant amplification when the dominant incoming wave frequency coincides with the natural frequency of the interconnected basin.

Dogan et al. (2019) present the results of the field survey conducted after the 20 July 2017 Bodrum-Kos earthquake and tsunami in the Mediterranean Sea, between the Bodrum Peninsula in Turkey, and Kos Island in Greece. A minor tsunami followed the earthquake, producing small runup around 1 m. Maximum runup and inundation distances were observed in Bodrum peninsula, with moderate damage observed in Kos port. Interviews conducted during the survey revealed extremely low tsunami awareness among the local population and the numerous tourists. As tsunamis are infrequent in this area, the study of minor tsunamis is important and it helps test local tsunami warning systems.

4. *Tsunami Source Models*

Ramírez-Herrera et al. (2019) analyse the source of the 27 August 1810 Loreto event in Baja

California. The main difference between this event and other events of the 19th and 20th Century is the fact that the tsunami followed the earthquake and produced significant flooding along the coast. Historical information on the tsunami is presented and used to constrain the source of the earthquake. The discrepancy between the maximum earthquake intensity (IX), the estimated magnitude (7.4) of the earthquake and the tsunami travel time to the coast suggest that the tsunami was triggered by a submarine landslide. The authors confirm this mechanism, based on numerical modelling studies, using various dislocation and landslide sources.

Pranantyo and Cummins (2019) revisit the source of the 1992 Flores earthquake and tsunami. They estimate the slip model of the Flores 1992 earthquake through a joint finite-fault source inversion technique, using all available datasets, teleseismic body and surface waveforms, co-seismic vertical deformation data, one tsunami waveform, and tsunami runup heights. The largest slip was found near the hypocenter, which had not been well resolved by previous models. They conclude that the earthquake ruptured a distinct segment of the back-arc thrust system in the eastern Sunda Arc, Indonesia.

Voronina et al. (2019) propose a method to identify an initial tsunami source from observed waveforms at DART buoys with the only additional information about the source spatial localization. They regularise an ill-posed inverse problem using a least square inversion with the truncated Singular Value Decomposition and r-solution methods. Their computational algorithm improves the inversion by selecting the most informative set of available observation stations, suppressing the instability of the numerical solution of this ill-posed problem. The method was successfully applied to the 16 September 2015 Illapel, Chile tsunami.

Yolsal-Çevikbilen et al. (2019) present finite-fault source rupture models and numerical modelling for the tsunamis generated by the 28 October 2012 Haida Gwaii, Canada (M_w 7.7) and 16 September 2012 Illapel, Chile (M_w 8.3) earthquakes. Fault parameters were estimated by analysing teleseismic long-period P- and SH-waves and broadband P-waveforms and using waveform inversion and hybrid back-projection methods. They reveal heterogeneous slip distributions on fault planes with long source durations (80 s and 150 s) and low stress drop values (10–15 bar). Further, they numerically model tsunami propagation for both events, using nonlinear long-wave models for both uniform and non-uniform slip distributions. They conclude that, in general, leading wave shapes and arrival times are adequately modelled with the heterogeneous slip distribution models. Moreover, in order to obtain a better comparison between numerical modelling and field data, they emphasise the need for joint inversions of GPS, tsunami, teleseismic and strong ground motion records, and for higher resolution of bathymetric data.

5. *Landslide Generated Tsunamis*

Paris et al. (2019) investigate the landslide-triggered tsunami event on 17 June 2017 in Karrat Fjord, Greenland, which flooded several villages, killing 4 people and destroying 11 houses at Nuugaatsiaq. They numerically model the landslide by considering it as a 50 million m^3 granular flow under gravity forces. They obtain amplitudes similar to those estimated from videos, three waves between 1 and 1.5 m arriving at Nuugaatsiaq with a period of roughly 3 min. Their amplitude estimates are also in agreement with the deconvolution of oscillations recorded on a horizontal seismogram operating at Nuugaatsiaq. Further, they study future potential subaerial landslides next to the one on 17 June 2017, threatening Karrat Fjord. They perform a sensitivity study with 2, 7, 14 and 38 million m^3 subaerial landslides and conclude that the shape of the waves is independent of volume, the relationship between the volume and the wave height is linear, and the orientation of the slide influences neither the period nor the shape of the generated wave.

The first probabilistic tsunami flooding maps, called probabilistic tsunami hazard assessment (PTHA), were developed by González et al. (2009), integrating tsunami inundation modelling with methods of probabilistic seismic hazard assessment (PSHA). Since then, PTHAs have been considered or performed in many tsunami-prone areas. Here, Schambach et al. (2019) present part of a tsunami hazard assessment study, which was carried out since 2010 to develop tsunami inundation maps for the US East Coast, under the auspices of the US National Tsunami Hazard Mitigation Program (NTHMP). In this study, submarine mass failures (SMFs) were considered as a form of potential hazard along the upper US East Coast and modelled using the three-dimensional non-hydrostatic model NHWAVE with two types of slump motion: (1) rigid, for which the geometry and law of motion are specified as bottom boundary conditions, and (2) a deforming slide, for which motion is modelled using a depth-integrated bottom layer of dense Newtonian fluid, fully coupled to the overlying fluid motion. They use the Boussinesq-type model FUNWAVE-TVD to evaluate tsunami propagation once the SMFs have stopped. The authors discuss the need to use a dispersive model to simulate for landslide tsunami generation, the model which depends on the relative size of the SMF and of the relative height of generated waves, as compared to the local depth. A non-dispersive model would have been adequate for the very large SFMs considered in this study. In addition, a dispersive model, such as a long-wave Boussinesq model, is needed to accurately simulate both far-field and near-field tsunami propagation; Schambach et al. (2019) also stress the need for sufficiently high-resolution model grids.

Salmanidou et al. (2019) investigate the tsunami hazard potential associated with submarine mass failures in the Indus Canyon in the Northwestern Indian Ocean, which featured numerous failures in the past. They use statistical emulation, training it with 60 slump scenarios parametrising slide thickness, slide width, travel distance, and submerged depth. Then, they use the trained emulator to predict 500,000 trial scenarios and to study probabilistically the tsunami hazard over the near field. Their results imply, most likely, 0.2–1.0 m and 2.5–13 m/s tsunami amplitudes and flow speeds respectively, which could potentially impact vessels and maritime facilities. Further, they discuss the advantage of using an emulator-based approach in probabilistic hazard analysis, allowing the generation of thousands of tsunami scenarios in a few seconds, as compared to several days on high performance computing facilities for a single run of the dispersive tsunami solver.

6. Tsunami Numerical Modelling

Sogut and Yalcıner (2019) present validation and model comparison studies for two numerical models, NAMI DANCE and FLOW-3D® (Flow Science 2002) for tsunami studies. While the former was developed in-house as a tsunami numerical model solving nonlinear shallow water-wave equations, the latter is a general purpose computational fluid dynamics software, which is based on the Volume-of-Fluid technique. Both models were tested against data for tsunami currents (Lynett et al. 2017), in addition to the benchmark results listed by Synolakis et al. (2008).

Periáñez and Cortés (2019) present a numerical study of the propagation of tsunamis generated by submarine earthquakes in the Caspian Sea. They apply their modelling to the 20 June 1990 Rudbar earthquake in north Iran showing results coherent with previous studies. In addition, they model the 8 July 1895 Krasnovodsk earthquake, which, previously, had not been simulated. Modelling results produced coastal flooding, as reported in literature, with a rather local effect since the tsunami was generated in very shallow waters. Further, three additional hypothetical scenarios were considered in the most seismically active regions of the Caspian Sea. Results showed that both the east and west coasts of the central Caspian Sea would be the most affected with potentially significant effects in such cities as Baku, Azerbaijan.

Torres et al. (2019) study electric currents induced by tsunami flow across the geomagnetic field both in the ocean and the ionosphere. These currents generate secondary magnetic fields, observable at the seafloor and on land. They model magnetic field disturbances, recorded at the Easter Island Observatory for the Chilean tsunamis of 27 February 2010 and 16 September 2015, by applying the Biot–Savart law to tsunami simulations, and emphasize that using Maxwell equations may present a significant improvement to their model. They calculate tsunami-induced perturbations in the vertical component of the geomagnetic field and obtain good agreement with the observations. They hypothesise that the magnetic effects may be produced by a combination of oceanic and ionospheric electric currents, and that even a relatively small tsunami might be monitored on magnetic records. Encouraged by the results, they decided to install a land-based magnetometer in San Vicente Bay, close to Concepción, Chile.

7. Tsunami Hazard Assessment and Warning

In some tsunami-prone regions, there is a lack of high-resolution digital elevation models to evaluate tsunami inundation hazard in order to mitigate its impact through numerical modelling. Nicolsky et al. (2019) present a methodology to approximate tsunami hazard zones in areas with poor topographic and bathymetric coverage, with specific application to many rural coastal communities in Alaska, from tsunamigenic earthquakes along the Alaska–Aleutian megathrust. They perform modelling of earthquake-generated tsunamis in the Andreanof Islands, and estimate the potential inundation zone for the Adak community for three scenarios, exploring a sensitivity of maximum runup with respect to the coseismic slip distribution. Hence, they define a safety factor, 1.3 in this specific case, validated against high-resolution modelling results. The authors assume that the safety factor is specific to local topographic/bathymetric features, e.g. the heads of bays might result in larger amplification, hence in larger safety factors.

Triantafyllou et al. (2019) applied a tsunami risk assessment model to the coast west Heraklion, the capital city of Crete Island, Greece. Tsunami risk assessment, which is considered as a convolution of tsunami hazard, vulnerability of the assets at risk and the economic value exposed, is an important component of the planning for risk reduction. Heraklion was hit by strong tectonic and volcanic tsunamis in the past. The Minoan tsunami, generated by pyroclastic flows during the eruption of Thera (17th century BC), was selected as the worst scenario. A numerical modelling study was performed using a Boussinesq-type model, which produced 1.2 km inundation and 14 m maximum water depth. Using their tsunami simulation results, building vulnerability was determined with the use of the DAMASCHE, empirical GIS tool, which is based on building damage data from the 2004 Indian Ocean tsunami. Results indicated that most buildings would suffer repairable damage, while 183 buildings would collapse partially or totally. The damage level was translated into monetary loss following building repair or reconstruction, based on the rates determined officially after the destructive Greek earthquakes of 2014.

Since it can be impractical to develop regional and global tsunami hazard analysis for some tsunami-prone communities, it is essential to use simplified and efficient methods to estimate the tsunami inundation height and its related uncertainty. One approach is to use an amplification factor, which describes the relation between offshore wave height and the maximum inundation height, as predicted by linearized plane wave models employed for incident waves with different wave characteristics. Glimsdal et al. (2019) present a new amplification factor method that takes into account the offshore bathymetry near the coastal site, and apply it to the North-Eastern Atlantic and Mediterranean (NEAM) region. They first quantify the uncertainty in inundation height by analysing its variability in more than 500 high-resolution inundation simulations at six different coastal sites with different earthquake sources, in order to produce various wave periods and polarities. They conclude that the probability density of the maximum inundation height can be modelled with a log-normal distribution, whose median is quite well predicted by the amplification factor, and that the associated maximum inundation height uncertainties are indeed significant and must be accounted for in tsunami hazard analysis. They apply their methodology to the recently developed Tsunami Hazard Maps for the NEAM (TSUMAPS-NEAM) region PTHA.

Over the past decade, the number of open-ocean gauges capable of parsing information about a passing tsunami has steadily increased, particularly through national cable networks and international buoyed systems such as the DART sensors (e,g. Rabinovich and Eblé 2015). However, most current warnings that incorporate direct tsunami observations are located at mid- and far-field localities. Williamson and Newman (2019) study the effectiveness of a localised near-field systems in four seismically active subduction zones, Cascadia, Japan, Chile, and Java, for their potential to facilitate local tsunami early warning. They assess which of them currently have instrumentation at the right locations for direct tsunami observations with enough lead time. Their primary findings indicate that while such regions as Chile are ill-suited in this respect, other localities, like Java could incorporate direct tsunami observations into their hazard forecasts with enough lead time for effective emergency response by coastal communities.

Greenslade et al. (2019) evaluate tsunami travel times (TTT) from scenario database T2 at the Joint Australian Tsunami Warning Centre (JATWC), and tsunami travel times from GEOWARE (2007), and compare them to observed arrival times at tide gauges and tsunameters from the National Geophysical Data Center/World Data Service (NGDCS/WDS), in order to evaluate current operational practice at the JATWC. They conclude that travel times from the T2 scenario database are at least as accurate as those from TTT and in most cases, more accurate. They also infer that the 'best' scenario-based method is Tp20—the time of arrival of the first peak that is at least 20% of the maximum amplitude at that location.

8. Conclusions

In conclusion, we reflect on Okal's (2015) evocation of Zeno of Elea's paradox in the context of the advancement of tsunami science and of efforts in mitigation: "Like Achilles, our communities seem well armed to constantly gain in wisdom and progress towards mitigation of a perceived level of tsunami hazard, but individual new events often bring an element of diversity which redefines the goal, and like the Tortoise, keeps pushing it forward, in an everlasting challenge". While the present collection of papers point to substantial development across the various fields of tsunami research in the past 25 years, our progress is, and will remain, a never-ending story, and we can trust that a continued string of similar special issues will appear in the future.

Acknowledgements

We thank Ms. Shamima Banu Rajesh and Mr. Sathish Srinivasan at the Journals Editorial Office of Springer for their timely editorial. We thank Francisco Azpilicueta, and the editors of the previous issue, Hermann M. Fritz and Eric Geist, for editing those papers which were transferred to this issue. We acknowledge Professor Kenji Satake (University of Tokyo, Japan) and Dr. Vasily Titov (PMEL/NOAA, Seattle, WA, USA) for their support. We thank the authors who contributed papers to these topical volumes. Finally, we would like to especially thank the reviewers who shared their time, effort, and expertise to maintain the scientific rigour of this volume. AR's contribution was partially supported by the Russian State Assignment of IORAS 0149-2019-0005.

References

Allgeyer, S., & Cummins, P. (2014). Numerical tsunami simulation including elastic loading and seawater density stratification. *Geophysical Research Letters, 41*(7), 2368–2375. https://doi.org/10.1002/2014gl059348.

Arcos, N. P., Dunbar, P. K., Stroker, K. J., & Kong, L. S. L. (2019). The impact of post-tsunami surveys on the NCEI/WDS global historical tsunami database. *Pure and Applied Geophysics*. https://doi.org/10.1007/s00024-019-02191-7. **(this issue)**.

Atwater, B. F., Musumi-Rokkaku, S., Satake, K., Tsuji, Y., Ueda, K., & Yamaguchi, D. K. (2005). *The orphan tsunami of 1700: Japanese clues to a parent earthquake in North America*. US Geological Survey. https://doi.org/10.3133/pp1707.

Bardet, J.-P., Synolakis, C. E., Davies, H. L., Imamura, F., & Okal, E. A. (2003). Landslide tsunamis: Recent findings and research directions. *Pure and Applied Geophysics, 160*(10–11), 1793–1809. https://doi.org/10.1007/s00024-003-2406-0.

Bernard, E., & Robinson, A. (Eds.) (2009). *Tsunamis*. In *The Sea, ideas and observations on progress in the study of the seas* (Vol. 15). Cambridge: Harvard University Press.

Braitenberg, C., & Rabinovich, A. B. (2017). *The Chile-2015 (Illapel) earthquake and tsunami* (p. 335). Basel: Birkhäuser/Springer. https://doi.org/10.1007/978-3-319-57822-4.

Clague, J. J., Bobrowsky, P. T., & Hutchinson, I. (2000). A review of geological records of large tsunamis at Vancouver Island, British Columbia, and implications for hazard. *Quaternary Science Reviews, 19,* 849–863.

Cummins, P. R., Kong, L. S. L., & Satake, K. (2008). Tsunami science four years after the 2004 Indian Ocean tsunami. Part I: Modelling and hazard assessment. *Pure and Applied Geophysics, 165*(11–12, topical issue).

Cummins, P. R., Kong, L. S. L., & Satake, K. (2009). Tsunami science four years after the 2004 Indian Ocean tsunami. Part II: Observation and data analysis. *Pure and Applied Geophysics, 166*(1–2, topical issue).

Dogan, G. G., Annunziato, A., Papadopoulos, G. A., et al. (2019). The 20th July 2017 Bodrum-Kos tsunami field survey. *Pure and Applied Geophysics.* https://doi.org/10.1007/s00024-019-02151-1. **(this issue)**.

Filloux, J. H. (1982). Tsunami recorded on the open ocean floor. *Geophysical Research Letters, 9*(1), 25–28. https://doi.org/10.1029/gl009i001p00025.

Flow Science. (2002). *FLOW-3D User's Manual.*

Fujii, Y., & Satake, K. (2007). Tsunami source of the 2004 Sumatra-Andaman earthquake inferred from tide gauge and satellite data. *Bulletin of the Seismological Society of America, 97*(1A), S192–S207. https://doi.org/10.1785/0120050613.

Geist, E. L., Fritz, H. M., Rabinovich, A. B., & Tanioka, Y. (2016). Global tsunami science: Past and future. Volume I. *Pure and Applied Geophysics, 173*(12, topical issue).

GEOWARE. (2007). TTT—A tsunami travel-time calculator. http://www.geoware-online.com.

Glimsdal, S., Løvholt, F., Harbitz, C. B., et al. (2019). A new approximate method for quantifying tsunami maximum inundation height probability. *Pure and Applied Geophysics.* https://doi.org/10.1007/s00024-019-02091-w. **(this issue)**.

Goldfinger, C., Nelson, C. H., & Morey, A. E., et al. (2012). Turbidite event history—Methods and implications for Holocene paleoseismicity of the Cascadia subduction zone. In R. Kayen (Ed.) (p. 170). US Geological Survey professional paper 1661-F. https://doi.org/10.3133/pp1661f.

González, F. I., Geist, E. L., Jaffe, B., et al. (2009). Probabilistic tsunami hazard assessment at Seaside, Oregon, for near- and far-field seismic sources. *Journal of Geophysical Research, 114,* C11023. https://doi.org/10.1029/2008jc005132.

Goto, K., Chagué-Goff, C., Goff, J., & Jaffe, B. (2012). The future of tsunami research following the 2011 Tohoku-oki event. *Sedimentary Geology, 282,* 1–13. https://doi.org/10.1016/j.sedgeo.2012.08.003.

Greenslade, D. J. M., Greenwood, R. I., & Allen, S. C. R. (2019). An evaluation of modelled tsunami arrival times. *Pure and Applied Geophysics.* https://doi.org/10.1007/s00024-018-2004-9. **(this issue)**.

Gusiakov, V. K., Dunbar, P. K., & Arcos, N. (2019). Twenty-five years (1992–2016) of global tsunamis: Statistical and analytical overview. *Pure and Applied Geophysics.* https://doi.org/10.1007/s00024-019-02113-7. **(this issue)**.

Imamura, F. (1996). Review of tsunami simulation with a finite difference method. In H. Yeh, P. Liu, & C. Synolakis (Eds.), *Long wave runup models* (pp. 25–42). Hackensack, N.J: World Scientific Publishing.

Kagan, Y. Y., & Jackson, D. D. (2013). Tohoku earthquake: A surprise? *Bulletin of the Seismological Society of America, 103*(2B), 1181–1194. https://doi.org/10.1785/0120120110.

Kanamori, H. (1972). Mechanism of tsunami earthquakes. *Physics of the Earth and Planetary Interiors, 6*(5), 346–359. https://doi.org/10.1016/0031-9201(72)90058-1.

Kânoğlu, U., & Synolakis, C. E. (1998). Long wave runup on piecewise linear topographies. *Journal of Fluid Mechanics, 374,* 1–28. https://doi.org/10.1017/s0022112098002468.

Kânoğlu, U., Titov, V., Bernard, E., & Synolakis, C. (2015). Tsunamis: Bridging science engineering and society. *Philosophical Transactions of the Royal Society A: Mathematical, Physical and Engineering Sciences, 373*(2053), 20140369. https://doi.org/10.1098/rsta.2014.0369.

Kong, L. (2011). Post-tsunami field surveys are essential for mitigating the next tsunami disaster. *Oceanography, 24*(2), 222–226. https://doi.org/10.5670/oceanog.2011.48.

Lynett, P. J., Gately, K., Wilson, R., et al. (2017). Inter-model analysis of tsunami-induced coastal currents. *Ocean Modelling, 114,* 14–32. https://doi.org/10.1016/j.ocemod.2017.04.003.

Martin, S. S., Li, L., Okal, E. A., et al. (2019). Reassessment of the 1907 Sumatra tsunami earthquake based on macroseismic seismological, and tsunami observations, and modeling. *Pure and Applied Geophysics.* https://doi.org/10.1007/s00024-019-02134-2. **(this issue)**.

McAdoo, B. G., Dengler, L., Prasetya, G., & Titov, V. V. (2006). Smong: How an oral history saved thousands on Indonesia's Simeulue Island during the December 2004 and March 2005 tsunamis. *Earthquake Spectra, 22*(S3), 661–669. https://doi.org/10.1193/1.2204966.

Mofjeld, H. O. (2009). Tsunami measurements. In A. Robinson & E. Bernard (Eds.), *The Sea, ideas and observations on progress in the study of the seas,* Vol. 15, Chap. 7. *Tsunamis,* (pp. 201–235), Cambridge: Harvard University Press.

Nicolsky, D. J., Suleimani, E. N., Koehler, R. D., & Salisbury, J. B. (2019). Developing an approximate tsunami hazard zone for areas with poor topographic coverage in Alaska. *Pure and Applied Geophysics.* https://doi.org/10.1007/s00024-019-02180-w. **(this issue)**.

Okal, E. A. (2015). The quest for wisdom: Lessons from 17 tsunamis 2004–2014. *Philosophical Transactions of the Royal Society A: Mathematical, Physical and Engineering Sciences, 373*(2053), 20140370. https://doi.org/10.1098/rsta.2014.0370.

Okal, E. A. (2019a). Twenty-five years of progress in the science of "geological" tsunamis following the 1992 Nicaragua and Flores events. *Pure and Applied Geophysics.* https://doi.org/10.1007/s00024-019-02244-x. **(this issue)**.

Okal, E. A. (2019b). The large Andaman Islands earthquake of 26 June 1941: Why no significant tsunami? *Pure and Applied Geophysics.* https://doi.org/10.1007/s00024-018-2082-8. **(this issue)**.

Okal, E. A., & Synolakis, C. E. (2004). Source discriminants for near-field tsunamis. *Geophysical Journal International, 158*(3), 899–912. https://doi.org/10.1111/j.1365-246x.2004.02347.x.

Paris, A., Okal, E. A., Guérin, C., Heinrich, P., Schindelé, F., & Hébert, H. (2019). Numerical modeling of the June 17, 2017 landslide and tsunami events in Karrat Fjord, West Greenland. *Pure and Applied Geophysics.* https://doi.org/10.1007/s00024-019-02123-5. **(this issue)**.

Periáñez, R., & Cortés, C. (2019). A modelling study on tsunami propagation in the Caspian Sea. *Pure and Applied Geophysics.* https://doi.org/10.1007/s00024-018-2057-9. **(this issue)**.

Pranantyo, I. R., & Cummins, P. R. (2019). Multi-data-type source estimation for the 1992 Flores earthquake and tsunami. *Pure and*

Applied Geophysics. https://doi.org/10.1007/s00024-018-2078-4. **(this issue)**.

Pugh, D., & Woodworth, P. (2014). *Sea-level science: Understanding tides, surges, tsunamis and mean sea-level changes*. Cambridge: Cambridge University Press.

Rabinovich, A. B., Borrero, J. C., & Fritz, H. M. (2014). Tsunamis in the Pacific Ocean: 2010–2011. *Pure and Applied Geophysics, 171*(12, topical issue).

Rabinovich, A. B., & Eblé, M. C. (2015). Deep ocean measurements of tsunami waves. *Pure and Applied Geophysics, 172*(12), 3281–3312. https://doi.org/10.1007/s00024-015-1058-1.

Rabinovich, A. B., Fritz, H. M., Tanioka, Y., & Geist, E. L. (2017b). Global tsunami science: Past and future, Volume II, *Pure and Applied Geophysics, 174*(8, topical issue).

Rabinovich, A. B., Fritz, H. M., Tanioka, Y., & Geist, E. L. (2018). Global tsunami science: Past and future, Volume III, *Pure and Applied Geophysics, 175*(4, topical issue).

Rabinovich, A. B., Geist, E. L., Fritz, H. M., & Borrero, J. C. (2015a). Tsunami science: Ten years after the 2004 Indian Ocean tsunami. Volume I. *Pure and Applied Geophysics, 172*(3–4, topical issue).

Rabinovich, A. B., Geist, E. L., Fritz, H. M., & Borrero, J. C. (2015b). Tsunami science: Ten years after the 2004 Indian Ocean tsunami. Volume II. *Pure and Applied Geophysics, 172*(12, topical issue).

Rabinovich, A. B., & Stephenson, F. E. (2004). Longwave measurements for the coast of British Columbia and improvements to the tsunami warning capability. *Natural Hazards, 32*(3), 313–343.

Rabinovich, A., Stroker, K., Thomson, R., & Davis, E. (2011). DARTs and CORK in Cascadia Basin: High-resolution observations of the 2004 Sumatra tsunami in the northeast Pacific. *Geophysical Research Letters, 38*(8), L08607. https://doi.org/10.1029/2011gl047026.

Rabinovich, A. B., Thomson, R. E., Krassovski, M. V., Stephenson, F. E., & Sinnott, D. C. (2019). Five great tsunamis of the 20th century as recorded on the coast of British Columbia. *Pure and Applied Geophysics*. https://doi.org/10.1007/s00024-019-02133-3. **(this issue)**.

Rabinovich, A. B., Thomson, R. E., & Stephenson, F. E. (2006). The Sumatra Tsunami of 26 December 2004 as observed in the North Pacific and North Atlantic Oceans. *Surveys In Geophysics, 27*, 647–677.

Rabinovich, A. B., Titov, V. V., Moore, C. W., & Eblé, M. C. (2017a). The 2004 sumatra tsunami in the Southeastern Pacific Ocean: New global insight from observations and modeling. *Journal of Geophysical Research: Oceans, 122*, 7992–8019. https://doi.org/10.1002/2017JC013078.

Ramírez-Herrera, M. T., Corona, N., & Castillo-Aja, R. (2019). Revealing the source of the 27 August 1810 Loreto, Baja California, tsunami from historical evidence and numerical modelling. *Pure and Applied Geophysics*. https://doi.org/10.1007/s00024-019-02161-z. **(this issue)**.

Salmanidou, D. M., Heidarzadeh, M., & Guillas, S. (2019). Probabilistic landslide-generated tsunamis in the Indus Canyon NW Indian Ocean, using statistical emulation. *Pure and Applied Geophysics*. https://doi.org/10.1007/s00024-019-02187-3. **(this issue)**.

Satake, K., Bourgeois, J., Abe, Ku, Abe, Ka, Tsuji, Y., Imamura, F., et al. (1993). Tsunami field survey of the 1992 Nicaragua earthquake. *EOS Transactions American Geophysical Union, 74*(13), 145–157. https://doi.org/10.1029/93eo00271.

Satake, K., & Imamura, F. (1995). Tsunamis: 1992–1994. *Pure and Applied Geophysics, 144*(3–4, topical issue).

Satake, K., Okal, E. A., & Borrero, J. C. (2007). Tsunami and its hazard in the Indian and Pacific oceans. *Pure and Applied Geophysics, 164*(2–3, topical issue).

Satake, K., Rabinovich, A. B., Dominey-Howes, D., & Borrero, J. C. (2013a). Historical and recent catastrophic tsunamis in the world: Past, present, and future. Volume I: The 2011 Tohoku tsunami. *Pure and Applied Geophysics, 170*(6–8, topical issue).

Satake, K., Rabinovich, A. B., Dominey-Howes, D., & Borrero, J. C. (2013b). Historical and recent catastrophic tsunamis in the world: Past, present, and future. Volume II: Tsunamis from 1755 to 2010. *Pure and Applied Geophysics, 170*(9–10, topical issue).

Satake, K., Rabinovich, A. B., Kânoğlu, U., & Tinti, S. (2011a). Tsunamis in the world ocean: past, present, and future. Volume I. *Pure and Applied Geophysics, 168*(6–7, topical issue).

Satake, K., Rabinovich, A. B., Kânoğlu, U., & Tinti, S. (2011b). Tsunamis in the World Ocean: Past, present, and future. Volume II. *Pure and Applied Geophysics, 168*(11, topical issue).

Satake, K., Shimazaki, K., Tsuji, Y., & Ueda, K. (1996). Time and size of a giant earthquake in Cascadia inferred from Japanese tsunami records of January 1700. *Nature, 379*, 246–249.

Schambach, L., Grilli, S. T., Kirby, J. T., & Shi, F. (2019). Landslide tsunami hazard along the upper US East Coast: Effects of slide deformation bottom friction, and frequency dispersion. *Pure and Applied Geophysics*. https://doi.org/10.1007/s00024-018-1978-7. **(this issue)**.

Sogut, D. V., & Yalcıner, A. C. (2019). Performance comparison of NAMI DANCE and FLOW-3D models in tsunami propagation inundation and currents using NTHMP benchmark problems. *Pure and Applied Geophysics*. https://doi.org/10.1007/s00024-018-1907-9. **(this issue)**.

Stein, S., & Okal, E. A. (2007). Ultralong period seismic study of the December 2004 Indian Ocean earthquake and implications for regional tectonics and the subduction process. *Bulletin of the Seismological Society of America, 97*(1A), S279–S295. https://doi.org/10.1785/0120050617.

Synolakis, C. E., Bernard, E. N., Titov, V. V., Kânoğlu, U., & González, F. I. (2008). Validation and verification of tsunami numerical models. *Pure and Applied Geophysics, 165*(11–12), 2197–2228. https://doi.org/10.1007/s00024-004-0427-y.

Synolakis, C., & Kânoğlu, U. (2015). The Fukushima accident was preventable. *Philosophical Transactions of the Royal Society A: Mathematical Physical and Engineering Sciences, 373*(2053), 20140379. https://doi.org/10.1098/rsta.2014.0379.

Synolakis, C. E., & Kong, L. (2006). Runup measurements of the December 2004 Indian Ocean tsunami. *Earthquake Spectra, 22*(S3), 67–91. https://doi.org/10.1193/1.2218371.

Synolakis, C. E., & Okal, E. A. (2005). 1992–2002: Perspective on a decade of post-tsunami surveys. In K. Satake (Ed.), *Tsunamis: Case studies and recent developments* (pp. 1–29), Dordrecht: Springer-Verlag. https://doi.org/10.1007/1-4020-3331-1_1.

Tappin, D. R. (2007). Sedimentary features of tsunami deposits—Their origin recognition and discrimination: An introduction. *Sedimentary Geology, 200*(3–4), 151–154. https://doi.org/10.1016/j.sedgeo.2007.01.001.

Thomson, R. E., Fine, I. V., Rabinovich, A. B., et al. (2011). Observations of the 2009 Samoa tsunami by the NEPTUNE-Canada cabled observatory: Test data for an operational regional tsunami forecast model. *Geophysical Research Letters, 38*, L11701. https://doi.org/10.1029/2011GL046728.

Titov, V. V. (2009). Tsunami forecasting. In A. Robinson & E. Bernard (Eds.), *The Sea, ideas and observations on progress in the study of the seas* Vol. 15. Chap. 12 *Tsunamis,* (pp. 371–400). Cambridge: Harvard University Press.

Titov, V. V., & González, F. I. (1997). Implementation and testing of the method of splitting tsunami (MOST) model, NOAA Tech. Memo., ERL PMEL-112 (p. 11).

Titov, V. V., González, F. I., Bernard, E. N., et al. (2005a). Real-time tsunami forecasting: Challenges and solutions. *Natural Hazards, 35*(1), 35–41. https://doi.org/10.1007/1-4020-3607-8_3.

Titov, V., Kânoğlu, U., & Synolakis, C. (2016). Development of MOST for real-time tsunami forecasting. *Journal of Waterway Port, Coastal, and Ocean Engineering, 142*(6), 03116004. https://doi.org/10.1061/(asce)ww.1943-5460.0000357.

Titov, V. V., Rabinovich, A. B., Mofjeld, H., Thomson, R. E., & González, F. I. (2005b). The global reach of the 26 December 2004 Sumatra tsunami. *Science, 309,* 2045–2048.

Titov, V. V., & Synolakis, C. E. (1998). Numerical modeling of tidal wave runup. *Journal of Waterway Port, Coastal, and Ocean Engineering, 124*(4), 157–171. https://doi.org/10.1061/(asce)0733-950x(1998)124:4(157).

Torres, C. E., Calisto, I., & Figueroa, D. (2019). Magnetic signals at Easter Island during the 2010 and 2015 Chilean tsunamis compared with numerical models. *Pure and Applied Geophysics.* https://doi.org/10.1007/s00024-018-2047-y. **(this issue)**.

Triantafyllou, I., Novikova, T., Charalampakis, M., Fokaefs, A., & Papadopoulos, G. A. (2019). Quantitative tsunami risk assessment in terms of building replacement cost based on tsunami modelling and GIS methods: The case of Crete Isl. Hellenic Arc. *Pure and Applied Geophysics.* https://doi.org/10.1007/s00024-018-1984-9. **(this issue)**.

UNESCO-IOC. (2014). The UNESCO Intergovernmental Oceanographic Commission International Tsunami Survey Team (ITST) Post-Tsunami Field Survey Guide, 2nd Edition. https://unesdoc.unesco.org/ark:/48223/pf0000229456.

Voronina, T. A., Voronin, V. V., & Cheverda, V. A. (2019). The 2015 Illapel tsunami source recovery by inversion of DART tsunami waveforms using the R-solution method. *Pure and Applied Geophysics.* https://doi.org/10.1007/s00024-019-02100-y. **(this issue)**.

Watada, S., Kusumoto, S., & Satake, K. (2014). Traveltime delay and initial phase reversal of distant tsunamis coupled with the self-gravitating elastic Earth. *Journal of Geophysical Research: Solid Earth, 119*(5), 4287–4310. https://doi.org/10.1002/2013jb010841.

Wei, Y., Bernard, E. N., Tang, L., et al. (2008). Real-time experimental forecast of the Peruvian tsunami of August 2007 for U.S. coastlines. *Geophysical Research Letters, 35,* L04609. https://doi.org/10.1029/2007gl032250.

Williamson, A. L., & Newman, A. V. (2019). Suitability of open-ocean instrumentation for use in near-field tsunami early warning along seismically active subduction zones. *Pure and Applied Geophysics.* https://doi.org/10.1007/s00024-018-1898-6. **(this issue)**.

Yeh, H., Liu, P. L.-F., Briggs, M., & Synolakis, C. (1994). Propagation and amplification of tsunamis at coastal boundaries. *Nature, 372,* 353–355. https://doi.org/10.1038/372353a0.

Yolsal-Çevikbilen, S., Ulutaş, E., & Taymaz, T. (2019). Source models of the 2012 Haida Gwaii (Canada) and 2015 Illapel (Chile) earthquakes and numerical simulations of related tsunamis. *Pure and Applied Geophysics.* https://doi.org/10.1007/s00024-018-1996-5. **(this issue)**.

(Received June 13, 2019, revised June 18, 2019, accepted June 25, 2019, Published online July 3, 2019)

Pure Appl. Geophys. 176 (2019), 2771–2793
© 2019 Springer Nature Switzerland AG
https://doi.org/10.1007/s00024-019-02244-x

Pure and Applied Geophysics

Twenty-Five Years of Progress in the Science of "Geological" Tsunamis Following the 1992 Nicaragua and Flores Events

EMILE A. OKAL[1]

Abstract—We review a set of 47 tsunamis of geological origin (triggered by earthquakes, landslides or volcanoes) which have occurred over the past 25 years and provided significant new insight into theoretical, experimental, field, or societal aspects of tsunami science. Among the principal developments in our command of various aspects of tsunamis, we earmark the development of the *W*-phase inversion for the low-frequency moment tensor of the parent earthquake; the abandonment of the concept of a maximum earthquake magnitude for a given subduction zone, controlled by simple plate properties; the development and implementation of computer codes simulating the interaction of tsunamis with initially dry land at beaches, thus introducing a quantitative component to realistic tsunami warning procedures; and the recent in situ investigation of current velocities, in addition to the field of surface displacements, during the interaction of tsunamis with harbors. Continued research remains warranted, notably in the field of the real time identification of "tsunami earthquakes" whose tsunamis are larger than expected from their seismic magnitudes, especially conventional ones. The recent tragedy during the 2018 Krakatau flank collapse, along a scenario which had been quantitatively forecast, also emphasizes the need for a continued effort in the education of the populations at risk.

Key words: Tsunamis, 1992 Nicaragua tsunami, 1992 Flores tsunami, 2004 Sumatra tsunami, 2011 Tohoku tsunami.

1. Introduction

It has now been slightly over a quarter century since two very different earthquakes literally re-awakened the field of observational tsunami science, which had been to some extent stagnant since the catastrophic events of the 1950s and early 1960s (Kamchatka, 1952; Aleutian, 1957, Chile, 1960; Alaska, 1964), despite significant tsunamis in 1976 in Mindanao, Philippines, and in 1983 in the Sea of Japan. While substantial progress was made in the 1970s and 1980s on the theoretical and experimental front (e.g., Hammack 1973; Ward 1980; Bernard and Milburn 1985), scientists still lacked the motivation provided by exceptional and intriguing field observations.

The two tsunamis of 02 September 1992 in Nicaragua and 12 December 1992 in Flores Island, Indonesia provided our community with a wealth of challenges which spawned a large number of new investigations covering the observational, experimental and theoretical aspects of our science. They were followed over the next 25 years by many more events (in particular by the catastrophic tsunamis of 2004 in Sumatra, and 2011 in Japan), many of which resulted in significant new scientific developments, which are the subject of this special issue.

We restrict our study to those "geological" tsunamis generated by earthquakes, landslides and volcanic eruptions. Whether their origin rests with strain accumulated during the interseismic cycle (and released during earthquakes), gravitational unbalance resulting from geological processes (landslides), or volcanic activity, their energy can be traced eventually to convective heat loss in the Earth's interior in the context of Plate Tectonics. By contrast, a separate paper (Rabinovich 2019) examines the case of "meteo-tsunamis" generated by coupling between atmospheric perturbations and the oceanic column. As such, the energy of meteo-tsunamis is drawn from the weather system, and eventually from the radiation received from the Sun.

In this paper, we present a detailed, chronological discussion of the unique properties of a selection of 47 among such geological tsunamis. Their sources

[1] Department of Earth and Planetary Sciences, Northwestern University, Evanston, IL 60208, USA. E-mail: emile@earth.northwestern.edu

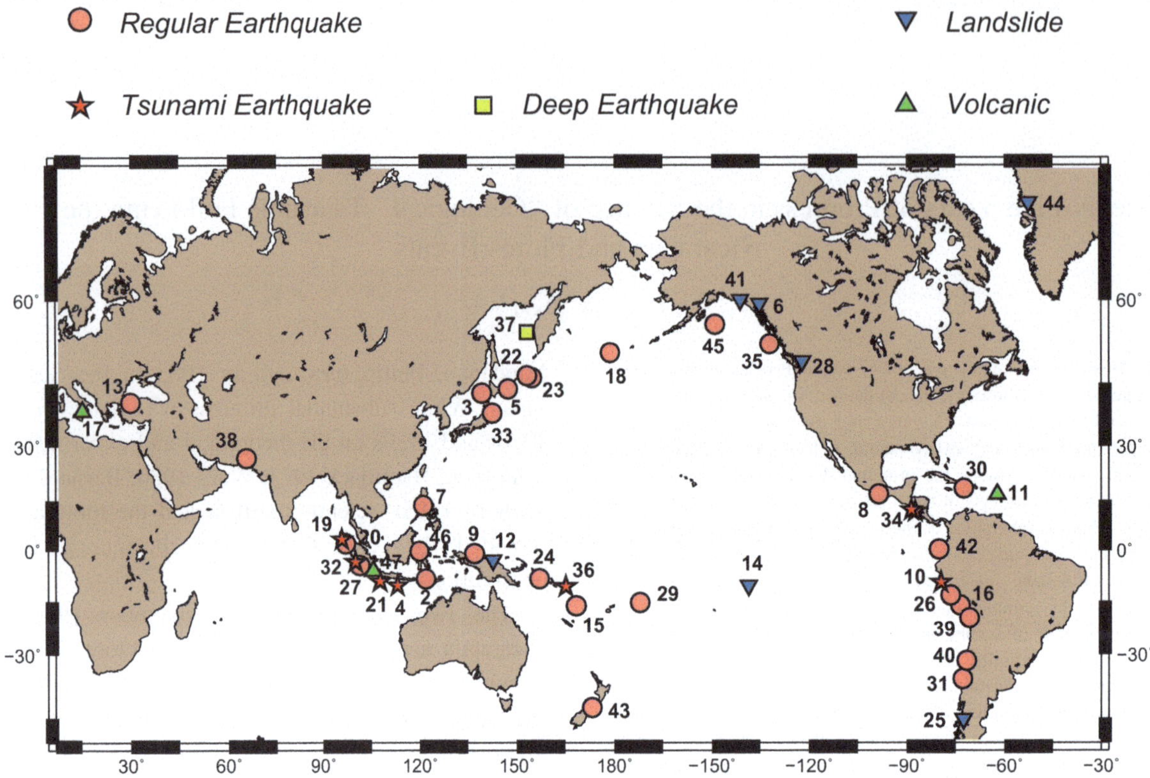

Figure 1
Map of the 47 events described in the text. Special symbols are used for tsunami earthquakes, for the deep Okhotsk earthquake (37), for tsunamis generated during volcanic events, and for those primarily due to major landslides. Event numbers are keyed to Table 1 and the text

are mapped on Fig. 1 and their fundamental parameters listed in Table 1. We emphasize that the list is not exhaustive, and represents a choice of events of significance in the development of tsunami science, out of the 320 events listed for that time window in NOAA's Global Tsunami database, 39 of which involving casualties, as reviewed in the companion paper by Gusiakov et al. (2019). Like all such choices, it may be subjective. Chronological plots of tsunami casualties and maximum run-up are given on Figs. 2 and 3. In this context, both Table 1 and Fig. 2 attempt to reflect tsunami casualties, not counting those directly attributable to the parent earthquake, which in some instances (e.g., İzmit, Event **13** or Pisco, Event **26**) were much heavier; we stress that in most cases, such a distinction remains a difficult exercise. Figure 4 shows energy-to-moment ratios (Newman and Okal 1998) for the 39 relevant earthquake sources.

2. *Significant Tsunamis, 1992–2018: A Timeline*

1. *Nicaragua, 02 September 1992*

This event was remarkable in many respects. First, it constituted the first *"tsunami earthquake"* to take place in the era of digital seismic instrumentation. We recall that the term "tsunami earthquake" was coined by Kanamori (1972) to characterize an event whose tsunami is significantly larger than expected from its magnitudes, especially conventional ones. Such events are particularly ominous as they can be felt, if at all, at a deceptively low level of shaking, while the ensuing tsunami can be catastrophic (170 fatalities in Nicaragua), thus negating the adage "The Shaking is the Warning", commonly used to sensitize populations at risk on tsunami-prone coastlines in the near field. While Kanamori (1972) was able to suggest that their large tsunamis resulted

Table 1

Significant tsunamis, 1992–2018

Number	Date D M (J) Y	Region	Seismic moment (10^{27} dyn × cm)	Seismic moment M_w	Death Toll	Maximum run-up (m)	References	Notes
1	02 SEP (246) 1992	Nicaragua	3.4	7.6	170	10	Abe et al. (1993)	Tsunami earthquake
2	12 DEC (347) 1992	Flores, Indonesia	5.1	7.7	1169	26	Tsuji et al. (1995)	Babi effect; Rangrioko landslide
3	12 JUL (193) 1993	Okushiri, Japan	4.7	7.7	198	31	Shuto and Matsutomi (1995)	"Tsuji Gulch"
4	02 JUN (153) 1994	East Java, Indonesia	5.3	7.8	238	16	Synolakis et al. (1995)	Tsunami earthquake
5	04 OCT (277) 1994	Shikotan, Kuriles	30	8.3	0	9	Yeh et al. (1995)	Slab tear event
6	03 NOV (307) 1994	Skagway, Alaska	–	–	1	9	Synolakis et al. (2002a)	Landslide tsunami; no earthquake detected
7	14 NOV (318) 1994	Mindoro, Philippines	0.51	7.1	81	7.3	Imamura et al. (1995b)	Strike-Slip event
8	09 OCT (282) 1995	Manzanillo, Mexico	11.5	8.0	1	11	Borrero et al. (1997)	Evidence of leading depression
9	17 FEB (048) 1996	Biak, Indonesia	24	8.2	110	7.7	Imamura et al. (1997)	Rare thrusting event at oblique boundary
10	21 FEB (052) 1996	Chimbote, Peru	2.2	7.5	12	5.1	Bourgeois et al. (1996)	Tsunami earthquake
11	26 DEC (361) 1997	Montserrat	–	–	0	3	Heinrich et al. (1998b)	Anticipated volcanic tsunami
12	17 JUL (198) 1998	Aitape, PNG	0.37	7.0	2200	15	Kawata et al. (1999)	Underwater landslide triggered by earthquake
13	17 AUG (229) 1999	İzmit, Turkey	2.9	7.6	155	3	Altınok et al. (2001)	Strike-slip earthquake
14	13 SEP (256) 1999	Fatu Hiva, Marquesas Is.	–	–	0	8	Okal et al. (2002a)	Subaerial landslide
15	26 NOV (330) 1999	Vanuatu	1.7	7.4	5	6.6	Caminade et al. (2000)	Successful evacuation at Baie Martelli
16	23 JUN (174) 2001	Camaná, Peru	47	8.4	86	8.8	Okal et al. (2002b)	First ionospheric detection
17	30 DEC (364) 2002	Stromboli, Italy	–	–	0	11	Tinti et al. (2005)	Volcanic Landslides
18	17 NOV (321) 2003	Aleutian Islands	5.3	7.7	0	0.26	Meining et al. (2005)	First alarm called off using DART
19	26 DEC (361) 2004	Sumatra-Andaman	1200	9.3	>200,000	51	[a]	Deadliest in history of mankind
20	28 MAR (087) 2005	Nias, Indonesia	100	8.6	8	4.2	Kerr (2005)	Deficient far-field tsunami from source in shallow waters
21	17 JUL (198) 2006	Central Java, Indonesia	4.6	7.5	802	21	Fritz et al. (2007)	Tsunami earthquake; similar to 1994 (4.)
22	15 NOV (319) 2006	Central Kuriles	35	8.3	0	22	Lobkovsky et al. (2009)	First (interplate thrust) of seismic doublet
23	13 JAN (013) 2007	Central Kuriles	18	8.1	0	16	Lobkovsky et al. (2009)	Second (normal, outer-rise) of seismic doublet
24	01 APR (091) 2007	Gizo, Solomon Is.	16	8.1	52	12.1	Fritz and Kalligeris (2008)	Rupture across triple junction
25	21 APR (111) 2007	Aysén, Chile	0.03	6.3	10	50	Barrientos et al. (2009)	Rockslide into fjord triggered by s.-slip earthquake
26	15 AUG (227) 2007	Pisco, Peru	11	8.0	3	10	Fritz et al. (2008)	Largely successful self-evacuation
27	12 SEP (255) 2007	Bengkulu, Indonesia	67	8.5	0	4	Borrero et al. (2009)	Only partial release of post-1833 strain
28	04 DEC (338) 2007	Chebalis Lake, B.C.	–	–	0	38	Wang et al. (2015)	Rock slide into closed lake
29	29 SEP (272) 2009	Samoa	17	8.1	192	22	Okal et al. (2010)	Composite seismic source
30	12 JAN (012) 2010	Haiti	0.44	7.0	7	3	Fritz et al. (2012)	Underwater landslides triggered by earthquake
31	27 FEB (058) 2010	Maule, Chile	187	8.8	156	29	Fritz et al. (2011b)	First transpacific tsunami in 46 years
32	25 OCT (298) 2010	Mentawai, Indonesia	6.8	7.8	431	17	Hill et al. (2012)	Tsunami earthquake

Table 1 *continued*

Number	Date	Region	Seismic moment		Death Toll	Maximum run-up	References	Notes
	D M (J) Y		(10^{27} dyn × cm)	M_w		(m)		
33	11 MAR (070) 2011	Tohoku, Japan	395	9.0	19,000	39	[a]	2nd deadliest in past 100 years; Fukushima disaster
34	27 AUG (240) 2012	El Salvador	1.3	7.3	0	6.3	Borrero et al. (2014)	Tsunami earthquake
35	28 OCT (302) 2012	Haida Gwaii, Canada	5.7	7.8	0	13	Fine et al. (2015)	Partitioned plate boundary
36	06 FEB (037) 2013	Santa Cruz, Solomon Is.	9	7.9	10	11	Fritz et al. (2013)	Tsunami earthquake
37	24 MAY (144) 2013	Sea of Okhotsk	39.5	8.3	0		Okal (2017)	Tsunami generated by deep earthquake
38	24 SEP (267) 2013	Gwadar, Pakistan	5.6	7.8	0	1	Heidarzadeh and Satake (2014)	Tsunami from onland earthquake
39	01 APR (091) 2014	Iquique, Chile	19	8.1	0	4.6	Catalán et al. (2015)	Major, successful evacuation
40	16 SEP (259) 2015	Illapel, Chile	32	8.3	8	13	Aránguiz et al. (2016)	In situ measurements of far-field currents
41	18 OCT (290) 2015	Taan Fjord, Alaska	–	–	0	193	Higman et al. (2018)	Rockslide into fjord
42	16 APR (107) 2016	Ecuador	5.9	7.8	0	0.1	Ye et al. (2016)	Large event with mediocre tsunami
43	13 NOV (318) 2016	Kaikōura, NewZealand	6.5	7.8	0	6.3	Power et al. (2017)	Unexpected event with complex rupture
44	17 JUN (168) 2017	Karrat Fjord, Greenland	–	–	4	50	Paris et al. (2019)	Rockslide into fjord
45	23 JAN (023) 2018	Gulf of Alaska	10	7.9	0	0.25	Lay et al. (2018)	Large strike-slip earthquake with benign tsunami
46	28 SEP (271) 2018	Palu, Indonesia	2.8	7.6	2200	11	Heidarzadeh et al. (2019)	Strike-slip event at bayhead
47	22 DEC (356) 2018	Anak Krakatau, Indonesia	–	–	437	30	Prasetya et al. (2019)	Volcanic flank collapse

[a]Being spoiled for choice, we give no reference for the 2004 Sumatra-Andaman and 2011 Tohoku events

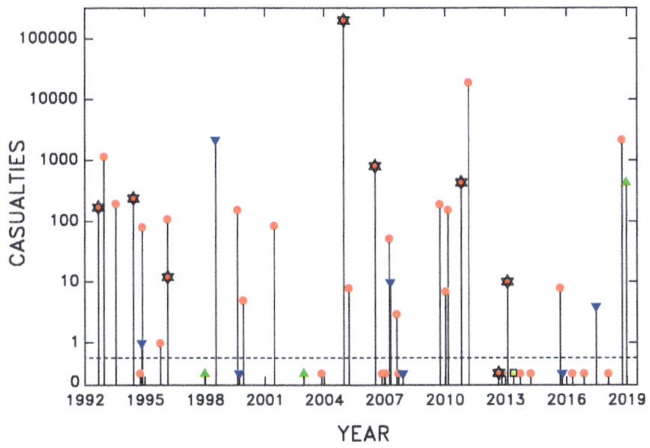

Figure 2
History of the 47 events (1992–2018). The vertical bars are scaled to the logarithm of the death toll, with events without known casualties plotted below the dashed line. Symbols are keyed to Fig. 1

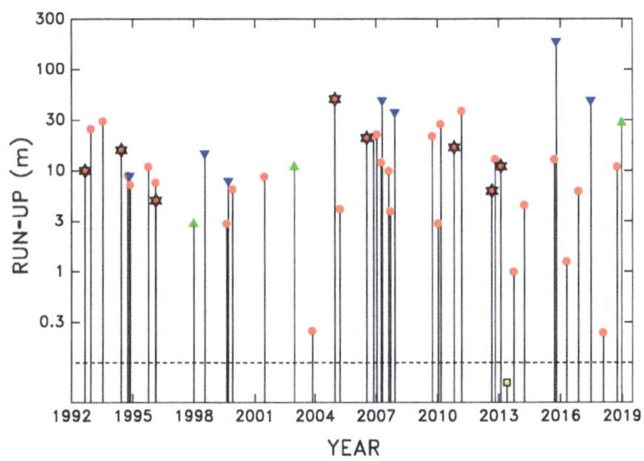

Figure 3
History of the 47 events (1992–2018). The vertical bars are scaled to the logarithm of the maximum observed run-up. The deep Okhotsk event (Number 37), for which no data is available, is plotted below the dashed line. Symbols are keyed to Fig. 1

from an anomalously slow seismic rupture, the charter tsunami earthquakes in his study, the 1896 Sanriku and 1946 Aleutian shocks, obviously predated digital seismology. It was at the time extremely challenging if not outright impossible to quantify the high-frequency part of the seismic spectrum from analog records. Following the deployment of broadband digital instruments (Hutt et al. 2002), Boatwright and Choy (1986) made this possible by developing an algorithm to compute radiated seismic energy, and the 1992 Nicaragua tsunami earthquake (together with similar ones in 1994 in Java and Chimbote, Peru in 1996) motivated Newman and Okal (1998) to introduce the source slowness

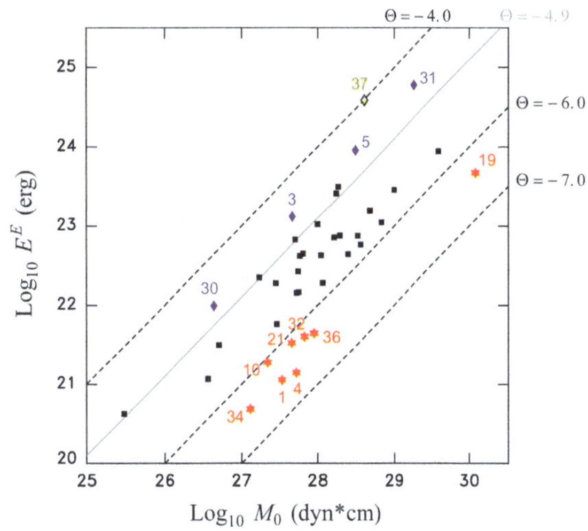

Figure 4
Energy-to-moment ratios for the 39 earthquake sources among the events studied (1992–2018). This plot is analogous to those published, e.g., by Newman and Okal (1998). The dashed lines identify constant values of Θ, with the grey solid line corresponding to the theoretical value (−4.90) expected from scaling laws. Tsunami earthquakes (featuring $\Theta \leq -5.8$) are shown as red stars and keyed to their number in Table 1. Events defined as "snappy" (Okal and Saloor 2017) ($\Theta \geq -4.7$) are similarly shown as purple diamonds, with the deep Sea of Okhotsk earthquake shown in yellow

parameter $\Theta = \log_{10}(E^E/M_0)$ where E^E is the estimated radiated seismic energy and M_0 the seismic moment, as a robust identifier of tsunami earthquakes.

Second, the availability of high-quality broadband data allowed Kanamori (1993) to discover, on records of the Nicaraguan earthquake, an ultra-long period, fast-propagating seismic wavetrain, which he named the *W* phase, with the potential for a rapid estimate of the long-period seismic moment controlling the generation of tsunamis. Despite some initial interest (Okal 1993), the concept had to wait for the full development of digital networks to allow its now routine use for rapid, full-fledge moment tensor inversion (Kanamori and Rivera 2008).

Next, the 1992 Nicaragua tsunami saw the first of comprehensive International Post-Tsunami Surveys (Abe et al. 1993), which were to become systematic following later significant events (Synolakis and Okal 2005), even though similar efforts had occasionally taken place, notably after the 1976 Mindanao tsunami (Wallace et al. 1977). The substantial values of run-up documented by the 1992 survey (8–10 m) could not be modeled using then standard simulation algorithms, which stopped the simulation at a shallow, but arbitrary, water depth (typically 5–10 m), and considered the coastline as a fully reflecting boundary (Imamura et al. 1993), thus underestimating the surveyed values by as much as one order of magnitude. This motivated the development of codes extending the computation over initially dry land (Titov and Synolakis 1993), which later matured into codes such as, e.g., MOST (Titov and Synolakis 1998; Titov et al. 2016), TUNAMI–2 (Goto et al. 1997), or COMCOT (Liu et al. 1994).

Finally, the 1992 Nicaragua event provided the first opportunity for space-based detection of a tsunami on the high seas, using an 8-cm signal on ERS-1 and TOPEX/POSEIDON radar altimeters (Okal et al. 1999).

2. *Flores, 12 December 1992*

This event took place on the Northern shore of Flores Island, Indonesia, where the Sunda plate subducts Southwards under the Australian plate. The tsunami was remarkable on two accounts. First, it gave rise to the intriguing "Babi effect": at Babi Island, located a few km North of the main island, the maximum run-up (7.2 m) occurred in the lee of the incident tsunami, as the waves interfered constructively after rounding this circular island (Yeh et al. 1993). This effect was later reproduced in the laboratory (Briggs et al. 1995) and modeled theoretically (Yeh et al. 1994; Liu et al. 1995; Imamura et al. 1995a; Tinti and Vannini 1995). It negated the conventional wisdom suggesting that island structures could protect their residents on the lee side of an incoming tsunami wave.

In addition, the 1992 Flores tsunami featured an isolated extreme run-up of 26 m at Rangrioko in Northeastern Flores (Tsuji et al. 1995), which underwater surveys later documented was caused by a submarine landslide triggered by the earthquake (Plafker 1997). This phenomenon contributed to raising awareness for this particular kind of hazard.

3. *Okushiri, 12 July 1993*

This event took place in the Sea of Japan and resulted in the destruction of the port village of Aonae, with a run-up of 10 m. A detailed field survey (Shuto and Matsutomi 1995) documented a run-up reaching 31 m due to amplification by a river valley at the so-called "Tsuji Gulch". Later simulation by Titov and Synolakis (1997) established the possibility of successfully modeling the highly non-linear interaction of tsunamis with originally dry land featuring extreme relief.

4. *East Java, 02 June 1994*

This was the second tsunami earthquake of the 1990s, running up as much as 16 m on beaches where the earthquake had not been felt, and resulting in 238 deaths (Synolakis et al. 1995).

5. *Shikotan, Kuriles, 04 October 1994*

This large earthquake (at the time the second largest in the CMT catalog) represented a tear in the slab below the Southern Kuriles. It resulted in significant local destruction, but the tsunami inflicted no casualties despite run-up of 5–9 m, as surveyed by Yeh et al. (1995). It nevertheless emphasized the need to include events other than interplate thrusts in the assessment of tsunami hazard at subduction zones, as already evidenced by such disasters as the 1933 Sanriku tsunami (Okal et al. 2016).

6. *Skagway, Alaska, 03 November 1994*

This local tsunami in the port of Skagway destroyed a wharf in a shipping terminal and killed one worker, in the absence of any detectable earthquake (Kulikov et al. 1996). Thomson et al. (2001) suggested that the failure of the wharf was responsible for the tsunami; however, the identification of an underwater slide down the fjord led Synolakis et al. (2002a) to a successful simulation of the failure of the wharf under a system of waves generated by that natural landslide.

7. *Mindoro, Philippines, 14 November 1994*

This was a very significant tsunami, which ran up as high as 7.3 m, killed 81 persons, and displaced a 4000-ton barge 1.6 km in a local estuary (Imamura et al. 1995b). Yet, the parent earthquake had a strike-slip mechanism, generally perceived as posing little if any tsunami risk. However, Tanioka and Satake (1996) showed that the lateral offset produced by a strike-slip fault at the edge of the island can generate a tsunami in the same fashion as a wavemaker paddle in the laboratory. The reverse situation (a marine strike-slip fault hitting the head of a bay) is thought to be responsible for the catastrophic 2018 tsunami at Palu, Indonesia (see **46.** below), even though ancillary landslides certainly played an additional role in its generation (Satake, pers. comm., 2019).

8. *Manzanillo, Mexico, 09 October 1995*

This was a typical tsunami generated by a major interplate thrust earthquake at the Mexican subduction zone. As part of the field survey, Borrero et al. (1997) obtained photographs from witnesses documenting a leading recession at Tencatita Bay. Although this phenomenon had been predicted theoretically by Tadepalli and Synolakis (1994, 1996), this observation verified that, in the most widely expected geometry of a shallow angle interplate thrust fault, the local beach will benefit from a natural warning to the population at risk, in the form of an initial down-draw; this may help explain the relatively low death toll, reported between 1 and 10.

9. *Biak, Indonesia, 17 February 1996*

This tsunami took place at a subduction zone featuring very oblique convergence, resulting in plate motion partitioning, and in a scarcity of megathrust earthquakes at the subduction interface, a situation to be repeated during the 2012 Haida Gwaii earthquake (see **35.** below). A remarkable aspect of the field survey is the estimation, for the first time, and based on an application of Bernoulli's theorem, of current velocities (Matsutomi and Iizuka 1998; Matsutomi et al. 2001).

10. *Chimbote, Peru, 21 February 1996*

Just 4 days after the Biak event, a new significant tsunami took place in Northern Peru, following the third "tsunami earthquake" of the decade. Investigations into the anomalous slowness of the earthquake (Ihmlé et al. 1998) led to successful tsunami simulation using for the first time a reduced rigidity μ for the seismic source material (Heinrich et al. 1998a).

11. *Montserrat, Lesser Antilles, 26 December 1997*

This tsunami resulted from the penetration of the ocean by a pyroclastic flow during the eruption of Montserrat. It ran up 1 m at Deshaies on the nearby island of Guadeloupe, in a geometry comparable to that of the 2018 Krakatau disaster (see **47.** below), but on a much reduced scale. Remarkably, this scenario had been accurately anticipated[1] by Heinrich et al. (1998b). It was essentially repeated at the same location on 13 July 2003 (Pelinovsky et al. 2004).

12. *Aitape, Papua New Guinea, 17 July 1998*

With about 2100 casualties, this tsunami was one of the deadliest of the past century. Yet, it followed a moderate earthquake whose moment ($M_0 = 3.7 \times 10^{26}$ dyn $\times$ cm) could not justify run-up reaching 15 m on a narrow stretch of coastline centered on Sissano Lagoon (Kawata et al. 1999). The combination of extreme amplitudes, limited lateral extent of the devastation, minimal amplitudes in the far field, and an apparent delay in the arrival of the wave, led Synolakis et al. (2002b) to propose as the origin of the tsunami a submarine landslide triggered 13 minutes after the seismic shock. The landslide was later identified during a survey involving seismic reflection and submarine ROV operations (Sweet and Silver 2003), and documented through its T phases throughout the Pacific Basin (Okal 2003).

The high death toll of the Papua New Guinea tsunami raised awareness among scientists and Civil Defense officials worldwide for hazards from potential landslides, notably in areas where offshore seismicity would have otherwise minimized tsunami risk, such as the Los Angeles Basin (Borrero et al. 2003).

13. *Kocaeli (İzmit), Turkey, 17 August 1999*

This tsunami added significant damage in the Eastern part of the Sea of Marmara to a catastrophic earthquake with a total death toll of 17,000. Its source was a combination of a strike-slip mechanism at the head of İzmit Bay (in retrospect comparable to the 2018 Palu geometry (see **46.** below)), vertical motion at pull-apart basins, and possible landslides in the Sea of Marmara, with inundation aggravated by local

---

[1] We refrain from using the word *"predicted"*, which would imply a precise date for the forecast. Note that the received date for the original manuscript (05 November 1997) is indeed anterior to the event.

subsidence, notably at Gölçuk (Altınok et al. 2001; Tinti et al. 2006).

This event is particularly significant, as it falls in the pattern of inexorable, if somewhat slow and certainly irregular, westward stress transfer along the Northern Anatolian Fault, initiated in Erzincan in 1939 (Stein et al. 1997). As such, it casts the ominous augury of the next event, expected to occur off İstanbul, a megalopolis with a population of 15 million.

14. *Fatu Hiva, Marquesas, 13 September 1999*

In the absence of any detectable seismic tremor, this tsunami was caused by the spontaneous collapse, into the sea, of a 4 million m^3 rock slide, part of the flank of an eroding 1.3 m.y. old volcano, following a 4-month drought throughout the Pacific, which acted to desiccate clays resulting from alteration of volcanic dykes (Okal et al. 2002a). It destroyed coastal infrastructure, including the wall of the school yard of the village of Omoa, and flooded the school where miraculously, 85 children escaped unharmed.

In a rare example of proactive development, the village school was relocated out of harm's way a few years later, about 1 km up the valley.

15. *Vanuatu, 26 November 1999*

The tsunami followed a moderate earthquake (Pelletier et al. 2000), and is noteworthy because of the successful evacuation of the village of Baie Martelli on the Southern coast of Pentecost Island (Caminade et al. 2000). The villagers had been sensitized to tsunami danger through a video based on the 1998 Fatu Hiva disaster, stressing the need to immediately self-evacuate low-lying areas upon feeling strong earthquake tremors, especially if accompanied by a recess of the sea. When the earthquake struck in the middle of the night, and a down-draw was reported, the village chief ordered an immediate full evacuation. The tsunami completely destroyed the village, but, out of the 300 residents, only three lost their lives: two elderly invalids who could not be evacuated and a drunken man, who refused to.

The lesson, if need be, from this event is simply: "Education works!"

16. *Camaná, Peru, 23 June 2001*

The earthquake was, at the time, the largest recorded worldwide in 36 years, and it generated a tsunami damaging 400 km of coastline, and killing 86 people (Okal et al. 2002b). A most remarkable aspect of the field survey was the concentration of the casualties on the river delta at Camaná, where farmers from the hinterland, poorly sensitized to tsunami hazards, were at work, while residents of fishing communities with a strong ancestral tradition had largely self-evacuated upon noticing the initial down-draw (Dengler et al. 2003). A much greater disaster was avoided by the occurrence of the tsunami on a beach deserted during the Winter season, which would have been packed with tourists during the Summer.

The Camaná event is also remarkable in that it provided the first detection of a tsunami through the oscillation of the ionosphere due to the continuation of its eigenfunction into the atmosphere (Artru et al. 2005), an idea originally suggested by Peltier and Hines (1976).

17. *Stromboli, Italy, 30 December 2002*

This volcanic tsunami was triggered by landslides, one submarine, the other aerial falling into the sea, at La Sciara on the flank of Stromboli (Tinti et al. 2005). It caused major damage in the nearby villages of Piscità and Ficogrande, but again a major disaster was avoided thanks to its occurrence during a season of low tourist activity.

A remarkable aspect of this event was the recording of the actual tsunami waves on horizontal seismometers located 20 km away on the island of Panarea (La Rocca et al. 2004), to our knowledge the first modern such instance, following Angenheister's (1920) observation in the far field after the Kuriles event of 07 September 1918.

18. *Aleutian Islands, 17 November 2003*

While this tsunami caused no damage or casualties, and had a maximum reported run-up of only 26 cm at Shemya, it is noteworthy because it constituted the first use of DART sensor records as part of a real-time accurate forecast of tsunami amplitudes inside a distant harbor, in this case 20 cm at Hilo, Hawaii

(Titov et al. 2005a). The alert in the Hawaiian Islands was canceled, thus preventing an unnecessary evacuation, and saving an estimated 68 million USD (Meining et al. 2005).

19. *Sumatra-Andaman, 26 December 2004*

With more than 200,000 casualties, this tsunami, the first in recorded history to have exported death and destruction across the Indian Ocean Basin, was probably the most lethal one in the history of mankind, considering the exponential growth of coastal populations in the past few centuries. The tsunami was the first in the era of modern technology to be detected worldwide (Rabinovich and Thomson 2007) and to be simulated on a full spherical model of the Earth's oceans (Titov et al. 2005b). It transformed the word "tsunami", used at the time only by a small community of specialists, into a worldwide household term.

It cannot be our purpose here to list every scientific result obtained in its wake, but rather we pinpoint fundamental observations which had a seminal effect on tsunami science, technology or societal response in the following years.

- From the operational standpoint, the failure to provide adequate warning to distant populations was traced to inexistent means of communication and unpreparedness in the coutries at risk, rather than to a systematic deficiency in scientific analysis (Marris 2005); at the time, the existing warning centers, in particular PTWC and ATWC, had no responsibility for the Indian Ocean, and hence no way to transmit a warning following their detection of the earthquake. The 15 years elapsed since the Sumatra disaster have seen a considerable investment into the worldwide development of a number of tsunami warning centers, notably in the Indian Ocean and in particular in Australia, Indonesia, and Oman. Only the trial of time will evaluate their performance for far-field warning.
- Perhaps the most important scientific result in the wake of the 2004 tsunami was the development of a new CMT inversion algorithm based on the *W* phase (Kanamori and Rivera 2008), following its observation at a record amplitude during the 2004 earthquake (Lockwood and Kanamori 2006). The algorithm has now been implemented worldwide at warning centers (Hayes et al. 2009), and has become the new de facto standard in long-period moment-tensor inversion.
- Another crucial development has been the abandonment of the paradigm linking plate age and convergence rate to a proposed maximum earthquake magnitude at subduction zones, originally proposed by Ruff and Kanamori (1980). Under this model, the Sumatra trench should not have hosted events bigger than $M_w \approx 8.2$, below the threshold for transoceanic tsunami hazard. Noting that a similar problem plagued the Cascadia subduction zone, where evidence for the 1700 megathrust event (Satake et al. 1996) came posterior to Ruff and Kanamori's (1980) study, Stein and Okal (2007) conducted a critical analysis of an updated dataset of tectonic and earthquake parameters, finding a significantly degraded correlation to support the paradigm; they also noted that Ruff's (1989) later suggestion to involve trench sediment thickness fared no better. In this context, a precautionary attitude is to consider that any sufficiently long subduction zone may be the site of a mega-event in the future, a conclusion shared by McCaffrey (2007).
- A large number of observations established the slowness of the seismic rupture of the 2004 event. They ranged from the growth of seismic moment with period (Stein and Okal 2005), seismic source tomography (e.g., Ishii et al. 2005), and the duration of high-frequency *P* waves (Ni et al. 2005) and hydroacoustic *T* waves (Guilbert et al. 2005), in addition to a clearly deficient Θ parameter (Stein and Okal 2007). In this context, and following Choy and Boatwright (2007), we classify the event as a "tsunami earthquake" on Figs. 1, 2, 3 and 4.

  In view of the slow precursor to the 1960 Chilean earthquake identified by Kanamori and Cipar (1974), and of the reassessment of the low-frequency moment of the 1964 Alaskan event (Nettles et al. 2005), this slowness in the source of the 2004 earthquake raises the legitimate question of whether all "super-mega" earthquakes ($M_0 > 10^{30}$ dyn $\times$ cm) feature a similar behavior, which would have significant implications for tsunami hazard in their aftermath (Okal 2013); the extreme scarcity of such events (a grand total of 3 have been analyzed) leaves the jury out at this time.

- The 2004 Sumatra earthquake was so large that its tsunami was recorded by a number of instruments which were not designed for it, generally as the result of coupling between very different media, such as the sea, the solid Earth and the atmosphere. This coupling would be expected to be very small, but in the presence of a monumental event, can yield tangible observations which were not only documented, but also often quantitatively forward- or inverse-modeled, occasionally all the way to an appropriate estimate of the moment of the earthquake. This was the case, for example of hydrophones of the IMS (Hanson and Bowman 2005; Okal et al. 2007), radar altimeters (Smith et al. 2005) leading to an explanation of the intriguing case of "tsunami shadows" (Godin et al. 2009), of the detection of the tsunami in the ionosphere from perturbation to GPS signals (Liu et al. 2006; Occhipinti et al. 2006), and systematically by horizontal seismometers on oceanic islands (Yuan et al. 2005) in a pattern that Okal (2007) showed could be interpreted as mirroring an Ocean-Bottom Seismometer record. Most such techniques were later applied regularly to smaller tsunamis.
  Finally, the tsunami was recorded by seismometers temporarily deployed on large icebergs floating off the coast of Antarctica, providing what remains to this day the only in situ recording of the horizontal motion of the ocean surface during the passage of a tsunami (Okal and MacAyeal 2006).
- Surveying in the near field benefited from technological advances such as the scientific decoding of video recordings to reconstruct a quantitative estimate of tsunami flow velocities (Fritz et al. 2006), which in particular documented their increase with time within a wavetrain, leaving essentially no chance for the population to escape.
- An unexpected observation compiled at several locations during the numerous field surveys covering the entire Indian Ocean Basin, was the occasional development of strong currents in distant harbors several hours after the arrival of the initial wave, and often after the all clear had been given by local authorities, e.g., in Toamasina, Madagascar (Okal et al. 2006a), Le Port, Réunion (Okal et al. 2006b) and Salallah, Oman (Okal et al. 2006c). This effect, which can lead to hazardous situations when major vessels break their moorings in ports, is explained as the setting up of harbor resonance by higher-frequency components of the tsunami traveling outside the shallow-water approximation. Since 2004, similar effects have been observed during other events. In this context, harbor communities must be educated to this form of hazard, and evacuation procedures, especially the issuance of all clear, adapted accordingly.

20. *Nias, Indonesia, 28 March 2005*

This "second" Sumatra event, remarkably anticipated a few days earlier by McCloskey et al. (2005) in the context of Coulomb stress transfer, was indeed gigantic ($M_0 = 1.05 \times 10^{29}$ dyn $\times$ cm), and took 600 lives in Nias and 100 in Simeulue. However, its tsunami was modest, running up to only 4.2 m, apparently without fatalities in the near field. It was also negligible in the far field, despite 10 casualties in traffic accidents due to panic generated by night-time evacuation (Okal et al. 2006a; Borrero et al. 2011). This intriguing situation is explained by a significant fraction of the deformation in the source area taking place either under large islands ("the earthquake moved a large amount of rock and not much water"), or at shallow depths, the tsunami then faltering under Green's (1838) law when reaching deep basins (Synolakis and Arcas, quoted by Kerr 2005).

In addition, the population of Simeulue was sensitized to tsunami hazard through the *Smong* tradition, possibly derived from the catastrophic "tsunami earthquake" of 1907 (Martin et al. 2019), and self-evacuated during both the 2004 and 2005 tsunamis (McAdoo et al. 2006).

21. *Central Java, 17 July 2006*

This "tsunami earthquake" was essentially a repeat of the 1994 event (see **4.** above), 580 km to the East. As such, it reopens the question of any possibly regional character to tsunami earthquakes. In an early study, Okal and Newman (2001) had found meager evidence for it in the Peruvian system. The 2006 event in Java shows a stronger case, supported also by the 2012 El Salvador and 2013 Santa Cruz events (see **34.** and **36.** below).

22. *Central Kuriles, 15 November 2006*
23. *Central Kuriles, 13 January 2007*

We discuss those two events together, since they occurred as a seismic doublet, in Winter, in a remote, unpopulated part of the island chain, and it took 9 months before a field survey could be arranged (MacInnes et al. 2009), which thus could not assign *a priori* the field evidence (up to 22 m of run-up) to one or the other. Tsunami records in the far field (at a minimum distance of 600 km) were obtained at tidal gauges across the Pacific and on DART sensors (Rabinovich et al. 2008), showing a smaller tsunami in 2007 at all locations.

The first event was a classical interplate thrust, followed 2 months later by a slightly smaller normal faulting event on the outer rise, probably triggered by stress transfer (Ammon 2008). As expected under such a scenario, source tomography studies (e.g., Lay et al. 2009) revealed that the smaller 2007 event featured larger slip distributed over a smaller surface than the 2006 mainshock. Using these source models, Rabinovich et al. (2008) carried out simulations which confirmed far-field observations, but suggested that the 2007 event could have featured larger amplitudes in deep water along the Central Kuriles than the 2006 one (on the order of 7 m as opposed to 5). Based on these results, and on full run-up calculations at a number of surveyed sites, MacInnes (2010) assigned some of her maximum run-up amplitudes to the 2007 event.

These two tsunamis emphasized the fundamental difference between near-field amplitudes, directly related to maximum slip on the fault, and far-field ones, controlled by the full integral of its distribution, i.e., by the seismic moment.

Perhaps because of the lack of local information, the warning centers for the Pacific Basin underestimated the 2006 event and canceled initial warnings, even though the tsunami was to cause one injury and damage in Hawaii, and especially in Crescent City, California, to the extent of 9.8 million USD (Dengler et al. 2009). The 2006 tsunami served as a particularly acute reminder of the vulnerability of that community to a wide range of tsunami sources in the Pacific, and resulted in the implementation of a more detailed, quantified set of warning criteria at the US Tsunami Warning Centers (Whitmore et al. 2008).

24. *Gizo, Solomon Islands, 01 April 2007*

This earthquake was remarkable in that it ruptured along the Solomon Islands trench, across the triple junction separating the Australian plate from the Woodlark mini-plate (Taylor et al. 2008), an important observation which shows that plate boundaries cannot be considered as barriers limiting the size of potentially tsunamigenic earthquakes in subduction zones. A similar scenario was proposed for the 1932 Manzanillo, Mexico earthquake by Okal and Borrero (2011).

The death toll (52) was contained despite widespread flooding and destruction, thanks to systematic self-evacuation in this region characterized by a high level of seismic and volcanic activity (Fritz and Kalligeris 2008). Very similar conclusions were reached after the tsunami of 03 January 2010, only 200 km to the ESE, which involved no casualties (Newman et al. 2011a).

25. *Aysén, Chile, 21 April 2007*

This tsunami resulted from a massive rock slide into Aysén Fjord (Barrientos et al. 2009), triggered by an earthquake on the nearby Liquiñe-Ofqui fault system (LOFS), which marked the culmination of an intense seismic swarm lasting 3 months (Legrand et al. 2011). The remoteness of the area helped contain the death toll, but splashes on Mentirosa Island reached ~ 50 m, in a scenario reminiscent of the 1958 Lituya Bay tsunami (Miller 1960), albeit on a smaller scale. The exact role of the LOFS in the local plate tectonic framework should be further investigated in the context of local tsunami hazard, especially since Kanamori et al. (2019) have recently suggested that it may have contributed a significant component to the great 1960 Chilean earthquake.

26. *Pisco, Peru, 15 August 2007*

Following heavy structural damage and casualties inflicted by the earthquake on this Central Peruvian city, the tsunami locally reached run-up heights of 10 m, but "sergeants" acting in a sense as lifeguards on beaches conducted an evacuation that helped contain the death toll from the tsunami to three casualties in the town of Lagunilla, which inexplicably was not part of the program (Fritz et al. 2008).

It is noteworthy that this event had been anticipated, albeit on a greater scale, by Okal et al. (2006d).

27. *Bengkulu, Indonesia, 12 September 2007*

This very large event produced a relatively contained tsunami, with run-up not exceeding 4 m (Borrero et al. 2009), and despite significant tsunami damage, no deaths were reported, thanks to successful self-evacuation.

The event is noteworthy, since it took place in a region highlighted for tectonic strain accumulated since the 1833 mega-earthquake, as well as under stress transfer in the wake of the 2004 Sumatra and 2005 Nias earthquakes (Nalbant et al. 2005; Pollitz et al. 2006). However, the 2007 Bengkulu earthquake failed to release all the slip accumulated since 1833, and thus the area remains a candidate for a significant tsunami in the near future (Okal et al. 2009).

28. *Chebalis Lake, B.C., 04 December 2007*

This event resulted from the fall of a 3-million m^3 rockslide into Lake Chebalis in Southwestern British Columbia. Fortunately, the local campsites were closed for the Winter, and no lives were lost, but a wave splashed 38 m onto the opposite bank of the 1-km wide lake. The event was successfully modeled by Wang et al. (2015).

29. *Samoa, 29 September 2009*

This large event at the Northern corner of the Tonga arc actually had a composite source, starting as a normal faulting shock, followed ~ 100 s later by a more regular interplate thrust one (Li et al. 2009; Lay et al. 2010). It raised a powerful tsunami, running up to 22 m on Tafahi, Tonga, and bringing death and destruction to American Samoa, Samoa, and Tonga, with a total of 192 casualties, mostly in [independent] Samoa.

Despite the complex mechanism, tsunami modeling based on the inversion of DART data assuming a regular shallow angle thrust provided acceptable fits to surveyed amplitudes, thus demonstrating the robustness of the approach (Okal et al. 2010). The field survey revealed a generally successful evacuation on American Samoa, reflecting a well-prepared community, as opposed to [independent] Samoa, where in particular the absence of signage may have contributed to a higher death toll (Fritz et al. 2011a).

30. *Haiti, 13 January 2010*

This is another example of a catastrophic strike-slip earthquake generating a minor tsunami, constrained to a few locations in Gonâve Bay and along the Southern coast (Fritz et al. 2012), with a run-up not exceeding 3 m, and a death toll of 7 (as opposed to perhaps 300,000 from the earthquake). At both locations, the tsunamis were due to underwater landslides triggered by the earthquake; note in particular that in this case, the Enriquillo strike-slip fault does not intersect the shoreline, preventing a "Mindoro"-type generation, which could have been much more damaging.

31. *Chile, 27 February 2010*

This megathrust event generated the first tsunami in 46 years with transpacific runup in excess of 1 m, in the Kuril Islands and at certain sites in California. In the near field, it caused considerable damage, with run-up reaching 29 m at Constitución, and 156 fatalities, many of them trapped on a campground on Orrego Island in the mouth of the Maule River (Fritz et al. 2011b). The real-time handling of the warning by Chilean authorities was generally regarded at best as deficient, at worst as inept, and in particular no warning was issued for Juan Fernández Island, where the loss of 18 lives could have been prevented, when the tsunami attacked the island 49 minutes after the earthquake. This led to a successful major restructuring of the Chilean Tsunami Warning Center (see **39.** below).

The 2010 Maule tsunami was also the first one to be detected as a perturbation of the geomagnetic field of the Earth (Manoj et al. 2011), following Tyler's (2005) model of induction currents caused by the displacement of conducting seawater in the Earth's field.

32. *Mentawai, Indonesia, 25 October 2010*

This "tsunami earthquake" occurred up-dip of the 2007 Bengkulu event (see **27.** above), in the typical configuration of an "Aftershock Tsunami Earthquake" (Okal and Saloor 2017). The high death toll (431) resulted from the combination of deceptively low shaking and of the occurrence, only a year before, of the intermediate-depth, intraplate Padang earthquake (Hill et al. 2012). The 2009 event had created a negligible tsunami (maximum run-up 27 cm), but had been strongly felt in the Mentawai Islands, instilling in the residents the false sense that a dangerous tsunami would require even greater shaking.

Incidentally, the anomalously slow nature of the earthquake had been documented 17 minutes after origin time (and thus before the tsunami could reach the Mentawai Islands), on the basis of the comparison of duration and energy of teleseismic *P* waves (Newman et al. 2011b), but there existed no way to transmit this information to Indonesia, let alone to the people at risk, many of whom were in communities lacking electrical power, where the local warning from the Indonesian Tsunami Warning Center was apparently not received.

33. *Tohoku, Japan, 11 March 2011*

The catastrophic 2011 Tohoku tsunami ranks only second the 2004 Sumatra one in terms of death toll in the past 100 years. Its run-up reached 6 m in the far field, and it wrought damage as far as the Chilean coast, and even in Antarctica, where it caused calving of a large iceberg (Brunt et al. 2011). A selection of the main points of scientific and societal interest includes:

- *Violation of Scaling Laws*
  Detailed seismic tomography of the earthquake source, later confirmed by GPS studies and modeling of DART records, identified a patch of extremely large seismic slip on the fault, reaching ~ 50 m over a length of no more than 100 km (Lay et al. 2011), thus clearly violating seismic scaling laws widely used in the assessment of tsunami risk.
- *Great Success in the Far Field*
  Despite large run-up amplitudes leading to some structural damage, adequate warnings were given—and generally well heeded—in the far field, resulting to our knowledge in only two fatalities (one in Crescent City, California, and one close to Jayapura, Indonesia), which were preventable since both occurred in violation of mandatory evacuation orders. This positive outcome may have been helped by the daytime occurrence of the tsunami in Japan, which allowed immediate worldwide TV coverage of the catastrophic magnitude of the event (as opposed for example, to the night-time occurrence of the Chilean tsunami the previous year).
- *Fukushima Nuclear Accident*
  The 2011 tsunami is of course known worldwide for the nuclear meltdown at the Fukushima Dai-Ichi nuclear plant, caused by cooling water starvation following the flooding of the emergency diesel power generators by the tsunami. As of 2019, and probably for decades to come, an exclusion zone of $\sim 300\ \mathrm{km}^2$ is maintained, affecting about 50,000 residents.
  Careful analyses of the origin of the disaster by Nöggerath et al. (2011) and Synolakis and Kânoğlu (2015) have evidenced a cascade of design, scientific, engineering, and managerial blunders in the construction and operation of the plant, which suggests that the disaster was preventable, the best proof being that the Oganawa nuclear power plant, located only 127 km to the North, in a stretch of coastline which suffered considerably higher run-up, was safely shut down, and after a lengthy upgrade, is now scheduled to be re-started in the next few months.
- *Reasonable Success in Evacuation*
  As offensive as this statement may sound in the wake of a 19,000 death toll, the evacuation process in the near field can be regarded as generally successful. It has been estimated that approximately 200,000 people were present in the areas totally eradicated by the tsunami, and thus that about 90% of them saved their lives by proper evacuation (Dengler, pers. comm., 2011; Fritz, pers. comm., 2011); this figure might even have been higher, but for horrific instances of orderly evacuation to shelters located at underestimated heights. It nevertheless constitutes a tribute to the awareness, education, and responsibility of the

people of Japan, once again stresses the value of education in the mitigation of natural disasters, and stands in blunt contrast to the ineptitude underlying the whole Fukushima tragedy.

34. *El Salvador, 27 August 2012*

This tsunami went largely unnoticed as it resulted in very little damage and no fatalities, mostly because it occurred in an unpopulated segment of the Pacific coast of Central America. Yet, it ran up to 6.3 m, and most significantly, was identified as a tsunami earthquake through its deficient energy-to-moment ($\Theta = -6.0$) and energy-to-duration ratios (Borrero et al. 2014).

This is important, since it brings some support to the concept that the Central American coast, site of the 1992 Nicaragua event only 150 km away (see **1.** above), could feature a regional preference for tsunami earthquakes. A rather obscure event on 26 February 1902 towards the Guatemalan border may also have shared some comparable properties (Cruz and Wyss 1983).

35. *Haida Gwaii, British Columbia, 28 October 2012*

This earthquake featured a rare shallow thrust fault mechanism at the Pacific-North America plate boundary, characterized in the area by extreme oblique convergence, and thus slip partitioning along the Queen Charlotte Fracture Zone (Lay et al. 2013). It featured locally substantial run-up values (up to 13 m), but took place in a seasonally unpopulated area (Leonard and Bednarski 2014). We dismiss the lone casualty, from a traffic accident in Hawaii caused by a drunken driver ramming into cars parked during the evacuation, as it is not a direct victim of the tsunami.

Like the 1996 Biak earthquake (see **9.** above), this event underscores the possibility of significant tsunami hazard from rare thrust components to oblique subduction along partitioned boundaries featuring mostly strike-slip events.

36. *Santa Cruz Islands, 06 February 2013*

This event, which qualifies as a "tsunami earthquake", triggered a tsunami running up to 11 m on Nendö Island, where it killed 10 people (Fritz et al. 2013). It is noteworthy that it shares its epicenter with another tsunami earthquake, on 21 July 1934, which led Okal and Saloor (2017) to identify systematic variations of coupling correlating with the slowness parameter Θ along the Solomon-Vanuatu arc; however the 1934 event happened as an aftershock of a regular earthquake, 100 km to the South, while the 2013 event was primary.

37. *Sea of Okhotsk, 24 May 2013*

This event, the largest deep earthquake ever recorded ($M_0 = 3.95 \times 10^{28}$ dyn $\times$ cm), did generate a tsunami recorded at millimetric amplitudes by two DART sensors off the Kuril Islands (Okal 2017). While it did not cause any damage, this unique detection raises the possibility of tsunamis observable at coastlines being generated by deep earthquakes, should even larger ones occur in the future.

38. *Gwadar, Pakistan, 24 September 2013*

This rather unusual scenario involved a tsunami in the Indian Ocean triggered by an earthquake occurring onland, about 200 km from the nearest shoreline. While of moderate size, the tsunami reached 1 m on the coast of Oman, and the process was accompanied by the emergence of a mud island off Gwadar, Pakistan. Heidarzadeh and Satake (2014) analyzed tidal gauge records of the event and concluded that the tsunami could not be caused by the emergence of the mud volcano, but rather by a probable underwater slump located about 70 km off the coast, at a total distance of $\sim$ 400 km from the epicenter of the earthquake. This event serves as a reminder, if need be, of the capability of inland earthquakes to generate tsunamis through the triggering of underwater landslides; the 400-km distance is indeed large, but far from a record, since the 1910 Rukwa earthquake near Lake Tanganyika, in present day Tanzania, severed telegraphic cables in the Mozambique channel, at a distance of 900 km (Ambraseys 1991).

39. *Iquique, Chile, 01 April 2014*

This large event took place in what was defined as the Northern Chilean seismic gap (Chlieh et al. 2011), where the last major earthquake and tsunami took place in 1877. However, despite its large magnitude, it did not release the full accumulated strain, and thus seismic and hence tsunamigenic potential remains to this day (Hayes et al. 2014).

A remarkable aspect of the 2014 Chilean tsunami is that it apparently did not cause any casualties. This reflects the orderly, professional evacuation of about 1 million people from coastal areas in less than an hour, in stark contrast to the case of the Maule tsunami just 4 years earlier, and following a complete revamping of the country's Tsunami Warning Center system.

40. *Illapel, Chile, 16 September 2015*

This new Chilean tsunami took place about 1000 km South of the 2014 epicenter. It produced much higher run-up reaching 13 m, and resulted in 8 fatalities (Aránguiz et al. 2016). As in the case of the 2014 event (see **39.** above), a successful warning and evacuation took place, again involving about 1 million residents.

A remarkable experiment conducted during this tsunami was the in situ measurement of currents in the harbor of Ventura, California. Taking advantage of the ~ 14-h travel time of the tsunami from Chile to California, Kalligeris et al. (2016) were able to deploy floaters equipped with GPS receivers, and documented current velocities of up to 1.5 m/s in turbulent flow. Because hazards to navigation in harbors depend crucially on current velocities, the collection of such data opens new opportunities for research critical to their further mitigation.

41. *Taan Fjord, Icy Bay, Alaska, 18 October 2015*

This remarkable event, fully described by Higman et al. (2018), shares many similarities with the classic 1958 Lituya Bay tsunami, 260 km to the SE, which splashed to a record height of 525 m (Miller 1960): it was generated by a massive landslide into a narrow fjord serving as estuary to an ice stream, and essentially faltered (because of extremely short wavelengths) upon reaching the open ocean. However, no detectable earthquake could be identified as its trigger, while the 1958 event followed a large shock ($M_s = 7.9$). The field survey documented run-up reaching 193 m on the opposite side of the fjord. Seismic waves from the landslide were detected worldwide and inverted to suggest a slide mass of 10^{11} kg.

42. *Ecuador, 16 April 2016*

This major earthquake occurred in a section of the South American subduction system characterized by diversity and complexity in the mechanism of strain release. The tsunami was only of marginal amplitude, reaching at most 10 cm on local tide gauges, and resulting in no fatalities. By contrast, the Colombia-Ecuador region was the site a mega-event in 1906, whose tsunami was recorded Pacific-wide and killed about 5000 people in the near field. It was followed by smaller shocks in 1942, 1958 and 1979, which released strain only over distinct fragments of the presumed 1906 rupture area. Only the 1979 event generated a substantial tsunami with about 600 victims. No mention can be found of a tsunami following the 1942 earthquake.

From the standpoint of its seismic source, and as detailed by Ye et al. (2016), the 2016 event may be a replica of the one in 1942, and the deceptive amplitude of their tsunamis explained by a relatively deep rupture, failing to reach the uppermost part of the subduction contact (Heidarzadeh et al. 2017). This brings the question of the nature of coupling at that shallow interface, which, if presently locked, could be ripening for a future event in the form of a tsunami earthquake. However, no such events are known in the historical record, which incidentally is very short in Southern Colombia and Northern Ecuador, with no events listed by Solov'ev and Go (1984) prior to 1906, despite Spanish presence going back to 1531.

43. *Kaikōura, New Zealand, 13 November 2016*

This earthquake nucleated on land, but propagated along a complex series of offset faults, both on land and at sea, in the general transition zone between the Alpine Fault system and the Hikurangi Trough (Power et al. 2017). The tsunami was destructive both in the southern part of the rupture area, with a maximum run-up of 6.3 m, and farther South in the Banks Peninsula, as a result of the extreme indentation of the latter's numerous bays. No comparable historical seismicity was known in the area, and thus the tsunami was unexpected. No casualties were attributed to the tsunami.

Modeling by Heidarzadeh and Satake (2017) has suggested a combined generation by elements of the complex dislocation and possible submarine landslides.

44. *Karrat Fjord, Greenland, 17 June 2017*

This tsunami was generated by the fall of an aerial landslide involving about 50 million m^3 into a fjord on the West coast of Greenland. At its source, the tsunami ran up to a height of 50 m across the fjord (Fritz et al. 2018). It attacked the village of Nuugaatsiaq, located 32 km away, running up 10 m and killing 4 persons. While no tidal gauges were available, a seismic station was being operated at Nuugaatsiaq, and its horizontal records were deconvolved by Paris et al. (2019) to reconstruct an equivalent tsunami amplitude time series in the channel facing the village. This documented a resonant oscillation of the channel comparable to observations in the Panama Canal by McNamara et al. (2011).

45. *Gulf of Alaska, 23 January 2018*

With a moment of only 1.0×10^{28} dyn $\times$ cm, this intraplate earthquake was still the largest one recorded in 2018. It was characterized by an extremely complex rupture, involving several subsources located on offset segments of a primarily strike-slip fault structure (Lay et al. 2018), in the general vicinity of the slightly smaller events of 30 November 1987 and 06 March 1988. Like these two events, incidentally the first ones recorded by a prototype DART sensor (González et al. 1991), the 2018 earthquake generated only a mediocre tsunami, barely reaching 25 cm at Kodiak, and in the far field at Crescent City and in the Marquesas Islands.

46. *Palu, Indonesia, 28 September 2018*

With more than 2200 casualties, this tsunami stands as the most lethal one since the great 2011 Tohoku event. At the time of writing, its mechanism remains unclear. The geometry of the event advocates a "Mindoro-type" bayhead effect (see **7.** above), where the strike-slip fault hits the coast of Sulawezi (Heidarzadeh et al. 2019); however visual evidence of underwater slides, in the form of surface eddies, suggests their contribution to the genesis of the tsunami (Harig et al. 2019).

47. *Anak Krakatau, Indonesia, 22 December 2018*

This catastrophic tsunami, killing upwards of 400 persons on the coasts of Java and Sumatra, resulted from the flank collapse of Anak Krakatau, as part of an eruption cycle which had started 6 months earlier. It is remarkable that the scenario had been forecast, at a quantitatively astonishing level of accuracy, by Giachetti et al. (2012). The flank collapse was well recorded seismically, and interpreted as a 0.3-km^3 slide, originating above sea level and propagating about 3 km under water (G. Ekström, pers. comm., 2019). In the absence of records from DART sensors (which did not trigger into higher sampling mode), horizontal seismograms at island stations can be deconvolved (Okal et al. 2019) to confirm amplitudes in the range of 0.1–1.5 mm on the high seas, due to the faltering, according to Green's (1838) law, of a tsunami generated in extremely shallow waters.

Preliminary results from surveys suggest flow depths on the order of 2–5 m along the coast of Java, and run-up values of 30 m in the source area (Prasetya et al. 2019).

The 2018 tsunami differs fundamentally from the monstrous, worldwide one in 1883, which Press and Harkrider (1966) showed was due to an acoustic-gravity wave following an atmospheric explosion which they estimated at between 100 and 150 Megatons. In 2019, infrasound signals at the 10 Pa level were recorded at the Cocos Island infrasound array (1160 km away) from an explosion following the landslide by 7 minutes, but had already fallen below noise level at Diego Garcia, 3440 km away (Okal et al. 2019).

3. Discussion and Conclusion

In conclusion of this review, the main developments in the past 25 years can be summarized as follows:

- We have acquired the capacity to obtain in quasi-real time (i.e., possibly before the tsunami reaches regional beaches) quantitative information on the earthquake source spectrum, both at low frequencies through W–phase inversions (Kanamori and Rivera 2008) and at high-frequency through the energy radiated by P waves, which can be compared either to seismic moment, or to source duration (Newman et al. 2011b).

- We have developed algorithms to successfully model run-up as the non-linear interaction of tsunamis with originally dry land at the beaches, both in the near and far fields. As a result, countless scenarios have been prepared to assist Civil Defense authorities in the advanced mitigation of tsunami hazards (e.g., Barberopoulou et al. 2011), the resulting inundation maps providing a considerable improvement for the issuance of tsunami evacuation orders under operational conditions.
- In the wake of the recent tsunami activity, and most particularly of the 2004 Sumatra event, the past 15 years have seen an explosion in the amount of high-quality data available under operational conditions to Tsunami Warning Centers, including broad-band seismic data and coastal and ocean-bottom sensors. We have developed real-time algorithms, such as SIFT (Gica et al. 2008), to successfully forecast tsunami amplitudes in the far field, introducing a quantitative component to the issuance of warnings for distant communities.
- Possibly as a result of this progress, we have seen, since the 2004 Sumatra disaster, a spectacular reduction in the number of casualties in the far field (i.e., 1000 km or more from the epicenter), which amounts to a grand total of 2 in the past 14 years (not counting traffic fatalities during evacuation, which are not direct effects of the tsunami), both of which occurred in violation of mandatory evacuation. This also reflects a better education of the lay public worldwide to tsunami hazards, obviously the result of the catastrophic tsunamis of 2004 and 2011.
- Elaborate techniques during post-tsunami surveys are starting to greatly improve our understanding of currents, i.e., particle flow velocities at beaches and harbors, which are the primary factors controlling momentum transfer, and therefore structural damage during inundation.
- The examples of Montserrat and Krakatau show that accurate quantitative forecasts can be made even for complex sources such as flank collapses during episodes of volcanic activity.

However, there remain significant challenges which need to be addressed in the forthcoming years:

- Paramount among them is the question of "tsunami earthquakes", which have not been recognized effectively by warning centers in the past 15 years (e.g., Java, 2006; Mentawai, 2010; El Salvador, 2012). While an approach such as Newman et al.'s (2011b) is promising in this respect, a major challenge will be to instill its fundamental ingredient (the disparity between the duration and amplitude of shaking) in the minds of the populations at risk. In this context, it is remarkable that some survivors of the Mentawai disaster have reported to the Post-Tsunami Survey Team that the shaking had been "weak and long" (Hill et al. 2012). An emphasis on the ominous character of such a harbinger must be part of the tsunami education of coastal communities.
- Finally, it is a bitter thought that the latest lethal tsunami took place during a catastrophic episode of a volcanic crisis which had taken months to build up, during which adequate seismic and geodetic monitoring could have been used to maintain awareness in the populations at risk. In hindsight, common sense would suggest that a closed building on an unprotected beach may not have been the best venue to hold a rock concert on the coast of Java, while Anak Krakatau was glowing red in the night sky, only 40 km away. Here again, a considerable effort of education is necessary in the communities at risk.

Acknowledgements

I am grateful to my many colleagues and students over the years, who helped me discover challenges in so many theoretical, experimental or field aspects of tsunami science; they are too numerous to list, but they know who they are. I thank Editor A.B. Rabinovich for motivating me to write this review, and for pointing out the importance of a few events which I had originally left aside. I am very grateful to

Paul Whitmore and Slava Gusiakov for constructive reviews. Figure 1 was drafted using the GMT software (Wessel and Smith 1991).

Publisher's Note Springer Nature remains neutral with regard to jurisdictional claims in published maps and institutional affiliations.

References

Abe, K., Abe, K., Tsuji, Y., Imamura, F., Katao, H., Iio, Y., et al. (1993). Field survey of the Nicaragua earthquake and tsunami of September 2, 1992. *Bulletin of the Earthquake Research Institute of Tokyo University*, *68*, 23–70.

Altınok, Y., Tinti, S., Alpar, V., Yalçıner, A. C., Ersoy, Ş., Bortolucci, E., et al. (2001). The Tsunami of August 17, 1999 in İzmit Bay. *Turkey, Natural Hazards*, *24*, 133–146.

Ambraseys, N. N. (1991). The Rukwa earthquake of 13 December 1910 in East Africa. *Terra Nova*, *3*, 202–211.

Ammon, C. J. (2008). A great earthquake doublet and seismic stress transfer cycle in the central Kuril Islands. *Nature*, *451*, 561–565.

Angenheister, G. (1920). Vier Erdbeben and Flutwellen im pazifischen Ozean, beobachtet am Samoa Observatorium, 1917–1919, *Nachrichten von der Gesellschaft der Wissenschaften zu Göttingen, Mathematische-physikalische Klasse*, 201–204.

Aránguiz, R., González, G., González, J., Catalán, P. A., Cienfuegos, R., Yagi, Y., et al. (2016). The 16 September 2015 Chile tsunami from the post-tsunami survey and numerical modeling perspectives. *Pure and Applied Geophysics*, *173*, 333–348.

Artru, J., Dučić, V., Kanamori, H., Lognonné, P., & Murakami, M. (2005). Ionospheric detection of gravity waves induced by tsunamis. *Geophysical Journal International*, *160*, 840–848.

Barberopoulou, A., Borrero, J. C., Uslu, B., Legg, M. R., & Synolakis, C. E. (2011). A second generation of tsunami inundation maps for the State of California. *Pure and Applied Geophysics*, *168*, 2133–2146.

Barrientos, S. E., Vigny, C., Ward, S. N., & Bataille, K. D. (2009). Earthquake-induced rockfall and tsunami in Southern Chile. *Eos, Transactions of the American Geophysical Union*, *90*(53), NH43B–1315 [abstract].

Bernard, E. N., & Milburn, H. B. (1985). Long-wave observations near the Galápagos Islands. *Journal of Geophysical Research*, *90*, 3361–3366.

Boatwright, J., & Choy, G. L. (1986). Teleseismic estimates of the energy radiated by shallow earthquakes. *Journal of Geophysical Research*, *91*, 2095–2112.

Borrero, J. C., Ortíz, M., Titov, V. V., & Synolakis, C. E. (1997). Field survey of Mexican tsunami produces new data, unusual photos, Eos. *Transactions of the American Geophysical Union*, *78*, 85 and 87–88.

Borrero, J. C., Yalçıner, A. C., Kânoğlu, U., Titov, V. V., McCarthy, D., & Synolakis, C. E. (2003). Producing tsunami inundation maps in California. In A. C. Yalçıner, E. N. Pelinovsky, E. A. Okal, & C. E. Synolakis (Eds.), *Submarine landslides and tsunamis* (pp. 315–329). Dordrecht: Kluwer Academic Publishers.

Borrero, J. C., Weiss, R., Okal, E. A., Hidayat, R., Suranto, A. D., & Titov, V. V. (2009). The tsunami of 12 September 2007, Bengkulu Province, Sumatra, Indonesia: Post-tsunami survey and numerical modeling. *Geophysical Journal International*, *178*, 180–194.

Borrero, J. C., McAdoo, B. G., Jaffe, B., Dengler, L., Gelfenbaum, G., Higman, B., et al. (2011). Field survey of the March 28, 2005 Nias-Simeulue earthquake and tsunami. *Pure and Applied Geophysics*, *168*, 1075–1088.

Borrero, J. C., Kalligeris, N., Lynett, P. J., Fritz, H. M., Newman, A. V., & Convers, J. A. (2014). Observations and modeling of the August 27, 2012 earthquake affecting El Salvador and Nicaragua. *Pure and Applied Geophysics*, *171*, 3421–3435.

Briggs, M. J., Synolakis, C. E., Harkins, G. S., & Green, D. R. (1995). Laboratory experiments of tsunami run-up on a circular island. *Pure and Applied Geophysics*, *144*, 569–594.

Brunt, K. M., Okal, E. A., & MacAyeal, D. R. (2011). Antarctic ice-shelf calving triggered by Honshu earthquake and tsunami, March 2011. *Journal of Glaciology*, *57*, 785–788.

Caminade, J.-P., Charlie, D., Kânoğlu, U., Koshimura, S., Matsutomi, H., Moore, A., et al. (2000). Vanuatu earthquake and tsunami cause much damage, few casualties, Eos. *Transactions of the American Geophysical Union*, *81*, 641 and 646–6470.

Catalán, P. A., Aránguiz, R., González, G., Tomita, T., Cienfuegos, R., González, J., et al. (2015). The 1 April 2014 Pisagua tsunami: Observations and modeling. *Geophysical Research Letters*, *42*, 2918–2925.

Chlieh, M., Perfettini, H., Tavera, H., Avouac, J.-P., Remy, D., Nocquet, J.-M., et al. (2011). Interseismic coupling and seismic potential along the Central Andes subduction zone. *Journal of Geophysical Research*, *116*, B12405, 21 pp.

Choy, G. L., & Boatwright, J. (2007). The energy radiated by the 26 December 2004 Sumatra-Andaman earthquake estimated from 10-minute *P*-wave windows. *Bulletin of the Seismological Society of America*, *97*, S18–S24.

Cruz, G., & Wyss, M. (1983). Large earthquakes, mean sea level, and tsunamis along the Pacific coast of Mexico and Central America. *Bulletin of the Seismological Society of America*, *73*, 553–570.

Dengler, L., Borrero, J., Gelfenbaum, G., Jaffe, B., Okal, E. A., Ortíz, M., et al. (2003). Tsunami. In A. Rodriguez-Marek & C. Edwards (Eds.), *Southern Peru earthquake of 23 June 2001, reconnaissance report* (Vol. 19, pp. 115–144). Oakland: Earthquake Spectra. Supp. A.

Dengler, L., Uslu, B., Barberopoulou, A., Yim, S. C., & Kelly, A. (2009). The November 15, 2006 Kuril Islands-generated tsunami in Crescent City. *California, Pure and Applied Geophysics*, *166*, 37–53.

Fine, I. V., Cherniawsky, J. Y., Thomson, R. E., Rabinovich, A. B., & Krassovski, M. V. (2015). Observations and numerical modeling of the 2012 Haida Gwaii tsunami off the coast of British Columbia. *Pure and Applied Geophysics*, *172*, 699–718.

Fritz, H. M., & Kalligeris, N. (2008). Ancestral heritage saves tribes during 1 April 2007 Solomon Islands tsunami. *Geophysical Research Letters*, *35*(1), L01607, 5 pp.

Fritz, H. M., Borrero, J. C., Synolakis, C. E., & Yoo, J. (2006). 2004 Indian Ocean tsunami flow velocity measurements from survivor videos. *Geophysical Research Letters*, *33*(24), L24605, 5 pp.

Fritz, H. M., Kongko, W., Moore, A., McAdoo, B., Goff, J., Harbits, C., et al. (2007). Extreme run-up from the 17 July 2006 Java tsunami. *Geophysical Research Letters, 34*(12), L12602.

Fritz, H. M., Kalligeris, N., Borrero, J. C., Broncano, P., & Ortega, E. (2008). The 15 August 2007 Peru tsunami run-up observations and modeling. *Geophysical Research Letters, 35*(10), L10604, 5 pp.

Fritz, H. M., Borrero, J. C., Synolakis, C. E., Okal, E. A., Weiss, R., Titov, V. V., et al. (2011a). Insights on the 2009 South Pacific tsunami in Samoa and Tonga from field surveys and numerical simulations. *Earth Science Reviews, 107*, 66–75.

Fritz, H. M., Petroff, C. M., Catalán, P. A., Cienfuegos, R., Winckler, P., Kalligeris, N., et al. (2011b). Field survey of the 27 February 2010 Chile tsunami. *Pure and Applied Geophysics, 168*, 1989–2010.

Fritz, H. M., Vilmond-Hillaire, J., Molière, E., Wei, Y., & Mohammed, F. (2012). Twin tsunamis triggered by the 12 January 2010 Haiti earthquake. *Pure and Applied Geophysics, 170*, 1463–1474.

Fritz, H. M., Papantoniou, A., Biuokoto, L., & Gilly, A. (2013). The Solomon Islands tsunami of 6 February 2013 field survey in the Santa Cruz Islands. *Eos, Transactions of the American Geophysical Union, 96*(53), NH41A-1696. [abstract].

Fritz, H. M., Giachetti, T., Anderson, S. A., & Gauthier, D. (2018). Field survey of the 17 June 2017 landslide-generated tsunami in Karrat Fjord, Greenland. *Geophysical Research Abstracts, 20*, EGU-2018-18345 [abstract].

Giachetti, T., Paris, R., Kelfoun, K., & Ontowirjo, B. (2012). Tsunami hazard related to a flank collapse of Anak Krakatau Volcano, Sunda Strait, Indonesia. *Geological Society of London, Special Publications, 361*, 79–90.

Gica, E., Spillane, M., Titov, V. V., Chamberlin, C., & Newman, J. C. (2008). Development of the forecast propagation database for NOAA's Short-term Inundation Forecast for Tsunamis (SIFT). *NOAA Tech. Memo. OAR PMEL, 139*, 89. pp.

Godin, O. A., Irisov, V. G., Leben, R. R., Hamlington, B. D., & Wick, G. A. (2009). Variations in sea surface roughness induced by the 2004 Sumatra-Andaman tsunami. *Natural Hazards Earth Science Systems, 9*, 1135–1147.

González, F. I., Mader, C. L., Eble, M. C., & Bernard, E. N. (1991). The 1987–88 Alaskan Bight tsunamis: Deep ocean data and model comparisons. *Natural Hazards, 4*, 119–139.

Goto, C., Ogawa, Y., Shuto, N., & Imamura, F. (1997). Numerical method of tsunami simulation with the leap-frog schemes. *IOC Manuals and Guides, 35*, 126. pp., UNESCO, Paris.

Green, G. (1838). On the motion of waves in a variable canal of small depth and width. *Transactions of the Cambridge Philosophical Society, 6*, 457–462.

Guilbert, J., Vergoz, J., Schisselé, E., Roueff, A., & Cansi, Y. (2005). Use of hydroacoustic and seismic arrays to observe rupture propagation and source extent of the $M_w = 9.0$ Sumatra earthquake. *Geophysical Research Letters, 32*(15), L15310, 5 pp.

Gusiakov, V. K., Dunbar, P., & Arcos, N. (2019). Twenty-five years (1992–2016) of global tsunamis: Statistical and analytical overview. *Pure and Applied Geophysics*. https://doi.org/10.1007/s00024-019-02113-7.

Hammack, J. L. (1973). A note on tsunamis: Their generation and propagation in an ocean of uniform depth. *Journal of Fluid Mechanics, 60*, 769–799.

Hanson, J. A., & Bowman, J. R. (2005). Dispersive and reflected tsunami signals from the 2004 Indian Ocean tsunami observed on hydrophone and seismic stations. *Geophysical Research Letters, 32*(17), L17606. 5 pp.

Harig, S., Androsov, A., & Rakowsky, N. (2019). Simulating landslide generated tsunamis in Palu Bay, Sulawesi, Indonesia. *Geophysical Research Abstracts, 21*, EGU2019-7094 [abstract].

Hayes, G. P., Rivera, L., & Kanamori, H. (2009). Source inversion of the *W* phase: Real-time implementation and extension to low magnitudes. *Seismological Research Letters, 80*, 817–822.

Hayes, G., Herman, M. W., Barnhart, W. D., Furlong, K. P., Riquelme, S., Benz, H. M., et al. (2014). Continuing megathrust earthquake potential in Chile after the 2014 Iquique earthquake. *Nature, 512*, 295–298.

Heidarzadeh, M., & Satake, K. (2014). Possible sources of the tsunami observed in the northwestern Indian Ocean following the 2013 September 24 $M_w = 7.7$ Pakistan inland earthquake. *Geophysical Journal International, 199*, 752–766.

Heidarzadeh, M., & Satake, K. (2017). Possible dual earthquake-landslide source of the 13 November 2016 Kaikoura, New Zealand tsunami. *Pure and Applied Geophysics, 174*, 3737–3749.

Heidarzadeh, M., Murotani, S., Satake, K., Takagawa, T., & Saito, T. (2017). Fault size and depth extent of the Ecuador earthquake ($M_w = 7.8$) of 16 April 2016 from teleseismic and tsunami data. *Geophysical Research Letters, 44*, 2211–2219.

Heidarzadeh, M., Muhari, A., & Wijanarto, A. B. (2019). Insights on the source of the 28 September 2018 Sulawesi tsunami, Indonesia based on spectral analyses and numerical simulations. *Pure and Applied Geophysics, 176*, 25–43.

Heinrich, P., Schindelé, F., Guibourg, S., & Ihmlé, P. (1998). Modeling of the 1996 Peruvian tsunami. *Geophysical Research Letters, 25*, 2687–2690.

Heinrich, P., Mangeney, A., Guibourg, S., Roche, R., Boudon, G., & Cheminée, J.-L. (1998). Simulation of water waves generated by a potential debris avalanche in Montserrat, Lesser Antilles. *Geophysical Research Letters, 25*, 3697–3700.

Higman, H., Shugar, D. H., Stark, C. P., Ekström, G., Koppes, M. N., Lynett, P., et al. (2018). The 2015 landslide and tsunami in Taan Fiord, Alaska. *Science Reports, 8*(1), 12993, 12 pp.

Hill, E. M., Borrero, J. C., Huang, Z., Qiu, Q., Banerjee, P., Natawidjaja, D. H., et al. (2012). The 2010 $M_w = 7.8$ Mentawai earthquake: Very shallow source of a rare tsunami earthquake determined from tsunami field survey and near-field GPS data. *Journal of Geophysical Research, Solid Earth, 117*(6), B06402, 21 pp.

Hutt, C. R., Bolton, H. F., & Holcomb, L. G. (2002). U.S. contribution to digital global seismograph network. In W. H. K. Lee, H. Kanamori, P. Jennings, & C. Kisslinger (Eds.), *International Handbook of Earthquake and Engineering Seismology* (pp. 319–332). New York: Academic Press.

Ihmlé, P., Gomez, J.-M., Heinrich, P., & Guibourg, S. (1998). The 1996 Peru tsunamigenic earthquake: Broadband source process. *Geophysical Research Letters, 25*, 2691–2694.

Imamura, F., Shuto, N., Ide, S., Yoshida, Y., & Abe, K. (1993). Estimate of the tsunami source of the 1992 Nicaraguan earthquake from tsunami data. *Geophysical Research Letters, 20*, 1515–1518.

Imamura, F., Gica, E., Takahashi, T., & Shuto, N. (1995). Numerical simulations of the 1992 Flores tsunami: Interpretation of tsunami phenomena in Northeastern Flores and damage at Babi Island. *Pure and Applied Geophysics, 144*, 555–568.

Imamura, F., Synolakis, C. E., Gica, E., Titov, V. V., Listanco, E., & Lee, H. J. (1995). Field survey of the 1994 Mindoro Island, Philippines tsunami. *Pure and Applied Geophysics, 144*, 875–890.

Imamura, F., Subandono, D., Watson, G., Moore, A., Takahashi, T., Matsutomi, H., et al. (1997). Irian Jaya earthquake and tsunami cause serious damage. *Eos, Transactions of the American Geophysical Union, 78*, 197 and 201.

Ishii, M., Shearer, P. M., Houston, H., & Vidale, J. E. (2005). Extent, duration and speed of the 2004 Sumatra-Andaman earthquake, imaged by the Hi-Net array. *Nature, 435*, 933–936.

Kalligeris, N., Skanavis, V., Tavakkol, S., Ayça, A., El Safty, H., Lynett, P. J., et al. (2016). Lagrangian flow measurements and observations of the 2015 Chilean tsunami in Ventura, CA. *Geophysical Research Letters, 43*, 5217–5224.

Kanamori, H. (1972). Mechanism of tsunami earthquakes. *Physics of the Earth and Planetary Interiors, 6*, 346–359.

Kanamori, H. (1993). *W* phase. *Geophysical Research Letters, 20*, 1691–1694.

Kanamori, H., & Cipar, J. J. (1974). Focal process of the great Chilean earthquake, May 22, 1960. *Physics of the Earth and Planetary Interiors, 9*, 128–136.

Kanamori, H., & Rivera, L. (2008). Source inversion of *W* phase: Speeding up tsunami warning. *Geophysical Journal International, 175*, 222–238.

Kanamori, H., Rivera, L., & Lamotte, S. (2019). Evidence for a large strike-slip component during the 1960 Chilean earthquake. *Geophysical Journal International, 218*, 1–32.

Kawata, Y., Benson, B. C., Borrero, J. C., Borrero, J. L., Davies, H. L., de Lange, W. P., et al. (1999). Tsunami in Papua New Guinea was as intense as first thought. *Eos, Transactions of the American Geophysical Union, 80*, 101–105.

Kerr, R. (2005). Model shows islands muted tsunami. *Science, 308*, 341.

Kulikov, E. A., Rabinovich, A. B., Thomson, R. E., & Bornhold, B. D. (1996). The landslide tsunami of November 3, 1994, Skagway Harbor, Alaska. *Journal of Geophysical Research, 101*, 6609–6615.

La Rocca, M., Galluzzo, D., Saccorotti, G., Tinti, S., Cimini, G. B., & Del Pezzo, E. (2004). Seismic signals associated with landslides and with a tsunami at Stromboli Volcano, Italy. *Bulletin of the Seismological Society of America, 94*, 1850–1867.

Lay, T., Kanamori, H., Ammon, C. J., Hutko, A. R., Furlong, K., & Rivera, L. (2009). The 2006–2007 Kuril Islands great earthquake sequence. *Journal of Geophysical Research, Solid Earth, 114*(B11), B11308, 31 pp.

Lay, T., Ammon, C. J., Kanamori, H., Rivera, L., Koper, K. D., & Hutko, A. R. (2010). The 2009 Samoa–Tonga great earthquake triggered doublet. *Nature, 466*, 964–968.

Lay, T., Ammon, C. J., Kanamori, H., Xue, L., & Kim, M. J. (2011). Possible large near-trench slip during the 2011 $M_w = 9.0$ off the Pacific coast of Tohoku earthquake. *Earth, Planets and Space, 63*, 687–692.

Lay, T., Ye, L., Kanamori, H., Yamazaki, Y., Cheug, K. F., Kwong, K., et al. (2013). The October 28, 2012 $M_w = 7.8$ Haida Gwaii underthrusting earthquake and tsunami: Slip partitioning along the Queen Charlotte Fault transpressional plate boundary. *Earth and Planetary Science Letters, 375*, 57–70.

Lay, T., Ye, L., Bai, Y., Cheung, K. F., & Kanamori, H. (2018). The 2018 $M_w = 7.9$ Gulf of Alaska earthquake: Multiple fault rupture in the Pacific plate. *Geophysical Research Letters, 45*, 9542–9551.

Legrand, D., Barrientos, S. E., Bataille, K., Cembrano, J., & Pavez, A. (2011). The fluid-driven tectonic swarm of Aysén Fjord, Chile (2007) associated with two earthquakes ($M_w = 6.1$ and $M_w = 6.2$) within the Liquiñe-Ofqui Fault Zone. *Continental Shelf Research, 31*, 154–161.

Leonard, L. J., & Bednarski, J. M. (2014). Field survey following the 28 October 2012 Haida Gwaii tsunami. *Pure and Applied Geophysics, 171*, 3467–3482.

Li, X., Shao, G., & Ji, C. (2009). Rupture process of the 2009 $M_w = 8.1$ Samoa earthquake constrained by joint inverting teleseismic body, surface waves and local strong motion. *Eos, Transactions of the American Geophysical Union, 91*(53), U21D-03 [abstract].

Liu, J. Y., Tsai, Y. B., Chen, S. W., Lee, C. P., Chen, Y. C., Yen, H. Y., et al. (2006). Giant ionospheric disturbances excited by the $M = 9.3$ Sumatra earthquake of 26 December 2004. *Geophysical Research Letters, 33*, L02103, 4 pp.

Liu, P. L.-F., Cho, Y.-S., Yoon, S. B., & Seo, S. N. (1994). Numerical simulations of the 1960 Chilean tsunami propagation and inundation at Hilo, Hawaii. In M. I. El-Sabh (Ed.), *Recent Developments in Tsunami Research* (pp. 99–115). Dordrecht: Kluwer.

Liu, P. L.-F., Cho, Y.-S., Briggs, M. J., Synolakis, C. E., & Kânoğlu, U. (1995). Run-up of solitary waves on a circular island. *Journal of Fluid Mechanics, 302*, 259–285.

Lobkovsky, L. I., Rabinovich, A. B., Kulikov, E. A., Ivashchenko, A. I., Fine, I. V., Thomson, R. E., et al. (2009). The Kuril earthquakes and tsunamis of November 15, 2006 and January 13, 2007: Observations, analysis and numerical modeling. *Oceanology, 49*, 166–181.

Lockwood, O. G., & Kanamori, H. (2006). Wavelet analysis of the seismograms of the 2004 Sumatra-Andaman earthquake and its application to tsunami early warning. *Geochemistry, Geophysics and Geosystems, 7*(9), GC001272, 10 pp.

MacInnes, B. T. (2010). Bridging seismology and geomorphology: Investigations of the 2006 and 2007 Kuril Islands tsunamis, *Ph.D. Dissertation,* 188 pp., University of Washington, Seattle.

MacInnes, B. T., Pinegina, T. K., Bourgeois, J., Razhigaeva, N. G., & Kaistrenko, V. M. (2009). Field survey and geological effects of the 15 November 2006 Kuril tsunami in the Middle Kuril Islands. *Pure and Applied Geophysics, 166*, 9–36.

Manoj, C., Maus, S., & Chulliat, A. (2011). Observation of magnetic fields generated by tsunamis. *Eos, Transactions of the American Geophysical Union, 92*, 13–14.

Marris, E. (2005). Inadequate warning system left Asia at the mercy of tsunami. *Nature, 433*, 3–5.

Martin, S. S., Li, L., Okal, E. A., Morin, J., Tetteroo, A., Switzer, A., et al. (2019). Reassessment of the 1907 Sumatra "tsunami earthquake" based on macroseismic, seismological, and tsunami observations, and modeling. *Pure and Applied Geophysics*. https://doi.org/10.1007/s00024-019-02134-2.

Matsutomi, H., & Iizuka, H. (1998). Tsunami current velocity on land and its simple estimation method. *Proceedings of Coastal Engineering, Japan Society of Coastal Engineers, 45*, 361–365. [in Japanese].

Matsutomi, H., Shuto, N., Imamura, F., & Takahashi, T. (2001). Field survey of the 1996 Irian Jaya earthquake tsunami in Biak Island. *Natural Hazards, 24*, 199–212.

McAdoo, B. G., Dengler, L., Prasetya, G., & Titov, V. V. (2006). Smong: How an oral history saved thousands on Indonesia's Simeulue Island during the December 2004 and March 2005 tsunamis. *Earthquake Spectra, 22*, S661–S669.

McCaffrey, R. (2007). The next great earthquake. *Science, 315*, 1675–1676.

McCloskey, J., Nalbant, S. S., & Steacy, S. (2005). Earthquake risk from co-seismic stress. *Nature, 434*, 291.

McNamara, D. E., Ringler, A. T., Hutt, C. R., & Gee, L. S. (2011). Seismically observed seiching in the Panama Canal. *Journal of Geophysical Research, 116*(B4), B04312, 12 pp.

Meining, C., Stalin, S. E., Nakamura, A. I., González, F. I., & Milburn, H. B. (2005). Technology developments in real-time tsunami measuring, monitoring and forecasting. *Proceedings of OCEANS 2005, MTS/IEEE, 93*, 1673–1679.

Miller, D. J. (1960). The Alaska earthquake of July 10, 1958: Giant wave in Lituya Bay. *Bulletin of the Seismological Society of America, 50*, 253–266.

Nalbant, S., Steacy, S., Sieh, K., Natawidjaja, D., & McCloskey, J. (2005). Updated earthquake hazard in Sumatra. *Nature, 435*, 756–757.

Nettles, M., Ekström, G., Dziewoński, A. M., & Maternovskaya, N. (2005). Source characteristics of the great Sumatra earthquake and its aftershocks. *Eos, Transactions of the American Geophysical Union, 86*(18), U43A–01 [abstract].

Newman, A. V., & Okal, E. A. (1998). Teleseismic estimates of radiated seismic energy: The E/M_0 discriminant for tsunami earthquakes. *Journal of Geophysical Research, 103*, 26885–26898.

Newman, A. V., Feng, L., Fritz, H. M., Lifton, Z. M., Kalligeris, N., & Wei, Y. (2011). The energetic 2010 $M_w = 7.1$ Solomon Islands tsunami earthquake. *Geophysical Journal International, 186*, 775–781.

Newman, A. V., Hayes, G., Wei, Y., & Convers, J. (2011). The 25 October 2010 Mentawai tsunami earthquake, from real-time discriminants, finite-fault rupture, and tsunami excitation. *Geophysical Research Letters, 38*(5), L05302, 10 pp.

Ni, S., Kanamori, H., & Helmberger, D. V. (2005). Energy radiation from the Sumatra earthquake. *Nature, 434*, 582.

Nöggerath, J., Geller, R. J., & Gusiakov, V. K. (2011). Fukushima: The myth of safety, reality of geoscience. *Bulletin of Atomic Scientists, 67*(6), 36–37.

Occhipinti, G., Lognonné, P., Kherani, A. A., & Hébert, H. (2006). Three-dimensional waveform modeling of ionospheric signature induced by the 2004 Sumatra tsunami. *Geophysical Research Letters, 33*, L20104.

Okal, E. A. (1993). WM_m: An extension of the concept of mantle magnitude to the W phase, with application to real-time assessment of the ultra-long component of the seismic source. *Eos, Transactions of the American Geophysical Union, 74*(33), 344 [abstract].

Okal, E. A. (2003). T waves from the 1998 Papua New Guinea earthquake and its aftershocks: Timing the tsunamigenic slump. *Pure and Applied Geophysics, 160*, 1843–1863.

Okal, E. A. (2007). Seismic records of the 2004 Sumatra and other tsunamis: A quantitative study. *Pure and Applied Geophysics, 164*, 325–353.

Okal, E. A. (2013). From 3-Hz P waves to ${}_0S_2$: No evidence of a slow component to the source of the 2011 Tohoku earthquake. *Pure and Applied Geophysics, 170*, 963–973.

Okal, E. A. (2017). The excitation of tsunamis by deep earthquakes. *Geophysical Journal International, 209*, 234–249.

Okal, E. A., & Borrero, J. C. (2011). The "tsunami earthquake" of 22 June 1932 in Manzanillo, Mexico: Seismological study and tsunami simulations. *Geophysical Journal International, 187*, 1443–1459.

Okal, E. A., & MacAyeal, D. R. (2006). Seismic recording on drifting icebergs: Catching seismic waves, tsunamis and storms from Sumatra and elsewhere. *Seismological Research Letters, 77*, 659–671.

Okal, E. A., & Newman, A. V. (2001). Tsunami earthquakes: The quest for a regional signal. *Physics of the Earth and Planetary Interiors, 124*, 45–70.

Okal, E. A., & Saloor, N. (2017). Historical tsunami earthquakes in the Southwest Pacific: An extension to $\Delta > 80°$ of the Energy-to-Moment parameter Θ. *Geophysical Journal International, 210*, 852–873.

Okal, E. A., Piatanesi, A., & Heinrich, P. (1999). Tsunami detection by satellite altimetry. *Journal of Geophysical Research, 104*, 599–615.

Okal, E. A., Fryer, G. J., Borrero, J. C., & Ruscher, C. (2002). The landslide and local tsunami of 13 September 1999 on Fatu-Hiva (Marquesas Islands; French Polynesia). *Bulletin de la Société Géologique de France, 173*, 359–367.

Okal, E. A., Dengler, L., Araya, S., Borrero, J. C., Gomer, B., Koshimura, S., et al. (2002). A field survey of the Camaná, Peru tsunami of June 23, 2001. *Seismological Research Letters, 73*, 904–917.

Okal, E. A., Fritz, H. M., Raveloson, R., Joelson, G., Pančošková, P., & Rambolamanana, G. (2006a). Madagascar field survey after the December 2004 Indian Ocean tsunami. *Earthquake Spectra, 22*, S263–S283.

Okal, E. A., Sladen, A., & Okal, E. A.-S. (2006b). Rodrigues, Mauritius and Réunion Islands field survey after the December 2004 Indian Ocean tsunami. *Earthquake Spectra, 22*, S241–S261.

Okal, E. A., Fritz, H. M., Raad, P. E., Synolakis, C. E., Al-Shijbi, Y., & Al-Saifi, M. (2006c). Oman field survey after the December 2004 Indian Ocean tsunami. *Earthquake Spectra, 22*, S203–S218.

Okal, E. A., Borrero, J. C., & Synolakis, C. E. (2006d). Evaluation of tsunami risk from regional earthquakes at Pisco, Peru. *Bulletin of the Seismological Society of America, 96*, 1634–1648.

Okal, E. A., Talandier, J., & Reymond, D. (2007). Quantification of hydrophone records of the 2004 Sumatra tsunami. *Pure and Applied Geophysics, 164*, 309–323.

Okal, E. A., Fritz, F. M., & Sladen, A. (2009). 2004 Sumatra tsunami surveys in the Comoro Islands and Tanzania, and regional tsunami hazard from future Sumatra events. *South African Journal of Geology, 112*, 343–358.

Okal, E. A., Fritz, H. M., Synolakis, C. E., Borrero, J. C., Weiss, R., Lynett, P. J., et al. (2010). Field survey of the Samoa tsunami of 29 September 2009. *Seismological Research Letters, 81*, 577–591.

Okal, E. A., Kirby, S. H., & Kalligeris, N. (2016). The Showa Sanriku earthquake of 1933 March 2: A global seismological reassessment. *Geophysical Journal International, 206*, 1492–1514.

Okal, E. A., Hyvernaud, O., Paris, A., Hébert, H., & Heinrich, P. (2019). Ancillary records of the Krakatau eruption and tsunami of 22 December 2018. *Geophysical Research Abstracts, 21*, EGU2019–11563 [abstract].

Paris, A., Okal, E. A., Guérin, C., Heinrich, P., Schindelé, F., & Hébert, H. (2019). Numerical modeling of the June 17, 2017 landslide and tsunami events in Karrat Fjord, West Greenland. *Pure and Applied Geophysics*. https://doi.org/10.1007/s00024-019-02123-.

Pelinovsky, E., Zahibo, N., Dunkley, P., Edmonds, M., Herd, R., Talipova, T., et al. (2004). Tsunami generated by the volcano eruption on July 12–13, 2003 at Montserrat, Lesser Antilles. *Science of Tsunami Hazards, 22*, 44–57.

Pelletier, B., Régnier, M., Calmant, S., Pillet, R., Cabioch, G., Lagabrielle, Y., et al. (2000). Le séisme d'Ambryn-Pentecôte du 26 novembre 1999 ($M_w = 7,5$): données préliminaires sur la séismicité, le tsunami et les déplacements associés. *Comptes-Rendus de l'Académie des Sciences (Paris), Série 2, 331*, 21–28.

Peltier, W. R., & Hines, C. O. (1976). On the possible detection of tsunamis by a monitoring of the ionosphere. *Journal of Geophysical Research, 81*, 1995–2000.

Plafker, G. L. (1997). Catastrophic tsunami generated by submarine slides and backarc thrusting during the 1992 earthquake on Eastern Flores I., Indonesia. *Geological Society of America, Abstracts with Program, 29*(5), 57 [abstract].

Pollitz, F., Banerjee, P., Bürgmann, R., Hashimoto, M., & Choosakul, N. (2006). Stress changes along the Sunda trench following the 26 December 2004 Sumatra-Andaman and 28 March 2005 Nias earthquakes. *Geophysical Research Letters, 33*(6), L06309, 4 pp.

Power, W., Clark, K., King, D. N., Borrero, J., Howarth, J., Lane, E. M., et al. (2017). Tsunami runup and tide-gauge observations from the 14 November 2016 $M = 7.8$ Kaikōura earthquake, New Zealand. *Pure and Applied Geophysics, 174*, 2457–2473.

Prasetya, G., Husrin, S., Kongko, W., Istyanto, D., Hidayat, R., Asvaliantina, V., et al. (2019). The 22nd December 2018 Anak Krakatau Tsunami in Sunda Straits, Indonesia. *Geophysical Research Abstracts, 21*, EGU2019–12276 [abstract].

Press, F., & Harkrider, D. G. (1966). Air-Sea waves from the explosion of Krakatoa. *Science, 154*, 1325–1327.

Rabinovich, A. B. (2019). Twenty-seven years of progress in the science of meteorological tsunamis following the 1992 Daytona Beach event. *Pure and Applied Geophysics*, submitted.

Rabinovich, A. B., & Thomson, R. E. (2007). The 26 December 2004 Sumatra tsunami: Analysis of tide gauge data from the world ocean, Part I. Indian Ocean and South Africa. *Pure and Applied Geophysics, 164*, 261–308.

Rabinovich, A. B., Lobkovsky, L. I., Fine, I. V., Thomson, R. E., Ivelskaya, T. N., & Kulikov, E. A. (2008). Near-source observations and modeling of the Kuril Islands tsunamis of 15 November 2006 and 13 January 2007. *Advances in Geosciences, 14*, 105–116.

Ruff, L. J. (1989). Do trench sediments affect great earthquake occurrence in subduction zones? *Pure and Applied Geophysics, 129*, 263–282.

Ruff, L. J., & Kanamori, H. (1980). Seismicity and the subduction process. *Physics of the Earth and Planetary Interiors, 23*, 240–252.

Satake, K., Shimazaki, K., Tsuji, Y., & Ueda, K. (1996). Time and size of a giant earthquake in Cascadia inferred from Japanese tsunami records of January 1700. *Nature, 379*, 246–249.

Shuto, N., & Matsutomi, H. (1995). Field survey of the 1993 Hokkaido-Nansei-Oki earthquake tsunami. *Pure and Applied Geophysics, 144*, 406–449.

Smith, W. H. F., Scharroo, R., Titov, V. V., Arcas, D., & Arbic, B. K. (2005). Satellite altimeters measure tsunami. *Oceanography, 18*, 11–13.

Solov'ev, S. L., & Go, C. N. (1984). Catalogue of tsunamis on the Eastern shore of the Pacific Ocean. *Canadian Translations. Fisheries and Aquatic Sciences, 5078*, 293. pp.

Stein, R. S., Barka, A. A., & Dieterich, J. H. (1997). Progressive failure on the North Anatolian fault since 1939 by earthquake stress triggering. *Geophysical Journal International, 128*, 594–604.

Stein, S., & Okal, E. A. (2005). Size and speed of the Sumatra earthquake. *Nature, 434*, 581–582.

Stein, S., & Okal, E. A. (2007). Ultra-long period seismic study of the December 2004 Indian Ocean earthquake and implications for regional tectonics and the subduction process. *Bulletin of the Seismological Society of America, 97*, S279–S295.

Sweet, S., & Silver, E. A. (2003). Tectonics and slumping in the source region of the 1998 Papua New Guinea tsunami from seismic reflection images. *Pure and Applied Geophysics, 160*, 1945–1968.

Synolakis, C. E., & Kânoğlu, U. (2015). The Fukushima accident was preventable. *Philosophical Transactions of the Royal Society (London), 373A*, 20140374. 23 pp.

Synolakis, C. E., & Okal, E. A. (2005). 1992–2002: Perspective on a decade of post-tsunami surveys. Advances in Natural and Technological Hazards. In K. Satake (Ed.), *Tsunamis: Case studies and recent developments* (Vol. 23, pp. 1–30). Berlin: Springer.

Synolakis, C. E., Imamura, F., Tsuji, Y., Matsutomi, H., Tinti, S., Cook, B., et al. (1995). Damage, conditions of East Java tsunami analyzed, Eos. *Transactions of the American Geophysical Union, 76*(26), 257–264.

Synolakis, C. E., Yalçıner, A. C., Borrero, J. C., & Plafker, G. L. (2002). Modeling of the November 3, 1994 Skagway, Alaska tsunami. Solutions to Coastal Disasters. In L. Wallendorf & L. Ewing (Eds.), *Proceedings of the American Society of Civil Engineers* (Vol. 78, pp. 915–927).

Synolakis, C. E., Bardet, J.-P., Borrero, J. C., Davies, H. L., Okal, E. A., Silver, E. A., et al. (2002). The slump origin of the 1998 Papua New Guinea tsunami. *Proceedings of the Royal Society (London), Series A, 458*, 763–789.

Tadepalli, S., & Synolakis, C. E. (1994). The run-up of N-waves. *Proceedings of the Royal Society (London), Series A, 445*, 99–112.

Tadepalli, S., & Synolakis, C. E. (1996). Model for the leading waves of tsunamis. *Physical Review Letters, 77*, 2141–2145.

Tanioka, Y., & Satake, K. (1996). Tsunami generation by horizontal displacement of ocean bottom. *Geophysical Research Letters, 23*, 861–864.

Taylor, F. W., Briggs, R. W., Frohlich, C., Brown, A., Hornback, M., Papabatu, A. K., et al. (2008). Rupture across arc segment and plate boundaries in the 1 April 2007 Solomons earthquake. *Nature Geoscience, 1*, 253–257.

Thomson, R. E., Rabinovich, A. B., Kulikov, E. A., Fine, I. V., & Bornhold, B. D. (2001). On numerical simulation of the landslide-generated tsunami of November 3, 1994 in Skagway Harbor, Alaska. In G. Hebenstreit (Ed.), *Tsunami research at the end of a critical decade* (pp. 243–282). Dordrecht: Kluwer Academic Publishers.

Tinti, S., & Vannini, C. (1995). Tsunami trapping near circular islands. *Pure and Applied Geophysics*, *144*, 595–620.

Tinti, S., Manucci, A., Pagnoni, G., Armigliato, A., & Zaniboni, F. (2005). The 30 December 2002 landslide-induced tsunamis in Stromboli: Sequence of the events reconstructed from the eyewitness accounts. *Natural Hazards Earth Systems Science*, *5*, 763–775.

Tinti, S., Armigliato, A., Manucci, A., Pagnoni, G., Zaniboni, F., Yalçıner, A. C., et al. (2006). The generating mechanisms of the August 17, 1999 İzmit Bay (Turkey) tsunami: Regional (tectonic) and local (mass instabilities) causes. *Marine Geology*, *225*, 311–330.

Titov, V. V., & Synolakis, C. E. (1993). A numerical study of wave run-up of the September 2, 1992 Nicaraguan tsunami. In Y. Tsuchiya & N. Shuto (Eds.), *Proceedings of the Tsunami Symposium, International Union of Geology and Geophysics Japan Society of Civil Engineers*, (pp. 627–635). Wakayama, Japan.

Titov, V. V., & Synolakis, C. E. (1997). Extreme inundation flow during the Hokkaido-Nansei-Oki tsunami. *Geophysical Research Letters*, *24*, 1315–1318.

Titov, V. V., & Synolakis, C. E. (1998). Numerical modeling of tidal wave runup. *Journal of Waterway, Port, Coastal and Ocean Engineering*, *124*, 157–171.

Titov, V. V., González, F. I., Bernard, E. N., Eble, M. C., Mofjeld, H. O., Newman, J. C., et al. (2005). Real-time tsunami forecasting: Challenges and solutions. *Natural Hazards*, *35*, 41–58.

Titov, V. V., Rabinovich, A. B., Mofjeld, H., Thomson, R. E., & González, F. I. (2005). The global reach of the 26 December 2004 Sumatra tsunami. *Science*, *309*, 2045–2048.

Titov, V. V., Kânoğlu, U., & Synolakis, C. E. (2016). Development of MOST for real-time tsunami forecasting. *Journal of Waterway, Port, Coastal and Ocean Engineering*, *142*, 03116004, 16.

Tsuji, Y., Matsutomi, H., Imamura, F., Takeo, M., Kawata, Y., Matsuyama, M., et al. (1995). Damage to coastal villages due to the 1992 Flores Island earthquake tsunami. *Pure and Applied Geophysics*, *144*, 481–524.

Tyler, R. H. (2005). A simple formula for estimating the magnetic fields generated by tsunami flow. *Geophysical Research Letters*, *32*(9), L09608, 4 pp.

Wallace, R. E., Pararas-Carayannis, G., Valenzuela, R., & Taggart, J. N. (1977). Earthquake and tsunamis of August 16, 1976, Mindanao, Philippines. *Geological Society of America Abstracts with Program*, *9*, 523 [abstract].

Ward, S. N. (1980). Relationships of tsunami generation and an earthquake source. *Journal of Physics of the Earth*, *28*, 441–474.

Wang, J., Ward, S. N., & Xiao, L. (2015). Numerical simulation of the December 4, 2007 landslide-generated tsunami in Chebalis Lake, Canada. *Geophysical Journal International*, *201*, 372–376.

Wessel, P., & Smith, W. H. F. (1991). Free software helps map and display data. *Eos, Transactions of the American Geophysical Union*, *72*, 441–446.

Whitmore, P., Benz, H., Bolton, M., Crawford, G., Dengler, L., Fryer, G., et al. (2008). NOAA/West Coast and Alaska Tsunami Warning Center Pacific Ocean response criteria. *Science of Tsunami Hazards*, *27*(2), 1–21.

Ye, L., Kanamori, H., Avouac, J.-P., Li, L., Cheung, K. F., & Lay, T. (2016). The 16 April 2016, $M_w = 7.8$ ($M_s = 7.5$) Ecuador earthquake: A quasi-repeat of the 1942 $M_s = 7.5$ earthquake and partial re-rupture of the 1906 $M_s = 8.6$ Colombia–Ecuador earthquake. *Earth and Planetary Science Letters*, *454*, 248–258.

Yeh, H., Imamura, F., Synolakis, C. E., Tsuji, Y., Liu, P. L.-F., & Shi, S. (1993). The Flores Island tsunami. *Eos, Transactions of the American Geophysical Union*, *74*, 369–373.

Yeh, H., Liu, P. L.-F., Briggs, M., & Synolakis, C. E. (1994). Propagation and amplification of tsunamis at coastal boundaries. *Nature*, *372*, 353–355.

Yeh, H., Titov, V. V., Gusiakov, V., Pelinovsky, E., Khramushin, V., & Kaistrenko, V. (1995). The 1994 Shikotan tsunamis. *Pure and Applied Geophysics*, *144*, 855–874.

Yuan, X., Kind, R., & Pedersen, H. (2005). Seismic monitoring of the Indian Ocean tsunami. *Geophysical Research Letters*, *32*(15), L15308, 4 pp.

(Received May 27, 2019, revised June 3, 2019, accepted June 5, 2019, Published online June 14, 2019)

Pure Appl. Geophys. 176 (2019), 2795–2807
© 2019 Springer Nature Switzerland AG
https://doi.org/10.1007/s00024-019-02113-7

Pure and Applied Geophysics

Twenty-Five Years (1992–2016) of Global Tsunamis: Statistical and Analytical Overview

Viacheslav K. Gusiakov,[1] Paula K. Dunbar,[2,3] and Nicolás Arcos[2,3]

Abstract—The paper presents the analysis of global tsunami occurrence from 1992 to 2016. During this 25-year period, 290 tsunamis in marine basins and tsunami-like events in inland waters were registered. While this period represents less than 1% of the total catalog duration (ca. 4000 years), it contains almost 12% of the total number of historical events and nearly 54% of the total run-up and inundation depth measurements. The latter is mainly due to the data obtained in the field surveys of two trans-oceanic mega-tsunamis that occurred within this period—the December 26, 2004 Indian Ocean tsunami (1625 total measurements) and the March 11, 2011 Tohoku tsunami (6271 total measurements). Among 290 registered tsunamis and tsunami-like events, there were 95 damaging or potentially-damaging events that had maximum run-up or inundation height > 1 m and 32 destructive tsunamis that resulted in human fatalities. As throughout history, the Pacific continues to be the main tsunamigenic region with 223 events (77%) that occurred within its basin; 26 events (9%) occurred in the Indian Ocean, 26 events (9%) in the Atlantic, and 15 events (5%) in the Mediterranean region. Six tsunami-like events occurred in internal basins (rivers and lakes). The complete statistics of the recent tsunamis in the form of their distribution over type of sources, validity, fatalities and damaging effects, intensity, and warning efficiency are presented. The paper also outlines and discusses the existing problems with data compilation, cataloging and distribution, and discusses incompleteness of run-up and wave-form data for a considerable number of non-damaging tsunamis, even those resulting from strong (magnitude higher than 7.5) submarine earthquakes.

Key words: Tsunami, tsunamigenic regions, historical catalogs, historical databases, post-event surveys.

1. Introduction

A crucial step in assessing tsunami risk is to understand what is known about the tsunami hazard and the uncertainties in the data. This paper presents a complete synthesis of global tsunami distribution, type of sources, validity, fatalities and damaging effects, and intensity, over the course of a 25-year period. While historical tsunami catalogs can span from approximately 2000 BC until present, this paper's analysis is focused on the time period 1992–2016. The reasons for focusing on this period are multifold.

Following the September 2, 1992 (UTC) earthquake and tsunami off the Pacific coast of Nicaragua, the first International Tsunami Survey Team (ITST) was formed to document the tsunami's impacts (Arcos et al. 2019). The ITST in Nicaragua used GPS, but the accuracy was limited, however, compact GPS receivers have since become a standard tool of the field measurements that has resulted in an unprecedented increase of accuracy in location of run-up and flow measurements. Field surveys benefited from a variety of technological improvements in the 1990s leading to new types of data and techniques. The improved coordination of ITSTs, most actively led by the International Tsunami Information Center (ITIC) and UNESCO-IOC, resulted in larger data sets. As a result, 1992 marked the beginning of a globally accepted standard of data collection in post-tsunami surveys; thereby, marking an appropriate time to begin our summary of recent global tsunamis.

This time period also saw the beginning of the Deep-ocean Assessment and Reporting of Tsunamis (DART®) type of deep ocean instrument deployment. The first modern Bottom Pressure Recorders (BPR) were installed in the NE Pacific in August of 1986 (Eblé and González 1991) and were eventually paired with a surface buoy for communication. The network of DART® instruments quickly grew, first in the Pacific under the NOAA's PacTOP Program, and later in the Indian Ocean and the Atlantic Ocean. The

[1] Institute of Computational Mathematics and Mathematical Geophysics (ICMMG), Novosibirsk, Russia. E-mail: gvk@sscc.ru
[2] Cooperative Institute for Research in Environmental Sciences (CIRES), University of Colorado at Boulder, Boulder, USA.
[3] NOAA National Centers for Environmental Information (NCEI), Asheville, USA.

advent of DART-type instruments has strengthened tsunami detection and forecasting, in addition it also has provided valuable data to validate tsunami models. These networks provide high quality digital deep-water records of almost all modern tsunamis including those that do not have any coastal manifestation. As a result, a higher level of confidence exists for having registered and detected more, weak and strong, tsunamis during this 25-year period (1992–2016) than previous eras.

During the same period, the transfer from analog to digital tsunami recording at tide gauges was basically completed. The digital recording along with real-time or near real-time transmission allows the data from sea-level gauges to be available immediately for operational decision making, processing, and distribution. As transmission and sampling rates improved, the ability to store and access data also improved which resulted in more robust and complete tsunami data.

Since the introduction of the first digital broadband seismic sensor (STS-1) in 1975 (Wielandt and Streckeisen 1982), instrumental measurement of moment magnitude (M_w) has become a routine operation for major seismic networks to better calculate and standardize the size of great earthquakes. The formula to calculate M_w includes the strength of the rock, area of fault, and amount of slip. The aforementioned factors are all important for understanding tsunami generation. More specifically, parameters that measure the ratio of seismic energy to moment magnitude were introduced as a valuable indicator of tsunami earthquakes (Kanamori 1972). As M_w has become a common measurement for medium to large earthquakes, our analysis of the 25-year time period can be done with a greater degree of confidence for tsunamis generated by earthquakes.

All the aforementioned scientific and technological advances resulted in an unprecedented increase in the quality and quantity of datasets (Arcos et al. 2019). The datasets for this 25-year period are more comprehensive than any previous period of historical tsunami observations. While this study's focus is on this 25-year period, comparisons spanning as far back as 1901 will be made to provide greater historical context. This statistical analysis supplies an understanding of completeness of tsunami catalogs and levels of uncertainty for users assessing tsunami risk.

2. *Global Historical Tsunami Databases*

Historical data on tsunami events and associated run-ups are critical to understanding tsunami generation, propagation and coastal impacts. Gusiakov (2009) provides a comprehensive account of historical tsunami databases including the distinct types of tsunami databases and their limitations. Currently, two global historical tsunami databases exist and are separately maintained by the NOAA National Centers for Environmental Information (NCEI) and the Novosibirsk Tsunami Laboratory of the Institute of Computational Mathematics and Mathematical Geophysics of Siberian Division of Russian Academy of Sciences (NTL/ICMMG SD RAS). In the paper, these databases are referenced as NCEI database (NCEI/WDS 2018) and GTDB (Global Tsunami Database) database (NTL/ICMMG GTDB 2018), respectively. For this 25-year period, the two databases are similar in content in terms of number of events and wave height observations (Fig. 1). The difference is mainly connected with event classification (e.g. type of sources, validity index). Another difference is that most of the events in the GTDB are provided with their tsunami intensity value (on Soloviev–Imamura scale) enabling analysis of the event recurrence. The NCEI database is used to determine the statistics on the occurrence of tsunamis and effects shown in Table 1 and in Figs. 2, 3, 4, 5, 6, 7, 8, and 9. The GTDB database is used to determine statistics related to tsunami intensity shown in Figs. 10 and 11.

In this analysis we consider the tsunamis and tsunami-like sudden level changes in coastal areas and inland water basins, thus excluding freak-type rogue waves (killer-waves) observed in the open sea. Although these waves sometimes manifest themselves on the coastline, their generation mechanism (non-linear redistribution of energy already existing in the dynamic ocean) is quite different from tsunami and tsunami-like phenomena (despite their big

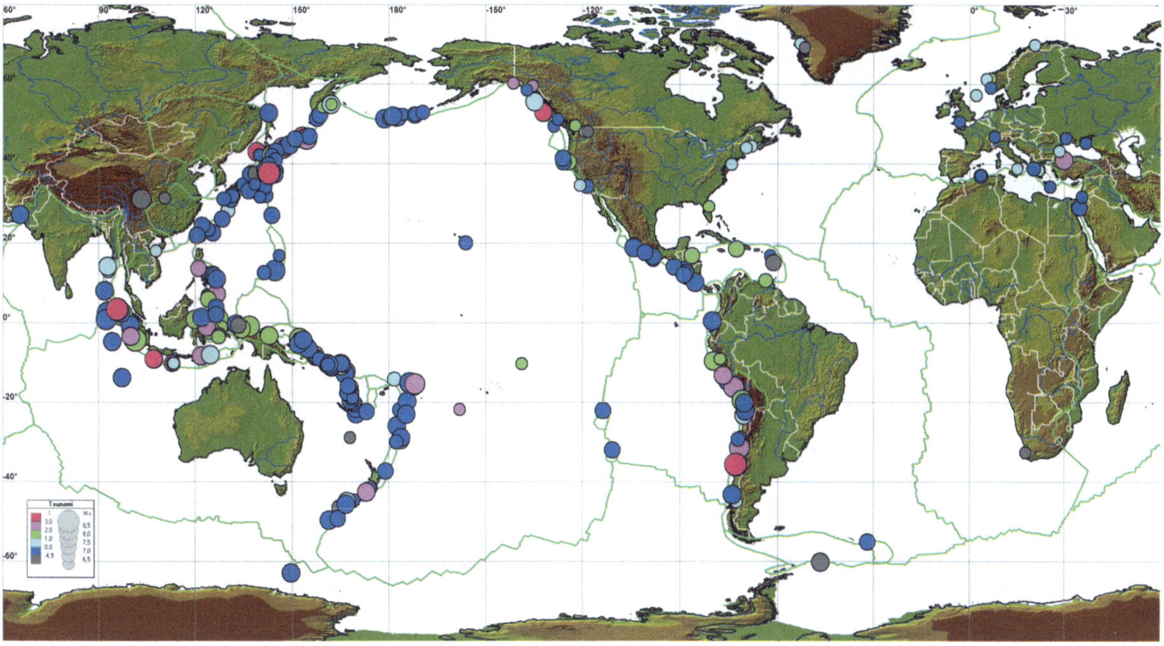

Figure 1
A source map of 290 tsunamigenic events in the World Ocean from 1992 to 2016. Size of circles is proportional to the source magnitude (for earthquake-induced tsunamis), color represents the tsunami intensity on Soloviev–Imamura scale. Grey color corresponds to the events whose intensity cannot be determined due to the lack of data on wave heights. Solid green lines are the main plate boundaries

Table 1

Top 15 tsunamis during the time period 1992-2016 by M_w magnitude, highest run-up (H_{max}), and number of fatalities (N_{FAT})

Top-15 tsunamis with highest M_w value, sorted by M_w			Top-15 tsunamis with highest run-up, sorted by H_{max}			Top-15 tsunamis with largest fatalities, sorted by N_{FAT}		
2011.03.11	9.1	Tohoku	2015.10.18	193 m	Icy Bay	2004.12.26	227,899	Sumatra
2004.12.26	9.1	Sumatra	2004.12.26	51 m	Sumatra	2011.03.11	18,454	Tohoku
2010.02.27	8.8	Chile	2000.11.21	50 m	Greenland	1998.07.17	1636	Papua
2012.04.11	8.6	Indonesia	2007.04.21	50 m	Chile	1992.12.12	1169	Indonesia
2005.03.28	8.6	Indonesia	2011.03.11	39 m	Tohoku	2006.07.17	802	Indonesia
2007.09.12	8.4	Indonesia	2007.12.04	38 m	Canada	2010.10.25	431	Indonesia
2001.06.23	8.4	Peru	1993.07.12	32 m	Okushiri Is.	1994.06.02	238	Indonesia
2015.09.16	8.3	Chile	1996.01.02	29 m	Kamchatka	1993.07.12	208	Okushiri Is.
2006.11.15	8.3	Kuriles	2010.02.27	29 m	Chile	2009.09.29	192	Samoa
2003.09.25	8.3	Japan	1992.12.12	26 m	Indonesia	1992.09.02	170	Nicaragua
1994.10.04	8.3	Shikotan	2010.04.13	23 m	Peru	2010.02.27	156	Chile
2014.04.01	8.2	Chile	2009.09.29	22 m	Samoa	1999.08.17	155	Turkey
1996.02.17	8.2	Indonesia	2006.11.15	22 m	Kuriles	1996.02.17	110	Indonesia
2012.04.11	8.2	Indonesia	2006.07.17	21 m	Indonesia	1994.11.14	81	Philippines
2009.09.29	8.1	Samoa	2010.10.25	17 m	Indonesia	2007.04.01	50	Solomon Is.

The dates of the events are given in the following format: yyyy.mm.dd

diversity in wave characteristics) that are always connected with additional energy input from an external source (earthquake, volcano, landslide or their combination).

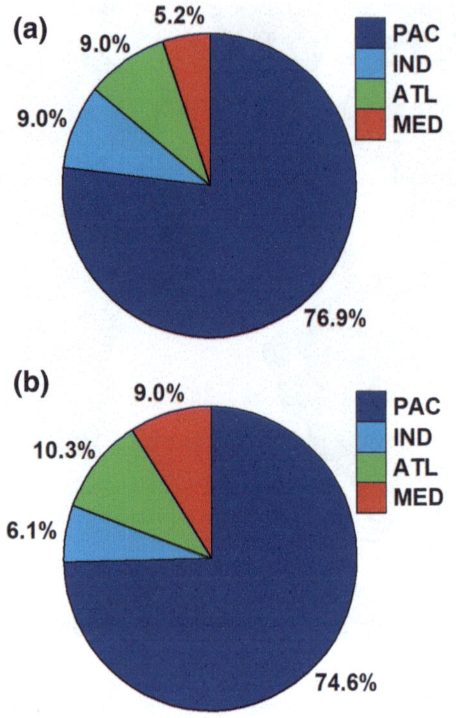

Figure 2
Distribution of tsunamigenic events by the main tsunamigenic regions of the World Ocean for 1992–2016 (**a**) and for 1901–2016 (**b**)

3. Data Analysis

3.1. Global Distribution of Tsunami Events from 1992 to 2016

The analysis of tsunami occurrence in the World Ocean for the period from 1992 to 2016 begins with a count of all events from the NCEI database. We included all tsunamis and tsunami-like events in the internal basins of all validities, except those that were considered to be erroneous events. A total of 290 tsunamis were observed during this period, which is approximately 116 events per decade. Their geographical distribution is shown in Fig. 1. There were 911 events for the time period 1901–1991 which is approximately 100 events per decade.

Among the events from 1992 to 2016, there were 32 destructive tsunamis that resulted in human fatalities, which included 2 trans-oceanic mega-tsunamis (2004 Indian Ocean and 2011 Tohoku). In addition to the 32 deadly tsunamis, there were 63 damaging or potentially damaging events that had maximum run-up $H_m > 1$ m, but with no fatalities. The NOAA Pacific Tsunami Warning Center's User's Guide (IOC 2014) categorizes $H_m > 1$ m to be a dangerous hazard that, and in most cases, would require a significant evacuation of the coast to protect lives. Thus, we consider events with maximum run-up $H_m > 1$ m to be potentially damaging.

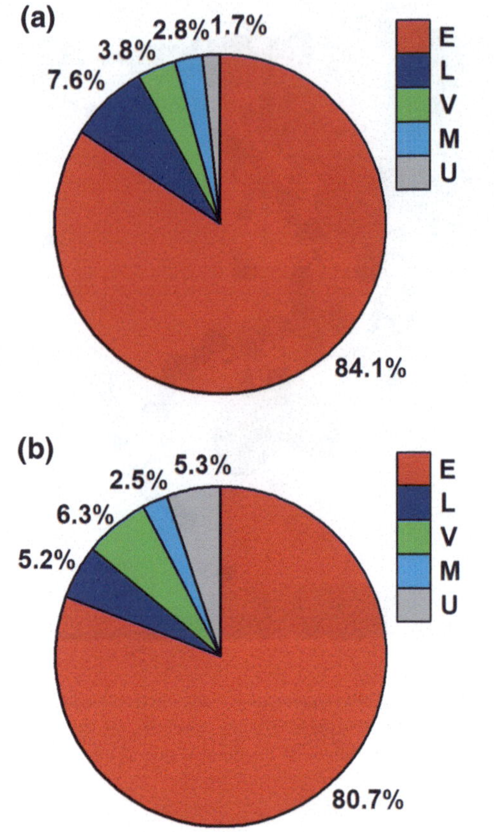

Figure 3
Distribution of tsunamigenic events by the type of sources for 1992–2016 (**a**) and for 1901–1991 (**b**). *E* earthquake, *L* landslide, *V* volcanic, *M* meteorological, *U* unknown

Table 1 provides the lists of the top fifteen tsunamis during the time period 1992 to 2016 by M_w magnitude, highest run-up, and number of fatalities. The December 26, 2004 Indian Ocean tsunami resulted in the historically largest number of fatalities (227,899 killed and missing people) from any tsunami in recorded history. The October 18, 2015 Icy Bay (Alaska) landslide tsunami ($H_m = 193$ m) gave the fourth largest run-up height among all the historical tsunamis.

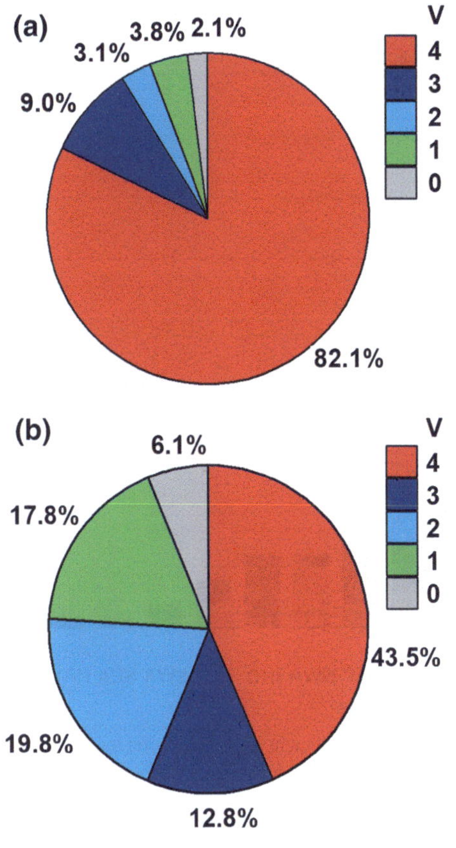

Figure 4
Distribution of tsunamigenic events by the NCEI validity index for 1992–2016 (**a**) and for 1901–1991 (**b**)

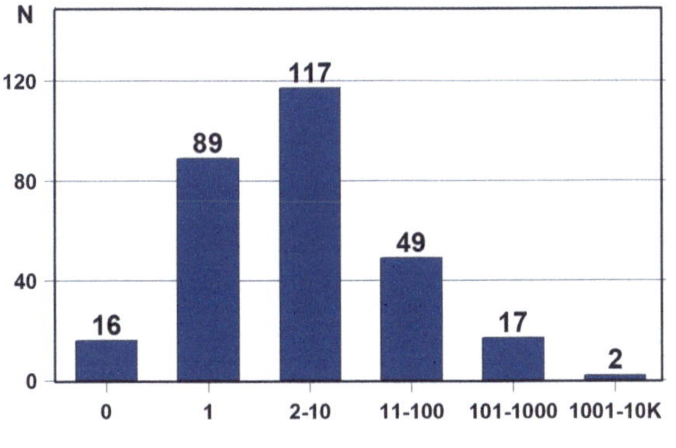

Figure 5
Distribution of 290 tsunamigenic events that occurred during 1992–2016 by the total number of run-up and height measurements available in the database

As one can expect, the three largest (by their M_w value) events, the December 26, 2004 Indian Ocean, the March 11, 2011 Tohoku and the February 27, 2010 Chile, appear in all three lists (M_w value, run-up, fatalities). However, the lowest event in the first list (M_w value), the M_w 8.1 September 29, 2009 Samoa tsunami, also appears in all three lists and the M_w 8.2 February 17, 1996 Indonesia tsunami appears in the third list (fatalities) as well. These low M_w 8 earthquakes generating destructive and fatal tsunamis show that M_w 8 events have the potential for large run-ups and significant loss of life.

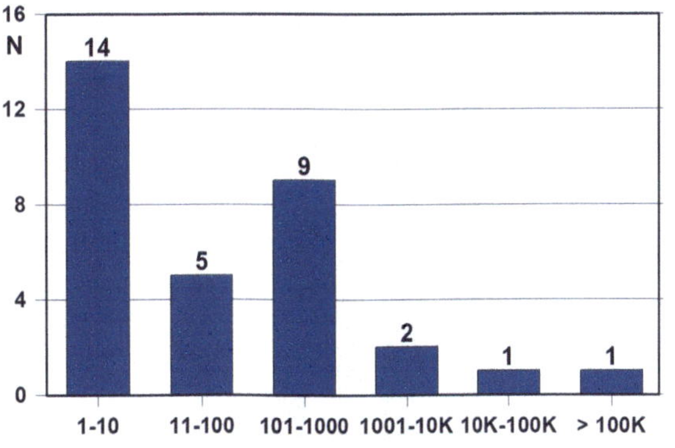

Figure 6
Distribution of 32 fatal events that occurred during 1992–2016 by the number of tsunami fatalities

3.2. Distribution of Tsunami Events in the World Ocean

We next examined the distribution of tsunami events in the World Ocean for 1992 to 2016. As throughout history, the Pacific continues to be the main tsunamigenic region with 223 events (76.9%) having occurred within its basin; 26 events (9.0%) occurred in the Indian Ocean, 26 events (9.0%) in the Atlantic (of these, 10 occurred in the Caribbean region), and 15 events (5.2%) in the Mediterranean region that also includes the Black and Azov seas (Fig. 2a). However, the 25-year period is too short to accurately reflect the level of tsunami activity in these large basins, especially taking into account the difference in seismic activity between, for instance, the Pacific and the Mediterranean basins. A similar analysis for 1901–2016 shows that there were 896 events (74.6%) that occurred within the Pacific basin; 73 events (6.1%) occurred in the Indian Ocean, 124 events (10.3%) in the Atlantic, and 108 events (9.0%) in the Mediterranean region (Fig. 2b). This ratio is a more accurate estimate of relative tsunami activity in the main tsunamigenic regions of the World Ocean.

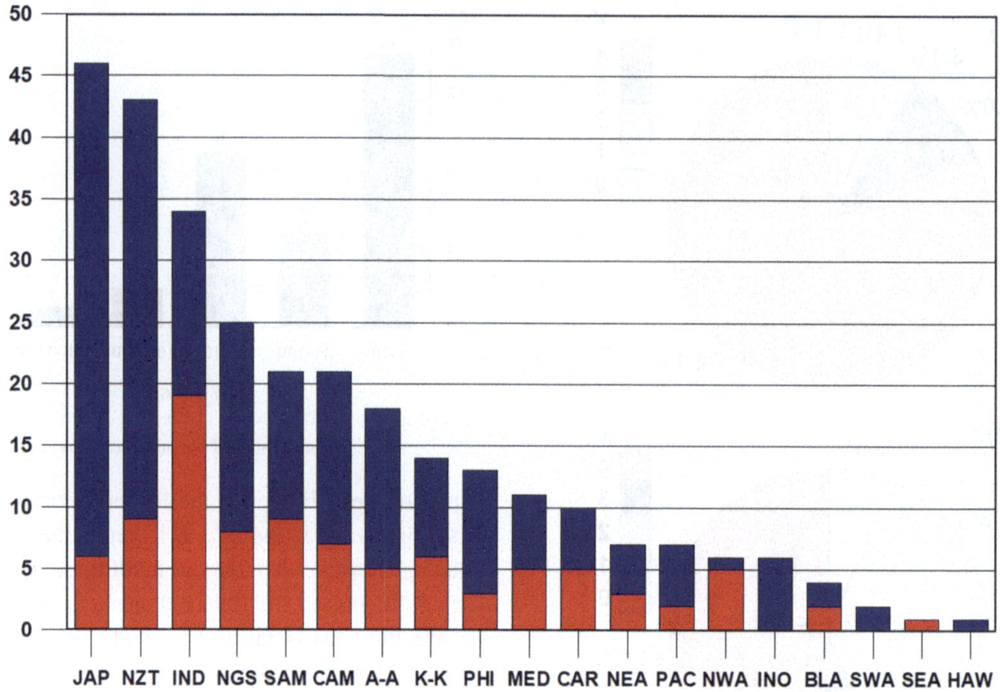

Figure 7
Distribution of 290 tsunamigenic events occurred in the World Ocean during 1992–2016 over the main tsunamigenic regions. Blue color—all events, red color—events with $H_m > 1.0$ m (damaging or potentially damaging events). *A–A* Alaska–Aleutians (including Canada west coast), *BLA* Black and Azov Seas, *CAM* Central America (including US west coast), *CAR* Caribbean, *IND* Indonesia, *INO* the rest of the Indian Ocean, *HAW* Hawaii, *JAP* Japan, *K–K* Kuril–Kamchatka, *MED* Mediterranean, *NEA* North-East Atlantic, *NWA* North-West Atlantic, *NZT* New Zealand and Tonga, *NGS* New Guinea and Solomon Is., *PAC* the rest of the Pacific Ocean, *PHI* Philippines, *SAM* South America, *SEA* South-East Atlantic, *SWA* South-West Atlantic

3.3. *Distribution of Tsunami Events by Source Types*

We then compared tsunami sources for the two time periods, 1992–2016 and 1901–1991. Since earthquake-induced movement of the ocean floor most often generates tsunamis, it is not surprising that for the period 1992–2016 there were 244 (84.1%) tsunamis generated by earthquakes and earthquake-generated landslides, and 22 landslide (7.6%), 11 volcanic (3.8%), 8 meteorological (2.8%), and 5 unknown (1.7%) sources (Fig. 3a).

The results for the time period 1901–1991 are fairly similar except that the fraction of events with unknown sources is much larger. There were 735 (80.7%) tsunamis generated by earthquakes and earthquake-generated landslides (Fig. 3b). This was followed by 47 landslide (5.2%), 58 volcanic (6.3%), 23 meteorological (2.5%), and 48 (5.3%) unknown sources (Fig. 3b).

3.4. *Distribution of Tsunami Events by Validity*

We next looked at the distribution of tsunamis by event validity. Each entry in the tsunami database has a validity score assigned by NCEI which ranges from − 1 for erroneous entries, 0 for seiches (standing wave oscillating in a partially or fully enclosed body of water), and 1–4 for questionable to definite tsunamis. For the period 1992–2016 the distribution is as follows: 6 (2.1%) validity 0, 11 (3.8%) validity 1, 9 (3.1%) validity 2, 26 (9.0%) validity 3, and 238 (82.1%) validity 4 (Fig. 4a). For the period 1901–1991 the distribution is as follows: 56 (6.1%) validity 0, 162 (17.8%) validity 1, 180 (19.8%) validity 2, 117 (12.8%) validity 3, and 396 (43.5%) validity 4 (Fig. 4b). There are 67 events with validity -1 for the period 1901–1991 and 5 events with validity − 1 for the period 1992–2016, but they were excluded from the analysis.

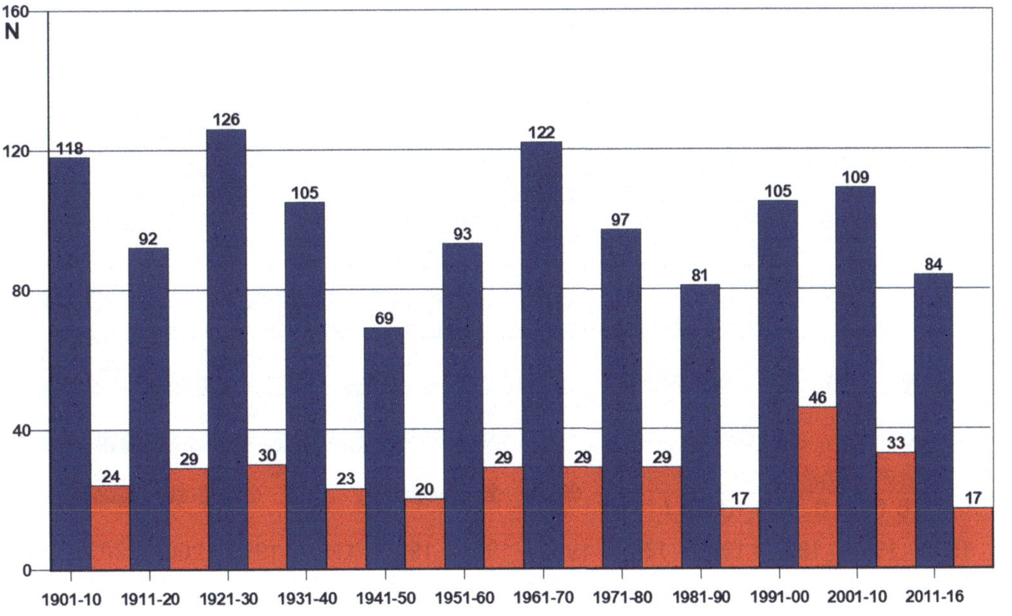

Figure 8
Number of tsunamigenic events (per complete decades except for 2011–2016) that occurred in the World Ocean from 1901 to 2016. Blue—all events, red—potentially damaging events with $H_m > 1.0$ m

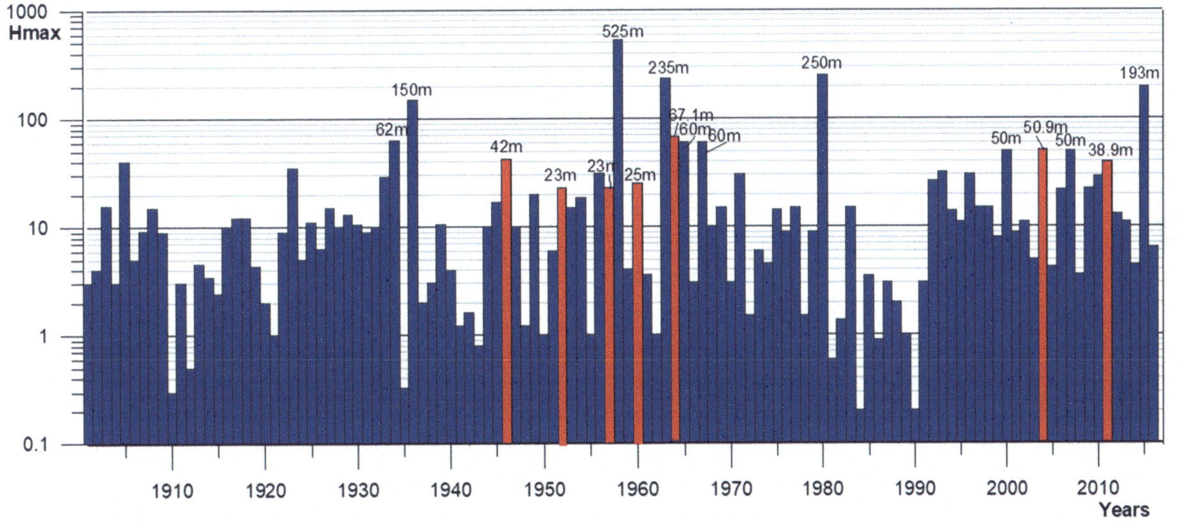

Figure 9
Maximum tsunami run-ups registered in the World Ocean during 1901–2016. Red color marks trans-oceanic mega-events resulting from M9 class subduction earthquakes. Top-10 run-up values and maximum run-ups related to 7 trans-oceanic mega-tsunamis are marked by their values

The percentage of tsunamis with event validities of 3 and 4 (91%) is much higher for 1992–2016 than for 1901–1991 (56%). As discussed in the Introduction, digital data and post-tsunami field surveys have improved data quality that in turn

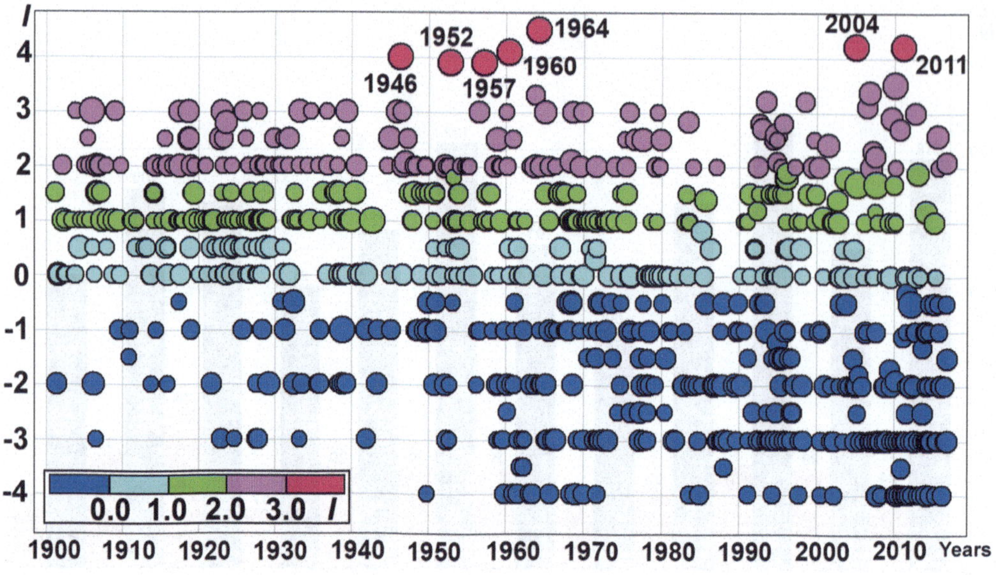

Figure 10
Tsunami occurrence in the World Ocean from 1901 to 2016. Events are shown as circles with the color determined by the tsunami intensity *I* and size proportional to the earthquake magnitude. Trans-oceanic mega-tsunamis are shown in red and marked by year of occurrence

increased the ability to register and validate tsunami manifestations.

3.5. Distribution of Tsunami Events by the Total Number of Run-up and Height Measurements

We next examined the distribution of tsunami events that occurred from 1992 to 2016 over the number of run-up and height measurements available for each event in the NCEI database (Fig. 5). The vast majority (195) of registered tsunamis and tsunami-like sudden level changes in coastal areas were weak non-damaging events and most were visible only on coastal mareograph or DART-type ocean-bottom instrumental records. These events have the average of 4.9 run-up and height measurements. Additionally, there are 15 events that are included in the database based on different sources of information, but they do not have any associated run-up, or height measurement, or eyewitness height estimates. For some of these events the run-up data is described in the tsunami event description, but it has not been entered into the run-up table. The events (49) with 11–100 run-up and height measurements reflect the increase in digital data available from sea level and DART-type instruments. Larger numbers (more than 100) of run-up and height measurements, are mainly the result of post-tsunami field surveys. Among all these events, only two trans-oceanic mega-tsunamis that resulted from M9 class subduction mega-earthquakes (2004 Sumatra and 2011 Tohoku) had more than 1000 run-up and height measurements.

3.6. Distribution of Tsunami Events by Number of Deaths

We then examined the distribution of tsunami events by the total number of fatalities that resulted from the tsunami (Fig. 6). The vast majority of events (258) did not result in any fatalities, so the percentage of deadly tsunamis (32) for this period is quite low (11%). Of the 32 fatal tsunamis the majority (28) caused less than 1000 deaths. There were a total of 251,869 deaths due to tsunamis during this period. The main input (97.8%) in the total tsunami death toll for this period (1992–2016) came from two trans-oceanic mega-tsunamis that occurred on December 26, 2004 in Indonesia (227,899) and on March 11, 2011 in Tohoku, Japan (18,454). Two additional events with more than 1000 tsunami fatalities were the July 17, 1998 Papua New Guinea tsunami (1636

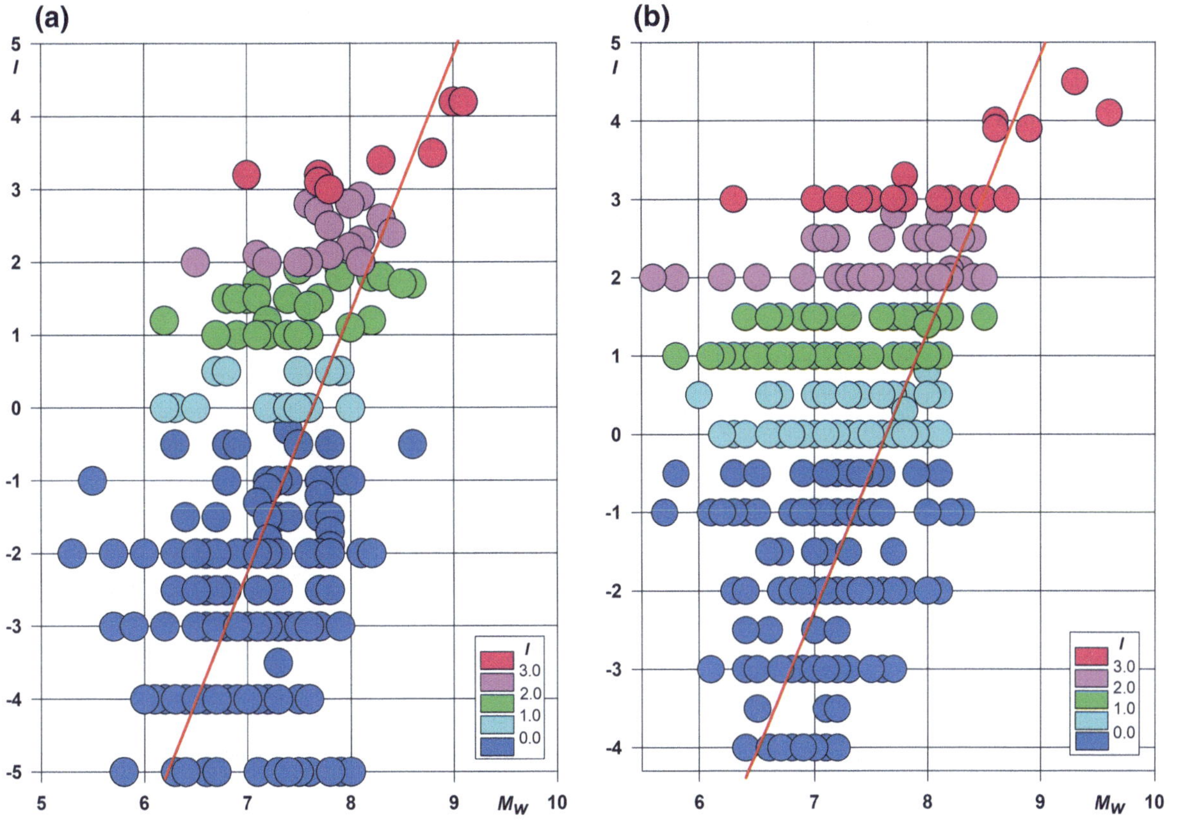

Figure 11
Dependence of tsunami intensity I on the source magnitude M_w for the period 1992–2016 (**a**) and for 1901–1991 (**b**). The solid red line shows the theoretical dependence $I = -2.71 + 3.55M_w$, obtained from Chubarov and Gusiakov (1985)

deaths) and the December 12, 1992 Flores tsunami in Indonesia (1169 deaths). All except two (November 4, 1994 Skagway, Alaska and April 13, 2010 Hualcan Lake, Peru) fatal tsunamigenic events from this period resulted from submarine earthquakes that occurred in an open sea. Among 32 deadly tsunamis, 18 events had the maximum wave height H_m more than 10 m, 7 events had H_m from 5 to 10 m and 7 events had H_m from 2.5 to 5 m.

3.7. Distribution of Tsunami Events over the Tsunamigenic Regions

The maximum wave height H_m is an important characteristic in determining the potential for a tsunami to cause damage or deaths. The NOAA Tsunami Warning Centers' guidance for a predicted H_m from 0.3 m to $\leq$ 1.0 m at a specific location indicates that it is likely a hazard for coastal recreation activities with potential for minor flooding. The guidance states that for a predicted $H_m >$ 1.0 m at a specific location there is likely a dangerous hazard or serious threat to coastal areas (IOC 2014). Following this, we subdivided all the tsunami events from the period from 1992 to 2016 into two groups: (1) the events with run-up heights $\leq$ 1.0 m and (2) the events with run-up heights $>$ 1.0 m. The latter are considered to be potentially damaging tsunamis.

The distribution of both groups of tsunami events was then determined for the main tsunamigenic regions of the World Ocean (Fig. 7). Of the 290 registered tsunamis and tsunami-like events, there were 95 damaging or potentially-damaging events that had maximum run-up or inundation height $>$ 1 m. The rest (195) were weak events and most were visible only on mareograph records. As expected, nine of the ten Pacific regions have the highest number of total tsunamis for this time period. Japan had the largest number of registered tsunamis (46) (Fig. 7). Of these, 6 tsunamis had $H_m > 1.0$ m. Five of these 6 events did indeed experience some level of damage from the tsunami. There were 43 tsunamis in New Zealand and the South Pacific islands on the Kermadec, Tonga, and New Hebrides Trenches. Of these, 9 tsunamis had $H_m > 1.0$ m and 6 had reports of damage. Indonesia and New Guinea-Solomon regions are in the third and fourth places, with 34 and 25 registered tsunamis, respectively. In total, these four regions produced 51% of all registered tsunamis and 44% of the potentially damaging tsunamis that occurred in the World Ocean during this period (1992–2016).

3.8. Distribution of Tsunami Events by Decade in the World Ocean

We next examined the total number of tsunami events per decade in the World Ocean during 1901–2016 (Fig. 8, blue column) and the total number of events per decade with $H_m > 1.0$ m (Fig. 8, red column). For the period from 1901 to 2010, there were on average 102 tsunami events per decade with 28 potentially damaging tsunamis ($H_m > 1.0$ m).

The fewest number of tsunamis was 69 events during the 1940s. This minimum is possibly the result of a lower reporting level during the Second World War. The decade of the 1980s gives the second clear minimum and it is especially clear for potentially damaging tsunamis (17 against 28 on average). Taking into account the likely lower catalog completeness during the 1940s, the 1980s possibly represents the decade minimum in tsunami activity for the entire XX century. From this summary, the 1990s appears to be a period of increased worldwide tsunami activity. The previous statement is corroborated by the number of potentially damaging tsunamis (46) that occurred in the 1990s.

3.9. Global Tsunami Occurrence in the World Ocean by Maximum Run-up Height

We then examined the maximum run-up and height measurement per year in the World Ocean during the time period 1901–2016 (Fig. 9). There are a total of 1201 tsunamigenic events during the time period and of those 399 did not have a measured or estimated run-up height, 476 had $H_m \leq 1.0$ m, 254 had $1.0 < H_m \leq 10.0$ m, and 72 had $H_m > 10.0$ m. The average one-year maximum run-up or height measurement during this 116-year period is 24.5 m. In consideration of this value, one should take into account that this number is largely affected by a few years with very high maximum run-ups, e.g. July 10, 1958 Alaska 525 m, May 18, 1980 Spirit Lake, Washington 250 m, October 9, 1963 Vajont Dam, Italy 235 m.

Seven trans-oceanic mega-tsunamis that occurred in the Pacific and Indian oceans during this period are marked in Fig. 9 by red color. These tsunamis had the largest run-up values in respective years with the average of $H_m = 38.6$ m per mega-event. However, the top-5 run-ups during this period (all with $H_m > 100$ m) resulted from much more local landslide-generated events that occurred in coastal bays and inlets and inland water reservoirs (lakes and rivers).

3.10. Tsunami Intensity and Global Tsunami Occurrence in the World Ocean

In the absence of practical methods of estimating the total tsunami energy, the intensity I on the Soloviev–Imamura scale is the best proxy for a tsunami magnitude and it most closely relates to the overall size of a tsunami. This scale was introduced by Soloviev (1972), who proposed to calculate the tsunami intensity according to the formula

$$I = 1/2 + \log_2 H_{av}, \quad (1)$$

where H_{av} is the *average* wave height along the nearest coast. The I scale is used in both NCEI and GTDB databases as the main parameter characterizing tsunami size. The initial determination of the tsunami intensity I for a large number of Pacific tsunamis was made by Soloviev during compilation of his catalogs (Soloviev and Go 1974, 1975; Soloviev 1978). Since 2003, the systematic determination of the tsunami intensity *I* has been made in the Novosibirsk Tsunami Laboratory of the ICMMG SD RAS for all the recent tsunamigenic events having a sufficient number (typically, more than 10) of run-up measurements on the basis of a procedure built into the WinITDB graphic shell (Gusiakov 2011).

Tsunami intensity *I* on the Soloviev–Imamura scale of each individual tsunami event versus time is presented (Fig. 10) to better understand global tsunami occurrence. In this diagram, all trans-oceanic mega-tsunamis (shown in red and marked by year of occurrence) are clearly plotted above all the other tsunamigenic events, even destructive ones (intensities 2–3). The obvious clustering of mega-events in the XX century (five destructive trans-oceanic mega-tsunamis during 18 years) with two recent mega-events (in 2004 and 2011) raises the question if this clustering is possible in the XXI century. The *I(t)* diagram further demonstrates the completeness of the global tsunami catalog in the second half of the XX century, as weak non-damaging tsunamis (registered only instrumentally) have almost regular occurrence.

3.11. Tsunami Intensity Versus Source Moment-Magnitude

One of the most important issues about seismogenic tsunamis is how their overall size or energy depends on the source magnitude of the parent seismic event. The relationship between the tsunami intensity on the Soloviev–Imamura scale and the source magnitude was first studied by Chubarov and Gusiakov (1985) and later, using more complete data by Gusiakov (2011). The main feature of this relationship is the large scattering of *I* values that exceeds six grades on the intensity scale, for events with $7.0 \leq M_w \leq 8.0$, where most tsunamigenic earthquakes occur.

The dependence $I(M_w)$ is shown in Fig. 11a for the period 1992–2016 and in Fig. 11b for the preceding period 1901–1991. As we can see, increasing the accuracy of intensity determination for recent tsunamis (in particular, for those that were followed by post-event field surveys) does not narrow the scattering in $I(M_w)$ dependence. In fact, looking at Fig. 11, one can conclude that this dependence does not actually exist. There is only a general tendency for an increase in the tsunami intensity with an increase in the source magnitude. This tendency pretty well fits the predicted dependence of *I* on M_w as discussed in Chubarov and Gusiakov (1985) and shown as red lines in Fig. 11.

The reasons for this large scattering are multifold and consistent with those described in Gusiakov (2011). Primary reasons include differences in the focus depth, source mechanism, source location, and the degree of involvement of secondary mechanisms, first of all, the involvement of submarine slides and slumps. There is some evidence that seismically induced slumping could greatly contribute to the increased run-up heights along some part of the Sanriku coast during the March 11, 2011 Tohoku tsunami (Tappin et al. 2014).

4. Discussion and Conclusions

Altogether, 290 tsunamis and tsunami-like events were observed in the World Ocean and inland water basins (lakes and rivers) during the 25-year period (1992–2016) marked by the beginning of post-tsunami surveys and wide distribution of digital recording of tsunami. Among them there were 32 destructive tsunamis that resulted in human fatalities which included 2 trans-oceanic mega-tsunamis (December 26, 2004 Indian Ocean and March 11, 2011 Tohoku). There were 63 damaging or potentially damaging tsunamis, having run-up or inundation height > 1 m, but with no fatalities. The rest (195) were weaker, non-damaging events usually visible only on instrumental records. Among them, 15 events were included in the database based on reports and publications, but they are still not associated with any run-up or water height values. As a result, global historical catalogs for tsunami events may be considered more complete for events than for run-ups.

This period (25 years) constitutes less than 1% of the total length of the historical tsunami catalog (from 2000 BC to present), but it contains 12% of the total number of events and 54% of the run-up and inundation heights. The latter is mainly due to the two trans-oceanic mega-tsunamis—the March 11, 2011

Tohoku and the December 26, 2004 Indian Ocean, that have 6,271 and 1,625 measured heights, respectively. In summary, they represent 54% of all the run-ups and 98% of all the tsunami fatalities reported during this period (1992–2016).

The Pacific remains the main tsunamigenic basin of the World Ocean where 77% of the 290 tsunamigenic events occurred, 9% occurred within the Indian Ocean basin, 9% within the Atlantic Ocean basin and 5% within the Mediterranean basin (including the Black and Azov seas). Among the main tsunamigenic regions, four regions: Japan, Indonesia, New Zealand-Tonga and New Guinea-Solomon, were the source for more than half (51%) of all the tsunamis that occurred in the World Ocean during this 25-year period. The frequency of tsunamis highlighted by the regional analysis presented herein helps quantify why tsunami research, risk assessment and mitigation activities have been more prevalent in these four regions.

For the last 25 years the global historical catalog can be considered as complete, at least, for potentially damaging tsunamis. Within this period, the average number of tsunamigenic events per decade is 116 (all events) and 38 (potentially damaging tsunamis). Based on this estimate, for the XX century (1901–2000) we can expect roughly 1160 tsunamigenic events in total with 380 potentially damaging tsunamis. In fact, the database contains 1008 events which includes 276 potentially damaging tsunamis. This demonstrates great potential for discovering new tsunami event information in existing data, particularly at the regional and local level.

Significant advances in tsunami instrumentation and data collection achieved in the recent years has led to improved quality of data, as evidenced by the increase in high validity events from 1992 to 2016. Despite these advances, our ability to register weak tsunamis remains inadequate in many regions of the World Ocean. It is seen from the fact that we do not have a single wave measurement for 7 shallow-focus coastal or submarine earthquakes with magnitude $M_w \geq 7.5$ (the threshold for issuing of a regional-wide tsunami warning) that occurred during these 25 years. Therefore, these potentially tsunamigenic events are not included in the historical database.

The main input in the total damage and the death toll came from two trans-oceanic mega-tsunamis that occurred on December 26, 2004 in Indonesia and on March 11, 2011 in Tohoku, Japan. These two events occurred in countries with quite different degrees of readiness to mitigate such disasters. While Indonesia at that time had not yet fully implemented its national tsunami warning system, Japan was at the highest degree of readiness among all the Pacific countries to counteract the tsunami attack, both technically and organizationally (Kong et al. 2015). Nevertheless, the number of resulted casualties (18,454) was the second largest [after the June 15, 1896 Sanriku tsunami resulted in 21,959 fatalities (JMA home page 2019)] throughout all Japanese history and the material damage reached 250 billion dollars. This fact clearly demonstrates the importance of considering the extreme rare mega-events, whose recurrence interval varies from 200–300 to 1000–1200 years, in the overall strategy of protection against natural disasters.

Acknowledgements

Some figures in the paper were drawn using procedures built in the PDM/TSU graphic shell (PDM/TSU 2018) developed in the NTL/ICMMG SD RAS as supporting software for GTDB maintenance. The authors wish to thank Fred Stephenson and Robert Geller for careful reading the manuscript and making numerous suggestions for its improvement as well as Tamara Kalashnikova for assistance in figures and tables preparation. The results presented in this paper were partly supported by the RFBR Grant 16-05-00450.

References

Arcos, N.P., Dunbar, P.K., Stroker, K.J., & Kong, L.S.L. (2019). Post-tsunami surveys impact on the NCEI/WDS global historical tsunami database. *Pure and Applied Geophysics, 176*, **(in press)**.

Chubarov, L. B., & Gusiakov, V. K. (1985). Tsunamis and earthquake mechanism in the island arc region. *Science of Tsunami Hazards, 3*(1), 3–21.

Eblé, M. C., & González, F. I. (1991). Deep-ocean bottom pressure measurements in the Northeast Pacific. *Journal of Atmospheric and Oceanic Technology, 8,* 221–233.

Gusiakov, V.K. (2009). Tsunami History: Recorded. In: Bernard, E. N., & Robinson, A. R. (Eds.), *The sea* (pp. 23–53), 15.

Gusiakov, V. K. (2011). Relationship of tsunami intensity to source earthquake magnitude as retrieved from historical data. *Pure and Applied Geophysics, 168*(11), 2033–2041. https://doi.org/10.1007/s00024-011-0286-2.

Intergovernmental Oceanographic Commission. (2014). Users guide for the Pacific Tsunami Warning Center enhanced products for the Pacific Tsunami Warning System, *IOC Technical Series, 105* (Revised ed.). Paris: UNESCO/IOC.

JMA (Japan Meteorological Agency) home page. 2019. https://www.data.jma.go.jp/svd/eqev/data/higai/higai-1995.html. Accessed 15 Jan 2019.

Kanamori, H. (1972). Mechanism of tsunami earthquakes. *Physics of the Earth and Planetary Interiors, 6,* 346–359.

Kong, L. S. L., Dunbar, P. K., & Arcos, N. (Eds.). (2015). *Pacific tsunami warning system: a half century of protecting the Pacific, 1965–2015.* Honolulu: International Tsunami Information Center.

NCEI/WDS Global Historical Tsunami Database, 2100 BC to Present (2018). https://doi.org/10.7289/v5pn93h7. Accessed 30 Nov 2018.

NTL/ICMMG GTDB (Novosibirsk Tsunami Laboratory of the Institute of Computational Mathematics and Mathematical Geophysics of Siberian Division of Russian Academy of Sciences) 2100 BC to Present (2018). http://tsun.sscc.ru/nh/tsunami.php. Accessed 15 Dec 2018.

PDM/TSU (Parametric Data Manager for Tsunami data management) graphic shell, NTL/ICMMG SD RAS. http://tsun.sscc.ru/PDM.htm. Accessed 15 Dec 2018.

Soloviev, S. L. (1978). *Basic data on tsunamis on the Pacific coast of the USSR, Izuchenie Tsunami v Otkrytom Okeane* (pp. 61–135). Moscow: "Nauka" Publishing House. **(in Russian)**.

Soloviev, S. L., & Go, C. N. (1974). *A catalogue of tsunamis on the western shore of the Pacific Ocean* (p. 308). Moscow: "Nauka" Publishing House. **(in Russian; English translation by the Canadian Institute for Scientific and Technical Information, No.5077, National Research Council, Ottawa, Canada, 1984, 439 pp.)**.

Soloviev, S. L., & Go, C. N. (1975). *A catalog of tsunamis on the eastern shore of the Pacific ocean.* Moscow: Academy of Science of the USSR, Nauka Publishing House. **(in Russian; English translation by the Canadian Institute for Science and Technical Information, No. 5078, National Research Council, Ottawa, Canada, 1984, 293 pp.)**.

Tappin, D. R., Grilli, S. T., Harris, J. S., et al. (2014). Did a submarine landslide contribute to the 2011 Tohoku tsunami? *Marine Geology, 357,* 344–361.

Wielandt, E., & Streckeisen, G. (1982). The leaf spring seismometer: Design and performance. *Bulletin of the Seismological Society of America, 72*(Part A), 2349–2367.

(Received December 18, 2018, revised January 18, 2019, accepted January 19, 2019, Published online February 11, 2019)

Pure Appl. Geophys. 176 (2019), 2809–2829

https://doi.org/10.1007/s00024-019-02191-7

Pure and Applied Geophysics

The Impact of Post-tsunami Surveys on the NCEI/WDS Global Historical Tsunami Database

Nicolas P. Arcos,[1,2] Paula K. Dunbar,[1,2] Kelly J. Stroker,[1] and Laura S. L. Kong[3]

Abstract—Since the late nineteenth century, post-tsunami survey teams have been collecting water height data (e.g. maximum inundation, runup, flow depth) as well as other types of data, including geologic and socio-economic effects. The NOAA National Centers for Environmental Information (NCEI) and co-located World Data Service (WDS) for Geophysics provide data management and access to global tsunami data. NCEI's Global Historical Tsunami Database includes information on over 2200 tsunami sources and over 26,000 runup points. The tsunami database has adapted to capture the advancements of post-tsunami survey data collection and distribution methods. The first International Tsunami Survey Team (ITST), which surveyed the impacts of the 02 September 1992 Nicaragua tsunami, spurred the formalization of standards and guidance for post-tsunami surveys. The post-tsunami survey data analyzed are a significant contribution, in quality and quantity, to the runup database. The 26 December 2004 Indonesia and 11 March 2011 Japan events alone account for over 25% of the total number of runup points in the database. Further, the analysis indicates that post-tsunami survey data capture a more complete representation of the tsunami that is being documented, not just the highest runup measurements. This range in post-tsunami survey runup heights for events, in the NCEI/WDS tsunami database, provides a more reliable historical tsunami dataset to test tsunami models against. This study also provides readers an understanding of the evolution of the tsunami database, particularly regarding the ongoing need to adapt to emerging scientific advances and standards.

Key words: Tsunami, database, survey, flow depth, runup, wave height, catalog.

[1] NOAA, NESDIS, National Centers for Environmental Information (NCEI), Boulder, CO 80305, USA. E-mail: nicolas.arcos@noaa.gov

[2] Cooperative Institute for Research in Environmental Sciences (CIRES), University of Colorado at Boulder, 216 UCB, Boulder, CO 80309, USA.

[3] UNESCO/IOC, and NOAA International Tsunami Information Center, Honolulu, HI 96818, USA.

1. Introduction

On 02 September 1992 (UTC) a deadly tsunami struck the Pacific coast of Nicaragua. Soon after, the first International Tsunami Survey Team (ITST) was formed to survey the affected area (Satake et al. 1993). Although this was the first recognized ITST, post-tsunami field surveys had taken place following the 1946 Aleutian Islands, 1964 Southern Alaska, 1983 Sea of Japan and 1985 Michoacán tsunami events, and many others. However, ITSTs are unique because of their international and interdisciplinary collaboration, as well as the standardization of survey methodology. Due in part to recognition of ITSTs, post-tsunami survey guidance and methodology has been adjusted and expanded over time (Kong 2011). Since their inception, ITSTs have emphasized standardized measurements in maximum runup, maximum water level and inundation extent (Synolakis and Okal 2005; Fig. 1). As a result, the general principles in standardization have allowed these measurements to be compared across continents, events and time. ITSTs have thus far documented a total of 33 tsunami events in the Pacific and Indian oceans and the Caribbean and Mediterranean seas (IOC 2014).

The NOAA National Centers for Environmental Information (NCEI, formerly National Geophysical Data Center) and co-located World Data Service for Geophysics (WDS) global tsunami archive (https://ngdc.noaa.gov/hazard/tsu.shtml) consists of the historical tsunami database, imagery, and raw and processed coastal tide gauge and DART ocean bottom pressure data. The NCEI/WDS historical tsunami database includes two related tables: global observations of tsunami sources and tsunami runup records (locations where tsunami waves were observed by eyewitnesses, field reconnaissance surveys, tide

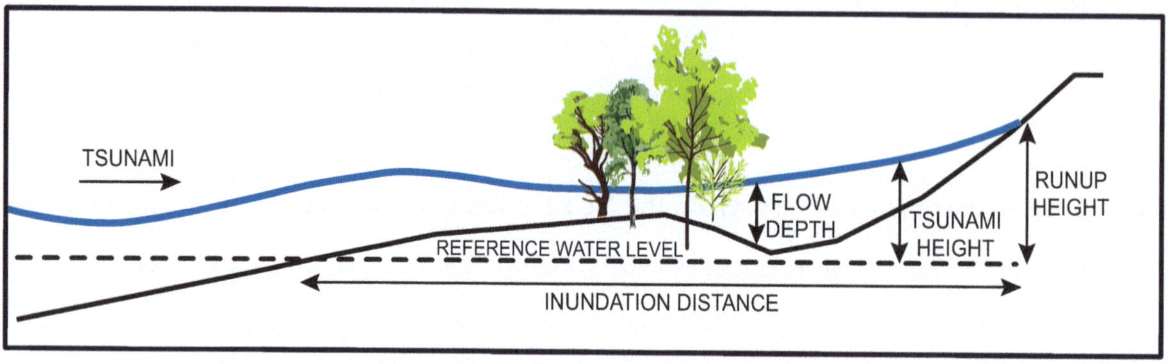

Figure 1
Tsunami hydrodynamic data terminology (NCEI/WDS)

gauges or deep-ocean sensors). The historical tsunami database incorporates, where available, maximum wave heights or runup data as well as inundation for each coastal location. The NCEI/WDS historical tsunami database has captured post-tsunami field surveys, pre- and post-1992, to enable user communities to access the information in a standardized format.

This study's intent is to assess the impact of post-tsunami surveys on the NCEI/WDS tsunami database with a focus on the pre- and ITST era. The post-tsunami survey data presented are not only a considerable contribution to the runup database, but also add information in locations where permanent data recorders (i.e. tide gauges) are not present. Simply put, post-tsunami surveys collect data in a wide variety of locations and environments to paint a more complete picture of the variation in runup and inundation along the coastline in a tsunami event. In addition, this study documents changes to the NCEI/WDS tsunami database classification of survey information through the same time period.

2. *Early Post-tsunami Surveys*

Arriving to impacted areas soon after a disaster, in this case a tsunami, is critical because as Quartenelli (1997) discusses, over time the collection of perishable data can be limited due to restricted access and limited cooperation among officials and victims. Furthermore, eyewitness accounts are not as reliable as time passes (Stallings 2007) and evidence of tsunami wave impact may grow faint with time, or is simply removed during post-disaster clean-up or reconstruction efforts. For this reason, post-disaster surveys have been an essential part of scientific data collection.

Tsunami survey teams have been collecting data since the late 1800s (Bourgeois 2009). Surveys are characterized by the systematic measurement and reporting of wave heights. For example, the 15 June 1896 Sanriku earthquake and tsunami resulted in approximately 27,000 deaths and was followed by two of the first post-tsunami surveys, one beginning 20 June and the other 28 July (Tsuji et al. 2014). The NCEI tsunami database contains 72 runup points for the tsunami, but none are currently classified as originating from a post-tsunami survey due to a lack of understanding of the methodology of the data collection at the time of entry into the database. Both post-tsunami surveys describe tsunami heights and socio-economic impacts (i.e. death, property damages) at surveyed locations. The changes in vegetation color, watermarks, debris lines and eyewitness accounts were utilized to measure tsunami heights much like a modern post-tsunami survey. Another early example, the 28 December 1908 Messina Straits tsunami survey included wave heights determined from tsunami traces and eyewitness testimonies. In his post-tsunami survey of the 1908 tsunami, Platania (1909) describes his post-event survey methodology which also includes socio-economic data, one of the first to do so.

Technological improvements have clearly improved aspects of surveys' precision. For the 1896 surveys, illustrative maps were used to convey location (Tsuji et al. 2014) not coordinates. Platania's survey used telescoping levels to determine heights (Platania 1909). The wide-spread use of GPS units and laser range finders in the ITST era have greatly increased the precision and volume of data collected. However, the underlying principles of what to measure (e.g. tsunami height traces, socio-economic data) in the early surveys were not considerably different from modern surveys.

3. *Historical Tsunami Measurements*

As post-tsunami surveys continued to take place in the early and mid-twentieth century, coordination of measurement standards among the scientific community was limited. Eaton et al. (1961) acknowledged a lack of clear guidance or standards in surveying techniques while reporting wave heights on the island of Hawaii following the deadly 22 May 1960 tsunami. Kajiura (1981) stated the challenges in the differing criteria for wave height measurements, noting the variation in datasets taken by different survey teams in the same area for the same tsunami. For instance, before terminology was standardized the phrase tsunami wave height used in pre-ITSTs surveys (e.g., Platania, 1909; Eaton et al. 1961; Tsuji et al. 2014) could refer to either height above sea-level or runup height (Fig. 1). In addition, the reference datum for tsunami wave heights varied. Tsuji et al. (2014) notes that the runup data from the 1896 survey were based on sea-level at the time of measurement. Yet, the Eaton et al. (1961) survey was based on the height above sea-level at the time of measurement, but later adjusted to mean lower low water. The confusion in terminology and the inconsistent reporting methodology could result in data providers and users misinterpreting the wave heights by many meters. This confusion is largely avoided in modern post-1992 surveys, due in large part to the guidance provided to ITSTs to correct water-level data for tidal conditions at the time of the tsunami (IOC 2014).

Runup is currently defined as the height of the land at the tsunami's farthest extent inland (inundation). Although instrumental observations from tide gauges (and DARTs) do not measure a runup on land, measurements from these instruments are listed as runup data in the tsunami database; thereby, enabling database users to more easily view and search across all tsunami wave data types. As such, all instrumental observations are listed as runup data in the database with a separate field to indicate the type of measurement (e.g. tide gauge, DART).

The NCEI/WDS historical tsunami database has endeavored to consistently utilize the terms defined in the Report of the International Measurements Workshop (Lander and Yeh 1995). However, the lack of clarity of datum utilized as well as the influence of tectonic uplift or subsidence are not always known from source documents. Dunbar et al. (2008) explain that despite a detailed data and literature review, the data manager may not be able to determine the survey methodology or datum utilized. However, Lander and Yeh (1995) and the subsequent Post-tsunami Survey Field Guide (IOC 1998) provided clarity, guidance and standardization to the post-tsunami survey community, thereby, creating more consistent datasets from that point going forward.

4. *Guides and ITSTs*

Collecting tsunami wave height or runup data throughout affected coastlines is critical to understanding the impact and cause of that particular event. Moreover, these records help contribute to a database that tsunami models can be reliably tested against. Although technological advances have been made, field work shortly after a hazardous event is still the most common and successful method of data collection (Tierney 2002). Thus, the establishment and updating of the Post-tsunami Survey Field Guide (IOC 1998) continues to be a critical aspect of establishing dependable databases for tsunami modeling and forecasting.

Efforts to standardize tsunami survey measurements began in the early 1980s (IOC 1998) but it was not until the early 1990s that the process began to be formalized by an internationally recognized body, what is now known as UNESCO's Intergovernmental Oceanographic Commission (IOC). The International Tsunami Information Center (ITIC) led the development of the 1998 Post-tsunami Field Guide (IOC 1998) that encouraged the minimum requirement of measuring runup height, water height and inundation extent, although it was the 1995 International Tsunami Measurements Workshop that initially determined the definitions and requirements of the aforementioned terms (Lander and Yeh 1995). The first ITST in Nicaragua was the basis for much of the guidance established at the 1995 workshop (Arcos et al. 2017). The workshop settled on the following definitions which were utilized by the 1998 field guide and persist with minimal modifications:

- Maximum runup is the difference between the elevation of maximum tsunami penetration and the elevation of the shoreline at the time of tsunami attack
- Maximum water level is the difference between the elevation of the highest local water mark and the elevation of the shoreline at the time of tsunami attack

Other data types highlighted for collection in the 1998 Guide included socio-economic impacts (damages, casualties, etc.) and geological effects. The tsunami database maintains socio-economic impacts with the following fields: deaths, missing, injuries, damages (in USD), and houses destroyed and damaged. A global standard is not applied across countries for reporting economic damages, as a result it is difficult to compare reported economic losses across time and places. However, the damages field provides users a broad understanding of damages sustained, particularly when the only damages reported are not captured by another existing field (e.g. houses destroyed, deaths, etc.). Information on the geological effects of tsunamis, such as erosion and sedimentation, have been documented in modern field surveys including prior to the 1998 field guide (Bourgeois 2009). Geological effects are documented in the historical tsunami database as well as in the NCEI/WDS Global Tsunami Deposits Database (Dunbar and McCullough 2012).

Following the ITST for the 29 September 2009 Samoa Islands tsunami, guidance and protocols were updated to provide more focus on improved coordination and data sharing (Kong 2011). This aimed to create better distribution of and access to data collected from post-tsunami surveys and gave guidance for combining all event data into a joint report. In 2014, the Post-tsunami Field Guide was updated to encourage the expansion of data collected on field surveys (IOC 2014). The most significant modification to the original 1998 document was the emphasis on measuring flow depth. Historically, flow depths had been measured on field surveys but perhaps not as extensively as following the 2004 Indian Ocean event (e.g. Borrero 2005b; Fritz et al. 2006b; Fritz and Borrero 2006). The increased frequency in flow depth measurements, in combination with the updated field guide, resulted in the addition of a flow depth classification in the NCEI/WDS historical tsunami database in 2010.

While many ITSTs did not have a direct impact on the tsunami database structure, they are worth highlighting for their unique scientific and data contributions. For instance, the 1992 Nicaragua post-tsunami survey recorded distinctive characteristics of "tsunami earthquakes" as eyewitnesses described the weak or soft shaking before the tsunami arrival (Kanamori 1972; Satake et al. 1993). The ITST for the 1993 tsunami in the Sea of Japan measured a 32-m maximum runup that is not consistent with the tsunami's other runup values. A steep valley location was the reason for this outlier runup value; thereby, emphasizing the need of high-resolution topography to determine runup potential (Shuto and Matsutomi 1995). Also, the first image of a leading depression of a tsunami at a local beach was collected by Borrero et al. (1997) during the ITST for the 1995 Manzanillo Mexico tsunami (Synolakis and Okal 2005). Following the 1998 Papua New Guinea tsunami, five marine expeditions took place to survey the offshore area off of the impacted coastline. This was the first large-scale international marine surveying effort to take place so soon after a tsunami (Kawata et al. 1999; Synolakis et al. 2002; Tappin et al. 2008). The marine investigations elucidated the presence of a landslide as the tsunami source as well as facilitated tsunami modeling efforts (Sweet and Silver 2003).

The ITSTs for the 2004 Indian Ocean tsunami had countless pioneering efforts, including the first analyses of post-tsunami satellite imagery enabling assessments of tsunami impacts (Borrero 2005a; Ramakrishnan et al. 2005). Clearly, ITSTs have contributed data to the tsunami database that are collected in a systematic and consistent manner, though more importantly new data types and data analyses captured in the NCEI/WDS historical tsunami database have preserved and furthered scientific advances.

As scientists develop new techniques to take advantage of emerging technologies, new data types and formats will emerge thus presenting challenges to data providers and managers. For instance, Fritz et al. (2006a) developed a procedure to analyze and determine tsunami flow velocity from amateur video of the 2004 Indian Ocean tsunami. The procedure, based on particle image velocimetry, took an evolutionary step as ground-based LiDAR measurements were incorporated to calibrate video of the 2011 Japan tsunami to determine tsunami flow velocities at Kesennuma Bay, Japan (Fritz et al. 2012). Most recently, Kalligeris et al. (2016) positioned GPS drifters and a camera to capture the arrival of the 16 September 2015 Chile tsunami at Ventura, CA. The data (i.e. flow velocity, current behavior) are important for developing guidance on engineering of structures (including vertical evacuation structures), maritime guidance, and evacuation planning. Consequently, it is critical for the scientific community to reach a consensus on guidance for collecting and distributing data based on video footage, particularly in an era where an increasing amount of amateur footage is available. NCEI/WDS and other data providers must then work with scientific community to determine the best manner for long-term archiving and access to the data for users attempting to validate inundation models or develop engineering guidance.

Following a tsunami when new research or anomalies are discovered, excerpts from published papers are often added to the "Comments" field of the particular tsunami event record. The related reference is also listed with the event. The comments field provides a more complete picture of the data and information from the tsunami and can be used to populate new data fields as they are added.

4.1. Challenges with Pre-ITST Surveys in the Database

Through the existence of the NCEI/WDS historical tsunami database, information fields have been added and definitions of fields modified to reflect the type of data being collected and scientific consensus on the definition of terms. Originally, the database had classified all wave observations as "runups", with no clear distinction among the types and quality of the observations. A significant effort was made to distinguish between instrumental (i.e. tide gauges) and non-instrumental (i.e. eyewitness, survey), as well as tsunami waves versus standing waves. The non-instrumental runup data were all assumed to be runup heights, including eyewitness measurements.

After the new post-tsunami survey classifications were added (e.g., flow depth, water height, runup height) the data points classified under runup height in the pre-ITST surveys were moved to the classification of eyewitness account. Consequently, the majority of pre-ITST observations derived from post-tsunami surveys prior to 1992 and added to the database before new classifications were added, are currently classified as "eyewitness measurements" due to lack of resources to revise the database and research whether the source was indeed an eyewitness or a post-tsunami survey. Efforts have been, and continue to be, made to review and correct pre-ITST event runup points to be labeled as post-tsunami survey measurements. This paper focuses on the 8 pre-ITST events with surveys contributing more than five runup points in the database, thereby excluding small survey efforts or events where the database may not be fully corrected. Currently, approximately 1100 runup points from 8 pre-ITST tsunamis have been assigned the classification of post-tsunami survey runup points.

The process of classifying pre-ITST data points as survey data is complicated by under reporting of methods and references. Early sources of tsunami data, such as scientific catalogues and technical reports, did not always provide details regarding their primary sources of data. Soloviev and Go (1974, 1975) are considered authoritative sources of tsunami information for the timespan in the Western and Eastern Pacific regions, respectively. However,

Soloviev and Go (1974) did not always differentiate between an eyewitness account and a post-tsunami survey. Thus, most non-instrumental observations are listed as eyewitness observations in the tsunami database. However, upon closer inspection, research may reveal that some of their listed observations did indeed originate from post-tsunami surveys. Although Soloviev and Go (1974) provide a list of references for each event summary, sources for each individual observation are not readily identifiable. Another consideration is that governments and political situations can delay or prevent the timely reporting of tsunami event data. For example, the USSR restricted the dissemination of data related to the damaging 1952 tsunami. Within the USSR the event did not show up in reports until after 1956 and was not widely released until the 1990s (Savarenskiy et al. 1958; Kaistrenko and Sedaeva 2001).

One example of the challenge of classifying data listed in Soloviev and Go (1974) is the 01 September 1923 earthquake and tsunami event. This event is the deadliest natural disaster in Japanese history, claiming over 140,000 lives primarily due to the ground shaking and subsequent fires. The largest tsunami wave was 13 m at Oshima Island with a total of 140 runup points recorded in the tsunami database. A total of 105 runup points are attributed to Soloviev and Go (1974) but a review of documents cited demonstrates that a post-event survey was conducted. Imamura (1924) discusses that the wave heights provided were collected by colleagues. Suda (1924) also confirms that at least some heights collected at Jogashima were based on traces left by the tsunami. Thus, confirming that at least some of the observations attributed to Soloviev and Go (1974) were not eyewitness accounts, as flagged in the tsunami database. The critical work of ensuring runup data exist in the database has been largely the focus of the quality control database effort, however, work remains to correct and/or refine the classification of the runup data.

5. Pre-ITST Surveys in the Database

The 1896 tsunami survey is among the earliest known post-tsunami surveys; however, this is not one of the eight pre-ITST events that has been corrected to reflect post-tsunami survey measurements. Thus, to compare the pre-ITST and post-1992 post-tsunami survey data, this analysis will be limited to tsunamis beginning in 1900. As of April 2019, only eight pre-ITST events have been revised to correct eyewitness runup points to post-tsunami survey runup points.

Of the 8 pre-ITST events analyzed, the 27 November 1945, 01 April 1946, and 09 July 1956 tsunami events include a significant number of runup points, 9 (out of 9), 89 (out of 416) and 68 (out of 68) respectively, collected decades after the events in the post-1992 era. For the three events, much of the post-tsunami survey runup data collected decades later were based on eyewitness testimonies with runup measurements taken using the post-1992 surveying methods (Okal et al. 2002b, 2009, 2015; Okal and Hébert 2007). Okal et al. (2003) did have some ability to map 1946 tsunami traces as they surveyed some areas that are unpopulated thus more able to preserve such markers. Additionally, the manner of reporting, as is common in modern surveys, is a table format that normally includes location name, coordinates (latitude, longitude), maximum wave height, and measurement type. The surveyors clearly and thoughtfully filtered out any testimonies that may be questionable, ensuring the runup data were accurate; thereby, the runup data are classified as post-tsunami survey runup points in the tsunami database.

The main difference between the tsunami survey methodology used for the post-1992 surveys of the 1945, 1946, and 1956 events and that of other post-1992 surveys conducted soon after the event, is the limited amount of perishable tsunami traces (e.g. water marks, flow direction indicators, wrack/debris lines). This methodology has allowed scientists to collect additional data for historical but underdocumented tsunami events and therefore the tsunami database continues to preserve eyewitness testimonies. In this paper's analysis, the 1945, 1946, and 1956 runup data collected post-1992 are grouped with pre-ITST events due to a combination of (1) the timing of the tsunami events and (2) the limited data centered on tsunami traces.

The post-tsunami surveys for the remaining five pre-ITST events were conducted soon after the event and were aided by tsunami traces. These five remaining pre-ITST events surveyed had a combination of descriptive and parametric components. Descriptive reports present challenges when converting to a parametric database as they require more interpretation from a database manager (Gusiakov 2009). For instance, the runup point locations are approximations. Moreover, wave heights are typically not calculated based on modern post-1992 reference datum.

Current IOC guidance (2014) insists that water-level data be corrected for tidal conditions at the time of the tsunami. However, post-tsunami survey runup data for the five remaining pre-ITST events did not make the aforementioned tidal corrections. For instance, runup data collected for the Hawaiian Islands was done relative to mean lower low water (Tsunami Technical Review Committee 2002). While the tsunami database notes the reference datum and cites the source of information, database users would not be able to easily or directly compare the runup data to post-tsunami survey runup data collected after 1992. Modern surveys report runup data in a table format that enables users to easily extract and interpret data, in comparison to the more easily misinterpreted descriptive format of pre-ITST surveys.

6. ITSTs in the Database

In response to the increase in post-tsunami scientific surveys in recent years, particularly following and including the 2004 Indian Ocean event, the tsunami database made adjustments to the classification of types of measurements to include post-tsunami survey measurements. The following post-tsunami survey classifications now exist within the runup field: runup height, water height, splash mark and flow depth.

The NCEI/WDS Global Historical Tsunami Database includes information on over 2200 tsunami sources and over 26,000 runup points (all measurement types). A total of 33 completed designated ITSTs (Fig. 2) have provided approximately 10,000 runup points to the tsunami database, which is approximately 40% of the tsunami runup table.

7. 2004 and 2011 Impact on the Database

Of the 33 ITSTs that have been completed, 2 recent events have had the most impact on the number of runup points in the database. The 26 December 2004 Indonesia and 11 March 2011 Japan tsunamis make up approximately 73% of the ITST post-tsunami survey runup points in the tsunami database. The data collected from these two devastating tsunami events have increased the quantity of runup points in the database and the advancement of the post-tsunami survey guidance allows an improved understanding of the data collected. The 2004 Indian Ocean tsunami event resulted in over 227,000 deaths and the wide-spread impact led to extensive media coverage (NCEI/WDS 2019). Hundreds of researchers and dozens of ITSTs surveyed 19 countries and territories to document the 2004 tsunami event. The 2004 Indian Ocean tsunami provided over 1400 post-tsunami survey runup points from ITSTs, approximately 15% of the ITST post-tsunami survey runup points. The 2011 Japan tsunami provided approximately 5700 post-tsunami survey runup points from ITSTs, approximately 59% of the ITST post-tsunami survey runup points. Approximately 22% of all tsunami runup points in the NCEI database are from the 2011 Japan tsunami post-tsunami surveys. Over 1000 post-tsunami survey runup points were $\geq$ 15 m (Fig. 3).

Most events see a decrease in the number of runup points between 10 and 15 m range to $\geq$ 15 m range, but Fig. 3 displays a curious increase in number runup points for the $\geq$ 15 m range for the 2004 and 2011 events. The 2004 Indian Ocean event has 456 post-tsunami survey runup points on the island of Sumatra, but 426 of those points are on a 250 km stretch of coastline on the north and northwestern part of Aceh province which happens to have some of the highest runup points (including the maximum runup of 51 m) (Fig. 4). The tsunami database shows that the post-tsunami survey data for this coastline was provided by 12 reports, each providing unique runup points. Therefore, it appears the hundreds of researchers and dozens of teams contributing post-tsunami survey runup data were attracted to this maximum runup area resulting in a large number of data points in a relatively small area. This clustering

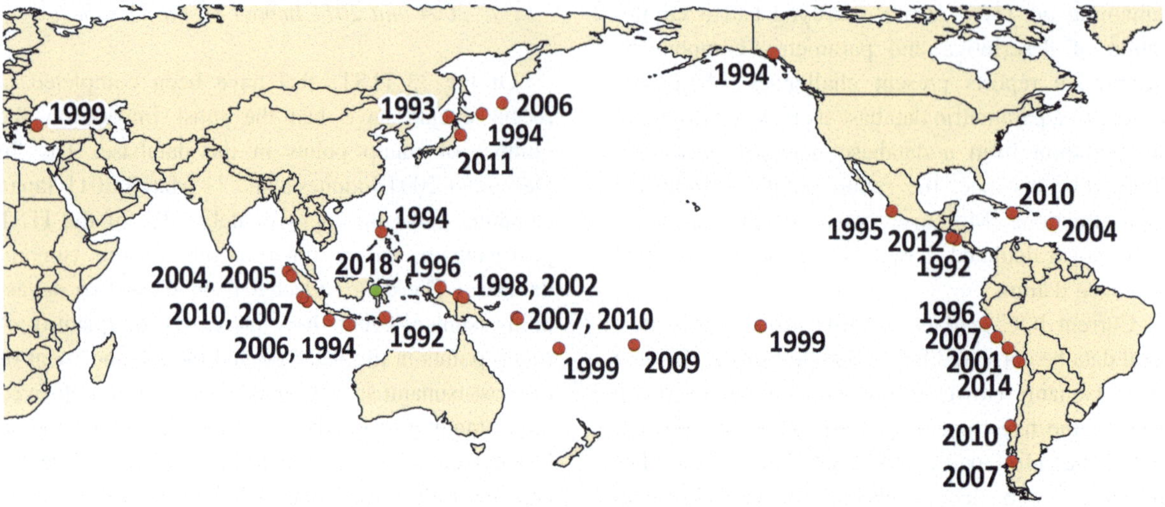

Figure 2
Spatial distribution and dates of ITSTs globally. Red circles (33) are completed ITSTs, green circle (1) is in progress ITST for 28 September 2018 Palu, Indonesia event (modified from Intergovernmental Oceanographic Commission 2014)

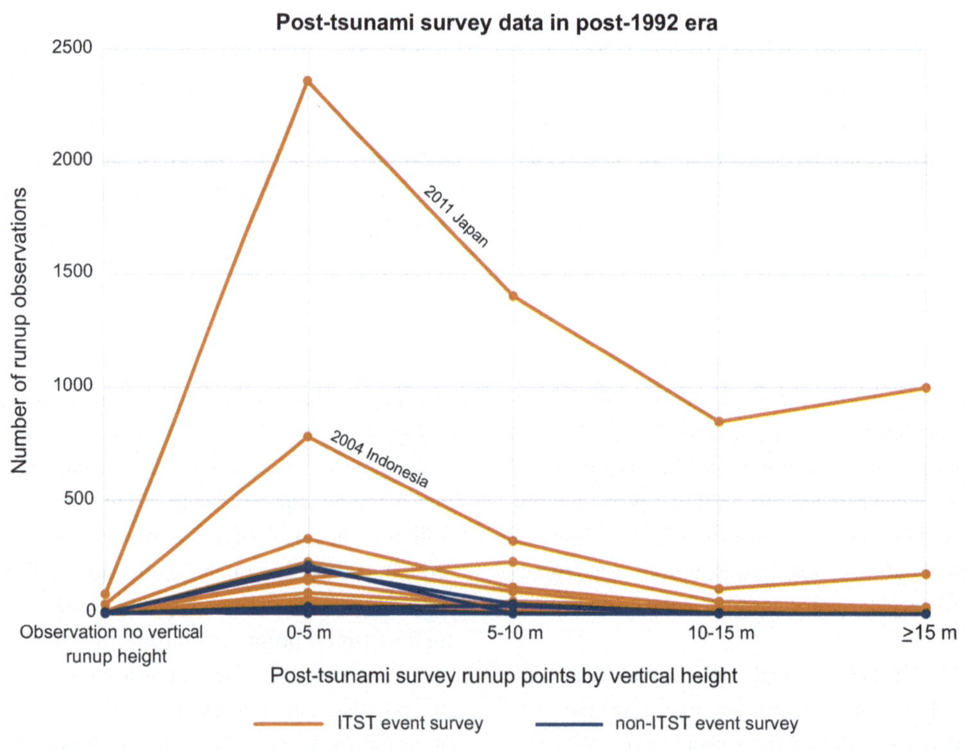

Figure 3
Post-tsunami survey data in post-1992 era in NCEI/WDS tsunami database. Each line represents a single tsunami surveyed by ITST (orange) or other post-tsunami survey team (blue). Runup points collected in the 0–5 m range are more abundant for all but three tsunami events (15 Nov 2006, 27 Feb 2010, 28 Oct 2012). The 2004 and 2011 tsunami events see an increase in number of runup points between 10 and 15 m range to $\geq$ 15 m, possibly due to the more focused post-tsunami survey effort near the highest runup points

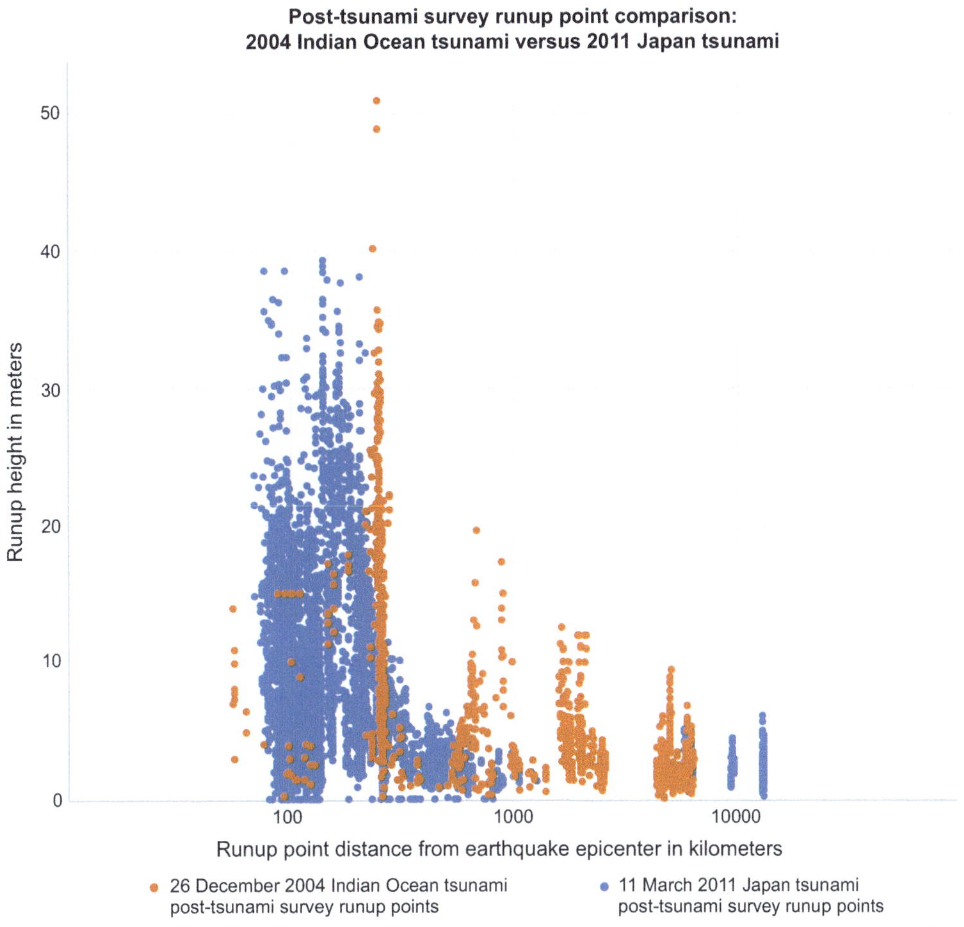

Figure 4
Post-tsunami survey runup point comparison. 2004 Indian Ocean (orange) and 2011 Japan (blue) post-tsunami survey rununp heights shown relative to distance from the earthquake epicenter. The 2004 and 2011 post-tsunami survey runup data both show a large amount of data collected at similar distances to the maximum runup, but more than 50% of the 2004 runup data are from the far-field (> 1000 km). The 2011 post-tsunami survey runup data has large clustering in near-field close to the maximum runup, but large amount of data still collected in far-field. Both events show a significant amount of runup data collected below 10 m height, in near- and far-field

of data points was largely restricted to Aceh, as post-tsunami survey data was collected from the event's 19 affected countries and territories. Figure 4 displays the distribution of runup data collected at distances far beyond the epicenter location and the maximum runup points.

Similar to the 2004 post-tsunami survey data, the 2011 Japan tsunami's increase in runup points from 10 to 15 m range to $\geq$ 15 m (Fig. 3) may be a result of focus on the most impacted area. A 450 km stretch of coastline on northeastern Honshu island is where almost all 10–15 m and all $\geq$ 15 m runup data are located despite the entirety of eastern Japan's coastlines being surveyed. As a result, it appears these unique tsunami events of 2004 and 2011 do in fact see more data collected near the highest runup points; however, as noted the importance of complete geographical coverage was not ignored (Fig. 4). Also, the 0–5 m range of post-tsunami survey runup points for both events is still significant for 2004 and 2011 tsunamis, 55% and 40% of total post-tsunami survey runup points respectively (Table 1).

Table 1

List of tsunami events with number of post-tsunami survey runup points by vertical runup height in meters in NCEI/WDS tsunami database

Date (UTC)	Source location	Survey	Observation with no vertical height		0–5 m		5–10 m		10–15 m		> 15 m	
			No.	No. near-field[a]	No.	No. near-field[a]	No.	No. near-field[a]	No.	No. near-field[a]	No.	No. near-field[a]
1945-11-27	Makran Coast, Pakistan[c]	E	0	0	4	4	4	4	1	1	0	0
1946-04-01	Unimak Is, Alaska, USA[c]	E	1	0	207	0	168	8	51	7	18	14
1952-11-04	Kamchatka, Russia	E	0	0	76	0	3	0	0	0	0	0
1956-07-09	Greece[c]	E	0	0	61	61	3	3	3	3	1	1
1957-03-09	Andreanof Is., Alaska, USA	E	0	0	144	0	29	3	3	1	4	3
1960-05-22	Central Chile	E	0	0	259	0	11	1	2	1	0	0
1964-03-28	Prince William Sound, Alaska, USA	E	0	0	56	0	1	0	0	0	1	1
1976-08-16	Moro Gulf, Philippines	E	11	9	7	6	2	2	0	0	0	0
1992-09-02	Nicaragua	I	0	0	22	22	22	22	0	0	0	0
1992-12-12	Flores Sea, Indonesia	I	2	2	68	54	8	8	4	3	4	2
1993-07-12	Sea of Japan	I	0	0	42	13	3	2	0	0	0	0
1994-06-02	Java, Indonesia	I	0	0	40	40	24	24	3	3	0	0
1994-10-04	Kuril Is, Russia	I	1	1	21	20	8	8	1	1	0	0
1994-11-04	Skagway, Alaska, USA	I	0	0	0	0	1	1	0	0	0	0
1994-11-14	Philippine Is	I	0	0	50	50	4	4	0	0	0	0
1995-10-09	Manzanillo, Mexico	I	0	0	16	16	2	2	1	1	0	0
1996-01-01	Sulawesi, Indonesia	P	0	0	16	16	0	0	0	0	0	0
1996-02-17	Biak, Indonesia	I	0	0	2	2	1	1	0	0	0	0
1996-02-21	Northern Peru	I	2	2	39	39	1	1	0	0	0	0
1998-07-17	Papua New Guinea	I	0	0	21	21	6	6	2	2	0	0
1999-08-17	Izmit Bay, Turkey	I	0	0	19	19	0	0	0	0	0	0
1999-09-13	French Polynesia	I	0	0	4	4	5	5	0	0	0	0
1999-11-26	Vanuatu Is	I	0	0	11	11	1	1	0	0	0	0
2001-06-23	Southern Peru	I	2	2	26	26	10	10	0	0	0	0
2002-09-08	Papua New Guinea	I	0	0	55	55	1	1	0	0	0	0
2003-09-25	Hokkaido Is, Japan	P	0	0	211	211	0	0	0	0	0	0
2004-11-21	Guadeloupe	I	0	0	1	1	0	0	0	0	0	0
2004-12-26	Sumatra, Indonesia	I	42	7	778	185	319	214	111	94	175	175
2005-03-28	Simeulue, Indonesia	I	0	0	38	38	0	0	0	0	0	0
2006-07-17	Java, Indonesia	I	0	0	92	92	38	38	14	14	16	16
2006-11-15	Kuril Is, Russia	I	0	0	7	7	55	55	33	33	19	19
2007-04-01	Solomon Is	I	0	0	146	146	28	28	1	1	0	0
2007-04-21	Southern Chile	I	0	0	0	0	1	1	1	1	1	1
2007-08-15	Southern Peru	I	0	0	59	59	7	7	1	1	0	0
2007-09-12	Sumatra, Indonesia	I	0	0	24	24	1	1	0	0	0	0
2009-09-29	Samoa Is	I	12	12	330	329	118	118	28	28	10	10
2010-01-03	Solomon Is	I	0	0	30	30	15	15	0	0	0	0
2010-01-12	Haiti	I	2	2	19	19	0	0	0	0	0	0
2010-02-27	Central Chile	I	1	0	158	96	230	230	55	55	31	31
2010-10-25	Mentawai, Indonesia	I	0	0	228	222	100	100	8	8	1	1

Table 1 *continued*

Date (UTC)	Source location	Survey	Observation with no vertical height		0–5 m		5–10 m		10–15 m		> 15 m	
			No.	No. near-field[a]	No.	No. near-field[a]	No.	No. near-field[a]	No.	No. near-field[a]	No.	No. near-field[a]
2010-12-25	Vanuatu Is	P	0	0	8	8	0	0	0	0	0	0
2011-03-11	Tohoku, Japan	I	87	0	2360	1943	1408	1397	852	852	1002	998
2012-08-27	El Salvador	I	1	1	22	22	4	4	0	0	0	0
2012-10-28	Haida Gwaii, Canada	P	0	0	26	26	35	35	1	1	0	0
2014-04-01	Northern Chile	I	0	0	51	51	0	0	0	0	0	0
2015-02-19	Vanuatu Is	P	5	4	0	0	1	1	0	0	0	0
2015-09-16	Central Chile	P	3	3	196	196	43	43	3	3	0	0
2016-11-13	New Zealand	P	0	0	10	10	1	1	0	0	0	0
2017-03-19	Iran[b]	P	3	3	11	11	0	0	0	0	0	0
2017-07-20	Bodrum, Turkey	P	0	0	28	28	0	0	0	0	0	0
2017-09-08	Chiapas, Mexico	P	0	0	8	8	0	0	0	0	0	0

E Post-tsunami survey data in pre-ITST era, *I* International Tsunami Survey Team, *P* Post-1992 (non-ITST) post-tsunami surveys with > 5 runups

[a]To classify as near-field, runup point is within 1000 km of source epicenter. Note: The point of origin for the 2017 Iran meteotsunami was assigned to the city of Bandar Dayyer

[b]Meteotsunami

[c]Include runup points collected post-1992

8. *Near-Field Versus Far-Field*

The broader comparisons in this study do not differentiate near-field versus far-field runup data. Far-field is defined here at a distance greater than 1000 km from the source epicenter, noting the complexities of comparisons of large rupture plane sources. Far-field runup points, in comparison to near-field, tend to be smaller values for the same tsunami event, although local conditions result in exceptions. One such exception is Crescent City, CA, which experiences amplified tsunami waves (Arcas and Uslu 2010). For example, the 1964 Southern Alaska tsunami resulted in a 4.8 m tsunami runup in Crescent City causing ten deaths (plus one death in nearby Klamath, CA) and significant damage. The next highest rununp point in California for this event was 3.7 m, while the more than 80% of the 57 non-instrumental (e.g. eyewitness accounts or post-tsunami survey) runup points in California were below 2 m.

When looking at the post-1992 tsunami survey runup data, the vast majority of the data is in the near-field and can be compared (Table 1). The two exceptions are the 1993 Sea of Japan tsunami and the 2004 Indian Ocean tsunami. In regard to the 1993 tsunami, the tsunami database contains over 40 runup points in the near-field provided by the Japan Meteorological Agency. However, the runup data only differentiates observation types between instrumental (i.e. tide gauge) and non-instrumental. As a result, the runup data are not easily identifiable as originating from a post-tsunami survey, though many are probably survey points. The 2004 Indian Ocean tsunami, as previously discussed, was an event with far reaching impacts thus the larger quantity (53%) of far-field runup data is not unexpected. In comparison, the 2011 Japan tsunami had only about 10% of runup points in the far-field; however, as this event is the most well documented event to date it had a considerable amount of post-tsunami survey runup data in the far-field, more than 500 runup points with some distances exceeding 13,000 km from the epicenter location (Fig. 4).

Unlike the post-1992 events, the eight pre-ITST events that are part of this study currently have more far-field runup data (Table 1). Reasons for this include the previously discussed post-1992 surveys of older events in far-field areas where data was lacking. Another significant reason is the importance of these events in the far-field, particularly in Hawaii. The Hawaii-focused Tsunami Technical Review Committee (2002) document is responsible for approximately 90% of the far-field post-tsunami survey runup points of the eight pre-ITST events. This document highlights the early post-tsunami surveys conducted in Hawaii beginning with the 1946 Alaska tsunami.

In regard to pre-ITST events, the tsunami database currently has a limited amount of runup data classified as being from post-tsunami surveys, particularly in the near-field. However, a broad comparison between pre-ITST and post-1992 runup data still gives the user an understanding of the data available. Meanwhile, comparing post-tsunami survey runup data within each era is appropriate as pre-ITST is largely far-field data and post-1992 is largely near-field data.

9. Database Differences Pre-ITST and Post-1992

International Tsunami Survey Teams (ITSTs) are IOC recognized, coordinated, interdisciplinary, international efforts that began with the 1992 Nicaragua tsunami survey. Although some post-tsunami surveys after 1992 Nicaragua tsunami may not formally be recognized by IOC as ITSTs, they often employed many of the same standards. Post-tsunami surveys prior to ITSTs may have had elements of an ITST, but ITSTs help ensure standardized data as well as increased confidence in the data reported. Additionally, ITSTs began in an era when precision measurement devices (e.g. GPS) and data sharing technologies were becoming widespread.

The tsunami database lists 519 confirmed tsunamis from 1900 to the ITST era. However, to limit the study to tsunamis potentially capable of impact to life and property, smaller tsunamis only detected instrumentally (e.g. tide gauge) were removed, resulting in fewer than 300 events. Approximately 3600 runup points are classified as eyewitness (i.e. not tide gauge, DART) for these events. This results in an average of almost 13 eyewitness runup points per event. Approximately 1100 pre-ITST runup points, from the 8 pre-ITST events, are classified as post-tsunami survey runup points, averaging to about 140 post-tsunami survey runup points per event. However, the 01 April 1946 Alaska tsunami accounts for approximately 40% of those post-tsunami survey runup points. When removing the 1946 event, the average of post-tsunami survey runup points per pre-ITST event drops to just under 100.

The tsunami database lists nearly 280 confirmed tsunamis in post-1992 era, beginning with the 1992 Nicaragua tsunami, this number decreases to 108 when removing events only recorded instrumentally. Approximately 700 runup points are classified as eyewitness, of a total of 94 post-1992 events that contain eyewitness runup data, averaging to over 7 eyewitness runup points per event. Although only 33 (plus the 2018 Palu post-tsunami survey in progress) post-tsunami surveys have been formally recognized by IOC/UNESCO as ITSTs, 50 post-1992 events in the database have runup points classified as post-tsunami survey runup points.

This paper focuses on the 39 post-1992 events with surveys contributing more than five runup points in the database, thereby excluding small survey efforts or events where the database may not be fully updated (Table 2). These 39 events have approximately 10,400 post-tsunami survey runup points. This results in an average of over 260 post-tsunami survey runup points for these events, a stark contrast to the 100 post-tsunami survey runup points from pre-ITST events. However, the 2004 and 2011 events make up nearly 70% of all the post-1992 post-tsunami survey runup points. When removing these two outlier events, the average of post-tsunami survey runup points per post-1992 event drops to approximately 82.

Comparing the number of collected post-tsunami survey runup points between pre-ITST and post-1992 cannot reveal trends given the limited number of corrected pre-ITST runup points in the database. However, the considerable number of points measured imparts that pre-ITST survey efforts were robust and not simply surveying points at, or nearest, the maximum runup height. Efforts to identify and correct more pre-ITST runup points from eyewitness to post-tsunami survey measurements must be made

Table 2

List of post-1992 tsunami events with > 5 post-tsunami survey runup points in NCEI/WDS tsunami database

Date (UTC)	Source location	Maximum runup (meters)	Number of post-tsunami survey runups	Select post-tsunami survey references[b]
1992-09-02	Nicaragua	9.9	44	Murty et al. (1993), Satake et al. (1993)
1992-12-12	Flores Sea, Indonesia	26.2	86	Tsuji et al. (1995a), Yeh et al. (1993)
1993-07-12	Sea of Japan	32.0	45	Shuto and Matsutomi (1995), Sigrist (1994)
1994-06-02	Java, Indonesia	13.9	67	Maramai and Tinti (1997), Tsuji et al. (1995b)
1994-10-04	Kuril Is, Russia	10.4	31	Gusiakov (1995), Yeh et al. (1995)
1994-11-14	Philippine Is	7.3	54	Imamura et al. (1995)
1995-10-09	Manzanillo, Mexico	11.0	19	Borrero et al. (1997), Trejo-Gómez et al. (2015)
1996-01-01	Sulawesi, Indonesia	3.4	16	Pelinovsky et al. (1997)
1996-02-21	Northern Peru	5.1	42	Bourgeois et al. (1999)
1998-07-17	Papua New Guinea	15.0	29	Lynett et al. (2003), Kawata et al. (1999)
1999-08-17	Izmit Bay, Turkey	2.5	19	Altınok et al. (2001), Yalçiner et al. (2000)
1999-09-13	Fatu Hiva, Marquesas, F.P.	8.0	9	Okal et al. (2002a)
1999-11-26	Vanuatu Is	6.6	12	Caminade et al. (2000)
2001-06-23	Southern Peru	8.8	38	Dengler et al. (2003)
2002-09-08	Papua New Guinea	5.5	56	Borrero et al. (2003)
2003-09-25	Hokkaido Is, Japan	4.4	211	Tanioka et al. (2004)
2004-12-26	Banda Aceh, Indonesia	50.9	1425	Borrero (2005b), Fritz (2006b), Fritz and Borrero (2006), Tsuji et al. (2006)
2005-03-28	Simeulue, Indonesia	4.2	38	Borrero et al. (2011)
2006-07-17	Java, Indonesia	20.9	160	Fritz et al. (2007)
2006-11-15	Kuril Is, Russia	21.9	114	MacInnes et al. (2009)
2007-04-01	Solomon Is	12.1	175	Fritz and Kalligeris (2008)
2007-08-15	Southern Peru	10.1	67	Fritz et al. (2008)
2007-09-12	Sumatra, Indonesia	5.0	25	Borrero et al. (2009)
2009-09-29	Samoa Is	22.4	498	Okal et al. (2010), Lamarche et al. (2010)
2010-01-03	Solomon Is	7.5	45	Newman et al. (2011)
2010-01-12	Haiti	3.2	21	Fritz et al. (2013)
2010-02-27	Central Chile	29.0	475	Fritz et al. (2011), Reymond et al. (2013)
2010-10-25	Mentawai, Indonesia	16.9	337	Hill et al. (2012), Satake et al. (2013)
2010-12-25	Vanuatu Is	4.1	8	Gibson (2015)
2011-03-11	Tohoku, Japan	38.9	5709	Mori et al. (2011), Satake et al. (2012), Suppasri et al. (2012)
2012-08-27	El Salvador	6.3	27	Borrero et al. (2014)
2012-10-28	Haida Gwaii, Canada	13.0	62	Leonard and Bednarski (2014)
2014-04-01	Northern Chile	4.6	51	Catalán et al. (2015)
2015-02-19	Vanuatu Is	8.0	6	Gibson (2015)
2015-09-16	Central Chile	13.6	245	Aránguiz et al. (2016), Contreras-López et al. (2016)
2016-11-13	New Zealand	6.9	11	Power et al. (2017)
2017-03-19	Iran[a]	3.0	14	Salaree et al. (2018)
2017-07-20	Bodrum, Turkey	1.9	28	Yalçıner et al. (2017)
2017-09-08	Chiapas, Mexico	2.7	8	Ramírez-Herrera et al. (2018)

[a]Meteotsunami

[b]Selection based on greatest number of post-tsunami survey runups attributed in the dataset analyzed in this study, noting that some references may be overview documents and not the original source of data. Complete listing of references available at NCEI/WDS and Intergovernmental Oceanographic Commission (2014)

Table 3

List of pre-ITST era tsunami events with > 5 post-tsunami survey runup points in NCEI/WDS tsunami database

Date (UTC)	Source location	Maximum runup (meters)	Number of post-tsunami survey runups	Select post-tsunami survey references[b]
1945-11-27	Makran Coast, Pakistan[a]	13.7	9	Okal et al. (2015)
1946-04-01	Unimak Is, Alaska, USA[a]	42.0	445	Okal et al. (2002b), Okal et al. (2003), Tsunami Technical Review Committee (2002)
1952-11-04	Kamchatka, Russia	23.0	79	Tsunami Technical Review Committee (2002)
1956-07-09	Greece[a]	30.0	68	Okal et al. (2009)
1957-03-09	Andreanof Is., Alaska, USA	22.8	180	Tsunami Technical Review Committee (2002)
1960-05-22	Central Chile	25.0	272	Tsunami Technical Review Committee (2002), Vitousek (1963)
1964-03-28	Prince William Sound, Alaska, USA	51.8	58	Lander et al. (1993), Leipold and Wood (1966)
1976-08-16	Moro Gulf, Philippines	9.0	20	Stratta and Haas (1977)

[a]Include runup points collected post-1992

[b]Complete listing of references available at NCEI/WDS

to have better understanding of pre-ITST survey data, as well as to provide database users with a more accurate account of the available data.

Currently, the post-1992 era appears to have resulted in fewer runup points collected in post-tsunami survey runup measurements, discounting outlier events. However, it should be noted that pre-ITST events' runup points identified as originating from post-tsunami surveys were mostly large tsunami events, with a median maximum runup of 24 m (Table 3). In comparison post-1992 events had a median maximum runup of 9 m, as a greater number of lower runup local and regional tsunamis were being more thoroughly investigated than in the pre-ITST era (Table 2).

10. Post-Tsunami Survey Runup Values

Typically, the larger the tsunami runup, the more data are collected for the event. Figure 5 validates the aforementioned statement, as there is generally a correlation between runup height and number of data points collected. Although the 1998 Papua New Guinea post-tsunami survey follows the aforementioned pattern, surveys for several other landslide-generated tsunamis are notable exceptions. The very localized extent of landslide tsunami impacts do not necessitate geographically extensive surveys resulting in a large number of data points despite the extreme runup measurements (e.g. 2007 Chile, 2015 Alaska, 2017 Greenland etc.) (Fig. 5). Other exceptions do exist and may be attributed to a variety of associated factors including, but not limited to, environmental conditions and stronger investment in scientific data collection due to socioeconomic reasons (e.g. high number of tsunami deaths, investment in scientific studies, etc.). The most striking examples of socioeconomic influence on the extent of post-tsunami surveys include the 2004 Indian Ocean event, resulting in over 227,000 deaths in 15 countries, and the 2011 Japan event, over $200 billion (USD) in damage and 18,454 deaths (NCEI/WDS 2019). Though such conditions are not investigated in this study, they should be considered for future work.

The collection of lesser runup values, it has been stated, are often overlooked by survey teams and, consequently, in tsunami databases (Fryer 2011). However, a closer analysis may prove otherwise. In the tsunami database, the eight pre-ITST events discussed with a 24 m median maximum runup have a range of 9–52 m (Table 3). However, the surveys demonstrate adequate coverage as most events show the majority of runup points collected to be between 0 and 5 m (Fig. 6 and Table 1). The 1976 Philippines tsunami is an anomaly since much of the post-tsunami survey data in the Stratta and Haas (1977) report, largely focused on structural

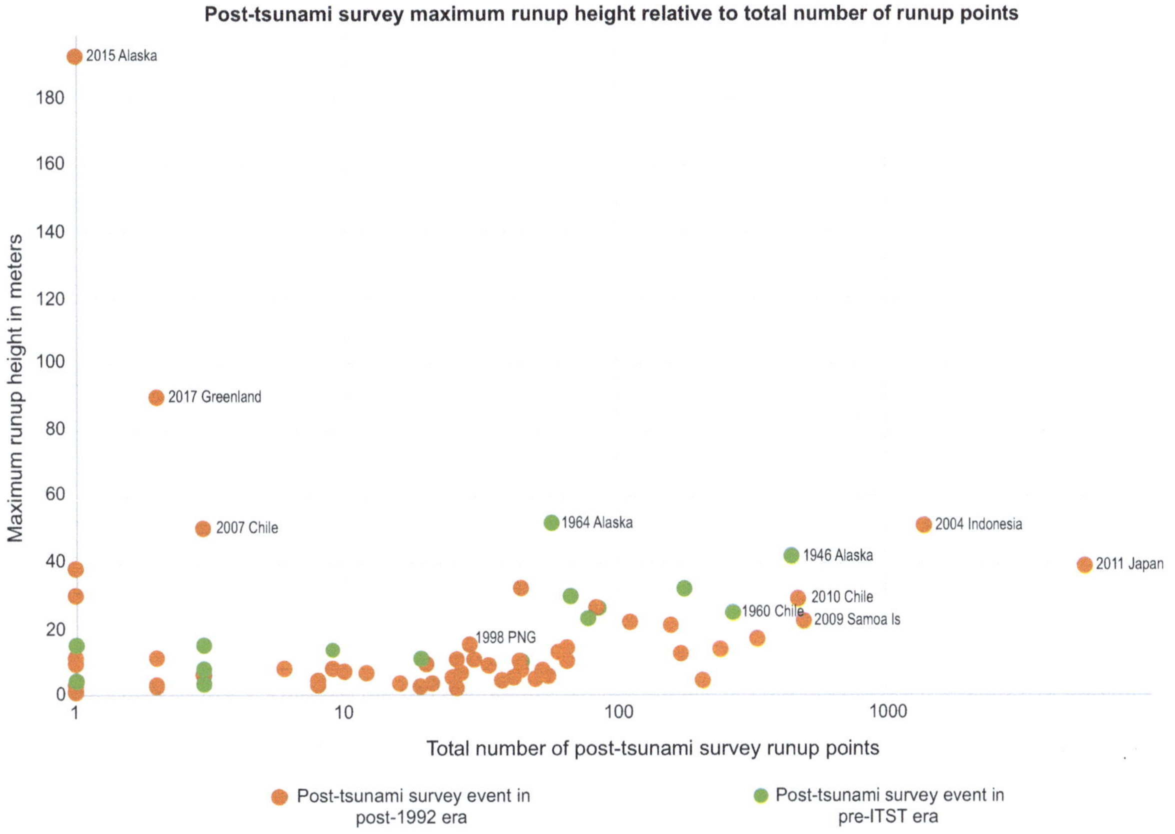

Figure 5
Post-tsunami survey maximum runup height relative to total number of post-tsunami survey runup points in NCEI/WDS tsunami database. Each circle represents a post-tsunami survey. Typically, the larger the tsunami runup, the more data are collected on that particular event. Although the 1998 Papua New Guinea post-tsunami survey follows the aforementioned pattern, surveys for several other landslide-generated tsunamis are notable exceptions. The very localized extent of landslide tsunami impacts do not necessitate geographically extensive surveys resulting in a large number of data points despite the extreme runup measurements (e.g. 2007 Chile, 2015 Alaska, 2017 Greenland etc.) (Note: 1946 Alaska includes runup points collected post-1992)

performance/damage, did not include runup heights, though some runup measurements were included in some other reports (e.g. Pararas-Carayannis 1976). Consequently, not only were a large quantity of data points collected in the eight pre-ITST events, but these survey efforts were not restricted to the highest runup points (Figs. 6 and 7). However, it should be noted that the post-tsunami survey runup points for the pre-ITST events were largely in the far-field, thus smaller values are expected (Fig. 6 and Table 1).

Similar to the pre-ITST events, post-1992 event runup points collected were more abundant in the 0–5 m range with three exceptions (15 Nov 2006, 27 Feb 2010, 28 Oct 2012 tsunamis). The three exceptions may be attributed to challenges in accessibility, consideration of other coastal processes, and clustering of data collection due to logistical considerations. The survey data from the 2009 Samoa Islands tsunami has a significantly higher number of runup points in the 0–5 m range; this may be attributable to the large number of scientists conducting the survey and improved coordination along a smaller coastline (Kong 2011) (Fig. 7 and Table 1). Further, even at epicentral distances where the runup points for 2004 and 2011 were well over 10 m, runup measurements were also recorded in the 0–5 m range (Fig. 4). If only considering ITSTs, approximately 50% of all ITST event runup observations were 0–5 m (Fig. 3 and Table 1).

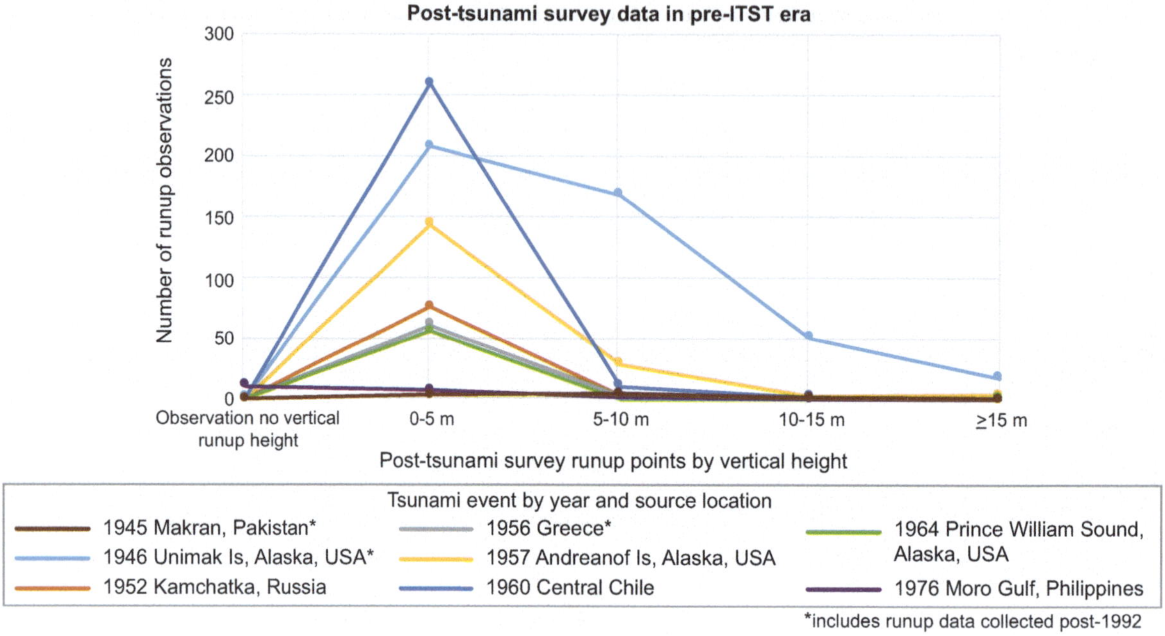

Figure 6
Post-tsunami survey data in pre-ITST era (1900–1992) with > 5 runup points in NCEI/WDS tsunami database. Each line represents a separate tsunami surveyed. Post-tsunami surveys in pre-ITST era collected a large quantity of data points, with the 0–5 m range being the most abundant range. The post-tsunami survey runup points collected for 1960 Chile tsunami were largely in the far-field, thus smaller values are expected. The 1976 Philippines tsunami post-tsunami survey runup data had limited inclusion of runup heights, as it was largely focused on structural performance/damage

11. Discussion and Conclusion

Post-tsunami surveys have contributed new data types, as well as increased quantity and quality of data. Additionally, new scientific fields have emerged due to post-tsunami surveys. A prominent example being the 1998 Papua New Guinea tsunami ITST survey data that allowed for modeling techniques that assist in determining the source of a tsunami (seismic versus landslide) (Okal and Synolakis 2004). Further, post-tsunami surveys contribution to the NCEI/WDS database is significant. The NCEI/WDS tsunami database includes information on tsunami events throughout the world that range in date from 1610 B.C. to present. However, post-tsunami survey measurements, ITSTs in particular, make up the majority of the runup data with the 2004 and 2011 events accounting for a disproportional percentage of runup points in the database.

The analysis presented in this paper reinforces that 1992 was a turning point in post-tsunami data collection, as a great number of runup data began to be collected systematically for local and regional tsunamis. The most important deductions from our analysis are the similarities between pre- and post-1992 runup heights, noting the limited runup data classified as post-tsunami survey for pre-ITST events. These similarities in measurement methodology impart a high level of confidence in pre-ITST surveys; however, the database must account for uncertainties in reference datum. Additionally, confidence in the extent of post-tsunami surveys is further strengthened as it is clear that best efforts have been made to capture a complete representation of a tsunami by documenting smaller runup values.

End users of tsunami data rely on consistent and comparable data between events and locations within a single event. For example, tsunami models studying sources or location-specific wave effects are validated using historical tsunami data, thus enhancements to tsunami data are vital to tsunami mitigation and forecasting. As a growing number of scientists are

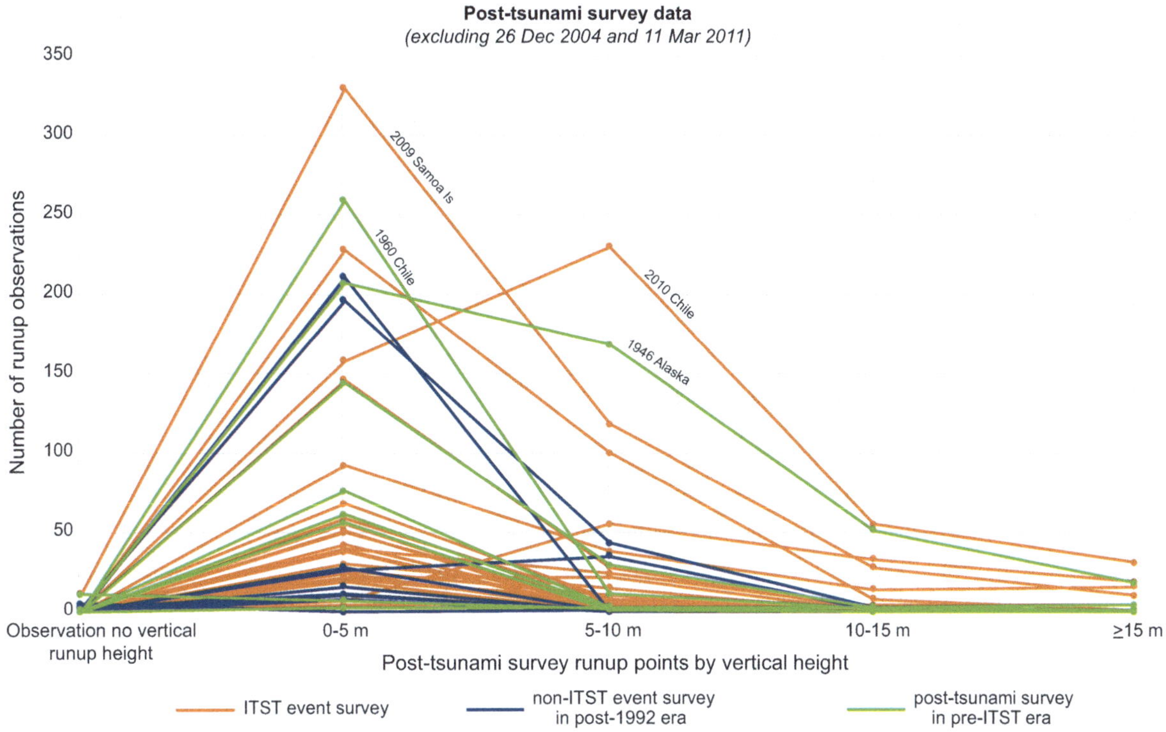

Figure 7
Post-tsunami survey data (excluding 26 December 2004 and 11 March 2011) with > 5 runup points in NCEI/WDS tsunami database. Each line represents a single tsunami surveyed by ITST (orange), post-tsunami survey team in post-1992 era (blue), or post-tsunami survey team in pre-ITST era (green). Regardless of era (pre- or post-1992) or ITST designation, runup points collected from post-tsunami surveys are more abundant in the 0–5 m range, with the three exceptions (15 Nov 2006, 27 Feb 2010, 28 Oct 2012 tsunamis). These exceptions may be attributed to challenges in accessibility, consideration of other coastal processes, and clustering of data collection due to logistical considerations

trained in surveying techniques, more data are likely to be collected. Continuing improvements to surveys, through standardization and technological advancements, result in larger datasets as well as improved data. This means that databases must keep up with the growing quantity of data and parameters.

The NCEI/WDS tsunami database reflects these advancements in data quality and quantity as evidenced by the addition of post-tsunami survey classifications. However, data collectors (i.e. ITST scientists) must continue to consider data providers, namely data centers, so that their data can be accessed. Data sharing helps save lives and protect property, but also advances the scientific fields of data collectors. For instance, ITST researchers also may be tsunami modelers that benefit from robust data sets accessible from data centers (e.g. NCEI/WDS). Thus, the responsibility is on both data centers and researchers to engage each other to ensure a reliable, robust, and continually adapting data center. ITSTs and their associated guidance have made significant progress to facilitate sharing of and consistency in tsunami data, and these efforts must continue to be nurtured.

The tsunami database must still assess pre-ITST era events to determine the origin of runup heights to better represent runup data. Recognizing the pre-ITST events may have employed a variety of survey methodologies (e.g. reference datum), the tsunami database may consider a distinct classification for pre-ITST post-tsunami survey data. The post-tsunami survey guidance developed since 1992 has encouraged consistency and cooperation in modern surveys. The tsunami database reflects these efforts and should continue to strive to better reflect pre-ITST post-tsunami surveys as well.

Acknowledgements

The manuscript was improved by valuable comments from Emile Okal and an anonymous reviewer. We thank the many colleagues at the NCEI and ITIC who provided us with the resources needed to conduct this study's analysis.

Appendix

NCEI (http://www.ngdc.noaa.gov/hazard) and the ITIC (http://itic.ioc-unesco.org) are committed to delivering data and information that enables informed decision-making and supports multiple uses of data. Most of the data in the NCEI archive are within the public domain and may be used without restriction, although users should cite data sources. These data can be searched via web maps or text interfaces by attribute (date, location, magnitude, etc.) and displayed as tables, reports, interactive maps, and imagery. Additional details of the archive are described by Dunbar et al. (2008).

References

Altınok, Y., Tinti, S., Alpar, V., Yalçıner, A., Ersoy, Ş., Bortolucci, E., et al. (2001). The Tsunami of August 17, 1999 in Izmit Bay, Turkey. *Natural Hazards, 24,* 133–146. https://doi.org/10.1023/A:1011863610289.

Aránguiz, R., González, G., Cienfuegos, R., Yagi, Y., Okuwaki, R., Urra, L., et al. (2016). The 16 September 2015 Chile tsunami from the post-tsunami survey and numerical modeling perspectives. *Pure and Applied Geophysics, 173*(2), 333–348. https://doi.org/10.1007/s00024-015-1225-4.

Arcas, D., & Uslu, B. (2010). A tsunami forecast model for Crescent City, California. *NOAA OAR Spec. Rep./PMEL Tsunami Forecast Ser. 2*, 112 pp., Gov. Print. Off., Seattle.

Arcos, N., Dunbar, P., Stroker, K., & Kong, L. (2017). The legacy of the 1992 Nicaragua tsunami. *Eos*. https://doi.org/10.1029/2017eo080845.

Borrero, J. C. (2005a). Field data and satellite imagery of tsunami effects in Banda Aceh. *Science, 308*(5728), 1596.

Borrero, J. C. (2005b). Field survey of northern Sumatra and Banda Aceh, Indonesia after the tsunami and earthquake of 26 December 2004. *Seismological Research Letters, 76*(3), 312–320. https://doi.org/10.1785/gssrl.76.3.312.

Borrero, J. C., Kalligeris, N., Lynett, P. J., Fritz, H. M., Newman, A. V., & Convers, J. A. (2014). Observations and modeling of the August 27, 2012 earthquake and tsunami affecting El Salvador and Nicaragua. *Pure and Applied Geophysics, 171*(12), 3421–3435.

Borrero, J. C., McAdoo, B., Jaffe, B., Dengler, L., Gelfenbaum, G., Higman, B., et al. (2011). Field survey of the March 28, 2005 Nias-Simeulue earthquake and tsunami. *Pure and Applied Geophysics, 168*(6–7), 1075–1088. https://doi.org/10.1007/s00024-010-0218-6.

Borrero, J., Ortíz, M., Titov, V., & Synolakis, C. (1997). Field survey of Mexican tsunami produces new data, unusual photos. *Eos, Transactions American Geophysical Union, 78*(8), 85–88.

Borrero, J., Saiang, J., Uslu, B., Freckman, J., Gomer, B., Okal, E., et al. (2003). Field survey and preliminary modeling of the Wewak, Papua New Guinea earthquake and Tsunami of 9 September 2002. *Seismological Research Letters, 74,* 393–405. https://doi.org/10.1785/gssrl.74.4.393.

Borrero, J. C., Weiss, R., Okal, E. A., Hidayat, R., Arcas, D., & Titov, V. (2009). The tsunami of 2007 September 12, Bengkulu province, Sumatra, Indonesia: post-tsunami field survey and numerical modelling. *Geophysical Journal International, 178,* 180–194. https://doi.org/10.1111/j.1365-246X.2008.04058.

Bourgeois, J. (2009). Geologic effects and records of tsunamis. In: E. N. Bernard & A. R. Robinson (eds.), The Sea, 15 (pp. 53-91).

Bourgeois, J., Petroff, C., Yeh, H., Titov, V., Synolakis, C., Benson, B., et al. (1999). Geologic setting, field survey and modeling of the Chimbote, Northern Peru, Tsunami of 21 February 1996. *Pure and Applied Geophysics, 154,* 513–540. https://doi.org/10.1007/s000240050242.

Caminade, P., Charlie, D., Kânoğlu, U., Koshimura, S., Matsutomi, H., Moore, A., et al. (2000). Vanuatu earthquake and tsunami cause much damage, few casualties. *Eos, 81,* 641–647. https://doi.org/10.1029/EO081i052p00641-02.

Catalán, P. A., Aránguiz, R., González, G., Tomita, T., Cienfuegos, R., González, J., et al. (2015). The 1 April 2014 Pisagua tsunami: observations and modeling. *Geophysical Research Letters, 42*(8), 2918–2925.

Contreras-López, M., Winckler, P., Sepúlveda, I., Andaur-Álvarez, A., Cortés-Molina, F., Guerrero, C. J., et al. (2016). Field survey of the 2015 Chile tsunami with emphasis on coastal wetland and conservation areas. *Pure and Applied Geophysics, 173*(2), 349–367. https://doi.org/10.1007/s00024-015-1235-2.

Dengler, L., Borrero, J., Gelfenbaum, G., Jaffe, B., Okal, E., Ortiz, M., et al. (2003). Tsunami. *Earthquake Spectra, 19,* 115–144. https://doi.org/10.1193/1.1737247.

Dunbar, P., & McCullough, H. (2012). Global tsunami deposits database. *Natural hazards, 63*(1), 267–278. https://doi.org/10.1007/s11069-011-9713-z.

Dunbar, P. K., Stroker, K. J., Brocko, V. R., Varner, J. D., McLean, S. J., Taylor, L. A., et al. (2008). Long-term tsunami data archive supports tsunami forecast, warning, research, and mitigation. *Pure and Applied Geophysics, 165,* 2275–2291. https://doi.org/10.1007/s00024-008-0419-4.

Eaton, J. P., Richter, D. H., & Ault, W. U. (1961). The tsunami of May 23, 1960, on the Island of Hawaii. *Bulletin of the Seismological Society of America, 51*(2), 135–157.

Fritz, H. M., & Borrero, J. C. (2006). Somalia field survey after the December 2004 Indian Ocean tsunami. *Earthquake Spectra, 22*(S3), 219–233. https://doi.org/10.1193/1.2201972.

Fritz, H. M., Borrero, J. C., Synolakis, C. E., & Yoo, J. (2006a). 2004 Indian Ocean tsunami flow velocity measurements from survivor videos. *Geophysical Research Letters*. https://doi.org/10.1029/2006GL026784.

Fritz, H. M., Hillaire, J. V., Molière, E., Wei, Y., & Mohammed, F. (2013). Twin tsunamis triggered by the 12 January 2010 Haiti earthquake. *Pure and Applied Geophysics, 170*(9–10), 1463–1474. https://doi.org/10.1007/s00024-012-0479-3.

Fritz, H., & Kalligeris, N. (2008). Ancestral heritage saves tribes during 1 April 2007 Solomon Islands tsunami. *Geophysical Research Letters, 35,* L01607. https://doi.org/10.1029/2007GL031654.

Fritz, H., Kalligeris, N., Borrero, J., Broncano, P., & Ortega, E. (2008). The 15 August 2007 Peru tsunami runup observations and modelling. *Geophysical Research Letters, 35,* L10604. https://doi.org/10.1029/2008GL033494.

Fritz, H. M., Kongko, W., Moore, A., McAdoo, B., Goff, J., Harbitz, C., et al. (2007). Extreme runup from the 17 July 2006 Java tsunami. *Geophysical Research Letters, 34,* L12602. https://doi.org/10.1029/2007GL029404.

Fritz, H. M., Petroff, C. M., Catalán, P. A., Cienfuegos, R., Winckler, P., Kalligeris, N., et al. (2011). Field survey of the 27 February 2010 Chile tsunami. *Pure and Applied Geophysics, 168*(11), 1989–2010. https://doi.org/10.1007/s00024-011-0283-5.

Fritz, H. M., Phillips, D. A., Okayasu, A., Shimozono, T., Liu, H., Mohammed, F., et al. (2012). The 2011 Japan tsunami current velocity measurements from survivor videos at Kesennuma Bay using LiDAR. *Geophysical Research Letters*. https://doi.org/10.1029/2011GL050686.

Fritz, H. M., Synolakis, C. E., & McAdoo, B. G. (2006b). Maldives field survey after the December 2004 Indian Ocean tsunami. *Earthquake Spectra, 22*(S3), 137–154. https://doi.org/10.1193/1.2201973.

Fryer, G. J. (2011). Walk the inundation limit: A suggestion for future post-tsunami surveys. *Earth-Science Reviews, 107*(1–2), 123–127. https://doi.org/10.1016/j.earscirev.2011.03.003.

Gibson, D. (2015). *National Report of Vanuatu (International Coordination Group for the Tsunami Warning System in the Pacific, XXVI Session)*. Paris: UNESCO Intergovernmental Oceanographic Commission.

Gusiakov, V. K. (1995). *The Shikotan, Kuril Islands, earthquake tsunami, october 4, 1994* (p. 12). Colorado: National Oceanic and Atmospheric Administration, National Geophysical Data Center, Boulder.

Gusiakov, V.K. (2009). Tsunami history: recorded. In: E. N. Bernard, & A. R. Robinson, (eds.), The Sea, 15, (pp. 23–53).

Hill, E. M., Borrero, J. C., Huang, Z., Qiu, Q., Banerjee, P., Natawidjaja, D. H., et al. (2012). The 2010 M_w 7.8 Mentawai earthquake: Very shallow source of a rare tsunami earthquake determined from tsunami field survey and near-field GPS data. *Journal of Geophysical Research: Solid Earth*. https://doi.org/10.1029/2012jb009159.

Imamura, A. (1924). Preliminary note on the great earthquake of southeastern Japan on September 1, 1923. *Bulletin of the Seismological Society of America, 14*(2), 136–149.

Imamura, F., Synolakis, C. E., Gica, E., Titov, V., Listanco, E., & Lee, H. J. (1995). Field survey of the 1994 Mindoro Island. *Philippines tsunami. Pure and applied geophysics, 144*(3–4), 875–890. https://doi.org/10.1007/BF00874399.

Intergovernmental Oceanographic Commission. (1998). Post-tsunami survey field guide. *IOC Manuals and Guides* No. 37, Paris.

Intergovernmental Oceanographic Commission. (2014). International Tsunami Survey Team (ITST) Post-Tsunami Survey Field Guide. 2nd Edition. *IOC Manuals and Guides* No. 37, Paris.

Kaistrenko, V., and Sedaeva, V. (2001), 1952 North Kuril tsunami: new data from archives, In: *Tsunami research at the end of a critical decade*, ed. by G.T. Hebenstreit, Adv. Natur. Technol. Res. 18, pp. 91–102, Kluver, Dordrecht, 2001.

Kajiura, K. (1981). Some statistics related to observed tsunami heights along the coast of Japan. *Tsunamis-their science and engineering*, 131–145.

Kalligeris, N., Skanavis, V., Tavakkol, S., Ayca, A., Safty, H. E., Lynett, P., et al. (2016). Lagrangian flow measurements and observations of the 2015 Chilean tsunami in Ventura, CA. *Geophysical Research Letters, 43*(10), 5217–5224.

Kanamori, H. (1972). Mechanism of tsunami earthquakes. *Physics of the Earth and Planetary Interiors, 6*(5), 346–359. https://doi.org/10.1016/0031-9201(72)90058-1.

Kawata, Y., Benson, B. C., Borrero, J. C., Borrero, J. L., Davies, H. L., de Lange, W. P., et al. (1999). Tsunami in Papua New Guinea was as intense as first thought. *Eos, Transactions American Geophysical Union, 80*(9), 101–105.

Kong, L. (2011). Post-tsunami field surveys are essential for mitigating the next tsunami disaster. *Oceanography, 24*(2), 222–226.

Lamarche, G., Pelletier, B., & Goff, J. (2010). Impact of the 29 september 2009 south pacific tsunami on Wallis and Futuna. *Marine Geology, 271*(3–4), 297–302. https://doi.org/10.1016/j.margeo.2010.02.012.

Lander, J., Lockridge, P.A., & Kozuch, J. (1993). Tsunamis affecting the west coast of the United States 1806–1992: NGDC Key to Geophysical Research Documentation No. 29. *NOAA., Boulder*, CO.

Lander, J. F., & Yeh, H. (1995). *Report of the international tsunami measurements workshop*. Colorado: Estes Park.

Leipold, L E., & Wood, F.J. (Eds.). (1966). *The prince william sound, Alaska, earthquake of 1964 and aftershocks* (Vol. 2). US Govt. Print. Off.

Leonard, L. J., & Bednarski, J. M. (2014). Field survey following the 28 October 2012 Haida Gwaii tsunami. *Pure and Applied Geophysics, 171*(12), 3467–3482. https://doi.org/10.1007/s00024-014-0792-0.

Lynett, P., Borrero, J., Liu, P., & Synolakis, C. (2003). Field survey and numerical simulations: A review of the 1998 Papua new Guinea Tsunami. *Pure and Applied Geophysics, 160,* 2119–2146. https://doi.org/10.1007/s00024-003-2422-0.

MacInnes, B. T., Pinegina, T. K., Bourgeois, J., Razhigaeva, N. G., Kaistrenko, V. M., & Kravchunovskaya, E. A. (2009). Field survey and geological effects of the 15 november 2006 kuril tsunami in the middle kuril islands. *Pure and Applied Geophysics, 166,* 9–36. https://doi.org/10.1007/s00024-008-0428-3.

Maramai, A., & Tinti, S. (1997). The 3 June 1994 Java tsunami: A post-event survey of the coastal effects. *Natural Hazards, 15*(1), 31–49. https://doi.org/10.1023/A:1007957224367.

Mori, N., Takahashi, T., Yasuda, T., & Yanagisawa, H. (2011). Survey of 2011 Tohoku earthquake tsunami inundation and runup. *Geophysical Research Letters, 38,* L00G14. https://doi.org/10.1029/2011gl049210.

Murty, T.S., Baptista, A.M., & Priest, G.R. (1993). Post-tsunami survey (Nov. 2–7, 1992) of run-up and inundation in the coast of Nicaragua. *Report to the Intergovernmental Oceanographic Commission (of UNESCO)*, Paris, France, 3.

National Centers for Environmental Information/World Data Service (NCEI/WDS). (2019). Global historical tsunami database. Natl. Cent. for Environ. Inf., Boulder, Colo., https://doi.org/10.7289/v5pn93h7. (Accessed 27 March 2019).

Newman, A., Feng, L., Fritz, H., Lifton, Z., Kalligeris, N., & Wei, Y. (2011). The energetic 2010 MW 7.1 Solomon Islands tsunami earthquake. *Geophysical Journal International, 186,* 775–781. https://doi.org/10.1111/j.1365-246X.2011.05057.x.

Okal, E. A., Fritz, H. M., Hamzeh, M. A., & Ghasemzadeh, J. (2015). Field survey of the 1945 Makran and 2004 Indian Ocean tsunamis in Baluchistan, Iran. *Pure and Applied Geophysics, 172*(12), 3343–3356.

Okal, E. A., Fritz, H. M., Synolakis, C. E., Borrero, J. C., Weiss, R., Lynett, P. J., et al. (2010). Field survey of the Samoa tsunami of 29 September 2009. *Seismological Research Letters, 81*(4), 577–591. https://doi.org/10.1785/gssrl.81.4.577.

Okal, E. A., Fryer, G. J., Borrero, J. C., & Ruscher, C. (2002a). The landslide and local tsunami of 13 September 1999 on Fatu Hiva (Marquesas islands; French Polynesia). *Bulletin de la Société géologique de France, 173*(4), 359–367.

Okal, E. A., & Hébert, H. (2007). Far-field simulation of the 1946 Aleutian tsunami. *Geophysical Journal International, 169*(3), 1229–1238.

Okal, E. A., Plafker, G., Synolakis, C. E., & Borrero, J. C. (2003). Near-field survey of the 1946 Aleutian tsunami on Unimak and Sanak Islands. *Bulletin of the Seismological Society of America, 93*(3), 1226–1234.

Okal, E. A., & Synolakis, C. E. (2004). Source discriminants for near-field tsunamis. *Geophysical Journal International, 158*(3), 899–912.

Okal, E. A., Synolakis, C. E., Fryer, G. J., Heinrich, P., Borrero, J. C., Ruscher, C., et al. (2002b). A field survey of the 1946 Aleutian tsunami in the far field. *Seismological Research Letters, 73*(4), 490–503.

Okal, E. A., Synolakis, C. E., Uslu, B., Kalligeris, N., & Voukouvalas, E. (2009). The 1956 earthquake and tsunami in Amorgos, Greece. *Geophysical Journal International, 178*(3), 1533–1554.

Pararas-Carayannis, G. (1976). Survey of Philippine earthquake and tsunami of August 16, 1976, ITIC Report 1976. *Tsunami Newsletter*, 9(3).

Pelinovsky, E., Yuliadi, D., Prasetya, G., & Hidayat, R. (1997). The 1996 Sulawesi tsunami. *Natural Hazards, 16*(1), 29–38. https://doi.org/10.1023/A:1007904610680.

Platania, G. (1909). *Il maremoto dello stretto di Messina del 28 dicembre 1908*. Societá tipografica modenese. **(In Italian)**.

Power, W., Clark, K., King, D. N., Borrero, J., Howarth, J., Lane, E. M., et al. (2017). Tsunami runup and tide-gauge observations from the 14 November 2016 M7. 8 Kaikōura earthquake, New Zealand. *Pure and Applied Geophysics, 174*(7), 2457–2473.

Quarantelli, E. L. (1997). The Disaster Center field studies of organized behavior in the crisis time period of disasters. *International Journal of Mass Emergencies and Disasters, 15*(1), 47–69.

Ramakrishnan, D., Ghosh, S. K., Raja, V. K. M., Chandran, R. V., & Jeyram, A. (2005). Trails of the killer tsunami: A preliminary assessment using satellite remote sensing technique. *Current Science, 88*(5), 709–711.

Ramírez-Herrera, M. T., Corona, N., Ruiz-Angulo, A., Melgar, D., & Zavala-Hidalgo, J. (2018). The 8 september 2017 tsunami triggered by the M_w 8.2 intraplate earthquake, Chiapas, Mexico. *Pure and Applied Geophysics, 175*(1), 25–34.

Reymond, D., Hyvernaud, O., & Okal, E. A. (2013). The 2010 and 2011 tsunamis in French polynesia: operational aspects and field surveys. *Pure and Applied Geophysics, 170*(6–8), 1169–1187. https://doi.org/10.1007/s00024-012-0485-5.

Salaree, A., Mansouri, R., & Okal, E. A. (2018). The intriguing tsunami of 19 March 2017 at Bandar Dayyer, Iran: field survey and simulations. *Natural Hazards, 90*(3), 1277–1307.

Satake, K., Bourgeois, J., Abe, K., Abe, K., Tsuji, Y., Imamura, F., et al. (1993). Tsunami field survey of the 1992 Nicaragua earthquake. *Eos Transaction AGU, 74*(13), 156–157. https://doi.org/10.1029/93eo00271.

Satake, K., Nishimura, Y., Putra, P. S., Gusman, A. R., Sunendar, H., Fujii, Y., et al. (2013). Tsunami source of the 2010 Mentawai, Indonesia earthquake inferred from tsunami field survey and waveform modeling. *Pure and Applied Geophysics, 170*(9–10), 1567–1582. https://doi.org/10.1007/s00024-012-0536-y.

Satake, K., Rabinovich, A.B., Dominey-Howes, D., & Borrero, J.C. (2012). Introduction to ''historical and recent catastrophic tsunamis in the world: Volume I. The 2011 Tohoku Tsunami''. *Pure and Applied Geophysics*, https://doi.org/10.1007/s00024-012-0615-0 [Special issue with a number of studies of the Tohoku Tsunami].

Savarenskiy, E.F., Tishchenko, V.G., Svyatlovskiy, A.E., Dobrovol'skiy, A.D., & Zhivago, A.V. (1958), *Tsunami 4-5 noyabrya 1952g*., Byulleten' sovieta po seismologii 4, 63 pp., Izdat. Akad. Nauk SSSR, Moskva **(in Russian)**.

Shuto, N., & Matsutomi, H. (1995). Field survey of the 1993 Hokkaido Nansei-Oki earthquake tsunami. *Pure and Applied Geophysics, 144*(3–4), 649–663. https://doi.org/10.1007/BF00874388.

Sigrist, D. J. (1994). July 12, 1993, Hokkaido Nansei-Oki Tsunami–Effects and observations in Japan, Russia, Democratic People's Republic of Korea, and Republic of Korea. In *Proceedings of the 3rd UJNR Tsunami Workshop* (No. 3315, p. 81). PWRI Technical Memorandum.

Soloviev, S. L., & Go, C. N. (1974). *A catalogue of tsunamis on the western shore of the Pacific Ocean* (p. 308). Moscow: ''Nauka'' Publishing House. (in Russian; English translation by the Canadian Institute for Scientific and Technical Information, No.5077, National Research Council, Ottawa, Canada, 1984, 439 pp.).

Soloviev, S. L., & Go, C. N. (1975). *A catalog of tsunamis on the eastern shore of the Pacific ocean*. Moscow: Academy of Science of the USSR, Nauka Publishing House. (in Russian; English translation by the Canadian Institute for Science and Technical Information, No. 5078, National Research Council, Ottawa, Canada, 1984, 293 pp.).

Stallings, R. A. (2007). Methodological issues. In H. Rodríguez, E. L. Quarantelli, & R. R. Dynes (Eds.), *Handbook of disaster research* (pp. 55–83). New York: Springer.

Stratta, J.L., & Haas, J.E. (Eds.). (1977). *Reconnaissance report, Mindanao, Philippines Earthquake, August 17, 1976*. EERI.

Suda, K. (1924). On the great Japanese earthquake of September 1, 1923, the observations of tsunami. *The Memoirs of the Imperial Marine Observatory, 1,* 4.

Suppasri, A., Koshimura, S., Imai, K., Mas, E., Gokon, H., Muhari, A., et al. (2012). Damage characteristic and field survey of the 2011 great east japan tsunami in Miyagi prefecture. *Coastal Engineering Journal*. https://doi.org/10.1142/s0578563412500052.

Sweet, S., & Silver, E. A. (2003). Tectonics and slumping in the source region of the 1998 Papua New Guinea tsunami from seismic reflection images. *Pure and Applied Geophysics, 160*(10–11), 1945–1968.

Synolakis, C. E., Bardet, J. P., Borrero, J. C., Davies, H. L., Okal, E. A., Silver, E. A., et al. (2002). The slump origin of the 1998 Papua New Guinea tsunami. *Proceedings of the Royal Society of London. Series A: Mathematical, Physical and Engineering Sciences, 458*(2020), 763–789.

Synolakis, C. E., & Okal, E. A. (2005). 1992–2002: perspective on a decade of post-tsunami surveys. *Tsunamis* (pp. 1–29). Dordrecht: Springer.

Tanioka, Y., Nishimura, Y., Hirakawa, K., Imamura, F., Abe, I., Abe, Y., et al. (2004). Tsunami run-up heights of the 2003 Tokachi-oki earthquake. *Earth, Planets and Space, 56*(3), 359–365.

Tappin, D. R., Watts, P., & Grilli, S. T. (2008). The Papua New Guinea tsunami of 17 July 1998: anatomy of a catastrophic event. *Natural Hazards and Earth System Science, 8*(2), 243–266.

Tierney, K. J. (2002). The field turns fifty: social change and the practice of disaster field work. In: R. A. Stallings (ed.). *Methods of Disaster Research* (pp. 349–374).

Trejo-Gómez, E., Ortiz, M., & Núñez-Cornú, F. J. (2015). Source model of the October 9, 1995 Jalisco-Colima tsunami as constrained by field survey reports, and on the numerical simulation of the tsunami. *Geofísica Internacional, 54*(2), 149–159.

Tsuji, Y., Imamura, F., Matsutomi, H., Synolakis, C.E., Nanang, P.T., Harada, S., Han, S.S., Arai, K.I., & Cook, B. (1995b). Field survey of the East Java earthquake and tsunami of June 3, 1994. In Tsunamis: 1992–1994, Kenji Satake and Fumihiko Imamura, eds., *Pure and applied geophysics*, *144*(3–4), 839–854.

Tsuji, Y., Matsutomi, H., Imamura, F., Takeo, M., Kawata, Y., Matsuyama, M., et al. (1995a). Damage to coastal villages due to the 1992 Flores Island earthquake tsunami. *Pure and Applied Geophysics, 144*(3–4), 481–524.

Tsuji, Y., Namegaya, Y., Matsumoto, H., Iwasaki, S. I., Kanbua, W., Sriwichai, M., et al. (2006). The 2004 Indian tsunami in Thailand: surveyed runup heights and tide gauge records. *Earth, Planets and Space, 58*(2), 223–232.

Tsuji, Y., Satake, K., Ishibe, T., Harada, T., Nishiyama, A., & Kusumoto, S. (2014). Tsunami heights along the pacific coast of northern Honshu recorded from the 2011 Tohoku and previous great earthquakes. *Pure and Applied Geophysics, 171*(12), 3183–3215.

Tsunami Technical Review Committee. (2002). *Field guide for measuring tsunami runups and inundations*. Civil Defense Division, Hawaii: State of Hawaii Department of Defense.

Vitousek, M. J. (1963). The tsunami of 22 may 1960 in French polynesia. *Bulletin of the Seismological Society of America, 53*(6), 1229–1236.

Yalçıner, A., Altınok, Y., Synolakis, C., Borrero, J., Imamura, F., Ersoy, Ş., et al. (2000). Tsunami waves in Izmit bay. *Earthquake Spectra, 16,* 55–62. https://doi.org/10.1193/1.1586146.

Yalçıner, A., Annunziato, A., Papadopoulos, G., Dogan, G.G., Guler, H.G., Cakir, T.E., Sozdinler, C.O., Ulutas, E., Arikawa, T., Suzen, L. & Kânoğlu, U. (2017). The 20th July 2017 (22: 31 UTC) Bodrum-Kos Earthquake and Tsunami: Post Tsunami Field Survey Report. *Online report at:* http://users.metu.edu.tr/Yalciner/July-21-2017-tsunami-report/Report-Field-Survey-of-July-20-2017-Bodrum-Kos-Tsunami.pdf.

Yeh, H., Imamura, F., Synolakis, C., Tsuji, Y., Liu, P., & Shi, S. (1993). The flores island tsunamis. *Eos, Transactions American Geophysical Union, 74*(33), 369–373.

Yeh, H., Titov, V., Gusiakov, V., Pelinovsky, E., Khramushin, V., & Kaistrenko, V. (1995). The 1994 Shikotan earthquake tsunamis. *Pure and Applied Geophysics, 144*(3–4), 855–874. https://doi.org/10.1007/BF00874398.

(Received October 19, 2018, revised April 6, 2019, accepted April 8, 2019, Published online April 12, 2019)

Pure Appl. Geophys. 176 (2019), 2831–2868
© 2019 Springer Nature Switzerland AG
https://doi.org/10.1007/s00024-019-02134-2

Pure and Applied Geophysics

Reassessment of the 1907 Sumatra "Tsunami Earthquake" Based on Macroseismic, Seismological, and Tsunami Observations, and Modeling

Stacey Servito Martin,[1] Linlin Li,[1] Emile A. Okal,[2] Julie Morin,[3] Alexander E. G. Tetteroo,[4] Adam D. Switzer,[1,5] and Kerry E. Sieh[1,5]

Abstract—On 4 January 1907, an earthquake occurred off the west coast of Sumatra, Indonesia, with an instrumental surface-wave magnitude (M_S) in the range of 7.5–8.0 at periods of ~ 40 s. The tsunami it generated was destructive on the islands of Nias and Simeulue, where it killed hundreds and gave rise to the legend of the S'mong. This tsunami was also observed in other parts of the Indian Ocean basin. Relative to its instrumented magnitude, the size of the tsunami was anomalous, qualifying the event as a "tsunami earthquake". However, unusually for a tsunami earthquake, the shaking on Nias was severe (7 EMS). We revisit the 1907 earthquake with a multidisciplinary approach by extracting evidence describing shaking effects or the tsunami from written documents and by acquiring new seismograms. Combining these, we discriminate two large earthquakes within an hour of each other with clear differences in seismological character. The first we interpret to be a tsunami earthquake with characteristic low levels of shaking, an estimated average seismic moment (M_0) of 2.5×10^{28} dyn cm ($M_W \approx 8.2$) in the frequency band 6–8 mHz, and an epicentral location close to the front of the Sunda Megathrust. The seismograms we analyzed also document a regular growth of moment with period, approaching $M_W \approx 8.4$ at the longest resolvable period (~ 170 s). For the second earthquake that caused damage on Nias, we estimate $M_S \approx 7$ based on seismograms and phase data. We also identify two $M_S \approx 6$ aftershocks within 24 h of the mainshock. Additionally, we present a dataset of 88 locations within the Indian Ocean basin where the tsunami was observed. Using a subset of these, we forward modeled the tsunami to propose a seismic rupture model extending along the Sunda Megathrust for about 220 km (~ 94.7°E to ~ 97°E) with a maximum modeled slip of ~ 21 m. Our new rupture model provides an acceptable fit to our new dataset of tsunami runup and inundation values from 88 local and far-field locations in the Indian Ocean basin. We also urge caution against an over-reliance on the S'mong legend for tsunami evacuation as its premise, that a tsunami will only follow an earthquake with very severe ground motions, is rendered ineffective for tsunami earthquakes.

Key words: Tsunami, earthquake, Simeulue, Nias, Sumatra, Indonesia, Indian Ocean, 1907.

Electronic supplementary material The online version of this article (https://doi.org/10.1007/s00024-019-02134-2) contains supplementary material, which is available to authorized users.

[1] Earth Observatory of Singapore, Nanyang Technological University, Singapore, Singapore. E-mail: 7point1@gmail.com; smartin@ntu.edu.sg
[2] Department of Earth and Planetary Sciences, Northwestern University, Evanston, USA.
[3] Laboratoire Magmas et Volcans, Université Clermont Auvergne, Clermont-Ferrand, France.
[4] Institute for History, Leiden University, Leiden, The Netherlands.
[5] Asian School of the Environment, Nanyang Technological University, Singapore, Singapore.

1. Introduction

Numerous large earthquakes have occurred off the west coast of Sumatra, Indonesia, during the historical and instrumental eras (e.g., Newcomb and McCann 1987), and in particular since 2000. These events include a tsunamigenic earthquake on 4 January 1907 near the islands of Simeulue and Nias (Fig. 1), which Newcomb and McCann (1987) associated with shaking of such severity that "people on Nias could not stand". As many as 370 people were killed on Nias, and at least 1818 lives were lost on Simeulue (Koloniaal Verslag van 1907; Openbaar Verbaal 1908), including 1205 in Tapah (Tapak in Dutch) District, 431 in Simeuloeë Rajou District, 130 in Salang District, and 52 in Leuköon District on Simeulue (Bataviaasch Nieuwsblad, 12 February 1907). It was feared that 1000 people were killed at Koela-Deh in Tapah alone (Utrechts Nieuwsblad, 14 February 1907). This led to the disaster being embodied in myth and legend on the island of Simeulue (e.g., McAdoo et al. 2006; Syafwina 2014; Rahman et al. 2017). As discussed herein, the 1907 tsunami was also recorded in the far field, as far away as the island of La Réunion (Bertho 1910), which gives the event a clearly anomalous character in the context of the comparatively low "Pasadena" magnitude (M_{PAS} = 7.6) assigned to the earthquake by

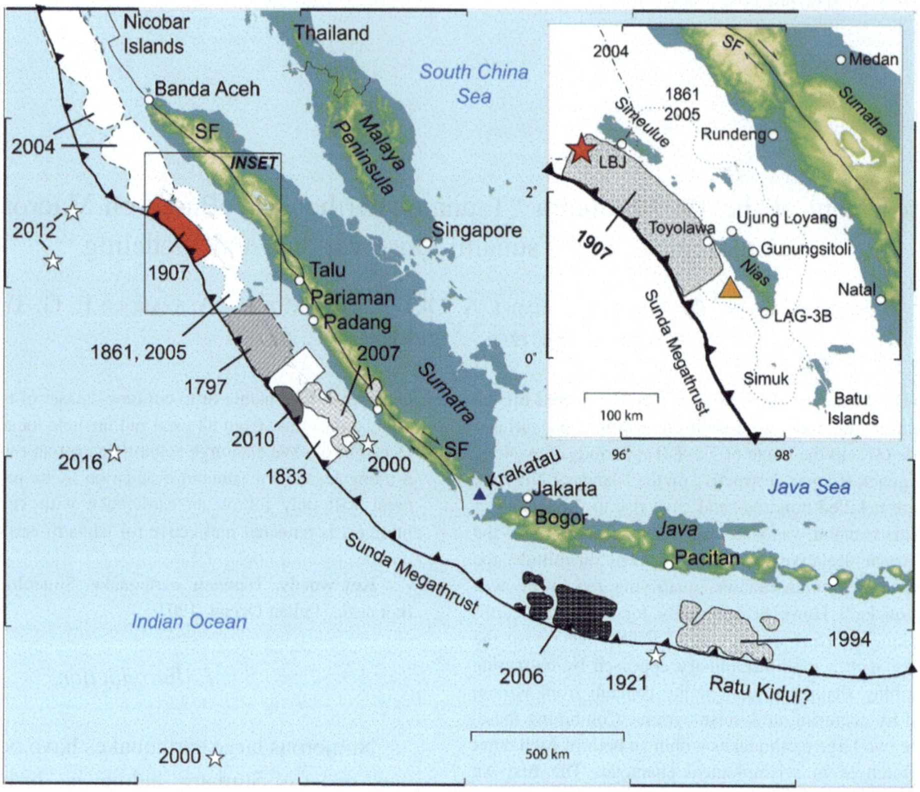

Figure 1
Generalized tectonic map of the Sunda Megathrust in Indonesia. First-order rupture of the 1907 earthquake (Event I, this study) indicated by a filled red polygon. Rupture areas for the 1797, 1833, 1861, 1994, 2004, 2006, 2007, and 2010 earthquakes are also shown (Bilek and Engdahl 2007; Chlieh et al. 2007; Konca et al. 2008; Hill et al. 2012). Location of the possible historical Ratu Kidul tsunami earthquake discussed in the text is labeled and appended by a question mark. Stars represent selected recent earthquakes. Inset box shows the islands of Simeulue and Nias with the first-order rupture of the 1907 earthquake (this study) indicated by a filled grey polygon, red star representing the epicenter of the 1907 earthquake (this study), and an orange triangle indicating the approximate location of Event II. Outlines of the 2004 (dot-dashed) and 2005 (dotted) lines are also displayed. Location of coral microatolls LBJ and LAG-3B from Meltzner et al. (2015) are also shown

Gutenberg and Richter (1954). Other estimates of conventional magnitudes for this earthquake have been presented by Abe and Noguchi (1983; M_S = 7.6), and recently the ISC (M_S = 7.8; Storchak et al. 2013). Duda (1965) proposed a figure of 7.8, but the nature of this magnitude was unspecified. More recently, Kanamori et al. (2010) conducted an extensive seismological study of the 1907 earthquake based on a number of historical seismograms. While they did not compute a seismic moment through waveform fitting, they measured a surface-wave magnitude $M_S = 7.8 \pm 0.25$ and estimated a moment magnitude $M_W = 7.8$ by scaling time-domain amplitudes of body and surface waves of the 1907 earthquake to those of nearby modern earthquakes with known moment tensors. These results were obtained in the period range 40–50 s, but Kanamori et al. (2010) stress that the source was obviously longer, and thus the moment should be larger at longer periods, suggestive of a "tsunami earthquake".

The term "tsunami earthquake" was first used by Kanamori (1972) to discuss the sources of the 1896 Meiji Sanriku and 1946 Unimak (Aleutian Islands) earthquakes, both of which resulted in anomalously large tsunamis with respect to their instrumental magnitudes. This type of event can be distinguished based on disproportionate relationships between surface-wave magnitude (M_S) and seismic moment (M_0), from the observation of longer than expected process times despite small rupture areas (Sykes 1971; Kanamori 1972; Pelayo and Wiens 1992; Polet and Kanamori 2000), and from lower than anticipated macroseismic intensities (Kanamori 1972; Fukao 1979; Bourgeois et al. 1999). The longer process times result in red-shifting of the source spectrum, and in inconsistencies between deficient seismic magnitudes measured at short to moderate periods (thus relevant to macroseismic effects) and enhanced ultralong-period seismic moments (controlling the generation of tsunamis). The ruptures of tsunami earthquakes have been observed to propagate toward the trench axis (Polet and Kanamori 2000) on very shallow dipping faults or on splays (Fukao 1979; Pelayo and Wiens 1990, 1992) located in weakly coupled regions of aseismic convergence (Pelayo and Wiens 1990, 1992; Bourgeois et al. 1999), or at the very top of the plate interface under conditions of sediment starvation leading to a jagged rupture (Tanioka et al. 1997). Rupture velocities for such earthquakes are also less than expected for typical earthquakes owing to the low rigidity of materials proximal to the trench axis (Fukao 1979; Pelayo and Wiens 1990, 1992; Heinrich et al. 1998; Ihmlé et al. 1998). In addition, tsunami earthquakes can occur as mainshocks, which Okal and Saloor (2017) qualified as "primary tsunami earthquakes" (PTEs), or as "aftershock tsunami earthquakes" (ATEs), following a larger, regular megathrust event.

With the exception of the 2010 Mentawai event (e.g., Newman et al. 2011a; Hill et al. 2012), no other tsunami earthquakes have been conclusively identified off the Sumatran coast during the modern or historical period. Furthermore, in stark contrast to other natural disasters in the Dutch East Indies such as the 1883 eruption of Krakatau for which detailed official reports were written (e.g., Verbeek 1885), locating an official scientific report for the 1907 disaster was futile despite exhaustive efforts on our part. Szirtes (1912a, p. 5) carries cursory mention of a "detailed study" (*eingehende Untersuchung* in German) by T.H. Staverman, including possibly a study of its epicentral location, but without a complete citation. The scientific bibliography relating to geology in the Dutch East Indies between 1907 and 1912 (e.g., Verbeek 1912) has no record of this document, and its whereabouts remain a mystery, including at Strasbourg where Siegmund Szirtes was based (L. Rivera, personal communication 2017). In light of missing colonial investigative reports, and despite the conclusion in more general terms by Kanamori et al. (2010) that the 1907 Sumatra earthquake bore all the hallmarks of a tsunami earthquake, key aspects of this event remain unaddressed, including (1) conclusive, quantitative evidence of its nature as a "tsunami earthquake" and (2) an estimate of the geometry and slip parameters of the source supporting the reported distribution of the tsunami. In addition, perplexing observations clearly in need of further study include (3) the anomalously violent ground motions (Newcomb and McCann 1987; Kanamori et al. 2010) in comparison with other tsunami earthquakes, (4) the lack of aftershocks, and (5) the lack of land level changes comparable to those identified for other large earthquakes in the Simeulue–Nias region (e.g., Meltzner et al. 2012, 2015).

In this article, we employ a multidisciplinary approach to tackle these points. Through scrutiny of original macroseismic reports and systematic analysis of a number of seismograms, we separate the mainshock (henceforth Event I) from a previously unsuspected large aftershock that occurred only $\sim$ 53 min later (henceforth Event II). Waveform analysis of digitized seismograms allows quantification of the mainshock at mantle wave periods up to 170 s and further demonstrates significant differences in source spectra between the mainshock and the large $\sim$ 06:12 GMT aftershock. We also compile a set of 88 qualitative and instrumental observations of the tsunami in the Indian Ocean basin (including nil reports) and utilize a subset of these to prepare a slip model for the 1907 mainshock.

2. *Reassessing Macroseismic Intensities and the Discovery of Event II*

The earliest macroseismic study of the 1907 Sumatra earthquake by Newcomb and McCann (1987) is a dual-zoned intensity map (Fig. 2a) utilizing the Modified Mercalli Intensity Scale (MMI) to distinguish between what they defined as weak (MMI I–IV) and strong shaking (MM V–VII), for what was assumed to be a single event. We use the 1998

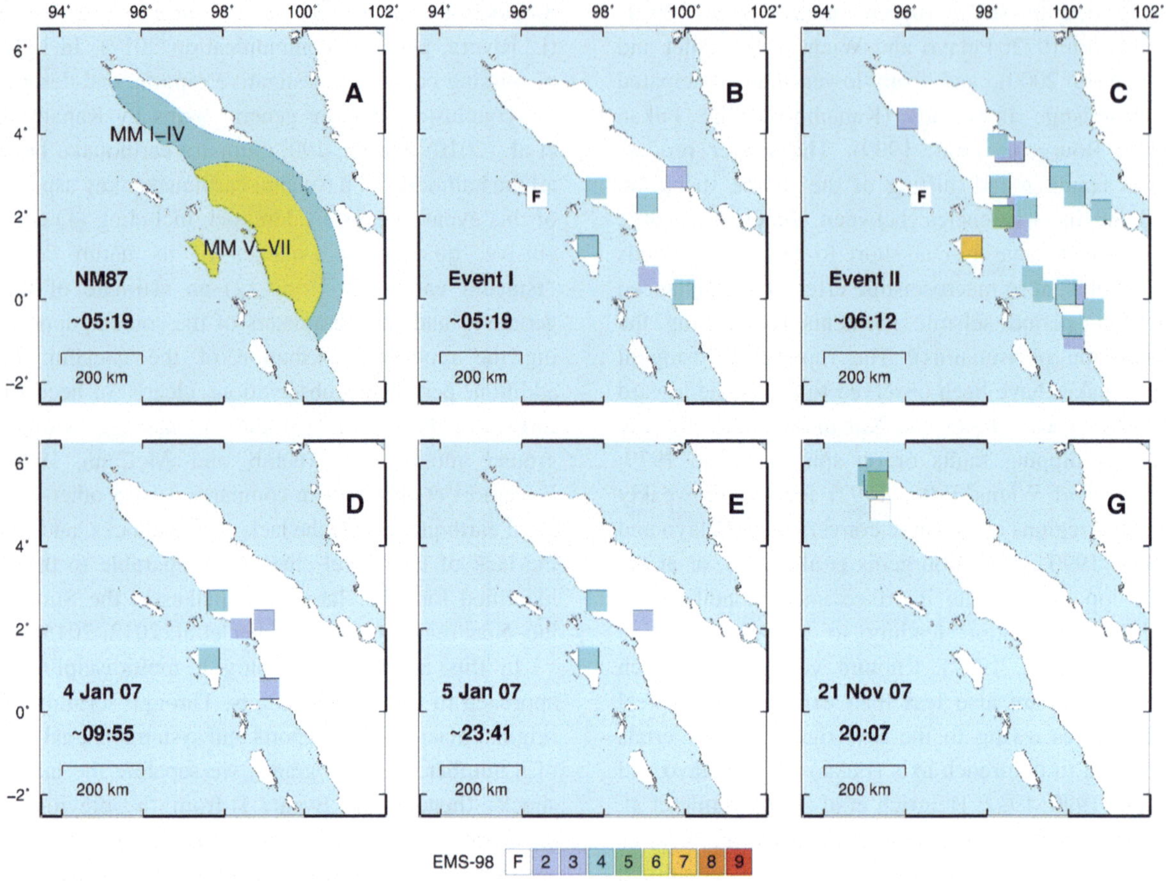

Figure 2
Intensity map (**a**) from Newcomb and McCann (1987) in comparison with intensities determined by our study for the mainshock at 05:19 GMT (**b**) and the largest aftershock at ~ 06:12 GMT (**c**). Intensity for the largest known aftershocks on 4 January (**d**) and 5 January (**e**) are also shown along with the earthquake of 21 November 1907 (**g**) near Banda Aceh. White boxes marked with an "F" indicate that an earthquake was felt but macroseismic data were insufficient to assign an intensity

European Macroseismic Scale (EMS-98; Grünthal 1998) to assess intensity with our choice of macroseismic intensity assessment adhering to modern practice (e.g., Toppozada and Real 1981; Ambraseys and Douglas 2004; Martin and Szeliga 2010). The EMS-98 scale is equivalent to MMI (Musson et al. 2010); its merits and its suitability for use outside Europe are discussed by Hough et al (2016), and Martin and Hough (2016). In this and subsequent sections, non-English text appears in parenthesis (bh: Bahasa Indonesia, nl: Dutch; fr: French; de: German) accompanying the modern reformed Indonesian or English spelling, e.g., earthquake (nl: *aardbeving*). The same applies for place names in use during the colonial Dutch period, e.g. Jakarta (nl: Batavia). Our reassessment of intensity focuses on an annual summary of felt earthquakes in the Dutch East Indies (Anonymous 1909) by the Royal Magnetic and Meteorological Observatory (nl: Koninklijk Magnetisch en Meteorologisch Observatorium, KMMO) that was culled from official correspondence. This was supplemented with further accounts extracted from colonial newspapers published in Indonesia and in the Netherlands (see Online Appendices A and B).

Accurate timekeeping is vital to accurately associate reports with individual earthquakes. In this respect, the KMMO summary was invaluable as it printed the standard time in Jakarta (known as Batavia Time or BT) for seismic disturbances recorded instrumentally at Jakarta or Bogor

Table 1

EMS-98 intensities for the 4 January 1907 mainshock (Event I)

Name	Lat.	Long.	EMS	Error	LT	BT	Account	References
Balige	2.333	99.083	4	1	11:45	+ 31	30 seconden. N-Z. Drie vrij sterke schokken.	Anonymous (1909)
Gunungsitoli	1.286	97.619	4–5	1	~ 12:00	+ 36	~ 60 seconden. O-W richting. Zwak golvende beweging. Gebouwen en boomen bewogen zichtbaar.	Anonymous (1909)
Natal	0.557	99.113	3	1	11:49	+ 30	1 se. ZO-NW richting. 2 zwakke schokken.	Anonymous (1909)
Ophir Districts	0.133	99.944	4	3	~ 12:00		Deze schokken zijn ook elders in de onderafdeeling Ophirdistricten meer of minder hevig gevoeld.	Het Nieuws van den Dag, 21 Mar 1907
Rundeng	2.685	97.834	4	4	11:55	+ 35	6 seconden. O-W. Drie sterke schokken	Anonymous (1909)
Sinabang	2.479	96.379	F	4	–	–	Den 4en dezer werden in den middag eenige elkander opvolgende schokken van aardbeving gevoeld, welke echter te Sinabang geene schade aanrichtten.	Het Nieuws van den Dag, 28 Jan 1907
Talu	0.225	99.976	4	1	12:15	+ 27	10 seconden. WZW–ONO. Vrij sterke schok. Waarnemer en diens schijver gevoelden zich onpasselijk door de beweging. Slingeren van lampen, bewegen van kasten, kraken van houtwerk in de kontroleurswoning, welke op een drassigen bodem in de Taloe-vallei staat en omringd in door sawah's.	Anonymous (1909)
Tanjung Balai	2.964	99.800	3	1	12:09	+ 28	5 seconden. O-W. Lichte horizontale schok.	Anonymous (1909)

Lat latitude (N°), *Long.* longitude (E°), *LT* local time, *BT* Batavia Time

(nl: Buitenzorg). BT was 7 h 7 min ahead of Greenwich Mean Time (GMT). It also published the difference in time at places outside Jakarta with respect to Batavia Time; For example, BT was 36 min ahead of local time at Gunungsitoli (nl: Goenoeng Sitoli) on Nias and 17 min behind local time at Pacitan (nl: Patjitan) in central Java. Thus, the local time was GMT + 06:31 at Gunungsitoli, GMT + 07:07 at Jakarta, and GMT + 07:24 at Pacitan.

A surprising result of our macroseismic reanalysis (Tables 1, 2) is the discovery of two distinctly separate earthquakes that were felt on Nias within an hour of each other on 4 January 1907 (Fig. 2b, c). A Milne seismograph in Jakarta (Anonymous 1909) recorded the first at 12:29 BT (05:19 GMT; Event I), and the magnetograph at Bogor detected the second at 13:24 BT (Event II). Engelbertus Schröder, the civil administrator (nl: kontrolleur) at Gunungsitoli on Nias (Online Appendix B), provided the strongest evidence with which to make this distinction; he described a weak but long-duration earthquake at about midday, followed by three "tidal waves" (nl: vloedgolven) at Toyolawa (nl: Tojolawa), which were later followed at 12:50 local time by a second violent earthquake at Gunungsitoli (Bataviaasch Nieuwsblad, 22 March 1907). His account appeared in several newspapers such as the Soerabaiasch Handelsblad (23 March 1907) and in the Nieuws van den Dag voor Nederlandsch-Indië (22 March 1907); the latter reported the time of the second earthquake to be "12:05". We believe this to be a typesetting error, as the identical account printed on the same day by the Bataviaasch Nieuwsblad (22 March 1907) states "12:50", and importantly 12:50 is also reported by Schröder (1917a) himself. The occurrence of two felt earthquakes separated by ~ 53 min is also corroborated by independent observers from Gunungsitoli (De Padanger, 26 January 1907), Natal, and Talu (nl: Taloe), as discussed in subsequent sections based on seismograms.

The weak, long-duration ground motions in the near field and the barely perceptible shaking at larger distances associated with Event I (Fig. 2b; Table 1) are characteristic of tsunami earthquakes (e.g., Kanamori 1972). In contrast, the descriptions of shaking and damage at Gunungsitoli from Event II

Table 2

EMS-98 intensities for the 4 January 1907 aftershock (Event II)

Name	Lat.	Long.	EMS	Error	LT	BT	Account	References
Airbangis	0.201	99.380	4	1	12:38	+ 29	~ 60 seconden. Hevige schok.	Anonymous (1909)
Bagan Si Api Api	2.157	100.816	4	1	12:45	+ 23	~ 60 seconden. O-W. Vrij hevige schommeling. De waarnemer had een gevoel als werd hij duizelig.	Anonymous (1909)
Barus	2.012	98.400	4–5	1	12:48	+ 33	~ 120 seconden. ZW-NO. Verscheidene kort op elkaar volgende vrij sterke schokken. De houten gebouwen kraakten. In de betonmuur van de verlaten benteng kwamen scheuren.	Anonymous (1909)
Gunungsitoli	1.286	97.619	7	1	12:50	+ 36	~ 60 seconden. O-W richtung. Zeer zware golvingen. Het was onmogelijk op de been te blijven. Alles schommelde sterk heen en weer. Allerwege groote schade aan gebouwen. Richting horizontale scheuren in cementen vloer douanekantoor, Noord 20 Oost, Noord en Noord 340 Oost. Cementen vloer gevangenis Noord 305 Oost.	Anonymous (1909)
Kuala Bëë	4.371	96.062	3	1	12:30	+ 42	45 seconden. N-Z.	Anonymous (1909)
Lambuhanbilik	2.520	100.165	4	1	12:45	+ 26	10 seconden. O-W. Twee vrij sterke schokken.	Anonymous (1909)
Langkat Regency	3.743	98.268	4	3	12:55		Een vrij sterke aardschock waargenomen in the richting Noord-Zuid.	Deli Courant, 4 Jan 1907
Lubuksikaping	0.137	100.167	3	1	~ 12:40	+ 26	~ 40 seconden. N-Z. Vele, zeer goed waarneembare golvingen. Een vrij zware golvende aardbeving waargenomen, die eenige seconden aanhield.	Anonymous (1909); Het Nieuws van den Dag, 21 Mar 1907
Medan	3.589	98.669	3	1	12:50		Hedenmiddag te 12.50 precies heeft hier terstede een duidelijk waarneembare aardbeving in noord-zuidelijke richting plaatsgehad. In het Medan-Hotel kwamen alle lampen, ook de zware in de groote zaal, in beweging, terwijl alle klokken er bleven stilstaan. Ook op het Postkantoor staakten de klokken den dienst. In den loop van den middag bereikten ons nog verschillende berichten, die met het bovenstaande overeenstemmen, en waaruit blijkt dat de aardbeving van vrij langen duur is geweest.	De Sumatra Post, 4 Jan 1907
Natal	0.557	99.113	4	1	12:42	+ 30	1 seconde. ZO-NW richting. 3 sterke schokken.	Anonymous (1909)
Padang Brahrang	3.588	98.435	3	1	12:50		Ook uit Padang Brahrang wordt ons omtrent een aardbeving gemeld. Ze is daar bedenmiddag eveneens 12:50 waargenomen en de richting was Oost-west. Het was een soort golvende beweging, die de lampen flink aan het slingeren bracht.	De Sumatra Post, 4 Jan 1907
Payakumbuh	–0.231	100.630	4	1	13:15	+23	10 seconden. O-W. Een lichte schok. Vrij hevige schok werd gevoeld.	Anonymous (1909); Het Nieuws van den Dag, 21 Mar 1907
Pulau Pandan	–0.949	100.140	3	1	12:55	+ 26	5 seconden. W-O. 5 lichte schokken.	Anonymous (1909)
Priaman	–0.628	100.116	4	1	13:02	+ 26	~30 seconden. O-W. Hevige horizonale bevingen.	Anonymous (1909)
Seribu Dolok	2.252	98.748	2	1	12:40	+ 32	Enkele seconden. 2 schokken.	Anonymous (1909)
Sibolga	1.739	98.783	3	1	12:35	+ 31	~ 54 seconden. NW-ZO. Zacht golvende beweging.	Anonymous (1909)
Siborongborong	2.210	98.972	4	1	~13:00	+ 30	10 seconden. Sterk.	Anonymous (1909)

Table 2 *continued*

Name	Lat.	Long.	EMS	Error	LT	BT	Account	References
Sidikalang	2.738	98.320	4	1	12:50	+ 33	10 seconden. NW-ZO. Zwakke schok. Beven der houten gebouwen, slingeren van lampen.	Anonymous (1909)
Talu	0.225	99.976	4	1	13:05	+ 27	30 seconden. WNW-OZO. Hevige schok. Waarnemer kreeg een onpasselijk gevoel. Slingeren van lampen, bewegen van kasten, kraken van houten deelen van het gebouw.	Anonymous (1909)
Tanjung Balai	2.964	99.800	4	1	12:53	+ 28	~120 seconden. O-W. Reeks opvolgende tamelijk sterke schokken. N-Z slingers klocken stopten. Lampen beschreven cirkels van 30 c.m. straal. Menschen kregen een gevoel van duizeligheid. Onmiddellijk na den schok brak een zware regenbui met donder los.	Anonymous (1909)

Lat. latitude (N°), *Long.* longitude (E°), *LT* local time, *BT* Batavia Time

(Fig. 2c; Table 2) lead us to believe that it was responsible for the violent shaking on Nias that was misconstrued as the mainshock by Newcomb and McCann (1987) and flagged as anomalous by Kanamori et al. (2010). Countless houses were destroyed on Nias by Event II, and even weeks later many people were living under temporary canopy shelters (bh: pondok) put up near their dwellings (Bataviaasch Nieuwsblad, 22 March 1907). Unusually, neither earthquake was reported in an official summary of news and affairs for January 1907 from Aceh (nl: Atjeh) i.e. "Atjeh-verslag over Januari" (Java Bode, 16 April 1907). These earthquakes also do not appear in catalogues of felt earthquakes in Thailand (Nutalaya et al. 1985), nor were they reported from anywhere else in Southeast Asia.

Our research underscores the importance of meticulously collated and analyzed noninstrumental evidence that allows the study of historical earthquakes for which instrumental data are lacking or severely limited. The serendipitous discovery of Event II was an unexpected outcome borne out of careful scrutiny of macroseismic data that led us to look for seismological evidence (discussed in subsequent sections) to support it. The absence of this event from earthquake catalogues (e.g., Gutenberg and Richter 1954; Storchak et al. 2013) is very conspicuous but not unusual, as with other recently discovered large early-instrumental earthquakes (Hough et al. 2005). This results from the limited number of seismograms at hand for early instrumental earthquakes, and the inadequacy (or simple inexistence) of proper algorithms to associate reported phase times, not to mention the adjustment of local times before implementation of standard time zones.

3. *Location of the Mainshock (Event I)*

The earliest published locations for Event I come from Turner et al. (1912) and Szirtes (1912a), who used arrival times from over 66 stations. As summarized by Kanamori et al. (2010), and not surprising for an event at the beginning of the 20th century, epicentral estimates are poorly constrained and significantly scattered (Fig. 3). Gutenberg and Richter's (1954) solution (2°N, 94½°E) is on the outer rise and slightly west of the solution (2°N, 95°E) by Turner et al. (1912), but both are very unlikely locations for a tsunami earthquake. By contrast, Szirtes (1912a) reports a location by T.H. Staverman (2°N, 96¼°E) in the immediate vicinity of the trench. In their recent compilation of relocations for early instrumental earthquakes, Storchak et al. (2013) locate the event at (1.87°N, 94.21°E), near the Turner et al. (1912) and Gutenberg and Richter (1954) epicenters, and with a remarkably small error ellipse (semi-axes: 28 km and 16 km). The most recent version (5.0) of the ISC-GEM catalog moves the epicenter north by 61 km, to

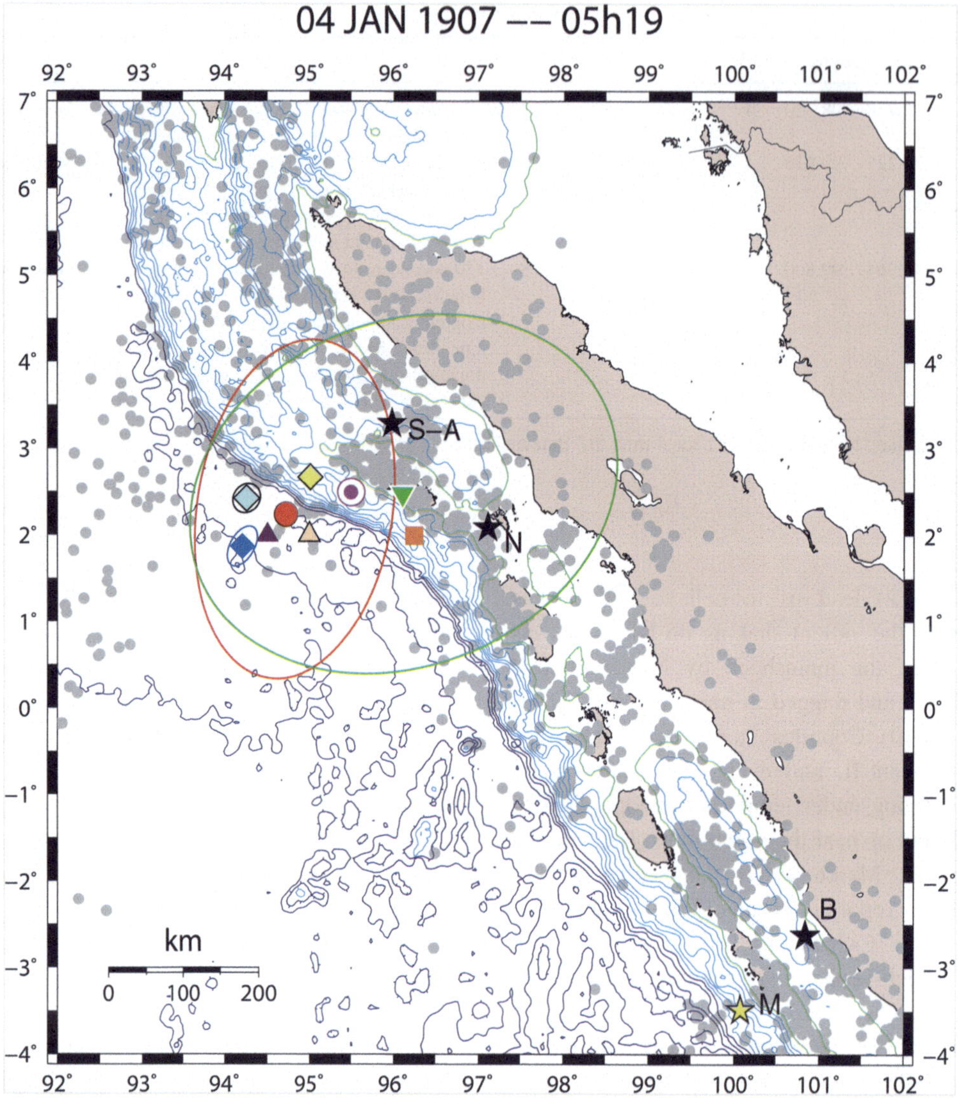

Figure 3

Relocations of the 1907 Sumatra event. The red dot shows our solution, with associated Monte Carlo ellipse. The green inverted triangle is the epicenter from Kanamori et al. (2010), with the confidence ellipse described in their online appendix shown in green. The dark-brown upward triangle is Gutenberg and Richter's (1954) epicenter, and the orange square T. Staveman's (Szirtes 1912a). The blue diamonds (with respective confidence ellipses) are the ISC solution by Storchak et al. (2013; solid, bright), and from the ISC-GEM (5.0) catalog (outlined, light blue); note that the ellipses do not intersect, suggesting that they are deceptively small. The yellow diamond is the best estimate from a grid search (D. Di Giacomo, pers. comm., 2017), and the preferred location is shown as the purple bull's eye symbol. The light-brown triangle is Turner et al.'s (1912) location. The black stars show the epicenters of the megathrust events of 2004 (Sumatra–Andaman, "S–A"), 2005 (Nias, "N"), and 2007 (Bengkulu, "B"), and the yellow star ("M") the 2010 Mentawai tsunami earthquake. Background seismicity (1970–2015; $h < 100$ km; $M \geq 5$) shown as light-grey dots. Isobaths are at 500-m intervals (green at 500 m; dark blue at 4000 m and deeper)

(2.422°N, 94.258°E). It is noteworthy that the very small confidence ellipse (20 km by 17 km) of the ISC-GEM solution does not overlap with the solution from Storchak et al. (2013) (Fig. 3). This casts doubt on the reliability of both locations, since they are supposed to be derived from the same algorithm. On the other hand, a systematic grid search of the same dataset confirms a large scatter of possible solutions but proposes a best-fitting epicenter at 2.67°N, 95.01°E (D. Di Giacomo, personal communication 2017), 125 km NE and across the plate boundary from the Storchak et al. (2013) solution.

Kanamori et al. (2010) relocated event I based on travel times listed in Gutenberg's personal notepads (Goodstein et al. 1980), with an emphasis on differential S – P times, to minimize errors due to clock uncertainties. Their solution (2.48°N, 96.11°E; Fig. 3) is located approximately 60 km closer to the trench than the 2004 and 2005 epicenters, at the seaward limit of the zone of interplate seismicity (Pesicek et al. 2010). This configuration is reminiscent of the 2010 Mentawai tsunami earthquake (Newman et al. 2011a; Hill et al. 2012). However, the 95 % confidence ellipse defined by Kanamori et al. (2010, p. 371) extends ~ 500 km, from the outer rise, beyond the Gutenberg and Richter (1954) and Storchak et al. (2013) locations, all the way to inland Sumatra.

We perform an independent relocation using the dataset listed by the ISC, and the interactive method of Wysession et al. (1991), which includes a Monte Carlo algorithm injecting Gaussian noise into the data. For an event in the 1900s, we give the noise a standard deviation $\sigma_G = 12.5$ s. Our solution converges to (2.24°N; 94.72°E: O.T. 05:19:13 GMT), a location on the outer rise, in the vicinity of the Storchak et al. (2013) and Gutenberg and Richter (1954) epicenters, but our confidence ellipse extends across the trench and grazes the solution from Kanamori et al. (2010); it is essentially contained inside these authors' ellipse. While all these relocations have very large uncertainties [we regard the Storchak et al. (2013) confidence ellipse as deceptively small], the emerging general pattern (Fig. 3) is that Event I was probably significantly displaced updip and trenchward with respect to the main seismogenic zone at the interplate contact on the Sunda Megathrust, to the general area located between the ISC grid-search solution and that of Staverman (in Szirtes 1912a), in a tectonic environment reminiscent of that of the Mentawai tsunami earthquake of 25 October 2010 (Newman et al. 2011a; Hill et al. 2012). This is our preferred location for the epicenter of Event I (2.5°N, 95.5°E) and is indicated by a red star on Fig. 4.

Our preferred location lies within an east-facing reentrant on the Sunda Megathrust (Franke et al. 2008) that indicates a zone of low seismic productivity between 1918 and 2007, and adjacent to, but not within, the rupture and aftershock zones of the 2004 and 2005 earthquakes (Chlieh et al. 2007; Engdahl et al. 2007; Konca et al. 2008; Pesicek et al. 2010). In this region, seismic reflection surveys (Franke et al. 2008) and three-dimensional (3-D) active source tomography (Tang et al. 2013) have identified a ridge of oceanic basement (Fig. 1b) that projects into the Wharton Basin, coinciding with a mapped fracture zone (Singh et al. 2011; Jacob et al. 2014). Onshore on Simeulue, this coincides with the Simeulue Saddle (Sieh et al. 2006), which has served as a persistent barrier to rupture in the past (Meltzner et al. 2012). Further downdip and beneath the Simeulue Saddle (Sieh et al. 2006), this morphological high can be associated with a region of strong coupling beneath central Simeulue (Tsang et al. 2015), and has been inferred as a structural control for local seismicity and the rupture dimensions of modern earthquakes under central Simeulue (Morgan et al. 2017). The dimensions of this morphological feature are debated (Franke et al. 2008; Tang et al. 2013; also see Fig. 4b in Morgan et al. 2017). Although our best-fit epicentral location lies slightly to the west of this feature, its uncertainty ellipse overlaps the western ramp of this morphological high, which we believe played a crucial role in rupture propagation and tsunami generation.

4. *Event II and Later Aftershocks*

Our study is the first to identify several aftershocks associated with the 1907 Sumatran earthquake. For the newly discovered large earthquake (Event II), waveforms were far fewer than for the mainshock, but we identified it in new records from Manila (Masó 1907) and Shimla (Patterson 1909), and the Osaka record scrutinized by Kanamori et al. (2010). Station bulletins from Manila (Masó 1907), Osaka (Anonymous 1931), and Shanghai (de Moidrey 1912) also list it. Masó (1907) interpreted the Manila records as requiring two events but pointed out the lack of evidence for Event II at Zikawei (Shanghai) and in Europe; we believe this to be partially incorrect, as Event II appears in the Zikawei bulletin (de Moidrey 1912). Anonymous (1909)

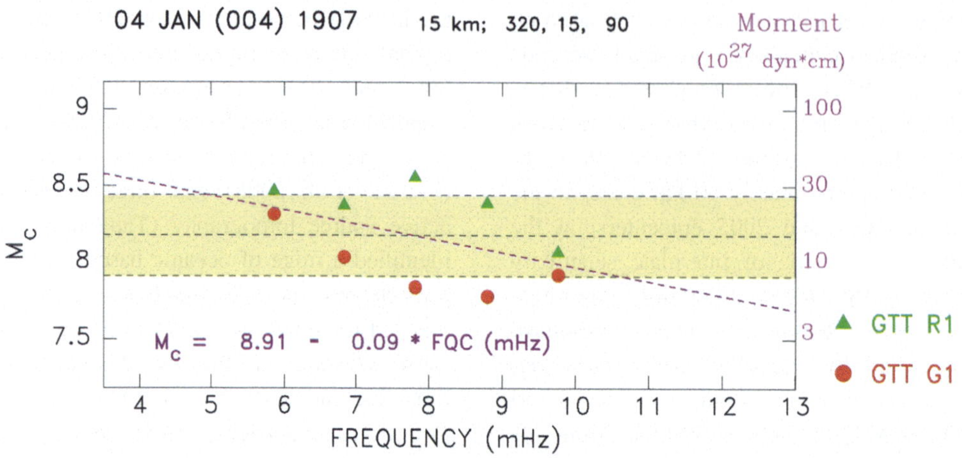

Figure 4

Spectral amplitudes of Rayleigh and Love waves at Göttingen, interpreted as corrected mantle magnitudes M_c (Okal and Talandier 1989), computed in the geometry $\phi = 320°$, $\delta = 15°$; $\lambda = 90°$, in the frequency band 6–10 mHz. The black dashed lines and yellow band show the average value and standard deviation of the full dataset, while the purple dashed line shows its best linear regression. Note the strong increase of moment with period

provides the time a magnetograph at Bogor detected this shock as 13:24 BT ($\sim$ 06:18 GMT), which gives an origin time of 06:15 GMT correcting for travel time to Bogor. We note, however, that the arrival time of the mainshock at this instrument is unknown. The difference in arrival time between the P waves from Event II and those of the mainshock at Osaka (Anonymous 1931), the difference in arrival time between the S waves from both events at Manila (Masó 1907) and Shanghai (de Moidrey 1912), and inspection of the waveform recorded at Manila (Masó 1907) suggest a $\sim$ 53 min interval between the two. We therefore estimate an origin time of $\sim$ 06:12 GMT, which is also supported by the first-hand account by Schröder (1917a) of two shocks $\sim$ 50 min apart.

The paucity of instrumental records prevents an instrumental location for Event II, but an epicentral distance (5420 km) to Osaka has been previously published (Anonymous 1931). Unfortunately, this datum is of little help, since Osaka is essentially equidistant (within 10 km) from the preferred locations of Events I and II. Finally, an epicenter in the region of Nias for Event II is further supported by the damage at Gunungsitoli (Table 2) and the destruction of buildings elsewhere on Nias (Bataviaasch Nieuwsblad, 22 March 1907). It is also likely, as discussed in more detail below, that uplifts inferred to have occurred in the Hinako Islands (Bataviaasch Nieuwsblad, 22 March 1907) and documented from a fossil coral microatoll on southern Nias (Meltzner et al. 2015) were related to Event II. This would suggest a source region in south–central Nias or off its south–central coast.

Two strongly felt aftershocks (Anonymous 1909) at $\sim$ 09:55 GMT on 4 January (Fig. 2d; Table 3) and $\sim$ 23:52 GMT on 5 January (Fig. 2e; Table 4) could be correlated with sparse regional and teleseismic recordings (Levitski 1909; Pechau 1907; Geiger 1909; Szirtes 1912b; Turner and Milne 1908a, b); origin times for both are corrected with respect to travel time to Jakarta from the Simeulue–Nias region. We estimated $M_S = 5.8$ for the 5 January event using phase data from Göttingen (Geiger 1909) and Jena (Pechau 1907). We also compute Milne magnitudes (M_M) of $\sim$ 6.1 to $\sim$ 6.2 for both events assuming an instrumental gain of 5 (Abe and Noguchi 1983); however, we emphasize that these values are very poorly constrained, and are associated with uncertainties of up to ± 0.5 (1σ).

In addition to the above, 19 felt aftershocks were counted on Simeulue on 4 January (Bataviaasch Nieuwsblad, 12 February 1907). Earthquakes were felt daily (Nieuws van den Dag voor Nederlandsch-Indië, 24 January 1907) with rarer aftershocks experienced after 15 January 1907 (De Sumatra Post, 1 February 1907). Earthquakes were also felt at Sinabang on Simeulue at 23:45 LT on 19 January 1907 (Bataviaasch Nieuwsblad, 12 February 1907) and on 8 July 1907 (Het Nieuws van den Dag, 20 September 1907), while on Nias,

Table 3

EMS-98 intensities for the 4 January 1907 aftershock at 9:55 GMT

Name	Lat.	Long.	EMS	Error	LT	BT	Account	References
Gunungsitoli	1.286	97.619	4	1	16:20	+ 36	NO-ZW. Meerdere sterke schokken. Slingeren van lampen	Anonymous (1909)
Natal	0.557	99.113	2	0	16:27	+ 30	1 second. ZO-NW. Zwakke schok.	Anonymous (1909)
Barus	2.012	98.400	3	0	16:27	+ 33	15-seconden. ZW-NO. Enkele kort op elkaar volgende lichte schokken.	Anonymous (1909)
Siborongborong	2.210	98.972	3	0	~ 17:00	+ 30	N-Z. Zwak.	Anonymous (1909)
Rundeng	2.685	97.834	4	0	16:25	+ 35	Sterke schok.	Anonymous (1909)

Lat. latitude (N°), *Long.* longitude (E°), *LT* local time, *BT* Batavia Time

Table 4

EMS-98 intensities for the 5 January 1907 aftershock at 23:52 GMT

Name	Lat.	Long.	EMS	Error	LT	BT	Account	References
Gunungsitoli	1.286	97.619	4	0	06:15	+ 36	2-seconden. Z-N. Korte sterke schok	Anonymous (1909)
Siborongborong	2.210	98.972	3	0	06:00	+ 30	N-Z. Zwak	Anonymous (1909)
Rundeng	2.685	97.834	4	0	06:00	+ 35	Een sterke schok	Anonymous (1909)

Lat. latitude (N°), *Long.* longitude (E°), *LT* local time, *BT* Batavia Time

aftershocks were felt continuously until 17 January (Bataviaasch Nieuwsblad, 29 January 1907). Aftershocks occuring later than 5 January 1907 were only reported from Nias (Anonymous 1909). Well-timed felt earthquakes are almost absent from Simeulue except for a shock that was quite heavy (nl: vrij hevig) at Sinabang around midday (~ 12:42 BT) on 4 January 1907 (Utrechts Nieuwsblad, 14 February 1907) without causing any damage (De Sumatra Post, 1 February 1907). We suspect this shock could be related to Event I or a strong local aftershock, neither of which we can definitively rule out. Another large earthquake was felt in the Banda Aceh region on 21 November 1907 (Fig. 2g; Anonymous 1909), but this was probably unrelated to Event I.

5. *Reassessment of Long-Period Seismic Moment and Source Slowness*

Kanamori et al. (2010) inferred a seismic moment (M_0) of about 6×10^{27} dyn cm for Event I at periods of ~ 50 s, through a comparison of time-domain amplitudes of body waves at Göttingen with those obtained at the nearby Black Forest Observatory (BFO) during the earthquake of 2 November 2002. This method, which assumes the same focal geometry for both events [strike (ϕ) = 297°, dip (δ) = 16°, slip angle (λ) = 73°], is confined to the time domain and, as such, tacitly assumes that Event I had a source spectrum similar to that of the reference event, and hence followed scaling laws. However, those are expected to be violated precisely by tsunami earthquakes, whose source spectrum is red-shifted towards lower frequencies with respect to a more traditional source, such as the 2002 event for which we have verified a totally regular energy-to-moment parameter ($\Theta = -4.92$). For this reason, it is necessary to further explore the source spectrum of Event I in the frequency domain to constrain its seismic moment at periods typical of mantle waves.

In addition to the waveforms interpreted by Kanamori et al. (2010), we revisit the record from Göttingen (GTT), and locate three previously unused waveforms from Pulkovo, Russia (Golitsyn 1908), Shimla, India (Patterson 1909), and Manila,

Philippines (Masó 1907). The Pulkovo record only features a partial, single component for Event I, and none for its aftershocks (Golitsyn 1908), while the Shimla seismogram clipped during Event I (Patterson 1909). The Manila record was written on a Vicentini mechanical seismometer with a period $T = 1.4$ s, which makes it the only short-period recording available to us of the 1907 Sumatra earthquakes; its crucial importance to our study is discussed in a subsequent section.

We use the first passages of Rayleigh (R_1) and Love (G_1) waves at Göttingen (GTT) for which precise metadata are available, with records hand-digitized at a sampling rate of 1 s. Our results are presented (Fig. 4) in the mantle range of frequencies (6–10 mHz), expressed as a mantle magnitude corrected for focal mechanism ($M_c = \log_{10} M_0 - 20$ with M_0 in dyn cm) in the formalism of Okal and Talandier (1989). We use a slightly adapted focal mechanism ($\phi = 320°$, $\delta = 15°$; $\lambda = 90°$), rotated only 8° in the formalism of Kagan (1991) from Kanamori et al.'s (2010), in order to reduce the scatter between Love and Rayleigh spectral amplitudes. Note that this new mechanism is also closer to the geometry of the nearby 2005 Nias megathrust earthquake ($\phi = 333°$, $\delta = 8°$; $\lambda = 118°$). Our results (Fig. 4) suggest an average moment $M_0 \approx 2.5 \times 10^{28}$ dyn cm ($M_c \approx 8.2$) at periods between 100 and 160 s; at the largest resolvable period ($\sim$ 170 s), the magnitude estimate approaches $M_c \approx 8.4$. This value of the seismic moment is about four times larger than proposed by Kanamori et al. (2010) from an estimate of time-domain amplitudes of S waves around 50 s and demonstrates slowness in the source of the 1907 mainshock, in clear agreement with its nature as a "tsunami earthquake". In addition, Fig. 4 documents a growth of moment with period, with a regression slope for M_C versus frequency of -0.09 logarithmic units per mHz. This number is comparable to values obtained for documented tsunami earthquakes (e.g., Java, 2006: -0.11; Mentawai, 2010: -0.08; Manzanillo, 22 June 1932: -0.14; Hikurangi, 1947: -0.07 and -0.08); by contrast, traditional subduction events feature lower slopes (in absolute value) that do not exceed -0.05 (e.g., Maule, 2010: -0.05; Illapel, 2015: -0.05; Manzanillo, 3 June 1932: -0.01; see Okal and Saloor 2017). This property provides a quantitative confirmation of the nature of Event I as a tsunami earthquake in conjunction with its anomalously low felt intensities.

6. *Ratio of Estimated Energies between Events I and II*

In the case of Event II, we recompute conventional magnitude estimates from those reports of ground amplitude for which associations could be made, using established formulæ and guidelines (Kárník et al. 1962; Geller and Kanamori 1977; Ambraseys and Douglas 2000). Our best estimate of M_S is $\sim$ 7.1 based on phase data from Osaka.

We also reproduce in Fig. 5 a copy of the Manila seismic record from Masó (1907). The seismogram in the top frame (Fig. 5a) was written on a Vicentini mechanical seismometer with a period $T = 1.4$ s. The distance to Manila is $\Delta = 29°$. Even though the orientation of this horizontal seismogram (NNW–SSE or 337°–157°) is pure transverse (the back-azimuth at Manila being $\beta = 247°$), this record clearly shows the P waves from two earthquakes separated by $\sim$ 53 min, corresponding to the two shocks identified on the basis of macroseismic data. We align the traces of the two events vertically to ease comparison of their waveforms (Fig. 5a, b). The body waves have a dominant recorded period of $\sim$ 4 s. The P waves of the mainshock are both much weaker and of longer duration than those of the aftershock. Note also that the mainshock P wavetrain consists of several subevents, lasting a total of about 2 min, while the aftershock waveform features two main packets, each lasting about 3 s. By contrast, we show in the lower frame a record (Fig. 5b) written on a longer-period Omori system with a natural period $T = 6.4$ s, oriented ENE–WSW, which this time is purely orthoradial. On that record, the amplitudes of the P waves from the two events become very comparable, and the main shock displays prominent Rayleigh waves featuring a period of $\sim$ 9 s at their maximum amplitude (probably reflecting the peaked response of the instrument), while the surface waves from the aftershock are much weaker.

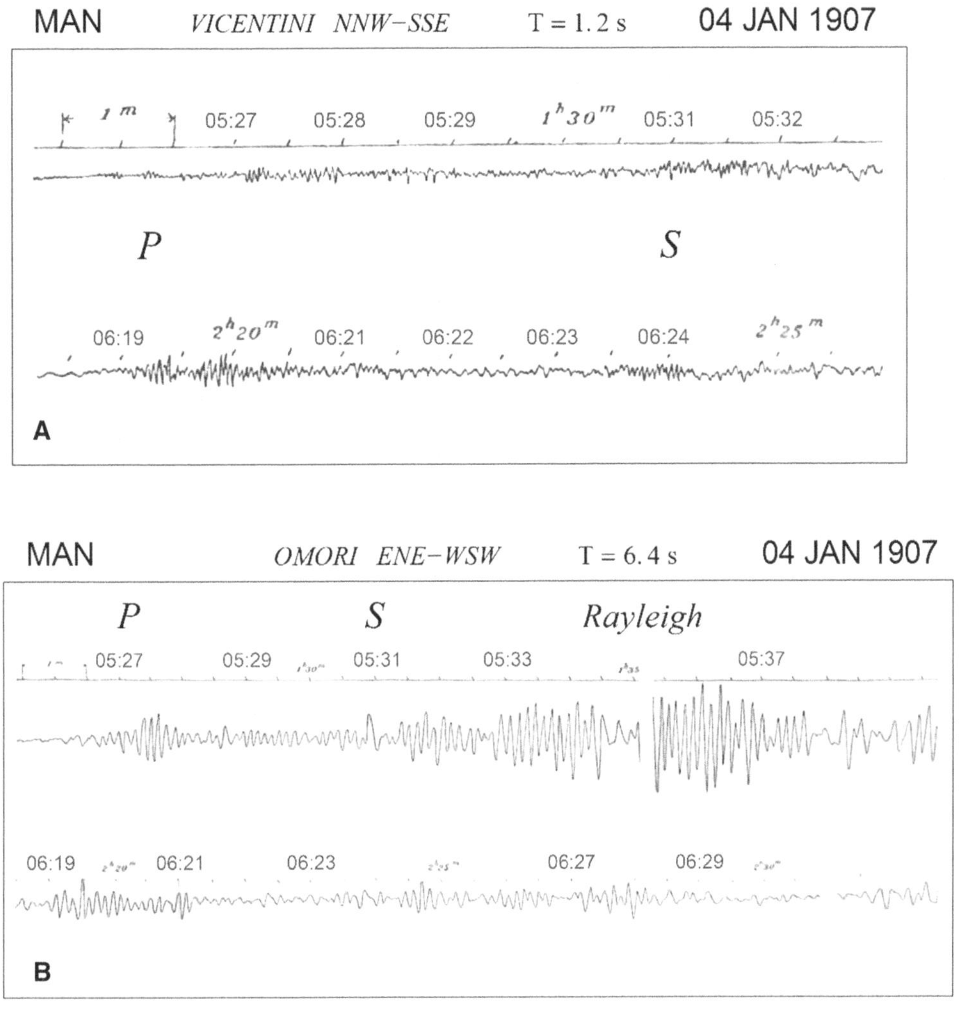

Figure 5
Seismic recordings at Manila (distance 29°; back-azimuth 247°), reproduced from Masó (1907), clearly showing the two events, separated by ~ 53 min. **a** Short-period Vicentini seismograph; **b** Omori seismograph. In both instances, the records are offset to align corresponding phases vertically. Time marks are at 30-s intervals, with original times given as local p.m. times (GMT + 8). Standard times (GMT) are overprinted in dark blue. These records dramatically illustrate the different characteristics of the two events

In an attempt to quantify these observations, we enlarged the Vicentini records (Fig. 5a) and digitized them at a time sampling $\delta t = 0.1$ s. Unfortunately, the Vicentini instrument was undamped (Masó 1907), which means that it would, at least theoretically, have an infinite response at its natural period. In this context, it would not be possible to formally compute an energy-to-moment ratio and a slowness parameter (Θ) as defined by Newman and Okal (1998). This is further compounded by the fact that the photographic magnification of the figure by Masó (1907) is unknown. Under these circumstances, we use the model of a very weakly damped instrument ($\varepsilon = 1.1$) and obtain a ratio of estimated energies between the mainshock (Event I) and aftershock (Event II) of $E_I/E_{II} \approx 0.2$. Furthermore, we obtain an estimate of $M_S \approx 7.1$ for the classical surface-wave magnitude of Event II as described in the previous section, which would correspond to a maximum moment of $\sim 4 \times 10^{26}$ dyn cm, assuming that it follows scaling laws. However, the strong intensities reported may suggest that its source is blue-shifted towards higher frequencies, in which case the moment could be lower. Given the long-period moment estimated above for the mainshock (2.5×10^{28} dyn cm), this would suggest a factor of 300 between the energy-to-moment ratios of Events I and II, amounting to a difference of ~ 2.5 units in their parameters Θ. While this value is not unreasonable, it comes close to the maximum range of Θ values that we have documented [see Okal and Saloor (2017) for an extensive dataset] and could be explained using both a very low value of $\Theta \approx -6.5$ for the mainshock, and a high value, $\Theta \approx -4$, for Event II. The former value is within the range of observed Θ for tsunami earthquakes [El Salvador 2012: −6.42; Hikurangi 1947 (II, 17 May): −6.51 (Okal and Saloor 2017)], while the latter would be typical of a "snappy" earthquake (Okal and Kirby 2002) occurring inside the overriding plate, such as the main normal faulting aftershock of the 2011 Tohoku earthquake (−4.12) or the 1939 Chillán earthquake (−4.04) (Okal and Kirby 2002; Okal et al. 2016). The jagged nature of the P waves of the mainshock at Manila would also be in line with observations during the 1992 Nicaragua event (Polet and Kanamori 2000) and the resulting model by Tanioka et al. (1997).

7. Indian Ocean Tsunami Observations

We were able to compile 88 observations of the tsunami generated by Event I across the Indian Ocean (Table 5; Fig. 6; Online Appendix B). They were obtained from individual accounts (e.g., Schröder 1917a), as well as from reports published in local newspapers. Most of the Dutch news reports were corroborated by later official Dutch correspondence (Openbaar Verbaal, 23 April 1908). First- and second-hand accounts from a few locales in the near and far field are available (e.g., Anonymous 1909; Visser 1922), and some were repeated by Solov'ev and Go (1974). We also extracted eyewitness or second-hand accounts appearing in contemporary anthropological (e.g., Baumwoll 2008; Rahman et al. 2017) and scientific studies (Yogaswara and Yulianto 2006; Whitlow 2008; Fujino et al. 2014).

The impact of the tsunami was greatest on the southern coast of Simeulue, and on the northern and western coasts of Nias (Fig. 6a, b). Without providing additional location details, news and official reports from Simeulue suggest the tsunami struck after Friday afternoon prayers, the sea flooding inland as much as 600–800 m with isolated inundations of 1200 m, or to the foot of the hills, sweeping coral boulders and marine fauna inland (Bataviaasch Nieuwsblad, 12 February 1907; Utrechts Nieuwsblad, 14 February 1907; Openbaar Verbaal, 23 April 1908; Yogaswara and Yulianto 2006). Better numerical constraint is forthcoming from Nias (Fig. 6a, c), for example, from Afulu (nl: Afoeloe), where the sea washed inland for over a kilometer (nl: ongeveer een paal)[1] as far as the foot of the hills, where it deposited debris that included human remains and large uprooted trees (Bataviaasch Nieuwsblad, 22 March 1907). Schröder (1917a) noted that a part of the bay at Afulu was filled with sand and other material (nl: Vermelding verdient de opvulling van een deel de baai van Afulu met zand en ander materiaal) at the time of his visit. Schröder (1917a) also reports that Hulo Uma near Afulu was "robbed of all life" (nl: van alle leven beroofd). Southwest of Toyolawa, only the west coast of the island of Pulau Mausi (nl: Ma'uso or Maoesi) was impacted by the tsunami (Schröder 1917a). However, to the south, the tsunami appears to have rapidly diminished in character at Lagundi (Fig. 6b; Bataviaasch Nieuwsblad, 22 March 1907).

The island of Simeulue Cut (nl: Simeuloeë Tjoet or Simaloer Tjoet) off the southern coast of Simeulue (Fig. 6a) "was lost" or "disappeared" (nl: is verdwenen) with only the hill at the center of the island above the water (Fig. 6b), suggestive of extreme tsunami erosion or, possibly, subsidence. West of Nias, on Pulau Wunga (Fig. 6b), the tsunami swept the island from the west (Schröder 1917a), destroying three-fourths of it including numerous coconut trees, with debris and victims swept into the lagoon at the center of the island (Bataviaasch Nieuwsblad, 22 March 1907). The height of the tsunami at Pulau Wunga is estimated at between 6 m and 15 m (Schröder 1917a; De Padanger, 25 February 1907), and in 2005, modern residents told one of the authors

[1] The old Dutch measure "een paal" used in colonial Indonesia is equivalent to 1507 m (Staring 1871).

Table 5

Tsunami observations for the 1907 tsunami extracted from newspaper reports and other publications from 88 locations (including nil reports) from the Indian Ocean basin

#	Name	CN	Lat.	Long.	HT (m)	IN (m)	Description
1	Aden	YE	12.802	45.018	–	–	"The tidal wave was not perceptible at this port" (Erskine 1909)
2	Afulu	ID	1.263	97.245	–	1507	Tsunami went inland about "een paal" (1507 m, *see text*) almost to the base of the hills where debris and skeletons of those killed were left; at least 50 killed (*Bataviaasch Nieuwsblad*, 22 Mar 1907)
3	Ayer	ID	2.591	95.968	–	–	Village completely "lost" (*Bataviaasch Nieuwsblad*, 12 Feb 1907)
4	Alafan	ID	2.824	95.756	> 8	–	Local history preserved by word of mouth recalls corpses left in the branches of durian trees, and make note to point out that the 2004 tsunami (that was 8-m high at this location) did not reach the height of the branches of durian trees (*Kompas*, 20 August 2014)
5	Angkë	ID	2.509	96.165	–	–	Village completely "lost" (*Bataviaasch Nieuwsblad*, 12 Feb 1907)
6	Apollo Bandar	IN	18.922	72.834	0.0508	–	"The effect of the tidal wave is noticeable between 7 P.M. on 4th, and 9 P.M. on 5th January, the oscillations of the pencil occurring only at or about the time of slack water at low and high tides. The greatest movement of the pencil out of the normal was 2 inches at 2-50 A. M. on 5th" (Erskine 1909)
7	Awe Ketjil	ID	2.436	96.252	–	–	Village completely "lost" (*Bataviaasch Nieuwsblad*, 12 Feb 1907)
8	Awe Sebal	ID	2.433	96.271	–	–	Village completely "lost" (*Bataviaasch Nieuwsblad*, 12 Feb 1907)
9	Bahoe	ID	–	–	–	–	Village completely "lost" (*Bataviaasch Nieuwsblad*, 12 Feb 1907)
10	Baie du Courrier	MG	−12.265	49.059	–	–	No record of the tsunami available (*see text*)
11	Barus	ID	2.012	98.339	–	–	"For the tidal wave, which destroyed the island of Simaloer on 4 January 1907, reports were received from Gunungsitoli, Nias, Pulau Wunga, Natal, Barus, Meulaboh, Kuala Bhee, Pulau Raja and Pulau Tello" (Anonymous 1909)
12	Boeman/Serie Boenon/Boenon Siri	ID	2.519	96.150	–	–	Village completely "lost" (*Bataviaasch Nieuwsblad*, 12 Feb 1907)
13	Cape Tolojawa	ID	1.404	97.062	–	–	Village almost completely destroyed; two killed; water leapt with great force landward but was broken by the coral reefs (*Bataviaasch Nieuwsblad*, 22 Mar 1907)
14	Colombo	SL	6.979	79.870	–	–	"It was observed as a disturbance, or tidal wave, at the mouth of the Kelani river, and by the villagers on the banks for some miles upstream" (*Ceylon Observer*, 10 January 1907)
15	Diego Suarez (Antsiranana)	MG	−12.289	49.302	–	–	No record of the tsunami available (*see text*)
16	Doemala	ID	1.171	97.304	–	–	Partially destroyed; the cape to the north–west broke the tsunami; three fatalities (*Bataviaasch Nieuwsblad*, 22 Mar 1907)

Table 5 *continued*

#	Name	CN	Lat.	Long.	HT (m)	IN (m)	Description
17	Galle	SL	6.031	80.231	–	–	"A great deal of excitement was caused in the Fort today when those assembled on the jetty noticed that the sea had receded in the harbor, the whole coast round the bay being high and dry some 25 or 35 feet. The lighters and other boats moored near the jetty were left on dry land. The hull of the wrecked "Dalswinton" was exposed to view. The coolies near the jetty secured a good deal of fish, lobsters, etc. The sea receded seven times at intervals of half-an-hour. After each recession the in-coming flow of water was very heavy. On the third occasion it was nearly level with the jetty. The sea was perfectly calm, and no waves were noticeable, but the current with the in-coming swell was very strong. The B. I. steamer "Loodiana" was in harbor discharging rice, and was affected by the movement of the seas, being moved backwards and forwards. Several lighters laden with rice were on their way from the steamer to the jetty, and were in danger of being dashed against the rocks owing to the heavy inflow" (*Ceylon Observer*, 10 January 1907; Also see Online Appendix B)
18	Guhanaga (Geni)	ID	4.683	95.686	–		An eyewitness claimed the Krueng Sabe river overflowed until Desa Geni i.e. modern Guhanaga (*Aceh Post*, 10 April 2013)
19	Gunungsitoli	ID	1.286	97.619	0.70	–	"At Gunungsitoli seawater, dark brown in colour, leapt against the beach to a height of about 70 cm (7 dM)" (*Bataviaasch Nieuwsblad*, 12 Feb 1907)
20	Hambantota	SL	6.123	81.122	–	–	"The first big wave was probably unrecorded and occurred, Mr. Barnard thinks, at about 1:30 p.m. The second great wave occurred at 3 p.m. and the third at 3:05 p.m. The waves are not always quite exact, which accounts for the time between the waves at which No. 7 wave occurred, and also at 4:05 p.m., when one was possibly unrecorded, or spoilt by the interfering waves. A wave after rushing up the beach recoiled and met the succeeding one, making its force diminished, and it might happen that one retiring wave just met an incoming small one and spoilt it...... between greatest rise and fall 14 feet measured at Jetty" (*Ceylon Observer*, 10 January 1907; Also see Online Appendix B)
21	Hulo Uma	ID	1.252	97.236	–	–	Hulo Uma near Afulu was "robbed of all life" (Schröder 1917a)
22	Île Amsterdam (*questionable*)	FR	−37.834	77.556	–	–	La même marée sismique a été observée aux Îles Saint-Paul, Amsterdam, Rodriguez, la Réunion, Maurice et enfin aux Seychelles (Bertho 1910)
23	Île Saint-Paul (*questionable*)	FR	−38.724	77.515	–	–	La même marée sismique a été observée aux Îles Saint-Paul, Amsterdam, Rodriguez, la Réunion, Maurice et enfin aux Seychelles (Bertho 1910)
24	Inoel/Inoer	ID	2.483	96.195	–	–	Village completely "lost" (*Bataviaasch Nieuwsblad*, 12 Feb 1907)
25	Karachi	PK	24.833	66.992	0.0762	–	"The disturbance commenced at 5 P.M. on 4th January, about the time of slack water at low tide and lasted till 10 A.M. on 5th. It was distinctly noticeable between 11 P.M. on 4th and e P.M. on 5th; at 1–15 A. M. at slack water at high tide on the 5th, the pencil showed an abnormal movement of the wave of 3 inches" (Erskine 1909)
26	Kayu Menang	ID	2.306	97.741	–	–	Possibly affected by flooding in 1907 causing residents to move to Singkil (Aceh Post, 10 April 2013)
27	Kidderpore	IN	22.548	88.320	–	–	"There is no trace of the tidal wave on the diagrams, at the river ports of Rangoon and Kidderpore" (Erskine 1909)

Table 5 *continued*

#	Name	CN	Lat.	Long.	HT (m)	IN (m)	Description
28	Koela Deh	ID	n/a	n/a	–	–	Village completely "lost" (*Bataviaasch Nieuwsblad*, 12 Feb 1907)
29	Kota Ladang	ID	n/a	n/a	–	–	Disappeared (*Utrechts Nieuwsblad*, 14 Feb 1907)
30	Kota Tinggi	ID	2.399	96.475	–	600–900	Village completely "lost" (*Bataviaasch Nieuwsblad*, 12 Feb 1907)
31	Kuala Bhee	ID	4.371	96.062	–	–	"For the tidal wave, which destroyed the island of Simaloer on 4 January 1907, reports were received from Gunungsitoli, Nias, Pulau Wunga, Natal, Barus, Meulaboh, Kuala Bhee, Pulau Raja and Pulau Tello" (Anonymous 1909)
32	La-eion	ID	2.547	96.108	–	–	Village completely "lost" (*Bataviaasch Nieuwsblad*, 12 Feb 1907)
33	Lahalos	ID	n/a	n/a	–	–	Village completely "lost" (*Bataviaasch Nieuwsblad*, 12 Feb 1907)
34	Lahoebang (Lakubang)	ID	2.586	95.989	–	2000–3000	Village completely "lost" (*Bataviaasch Nieuwsblad*, 12 Feb 1907). Dolphins (bh: ikan lumba-lumba) and whales (bh: ikan paus) carried inland up to distances of 2–3 km from the coast (Yogaswara and Yulianto 2006)
35	Lafu	ID	1.401	97.213	–	~ 100	"A few houses damaged; tidal wave ran about 100 m inland; damage minor" (*Bataviaasch Nieuwsblad*, 22 Mar 1907)
36	Lagundi	ID	0.582	97.737	–	–	"Weak tsunami" (*Bataviaasch Nieuwsblad*, 22 Mar 1907)
37	Lamajan (Cape)	ID	2.572	95.996	–	–	Village completely "lost" (*Bataviaasch Nieuwsblad*, 12 Feb 1907)
38	Laokeh	ID	n/a	n/a	–	–	Village completely "lost" (*Bataviaasch Nieuwsblad*, 12 Feb 1907)
39	Lasingaloe	ID	2.442	96.232	–	–	Village completely "lost" (*Bataviaasch Nieuwsblad*, 12 Feb 1907)
40	Lessihan	ID	2.407	96.324	–	–	"Disappeared" (*Utrechts Nieuwsblad*, 14 February 1907)
41	Lebang	ID	2.433	96.283	–	–	Village completely "lost" (*Bataviaasch Nieuwsblad*, 12 Feb 1907)
42	Leukoën (Lukon)	ID	2.609	95.867	–	–	65 killed (Utrechts Nieuwsblad, 14 Feb 1907). On the (small) islands of Simeuloee Tjoet, Linggang and Leukoën most the houses were destroyed. (*Bataviaasch Nieuwsblad*, 12 Feb 1907)
43	Linggang	ID	2.610	95.868	–	–	On the (small) islands of Simeuloee Tjoet, Linggang and Leukoën most the houses were destroyed. (*Bataviaasch Nieuwsblad*, 12 Feb 1907)
44	Madras (Chennai)	IN	13.091	80.298	–	–	"Oscillations due to the tidal wave are traceable on the tidal diagram between midnight of 4th and midnight of 5th January. They are insignificant" (Erskine 1909)
45	Matara	SL	5.942	80.543	–	–	"The sea receded as far back that the sunken ridge of rocks a quarter of a mile from the shore became visible. This happened two or three times" (*Ceylon Observer*, 10 January 1907)
46	Mauritius	FR	−20.285	57.573	–	–	La même marée sismique a été observée aux Îles Saint-Paul, Amsterdam, Rodriguez, la Réunion, Maurice et enfin aux Seychelles (Bertho 1910)
47	Meulaboh (Analabu)	ID	4.143	96.128	–	–	"For the tidal wave, which destroyed the island of Simaloer on 4 January 1907, reports were received from Gunungsitoli, Nias, Pulau Wunga, Natal, Barus, Meulaboh, Kuala Bhee, Pulau Raja and Pulau Tello" (Anonymous 1909)
48	Moedil	ID	2.466	96.217	–	–	Village completely "lost" (*Bataviaasch Nieuwsblad*, 12 Feb 1907)
49	Nantjalla	ID	2.450	96.227	–	–	Village completely "lost" (*Bataviaasch Nieuwsblad*, 12 Feb 1907)

Table 5 *continued*

#	Name	CN	Lat.	Long.	HT (m)	IN (m)	Description
50	Naibos	ID	2.489	96.187	8–10	~ 1000	Village completely "lost" (*Bataviaasch Nieuwsblad*, 12 Feb 1907); Sharks (bh: ikan hiu), whales (bh: ikan paus), and coral boulders were transported inland; marine fauna was swept inland as much as a kilometer at Naibos in south Simeulue where local narratives also allege human corpses and dead livestock such as buffalo were found in the tops of coconut trees at heights of 8–10 m (Yogaswara and Yulianto 2006)
51	Natal	ID	0.557	99.113	–	–	"For the tidal wave, which destroyed the island of Simaloer on 4 January 1907, reports were received from Gunungsitoli, Nias, Pulau Wunga, Natal, Barus, Meulaboh, Kuala Bhee, Pulau Raja and Pulau Tello" (Anonymous 1909)
52	Oejong	ID	2.688	95.776	–	–	87 killed (*Utrechts Nieuwsblad*, 14 Feb 1907)
53	Oedjoeng Lojang	ID	1.530	97.352	–	–	"All houses except one flattened to the ground" (*Bataviaasch Nieuwsblad*, 29 Jan 1907). *Note*: Uncertain if damage is from the earthquake of the tsunami
54	Oedjoeng Oesoej (Ujung Gossong?)	ID	–	–	–	–	Village completely "lost" (*Bataviaasch Nieuwsblad*, 12 Feb 1907)
55	Padang	ID	−1.003	100.379		–	Recorded by the automatic tide gauge in Teluk Bayur (nl: Emmahaven) (Anonymous 1909; also see text)
56	Pointe des Galets	FR	−20.940	55.305	–	–	"Le 4 janvier 1907, à 4h20 m de l'après-midi, au moment de la pleine mer, on vit l'eau monter assez rapidement dans le port, atteignant 60 mm à 70 mm au-dessus du niveau des marées de syzygies, puis tout à coup se retirer lentement bien au-dessous du niveau des plus basses mers. Cette élévation et cet abaissement se répétèrent avec un rythme parfait, de 10 en 10 min, ainsi que l'a bien indiqué le marégraphe. La mer, au large, était calme et belle, sans houle. La dénivellation, toutes les 10 min, était de lm, 70. Le phénomène continua jusqu'au 6 janvier au soir, avec des recrudescences d'intensité" (Bertho 1910)
57	Port Blair	IN	11.675	92.761	0.127	–	"The first disturbance appeared to have commenced at 1-45 P.M., on 4th January 1907; the oscillations of the pencil due to the tidal wave, were slight up to 2-40 P.M., after which they increased in frequency and in height up to 6-30 P. M., the time of slack water at low tide, when the wave was greatest, the height being 5 inches. After this the curve showed a diminishing of the wave until it ceased at 10-20 P.M. on 6th instant. The oscillations were most marked at each slack water at low and high tides" (Erskine 1909)
58	Pulau Mausi	ID	1.355	97.101	–	–	The west coast of the island of Pulau Mausi was impacted by the tsunami (Schröder 1917a)
59	Pulau Raja	ID	4.869	95.382	–	–	"For the tidal wave, which destroyed the island of Simaloer on 4 January 1907, reports were received from Gunungsitoli, Nias, Pulau Wunga, Natal, Barus, Meulaboh, Kuala Bhee, Pulau Raja and Pulau Tello" (Anonymous 1909)
60	Pulau Tello	ID	−0.066	98.264	–	–	"For the tidal wave, which destroyed the island of Simaloer on 4 January 1907, reports were received from Gunungsitoli, Nias, Pulau Wunga, Natal, Barus, Meulaboh, Kuala Bhee, Pulau Raja and Pulau Tello" (Anonymous 1909)
61	Pulau Woenga (Pulau Wunga)	ID	1.207	97.091	7–5	–	All houses washed into the lagoon with at least 100 killed. 120 coconut trees destroyed; estimated to have been 7 m–15 m high (Schroeder, 1912a; *De Padanger*, 25 Feb 1907; *Bataviaasch Nieuwsblad*, 22 Mar 1907)

Table 5 *continued*

#	Name	CN	Lat.	Long.	HT (m)	IN (m)	Description
62	Rangoon	MY	16.768	96.159	–	–	"There is no trace of the tidal wave on the diagrams, at the river ports of Rangoon and Kidderpore" (Erskine 1909)
63	Rodrigues Island	FR	−19.711	63.448	–	–	"The officer in charge of the Eastern Extension Telegraph office, Barrack Street, stated last night that at 8:35 p.m. the Company's office at Cocos Island had been informed by the Rodrigues manager that the sea was boiling, and that small tidal waves were being experienced. At about 9 o'clock information came to hand that the tide, which should have been at its lowest ebb, was rising rapidly, and that a small reef which was ordinarily visible standing out of the water was being covered. The tide was higher than it had ever been before. Half an hour later the tide was reported to be still rising, and the residents were hauling their boats up out of danger. Shortly before 11 p.m. the Rodrigues station reported that the tide was falling" (*The West Australian*, 5 January 1907)
64	Salang	ID	2.637	95.864	–	–	Village completely "lost" (*Bataviaasch Nieuwsblad*, 12 Feb 1907)
65	Saloer Ilir	ID	2.443	96.238	–	–	Village completely "lost" (*Bataviaasch Nieuwsblad*, 12 Feb 1907)
66	Saint-Benoît	FR	−21.044	55.717	–	–	"Le phénomène était bien marqué, et complètement visible de la « Batterie » , où beaucoup de personnes ont l'habitude de prendre le frais dans les journées chaudes vers le soir. Il était 4 h. ½. Tout à coup, les lames qui de tout temps déferlent à cet endroit, disparaissent, la mer se retire calme, laissant à sec un fond d'une centaine de mètres de large environ. Les eaux se retirent derrière la chaine de Caps, qui forme le récif de St-Benoit, caps qui n'émergeaient jamais quelque temps qu'il fît, mais que la retraite de la mer laisse entièrement à nu. Parmi les spectateurs du phénomène, des jeunes garçons, attirés par l'appât, se précipitent sur les poissons surpris, que l'eau a abandonnés. Mais à peine quelques minutes s'étaient écoulées, que les gens restés sur le rivage signalent à ces jeunes imprudents, une houle d'une hauteur extraordinaire venant de très loin. Ceux-ci n'eurent que le temps d'abandonner leur butin et de gagner la côte. Bien leur en prit : la houle est venue déferler et envahir la terre ferme à une limite que n'atteint pas d'ordinaire le niveau de la mer, le ressac balayant et entraînant tout : poissons et galets. Plusieurs vagues semblables se sont succédé à différents intervalles. A neuf du soir la mer a commencé à reprendre son niveau ordinaire, puis tout est rentré dans l'ordre normal" (*Journal de l'île de La Réunion*, 8 January 1907)
67	Saint-Gilles-de-Bain	FR	−21.053	55.224	–	–	"Peut-être quelques uns de nos érudits nous donneront-ils l'explication de ce phénomène qui a fortement impressionné la population saint gilloise et même les quelques baigneurs en villégiature" (*Journal de l'île de La Réunion*, 8 January 1907)
68	Saint-Marie	FR	−20.893	55.508	–	–	"….et qu'à Ste Marie des rochers qui n'émergent jamais, ont été complètement mis à découvert" (*La Patrie Créole*, 8 January 1907)
69	Saint-Pierre	FR	−21.334	55.471	–	–	"Il paraît que dans le port de St-Pierre, il a été relevé une différence de niveau de près de deux mètres, produisant, au retrait de l'eau, de véritables cascades descendant des hauts fonds non dragués" (*La Patrie Créole*, 8 January 1907)

Table 5 *continued*

#	Name	CN	Lat.	Long.	HT (m)	IN (m)	Description
70	Seychelles	SC	−4.712	55.491	–	–	La même marée sismique a été observée aux Îles Saint-Paul, Amsterdam, Rodriguez, la Réunion, Maurice et enfin aux Seychelles (Bertho 1910)
71	Si Linggas	ID	–	–	–	–	Village completely “lost” (*Bataviaasch Nieuwsblad*, 12 Feb 1907)
72	Si Toeboeh	ID	–	–	–	–	Village completely “lost” (*Bataviaasch Nieuwsblad*, 12 Feb 1907)
73	Sifahandro	ID	1.504	97.353	–	–	A few houses destroyed (*Bataviaasch Nieuwsblad*, 22 Mar 1907)
74	Sibolga (*questionable*)	ID	1.731	98.782	–	–	“It is worth mentioning that in the past 50 years, (at the time) when a strong earthquake destroyed Hinako (Nias), the sea at Sibolga rushed forth and a tidal wave was formed, the city was flooded. Now when people see the sea ebb and flow, they fear a tidal wave (report from the controller of Sibolga). More than likely that this report is related to the earthquake of 4 January 1907” (Visser 1931)
75	Simueloee Tjoet (Simuelue Cut)	ID	2.531	95.940	–	–	“Half lost” (*Bataviaasch Niewsblad*, 23 Jan 1907) or “partially lost in the sea” (De Sumatra Post, 1 Feb 1907). Barely the (little) top of the (little) hill was visible out of the water (*Utrechts Nieuwsblad*, 14 Feb 1907; *De Padanger*, 25 Feb 1907; Most of the houses destroyed (*Bataviaasch Niewsblad*, 12 Feb 1907). Also see main text
76	Sinabang	ID	2.479	96.379	0.5	–	0.5 m oscillations of harbor (*Utrechts Nieuwsblad*, 14 February 1907)
77	Siriwaoe/Saroeng Baroeng	ID	1.700	97.448	–	–	“Flooded” (*Bataviaasch Nieuwsblad*, 10 Jan 1907). Temporarily underwater (*Bataviaasch Nieuwsblad*, 29 Jan 1907)
78	Sirombu	ID	0.951	97.418	–	–	Deaths reported (*Bataviaasch Nieuwsblad*, 22 Mar 1907)
79	Sosoh	ID	3.721	96.825	–	–	“Went under water” (*Bataviaasch Nieuwsblad*, 12 Mar 1907)
80	Soea-Soea	ID	2.421	96.314	–	–	Village completely “lost” (*Bataviaasch Nieuwsblad*, 12 Feb 1907)
81	Taloe Siaba	ID	1.510	97.378	–	–	“Lost” (*De Padanger*, 26 Jan 1907)
82	Tamatave (Toamasina)	MG			–	–	No record of the tsunami available (*see text*)
83	Tanjung Gossong	ID	2.395	96.336	–	–	The newly established field or agricultural land (bh: ladang) was destroyed (*Bataviaasch Nieuwsblad*, 12 Feb 1907); Coral boulder carried inland supposedly in 1907 (Fujino et al. 2014)
84	Tangalle	SL	6.022	80.801	–	–	“The sea ebbed and flowed in the same manner, exposing the sea bottom as far as the rock which lies to the east of the Resthouse…”. (*Ceylon Observer*, 17 June 1909)
85	Tapah (island)	ID	2.355	96.240	–	–	“As good as without coconut trees” and on this island about 200 people were killed (*Utrechts Nieuwsblad*, 14 Feb 1907); One account says 1000 people were killed at Kuala Deh (on Tapah) alone (*Utrechts Nieuwsblad*, 14 Feb 1907); All villages completely “lost” (*Bataviaasch Nieuwsblad*, 12 Feb 1907)
86	Tapak Tuan	ID	2.357	97.180	–	–	“Ebb and flood observed thrice” (*Bataviaasch Nieuwsblad*, 12 Feb 1907)
87	Titi Kapoer	ID	–	–	–	–	Village completely “lost” (*Bataviaasch Nieuwsblad*, 12 Feb 1907)
88	Toerelaoeja	ID	1.532	97.532	–	–	“A fisherman was killed” (*Bataviaasch Nieuwsblad*, 22 Mar 1907)

See Online Appendix B for verbatim newspaper reports. Country name code (CN): France (FR), India (IN), Indonesia (ID), Madagascar (MG), Pakistan (PK), Seychelles (SC), Sri Lanka (SL), and Yemen (YE); Lat. = latitude, Long. = longitude, height (HT), and inundation distance (IN) of tsunami in meters (m) as best as can be estimated from available material

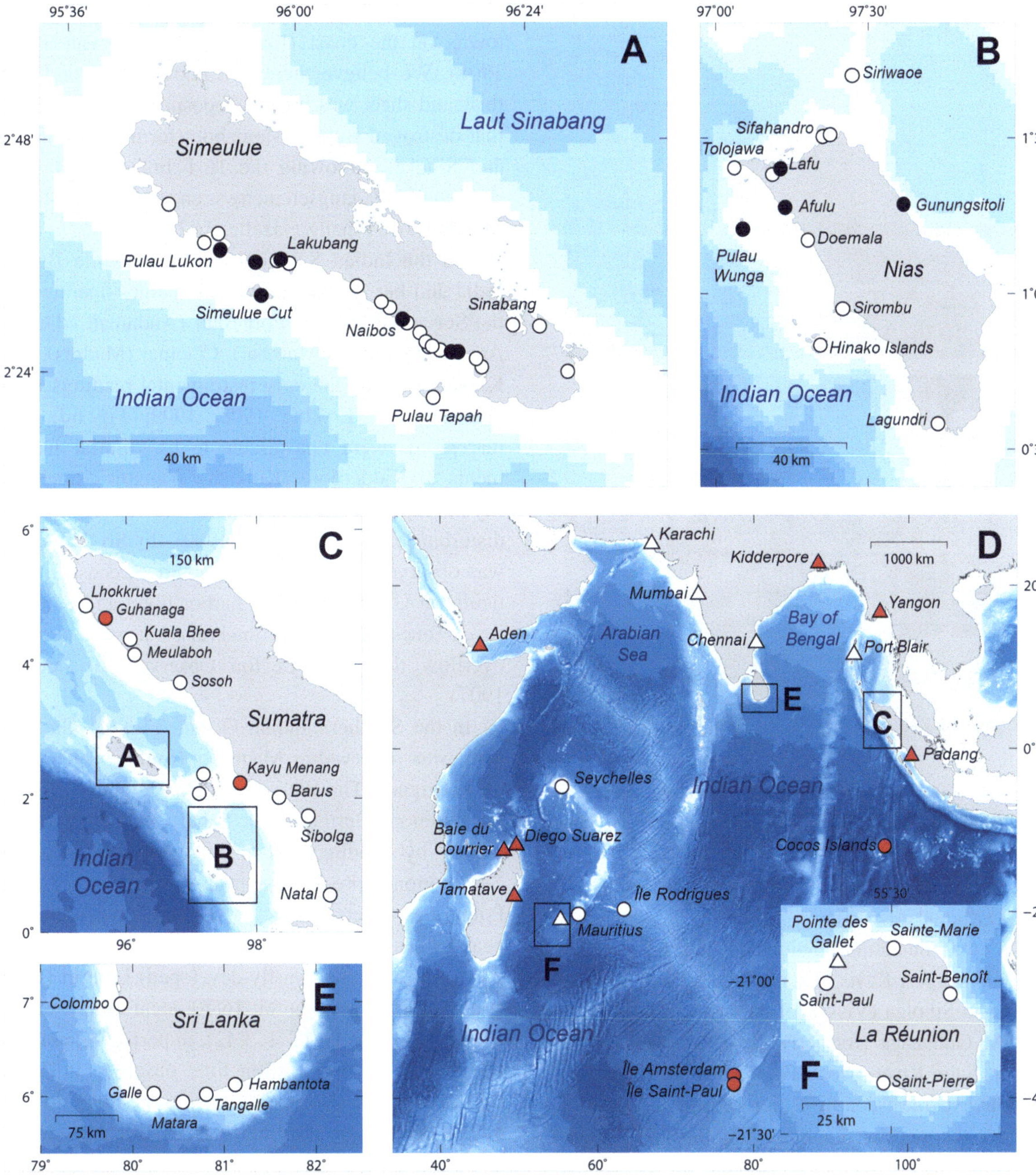

Figure 6

Tsunami observations on Simeulue (**a**), Nias (**b**), and northern Sumatra (**c**), and in the Indian Ocean basin (**d**) including Sri Lanka (**e**) and La Réunion (**f**), also tabulated in Table 5. White circles show locations where the tsunami was observed, and black circles indicate sites used to forward model the tsunami. Red circles mark questionable or false reports. White triangles show tide gauges where readings were available, and red triangles show tide gauges in operation in 1907 where either data were unavailable, the tsunami was unrecorded, or records were incomplete

(K.S.) that corpses were stranded in coconut trees in 1907. The only photographic evidence of this disaster that we found is from Pulau Wunga (Schröder 1917b; Fig. 7a, b); how representative these are of damage elsewhere on the island is unknown, but the similarity of the post-tsunami landscape to that of Pulau Sibigau

Figure 7
The northern (**a**) and western part (**b**) of the island of Pulau Wunga (Schröder 1917b)

(Hill et al. 2012) following the 2010 Mentawai tsunami earthquake is striking.

On the Sumatran mainland (Fig. 6a), vague accounts suggest Kayu Menang (Aceh Post, 10 June 2013) and Sibolga (Visser 1931) were possibly flooded. At Guhanaga (formerly Geni), a flood in the Krueng Sabe River (Aceh Post, 10 June 2013) was attributed to the tsunami by an eyewitness, but we consider this account cautiously as the date associated with it is unreliable. The tsunami was reportedly recorded by a tide gauge in the harbor at Padang (Anonymous 1909), but this record has been lost. Monecke et al. (2008) suggest that a sand sheet (unit A) deposited on the Sumatran coast to the north of Meulaboh was related to the 1907 tsunami. While accounts at our disposal from the west coast of Sumatra are sparse, we doubt a tsunami that flooded several kilometers inland would go unnoticed, and despite a heavy colonial military presence in the region at the time, there is no record of damage to towns on the coast (Koloniaal Verslag van 1907; 1908). We believe there is a higher likelihood that this sand sheet was instead deposited by the tsunami that destroyed Bubon, Meulaboh (formerly Analabu) and Singkel, following the 16 February 1861 Nias earthquake (Padangsch nieuws- en advertentie-blad, 2 March 1861; Pinang Gazette, 16 March 1861).

On the Indian Subcontinent in erstwhile British India and her dependencies, tide gauges operated by the Survey of India at Port Blair (Andaman Islands), Apollo Bandar (in Mumbai), Chennai (Madras), and Karachi provide the only instrumental readings from this tsunami (Erskine 1908, 1909; Fig. 6d), but maregrams were unavailable to us. Similar instruments at Kidderpore (Kolkata), Yangon (Rangoon, Myanmar), and Aden (Yemen) did not record the disturbance (Erskine 1908, 1909). In Sri Lanka, it was observed at points on the southern coast in particular at Galle, where the harbor filled and emptied seven times at half-hour intervals, but water did not overflow the jetty (Ceylon Observer, 10 January 1907).

In the Southern Indian Ocean (Fig. 6f), the tsunami was observed at Mauritius (Erskine 1909) and on Rodrigues (The West Australian, 5 January 1907). La Réunion (Bertho 1910; Sahal et al. 2011), supported by readings on a maregraph, reports an "oscillation" (presumably peak-to-peak) of 1.7 m at Pointe des Galets, about 1/3 (if meant to represent a peak-to-peak amplitude) of the runup observed in 2004 at nearby (and vastly developed) Le Port (Okal et al. 2006a), starting at 16:20 local time (which before 1911 was GMT + 3:42), in perfect agreement with the 7.5 h propagation time observed in 2004. Bertho (1910) also mentions observations in the Seychelles and on the French islands of Amsterdam and Saint Paul; however, the remoteness of the latter two islands, whose settlement at the time is not proven beyond doubt, lead us to question the veracity of the corresponding reports. Curiously, this event is missing from the French national tsunami database (Lambert and Terrier 2011).

As for Madagascar, historical tide gauge data available from Service hydrographique et océanographique de la Marine (SHOM) for Diego-Suarez (now Antsiranana; SHOM Shelfmark: SHD R

7JJ2384) have no entries for 3–6 January 1907 (Garnier-Loussaut, personal communication 2014). Observations of high and low water levels are missing for January 1907 from Tamatave (now Toamasina), and diurnal visual observations made with a tide staff are only available for 15 February until 2 March 1907 from Baie du Courrier (Ferret, personal communication 2014). No reports could be found in the several French-language newspapers published in Madagascar at the time.

8. *Tsunami Modeling Methodology*

We choose a subset of 12 locations with reliable geographic coordinates on the islands of Simeulue and Nias (Fig. 6a, b), at which either runup, wave heights, or inundation distances could be determined (Table 5). These also include islands such as Lukon, Pulau Wunga, Simeulue Cut, and Tapah that were completely or partially overtopped (Table 5). These sites extend approximately parallel to the northern Sunda Trench, along strike from Simeulue to Nias, for a distance of about 220 km. Their spatial distribution and their proximity to the Sunda Trench (60–80 km), along with the effect of directivity (e.g., Miyoshi 1955; Kajiura 1972) then allow a causal relationship between field observations and the parameters of the rupture, with each site serving as a control point to estimate and constrain slip on a corresponding segment of a model of the parent seismic source.

Our tsunami simulations use the Cornell Multi-grid Coupled Tsunami Model (COMCOT), a fully validated finite-difference algorithm solving the nonlinear shallow-water approximation of hydrodynamics (Liu et al. 1998). COMCOT simulates tsunami wave propagation in the deep ocean by solving linear, shallow-water equations in a spherical coordinate system (Wang 2009), while depth-averaged nonlinear, shallow-water equations are used to simulate near-shore wave propagation and onshore inundation (Wang 2009). It has been used to investigate both historical and modern tsunamis (e.g., Li et al. 2015; Wang and Liu 2006). We use three nested layers or grids (Grid 01–Grid 03) to simulate the tsunami at different scales, with the resolution of the grids varying from $\sim$ 2000 m in the source region near the Sunda Trench to 100 m in the coastal areas of Simeulue and Nias. Our setup of computational grids is shown in Fig. 8. A 30 arc-second grid (ca. 925 m) from the General Bathymetric Chart of the Oceans (GEBCO) digital bathymetry dataset (see "Data and Resources") is used to create the computational Grids 01 and 02. Three sublayers, each with a resolution of 100 m, are used for Grid 03 to cover the 12 selected sites on Simeulue and Nias. The topography in the innermost Grid 03 is the "bare-earth" digital elevation model (DEM), in which vegetation biases are removed from Shuttle Radar Topography Mission (SRTM) data using a multisensor approach (O'Loughlin et al. 2016). We derived nearshore bathymetry and topography by combining depth contours digitized by us from hard-copy maps, with digitally available topographic contours and the 30 arc-second GEBCO dataset (see "Data and Resources"). We also specify a constant Manning's roughness of 0.013 for the ocean floor and 0.025 for the inland area in the innermost grids.

The extremely limited nature and quality of the available seismological data preclude the use of commonplace modern methods (e.g., Kikuchi and Kanamori 1982, 1986, 1991) to constrain the parameters (fault length and width, seismic slip) necessary to compute a field of initial conditions for the simulation of the tsunami. As documented in a recent study (Ebel and Chambers 2016) seismicity occurring in the vicinity of past large ruptures in eastern North America and California shows a tendency to concentrate its largest events at the edges of past ruptures. They went on to show that this geographical trend can be used to map the rupture area of historical events predating the development of modern (and especially digital) data necessary to conduct detailed source tomography. In this context, and assuming that Ebel and Chambers' (2016) paradigm can be extended to the case of large subduction interfaces, we use our preferred epicentral location, and catalogues of relocated seismicity before and after the 2004 Sumatra–Andaman and 2005 Nias earthquakes (Engdahl et al. 2007; Pesicek et al. 2010), to define a zone of presently low seismic productivity in the uppermost part of the interface of the Sunda Megathrust, extending roughly 250 km

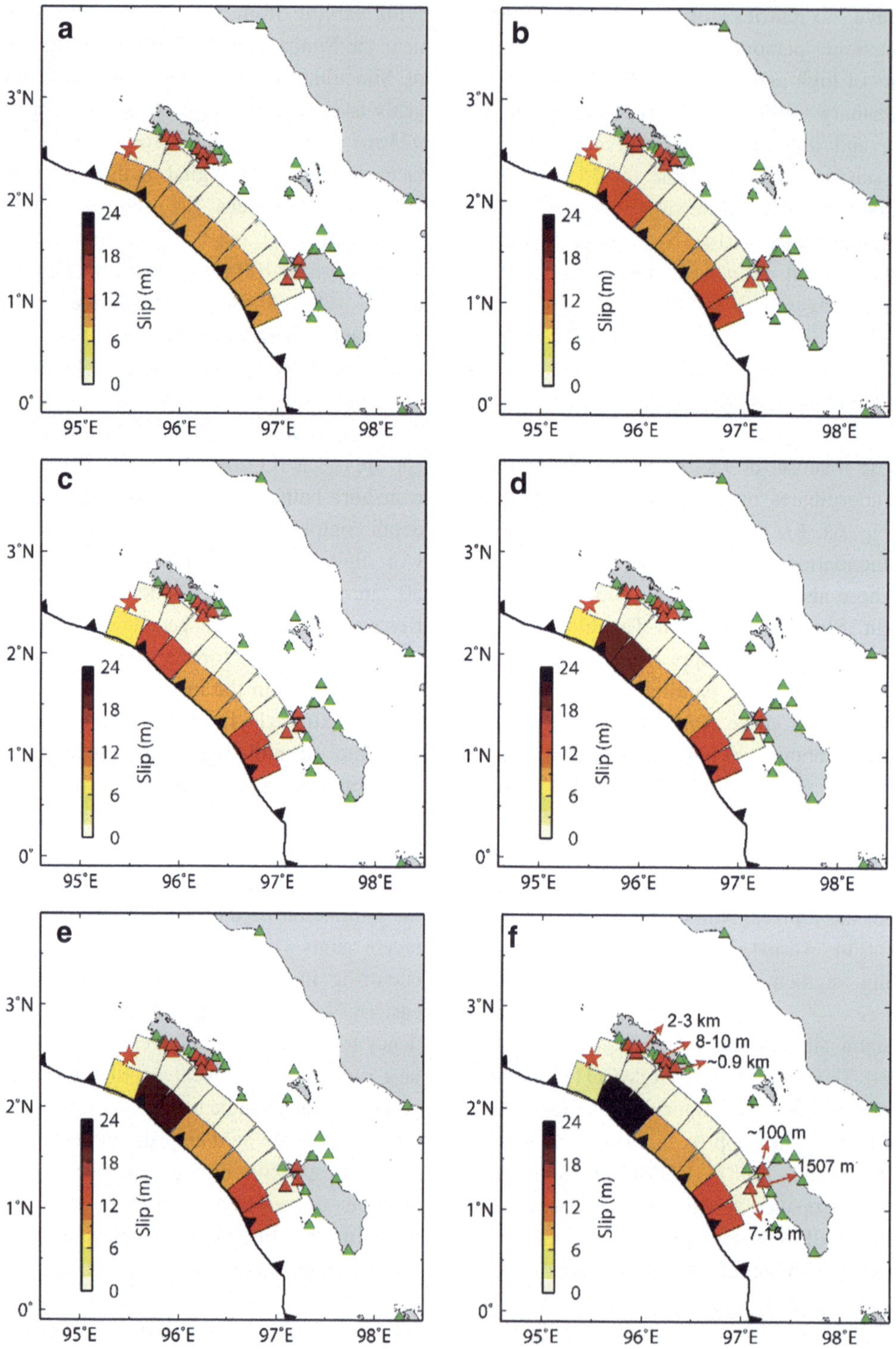

Figure 8
Hypothetical slip models utilized by us to reproduce the 1907 tsunami. Slip is initially assumed to be 9 m (**a**), but we increase this value incrementally to 24 m (**b–f**). Green triangles represent locations where the tsunami was reported, and red triangles represent locations used in the tsunami model (Table 5). Inundation distances estimated from documentary evidence are indicated in **f** with updip segments along the trench (from west to east) are **A** (light yellow), **B** (black), **C** (orange) and **D** (red)

between 94.7°E and 97°E. In this general area, we then envision a number of scenarios for seismic rupture along the shallow updip plate interface of the Sunda Megathrust, which we divide into 16 elementary subfaults, themselves regrouped into four segments labeled A to D. The 16 patches, each measuring 30 km × 40 km, are distributed in updip–downdip pairs (Fig. 8), their exact geometry following Slab 1.0 (Hayes et al. 2012).

Given the slowness we have documented in the source of the 1907 earthquake, we use slip estimates departing from the classical scaling laws applicable to subduction events (Geller 1976; Blaser et al. 2010) and consider models with variable rigidity (μ). We begin by assuming that the rupture was limited to the shallowest portion of the interface; conventional crustal material would feature a typical value $\mu = 3 \times 10^{11}$ dyn/cm^2, which using the moment $M_0 = 2.5 \times 10^{28}$ dyn cm obtained from mantle waves, would result in a slip (Δu) $\approx$ 7 m on each of the updip subfaults. We then model a tsunami earthquake by varying the rigidity on each segment between 0.7 and 2.0 × 10^{11} dyn/cm^2, which increases the slip to between 9 and 24 m (Fig. 8; S1–S6) in 3-m increments. These values of μ fall in the range proposed in the context of tsunami earthquakes by a number of investigators (Ide et al. 1993; Satake 1995; Bilek and Lay 1999; Geist and Bilek 2001; Hill et al. 2012; Li et al. 2015). For each of these models, we then compare the simulated tsunami amplitudes with those estimated from documentary material (Fig. 6).

Incidentally, the largest inferred slip in 1907 on the interface of the Sunda Megathrust coincides with a morphological high. As discussed previously, this feature has been associated with a region of strong coupling beneath central Simeulue (Tsang et al. 2015), and inferred as a structural control for local seismicity and for the rupture dimensions of modern earthquakes under central Simeulue (Sieh et al. 2006; Morgan et al. 2017). This pattern is very similar to that of other tsunami earthquakes and associates both event nucleation and the largest slip with the subduction of subsurface topography along strongly coupled interfaces (e.g., Tanioka et al. 1997; Abercrombie et al. 2001; Newman et al. 2011b; Bell et al. 2014). This remark is also supported by the jagged nature of the Omori waveform at Manila discussed above. Another morphological feature to the south in the vicinity of the Batu Islands (Natawidjaja et al. 2006; Fig. 1b) represents another barrier to rupture (Philibosian et al. 2014) and demarcates an important segment boundary on the Sunda Megathrust near the Batu Islands (Meltzner et al. 2015). The weaker tsunami amplitudes on southern Nias suggest that the 1907 rupture did not extend to the Batu Islands segment barrier or terminated in the neighborhood of this southern segment boundary near the likely rupture of a deadly tsunamigenic earthquake near the island of Simuk (Fig. 1b) to the south of Nias on 9 March 1861 (Reiche 1863). Since our rupture is only constrained by tsunami observations, further research would also be required to refine the western limit of the 1907 rupture in the region of the morphological high identified by Franke et al. (2008) and Tang et al. (2013), to determine whether the rupture stopped at, or propagated through, this segment boundary.

9. *Observed and Modeled Tsunami Impacts*

The tsunami impacts documented on Simeulue (Fig. 6a) were best replicated by more than 21 m of slip on segment B, which is our preferred rupture model (Fig. 9); For example, at Lakubang, the inundation distance is thought to have approached 3 km (Yogaswara and Yulianto 2006; Fig. 9b). At this location, despite slip greater than 21 m on segment B, the inundation remains underpredicted and we suspect that the inundation distance reported by Yogaswara and Yulianto (2006) is strongly controlled by local drainage that is known to allow tsunamis to travel much farther inland following the channels of natural and manmade drainage features (e.g., Mori et al. 2011). In the region of Nias to the south (Fig. 9c), we find that a slip of 15 m is required on segment D to reproduce the wave heights reported from Pulau Wunga (Fig. 9c). Our constraints on the amount of slip on segments A and C (Fig. 9) are weaker due to the lack of quantitative records from northern Simeulue and the Banyak Islands, respectively (Fig. 6a–c). However, we infer that the noticeable tsunami impacts on the west coast of Sumatra (Fig. 6c) such as the flooding at Susoh (Bataviaasch Nieuwsblad, 12 February 1907),

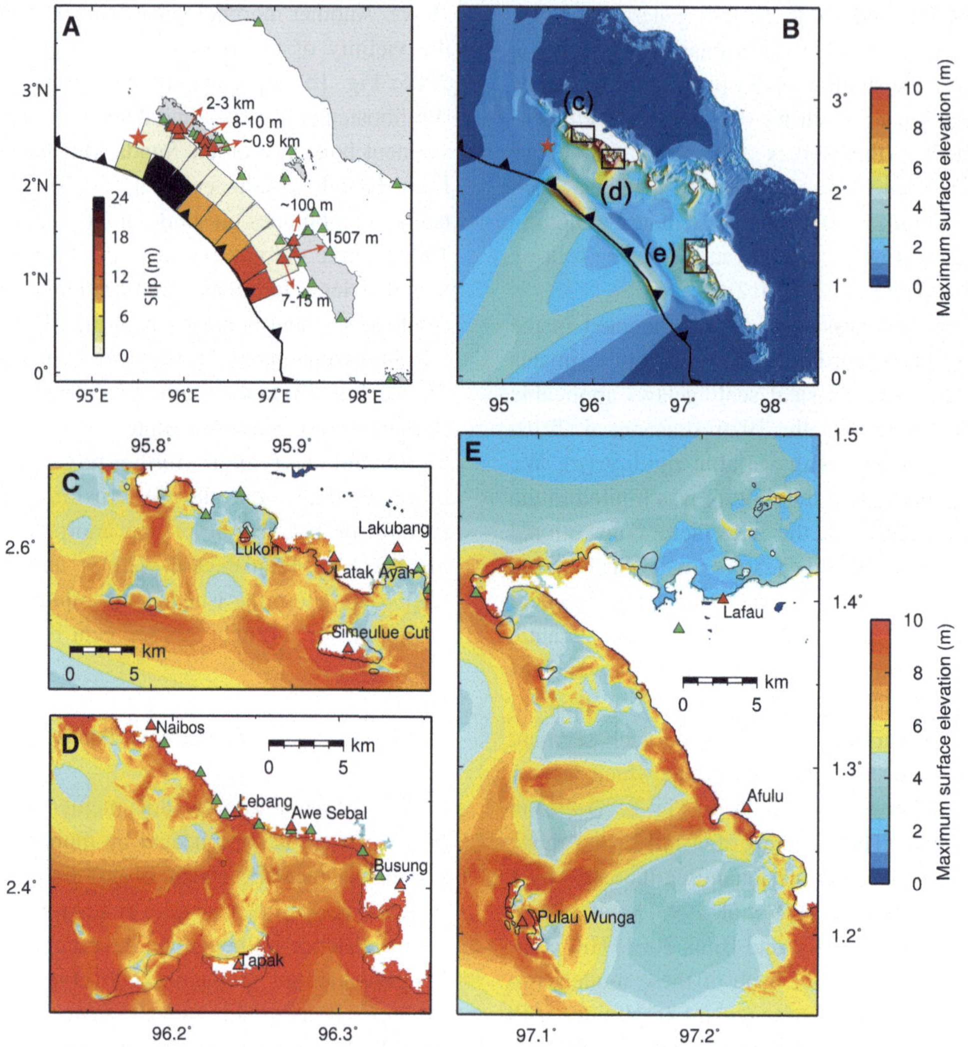

Figure 9
Hypothetical rupture geometry and slip parameters for the 1907 mainshock (**a**) estimated in this study off Simeulue (**c**, **d**) and Nias (**e**). Green triangles represent locations where the tsunami was reported (Table 5), and red triangles represent locations used in the tsunami model

Tapaktuan (Anonymous 1909), and possibly Kayu Menang (*Aceh Post*, 10 June 2013), are probably indicative of slip of 6–9 m on those segments.

In conclusion, our simulation experiments indicate that the values of runup and inundation estimated from the documentary material require seismic slips varying from 6–9 m on segments A and C to as much as 15 m on segment D, and probably more than 21 m on segment B. These values are greater, by factors ranging from 1.3 to 5, than expected under scaling laws (e.g., Geller 1976) for an earthquake of $M_0 = 2.5 \times 10^{28}$ dyn cm. This provides an additional argument for the anomalous properties of the 1907 event, and quantifies its character as a "tsunami earthquake", as defined in Kanamori's (1972) landmark study.

Our preferred rupture model satisfactorily predicts tsunami heights at all but one location, i.e., Afulu on Nias (Figs. 6b, 9b). We note that the modern village is located near a semicircular bay (Telok Afulu) opening to the south and sheltered on the west by a small island called Hulo Uma (1.251°N, 97.236°E); further, a review of topographic data (see "Data and Resources") indicates steep relief (≥ 20 m) close to

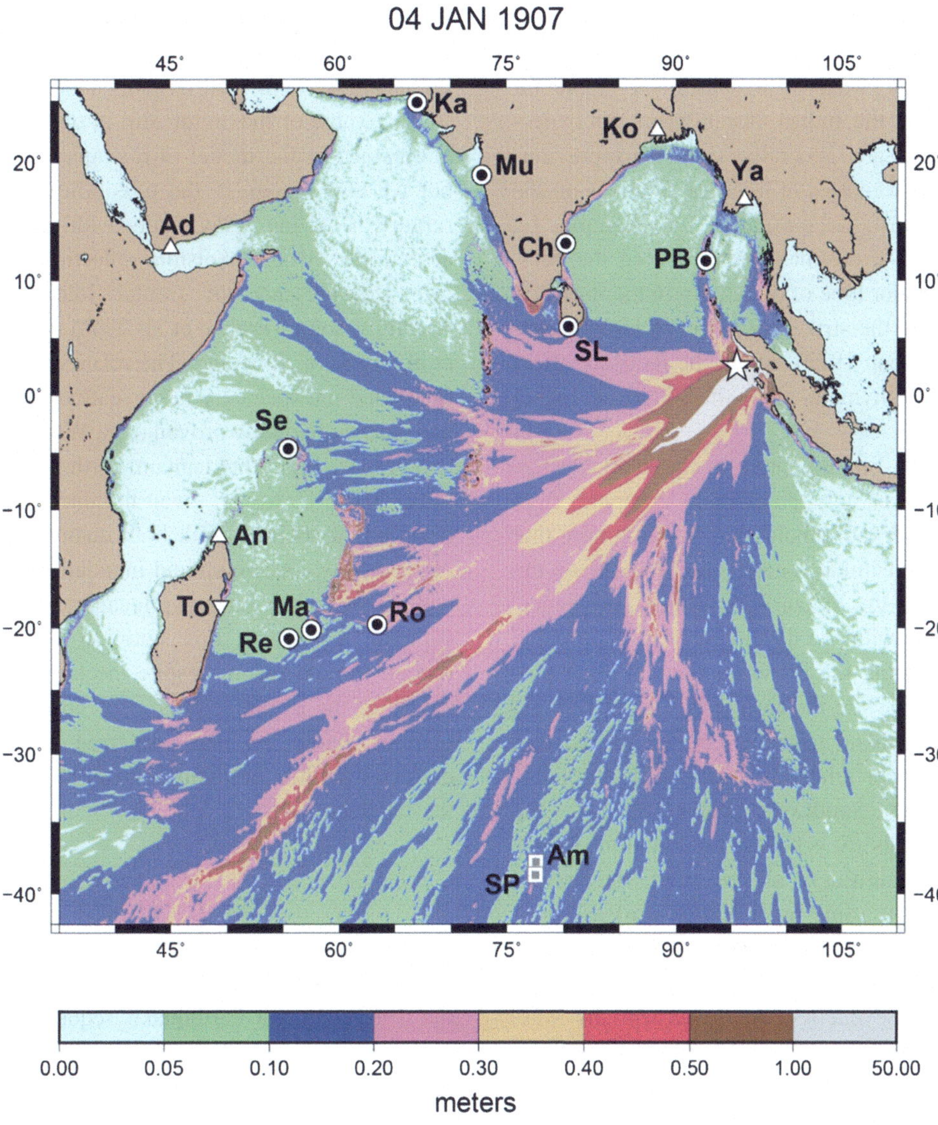

Figure 10
Maximum tsunami amplitude simulated in deep water using our preferred model of rupture for the 1907 mainshock, whose epicenter is shown as the large star. Locations with tsunami effects either instrumentally recorded or simply reported, are shown as black bull's eye symbols (Ka: Karachi; Mu: Mumbai; Ch: Chennai; SL: Sri Lanka; PB: Port Blair; Ro: Rodrigues; Se: Seychelles; Ma: Mauritius; Re: La Réunion). Instrumented locations not reporting the event are shown as open triangles (Ad: Aden; Ko: Kolkata; Ya: Yangon; An: Antsiranana/Diego Suarez). The port of Toamasina (Tamatave), Madagascar (To), discussed in the text, is shown as an inverted triangle. Amsterdam (Am) and Saint Paul (SP) islands, sites of questionable reports, are shown as grey-centered squares

the shore, and a drainage channel extending inland for ∼ 1 km from the northwestern shore of Telok Afulu. Such features can cause highly localized amplification of tsunami waves (e.g., Shimozono et al. 2014), while a small island can also amplify tsunami waves in its lee, a scenario known as the "Babi effect" (e.g. Briggs et al. 1995). Localized extreme inundation could also be due to an aerial or subaerial landslide, possibly triggered by the snappier Event II (e.g., Tsuji et al. 1995; Pflaker 1997). We also cannot rule out the possibility of a separate, local tsunami caused by Event II, especially given the lack of arrival times in the accounts of the tsunami at Afulu.

Figure 10 contours the maximum amplitudes simulated in deep water (i.e., not including the interaction of the wave with coastlines) in the far field, throughout the Indian Ocean basin. As expected, our results are generally similar to previous simulations for scenarios of large earthquakes on the Sunda Megathrust, as presented, for example, by Okal et al. (2009), the main pattern being the combination of a major lobe of directivity in the direction perpendicular to the strike of the trench (Ben-Menahem and Rosenman 1972), and of the focusing effect of the shallow bathymetry along the Southwest Indian Ocean Ridge.

Owing to the limited nature of the available far-field reports, it is only possible to compare them qualitatively with our simulations. We first note that reports of the tsunami at Karachi and Mumbai, on the western coast of the Indian Subcontinent, are generally compatible with the development of a probable edge wave hugging the broad continental shelf present in the area, while the river port sites at Kolkata and Yangon, with no tsunami reported on existing tide gauges, appear relatively sheltered in our simulations. In the Southwest Indian Ocean, the Mascarene Islands (Seychelles, La Réunion, Rodrigues, and Mauritius) all feature deep-water amplitudes on the same order (20–30 cm) as simulated for the 2004 tsunami (Okal et al. 2006a). On the other hand, the Mascarene Plateau, arcing from La Réunion to the Seychelles, clearly screens the Mascarene Basin in its lee, thus protecting the northern half of Madagascar, and explaining the lack of tsunami reports from the maregraph at Diego-Suarez. This would also explain the lack of observations at Tamatave, which was already the main trading port of the island and whose activity was at the time well described in the local press. By contrast, stronger amplitudes, potentially comparable to those at the Mascarene Islands, would be expected on the southern part of Madagascar, the site of the highest runup values in 2004 (Okal et al. 2006b); the absence of reports in the press may reflect poor communications and possibly an ongoing state of insurgency. Finally, regarding the islands of Amsterdam and Saint Paul, deep-water amplitudes in the range of 0.2 m and an individual simulated value reaching 0.9 m within 5 km of the coast of Saint Paul would give some veracity to the report of an observable wave (Bertho 1910), but the presence of a settlement at the time on the island of Saint Paul remains unproven.

The value of maximum slip in our rupture model is obtained in the model of a simple planar rupture, but we also recognize the possibility of a complex rupture involving splay faults within the overlying accretionary wedge. Seismic reflection profiles indicate the presence of frontal folds and pop-up structures south (Franke et al. 2008) and west (Singh et al. 2008) of Simeulue. The relation between splay faults and unexpectedly large seafloor displacements has been numerically proven (e.g., Wendt et al. 2009) and proposed for other tsunami earthquakes (e.g., Fan et al. 2017). However, given the uncertainties in the near field associated with both quantifiable numeric observations, i.e., runup and inundation distances, as well as the largely unknown local arrival times of the tsunami, we restrict this study to a simple planar model. In conclusion, despite the largely qualitative nature of the available dataset of observations, simulations based on our preferred rupture model correctly predict the main features of the tsunami in the far field.

10. Discussion

Our study resolves many unaddressed aspects of the 1907 Sumatra earthquake sequence but leaves unanswered the question of whether any subsidence was associated with this sequence. In this context, we note Hodgson's (1934) remark that "the south coast was partially submerged by an earthquake", which could be interpreted as coseismic subsidence (Meltzner et al. 2010, 2015) or as tsunami inundation, in the semantics of the 1930s. Reports from the southern coast of Simeulue are limited, and none of the available Dutch accounts mention land level changes associated with the 1907 earthquake, except in brief descriptions from Simeulue Cut (Haagsche Courant, 14 February 1907; Fig. 6a) and the Hinako Islands (Bataviaasch Nieuwsblad, 22 March 1907; Fig. 6b). We model subsidence (Figs. S7, S8) of $\leq$ 50 cm on Simeulue Cut and at the south coast of the island of Simeulue due to a dislocation in an elastic half-space (Okada 1985), and calculate even smaller values of subsidence ($\leq$ 15 cm) at the Hinako Islands. The accounts at our disposal do not state whether the writer visited the island of Simeulue Cut, viewed it from a distance, interviewed direct witnesses, or simply derived this information second-hand. If taken at face value, the description from Simeulue Cut might be inferred as coseismic subsidence or extreme tsunami erosion. The other perplexing detail comes from the Hinako Islands (Fig. 6b), where the seabed was reportedly exposed (uplifted?) for up to 2 h (Bataviaasch Nieuwsblad, 22 March 1907). Coral microatolls have been well studied on Simeulue and have contributed to our understanding of the rupture extents of great medieval earthquakes (Meltzner et al. 2015), as they are natural long-term indicators of changes in relative sea level, but these are often subject to preservation issues that can undermine temporal completeness (Meltzner et al. 2012). In this regard, and pertinent to our study, Meltzner et al. (2012) investigated a coral microatoll at Ujung Lambajo (LBJ; Fig. 1b) at a distance of $\sim$ 6 km northeast of Simeulue Cut but found that the paleorecord at this site only extended to 1955. Much further south, Meltzner et al. (2015) record the death of a microatoll (LAG-3B; Fig. 1b) near Lagundri on Nias and suggest this could either be related to the 1907 earthquake or had "nothing to do with tectonics or relative sea level (RSL) change". Unlike the locations of many tsunami earthquakes that are devoid of offshore islands, Simeulue and Nias lie within 100 km of the trench axis and therefore increase the likelihood that the coseismic signature of land level changes from tsunami earthquakes could be recorded in the paleorecord of microatolls. It is also entirely plausible that the inferred land level changes on Simeulue Cut and the Hinako Islands are indicative of a rupture that extended further downdip than in our first-order model. Also, they could have been, both or independently, associated with the aftershock, or with postseismic after-slip; the alleged Hinako uplift (Bataviaasch Nieuwsblad, 22 March 1907) could have occurred concurrently with the death of a coral microatoll (LAG-3B) previously documented by Meltzner et al. (2015), and thus could possibly represent coseismic uplift from Event II. Bearing in mind the uncertainties in all of the above, we do not incorporate these in our rupture model but suggest that further work is required in this unique physiographic environment to uncover the paleorecord of previous tsunami earthquakes near Simeulue.

Owing to its size and location, the identification of the 1907 Sumatra earthquake as a tsunami earthquake has a number of implications which must be discussed in a global perspective. Table 6 compiles characteristics of 21 known such events, including estimates of seismic moments and, when available, of the slowness parameter Θ introduced by Newman and Okal (1998). At $M_0 = 2.5 \times 10^{28}$ dyn cm, the 1907 earthquake ranks as the third largest documented "tsunami earthquake", after the 2004 Sumatra–Andaman earthquake and the 1946 Aleutian Islands earthquake. Note that the list in Table 6 may not be exhaustive since the origin of some historical destructive waves remains debated, such as the tsunami of 26 February 1902 in El Salvador (Cruz and Wyss 1983; Ambraseys and Adams 1996). Some evidence would also characterize the 1700 Cascadia megathrust event as a tsunami earthquake (Obermeier and Dickenson 2000; Okal 2011). We also speculate that the 16th century disaster on the south coast of Java relating to the Javanese and Sundanese legend of Ratu Kidul and recently interpreted as a tsunami based on surviving indigenous Babads and written Dutch sources (Reid 2014) could have been another tsunami earthquake, owing to the lack of reported shaking effects preserved in these materials. However, we emphasize that the brief nature of these documents warrants further work to confirm our assertions, as these could also be related to traditional earthquakes, seismically or gravitationally triggered turbidity currents, or meteorological phenomena. Paramount among remaining questions of primary importance is the relationship between PTEs and ATEs as defined by Okal and Saloor (2017) with traditional megathrust earthquakes, namely: (1) Does the distribution of PTEs bear a regional signal? (2) Does the occurrence of PTEs along a subduction zone rule out regular (and possibly larger) megathrust earthquakes? (3) Can a subduction zone be considered immune to tsunami earthquakes, either PTEs or ATEs?

A systematic study by Okal and Newman (2001) in the regions of the three 1990s PTEs (Nicaragua

Table 6

List of documented tsunami earthquakes since 1896

#	Date (D M Y)	Region	Type*	Moment (10^{27} dyn cm)	Θ	Notes
1	15 June 1896	Meiji Sanriku, Japan[a]	PTE	~ 12	NA	Tanioka and Satake (1996)
2	4 January 1907	Northern Sumatra	PTE	~ 25	−6.5?	This study
3	13 April 1923	Kamchatka	ATE (L)	1.2		Saloor and Okal (2018)
4	23 June 1932	Manzanillo, Mexico	ATE	5.2	−6.18	Okal and Borrero (2011)
5	21 July 1934	Santa Cruz Islands	ATE	10	−6.10	Okal and Saloor (2017)
6	1 April 1946	Aleutian Islands[a]	PTE (L)	85	−7.03	López and Okal (2006)
7	25 March 1947	Gisborne, New Zealand	PTE	4	−5.94	Okal and Saloor (2017)
8	17 May 1947	Gisborne, New Zealand	[b]	3	−6.51	Okal and Saloor (2017)
9	20 November 1960	Northern Peru	PTE	2.7	−6.13	Okal and Newman (2001)
10	20 October 1963	Kuril Islands	ATE	7.5	−6.42	Fukao (1979)
11	13 August 1965	Vanuatu	ATE	2	−5.88	Okal and Saloor (2017)
12	10 June 1975	Nemuro-Oki, Japan	ATE	0.8	−6.43	Fukao (1979)
13	19 December 1982	Tonga	PTE	2.0	−5.76	Newman and Okal (1998)
14	2 September 1992	Masachapa, Nicaragua	PTE	3.4	−6.30	Newman and Okal (1998)
15	2 June 1994	South of East Java	PTE	5.3	−6.01	Newman and Okal (1998)
16	21 February 1996	Chimbote, Peru	PTE	2.2	−5.94	Newman and Okal (1998)
17	26 December 2004	Sumatra–Andaman	PTE	1200	−6.40	Stein and Okal (2007), Choy and Boatwright (2007) and Okal (2011)
18	17 July 2006	South of Central Java	PTE	4.6	−6.13	Saloor and Okal (2018)
19	25 October 2010	Mentawai Islands	ATE	6.8	−6.22	Saloor and Okal (2018)
20	27 August 2012	El Salvador	PTE	1.3	−6.42	Saloor and Okal (2018)
21	6 February 2013	Santa Cruz Islands	PTE	9.4	−5.94	Okal and Saloor (2017)

**PTE* primary tsunami earthquake, *ATE* aftershock tsunami earthquake (Okal and Saloor 2017), *L* tsunami enhanced by ancillary landslide

[a]Charter events from Kanamori (1972)

[b]While #8 is smaller than its predecessor (#7), the moment ratio (3/4) is too close to qualify the second event as an aftershock of #7

1992, Java 1994, and northern Peru 1996) found no systematic trend for slowness in background seismicity but identified an older event in northern Peru in 1960 as a PTE, 340 km from the 1996 Chimbote source, thus suggesting a lateral coherence in the updip properties of the interface governing the occurrence of PTEs. Following their study, the 2006 Java earthquake provided a confirmation of this trend, occurring 580 km from the 1994 epicenter. Similarly, the 2012 El Salvador PTE occurred only 175 km from the site of the 1992 Nicaragua earthquake; should the 1902 event (Cruz and Wyss 1983; Ambraseys and Adams 1996) be confirmed as a PTE, then it would earmark a 350-km section of the Central American subduction zone as prone to PTEs. In this context, the 1907 event provides some limited new insight. It was indeed followed, 98 years later, by a slow megathrust earthquake, but the origin of their slow character may not be comparable: based on its location near the trench, the 1907 event's slowness was rooted in its shallow source, whereas in 2004, it emanated from a relatively slow and jagged propagation along its exceptionally long fault length (Ishii et al. 2005; de Groot-Hedlin 2005; Tolstoy and Bohnenstiehl 2005). More comparable are the events of 1907 and 2010, both of which took place along the shallowest, probably structurally similar, portions of the plate interface, and were separated laterally by 900 km. However, the 2010 Mentawai earthquake was clearly an ATE of the 2007 Bengkulu megathrust event, while the 1907 earthquake was primary: there is no concrete evidence in the known historical record for an immediately preceding mainshock which would have been significantly larger (Newcomb and McCann 1987).

Whether or not the occurrence of PTEs along a subduction zone rules out regular (and possibly larger) megathrust earthquakes is particularly critical along the south coast of Java, where two PTEs featuring very similar source characteristics took place in 1994 and 2006, and where no truly gigantic earthquake is known in the admittedly short historical record (Newcomb and McCann 1987). Our assertion would also apply to the Hikurangi subduction zone, off the North Island of New Zealand, where the historical record is even shorter and where the only large interplate thrust earthquakes since 1917 are the 1947 tsunami earthquakes (Doser and Webb 2003; Okal and Saloor 2017), although Cochran et al. (2006) inferred prehistoric tsunamis from an analysis of wetland sediment cores. Our assertion would also extend conceivably to Cascadia if indeed the 1700 event had a slow source. In this context, the proximity of the 1907 and 2005 epicenters would suggest a negative answer, namely that PTEs and regular megathrust events may occur along neighboring subduction segments (or even along the same one at different depths along the interface). Note that this conclusion might be supported by the case of the 1946 Aleutian PTE, in comparison to the nearby 1938 and 1957 megathrust earthquakes which did not exhibit source slowness, the former even generating a deceptively small tsunami (Stover and Coffmann 1993); however, this analogy remains tentative in the presence of the Shumagin gap between the 1946 and 1938 ruptures, where no large events are known since at least 1787 (Davies et al. 1981), which leaves the question of its potential for a PTE wide open until the next large earthquake fills it.

Tsunami earthquakes have not yet been documented in a number of subduction zones having produced large tsunamigenic earthquakes during the era of instrumental seismology (e.g., Southern Peru, Northern Chile, Makran), or during preinstrumental, historical times (e.g., Hellenic Arc, Ryukyu). A small-scale study in Vanuatu by Okal and Saloor (2017) documents a lateral, along-strike, variation of the slowness parameter Θ, in conjunction with the subduction of submarine structures, on a scale of only a few hundred kilometers. The underestimation of the magnitudes of great earthquakes prior to instrumentation (Hough 2013), the largely random nature of the fragmentation of rupture at subduction zones (Ando 1975), and the gross temporal undersampling of tectonic cycles by modern seismology which resulted in unexpectedly large events such as the 2004 Sumatra–Andaman and 2011 Tohoku earthquakes, have led to the demise of otherwise promising paradigms (Stein and Okal 2007; McCaffrey 2007). Thus the only prudent answer to this question must remain negative. The 1907 earthquake provides no new insight, however, on whether any subduction zone can be considered immune to tsunami earthquakes, either PTEs or ATEs.

In the absence of definitive evidence of aseismic slip, tremor, or slow-slip earthquakes updip (Feng et al. 2015; Tsang et al. 2015), and based on newly modeled estimates of convergence of 46 mm year^{-1} at 1°N, 96°E (Bradley et al. 2017), we believe that the rupture patch identified in this study could host a future $M_W \geq 7.7$ tsunami earthquake, updip and trenchward between the rupture patches of the 2004 and 2005 earthquakes with dimensions and physical properties similar to our preferred model (Fig. 8), and with lower than anticipated ground motions to elicit self-evacuation. We recall that, on Simeulue, the local legend of the S'mong associates strong earthquakes with tsunami hazard and elicits self-evacuation to high ground upon feeling strong shaking (McAdoo et al. 2006; Baumwoll 2008; Syafwina 2014), which turned out particularly effective in preventing heavy loss of life on Simeulue during the 2004 Sumatra–Andaman and 2005 Nias tsunamis, both preceded by damaging shaking (Martin 2005; McAdoo et al. 2006). It has been proposed (e.g., Syafwina 2014) that the S'mong is based on memories from 1907 preserved in verse or in stories passed down by older generations [we note, however, that Yogaswara and Yulianto (2006) record versions of the legend, both with and without severe shaking]. If so, the exact role played, in the S'mong legend, by the weak shaking due to Event I, and the presumably stronger one from Event II, remains unclear; at any rate, shaking from Event II, estimated to lag Event I by $\sim$ 53 min, may not have occurred before the tsunami attacked Simeulue; this is further suggested by E. Schröder's account of the second shock felt on Nias (located farther from Event I than Simeulue) after flooding by the tsunami (Bataviaasch Nieuwsblad, 22 March 1907). While the S'mong tradition may have saved many lives in 2004 and 2005, it is clear that it is not adapted to the particular challenge of tsunami earthquakes, as illustrated by the case of the 2010 Mentawai tsunami earthquake, 600 km southeast of Simeulue (Figs. 1, 3), a typical ATE following the 2007 Bengkulu megathrust earthquake. Some interviewed witnesses reported that the 2010 shaking on the Mentawai Islands was both weak and long, lasting as much as "several minutes" or was even not felt at some locations in the Pagai islands (Hill et al. 2012). This is an important observation, since source slowness is expected to both decrease high-frequency ground motion (and hence the level of shaking) and increase the duration of seismic wavetrains; the comparison of these two properties has been earmarked by Convers and Newman (2013) and Okal (2013) as a means of identifying tsunami earthquakes from the analysis of first-arriving P wave shapes. Yet, the ensuing 2010 tsunami attained runup heights of 17 m and killed more than 500 people. In this case, local villagers did not immediately evacuate as the shaking was weaker than during the seismologically much larger 2007 Bengkulu mainshock (Borrero et al. 2009) and during the 2009 intraplate Padang earthquake (McCloskey et al. 2010), both of which did not produce significant tsunamis in the Mentawai Islands. Accordingly, we end by reiterating the recommendation made by Hill et al. (2012) to include the duration of felt shaking as a warning for spontaneous evacuation, in a sense to implement a variant of Convers and Newman's (2013) algorithm based on human perception. This constitutes a formidable challenge, since our experience gathered from a large number of post-tsunami surveys (e.g., Synolakis and Okal 2005) is that the perception of relatively short time durations (seconds, minutes) by lay individuals in a situation of emergency and panic is even less accurate than that of distance.

11. Conclusions

The main conclusion of our study is the confirmation and quantification of the slowness of the 1907 mainshock (Event I), which we achieve from a broad range of seismological observations. By computing its seismic moment from the spectral amplitude of surface waves measured at mantle periods (up to 170 s), we obtain a value significantly greater (by a factor of about 4) than previously derived around 50 s by Kanamori et al. (2010). The identification of Event II, occurring only 53 min after the mainshock, clearly resolves the paradox of an event bearing many hallmarks of a "tsunami earthquake" but being felt at surprisingly high intensities. Although it must remain qualitative in the absence of instrumental metadata, the comparison of amplitudes and duration of body and surface waves at Manila clearly documents the difference in source properties between Events I and II, hence the slowness of the mainshock. We also propose a seismic rupture model which provides an acceptable fit to a new dataset of tsunami runup and inundation values from local and far-field locations in the Indian Ocean basin. Our preferred model extends from $\sim$ 94.7°E to $\sim$ 97°E along the Sunda Megathrust with as much as 21 m of slip. We also note that while the S'mong tradition may have prevented heavy loss of life on Simeulue in 2004 and 2005, both of which were preceded by severe shaking, it will be ineffective for future tsunami earthquakes that do not generate shaking violent enough to elicit societal concern, and in turn, to warrant spontaneous evacuations.

12. Data and Resources

We present a list of all newspaper titles and dates consulted including libraries and online databases (free and subscription-based) where these are currently available (Online Appendix A) with transcriptions of selected newspaper reports in Online Appendix B. We consulted official mail reports (nos. 88, 91, 179, 202, 258, and 1134) which include telegrams sent by the Governor of Atjeh and the Resident at Tapanoeli to the Governor-General in Batavia (Jakarta). These are stored in the Openbaar Verbaal (23-05-1908, no. 11) for "Atjeh en Onderhorigheden" belonging to the collection "Ministerie van Koloniën: Politieke Verslagen en Berichten uit de Buitengewesten" (2.10.52.01), inventory number 16 at the National Archives of the Netherlands, The Hague. "Handelingen der Staten-Generaal" which contains the "Koloniaal Verslag" for 1907 and 1908 can be found via the Staten-Generaal Digitaal (https://www.statengeneraaldigital.nl/). Figures were created using freely available Generic Mapping Tools (Wessel and Smith 1991) and QGIS software (http://download.qgis.org/). Bathymetry was digitized for Nias and Simeulue from AMS T503 series maps (1:250,000; sheets NA 47-5, NA 47-8, NA 47-9, NA 47-13, NA 47-14, NA 46-8) available via the Perry-Castañeda Map Collection at the University of Texas at Austin (http://www.lib.utexas.edu/maps/ams/indonesia/, last accessed July 2017). We also used the General Bathymetric Chart of the Oceans (GEBCO) digital bathymetry dataset (http://www.gebco.net/, last accessed in October 2014). Digital topographic contours (1:5000) are available from Badan Informasi Geospasial (http://tanahair.indonesia.go.id; last accessed 27 November 2017). The topographic data used in the tsunami simulations are bare-ground Shuttle Radar Topography Mission (SRTM) data downloaded from https://data.bris.ac.uk/data/dataset/10tv0p32gizt01nh9edcjzd6wa (last accessed, April 2018). For some key locations including Lakubang, Lukon, and Simeulue Cut, we replace the SRTM data with 4-m-resolution digital terrain data purchased from the Nippon Telegraph and Telephone (NTT) Data Corporation. Topographic maps (1:50,000) prepared by the 653rd Topographic Engineering Battalion of the United States Air Force for Simeulue (Series HIND 605 Sheet 5; NLA shelf mark: MAP G8082.S5 s50) and by the British War Office for Nias (Series HIND 614 Sheet 9; NLA shelf mark: MAP G8082.N5 s50) were procured from the National Library of Australia (NLA).

Acknowledgements

This article has benefited from interactions with Alicia Schriker, Anthony Reid, Aron Meltzner, Christina Widiwijayanti, Euan Smith, Lauriane Chardot, Hiroo Kanamori, Luis Rivera, Priyamvada Nanjundiah, Qiang Qiu, Shengji Wei, and Susan Bilek. We are also grateful for the constructive reviews of two anonymous reviewers and the comments of Guest Editor Yuichiro Tanioka. We would like to thank H.P. Shukla, Isabelle Garnier-Loussaut, Susanne Sargeant, Vineet Gahalaut, and Yann Ferret for assistance rendered in our efforts to locate additional seismograms and station bulletins for the earthquake, and maregrams. Hiroo Kanamori provided Japanese seismograms from Hongo and Osaka. Dominico Di Giacomo provided a customized grid-search relocation. Robert Engdahl provided relocated seismicity from Engdahl et al. (2007), and Alexandre Sahal provided French newspaper reports from La Réunion. Çağıl Karakaş, Iwan Hermawan, Koen van Noten, and Mari Hamahashi assisted with, or verified selected translations. Aron Meltzner provided field notes and photos from the 2005–2007 field season. Isabelle Autissier and Gracieuse Delépine provided insight on the conditions at Saint Paul and Amsterdam Islands. We also acknowledge the services rendered by the staff at libraries in The Hague (Koninklijk Bibliotheek and Nationaal Archief), Berkeley (National Information Service for Earthquake Engineering, University of California, Berkeley), Canberra (National Library of Australia), Jakarta (UNESCO), London (The British Library), Los Angeles (University of California, Los Angeles), Paris (Bibliothèque Nationale de France), and Singapore (National Library of Singapore). E.O. was partially supported at Northwestern University by the National Science Foundation, under subcontract from the University of Pittsburgh's Hazards SEES Grant Number OCE-1331463. This paper is a contribution to IGCP Project 639 Sea-Level Changes from Minutes to Millennia. This research is partly supported by the National Research Foundation Singapore and the Singapore Ministry of Education under the Research Centres of Excellence initiative. This work comprises Earth Observatory of Singapore contribution number 218.

References

Abe, K., & Noguchi, S. (1983). Determination of magnitude for large shallow earthquakes 1898–1917. *Physics of the Earth and Planetary Interiors, 32,* 45–59.

Abercrombie, R. E., Antolik, M., Felzer, K., & Ekström, G. (2001). The 1994 Java tsunami earthquake: Slip over a subducting seamount. *Journal of Geophysical Research, 106*(B4), 6595–6607. https://doi.org/10.1029/2000jb900403.

Ambraseys, N., & Adams, R. D. (1996). Large magnitude Central American earthquakes, 1898–1994. *Geophysical Journal International, 127,* 665–692.

Ambraseys, N., & Douglas, J. (2000). Reappraisal of surface wave magnitudes in the Eastern Mediterranean region and the Middle East. *Geophysical Journal International, 141*(2), 357–373. https://doi.org/10.1046/j.1365-246x.2000.00084.x.

Ambraseys, N., & Douglas, J. (2004). Magnitude calibration of north Indian earthquakes. *Geophysical Journal International, 159*(1), 165–206. https://doi.org/10.1111/j.1365-246x.2004.02323.x.

Ando, M. (1975). Source mechanism and tectonic significance of historical earthquakes along the Nankai Trough, Japan. *Tectonophysics, 27,* 119–140.

Anonymous. (1909). Vulkanische verschijnselen en aardbevingen in den Oost-Indischen Archipel waargenomen gedurende het jaar 1907. *Natuurkundig tijdschrift voor Nederlandsch Indië, 68*(2), 117–201.

Anonymous. (1931). *The seismological bulletin in Osaka (1882–1929)* (p. 132). Osaka: Osaka Meteorological Observatory.

Baumwoll, J. (2008). *The value of indigenous knowledge for disaster risk reduction: A unique assessment tool for reducing community vulnerability to natural disasters.* Webster University, Vienna, Master of Arts in International Relations thesis, p 153.

Bell, R., Holden, C., Power, W., Wang, X., & Downes, G. (2014). Hikurangi margin tsunami earthquake generated by slow seismic rupture over a subducted seamount. *Earth and Planetary Science Letters, 397,* 1–9. https://doi.org/10.1016/j.epsl.2014.04.005.

Ben-Menahem, A., & Rosenman, M. (1972). Amplitude patterns of tsunami waves from submarine earthquakes. *Journal of Geophysical Research, 77,* 3097–3128.

Bertho, J. (1910). Tremblements de terre, *Bulletin de la Société Astronomique de France et revue mensuelle d'Astronomie, de météorologie et de Physique du Globe*, Séance du 5 Octobre 1910, Communications écrites, p. 481, Paris.

Bilek, S. L., & Engdahl, E. R. (2007). Rupture characterization and relocation of aftershocks for the 1994 and 2006 tsunami earthquakes in the Java subduction zone. *Geophysical Research Letters, 34,* L20311. https://doi.org/10.1029/gl031357.

Bilek, S. L., & Lay, T. (1999). Rigidity variations with depth along interplate megathrust faults in subduction zones. *Nature, 400,* 443–446.

Blaser, L. F., Krüger, M. Ohrnberger, & Scherbaum, F. (2010). Scaling relations of earthquake source parameter estimates with special focus on subduction environment. *Bulletin of the Seismological Society of America, 100,* 2914–2926. https://doi.org/10.1785/0120100111.

Borrero, J. C., Weiss, R., Okal, E. A., Hidayat, R., Suranto, D. Arcas, & Titov, V. V. (2009). The tsunami of 12 September 2007, Bengkulu province, Sumatra, Indonesia: Post-tsunami survey and numerical modelling. *Geophysical Journal International, 178,* 180–194.

Bourgeois, J., Petroff, C., Yeh, H., Titov, V., Synolakis, C. E., Benson, B., et al. (1999). Geologic setting, field survey and modeling of the Chimbote, Northern Peru, tsunami of 21 February 1996. *Pure and Applied Geophysics, 154*(3), 513–540. https://doi.org/10.1007/s000240050242.

Bradley, K. E., Feng, L., Hill, E. M., Natawidjaja, D. H., & Sieh, K. (2017). Implications of the diffuse deformation of the Indian Ocean lithosphere for slip partitioning of oblique plate convergence in Sumatra. *Journal of Geophysical Research Solid Earth, 122,* 572–591. https://doi.org/10.1002/2016JB013549.

Briggs, M. J., Synolakis, C. E., Harkins, G. S., & Green, D. R. (1995). Laboratory experiments of tsunami run-up on a circular island. *Pure and Applied Geophysics, 144,* 569–594.

Chlieh, M., Avouac, J.-P., Hjörleifsdóttir, V., Song, T.-R. A., Ji, C., Sieh, K., et al. (2007). Coseismic slip and afterslip of the great M 9.15 Sumatra-Andaman earthquake of 2004. *Bulletin of the Seismological Society of America, 97,* S152–S173. https://doi.org/10.1785/0120050631.

Choy, G. L., & Boatwright, J. (2007). The energy radiated by the 26 December 2004 Sumatra-Andaman earthquake estimated from 10-minute P-wave windows. *Bulletin of the Seismological Society of America, 97,* S18–S24.

Cochran, U., Berryman, K., Zachariasen, J., Mildenhall, D., Hayward, B., Southall, K., et al. (2006). Paleoecological insights into subduction zone earthquake occurrence, eastern North Island, New Zealand. *Geological Society of America Bulletin, 118*(9–10), 1051–1074. https://doi.org/10.1130/B25761.1.

Convers, J. A., & Newman, A. V. (2013). Rapid earthquake rupture duration estimates from teleseismic energy rates, with application to real-time warning. *Geophysical Research Letters, 40,* 5844–5848.

Cruz, G., & Wyss, M. (1983). Large earthquakes, mean sea level, and tsunamis along the Pacific Coast of Mexico and Central America. *Bulletin of the Seismological Society of America, 73,* 553–570.

Davies, J., Sykes, L., House, L., & Jacob, K. (1981). Shumagin Seismic Gap, Alaska Peninsula: History of great earthquakes, tectonic setting, and evidence for high seismic potential. *Journal of Geophysical Research, 86,* 3821–3855.

de Groot-Hedlin, C. D. (2005). Estimation of the rupture length and velocity of the Great Sumatran earthquake of December 26, 2004 using hydroacoustic signals. *Geophysical Research Letters, 32,* 11.

de Moidrey, J. T. (1912). Bulletin des observations—Année 1907: Observatoire magnétique météorologique et sismologique de Zi-Ka-Wei (Chine), Fascicule C Sismologie, *La Mission Catholique (Les missionnaires de la Compagnie de Jésus). Shanghai, 33,* 12.

Doser, D. I., & Webb, T. H. (2003). Source parameters of large historical (1917–1961) earthquakes, North Island, New Zealand. *Geophysical Journal International, 152,* 795–832. https://doi.org/10.1046/j.1365-246x.2003.01895.x.

Duda, S. J. (1965). Secular seismic energy release in the circum-Pacific belt. *Tectonophysics, 2,* 409–452.

Ebel, J. E., & Chambers, D. W. (2016). Using the locations of the $M > 4$ earthquakes to delineate the extents of the ruptures of past major earthquakes. *Geophysical Journal International, 207,* 862–875.

Engdahl, E. R., Villaseñor, A., DeShon, H. R., & Thurber, C. H. (2007). Teleseismic relocation and assessment of seismicity (1918–2005) in the region of the 2004 M_W 9.0 Sumatra-Andaman and 2005 M_W 8.6 Nias Island great earthquakes. *Bulletin of the Seismological Society of America, 97,* S43–S61.

Erskine, C.F. (1908). The operations of the Field Parties—Tidal and Levelling. In: F.B. Longe (ed) *General report on the operations of the Survey of India administered under the Government of India during 1906–07*, Calcutta, p 28.

Erskine, C. F. (1909). Tidal and levelling operations, in *Extracts from narrative reports of officers of the Survey of India for the season 1906–07* (ed. S.G. Burrard), Calcutta, pp. 95–96.

Fan, W., Bassett, W. D., Jiang, J., Shearer, P. M., & Ji, C. (2017). Rupture evolution of the 2006 Java tsunami earthquake and the possible role of splay faults. *Tectonophysics, 721,* 143–150. https://doi.org/10.1016/j.tecto.2017.

Feng, L., Hill, E. M., Banerjee, P., Hermawan, I., Tsang, L. L. H., Natawidjaja, D. H., et al. (2015). A unified GPS-based earthquake catalog for the Sumatran plate boundary between 2002 and 2013. *Journal of Geophysical Research Solid Earth, 120,* 3566–3598. https://doi.org/10.1002/2014jb011661.

Franke, D., Schnabel, M., Ladage, S., Tappin, D. R., Neben, S., Djajadihardja, Y. S., et al. (2008). The great Sumatra-Andaman earthquakes—Imaging the boundary between the ruptures of the great 2004 and 2005 earthquakes. *Earth and Planetary Science Letters, 269,* 118–130. https://doi.org/10.1016/j.epsl.2008.01.047.

Fujino, S., Sieh, K., Meltzner, A. J., Yulianto, E., & Chiang, H.-W. (2014). Ambiguous correlation of precisely dated coral detritus with the tsunamis of 1861 and 1907 at Simeulue Island, Aceh Province, Indonesia. *Marine Geology, 357,* 384–391. https://doi.org/10.1016/j.margeo.2014.09.047.

Fukao, Y. (1979). Tsunami earthquakes and subduction processes near deep-sea trenches. *Journal of Geophysical Research, 84*(B5), 2303–2314.

Geiger, L. (1909). Seismische Registrierungen in Göttingen im Jahre 1907 mit einem Vorwort über die Bearbeitung der Erdbebendiagramme. *Nachrichten von der Königlichen Gesellschaft der Wissenschaften zu Göttingen* (pp. 1–44). Göttingen: Mathematisch-physikalische Klasse.

Geist, E. L., & Bilek, S. L. (2001). Effect of depth-dependent shear modulus on tsunami generation along subduction zones. *Geophysical Research Letters, 28,* 1315–1318. https://doi.org/10.1029/2000gl012385.

Geller, R. J. (1976). Scaling relations for earthquake source parameters and magnitudes. *Bulletin of the Seismological Society of America, 66,* 1501–1523.

Geller, R. J., & Kanamori, H. (1977). Magnitudes of great shallow earthquakes from 1904 to 1952. *Bulletin of the Seismological Society of America, 67*(3), 587–598.

Golitsyn, B. B. (1908). Seismometriche Beobachtungen in Pulkowa. *Académie impériale des Sciences, Commission sismique permanente, 3*(1), 117–172.

Goodstein, J. R., Kanamori, H., & Lee, W. H. K. (1980). Seismology microfiche publications from the Caltech archives. *Bulletin of the Seismological Society of America, 70,* 657–658.

Grünthal, G. (Ed.). (1998). *The European Macroseismic Scale EMS-98, Conseil de l'Europe* (Vol. 15, p. 101). Luxembourg: Cahiers du Centre Européen de Géodynamique et de Séismologie.

Gutenberg, B., & Richter, C. F. (1954). *Seismicity of the Earth and associated phenomena* (p. 310). Princeton: Princeton.

Hayes, G. P., Wald, D. J., & Johnson, R. L. (2012). Slab1.0: A three-dimensional model of global subduction zone geometries. *Journal of Geophysical Research, 117,* B01302. https://doi.org/10.1029/2011jb008524.

Heinrich, P., Schindelé, F., Guibourg, S., & Ihmlé, P. F. (1998). Modelling of the February 1996 Peruvian tsunami. *Geophysical Research Letters, 25*(14), 2687–2690. https://doi.org/10.1029/98gl01780.

Hill, E. M., Borrero, J. C., Huang, Z., Qiu, Q., Banerjee, P., Natawidjaja, D. H., et al. (2012). The 2010 M_W 7.8 Mentawai earthquake: Very shallow source of a rare tsunami earthquake determined from tsunami field survey and near-field GPS data. *Journal of Geophysical Research, 117,* B06402. https://doi.org/10.1029/2012jb009159.

Hodgson, F. N. (1934). *Malacca Strait Pilot, comprising Malacca Strait and its Northern Approaches* (p. 430). London: Singapore Strait and the West Coast of Sumatra, Hydrographic Department -Admiralty.

Hough, S. E. (2013). Missing great earthquakes. *Journal of Geophysical Research Solid Earth, 118,* 1098–1108. https://doi.org/10.1002/jgrb.50083.

Hough, S. E., Bilham, R., Ambraseys, N., & Feldl, N. (2005). Revisiting the 1897 Shillong and 1905 Kangra earthquakes in northern India: Site response, Moho reflections and a triggered earthquake. *Current Science, 88*(10), 1632–1638.

Hough, S. E., Martin, S. S., Gahalaut, V., Joshi, A., Landes, M., Bossu, R. (2016). A comparison of observed and predicted ground motions from the 2015 MW7.8 Gorkha, Nepal, earthquake. *Natural Hazards, 84*(3):1661–1684

Ide, S., Imamura, F., Yoshida, Y., & Abe, K. (1993). Source characteristics of the Nicaraguan tsunami earthquake of September 2, 1992. *Geophysical Research Letters, 20,* 863–866.

Ihmlé, P. F., Gomez, J.-M., Heinrich, P., & Guibourg, S. (1998). The 1996 Peru tsunamigenic earthquake: Broadband source process. *Geophysical Research Letters, 25*(14), 2691–2694.

Ishii, M., Shearer, P. M., Houston, H., & Vidale, J. E. (2005). Extent, duration and speed of the 2004 Sumatra-Andaman earthquake imaged by the Hi-Net array. *Nature, 435,* 933–936.

Jacob, J., Dyment, J., & Yatheesh, V. (2014). Revisiting the structure, age, and evolution of the Wharton Basin to better understand subduction under Indonesia. *Journal of Geophysical Research, 119,* 169–190. https://doi.org/10.1002/2013jb010285.

Kagan, Y. Y. (1991). 3-D rotation of double-couple earthquake sources. *Geophysical Journal International, 106,* 709–716.

Kajiura, K. (1972). The directivity of energy radiation of the tsunami generated in the vicinity of a continental shelf. *Journal of Oceanography, 28*(6), 260–277. https://doi.org/10.1007/bf02109296.

Kanamori, H. (1972). Mechanism of tsunami earthquakes. *Physics of the Earth and Planetary Interiors, 6,* 346–359.

Kanamori, H., Rivera, L., & Lee, W. H. K. (2010). Historical seismograms for unravelling a mysterious earthquake: The 1907 Sumatra earthquake. *Geophysical Journal International, 183,* 358–374. https://doi.org/10.1111/j.1365-246x.2010.04731.x.

Kárník, V., Kondorskaya, N. V., Ju, V., Riznitchenko, E. F., Savarensky, S. L., Soloviev, N. V., et al. (1962). Standardization of the earthquake magnitude scale. *Studia Geophysica et Geodaetica, 6*(1), 41–48.

Kikuchi, M., & Kanamori, H. (1982). Inversion of complex body waves. *Bulletin of the Seismological Society of America, 72*(2), 491–506.

Kikuchi, M., & Kanamori, H. (1986). Inversion of complex body waves—II. *Physics of the Earth and Planetary Interiors, 43*(3), 205–222.

Kikuchi, M., & Kanamori, H. (1991). Inversion of complex body waves—III. *Bulletin of the Seismological Society of America, 81*(6), 2335–2350.

Konca, A. O., Avouac, J.-P., Sladen, A., Meltzner, A. J., Sieh, K., Fang, P., et al. (2008). Partial rupture of a locked patch of the

Sumatra megathrust during the 2007 earthquake sequence. *Nature, 456,* 631–635. https://doi.org/10.1038/nature07572.

Lambert, J., & Terrier, M. (2011). Historical tsunami database for France and its overseas territories. *Natural Hazards and Earth System Science, 11,* 1037–1046. https://doi.org/10.5194/nhess-11-1037-2011.

Levitski, G.V. (1909). Январь - июня 1907 года, Известия постоянной центральной сейсмической комиссии, Императорская академия наук (January to June 1907, Report of the Permanent Central Seismic Commission, Imperial Academy of Sciences, 3(2), 17–21, Saint Petersburg

Li, L., Switzer, A. D., Wang, Y., Weiss, R., Qiu, Q., Chan, C.-H., et al. (2015). What caused the mysterious eighteenth century tsunami that struck the southwest Taiwan coast? *Geophysical Research Letters, 42,* 8498–8506. https://doi.org/10.1002/2015gl065567.

Liu, P.L.F., Woo, S.B., & Cho, Y.S. (1998). *Computer programs for tsunami propagation and inundation*. Technical report, Cornell University Ithaca, New York.

López, A. M., & Okal, E. A. (2006). A seismological reassessment of the 1946 Aleutian "tsunami" earthquake. *Geophysical Journal International, 165,* 835–849.

Martin, S. S. (2005). Intensity Distribution from the 2004 M 9.0 Sumatra-Andaman Earthquake. *Seismological Research Letters, 76*(3):321–330.

Martin, S. & Szeliga, W. (2010). A Catalog of Felt Intensity Data for 570 Earthquakes in India from 1636 to 2009. *Bulletin of the Seismological Society of America, 100*(2):562–569.

Martin, S. S., & Hough, S. E. (2016). Reply to, "Comment on 'Ground Motions from the 2015 M_W 7.8 Gorkha, Nepal, Earthquake Constrained by a Detailed Assessment of Macroseismic Data' by Stacey Martin, Susan E. Hough and Charleen Hung" by Andrea Tertulliani, Laura Graziani, Corrado Castellano, Alessandra Maramai, Antonio Rossi. *Seismological Research Letters, 87*(4), 369–377. https://doi.org/10.1785/0220160061.

Masó, M. S. (1907). *Seismological Bulletin for January 1907, Monthly Bulletin 1907, Department of the Interior—Weather Bureau* (pp. 37–39). Manila: Manila Central Observatory.

McAdoo, B. G., Dengler, L., Prasetya, G., & Titov, V. (2006). Smong: How an oral history saved thousands on Indonesia's Simeulue island during the December 2004 and March 2005 tsunamis. *Earthquake Spectra, 22*(S3), 661–669.

McCaffrey, R. (2007). The next great earthquake. *Science, 315,* 1675–1676.

McCloskey, J., Lange, D., Tilman, F., Nalbant, S. S., Bell, A. F., Natawidjaja, D. H., et al. (2010). The September 2009 Padang earthquake. *Nature Geoscience, 3,* 70–71.

Meltzner, A. J., Sieh, K., Chiang, H.-W., Shen, C.-C., Suwargadi, B. W., Natawidjaja, D. H., et al. (2012). Persistent termini of 2004- and 2005-like ruptures of the Sunda megathrust. *Journal of Geophysical Research, 117,* B04405. https://doi.org/10.1029/2011jb008888.

Meltzner, A. J., Sieh, K., Chiang, H.-W., Shen, C.-C., Suwargadi, B. W., Natawidjaja, D. H., et al. (2010). Coral evidence for earthquake recurrence and an AD 1390–1455 cluster at the south end of the 2004 Aceh-Andaman rupture. *Journal of Geophysical Research, 115,* B10. https://doi.org/10.1029/2010jb007499.

Meltzner, A. J., Sieh, K., Chiang, H.-W., Wu, C.-C., Tsang, L. L. H., Shen, C.-C., et al. (2015). Time-varying interseismic strain rates and similar seismic ruptures on the Nias-Simeulue patch of the Sunda megathrust. *Quaternary Science Reviews, 122,* 258–281. https://doi.org/10.1016/j.quascirev.2015.06.003.

Miyoshi, H. (1955). Directivity of the recent tsunamis. *Journal of the Oceanographic Society of Japan, 11*(4), 151–156.

Monecke, K., Finger, W., Klarer, D., Kongko, W., McAdoo, B., Moore, A. L., et al. (2008). A 1,000-year sediment record of tsunami recurrence in northern Sumatra. *Nature, 455,* 1232–1234.

Morgan, P. M., Feng, L., Meltzner, A. J., Lindsey, E. O., Tsang, L. L. H., & Hill, E. M. (2017). Sibling earthquakes generated within a persistent rupture barrier on the Sunda megathrust under Simeulue Island. *Geophysical Research Letters, 44,* 2159–2166. https://doi.org/10.1002/2016gl071901.

Mori, N., Takahashi, T., Yasuda, T., & Yanagisawa, H. (2011). Survey of 2011 Tohoku earthquake tsunami inundation and run-up. *Geophysical Research Letters, 38,* L00G14. https://doi.org/10.1029/2011gl049210.

Musson, R. M. W., Grunthal, G., & Stucchi, M. (2010). The comparison of macroseismic intensity scales. *Journal of Seismology, 14,* 413–428. https://doi.org/10.1007/s10950-009-9172-0.

Natawidjaja, D. H., Sieh, K., Chlieh, M., Galetzka, J., Suwargadi, B. W., Cheng, H., et al. (2006). Source parameters of the great Sumatran megathrust earthquakes of 1797 and 1833 inferred from coral microatolls. *Journal of Geophysical Research, 111,* B06403. https://doi.org/10.1029/2005jb004025.

Newcomb, K. R., & McCann, W. R. (1987). Seismic history and seismotectonics of the Sunda Arc. *Journal of Geophysical Research, 92*(B1), 421–439. https://doi.org/10.1029/jb092ib01p00421.

Newman, A. V., Feng, L., Fritz, H. M., Lifton, Z. M., Kalligeris, N., & Wei, Y. (2011a). The energetic 2010 M_W 7.1 Solomon Islands tsunami earthquake. *Geophysical Journal International, 186,* 775–781. https://doi.org/10.1111/j.1365-246x.2011.05057.x.

Newman, A. V., Hayes, G., Wei, Y., & Convers, J. (2011b). The 25 October 2010 Mentawai tsunami earthquake, from real-time discriminants, finite-fault rupture, and tsunami excitation. *Geophysical Research Letters, 38*(5), L05302. https://doi.org/10.1029/2010gl046498.

Newman, A. V., & Okal, E. A. (1998). Teleseismic estimates of radiated seismic energy: The E/M_0 discriminant for tsunami earthquakes. *Journal of Geophysical Research, 103,* 26885–26898.

Nutalaya, P., Sodsri, S., & Arnold, E. P. (1985). Series on seismology, Thailand. *Southeast Asia Association of Seismology and Earthquake Engineering, 2,* 409.

Obermeier, S. F., & Dickenson, S. E. (2000). Liquefaction evidence for the strength of ground motions resulting from Late Holocene Cascadia subduction earthquakes, with emphasis on the event of 1700 A.D. *Bulletin of the Seismological Society of America, 90,* 876–896.

Okada, Y. (1985). Surface deformation due to shear and tensile faults in a half-space. *Bulletin of the Seismological Society of America, 75*(4), 1135–1154.

Okal, E. A. (2011). Tsunamigenic earthquakes: Past and present milestones. *Pure and Applied Geophysics, 168,* 969–996.

Okal, E. A. (2013). From 3-Hz P waves to $_0S_2$: No evidence of a slow component to the source of the 2011 Tohoku earthquake. *Pure and Applied Geophysics, 170,* 963–973.

Okal, E. A., & Borrero, J. C. (2011). The "tsunami earthquake" of 22 June 1932 in Manzanillo, Mexico: Seismological study and tsunami simulation. *Geophysical Journal International, 187,* 1443–1459.

Okal, E. A., Fritz, H. M., Raveloson, R., Joelson, G., Pančošková, P., & Rambolamanana, G. (2006a). Madagascar field survey after the December 2004 Indian Ocean tsunami. *Earthquake Spectra, 22,* S263–S283.

Okal, E. A., Fritz, H. M., & Sladen, A. (2009). 2004 Sumatra tsunami surveys in the Comoro Islands and Tanzania and regional tsunami hazard from future Sumatra events. *South African Journal of Geology, 112,* 343–358.

Okal, E.A., & Kirby, S.H. (2002). Energy-to-moment ratios for damaging intraslab earthquakes: Preliminary results on a few case studies. *USGS Open File Report*, 02-328, pp. 127–131.

Okal, E. A., Kirby, S. H., & Kalligeris, N. (2016). The Showa Sanriku earthquake of 1933 March 2: A global seismological reassessment. *Geophysical Journal International, 206,* 1492–1514.

Okal, E. A., & Newman, A. V. (2001). Tsunami earthquakes: The quest for a regional signal. *Physics of the Earth and Planetary Interiors, 124,* 45–70.

Okal, E. A., & Saloor, N. (2017). Historical tsunami earthquakes in the Southwestern Pacific: An extension to $\Delta > 80°$ of the energy-to-moment parameter Θ. *Geophysical Journal International, 210,* 852–873.

Okal, E. A., Sladen, A., & Okal, E. A.-S. (2006b). Rodrigues, Mauritius and Réunion Islands field survey after the December 2004 Indian Ocean tsunami. *Earthquake Spectra, 22,* S241–S261.

Okal, E. A., & Talandier, J. (1989). M_m: A variable period mantle magnitude. *Journal of Geophysical Research, 94,* 4169–4193.

O'Loughlin, F. E., Paiva, R. C. D., Durand, M., Alsdorf, D. E., & Bates, P. D. (2016). A multi-sensor approach towards a global vegetation corrected SRTM DEM product. *RSEnv, 182*(Supplement C), 49–59. https://doi.org/10.1016/j.rse.2016.04.018.

Patterson, J. (1909). The Simla seismograms obtained between June 1905 and November 1908. *Memoirs of the Indian Meteorological Department, 20*(3), 97–143.

Pechau, W. (1907). *Monatliche Erdbebenberichte der Seismischen Station zu Jena—1907, Ant* (p. 28). Jena: Kämpfe, Buchdruckerei.

Pelayo, A. M., & Wiens, D. A. (1990). The November 20, 1960 Peru tsunami earthquake: Source mechanism of a slow event. *Geophysical Research Letters, 17*(6), 661–664.

Pelayo, A. M., & Wiens, D. A. (1992). Tsunami earthquakes: Slow thrust-faulting events in the accretionary wedge. *Journal of Geophysical Research, 97*(B11), 15321–15337.

Pesicek, J. D., Thurber, C. H., Zhang, H., DeShon, H. R., Engdahl, E. R., & Widiyantoro, S. (2010). Teleseismic double-difference relocation of earthquakes along the Sumatra-Andaman subduction zone using a 3-D model. *Journal of Geophysical Research, 115,* B10303. https://doi.org/10.1029/2010jb007443.

Pflaker, G. L. (1997). Catastrophic tsunami generated by submarine slides and backarc thrusting during the 1992 earthquake on Eastern Flores Island, Indonesia. *Geological Society of America (abstracts with program), 29*(5), 57.

Philibosian, B., Sieh, K., Avouac, J.-P., Natawidjaja, D. H., Chiang, H.-W., Wu, C.-C., et al. (2014). Rupture and variable coupling behavior of the Mentawai segment of the Sunda megathrust during the supercycle culmination of 1797 to 1833. *Journal of Geophysical Research, 119,* 7258–7287. https://doi.org/10.1002/2014jb011200.

Polet, J., & Kanamori, H. (2000). Shallow subduction zone earthquakes and their tsunamigenic potential. *Geophysical Journal International, 142,* 684–702. https://doi.org/10.1046/j.1365-246x.2000.00205.x.

Rahman, A., Sakurai, A., & Munadi, K. (2017). The analysis of the development of the Smong story on the 1907 and 2004 Indian Ocean tsunamis in strengthening the Simeulue island community's resilience. *International Journal of Disaster Risk Reduction, 29,* 13–24. https://doi.org/10.1016/j.ijdrr.2017.07.015.

Reiche, M Th. (1863). Aanteekeningen omtrent aardbevingen en andere natuurverschijnselen, waargenomen in den Indischen Archipel, gedurende het laaste gedeelte van 1860 en het jaar 1861, verzameld. *Natuurkundig tijdschrift voor Nederlandsch Indië, 25*(1), 108–137.

Reid, A. (2014). Two hitherto unknown Indonesian tsunamis of the seventeenth century: Probabilities and context. *Journal of Southeast Asian Studies, 47*(1), 88–108. https://doi.org/10.1017/s002246341500048x.

Sahal, A., Morin, J., Schindelé, F., & Lavigne, F. (2011). A catalog of tsunamis in La Réunion Island (France) from 1883 to 2010. *Science of Tsunami Hazards, 30*(3), 178–190.

Saloor, N., & Okal, E. A. (2018). Extension of the energy-to-moment parameter Θ to intermediate and deep earthquakes. *Physics of the Earth and Planetary Interiors, 274,* 37–48.

Satake, K. (1995). Linear and non-linear computations of the 1992 Nicaragua earthquake tsunami. *Pure and Applied Geophysics, 144,* 455–470.

Schröder, E.E.W.G. (1917a). *Nias. Ethnographische, geographische en historische aanteekeningen en studiën* (vol 1), E.J. Brill, Leiden, p. 866.

Schröder, E.E.W.G. (1917b). *Nias. Ethnographische, geographische en historische aanteekeningen en studiën: Platen en Kaarten* (vol 2), Plate 128, Images 241 and 242, E.J. Brill, Leiden.

Shimozono, T., Cui, H., Pietrzak, J. D., Hermann, F., Okayasu, A., & Hooper, A. J. (2014). Short wave amplification and extreme runup by the 2011 Tohoku tsunami. *Pure and Applied Geophysics, 171*(12), 3217–3228. https://doi.org/10.1007/s00024-014-0803-1.

Sieh, K., Briggs, R., Meltzner, A. J., Natawidjaja, D., Hananto, N., & Suprihanto, I. (2006). The Simeulue saddle: Evidence for a barrier to rupture for both the 2004 and 2005 Sumatran megathrust failures. *EOS, Transactions of the American Geophysical Union*, vol 87, p. 52, U14A-01 (Abstract).

Singh, S. C., Carton, H., Chauhan, A. S., Androvandi, S., Davaille, A., Dyment, J., et al. (2011). Extremely thin crust in the Indian Ocean possibly resulting from plume-ridge interaction. *Geophysics Journal International, 184*(1), 29–42. https://doi.org/10.1111/j.1365-246x.2010.04823.x.

Singh, S. C., Carton, H., Tapponnier, P., Hananto, N. D., Chauhan, A. P. S., Haryoto, D., et al. (2008). Seismic evidence for broken oceanic crust in the 2004 Sumatra earthquake epicentral region. *Nature Geoscience, 1,* 777–781.

Solov'iev, S.L., & Go, Ch. (1974). Каталог цунами на западном побережье тихого океана (Catalogue of tsunamis on the western shores of the Pacific Ocean), Наука (Nauka), p. 310. (English translation by Sidney O. Wigen, Institute of Ocean Sciences, Department of Fisheries and Oceans, #5077, Sidney, Canada, 1984).

Staring, W. C. H. (1871). *De binnen- en buitenlandsche maten, gewichten en munten van vroeger en tegenwoordig met hunne onderlinge vergelijkingen en herleidingen benevens vele andere, dagelijks te pas komende opgaven en berekeningen, 4th edition (1980 reprint)*. Arnhem: Gybers & Van Loon, p 196.

Stein, S., & Okal, E. A. (2007). Ultra-long period seismic study of the December 2004 Indian Ocean earthquake and implications for regional tectonics and the subduction process. *Bulletin of the Seismological Society of America, 97,* S279–S295.

Storchak, D. A., Giacomo, D. D., Bondár, I., Engdahl, E. R., Harris, J., Lee, W. H. K., et al. (2013). Public release of the ISC-GEM global instrumental earthquake catalogue (1900–2009). *Seismological Research Letters, 84*(5), 810–815.

Stover, C. W., & Coffmann, J. L. (1993). Seismicity of the United States, 1568–1989 (Revised). *United States Geological Survey Professional Paper*, vol 1527, p. 418.

Syafwina, S. (2014). Recognizing indigenous knowledge for disaster management: Smong, early warning system from Simeulue Island, Aceh. *Procedia Environmental Sciences, 20,* 572–582.

Sykes, L. (1971). Aftershock zones of great earthquakes seismicity gaps, and earthquake prediction for Alaska and the Aleutians. *Journal of Geophysical Research, 76,* 8021–8041.

Synolakis, C. E., & Okal, E. A. (2005). 1992–2002: Perspective on a decade of post-tsunami surveys. In K. Satake (Ed.), *Tsunamis: Case studies and recent developments. Advances in Natural and Technological Hazards* (pp. 1–30). Berlin: Springer.

Szirtes, S. (1912a). Registerierungen der besser ausgeprägten seismischen Störungen des Jahres 1907, Erganzung mit mikroseismischen Katalog, *Veröffentlichungen des Zentral-bureaus der Internationalen Seismologischen Assoziation* (Publications du bureau central de l'association internationale de séismologie), Strassburg, p. 111.

Szirtes, S. (1912b). Katalog der im Jahr 1907 registierten seismischen Störungen, *Veröffentlichungen des Zentral-bureaus der Internationalen Seismologischen Assoziation* (Publications du bureau central de l'association internationale de séismologie), Strassburg, p. 120.

Tang, G., Barton, P. J., McNeill, L. C., Henstock, T. J., Tilmann, F., Dean, S. M., et al. (2013). 3-D active source tomography around Simeulue Island offshore Sumatra: Thick crustal zone responsible for earthquake segment boundary. *Geophysical Research Letters, 40,* 48–53. https://doi.org/10.1029/2012gl054148.

Tanioka, Y., Ruff, L., & Satake, K. (1997). What controls the lateral variation of large earthquake occurrence along the Japan trench? *Island Arc, 6,* 261–266. https://doi.org/10.1111/j.1440-1738.1997.tb00176.x.

Tanioka, Y., & Satake, K. (1996). Fault parameters of the 1896 Sanriku tsunami earthquake estimated from tsunami modelling. *Geophysical Research Letters, 23,* 1549–1552.

Tolstoy, M., & Bohnenstiehl, D. R. (2005). Hydroacoustic constraints on the rupture duration, length, and speed of the great Sumatra-Andaman earthquake. *Seismological Research Letters, 76,* 419–425.

Toppozada, T. R., & Real, C. R. (1981). Preparation of isoseismal maps and summaries of reported effects for pre-1900 California earthquakes. *U.S. Geological Survey Open File Report, 1981–262,* 7–11.

Tsang, L. L. H., Meltzner, A. J., Philibosian, B., Hill, E. M., Freymueller, J. T., & Sieh, K. (2015). A 15-year slow-slip event on the Sunda megathrust offshore Sumatra. *Geophysical Research Letters, 42,* 6630–6638. https://doi.org/10.1002/2015gl064928.

Tsuji, Y., Matsutomi, H., Imamura, F., Takeo, M., Kawata, Y., Matsuyama, M., et al. (1995). Damage to coastal villages due to the 1992 Flores Island earthquake tsunami. *Pure and Applied Geophysics, 144,* 481–524.

Turner, H. H., & Milne, J. (1908a). Circular No. 15 issued by the Seismological Committee. *British Association for the Advancement of Science* (Shide Circulars), Newport, pp. 105–152.

Turner, H. H., & Milne, J. (1908b). Circular No. 16 issued by the Seismological Committee. *British Association for the Advancement of Science* (Shide Circulars), Newport, pp. 153–186.

Turner, H. H., Milne, J., Boys, C. V., Darwin, G., Darwin, H., Glazerbrook, R. T., Gray, M. H., Gray, R. K., Judd, J. W., Knott, C. G., Meldola, R., Oldham, R. D., Perry, J., Plummer, W. E., Sampson, R. A., & Schuster, A. (1912). Seismic activity, 1904 to 1910 inclusive with map of origins, Seismological Investigations. In: *17th report of the Committee, British Association for the Advancement of Science (BAAS)*, pp. 70–87.

Verbeek, R. D. M. (1885). Krakatau, Landsrukkerij, Batavia, p. 624.

Verbeek, R. D. M. (1912). Opgave van geschriften over geologie en mijnbouw van Nederlandsch Oost-Indië. *Geologisch-Mijnbouwkundig Genootschap voor Nederland en Koloniën, Geologische serie, 1,* 31–284.

Visser, S. W. (1922). *Inland and submarine epicentra of Sumatra and Java earthquakes, Verhandelingen 9 (complement to Verhandelingen 7)* (pp. 1–14). Batavia: Koninklijk Magnetisch en Meteorologisch Observatorium te Batavia.

Visser, S. W. (1931). Aardbevingen en getijden. *Natuurkundig tijdschrift voor Nederlandsch Indië, 91*(2), 153–166.

Wang, X. (2009). User manual for Cornell Multi-grid COupled Tsunami model-COMCOT V1.7, edited, http://ceeserver.cee.cornell.edu/pll-group/doc/COMCOT_User_Manual_v1_7.pdf. (last accessed, 1 October 2018).

Wang, X., & Liu, P. L. F. (2006). An analysis of 2004 Sumatra earthquake fault plane mechanisms and Indian Ocean tsunami. *Journal of Hydraulic Research, 44*(2), 147–154.

Wendt, J., Oglesby, D. D., & Geist, E. L. (2009). Tsunamis and splay fault dynamics. *Geophysical Research Letters, 36,* L15303. https://doi.org/10.1029/2009gl038295.

Wessel, P., & Smith, W. H. F. (1991). Free software helps map and display data. *Eos Transactions AGU, 72,* 441. https://doi.org/10.1029/90eo00319.

Whitlow, K. F. (2008). *The 2004 and 1861 tsunami deposits on Simeulue Island, western Sumatra.* Central Washington University, Master of Science in Geology thesis, p. 75.

Wysession, M. E., Okal, E. A., & Miller, K. L. (1991). Intraplate seismicity of the Pacific Basin, 1913–1988. *Pure and Applied Geophysics, 135,* 261–359.

Yogaswara, H., & Yulianto, E. (2006). *Pengetahuan local tenTang tsunami pada masyarakat Simeulue, Nanggroe Aceh Darussalam, Assessing and recognising community preparedness in natural disasters in Indonesia*, LIPI–UNESCO/ISDR, p. 69.

(Received October 6, 2018, revised February 7, 2019, accepted February 11, 2019, Published online March 26, 2019)

Pure Appl. Geophys. 176 (2019), 2869–2886

https://doi.org/10.1007/s00024-018-2082-8

Pure and Applied Geophysics

The Large Andaman Islands Earthquake of 26 June 1941: Why No Significant Tsunami?

EMILE A. OKAL[1]

Abstract—We present a modern seismological study of the earthquake of 26 June 1941 in the Andaman Islands, the largest pre-2004 event along that section of the India-Burma plate boundary. Despite a large conventional magnitude ($M_{PAS} = 8.1$), it generated at best a mediocre tsunami for which no definitive quantitative reports are available. We show that the 1941 earthquake took place under the Andaman accretionary prism and consisted of a composite event, whose nucleating phase had a strike-slip mechanism incompatible with a data set of spectral amplitudes of mantle Rayleigh and Love waves. Combining this initial phase with a larger normal faulting mechanism can reconcile them with P-wave first motions, reports of subsidence on the eastern coast of the Andaman Islands and the small amplitudes of any putative tsunami. The small tsunami results from a combination of that mechanism and of a source located under the islands themselves and in shallow water, implying a reduction in amplitude under Green's law when transitioning to a deeper basin.

Key words: Tsunami, Andaman Islands, historical earthquakes.

1. *Introduction*

The purpose of this article is to conduct a modern reassessment of the large 1941 earthquake in the Andaman Islands, which featured a surprisingly low tsunami (Rajendran 2013), given its conventional magnitude, $M_{PAS} = 8.1$ (Gutenberg and Richter 1954).

The Andaman Islands, located in the forearc of the northern extension of the Sumatra Trench (Cochran 2010), constitute the emerged part of a large accretionary prism approximately 400 by 100 km, with basement rocks formed in the Cretaceous and uplifted starting in the Eocene as a result of the Indian-Eurasian collision (Curray 2005).

The earthquake took place at 11:52 GMT on 26 June 1941, with an epicenter in the immediate vicinity of the Andaman Islands. In this area, the convergence between the Indian plate and the Sunda block, as determined, e.g., by the REVEL model, is extremely oblique, striking N14°E at 6.7 cm/year (Sella et al. 2002); furthermore, this motion is obviously partitioned given the abundant seismicity of the Andaman Sea, where an active spreading center is present (Fitch 1972), thus defining a "Burma" or "Andaman-Sumatra" plate sliver (Fig. 1), delineated by the northern extension of the Sumatra Fault System, which evolves into the Andaman Sea spreading center (Diehl et al. 2013).

The 1941 earthquake was originally given a "Pasadena" magnitude $M_{PAS} = 8.1$ by Gutenberg and Richter (1954). This value can be confirmed by an examination of B. Gutenberg's notepads (Goodstein et al. 1980). Richter (1958, p. 712) later proposed a significantly larger magnitude of 8.7, this discrepancy having the hallmarks of a departure from scaling laws, and suggesting that the 1941 earthquake may not be a traditional subduction event, by analogy with the case of the great 1933 Sanriku earthquake, as detailed by Okal et al. (2016, p. 1497). Abe (1981) recomputed a body-wave magnitude $m_B = 8.0$, using relatively long-period waves (8 s), and a surface-wave magnitude $M_s = 7.7$. Brune and Engen (1969) used Love wave spectral amplitudes to obtain a 100-s magnitude $M_{100} = 8.0$, corresponding to a moment $M_0 = 10^{28}$ dyn cm, according to Brune (1968, Fig. 1). Any of these values makes the 1941 event the largest earthquake along the India-Burma plate boundary since the dawn of instrumental seismology and prior to the 2004 Sumatra-Andaman mega-event. Kanamori (1977) proposed a

[1] Department of Earth and Planetary Sciences, Northwestern University, Evanston, IL 60208, USA. E-mail: emile@earth.northwestern.edu

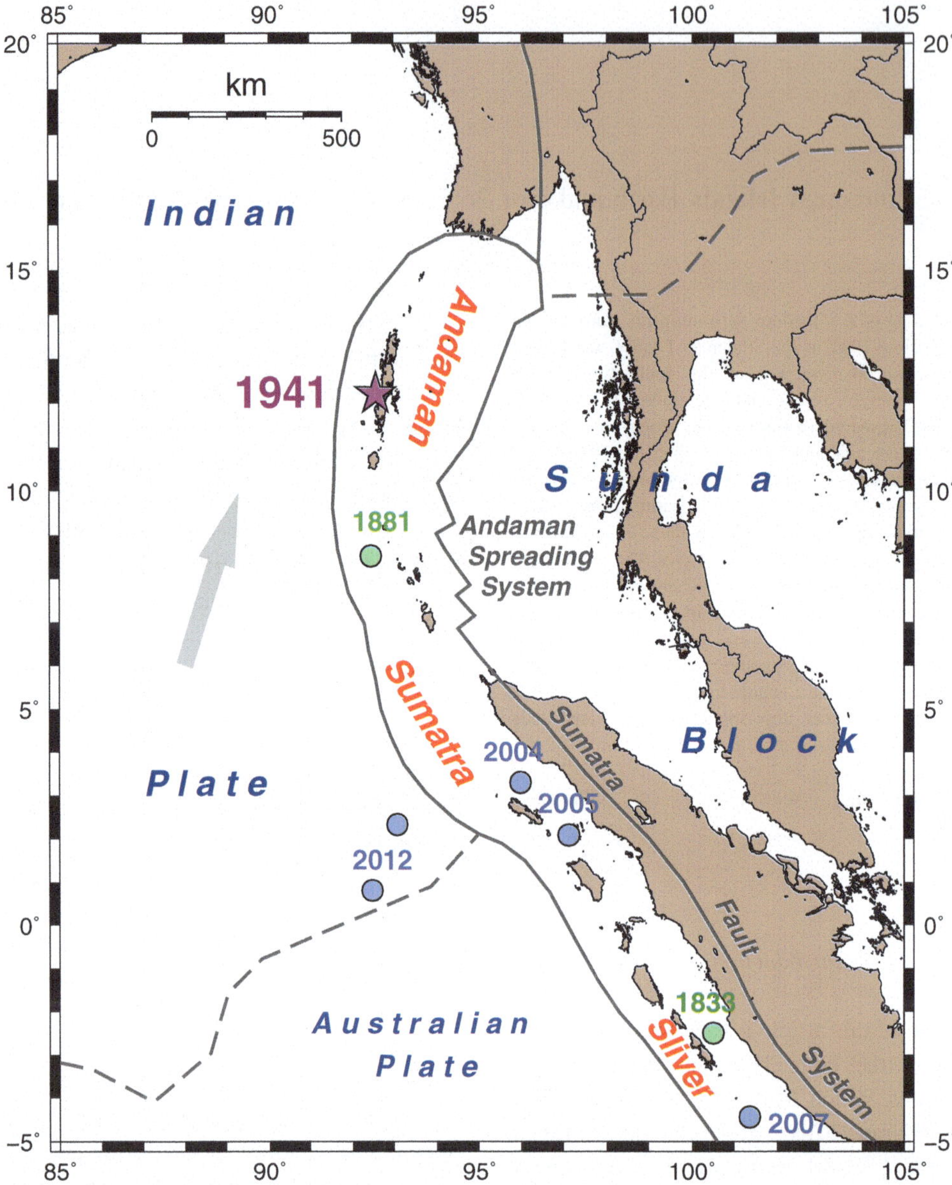

Figure 1
Sketch of the plate tectonics framework of the study area. The solid lines are plate boundaries, adapted from Bird's (2003) model. Dashed lines represent diffuse or uncertain boundaries. The large star shows the epicenter of the 1941 event and the circles those of recent (blue) and pre-instrumental (green) mega-earthquakes

significantly smaller seismic moment $M_0 = 3 \times 10^{27}$ dyn cm "from the 100-s magnitude."

A classical interpretation of the 1941 earthquake as expressing the subduction of the Indian plate under the Andaman plate (e.g., Bilham et al. 2005) gives rise to a significant problem in that the 2004 rupture propagated over its epicentral area, where it resulted in substantial seismic slip release; for example, Tsai et al.'s (2005) composite solution for the 2004 event positions a source of moment 10^{29} dyn cm only 100 km south of the 1941 event, with a slightly smaller one (8×10^{28} dyn cm) off North Andaman Island. Similarly, Ishii et al.'s (2005; Fig. 4) reverse imaging source tomography features a density of seismic release (per unit area) over the Andaman Islands of 65–70% of its maximum along the whole rupture. These results were confirmed, e.g., by the inversion of GPS data, which suggested slips on the order of 7–10 m (e.g., Chlieh et al. 2007) and by a combined inversion of GPS, tide gauge and satellite altimetry data, requiring slips of 7–12 m in the same area (Lorito et al. 2010). All these estimates of displacement in 2004 are at least twice the slip accumulated during the 63 years separating the 1941 and 2004 events, even under the assumption of a non-partitioned boundary taking up the full convergence between the Indian and Sunda plates; they suggest that the 1941 event was not an interplate thrust earthquake.

The tsunami generated by the 1941 Andaman earthquake constitutes another, highly controversial issue. Murty (1984) reported the observation of a tsunami with an amplitude of 0.75–1.25 m on the eastern coast of India (from unspecified sources), and later Murty and Rafiq (1991) suggested that it resulted in 5000 deaths in India. However, Ortíz and Bilham (2003) and Bilham et al. (2005) have failed to unravel any scientific or press evidence to support this assertion (in contrast, e.g., to the Car Nicobar event of 1881), casting strong doubt on its veracity: it is extremely unlikely that a 1-m tsunami would result in such a death toll. It would make it the most lethal tsunami in the far field in the twentieth century and the second or third most lethal one regardless of distance, surpassed clearly only by the 1952 Kamchatka event, a seismologically much larger event (Smyshlyaev 2003), and perhaps comparable to the 1976 Mindanao event for which the repartition of the 5000 to 8000 casualties between earthquake and tsunami is unclear. In particular, a human disaster of this scope should have been widely reported in the press, which was not the case, as forcefully expressed by Rajendran (2013). Further, it would have required catastrophic waves at the epicenter, which have not been documented in the Andaman Islands (which in 1941 were not yet under Japanese occupation). Jhingran (1953) gives a detailed account of the macroseismic effects of the 1941 earthquake, but makes only a vague mention of a tsunami ("there is no doubt that extensive waves were forced at least around the Andaman Islands"), with no specific description at any location. Following Rajendran (2013), we conclude that the 1941 earthquake was probably not accompanied by a substantial tsunami either in the vicinity of its epicenter or a fortiori on the coast of India.

These questions motivate a fresh look at the 1941 Andaman earthquake, which is the subject of this article. We will show that the event has a composite mechanism and does not represent an episode of subduction of the Indian plate.

2. *Relocation*

The 1941 earthquake was originally located by the International Seismological Summary (ISS) at (12.4°N; 92.5°E), an epicenter simply rounded to the nearest half-degree (12.5°N; 92.5°E) by Gutenberg and Richter (1954); by contrast, the USGS solution (12.8°N; 92.7°E) is about 50 km to the north. Modern relocations include the Centennial Catalog solution (Engdahl and Villaseñor 2002) (12.149°N; 92.478°E), Bilham et al.'s (2005) (12.13°N; 92.49°E) and the ISC-GEM solution (Storchak et al. 2015) (12.017°N; 92.575°E).

We conducted our own relocation, based on the ISS data set and using the interactive iterative algorithm of Wysession et al. (1991), which yielded an epicenter at (12.21°N; 92.55°E), only 10 km from the Centennial Catalog's and 22 km from the ISC-GEM source. All these relocations are summarized in Fig. 2, which also shows our Monte Carlo confidence ellipse, obtained by injecting Gaussian noise into the data set with a standard deviation $\sigma_G = 4.5$ s, adequate for an event in the early 1940s.

We found that the ISS data set of arrival times cannot constrain depth, a conclusion shared by Storchak et al. (2015), who used a fixed depth (20 km); by contrast, Engdahl and Villaseñor (2002) proposed a floated depth of 49 km. In Fig. 3a, and following Rees and Okal (1987), we examine the r.m.s. residuals for constrained-depth relocations, as a function of source depth; we find a very weak increase with depth, suggestive of a shallow source, but with no resolvable trend in the first $\sim$40 km. The minimal moveout of the resulting epicenters is shown by the small open circles in Fig. 2 and in cross section in

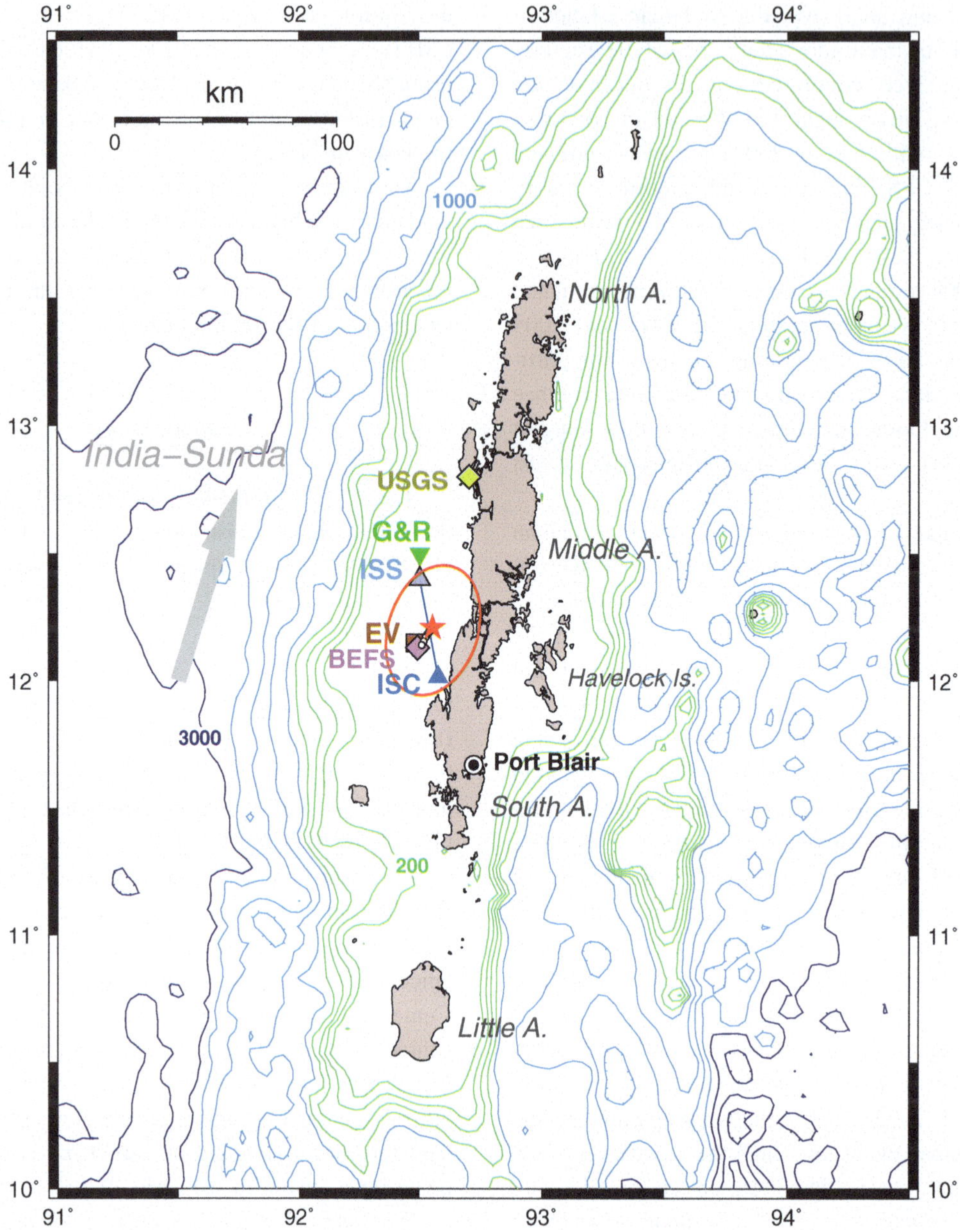

Figure 2
Relocation of the 1941 mainshock. Our relocation is shown as the red star, together with its Monte Carlo ellipse. The small open circles show the minimal moveout of the epicenter as a function of constrained source depth, from 20 to 150 km. ISS (light blue) and relocated ISC-GEM (dark blue) solutions shown as triangles, Gutenberg and Richter's (1954) as an inverted green triangle, Engdahl and Villaseñor's (2002) as a brown square and Bilham et al.'s (2005) as a magenta diamond; the USGS location is the yellow diamond. Isobaths every 200 m (green), then from 1000 m every 500 m (blue). The large gray arrow on the left shows the motion of India with respect to Sunda in the REVEL model (Sella et al. 2002)

Fig. 3b against a background of unrelocated seismicity (1963–2015; 11°N–14°N; $M > 5$), defining a seismogenic zone extending to ~ 40 km; in this context, the 49-km depth proposed by Engdahl and Villaseñor (2002) may be acceptable, but could also be an artifact of interpreting secondary arrivals as

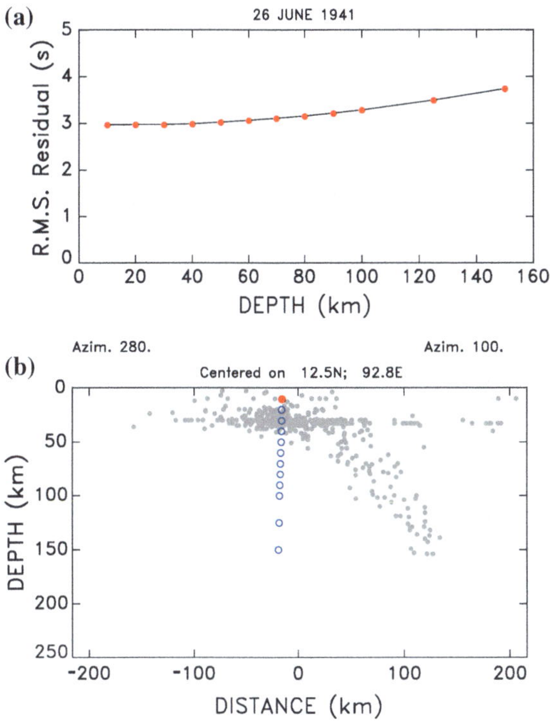

Figure 3
a Variation of r.m.s residual as a function of depth for constrained-depth relocations of the 1941 mainshock. Note lack of resolution above 40 km. **b** Cross section of epicentral area along the azimuth 280°–100°. The red dot is the preferred relocation, the open blue circles the hypocentral solutions obtained at constrained depths from 20 to 150 km. Background seismicity from NEIC catalog (1970–2016) in gray, suggesting a depth less than 40 km

depth phases because of the source complexity which will be documented below. At any rate, the lack of resolution of hypocentral depth prevents a definitive positioning of the source with respect to the plate interface.

It is remarkable that all proposed epicenters (and our confidence ellipse) are located on the accretionary shelf, as defined in Fig. 2 by the 200-m contour, and mostly on its submerged part, below sea level.

2.1. *Aftershocks*

Table 1 and Fig. 4 detail our relocation solutions for the mainshock and 16 aftershocks occurring within the remainder of 1941 from travel-time data listed by the ISS. We label them as "post-shocks" (Code P), as opposed to genuine aftershocks (Code A), events relocating outside the cluster of confidence ellipses and interpreted as triggered by stress transfer outside the zone of rupture of the mainshock; note in particular that Event P3 is the only one (with P15, about 200 km away) relocating outside the Andaman ridge. Only two genuine aftershocks (A11 and A13) were assigned magnitudes (in both instances, $M_{PAS} = 6.0$); they are shown in blue. They can be assumed to be the largest aftershocks; significantly, they seem to occur at the southern boundary of the fault area, thus supporting, in the case of this large event, the model of Ebel and Chambers (2016) obtained for smaller intraplate earthquakes.

Figure 4 suggests a fault area extending approximately 150 km × 80 km, which would correspond to a moment on the order of 1.6×10^{28} dyn cm assuming that the earthquake follows scaling laws (Geller 1976).

3. *Focal Mechanism*

A focal mechanism of the 1941 earthquake (hereafter "RV") was compiled by Ritsema and Veldkamp (1960) using polarities of *P* and *S* waves, based on questionnaires sent to worldwide stations; a modern description of its geometry features a very shallow angle thrust: ($\phi = 56°$, $\delta = 5°$, $\lambda = 90°$). However, the authors comment that the mechanism is poorly constrained, with "great possible variation" in the null axis and one of the planes. Wickens and Hodgson (1967, p. 43) later optimized this solution and proposed comparable mechanisms ("WH1", $\phi = 81°$, $\delta = 1°$, $\lambda = 117°$; "WH2", $\phi = 64°$, $\delta = 17°$, $\lambda = 100°$). For the purpose of tsunami simulations, Ramana Murthy et al. (2011) and Srivastava et al. (2012) proposed low-angle thrust mechanisms ("RM", $\phi = 20°$, $\delta = 20°$, $\lambda = 90°$; "SR", $\phi = 356°$, $\delta = 12°$, $\lambda = 90°$, respectively), but did not detail their sources.

We were able to read a data set of 15 *P*-wave first motions on copies of records at teleseismic stations, complemented by three polarities reported to the ISS for which records could not be inspected, with all pertinent information listed in Table 2. As shown in Fig. 5, this data set constrains the solution from this study ("TS") to a strike-slip mechanism ($\phi = 290°$,

Table 1

Events relocated in this study

Code of event	Date D M (J) Y	Origin Time (GMT)	Latitude (°N)	Longitude (°E)	Arrival times		r.m.s (s)	Magnitude M_{PAS}
					Read	Kept		
M1	**26 JUN (177) 1941**	**11:52:00.5**	**12.21**	**92.55**	**96**	**95**	**2.97**	**8.1**
A2	27 JUN (178) 1941	07:32:46.2	12.11	92.56	17	14	4.58	
P3	27 JUN (178) 1941	08:32:19.2	11.87	91.93	24	22	4.86	
A4	27 JUN (178) 1941	19:04:04.8	11.73	92.73	15	11	4.47	
A5	28 JUN (179) 1941	17:55:22.3	12.21	93.40	15	14	3.47	
A6	28 JUN (179) 1941	23:07:17.2	9.62	90.00	10	9	4.00	
A7	30 JUN (181) 1941	03:13:42.9	12.78	92.72	9	9	5.47	
A8	30 JUN (181) 1941	18:23:32.6	12.04	92.69	34	34	3.92	
A9	02 JUL (183) 1941	02:42:03.4	11.99	93.03	18	18	3.57	
A10	09 JUL (190) 1941	00:39:10.0	12.23	93.16	13	13	4.15	
A11	14 JUL (195) 1941	02:02:26.7	11.88	92.79	24	23	4.26	6.0
A12	21 JUL (202) 1941	20:19:30.0	12.47	92.63	13	11	2.73	
A13	09 AUG (221) 1941	22:17:37.7	12.06	92.79	37	35	5.21	6.0
A14	19 AUG (231) 1941	16:19:28.9	8.41	93.36	34	33	5.51	
P15	30 AUG (242) 1941	16:44:33.7	10.73	94.35	26	21	2.82	
A16	21 SEP (264) 1941	18:53:32.0	12.68	92.77	14	12	3.90	
A17	23 OCT (296) 1941	21:02:35.9	12.61	93.27	15	12	3.46	

$\delta = 80°$, $\lambda = 171°$), rotated 89° from Ritsema and Veldkamp's (1960) in the formalism of Kagan (1991) and similarly 92°, 87°, 99° and 105° from mechanisms WH1, WH2, RM and SR, respectively (out of a maximum of 120°). This very significant difference from previously published mechanisms must be considered robust, given the adequate distribution of stations on the focal sphere (see Fig. 5), whose overwhelming majority were read individually as part of this study.

In anticipation of the results of Sect. 4, we emphasize that the TS mechanism is derived from first-motion *P* waves and as such characterizes only the nucleation of the event, which will prove to be a complex earthquake; in particular, it will be found to be incompatible with long-period surface wave spectral amplitudes.

4. Estimation of Long-Period Seismic Moment and Low-Frequency Mechanism

Table 3 lists mantle waves for which we obtained scans of records, which we digitized and processed through the M_m mantle magnitude algorithm (Okal and Talandier 1989). Results are shown in Fig. 6 in terms of the magnitude M_c corrected for the focal mechanism TS, obtained from the first motions in Sect. 3. That focal geometry clearly misfits the distribution of spectral amplitude among stations, with, for example, Rayleigh and Love waves at COL requiring moments differing by ~2.5 orders of magnitude, as do Love waves at HON and TUC.

We are thus drawn to envision that the 1941 event must have featured a composite focal mechanism, with the first arrivals controlled by an initial strike-slip mechanism, followed by a main event with a different geometry controlling the spectral amplitude of mantle waves. We attempted to invert the mantle magnitude data set using the PDFM method (Reymond and Okal 2000), which we have successfully applied to a number of historical events (e.g., Okal 2005), but, unfortunately, we could not obtain a stable inversion. In this context, we resorted to a trial-and-error method to constrain the low-frequency mechanism of the 1941 earthquake.

In the introduction, we pointed out that significant moment release over the Andaman Islands during the 2004 event suggests that the 1941 earthquake was not a typical low-angle thrust event at the subduction interface and thus the main non-strike-slip component of its mechanism cannot be inferred simply by the plate tectonic context. We verified in particular that the focal geometry of the great 2004 earthquake does

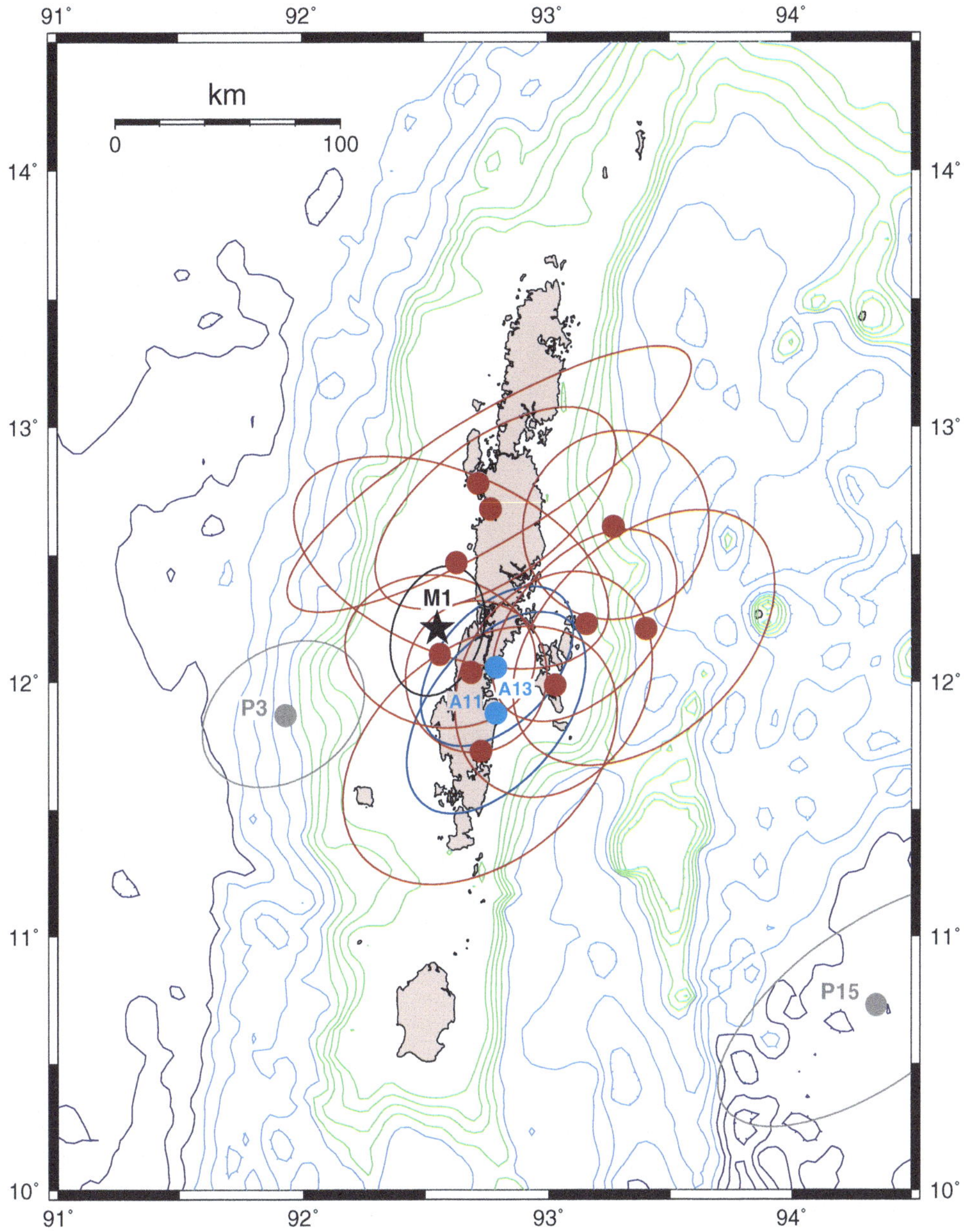

Figure 4
Map of relocations for the 17 events listed in Table 1. Each event is shown as its preferred epicenter, surrounded by its Monte Carlo ellipse. Genuine aftershocks are shown in dark red, the two large ones with a published magnitude in blue and the "postshocks" triggered by stress outside of the presumed rupture zone in gray; the mainshock is shown as the black star. See text for details

not fit our spectral amplitudes, with the Love waves at CTO and COL requiring moments varying by 2.5 orders of magnitude.

Faced with this condition, we explored the distribution of focal mechanisms in the background seismicity of the epicentral area. For this purpose, we extracted from the Global CMT catalog (Dziewonski et al. 1981; Ekström et al. 2012) all 64 shallow solutions ($h \leq 100$ km) in a box delimited by latitudes 11°N–14°N and longitudes 92°E–93.5°E; we

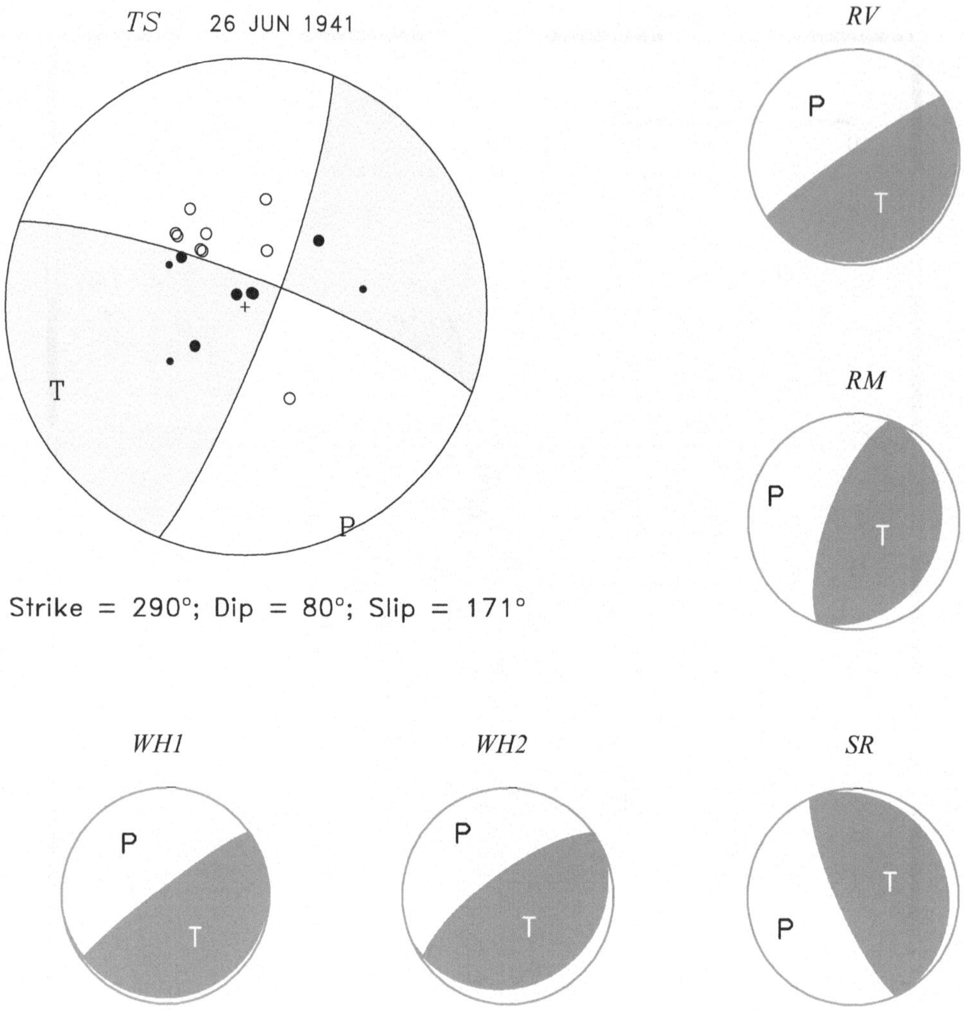

Figure 5
Top left (TS): focal mechanism obtained in this study from *P*-wave first motions. Solid dots represent upwards (anaseismic) first motions, open dots downwards (kataseismic) ones; smaller symbols relate to data retrieved unverified from the ISS. Bottom and right: mechanisms proposed in previous studies. See text for details

regard them as representative of stresses existing in the vicinity of the 1941 epicenter. We note that an overwhelming majority of these events (46 out of 64) took place in the aftermath of the great 2004 earthquake, presumably as the result of stress transfer. We further classify them as strike-slip (15 events), normal (13) or thrust (18) if their null, compressional or tensional axes, respectively, dip more than 50°; the remaining 18 events are considered hybrid. This classification is inspired by the formalism of ternary diagrams introduced by Frohlich and Apperson (1992). In each category, we further define an average mechanism as the best double-couple of the sum of the relevant moment tensors, unweighted by scalar moment, and explore the coherence of the group by analyzing the distribution of "Kagan" angles of solid rotation (Kagan 1991) between individual events and the average mechanism; we recall that the theoretical maximum value for a Kagan angle is 120°.

Results, listed in Table 4, indicate that normal faulting events (all but one post-Sumatra) are reasonably well grouped around their average mechanism, being rotated from it an average Kagan angle of 28° ± 10°. So are thrust events, but in their

Table 2

First-motion data set used in this study

Station Code	Location[a]	Distance (°)	Azimuth (°)	First motion[b]
Read in this study				
TAS	Tashkent, Uzbekistan	35.5	329	d
IRK	Irkutsk, Russia	41.1	11	d
BAK	Baku, Azerbaijan	46.9	315	d
MIZ	Mizusawa, Japan	49.0	252	u
PER	Perth, Western Australia	49.4	154	d
TIF	Tbilisi, Georgia	50.9	314	d
PUL	Pulkovo, Russia	65.7	331	d
ATU	Athens, Greece	65.9	306	u
GTT	Göttingen, Germany	75.9	320	d
DBN	De Bilt, The Netherlands	78.8	321	d
CTO	Cape Town, South Africa	84.0	234	u
COL	College, Alaska	90.9	22	d
PAS	Pasadena, California	125.3 (PKP)	31	u
TUC	Tucson, Arizona	130.3 (PKP)	26	u
SJG	San Juan, Puerto Rico	143.2 (PKP)	325	u
Transcribed from ISS				
MAN	Manila, Philippines	27.8	82	u
TAN	Antananarivo, Madagascar	54.1	236	u
HLW	Helwan, Egypt	59.1	298	u

[a]Locations and countries reflect present political boundaries

[b]u: anaseismic ("up"); d: kataseismic ("down")

case, the average mechanism differs significantly (by a 57° Kagan angle) from that of the 2004 mega-earthquake, supporting the idea that these earthquakes may not be located at the subduction interface; we note also that many (11 out of 18) occurred prior to the 2004 mega-earthquake. In both instances, the average mechanism gives a fair representation of the population of geometries of strain release in the epicentral area. By contrast, the strike-slip events (all of which postdate the 2004 mega-earthquake) are more scattered in geometry, with a mean Kagan angle of 40° with respect to their average mechanism; the latter is also 70° away from the mechanism (TS) inferred from the first motions in Sect. 3. A similar conclusion is reached for the hybrid group.

In this context, we select the two average mechanisms for normal and thrust faulting events (with moment tensors $\mathbf{M_N}$ and $\mathbf{M_T}$, respectively) and recompute corrected mantle magnitudes M_c for composite mechanisms obtained by adding the moment tensor of the first motion ($\mathbf{M_{TS}}$, as derived in Sect. 3), weighted by a variable coefficient α

$$\mathbf{M}_{composite} = \alpha\,\mathbf{M_{TS}} + (1 - \alpha)\,\mathbf{M_{N,T}}. \quad (1)$$

Finally, we optimize α by trial and error, so as to minimize the scatter in the resulting M_c values. As shown in Fig. 7, the best results are obtained with the normal faulting average solution and a parameter $\alpha = 0.20$; an acceptable but less satisfactory fit is also obtained with the thrusting mechanism ($\alpha = 0.21$).

We note in Fig. 7 that neither of the two solutions features a significant increase of moment with periods, the (negative) slope of the regression of M_c with frequency being 0.04 (normal faulting) and 0.05 (thrust) logarithmic units per mHz, well within the range of typical subduction zone events; we recall (Okal and Saloor 2017; Salaree and Okal 2018) that

Table 3

Mantle waves used in the moment determination

Station Code	Location	Distance (°)	Azimuth (°)	Intrument type	Phases used
DBN	De Bilt, The Netherlands	78.8	321	Golitsyn	R_1, G_1, G_3
CTO	Cape Town, South Africa	84.0	234	Milne Shaw	G_1
COL	College, Alaska	90.9	22	McRomberg	R_1, G_1
AUC	Auckland, New Zealand	91.2	127	Milne-Shaw	G_1
HON	Honolulu, Hawaii	103.1	64	Milne-Shaw	R_1, G_1
PAS	Pasadena, California	125.3	31	Benioff 1–90	G_3
TUC	Tucson, Arizona	130.3	26	Benioff 1–77	G_1
SJG	San Juan, Puerto Rico	143.2	325	Wenner	G_1

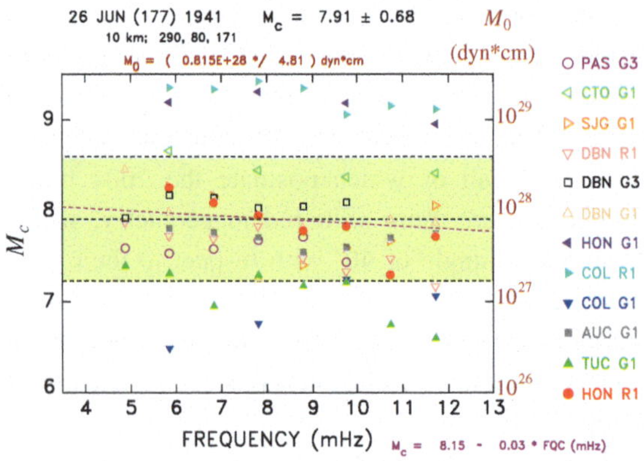

Figure 6

Mantle magnitudes M_c (Okal and Talandier 1989) corrected for the strike-slip focal mechanism obtained from first motions. The horizontal dashed line is the value averaged over all measurements and the colored band the 1–σ range. Note the widely scattered distribution of results, in particular the incompatibility between G_1 values at HON and TUC or COL

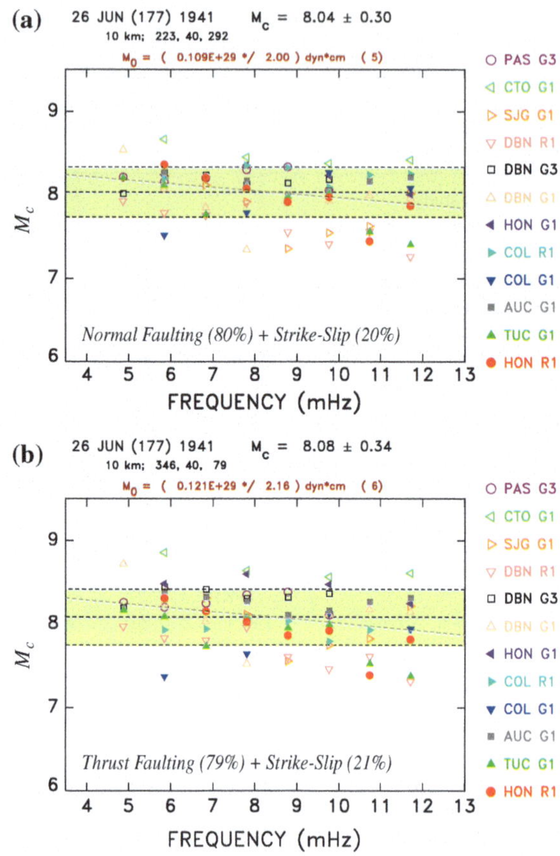

Figure 7

a Same as Fig. 6 for the preferred mechanism obtained by combining the average normal faulting mechanism ($\phi = 223°$; $\delta = 40°$; $\lambda = -68°$; weighted 80%) with the first-motion strike-slip mechanism, weighted 20%. Note the much reduced scatter. **b** Same as **a** for the average thrust faulting mechanism. See text for details

Table 4

Characteristics of background global CMT seismicity in the Andaman Islands, 1976–2018

Type	Number of events		Average mechanism			Kagan angle
	Total	Pre-2004	ϕ (°)	δ (°)	λ (°)	Average ± r.m.s. (°)
Normal	13	1	223	40	− 68	28 ± 10
Thrust	18	11	346	40	79	30 ± 10
Strike-slip	15	0	268	75	− 16	40 ± 18
Hybrid	18	5	176	84	125	46 ± 29

slow events, such as tsunami earthquakes, feature an absolute slope in excess of 0.07 logarithmic units per mHz.

In conclusion, our preferred solution is a composite focal mechanism ("NO") in which the strike-slip nucleation features the TS mechanism ($\phi = 290°$, $\delta = 80°$, $\lambda = 171°$) with a moment $M_0^{TS} = 2.2 \times 10^{27}$ dyn cm followed by a normal faulting event ($\phi = 214°$, $\delta = 54°$, $\lambda = -68°$) with a moment $M_0^{NF} = 8.7 \times 10^{27}$ dyn cm. An alternative solution ("TH") could feature a moment $M_0^{TS} = 2.5 \times 10^{27}$ dyn cm for the *TS* nucleation, followed by a thrust faulting event ($\phi = 346°$, $\delta = 40°$, $\lambda = 79°$) with a moment $M_0^{TF} = 9.6 \times 10^{27}$ dyn cm. Either of these combinations reconciles the data set of the first motions obtained in Sect. 3 with the spectral amplitudes of mantle waves at the eight stations listed in Table 3. Figure 8 illustrates the combination of mechanisms involved in the NO and TH mechanisms. We emphasize, however, that the final "beachballs" for the composite mechanisms do not represent the distribution of first motions on the focal sphere, since the main non-strike-slip component of the source is expected to be delayed.

While the data set of spectral amplitudes of mantle waves cannot discriminate between the mainly normal ("NO") and thrust ("TH") mechanisms, we turn our attention to the static offsets described by Jhingran (1953), who reported

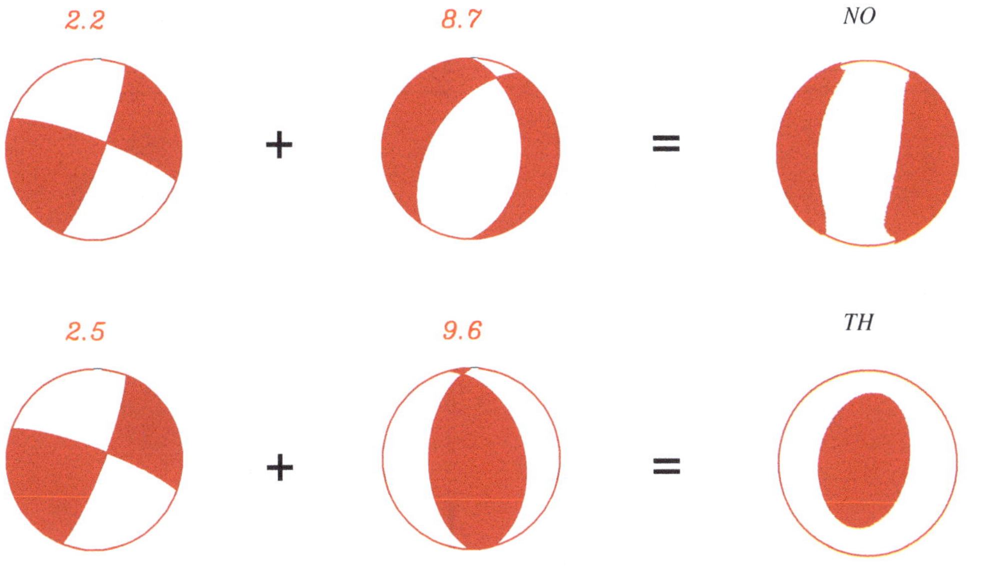

Figure 8
Top: illustration of the combination of the nucleating strike-slip mechanism (left) and of the average background normal faulting double-couple (center) resulting in the preferred "NO" composite mechanism (right). Bottom: same for the "TH" mechanism. The small numbers above the beachballs represent the seismic moments (in units of 10^{27} dyn cm) used in the various combinations. See text for details

subsidence of $\sim$70 cm ("$2 - 2\frac{1}{2}$ ft.") at Port Blair and farther north along the eastern coast of the main island group at a number of locations that we have identified as lying between latitudes 12.10°N and 12.40°N. By contrast, and despite a report of more extensive damage, Jhingran (1953) mentions no vertical motion on the much less populated west coast.

In Fig. 9, we compute static displacements due to the "TH" and "NO" mechanisms using the formalism of Mansinha and Smylie (1971). The computations are performed by superimposing the fields produced by the two components to the source, as described by Eq. (1). An additional complexity stems from a possible lateral offset between the strike-slip precursor and the mainshock; in this context, Fig. 9 represents the results of a trial-and-error search optimizing fits to both the above reports of static motion and the tsunami simulations in Sect. 5 for which they serve as initial conditions.

A robust property of the mechanisms tested is that, regardless of the lateral offset between the two components, a "TH" mechanism produces a strong uplift flanked by areas of marginal subsidence (Fig. 9b), while an "NO" mechanism results in a strong subsidence flanked by minor uplift. In this context, the reports of systematic subsidence extending from Port Blair to the north clearly favor the "NO" geometry and cannot be reconciled with a "TH" mechanism. The preferred mechanism shown in Fig. 9a features a normal component offset 40 km in the azimuth N135°E from the nucleation epicenter. It does replicate the subsidence reports and concentrates most of the vertical deformation on the Andaman ridge (loosely defined as the 200-m isobath, in red in Fig. 9, which will play an important role in the inefficient generation of the tsunami (see Sect. 5 below).

5. *Tsunami Simulations*

In this section, we use the composite mechanisms derived in Sect. 4 to simulate the generation of a tsunami in the Bay of Bengal and the Andaman Sea, and its propagation in the Indian Ocean, seeking in particular to explain the absence of confirmed reports of significant inundation during the 1941 earthquake.

Our simulations use the MOST algorithm (Titov and Synolakis 1998; Titov et al. 2016), which solves

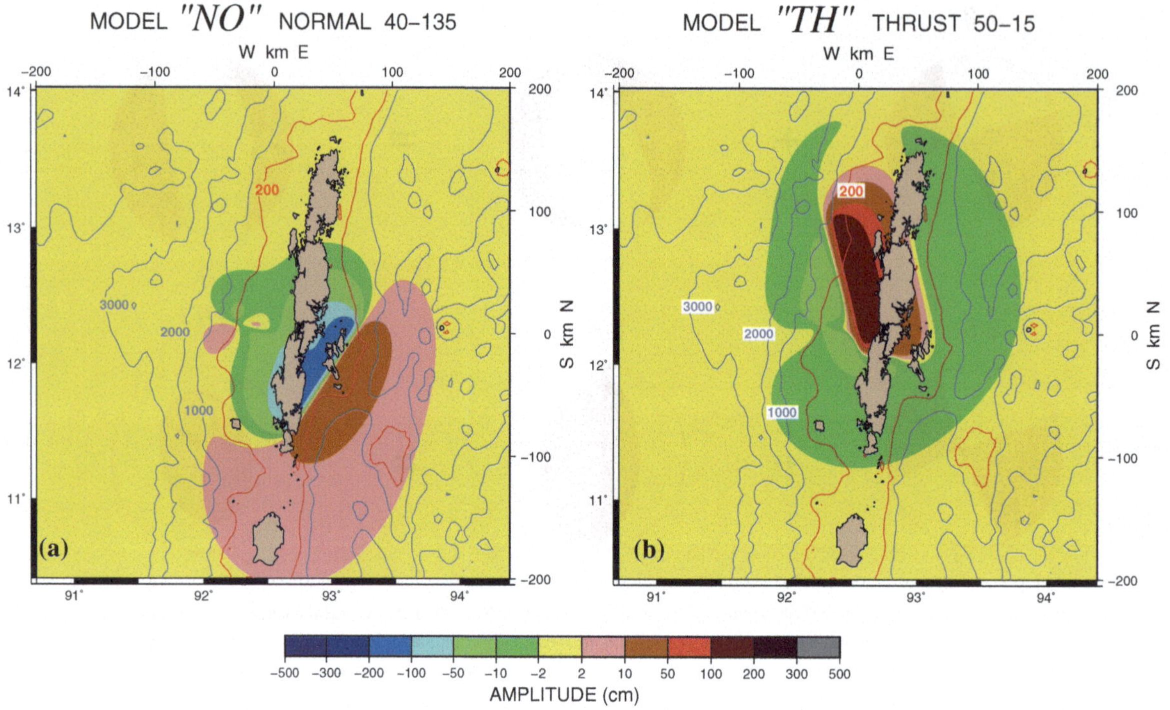

Figure 9

a Field of static displacements $\eta\,(x,\,y;\,0_+)$ for the composite mechanism "NO", featuring a normal faulting mainshock offset 40 km in the azimuth N135°E from the strike-slip nucleator, superimposed on morphologic and bathymetric data. The red contour outlines the 200-m isobath, the dark blue ones the 1000-m, 2000-m and 3000-m ones. Note that most of the deformation occurs within the range of the 200-m isobath. **b** Same as **a** for the thrust model "TH". See text for details

the full non-linear equations of hydrodynamics under the shallow-water approximation, by finite differences and through the method of alternate steps (Godunov 1959). MOST has been extensively validated through comparisons with laboratory and field data, per standard international protocols (Synolakis et al. 2008); full details can be found in Synolakis (2003).

As initial conditions for the vertical displacements of the sea surface, $\eta\,(t = 0_+)$, we use the field of static deformations resulting from the seismic dislocation, computed through the algorithm of Mansinha and Smylie (1971) in the geometry of a homogeneous half-space. This approximation is appropriate as the rise time of an earthquake is always much shorter than the characteristic time needed by a tsunami wave to flush the displaced water out of the generation area (e.g., Saito and Furumura 2009; Dalrymple and Derakhti 2018). For the composite mechanisms used in the present study, we superimpose the fields of deformation of the two individual subevents. In the absence of information on any possible spatial offset between the two subevents, we start by simply locating both at our preferred epicenter (12.21°N; 92.55°E). The simulation uses a 5-arcmin grid and a 10-s time step, satisfying the CFL stability condition (Courant et al. 1928); it is carried on for 5000 s and is stopped at the 20-m isobath in the vicinity of coastlines.

Figure 10 shows the results of our simulation for the normal faulting composite mechanism in the form of the field of maximum amplitudes $\eta_{\max}$ of vertical water displacement during the simulated 5000-s window. The tsunami remains very small ($\eta_{\max} < 3$ cm) over most of the Bay of Bengal and does not leak into the main Indian Ocean Basin. It remains smaller than 10 cm at the 20-m isobath where the computation is stopped, off most of the coastlines across the Bay of Bengal. In the Andaman Islands, it predictably affects mostly the western

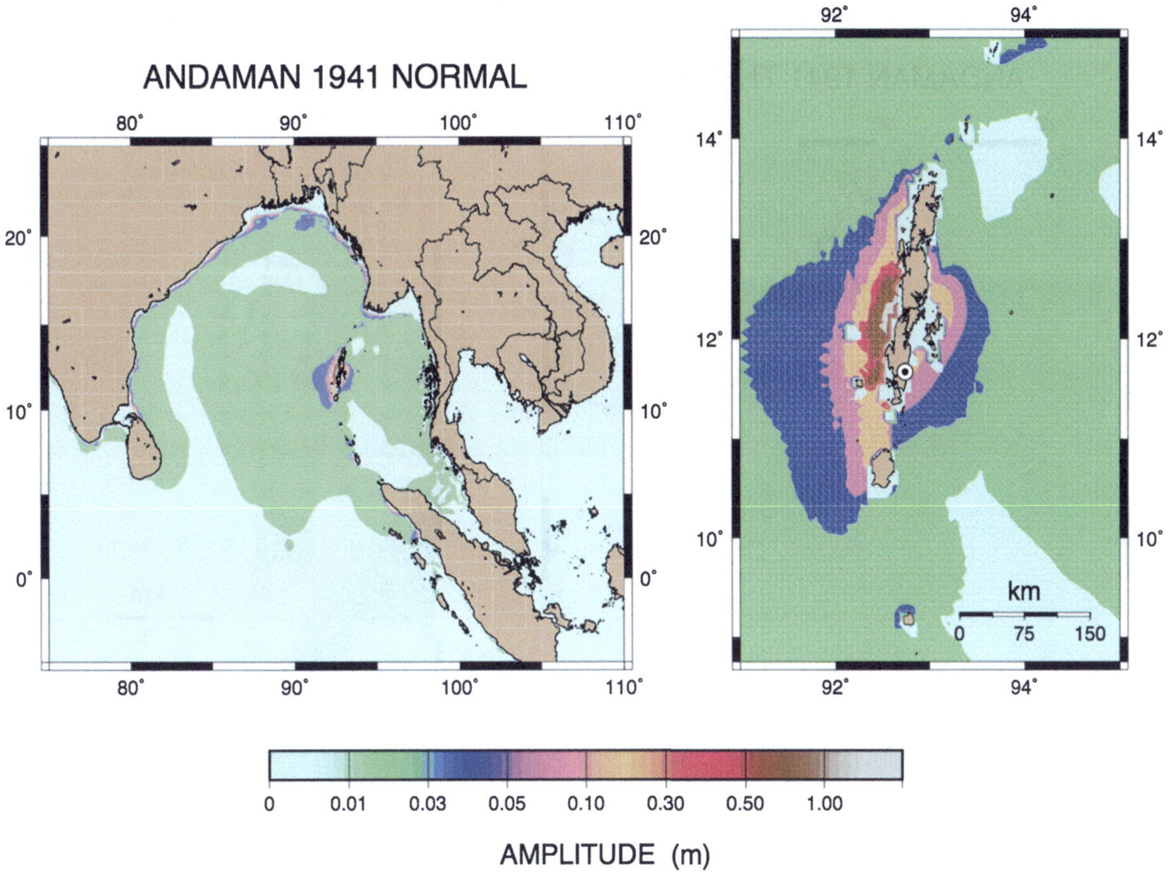

Figure 10

Maps of the field of maximum amplitudes η_{max} for the composite normal faulting mechanism. Left: Bay of Bengal and adjoining areas. Right: close-up around the Andaman Islands. The computation is stopped at the 20-m isobath. The bull's eye symbol shows the location of the capital and main population center, Port Blair

coasts, which in 1941 were far less populated that the eastern ones, and remain to this day. The tsunami amplitude on the high seas off the main city of Port Blair remains centimetric.

Similarly, Fig. 11 presents the field η_{max} for the thrust composite mechanism. Amplitudes along the shorelines of the Bay of Bengal reach observable levels (~ 15 cm at the 20-m isobath) along some eastern shores of India. In the Andaman Islands, the maxima are generally smaller on the eastern coasts and larger on the western ones.

In Fig. 12, we consider the preferred composite source "NO" described in Sect. 4, which features a spatial offset of 40 km in the N135°E azimuth between the nucleating strike-slip mechanism and the main normal faulting component. Under this scenario, the amplitudes at the 20-m isobath along the Indian and Western Burmese coasts are reduced to about 2 cm, while they are slightly increased, to about 5 cm, along the eastern shores of the Andaman Sea, in Eastern Burma and Western Thailand. In the Andaman Islands, the distribution of η_{max} is also altered, with larger amplitudes on the eastern shores and reduced ones in the west. On the eastern shores, where the population (and hence the potential for observation) was highest, e.g., in Port Blair, the 20-m amplitudes reach about 30 cm. In Fig. 13, we move the thrust fault component of the "TH" mechanism 50 km in the N15°E direction. Again, the general pattern of η_{max} is retained, with this time smaller amplitudes on the shores of the Bay of Bengal and

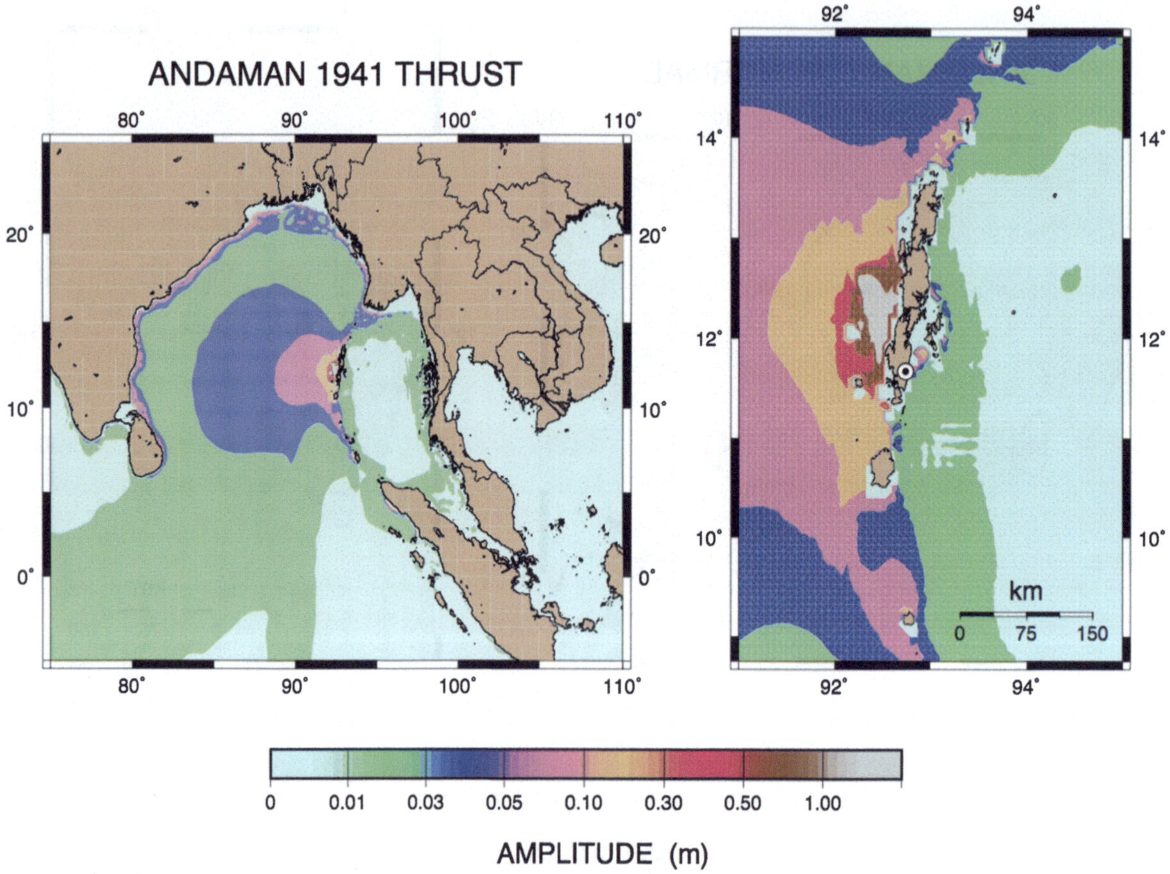

Figure 11
Same as Fig. 10 for the composite thrust mechanism

especially on the eastern coasts of the Andaman Islands.

In very general terms, all our simulations predict rather deceptive amplitudes for an earthquake with a total moment of 10^{28} dyn cm. In physical terms, we can identify several reasons contributing to a reduction of $\eta_{\max}$, in both the near field (on the islands) and far field, across the Bay of Bengal. The mere presence of the Andaman Islands significantly reduces the area of ocean floor where water is displaced coseismically; in simple terms, the earthquake moves more rock and less water. This situation is strikingly reminiscent of the geometry of the 2005 Nias earthquake, whose domain of rupture was significantly overlain by emerged land, namely the Nias and Simeulue Islands. Because of its much larger size, the 2005 earthquake did generate a significant tsunami; however, it did not reach the catastrophic proportions that its moment [1.4×10^{29} dyn cm (Okal and Stein 2009)] would have suggested, especially in the near field where considerable damage was wrought by the earthquake (Boen 2006).

In addition, the remainder of the source area of the 1941 earthquake is located under very shallow water, most of it < 200 m deep (Fig. 9). As a result, the tsunami falters according to Green's (1837) law, when it transitions to the deeper waters of the Bay of Bengal. This situation is again reminiscent of the 2005 Nias event, which produced a negligible tsunami in the far field, despite its large seismic moment (Okal and Synolakis 2008). Finally, the composite nature of the focal mechanism, involving a strike-slip component of about 1/5 of the total moment, reduces the vertical deformation in the epicentral area, and hence the amplitude of the tsunami, especially in the near field.

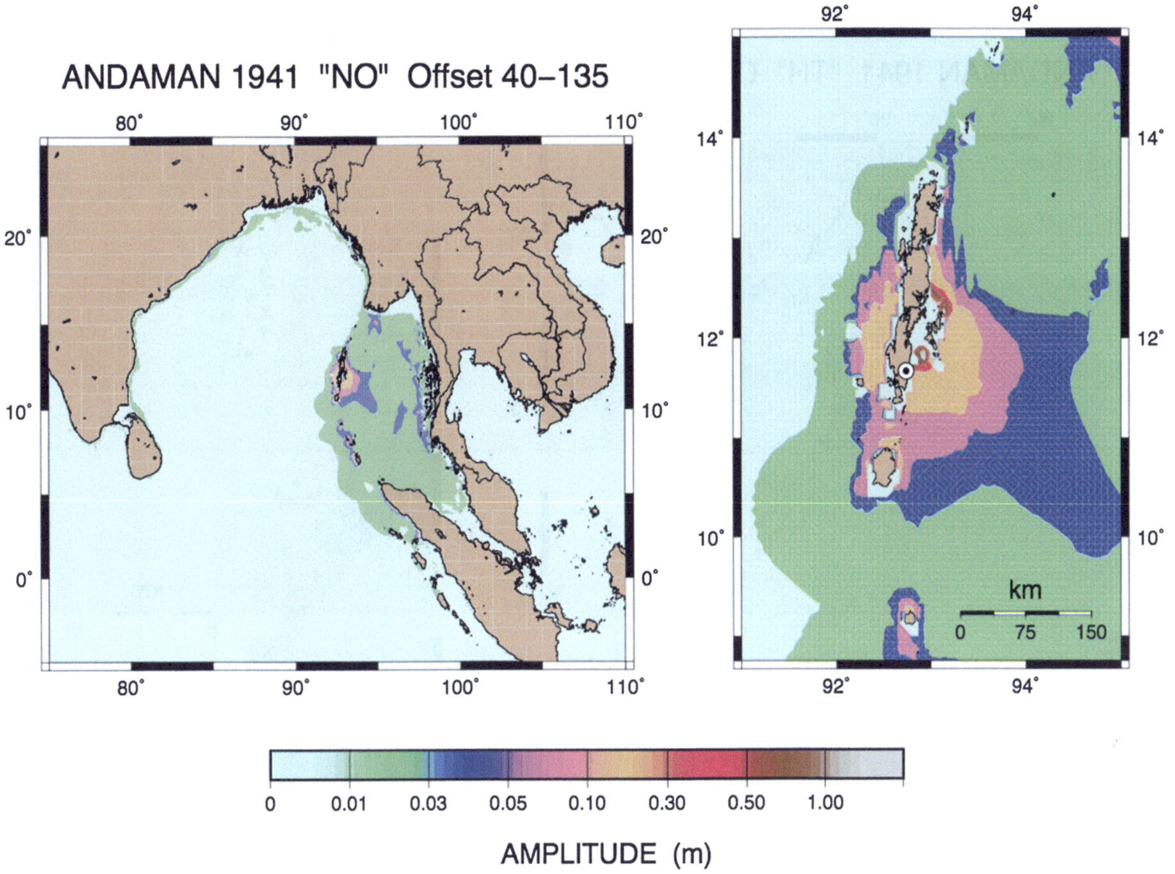

Figure 12
Same as Fig. 10 for a composite normal mechanism, where the normal component is moved 40 km in the N135°E direction

From the observational standpoint, and for all the models envisioned, the tsunami amplitudes at the 20-m depth where the computations are terminated are larger on the western shores of the Andaman Islands, away from the populated centers including the main city of Port Blair, which are located on the eastern shores.

6. Conclusion

We have revisited the 1941 Andaman earthquake, the largest in the 20th century in that part of the India-Burma plate boundary. We conclude that it cannot be interpreted as an interplate thrust event expressing the subduction of the Indian plate. The distribution of its teleseismic first motions requires nucleation as a strike-slip mechanism, which is however incompatible with the spectral amplitudes of Rayleigh and Love waves at mantle periods, thus requiring a composite mechanism. We prefer a normal faulting geometry for the main part of the source, as it can fit the coseismic subsidence reported on the eastern coast of the islands. That model predicts small tsunami amplitudes along the coasts of the Bay of Bengal, supporting the results of Ortíz and Bilham (2003), who failed to document any scientific or press reports of a substantial tsunami in India; it is also generally compatible with Jhingran's (1953) qualitative statement about "extensive waves…around the Andaman Islands."

We emphasize again that our models are by no means unique, and we do not pretend that our preferred one ("NO"; Figs. 9a, 12) is necessarily representative of the true mechanism of the event. Our purpose is merely to show that it is possible to

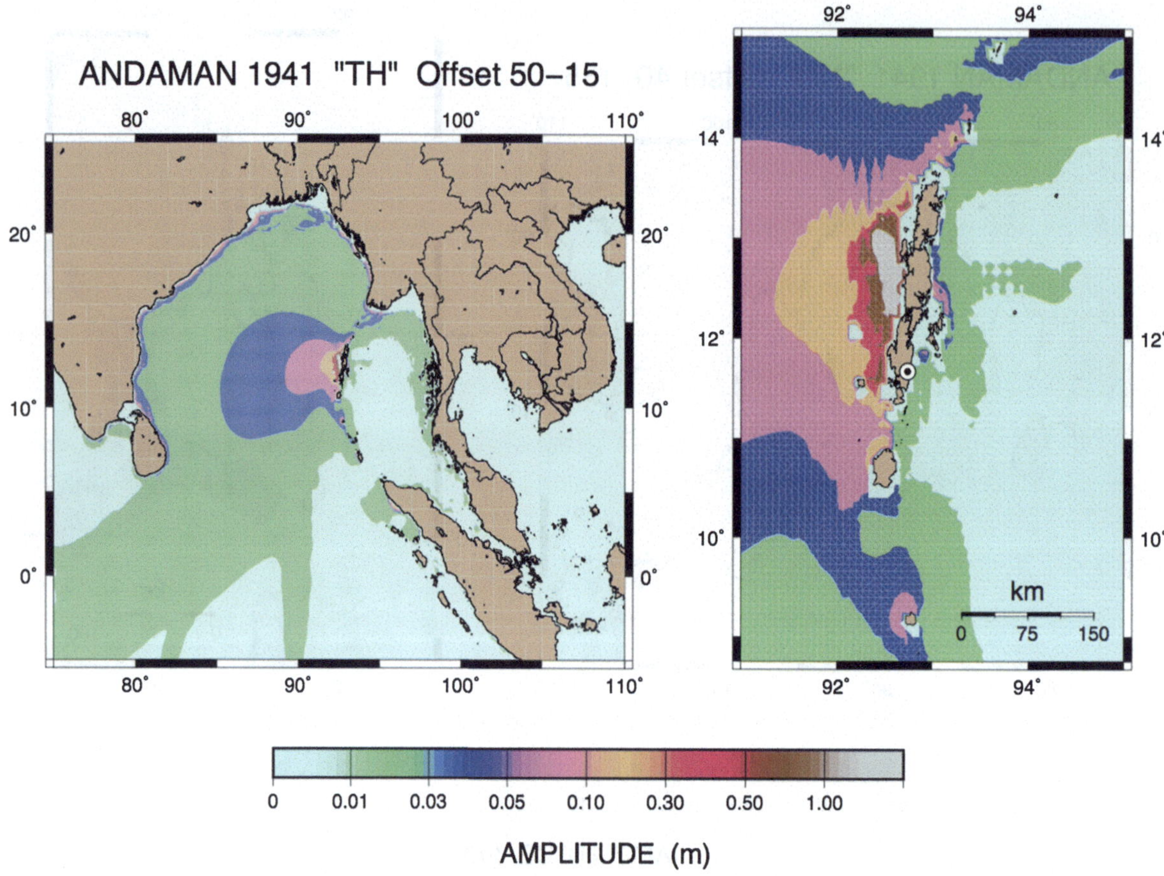

Figure 13
Same as Fig. 11 for a composite thrust mechanism, where the thrust component is moved 50 km in the N15°E direction

build one (or several) models that can reconcile the seismological data (first motions of body waves and spectral amplitudes of surface waves at mantle periods), Jhingran's (1953) reports of subsidence on the eastern shore of the Andaman Islands and the mediocre amplitudes of any putative tsunami.

We stress that the existence of major earthquakes deviating from an interplate thrust geometry is a common occurrence at subduction zones, especially those featuring oblique subduction. A further example would be the West Aleutian subduction zone, where mechanisms of all kinds (thrust, normal, strike-slip, hybrid) are well documented in modern-day seismicity.

Finally, this study reaffirms, if need be, the value of historical records in sharpening our understanding of tectonic processes at major plate boundaries, where the recurrence times of significant earthquakes can be much longer than the present extent of modern digital networks, even when combined with their analog predecessor, the World-Wide Standardized Seismic Network (Okal 2015).

Acknowledgements

We are grateful to Ian Saunders (Pretoria), Bernard Dost (De Bilt) and Brian Ferris (Lower Hutt) for access to the excellent seismological archives at their respective facilities. Some figures were drawn using the GMT software (Wessel and Smith 1991). The paper benefited from the constructive comments of two anonymous reviewers.

Publisher's Note Springer Nature remains neutral with regard to jurisdictional claims in published maps and institutional affiliations.

References

Abe, K. (1981). Magnitudes of large shallow earthquakes from 1904 to 1980. *Physics of the Earth and Planetary Interiors, 27*, 72–92.

Bilham, R., Engdahl, E. R., Feldl, N., & Satyabala, S. P. (2005). Partial and complete rupture of the Indo-Andaman plate boundary, 1847–2004. *Seismological Research Letters, 76*, 299–311.

Bird, P. (2003). An updated digital model of plate boundaries. *Geochemistry, Geophysics, Geosystems, 4*(3), GC000252.

Boen, T. (2006). Sructural damage in the March 2005 Nias-Simeulue earthquake. *Earthquake Spectra, 22*, S419–S434.

Brune, J. N. (1968). Seismic moment, seismicity, and rate of slip along major fault zones. *Journal of Geophysical Research, 73*, 777–784.

Brune, J. N., & Engen, G. R. (1969). Excitation of mantle Love waves and definition of mantle wave magnitude. *Bulletin of the Seismological Society of America, 59*, 923–933.

Chlieh, M., Avouac, J.-P., Hjörleifsdóttir, V., Song, T.-R. A., Ji, C., Sieh, K., et al. (2007). Coseismic slip and afterslip of the great $M_w = 9.15$ Sumatra-Andaman earthquake of 2004. *Bulletin of the Seismological Society of America, 97*, S152–S173.

Cochran, J. R. (2010). Morphology and tectonics of the Andaman Forearc, Northeastern Indian Ocean. *Geophysical Journal International, 182*, 631–651.

Courant, R., Friedrichs, K., & Lewy, H. (1928). Über die partiellen Differenzengleichungen der mathematischen Physik. *Mathematische Annalen, 100*, 32–74.

Curray, J. R. (2005). Tectonics and history of the Andaman Sea region. *Journal of Asian Earth Sciences, 25*, 187–232.

Dalrymple, R.A., & Derakhti, M. (2018). Tsunami generation by earthquakes: Seabed topography and inertial effects. In: *Proceedings 36th international conference on coastal engineering*, 1605, Baltimore [abstract].

Diehl, T., Waldhauser, F., Cochran, J. R., Kamesh Raju, K. A., Seeber, L., Schaff, D., et al. (2013). Back-arc extension in the Andaman Sea: Tectonic and magmatic processes imaged by high-precision teleseismic double-difference earthquake relocation. *Journal of Geophysical Research: Solid Earth, 118*, 2206–2224.

Dziewonski, A. M., Chou, T. A., & Woodhouse, J. H. (1981). Determination of earthquake source parameters from waveform data for studies of global and regional seismicity. *Journal of Geophysical Research, 86*, 2825–2852.

Ebel, J. E., & Chambers, D. W. (2016). Using the locations of $M \geq 4$ earthquakes to delineate the extents of the ruptures of past major earthquakes. *Geophysical Journal International, 207*, 862–875.

Ekström, G., Nettles, M., & Dziewoński, A. M. (2012). The global CMT project, 2004–2010: Centroid moment tensors for 13,107 earthquakes. *Physics of the Earth and Planetary Interiors, 200*, 1–9.

Engdahl, E. R., & Villaseñor, A. (2002). Global seismicity: 1900–1999. In *International earthquake and engineering seismology Part A* (pp. 665–690). New York: Elsevier.

Fitch, T. J. (1972). Plate convergence, transcurrent faults, and internal deformation adjacent to Southeast Asia and Western Pacific. *Journal of Geophysical Research, 77*, 4442–4460.

Frohlich, C., & Apperson, K. D. (1992). Earthquake focal mechanisms, moment tensors, and the consistency of seismic activity near plate boundaries. *Tectonics, 11*, 279–296.

Geller, R. J. (1976). Scaling relations for earthquake source parameters and magnitudes. *Bulletin of the Seismological Society of America, 66*, 1501–1523.

Godunov, S. K. (1959). Finite difference methods for numerical computations of discontinuous solutions of the equations of fluid dynamics. *Matematicheskiĭ Sbornik, 47*, 271–295.

Goodstein, J. R., Kanamori, H., & Lee, W. H. K. (1980). Seismology microfiche publications from the Caltech archives. *Bulletin of the Seismological Society of America, 70*, 657–658.

Green, G. (1837). On the motion of waves in a canal of variable depth. *Cambridge Philosophical Transactions, 6*, 457–462.

Gutenberg, B., & Richter, C. F. (1954). *Seismicity of the Earth and associated phenomena*. Princeton: Princeton University Press.

Ishii, M., Shearer, P. M., Houston, H., & Vidale, J. E. (2005). Extent, duration and speed of the 2004 Sumatra–Andaman earthquake imaged by the Hi-Net array. *Nature, 435*, 933–936.

Jhingran, A. G. (1953). A note on the earthquake in the Andaman Islands (26 June 1941). *Records of the Geological Survey of India, 82*, 300–307.

Kagan, Y. Y. (1991). 3-D rotation of double-couple earthquake sources. *Geophysical Journal International, 106*, 709–716.

Kanamori, H. (1977). The energy release in great earthquakes. *Journal of Geophysical Research, 82*, 2981–2987.

Lorito, S., Piatanesi, A., Cannelli, V., Romano, F., & Melini, D. (2010). Kinematics and source zone properties of the 2004 Sumatra-Andaman earthquake and tsunami: Nonlinear joint inversion of tide gauge, satellite altimetry, and GPS data. *Journal of Geophysical Research, 115*(B2), B02304.

Mansinha, L., & Smylie, D. E. (1971). The displacement fields of inclined faults. *Bulletin of the Seismological Society of America, 61*, 1433–1440.

Murty, T.S. (1984). Storm surges: Meteorological ocean tides, Can. Bull. Fisheries Aquat. Sci., 212, Dept. Fisheries Oceans, Ottawa.

Murty, T. S., & Rafiq, M. (1991). A tentative list of tsunamis in the Indian Ocean. *Natural Hazards, 4*, 81–83.

Okal, E. A. (2005). A re-evaluation of the great Aleutian and Chilean earthquakes of 1906 August 17. *Geophysical Journal International, 161*, 268–282.

Okal, E. A. (2015). Historical seismograms: Preserving an endangered species. *GeoResJ, 6*, 53–64.

Okal, E. A., Kirby, S. H., & Kalligeris, N. (2016). The Showa Sanriku earthquake of 1933 March 2: A global seismological reassessment. *Geophysical Journal International, 206*, 1492–1514.

Okal, E. A., & Saloor, N. (2017). Historical tsunami earthquakes in the Southwest Pacific: An extension to $\Delta > 80°$ of the Energy-to-Moment parameter Θ. *Geophysical Journal International, 210*, 852–873.

Okal, E. A., & Stein, S. (2009). Observations of ultra-long period normal modes from the 2004 Sumatra-Andaman earthquake. *Physics of the Earth and Planetary Interiors, 175*, 53–62.

Okal, E. A., & Synolakis, C. E. (2008). Far-field tsunami hazard from mega-thrust earthquakes in the Indian Ocean. *Geophysical Journal International, 172*, 995–1015.

Okal, E. A., & Talandier, J. (1989). M_m: A variable period mantle magnitude. *Journal of Geophysical Research, 94*, 4169–4193.

Ortíz, M., & Bilham, R. (2003). Source area and rupture parameters of the 31 December 1881 $M_w = 7.9$ Car Nicobar earthquake estimated from tsunamis recorded in the Bay of Bengal. *Journal of Geophysical Research, 108*(B4), 2215.

Rajendran, C. P. (2013). Was the 1941 Andaman earthquake tsunamigenic? Comments on "Inundation studies for Nagapattinam region on the East coast of India due to tsunamigenic earthquakes from the Andaman region" by Srivastava et al. *Natural Hazards, 65*, 981–984.

Ramana Murthy, M. V., Usha, T., Pari, Y., & Reddy, N. T. (2011). Tsunami vulnerability assessment of Cuddalore using numerical model and GIS. *Marine Geodesy, 34*, 16–28.

Rees, B. A., & Okal, E. A. (1987). The depth of the deepest historical earthquakes. *Pure and Applied Geophysics, 125*, 699–715.

Reymond, D., & Okal, E. A. (2000). Preliminary determination of focal mechanisms from the inversion of spectral amplitudes of mantle waves. *Physics of the Earth and Planetary Interiors, 121*, 249–271.

Richter, C. F. (1958). *Elementary seismology*. San Francisco: Freeman.

Ritsema, A. R., & Veldkamp, J. (1960). Fault plane mechanisms of Southeast Asian earthquakes. *Publ. Koninkl. Nederl. Meteorolog. Inst., 72*, 63. (De Bilt).

Saito, T., & Furumura, T. (2009). Three-dimensional tsunami generation simulation due to sea-bottom deformation and its interpretation based on the linear theory. *Geophysical Journal International, 178*, 877–888.

Salaree, A., & Okal, E. A. (2018). The "tsunami earthquake" of 13 April 1923 in Northern Kamchatka: Seismological and hydrodynamic investigations. *Pure and Applied Geophysics, 175*, 1257–1285.

Sella, G. F., Dixon, T. H., & Mao, A. (2002). REVEL: A model for Recent plate velocities from space geodesy. *Journal of Geophysical Research, 107*(B4), ETG-11.

Smyshlyaev, A. A. (2003). *Vremya krasnoĭ ryby*, pp. 310–315. Petropavlovsk-Kamchatskiĭ: Novaya Kniga [in Russian].

Srivastava, K., Krishna Kumar, R., Swapna, M., Swaroopa Rani, V., & Dimri, V. P. (2012). Inundation studies for Nagapattinam region on the East coast of India due to tsunamigenic earthquakes from the Andaman region. *Natural Hazards, 63*, 211–221.

Storchak, D. A., Di Giacomo, D., Engdahl, E. R., Harris, J., Bondár, I., Lee, W. H. K., et al. (2015). The ISC-GEM global instrumental earthquake catalogue (1900–2009): Introduction. *Physics of the Earth and Planetary Interiors, 239*, 48–63.

Synolakis, C. E. (2003). Tsunami and seiche. In W.-F. Chen & C. Scawthron (Eds.), *Earthquake engineering handbook* (pp. 9_1–9_90). Boca Raton: CRC Press.

Synolakis, C., Bernard, E., Titov, V., Kânoğlu, U., & González, F. (2008). Validation and verification of tsunami numerical models. *Pure and Applied Geophysics, 165*, 2197–2228.

Titov, V., Kânoğlu, U., & Synolakis, C. (2016). Development of MOST for real-time tsunami forecasting. *Journal of Waterway Port Coastal and Ocean Engineering, 142*(6), 03116004.

Titov, V. V., & Synolakis, C. E. (1998). Numerical modeling of tidal wave runup. *Journal of Waterway Port Coastal and Ocean Engineering, 124*, 157–171.

Tsai, V. C., Nettles, M., Ekström, G., & Dziewonski, A. M. (2005). Multiple CMT source analysis of the 2004 Sumatra earthquake. *Geophysical Research Letters, 32*(17), L17304.

Wessel, P., & Smith, W. H. F. (1991). Free software helps map and display data. *Eos Transactions American Geophysical Union, 72*, 441 and 445–446.

Wickens, A.J., & Hodgson, J.H. (1967). *Computer re-evaluation of earthquake mechanism solutions, 1922–1962*. Publications of the Dominion Observatory, Ottawa (vol. 33, no. 1). Ottawa: Dept. Energy, Mines, Res.

Wysession, M. E., Okal, E. A., & Miller, K. L. (1991). Intraplate seismicity of the Pacific Basin, 1913–1988. *Pure and Applied Geophysics, 135*, 261–359.

(Received October 16, 2018, revised December 18, 2018, accepted December 19, 2018, Published online January 17, 2019)

Pure Appl. Geophys. 176 (2019), 2887–2924

https://doi.org/10.1007/s00024-019-02133-3

Pure and Applied Geophysics

Five Great Tsunamis of the 20th Century as Recorded on the Coast of British Columbia

Alexander B. Rabinovich,[1,2] Richard E. Thomson,[1] Maxim V. Krassovski,[1] Fred E. Stephenson,[1] and Denny C. Sinnott[1]

Abstract—The five great trans-Pacific tsunamis of the 20th century that occurred in 1946, 1952, 1957, 1960 and 1964 were accurately recorded by analogue tide gauges on the coast of British Columbia. All available pen-and-paper records of these events were collected, digitised, de-tided and analysed. The 1946 Aleutian Islands event was recorded at two stations, Tofino and Victoria, where maximum trough-to-crest tsunami wave heights were 55 and 27 cm, respectively. These two gauged stations, as well as Prince Rupert, Alert Bay and Kitimat, also recorded the 1952 Kamchatka tsunami, which generated a maximum wave height of 77 cm at Tofino. The 1957 Andreanof Islands tsunami was recorded at six primary stations and the 1960 Chile tsunami by 17 primary and temporary tide gauges. For both of these events, the maximum tsunami wave heights also occurred at Tofino: 48 cm (1957) and 132 cm (1960). The 1964 Alaska tsunami remains the strongest tsunami yet instrumentally recorded on the coast of British Columbia. Our examination of 16 records from this event shows that maximum wave heights at eight stations were higher than 1 m, including Port Alberni (770 cm), Ocean Falls (376 cm), Tofino (237 cm) and Alert Bay (222 cm). We also find that the maximum wave at all stations for this event was among the first three waves. Subsequent wave heights rapidly attenuated following this group of waves. Frequency-time (*f–t*) analyses of the tsunami waveforms reveal that, for each station, the dominant frequencies of the waves and their evolution with time were very similar for different tsunamis, but differed considerably among sites for a particular tsunami, indicating the strong influence of local/regional topography on the incoming waves. From the latter point of view, the 1964 tsunami was exceptional. The epicentre of the 1964 Alaska earthquake was located much closer to the BC coast than for the other events and, therefore, the influence of the source was much stronger. The "ringing" of this tsunami was substantially shorter (< 1.5 days), the energy decay much faster, and the *f–t* diagrams at all sites more similar than for the other major events. During the 1964 event, energy associated with the dominant period of 2 h rapidly decayed. All of the tsunamis examined penetrated deep inside the narrow channels, fjords and inlets, typical of the BC coast. Waves also propagated far into some of the rivers. The 1952 Kamchatka tsunami was recorded at Kitimat, located at the head of Kitimat Arm about 80 km from Hecate Strait, while the 1957 Andreanof Islands tsunami was measured at Bella Coola, located at the remotest part of Burke Channel, a distance of ~110 km from the open ocean (Queen Charlotte Sound). The 1960 Chile tsunami was recorded at five stations located in the complex network of inlets and channels that make up the Seymour–Belize Inlet system on the central mainland coast of British Columbia, while the 1964 Alaska tsunami was observed at many stations in the Fraser River, including Pitt Lake that is connected to the Strait of Georgia through a 64-km route upstream in the Fraser and Pitt rivers. Our results further show that occasionally the fjords and inlets of the BC coast not only do not hinder the incoming waves but can strongly enhance them through resonant amplification. This was the case for the 1964 tsunami waves in Alberni Inlet, where the resonant response to the incoming waves resulted in severe damage to Port Alberni at the head of the inlet. The results of our analyses make it possible to isolate the tsunami signal and evaluate the principal parameters of tsunami waves. This, in turn, is of considerable value for three major problems: (1) Cataloguing of tsunami events; (2) estimation of the potential tsunami risk to the BC coast; and (3) verification and calibration of numerical tsunami models.

Key words: 1946 Aleutian tsunami, 1952 Kamchatka tsunami, 1957 Andreanof Islands tsunami, 1960 Great Chile tsunami, 1964 Alaska earthquake and tsunami, tide gauge tsunami records, wavelet analysis.

1. Introduction

The highly destructive and deadly tsunamis originating off Sumatra (2004), Chile (2010) and Japan (2011) once again emphasise the serious threat that major seismically generated tsunamis pose to coastal communities within the world oceans. The 2004 Sumatra tsunami killed about 230,000 people in 14 countries bordering the Indian Ocean and is likely the deadliest event of its kind in recorded history. The 2011 Tohoku (Great East Japan) tsunami killed almost 20,000 people, destroyed the Fukushima Daiichi nuclear power plant and caused considerable destruction along the

[1] Fisheries and Oceans Canada, Institute of Ocean Sciences, 9860 West Saanich Road, Sidney, BC V8L 4B2, Canada. E-mail: A.B.Rabinovich@gmail.com

[2] Russian Academy of Sciences, P.P. Shirshov Institute of Oceanology, 36 Nakhimovsky Prosp., Moscow 117997, Russia.

coast of Japan. The 2010 Chile tsunami caused hundreds of deaths and severe damage on the Pacific coast of South America.

In contrast to most other natural hazards, such as earthquakes, volcanic explosions, floods, landslides, and storm surges, which have mainly a regional character, tsunamis generated by the strongest megathrust earthquakes ($M_w \geq 8.5$–9.0) are global-scale events that can have devastating effects thousands of km from the source area (Gusiakov 2014). Impact from the Sumatra tsunami of 26 December 2004, induced by an earthquake with an estimated magnitude M_w 9.1–9.3, extended all the way to South Africa, roughly 8000 km from the earthquake epicentre (Rabinovich and Thomson 2007). Global tsunami propagation models (Titov et al. 2005; Kowalik et al. 2007) and observational data confirm the world-wide reach of this tsunami, including far distant regions of the North Pacific and North Atlantic, approximately 26,000 km from the source (Rabinovich et al. 2006). Similarly, the 2011 Tohoku tsunami, produced by a M_w 9.0 earthquake, propagated throughout the entire Pacific Ocean and even penetrated into the Atlantic Ocean, while the 2010 Chile earthquake (M_w 8.8) generated a tsunami, with > 2 m wave heights, on the coasts of California, Japan and the Kuril Islands.

The three major events listed above occurred during the "digital era" and were clearly recorded by hundreds of coastal and open-ocean high-quality gauges. The precise measurement of tsunamis at widely separated locations has enabled scientists to examine the events in detail and has renewed interest in major tsunamis that occurred prior to the establishment of permanent networks of digital tsunami gauges. Tsunami waves recorded by analogue ("pen-and-paper") coastal tide gauges during the pre-digital era were frequently "chopped off" or "blurred" by instrumental or environmental noise. Using the original records, it was difficult to recognise and isolate the tsunami signal that was superimposed on tides and other longwave processes. Nevertheless, even such sparse information on major events is valuable, especially for the estimation of the tsunami risk and long-term tsunami forecast for coastal areas (cf. Leonard et al. 2014).

Five great trans-oceanic tsunamis were generated by earthquakes in the Pacific Ocean during the pre-digital era of the 20th century: the 1946 Aleutian Islands tsunami generated by a M_w 8.6 earthquake; the 1952 Kamchatka tsunami (M_w 9.0); the 1957 Andreanof Islands (Aleutian) tsunami (M_w 8.6); the 1960 Great Chile (Valdivia) tsunami (M_w 9.5); and the 1964 Alaska (Prince William Sound) tsunami (M_w 9.2) (Abe 1979; Satake and Atwater 2007; Gusiakov 2014). The source areas of the great tsunamis were located along the main subduction zones of the Pacific (Fig. 1). These particular tsunamis played key roles in the establishment of the present International Tsunami Warning System (cf. Kong et al. 2015) and created the foundation for modern tsunami science. All five tsunamis were clearly recorded on the coast of British Columbia (BC). Moreover, prior to the 2011 Tohoku tsunami, these five events were the strongest seismically generated tsunamis ever measured on the BC coast (Fig. 2). The 1964 Alaska tsunami was the most hazardous tsunami ever to impact the west coast of Canada, causing multi-million-dollar damage at Port Alberni and Prince Rupert (White 1966; Murty and Boilard 1970; Murty 1992; Stephenson et al. 2007).

Large segments of the British Columbia coastline are susceptible to tsunamis generated within the Pacific Ocean. Destructive tsunamis have occurred on this coast in the past and will occur again in the future (cf. Clague 2001; Clague et al. 2003). Estimation of the potential tsunami risk to the British Columbia coast is vitally important, especially for areas being considered for new construction or major renovations (Leonard et al. 2014). A thorough examination of the 1946, 1952, 1957, 1960 and 1964 tsunamis is, therefore, highly warranted, but for some reason, has never been done before. In fact, there are only two short reports describing the effects of the 1960 Chile (Wigen 1960) and 1964 Alaska (Wigen and White 1964) tsunamis on the BC coast and very limited information on the other three events (Wigen 1983; Stephenson et al. 2007). One possible reason for this omission is the considerable preliminary effort required to digitise the analogue traces of these tsunamis in the tide gauge records. This work was undertaken as part of the present study.

The purpose of this paper is to provide estimates of the main statistical parameters and physical properties of the recorded waves. The information arising

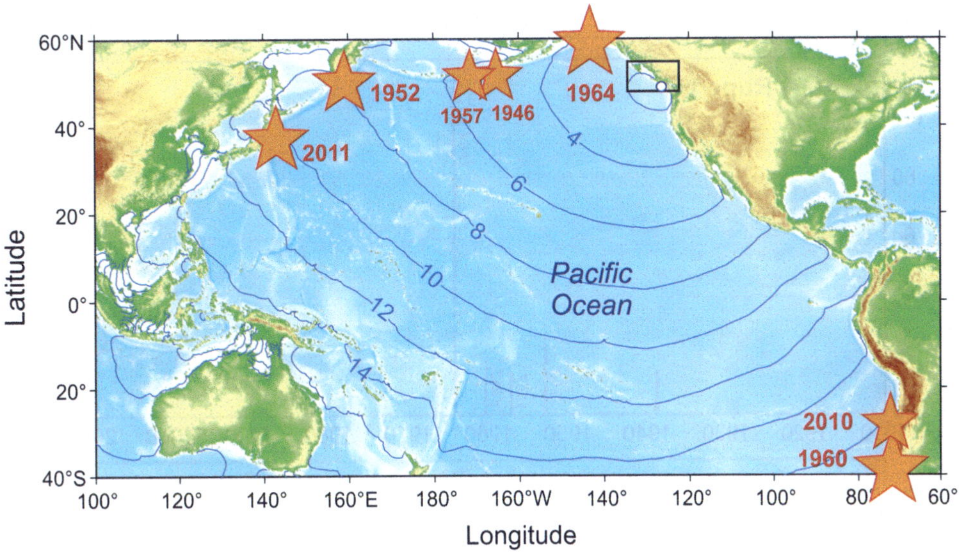

Figure 1
Epicentres (orange stars) of the five great earthquakes of the 20th century (1946 Aleutian Islands, 1952 Kamchatka, 1957 Andreanof Islands., 1960 Chile and 1964 Alaska) and the two of the 21st century (2010 Chile and 2011 Tohoku) in the Pacific Ocean that produced major trans-oceanic tsunamis. The size of the star is proportional to the earthquake magnitude (M_w). The solid blue lines are inverted isochrones of the tsunami travel time (in h) from various parts in the Pacific Ocean to Tofino, British Columbia (denoted by the white circle in the black box). The black box encloses the coast of British Columbia, the main study region

from this study can be effectively used to evaluate the risk from major tsunamis arriving at the coast of British Columbia and, more generally, at the Pacific coast of North America. This information will also help to identify the potential sources of tsunami waves that are the greatest threat to the BC coast and to provide background information in support of the future numerical modelling of tsunami waves for this region.

2. Data

Sea level measurements on the coast of British Columbia have been performed by the Canadian Hydrographic Service (CHS) for more than 110 years. The first tide gauge stations to operate on a permanent basis in British Columbia began in 1906–1909 at Victoria (Juan de Fuca Strait, the southeastern coast of Vancouver Island), Tofino (oceanic, southwestern coast of Vancouver Island), Prince Rupert (the head of Dixon Entrance, the northern mainland coast), Caulfeild/Point Atkinson (the Strait of Georgia, near the city of West Vancouver) and Vancouver (within Burrard Inlet). These five stations (Fig. 3) have the longest records for the Pacific coast of Canada and, in general, among the longest records in North America. Prince Rupert, Tofino and Victoria (Fig. 2) have the longest span of recorded tsunami events on the west coast of Canada. An additional tide gauge was installed at Alert Bay (the northeastern coast of Vancouver Island) in 1948 and in Fulford Harbour (southern coast of Salt Spring Island, the Strait of Georgia) at the end of November 1952 (3 weeks after the 1952 Kamchatka tsunami). The available records from these particular permanent instruments (Fig. 3) were used to examine the five great tsunamis. One more permanent tide gauge was installed in 1961 in Bella Bella and was one of the instruments that most clearly recorded the 1964 tsunami.

In addition to the permanent CHS network of tide gauges on the coast of British Columbia, there were also temporary gauge deployments, whose purpose was to measure the tidal regime at the respective site. Some of these measurements were conducted during the tsunami events. For example, the 1952 Kamchatka tsunami was recorded by a temporary gauge at

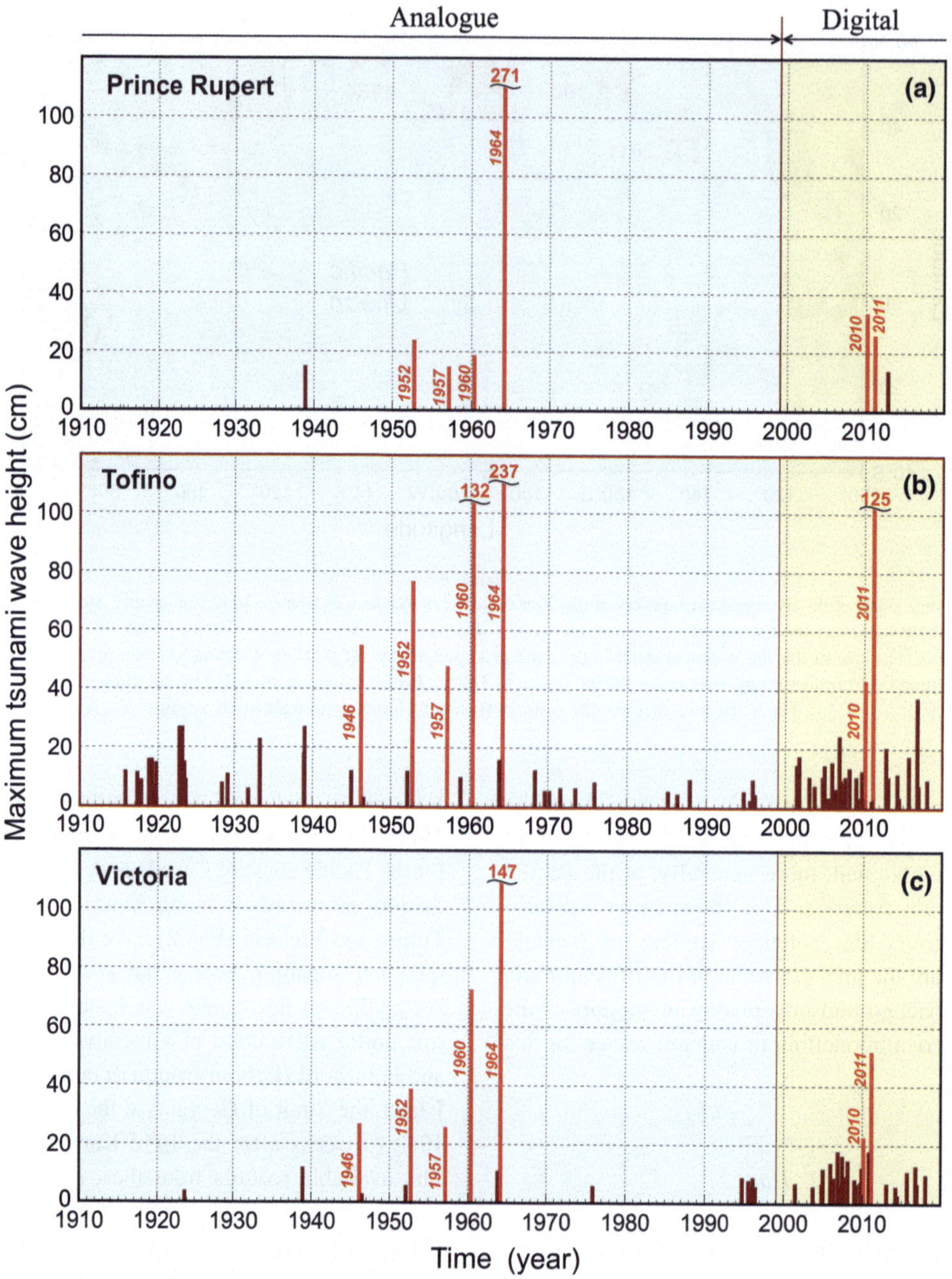

Figure 2

Maximum wave heights of tsunamis recorded at **a** Prince Rupert, **b** Tofino and **c** Victoria during the period 1910–2018. Up to 1998, measurements were made by analogue tide gauges; since 1998, digital tide gauges have been used. Red bars indicate tsunami wave heights for the five great tsunamis of the 20th century (1946, 1952, 1957, 1960 and 1964) and two of the 21st century (2010 and 2011); all the others are indicated by brown bars

Kitimat, while the 1957 Aleutian Islands tsunami was recorded by a temporary gauge at Bella Coola. A large number of temporary tide gauges were operating on the coast of British Columbia in May 1960, when the 1960 Chilean tsunami occurred, specifically McKenney Islands, Griffin Pass, Klemtu, Copper Island, Cape St. James and Barkley Sound. At the time of this event, the CHS was conducting

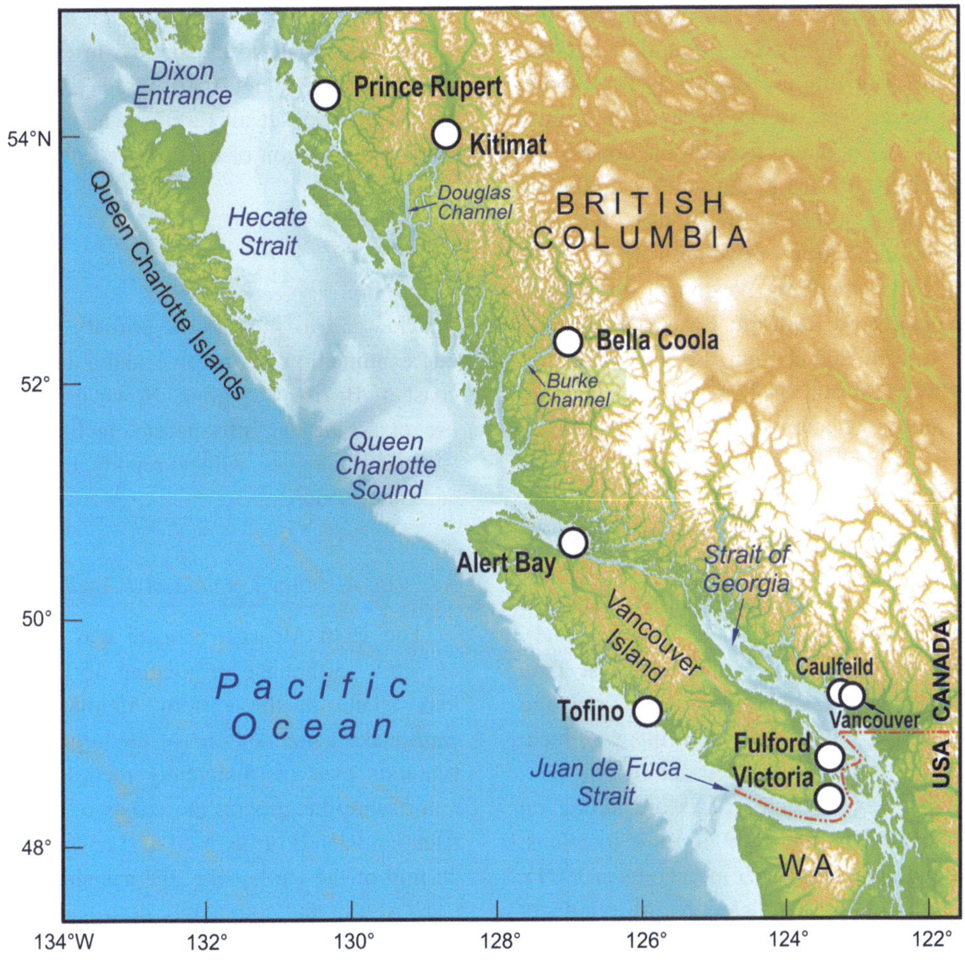

Figure 3
Locations of CHS permanent and temporary tide gauges operating on the coast of British Columbia before 1960 (white circles). Five gauges—Tofino, Victoria, Caulfeild, Vancouver and Prince Rupert—were working during three events: 1946, 1952 and 1957. The Tofino and Victoria tide gauges recorded all three tsunamis, the Prince Rupert tide gauge two of them (1952 and 1957), while Caulfeild (Point Atkinson) and Vancouver, located inside the Strait of Georgia and sheltered from incoming waves, did not record any of these tsunamis. The Alert Bay tide gauge on the northeastern coast of Vancouver Island was in operation during the 1952 and 1957 tsunamis and clearly recorded both events. The Kitimat temporary gauge detected the 1952 tsunami; the Fulford Harbour and Bella Coola instruments recorded the 1957 tsunami. The Queen Charlotte Islands are now officially known as "Haida Gwaii"

hydrographic surveys in the "Seymour-Belize Inlet system" on the central BC coast. Eight temporary tide gauges were deployed in this region and five of them—Johnson Point, Mereworth Sound, Belize Inlet, Nugent Sound and Seymour Inlet—clearly recorded tsunami waves from the event. The three other gauges—Bamford Lagoon, Frederick Sound and Allison Sound—were too well sheltered from the incoming waves to record the tsunami (Wigen 1960). As a result, a total of 17 tide gauges on the BC coast recorded the 1960 tsunami (Wigen 1960; Stephenson et al. 2007)[1] and we were able to use these records for our analysis of the event. Likewise, several temporary tide gauges were in operation during the 1964 Alaska tsunami: Tasu Sound, Port Alberni, Ballenas Islands and Port Moody. All of these gauges, as well as all permanent gauges, both those on the outer coast (Prince Rupert, Bella Bella, Alert Bay, Tofino and Victoria) and in the Strait of Georgia (Fulford

[1] See a map with location of all these stations in Sect. 3.4 describing the 1960 event.

Harbour, Point Atkinson and Vancouver), recorded the 1964 tsunami. Moreover, several additional records of this tsunami are available from other sources. The record from Ocean Falls on the central mainland coast was obtained through the courtesy of Crown Zellerbach Ltd.; the gauge was located at the power house about 300 m upstream from the river mouth. Four records from the Fraser River delta (North Arm, Steveston, New Westminster and Deas Island) and one record from Pitt Lake (a freshwater tidal lake, ~60 km inland from the sea) were provided by the Water Resources Branch, Department of Northern Affairs and National Resources. Thus, there are a total number of 18 records available for this event (Wigen and White 1964).[2]

Most records of the five great tsunamis on the BC coast are from float-actuated gauges, employing a stilling well with an inlet hole to suppress high-frequency oscillations (Wigen and White 1964). However, the records of the 1964 tsunami at Tasu Sound and Ballenas Islands were from pressure gauges (Wigen and White 1964). The main problem with all these instruments was their analogue ("pen-and-paper") recording system. Tides are strong on the coast of British Columbia and, in some locations, such as Prince Rupert, exceed 6 m (Thomson 1981). Such large tides make the detection of relatively weak tsunamis from paper records almost impossible. Moreover, this makes problematic the reliable direct estimation of tsunami parameters even for great tsunamis, especially for those that have long-period signals. Sydney Wigen, who collected and examined all available data from the 1960 and 1964 events (Wigen 1960; Wigen and White 1964) and the Tofino records for the entire observational period of 1909–1980 (Wigen 1983), concluded that the minimum tsunami wave height recognisable in these records is about 6 cm.

To estimate the wave parameters of the five major historical events, we digitised all available tsunami records and resampled them to a 1-min time step. The astronomical tides were calculated for all sites and subtracted from the original records; the resulting "residual records" were statistically analysed. The estimated wave parameters for these historical tsunamis were found to be slightly different from those presented by Wigen (1960), Wigen and White (1964) and Stephenson et al. (2007), which were based on direct examination of the paper records.

[2] A map showing locations of all these stations is presented in Sect. 3.5, describing the 1964 event.

3. Observations

Tide gauges provide the primary records used in our examination of the five major tsunamis on the coast of British Columbia. However, we also make use of additional information, including that from tsunami catalogues, witness reports, local newspapers and other sources.

3.1. The Aleutian Tsunami of 1 April 1946

The 1946 Aleutian tsunami was generated by a $M_w = 8.6$ earthquake originating 150 km from Unimak Island in the eastern Aleutian Islands. The earthquake generated one of the largest trans-Pacific tsunamis ever instrumentally recorded, with maximum runup heights on the coasts of Alaska and the Aleutian Islands of up to 40 m (Lander 1996). Within 48 min of the earthquake, the tsunami struck Scotch Cap, Alaska and completely destroyed a newly built US Coast Guard Lighthouse, surging over the coastal cliff to a height of 35 m above mean sea level (Lander 1996). Then, 4.5 h after the main shock, destructive waves reached the coasts of the Hawaiian Islands. The waves, which were totally unexpected by the island residents, reached a maximum runup of 17 m at Hilo on the Big Island and 10 and 11 m, on the islands of Maui and Oahu, respectively. The tsunami was also recorded along the entire west coast of the United States and crossed the Pacific Ocean to produce 18–20 m runup at locations in the Marquesas Islands, French Polynesia (Okal et al. 2002) and extended southeastwards, damaging boats in Chile. Altogether, the tsunami killed 168 people, most of them in Hawaii (159), but also at the Scotch Cap lighthouse (5), Marquesas (2), Peru (1), and California (1) (Lander and Lockridge 1989). The property damage in the Hawaiian Islands amounted to $25 million (1946 dollars) (Weeks and Studds 1953). A distinguishing feature of the 1946 tsunami was its

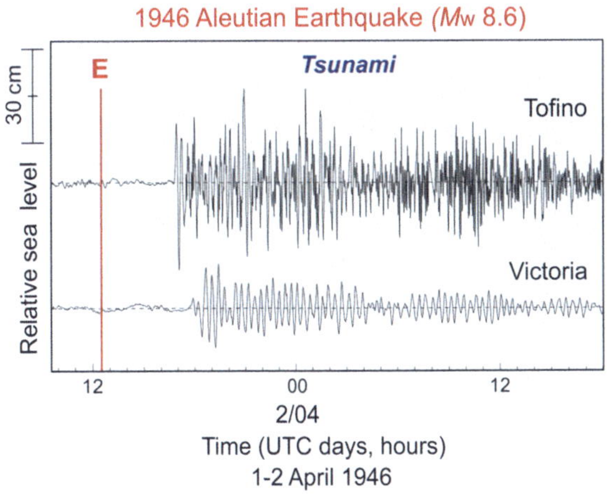

Figure 4
The 1 April 1946 Aleutian tsunami recorded at Tofino and Victoria. The solid vertical red line labelled "E" denotes the time of the earthquake

pronounced directivity, consisting of a narrow energy beam much like a "spotlight" radiating energy towards the Hawaiian and Marquesas islands, and onward to Antarctica (Fuchs 1982; Gusiakov 2014).

The tsunami was unusually powerful for the magnitude of the earthquake. Due to the large discrepancy between the tsunami magnitude, $M_t = 9.3$ (Abe 1979), and the surface wave magnitude, $M_s = 7.4$, the 1946 earthquake was identified as a "*tsunami earthquake*" (Kanamori 1972; Abe 1973), i.e., as an earthquake generating a tsunami that is much larger than it should be based on its magnitude. The severe destruction and large number of casualties in the Hawaiian Islands prompted the United States to create, in Honolulu, Hawaii, the Seismic Sea Wave Warning System (SSWWS), which, following the 1960 Great Chile tsunami, became the Pacific Tsunami Warning Center (PTWC) (Kong et al. 2015).

The 1946 tsunami was recorded at a number of stations in the Pacific Ocean (Green 1946; Sheppard et al. 1950). Of the five CHS tide gauges in operation at the time of the event (Prince Rupert, Tofino, Victoria, Caufeild and Vancouver; see Fig. 3), the tsunami signal was evident at two: Tofino and Victoria (cf. Stephenson et al. 2007). These two records were digitised and the resulting records de-tided. Significant tsunami waves were recorded at both stations (Fig. 4). The Prince Rupert record was also digitised and de-tided, but no tsunami was detected in the record. The question then is: *Why would the 1946 tsunami reach Tofino and Victoria, but not Prince Rupert?* This is especially intriguing considering that the four other great tsunamis (1952, 1957, 1960 and 1964) were clearly recorded at this station. The most probable explanation is the narrow focus of the tsunami "beam", with little energy propagating eastward toward Prince Rupert. The 1946 tsunami was also not recorded at two stations, Caulfeild (Point Atkinson) and Vancouver, located deep inside the Strait of Georgia, and therefore well sheltered from arriving tsunami waves; neither the 1952 nor the 1957 tsunamis were observed at these stations (Sects. 3.2 and 3.3). Therefore, our analysis focuses mainly on the Tofino and Victoria records (Fig. 4). The main parameters of the observed waves are shown in Table 1.

The tsunami waves arrived at Tofino at 16:48 UTC, precisely 4 h and 19 min after the earthquake. The first positive wave (wave crest) had a well-

Table 1

Parameters of the Aleutian tsunami of 1 April 1946 recorded by CHS analogue tide gauges on the coast of British Columbia (main shock, $M_w = 8.6$ at 12:29 UTC)

Station	First wave			Maximum wave			Visible period (min)
	Arrival time (UTC)	Travel time (hh:mm)	Amplitude (cm) and sign	Amplitude (cm)	Time (UTC) of max amplitude	Height (cm)	
Tofino	16:48	4:19	+ 18	34	20:55	55	18, 4.5
Victoria	17:48	5:19	+ 4	16	19:23	27	22

The analogue records of this tsunami were digitised at 1 min sampling

The tide gauge record at Prince Rupert was digitised, but no evidence for the 1946 tsunami was detected

"Visible period" (last column) is the period that can detected visually in the record

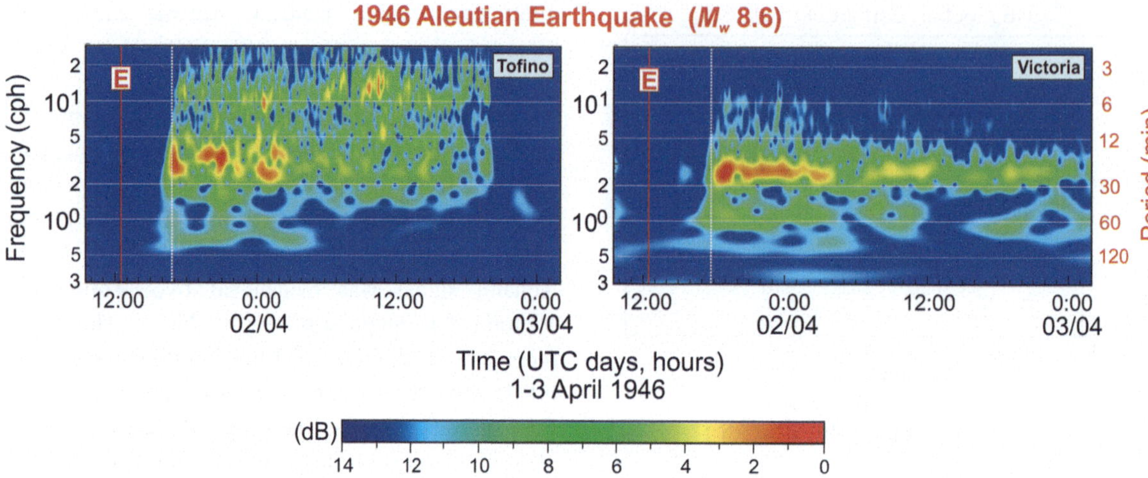

Figure 5
Frequency–time plots (*f–t* diagrams) for the 1 April 1946 Aleutian tsunami records at Tofino and Victoria. The solid vertical red line labelled "E" denotes the time of the earthquake; the dashed white line indicates the tsunami arrival

defined amplitude of 18 cm. Remarkably, it was not until more than four hours after the first wave arrival that the maximum wave, with an amplitude of 34 cm, arrived at the Tofino tide gauge (Fig. 4). Two wave periods prevail in this record: 18 and 4.5 min. Exactly one hour after the first wave arrival at Tofino, the first wave arrived at the Victoria tide gauge. The arrival time of 5 h 19 min after the main earthquake shock was, in good agreement with the theoretical travel time and with earlier estimates of arrival times (cf. Stephenson et al. 2007). Based on a paper trace record, Sheppard et al. (1950) estimated the maximum tsunami wave height at Victoria to be 20 cm, while according to our estimates, the maximum wave height was 27 cm. Our estimate is approximately half that of the maximum wave height at Tofino (55 cm). The observed tsunami waves were fairly monochromatic, with a very consistent period of ~22 min (Fig. 4).

To examine temporal variations of the observed 1946 Aleutian tsunami at Tofino and Victoria in the frequency domain, we used a multiple-filter method, which is similar to wavelet analysis (cf. Thomson and Emery 2014; Rabinovich et al. 2006, 2013). The method shows changes of tsunami waves as a function of frequency, *f*, and time, *t*, termed "*f–t* diagrams". The plots for the 1946 event (Fig. 5) reveal distinct characteristics of tsunami waves at these two stations. Specifically, the arrival times are well-defined and mutually consistent (taking into account the one hour time delay of Victoria relative to Tofino). At Tofino, a broad frequency band of 0.7–20 cph (periods of 85–3 min) of tsunami wave oscillations was induced by the incoming waves, with oscillations with peak periods of 18 and 4.5 min clearly dominant. It is noteworthy that the lower frequency waves (18 min) prevail during the first 12 h following the tsunami arrival at this station, while the higher frequency (4.5 min) dominated afterwards. This abrupt change in the character of the Tofino tsunami oscillations is evident in the record (Fig. 4).

During the 1946 event the tsunami waves at Victoria were substantially different than those at Tofino. As illustrated by the *f–t* diagram for Victoria (Fig. 5), tsunami oscillations at this station were characterised by a narrow and persistent frequency band of enhanced energy with a peak period of 22 min, compared to 18 min for Tofino. There is also another amplified frequency band with a period of ~55 min at Victoria, which is not noticeable in the record but obvious in the corresponding *f–t* diagram. The same two spectral peaks for Victoria were obtained by Rabinovich and Stephenson (2004) based on analysis of 9-day background oscillations at this site and by Rabinovich et al. (2018) based on examination of several tsunamis detected at this site. The first period (22 min) is likely associated with the fundamental eigen period of Victoria's Inner Harbour, while the second period (55 min) appears to be related to the resonant properties of the outer basin and Juan de Fuca Strait (Rabinovich et al. 2018).

3.2. The Kamchatka Tsunami of 4 November 1952

The 1952 tsunami was generated by a $M_w = 9.0$ earthquake originating near the southern end of the Kamchatka Peninsula, Russia. In the near-field region, the tsunami attained wave heights > 18 m and killed several thousand people in Severo-Kurilsk (Paramushir Island, Northern Kuril Islands) and southern Kamchatka, making it the most devastating tsunami in Russian history (Gusiakov 2014). Six hours after the initial earthquake, 6–8 m waves struck the Hawaiian Islands, reaching a maximum height of 9.1 m on the easternmost coast of Oahu Island (Lander and Lockridge 1989). However, no fatalities were reported from the far-field areas affected by this tsunami and the property damage in the Hawaiian Islands, estimated at $0.8–1.0 million (in 1952 dollars), was significantly smaller than in 1946. The fact that no lives were lost and that the damage was relatively small, is mainly attributed to the seismic sea wave warning system organised on the Hawaiian Islands after the 1946 catastrophic event (Weeks and Studds 1953).

The 1952 tsunami was recorded by many coastal tide gauges around the Pacific Ocean, including gauges in Peru, Chile and numerous islands in the Pacific. As noted by Weeks and Studds (1953), "*This <1952> tsunami was probably recorded by more tide gauges than any other before*". The authors of this report also indicate that "*Even a few stations up tidal tributaries from the sea recorded it*". As an example, they referred to Kitimat, BC, located 80 km up Douglas Channel (Fig. 3).

The 1952 Kamchatka tsunami was recorded at several locations along the coast of British Columbia. According to Weeks and Studds (1953), Wigen (1983) and Stephenson et al. (2007), the maximum wave height at Tofino was 58 cm, at Victoria and Alert Bay ~40 cm and at Kitimat noticeable, but small. We digitised and re-examined all five records. The de-tided and high-pass filtered tsunami records are shown in Fig. 6

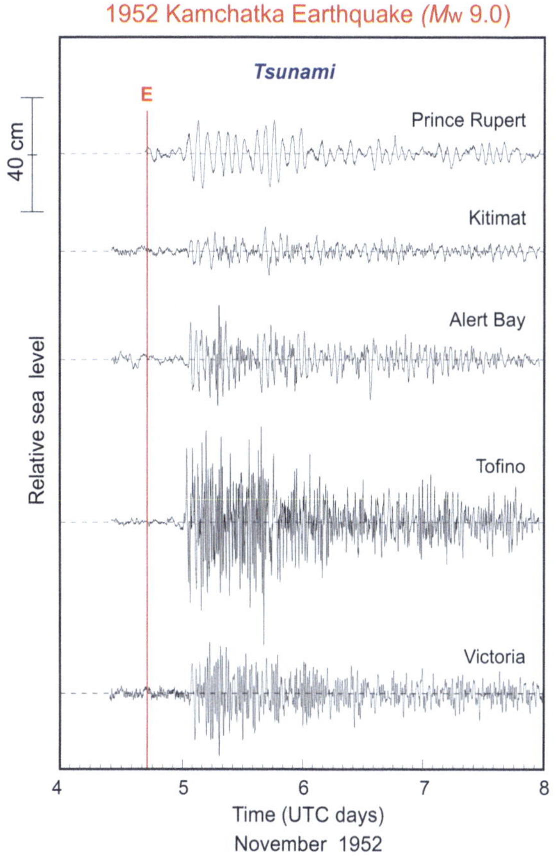

Figure 6
The 4 November 1952 Kamchatka tsunami recorded at Prince Rupert, Kitimat, Alert Bay, Tofino and Victoria on the coast of British Columbia. The solid vertical red line labelled "E" denotes the time of the earthquake

and the estimated parameters of the measured tsunami waves presented in Table 2.

Well-defined tsunami waves, including the first wave arrival, were observed at all five stations, including Kitimat. The tsunami first arrived at Tofino at 00:30 UTC on 5 November 1952 (7 h 32 min after the earthquake), 13 min later at Prince Rupert (00:43 UTC), then at 01:17–01:19 UTC at Kitimat and Alert Bay, and finally, at 01:40 UTC at Victoria. The maximum observed wave heights were approximately 77 cm (Tofino), 39 cm (Victoria), 35 cm (Alert Bay), 24 cm (Prince Rupert) and 17 cm (Kitimat). Results from the digitised records (see Table 2) can be compared with those previously estimated by Weeks and Studds (1953) and Wigen (1983) directly from the analogue tide gauge records. The arrival times for Tofino, Prince Rupert, Alert Bay

Table 2

Parameters of the Kamchatka tsunami of 4 November 1952 recorded by five analogue CHS tide gauges on the coast of British Columbia (main shock, $M_w = 9.0$ at 16:58 UTC)

Station	First wave			Maximum wave			Visible period (min)
	Arrival time (UTC)	Travel time (hh:mm)	Amplitude (cm) and sign	Amplitude (cm)	Time (UTC) of max amplitude	Height (cm)	
Victoria	1:40	8:42	+ 6.0	18.0	5:30	38.8	25
Tofino	0:30	7:32	+ 26.2	33.7	15:38	77.0	22
Alert Bay	1:19	8:21	+ 14.7	19.2	7:16	35.3	25, 70
Kitimat	1:17	8:19	+ 5.5	8.5	16:26	17.0	30, 100
Prince Rupert	0:43	7:45	+ 7.0	11.8	18:13	23.7	110

The analogue records of these tsunamis were digitised at 1 min sampling

Arrival and maximum wave times are for 5 November 1952

and Victoria were indicated to be: 00:42, 02:20, 01:18 and 01:40 UTC. Thus, some of the earlier estimates were quite precise, while others were not. In particular, the Kitimat record could not be examined for the earlier study, while for Prince Rupert the only estimated parameter was the arrival time and the error in this parameter was 1 h 37 min. The main reason for the latter discrepancy is that the tsunami signal at Prince Rupert has a very low-frequency (Fig. 6) and is difficult to discern in the original analogue record.

In general, the 1952 tsunami parameters at the five stations, estimated from the digitised and de-tided tide gauge records (Table 2), significantly correct the results of earlier studies. In particular, the 77-cm tsunami wave height measured at Tofino was higher than any other tsunami earlier recorded on the coast of British Columbia (Fig. 2). Remarkably, this wave occurred 15.1 h after the first tsunami arrival (Table 2, Fig. 6), consistent with the very long (several days) "ringing" and slow energy decay of the tsunami waves observed at all BC stations (Fig. 6). This property was noted by Weeks and Studds (1953), who examined the analogue records at many stations in the Pacific Ocean and found that the typical duration of the 1952 tsunami oscillations was from 3.5 to 5.5 days. At most of the BC stations, as well as at many other stations (Weeks and Studds 1953), maximum waves were observed several hours after the first wave arrival.

Although the maximum wave height recorded at Victoria, ~39 cm, was about one half that at Tofino, it was higher than any tsunami recorded at this site prior to 1952 (Fig. 2). Tsunami waves were noticed by shipping in Victoria Harbour, but there was no damage (Stephenson et al. 2007). The District Engineer (CHS, Victoria) reported "*This wave was of somewhat greater amplitude on our coast than the one of 1946 and continued to oscillate for a considerably longer time…*" (Weeks and Studds 1953).

The 1952 event was the first tsunami ever recorded or reported for Alert Bay and Kitimat. The fact that this wave was so clearly recorded at Kitimat is especially remarkable considering that the station is located at the head of Kitimat Arm, the innermost inlet at the end of the long and narrow Douglas Channel, roughly 80 km from Hecate Strait (Fig. 3). The tsunami parameters, which were unrecognisable in the original paper record (cf. Weeks and Studds 1953; Stephenson et al. 2007), were easily determined from the digitised and de-tided record (Table 2, Fig. 6).

Tsunami waves were not detected at Prince Rupert during the 1946 Aleutian event but were clearly evident in the 1952 Kamchatka tsunami record (Fig. 6). The waves measured at Prince Rupert had much longer periods than those at Tofino and Victoria (Fig. 6). Wave periods at Kitimat and Alert Bay were intermediate between these sites in that both low and high-frequency waves prevailed. These

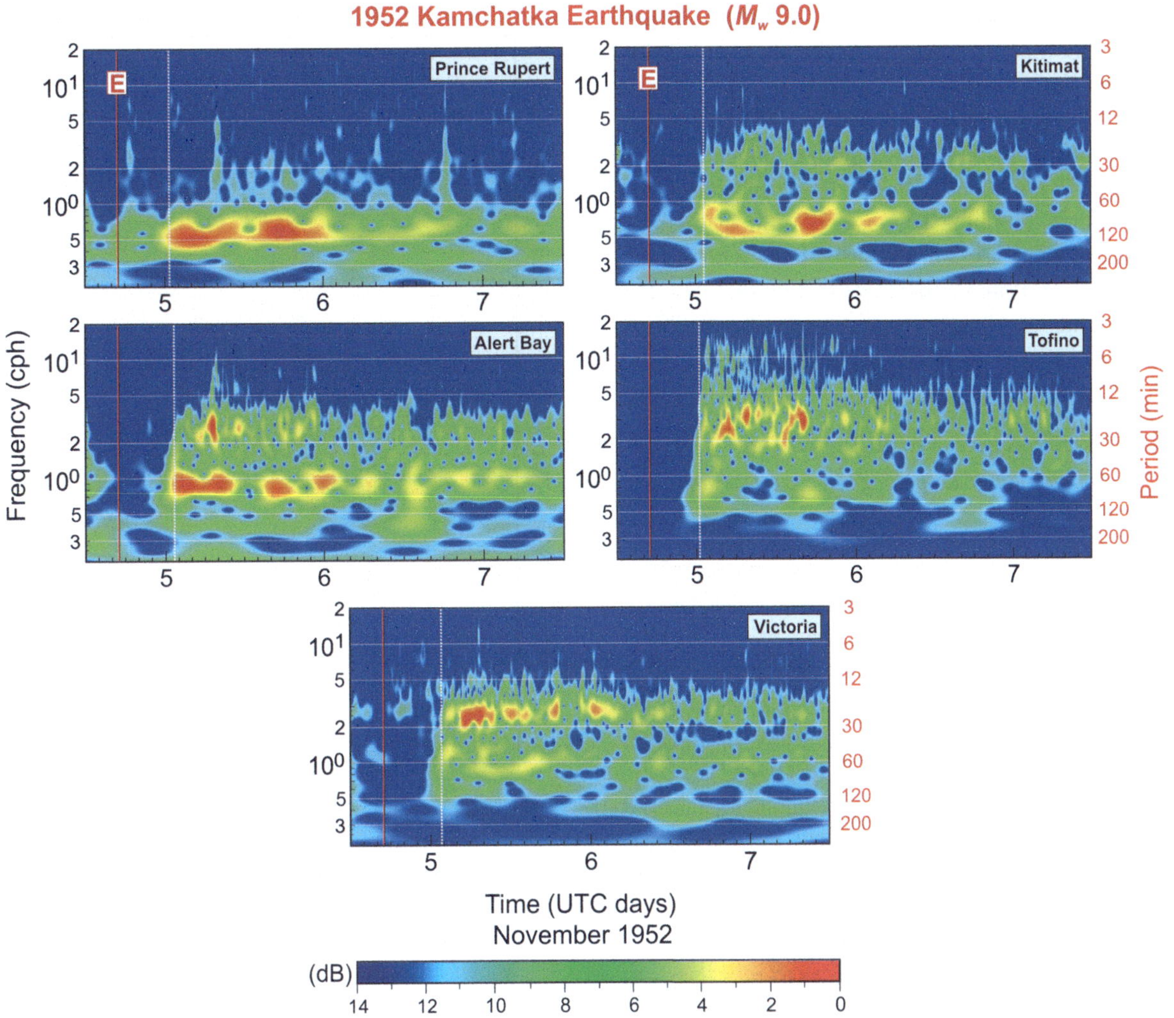

Figure 7

Frequency-time plots (*f–t* diagrams) for the 4 November 1952 Kamchatka tsunami records at five tide gauge stations on the coast of British Columbia. The solid vertical red line labelled "E" denotes the time of the earthquake; the dashed white line indicates the tsunami arrival

particular features of the tsunami oscillations are visible in the *f–t* plots of the records shown in Fig. 7.

The *f–t* diagram for Prince Rupert is characterised by a dominant frequency band with a peak period of ~110 min. There are no high-frequency oscillations with $f > 0.8$ cph (i.e., periods $T < 75$ min). As observed in previous studies (cf. Rabinovich and Stephenson 2004; Rabinovich et al. 2018), tsunamis generate relatively high wave energy at low frequencies (with peak frequencies at 0.5–0.6 cph; periods of 100–120 min) but little wave energy at high-frequencies. It appears that Dixon Entrance and the shelf adjacent to Prince Rupert act as an effective low-pass filter that strongly suppresses high-frequency waves, incoming to this region (this effect for that region was discussed in detail by Rabinovich et al. 2018).

The *f–t* diagrams at Kitimat and Alert Bay are similar. Low-frequency motions are predominant at both sites (with a peak period of about 100 min at Kitimat and 70 min at Alert Bay), but as are relatively high-frequency oscillations with periods of 25–30 min (Fig. 7). No oscillations with periods shorter than 15 min were observed at either Kitimat or Alert Bay, indicating that these oscillations are strongly suppressed by the long channels and straits of the region.

The *f–t* diagrams for the 1952 tsunami at Tofino and Victoria are similar, as are the *f–t* diagrams for the 1946 tsunami (Fig. 5). At both stations, arriving tsunami waves induce broad-band (0.5–5 cph) oscillations with two frequency bands of enhanced wave energy: (1) a predominant, shorter-period band with periods of ~22 min at Tofino and 25 min at Victoria; and (2) a secondary, longer-period band with periods of 70 min at Tofino and 55–60 min at Victoria. The latter bands are clearly evident in the *f–t* diagrams, but not distinguishable in the tsunami records themselves (Fig. 6).

3.3. The Andreanof Islands (Aleutian) Tsunami of 9 March 1957

The tsunami of 9 March 1957 was generated by a M_w =8.6 earthquake originating south of the Andreanof Islands group, in the central part of the Aleutian Islands (Fig. 1). Based on the aftershock distribution, Johnson et al. (1994) estimated the source area of the earthquake as one of the longest of any earthquakes instrumentally recorded, stretching almost 1200 km along the Aleutian Trench. The earthquake generated a major tsunami that was recorded by tide gauges throughout the entire Pacific Ocean (Salsman 1959). The tsunami wave heights on the coasts of some of the unpopulated Aleutian Islands were estimated to be more than 20 m high, with a maximum observed wave height of 22.8 m at Umnak Island (Lander 1996; https://www.drgeorgepc.com/Tsunami1957.html). The north coast of Kauai Island (Hawaii) was hit by waves up to 16 m high. However, the maximum recorded wave height was only 3.6 m—Kahului on the Island of Maui in Hawaii. The Hawaiian Islands suffered the most, with infrastructure damage of approximately $5 million (in 1957 US dollars). Timely tsunami warnings prevented any direct fatalities in the far-field areas affected (Lander and Lockridge 1989).

The effects of the 1957 event were poorly documented for the coast of British Columbia. An analysis of the analogue records by Wigen (1983) indicated that the maximum wave height at Tofino was 52 cm, but he did not examine other records, probably because the tsunami signal was weak and could not be readily extracted from tidal analogue

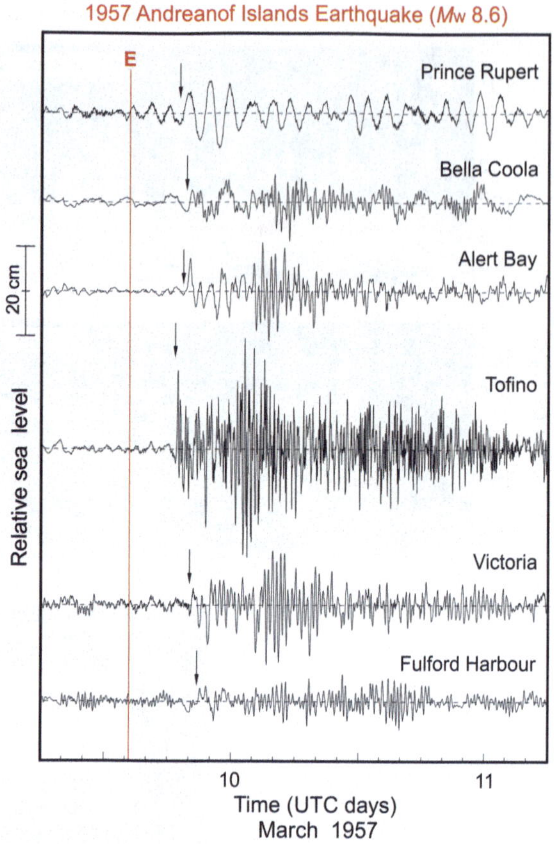

Figure 8
The 9 March 1957 Andreanof Islands tsunami recorded at six stations on the coast of British Columbia. The solid vertical red line labelled "E" denotes the time of the earthquake; the arrows indicate the tsunami arrival

records. For the present study, we digitised all available paper records of this event, except for Port San Juan (Port Renfrew), which was too noisy. We found that the tsunami was recorded at six stations: at five permanent stations—Prince Rupert, Alert Bay, Tofino, Victoria and Fulford Harbour—and at a temporary station at Bella Coola (see Fig. 3 for the location of these stations). The de-tided and high-pass filtered tsunami records for these six stations are shown in Fig. 8 and the estimated parameters of the observed tsunami waves are given in Table 3.

As with the 1952 tsunami event, Tofino was the first station on the BC coast to record the 1957 tsunami. The leading wave arrived at Tofino at 18:46 UTC on 9 March 1957 (4 h 23 min after the earthquake) and, 21 min later, at Prince Rupert (19:07 UTC). Subsequent recorded arrival times

Table 3

Parameters of the Andreanof Islands tsunami of 9 March 1957 recorded by six analogue CHS tide gauges on the coast of British Columbia (main shock, $M_w = 8.0$ at 14:23 UTC)

Station	First wave			Maximum wave			Visible period (min)
	Arrival time (UTC)	Travel time (hh:mm)	Amplitude (cm) and sign	Amplitude (cm)	Time (UTC) of max amplitude	Height (cm)	
Prince Rupert	19:07	4:44	+ 6.70	7.2	21:43	15.0	110
Bella Coola	19:44	5:21	+ 4.8	6.3	4:00*	15.0	24
Alert Bay	19:23	5:00	+ 8.3	11.0	2:51*	22.2	24
Tofino	18:46	4:23	+ 20.2	24.9	1:14*	48.2	24
Victoria	20:05	5:42	+ 6.3	12.2	3:55*	25.7	22
Fulford Harbour	20:50	6:27	+ 3.5	6.1	10:30*	11.2	25

The analogue records for this tsunami were digitised at 1 min sampling

Maximum wave times indicated by "*" are on 10 March 1957

Tide gauge record at Port San Juan (Port Renfrew) was very noisy and no tsunami signal was recognisable

were 19:23 UTC at Alert Bay, 19:44 UTC at Bella Coola, 20:05 UTC at Victoria and lastly 20:50 UTC at Fulford Harbour. Once again, the maximum wave height among all stations was recorded at Tofino: ~48 cm. The observed wave heights at other stations were significantly smaller, ranging from 11 cm at Fulford Harbour to 26 cm at Victoria (Table 3). Perhaps the most remarkable feature of the measured waves on the BC coast is that the 1957 tsunami was recorded at Bella Coola. This station is located at the head of North Bentinck Arm, which is the remotest part of the long and narrow Burke Channel. The distance from the open ocean (Queen Charlotte Sound) to this station is ~110 km (Fig. 3). Nevertheless, waves with a maximum height of 15 cm were recorded at this station less than one hour after they arrived at Tofino. A notable property of the 1957 tsunami is the consistency of the visible period of 22–25 min observed at all stations, except Prince Rupert, where this period is 110 min (Table 3), in agreement to what was observed at this station during the 1946 and 1952 events. Additional details of the recorded tsunami oscillations at various stations are seen in their *f–t* diagrams (Fig. 9).

The specific attribute of all *f–t* diagrams, except Prince Rupert (i.e., for Bella Coola, Alert Bay, Tofino, Victoria and Fulford Harbour), is their marked similarity.

(1) The main energy of observed tsunami waves spans the frequency band of 1.0–5.0 cph (i.e., periods of 60–12 min).
(2) The dominant frequency at the five stations is almost the same: ~ 2.5 cph ($T \sim 24$ min).
(3) The maximum tsunami waves were associated, not with the leading waves, but with a train of waves that arrived 6–8 h later. This aspect of the incoming tsunami waves is visible in the records themselves (Fig. 8).

A particular feature of the Alert Bay and Victoria *f–t* diagrams is the presence of a secondary frequency band of amplified-energy at frequencies 0.9–1.1 cph ($T \sim 55$–65 min). For these two stations, approximately the same amplified frequency bands (~ 2.5 cph and ~ 0.9–1.1 cph) were also evident in the *f–t* diagrams for the 1952 tsunami (Fig. 7); however, for Alert Bay in 1952, low-frequency oscillations (~ 0.9 cph) prevailed, while higher-frequency oscillations (~ 2.5 cph) were secondary.

The *f–t* diagram for the 1957 tsunami at Prince Rupert is almost identical to that for the 1952 tsunami (Fig. 7). Specifically, there is one well-defined frequency band with a peak period of ~110 min and no oscillations at frequencies higher than 0.85–0.9 cph (i.e., at periods shorter than 65–70 min).

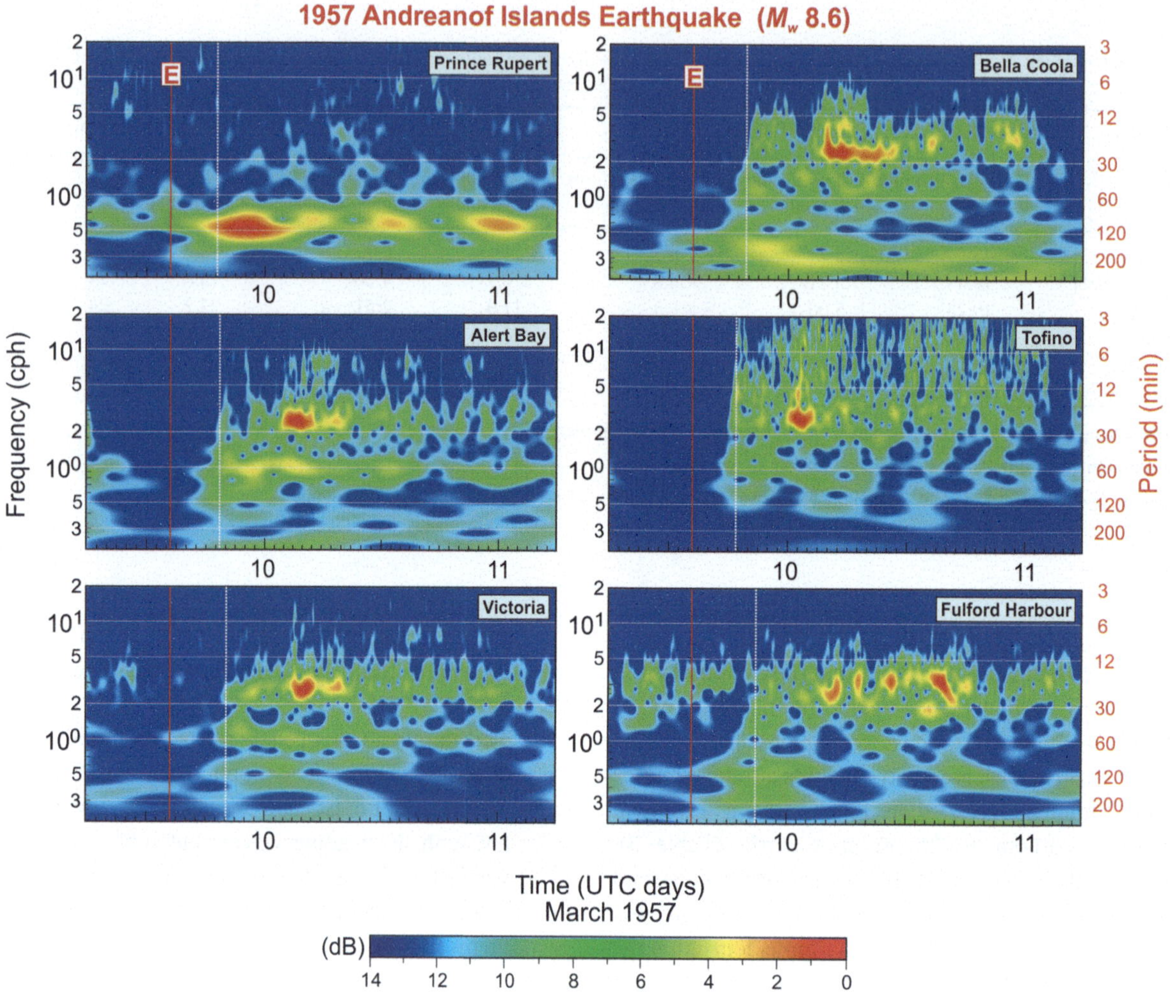

Figure 9
Frequency-time plots (*f–t* diagrams) for the 9 March 1957 Andreanof Islands tsunami records at six tide gauge stations on the coast of British Columbia. The solid vertical red line labelled "E" denotes the time of the earthquake; the dashed white line indicates the tsunami arrival

3.4. The Chilean Tsunami of 22 May 1960

The Chilean tsunami of 22 May 1960 was generated by the $M_w = 9.5$ Great Chilean (Valdivia) Earthquake, the strongest earthquake ever instrumentally recorded on land or in the ocean. The main shock of the earthquake occurred at 19:11 UTC; the earthquake lasted for about 10 min and the rupture length was approximately 900 km (Satake and Atwater 2007). The tsunami from this event was the largest of the 20th century. In southern Chile, about 1655 people were killed, 3000 injured and two million displaced; 61 people lost their lives on the coasts of the Hawaiian Islands, 142 in Japan, 32 in the Philippines and 2 on the west coast of the USA (Kong et al. 2015). The high degree of destruction and loss of life in a number of the Pacific countries located far from the source area was the impetus for international cooperation in tsunami research and mitigation, and resulted in the establishment of the International Tsunami Warning System in the Pacific (ITSU) and the Pacific Tsunami Warning Center (PTWC) in Honolulu, Hawaii, based on the former Seismic Sea Wave Warning System (SSWWS) (Kong et al. 2015).

The 1960 Chilean tsunami propagated throughout the entire Pacific Ocean and even penetrated into the Atlantic and Indian oceans (Van Dorn 1984). The

Figure 10

Location of CHS tide gauges on the coast of British Columbia that recorded the 1960 Chile tsunami. The white circles denote permanent CHS stations, the yellow circles are temporary stations. The inset shows five stations deployed in narrow mainland inlets and channels of the central BC coast (Seymour–Belize inlet system) as part of the CHS hydrographic study of this region; "QC Str." denotes Queen Charlotte Strait. The Queen Charlotte Islands are now officially known as "Haida Gwaii"

maximum observed far-field wave heights were 10.5 m on the Hawaiian Islands, 6–9 m on other Pacific Islands and in Japan, and 4–6 m in Russia. The tsunami was recorded by an unprecedented number of about 250 tide gauges, including far-field sites at the Aleutian Islands and in the Sea of Okhotsk, located 16–18 thousand km from the source area (Berkman and Symons 1960; Takahashi and Hatori 1961). Strong oscillations with trough-to-crest wave heights of more than 3–5 m were recorded at a number of stations on the coast of Alaska (Lander 1996), at sites relatively close to Prince Rupert. At many tide gauge stations around the Pacific, extreme positive or negative tsunami wave heights were "chopped off"; the instruments were simply not built for such strong oscillations (Berkman and Symons 1960).

Altogether, there were 21 CHS tide gauges in operation on the coast of British Columbia in May 1960. Of the 7 permanent and 14 temporary gauges, 17 of them (6 permanent and 11 temporary; Fig. 10) recorded the 1960 Chilean tsunami (Wigen 1960). The exceptions were the permanent tide gauge at Vancouver and three sheltered temporary gauges installed in the "Seymour-Belize Inlet system" (see Sect. 2 for details). There were also numerous eye-witness reports of strong oscillations along the BC coast. In particular, at Shields Bay located on the west coast of Graham Island, the Queen Charlotte Islands (the bay is indicated in Fig. 10), loggers reported that the water poured in and out of the bay with a violence resembling Seymour Narrows (where maximum tidal currents are 14–16 knots); from photos made in the bay during these disturbances it appeared that the maximum oscillation heights were more than seven feet (2.1 m). One of the most striking features of the 1960 tsunami was the transmission of waves through fast running narrow passages, like Nakwakto Rapids, connecting the Seymour Inlet area to the ocean (see the inset in Fig. 10). No damage or loss of lives caused by the tsunami was reported, but an Indian village on the west coast of Vancouver Island was partly inundated and moored log booms were torn adrift along this coast (Wigen 1960).

From an observational perspective, the entire situation with the 1960 tsunami tide gauge measurements was completely different from that for the 1946, 1952 and 1957 events. While tsunami waves for the previous events had been recorded at only a few BC stations, the 1960 tsunami was observed at 17 stations, many of them located in regions where tsunami waves had never been observed before. Moreover, the 1960 tsunami was much stronger than any of the previous events recorded on the BC coast (cf. Fig. 2). All 17 existing records were examined by Wigen (1960) using the original paper records, in which tsunami oscillations were superimposed on tides. Thus, we could compare the main characteristics of the observed tsunami waves estimated from digitised and de-tided records with those estimated by Wigen (1960).

One-day records of the 1960 tsunami at the 17 BC tide gauges were digitised and examined in the same manner as for the 1946, 1952 and 1957 events; the evaluated statistical parameters of this tsunami are given in Table 4, the de-tided and high-pass filtered records are shown in Figs. 11 and 12. The arrival times and maximum tsunami wave heights from Wigen (1960) are presented in the respective columns in Table 4 in brackets.[3]

[3] Arrival times in Wigen (1960) are given in the Pacific Standard Time (PST), i.e. in UTC—8 h; tsunami wave heights are in feet with decimal fractions (one digit). In Table 4 these values are transferred into UTC and cm, respectively.

Comparison of Wigen's (1960) estimates with those from our study indicates that in most cases they agree quite well. The corrections in Wigen's arrival times are small: from 0 min (Alert Bay) to 8 min (Griffin Pass). There is only one exception—Belize Inlet—where the arrival time was corrected by 25 min. Similarly, maximum wave heights differed mainly by a few cm, except at the Barkley Sound site, which was corrected by 27 cm (Table 4). The problem, however, was that Wigen (1960) could not identify the tsunami arrival times at four stations (Prince Rupert, Seymour Inlet, Fulford Harbour and Caulfeild) and maximum wave heights at two stations (Prince Rupert and Caulfeild). In the newly digitised and de-tided records, the tsunami arrivals and maximum wave heights at all stations were clearly observed and these parameters were estimated (Table 4).

Cape St. James, located at the southernmost end of the Queen Charlotte Islands (Fig. 10), was the first station on the coast of British Columbia to detect the 1960 Chile tsunami. The wave arrived at Cape St. James at 12:17 UTC, corresponding to a travel time from the source area near southern Chile of 17 h and 6 min, in good agreement with the theoretical travel time (Fig. 1). Roughly 6–7 min later, the tsunami arrived at Tofino and Barkley Sound, and then at other stations on the outer coast (Table 4). At 12:57 UTC (34 min later than at Tofino), the tsunami arrived at Victoria. The wave then travelled through the Strait of Georgia to Fulford Harbour, arriving at 13:31 UTC, and then to Caulfeild, arriving at 14:02 UTC. The 1960 tsunami also entered Queen Charlotte Sound and propagated northward across the wide and relatively shallow waters of Hecate Strait (Fig. 10), eventually reaching Copper Island (12:42 UTC), McKenney Islands (13:02 UTC), Klemtu (13:13 UTC) and Griffin Passage (13:28 UTC) (Table 4). The wave then reached Prince Rupert at 14:45 UTC. It is not known, without high-resolution numerical modelling, if the first tsunami wave propagated from the south through Hecate Strait or from the west through Dixon Entrance (Fig. 10). The results of numerical simulation of the 2010 Chilean tsunami (see Fig. 5 in Rabinovich et al. 2013), which had the source approximately at the same area as the 1960

Table 4

Parameters of the Great Chile tsunami of 22 May 1960 recorded by CHS analogue tide gauges on the coast of British Columbia (main shock, $M_w = 9.5$ at 19:11 UTC)

Station	First wave			Maximum wave			Visible period (min)
	Arrival time (UTC)	Travel time (hh:mm)	Amplitude (cm) and sign	Amplitude (cm)	Time (UTC) of max amplitude	Height (cm)	
Prince Rupert	14:45 (Indef)	19:34	+ 6	12	21:56	19 (Indef)	100
Griffin Passage	13:28 (13:36)	18:17	+ 36	36	14:44	47 (43)	30
Klemtu	13:13 (13:20)	18:02	+ 32	32	14:25	42 (40)	20, 50
McKenney Is.	13:02 (13:03)	17:51	+ 16	25	17:49	49 (55)	14, 25
Copper I.	12:42 (12:47)	17:31	+ 11	18	20:29	27 (21)	30
C. St. James	12:17 (12:21)	17:06	+ 16	22	15:57	45 (43)	10, 35
Alert Bay	13:45 (13:45)	18:34	+ 22	12	17:14	51 (46)	30
Tofino	12:23 (12:21)	17:12	+ 35	60	19:48	132 (134)	22
Barkley Sound	12:24 (12:22)	17:13	+ 50	60	16:04	131 (104)	30
Johnson Pt.	15:42 (15:40)	20:31	+ 3	10	18:06	16 (15)	15, 25, 40
Nugent Sd.	16:00 (16:03)	20:49	+ 5	11	17:48	19 (15)	30
Belize Inlet	16:38 (16:13)	21:27	+ 2.5	4	18:24	6 (6)	20
Mereworth Sd.	16:44 (16:40)	21:33	+ 3	8	18:06	16 (15)	22
Seymour Inlet	16:05 (Indef)	20:54	+ 5	4	16:54	6 (3)	27, 120
Victoria	12:57 (12:58)	17:46	+ 40	41	19:48	73 (73)	22
Fulford Hbr.	13:31 (Indef)	18:20	+ 12	17	22:56	34 (30)	30
Caulfeild	14:02 (Indef)	18:51	+ 12	12	15:47	20 (Indef)	180

Parameters of the tsunami waves were estimated based on digitised analogue tide gauge records and re-sampling to 1-min

Temporary CHS tide gauge stations are indicated by italics

Tsunami arrival times and maximum wave heights estimated by Wigen (1960) are given in brackets; "Indef" means that the parameter estimates were indefinite

Typical feature of most of the records is a small negative wave ("negative phase") preceding the main frontal tsunami wave. A similar feature was discovered for the 27 February 2010 Chile tsunami (Rabinovich et al. 2013). As shown by Watada et al. (2014), this negative phase is mainly related to the elasticity of the Earth. Eblé et al. (2015) showed that this is a typical feature of both open-ocean and coastal records of major distant tsunamis. This negative wave was ignored to evaluate the true parameters of the 1960 tsunami

At most stations, the first (frontal) wave appears in the record as an extensive (~ 2 h) "hump" followed by higher frequency oscillations

tsunami, showed the tsunami waves reached Prince Rupert almost simultaneously by both routes.

Another branch of the 1960 tsunami in Queen Charlotte Sound went around the northern coast of Vancouver Island and at 13:45 UTC was recorded at Alert Bay, on the northeastern side of the island. The tsunami waves also reached the mainland coast and penetrated into the narrow fjords and channels, typical of this coast. After passing through Nakwakto Rapids (see the inset in Fig. 10), the waves entered the Seymour–Belize Inlet System, where they were measured at five stations: Johnson Point, Nugent Sound, Seymour Inlet, Belize Inlet and Mereworth Sound. The last arrival time was at Mereworth Sound at 16:44 UTC, i.e., 21 h 33 min after the main earthquake shock and 4 h 27 min after arrival at Cape St. James, which, as noted earlier, was the first station on the BC coast to record the 1960 tsunami. Thus, in approximately 4.5 h, the 1960 Chilean tsunami spread along the entire coast of British Columbia, even propagating into the innermost inlets, fjords and channels of the mainland coast.

At two stations, Tofino and Barkley Sound, the maximum observed tsunami wave heights were 132 and 131 cm, respectively. Once again, the highest tsunami wave was recorded at Tofino. At six other stations, including Victoria (73 cm), it was more than 40 cm (Table 4). For most stations, the 1960 tsunami

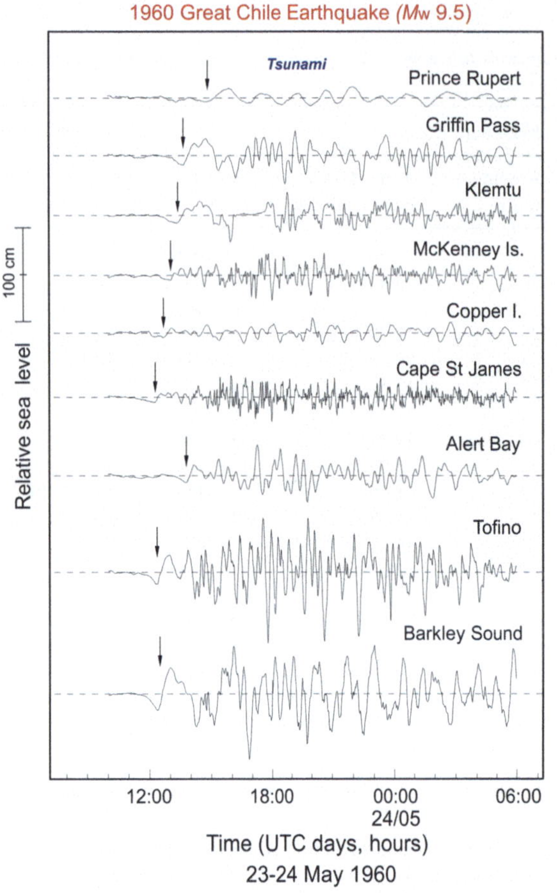

Figure 11
The 22 May 1960 Great Chile Tsunami recorded at nine stations located on the northern mainland coast of British Columbia and on the outer coast of Vancouver Island. The arrows indicate the tsunami arrival. The negative denivelation ("trough wave") preceding the arrival of the frontal tsunami wave is related to the elasticity of the Earth's core deflecting under the weight of the propagating tsunami wave (see details in Watada et al. 2014; Eblé et al. 2015)

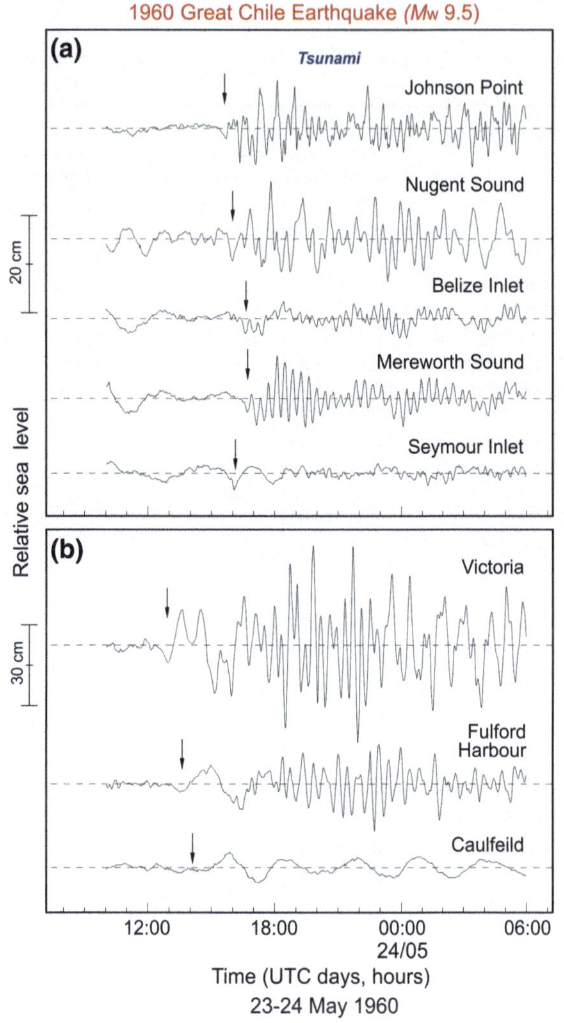

Figure 12
The same as in Fig. 11, but for **a** five stations located in inlets and channels of the Belize-Seymour Inlet system on the mainland coast of British Columbia, and **b** a group of southern BC stations consisting of Victoria (Juan de Fuca Strait) and Fulford Harbour and Caulfeild (the Strait of Georgia)

was the first tsunami ever recorded, while for Tofino, Victoria and Alert Bay, this was the highest recorded tsunami. An important feature of the 1960 tsunami, which was observed not only in British Columbia but throughout the entire Pacific Ocean, was very long ringing (> 4–5 days) and slow energy decay (cf. Berkman and Symons 1960; Takahashi and Hatori 1961; Miller et al. 1962). The maximum tsunami amplitudes were commonly observed within several hours after the arrival of the first wave. For example, at Victoria, the first waves were weak and had a period of $T \sim 55$ min; however, roughly four hours after the first wave arrival, a train of prominent waves with much shorter periods ($T \sim 22$ min) arrived. The later wave trains appear to be due to wave dispersion during the long oceanic trajectory of the tsunami. The maximum trough-to-crest height of 73 cm at Victoria was associated with this train of waves, with the maximum wave arrival at 19:48 UTC (Table 4), almost 7 h after the leading wave.

Pronounced sea level oscillations in several harbours were accompanied by strong currents. This once again demonstrates the prolonged threat of tsunami waves and associated tsunami-induced currents for coastal communities and infrastructure. The typical visual periods of the recorded waves were 20–30 min, but at some BC coastal stations, such as Prince Rupert and Caulfeild, much longer periods were observed (Table 4).

Based on the characteristics of the 1960 tsunami waves, evident in the *f–t* diagrams for BC coastal sites (Figs. 13 and 14), we can divide all stations into three distinct groups:

(1) *Low-frequency* (Prince Rupert, Caulfeild and Seymour Inlet). The tsunami waves recorded at Prince Rupert and Caulfeild are characterised by prolonged monochromatic oscillations with a distinct long period of ~ 100 and 180 min, respectively. At Seymour Inlet, long-period oscillations (~ 120 min) are also strongly predominant, but they continue for only a few hours; weak multi-frequency oscillations arrive later.

(2) *Moderate-frequency* (Belize Inlet, Mereworth Sound and Fulford Harbour). Tsunami oscillations at all three stations were monochromatic, with one predominant period, corresponding to 22 min at Belize and Mereworth and 30 min at Fulford.

(3) *Polychromatic* (other 11 stations). At all other stations, at least two and up to three, frequencies prevail with typical periods between 10 and 90 min.

An important feature of the *f–t* diagrams for Tofino, Victoria, Prince Rupert and Alert Bay (i.e., for those stations that measured other great tsunamis) is the similarity of the plots at each station for different events. In fact, it is well known that the main spectral peaks for different tsunamis at the same site are very similar, while the spectral peaks, from the same tsunami event, even at neighbouring sites are substantially distinct (cf. Honda et al. 1908). This property was considered in detail by Rabinovich (1997) (see also Fig. 8 and the corresponding discussion in Zaytsev et al. 2017). The similarity of the *f-t* diagrams at the same site for different events demonstrates the dominant influence of the local topography over the individual characteristics of the tsunami source.

3.5. *The 1964 Alaska ("Good Friday") Tsunami of 28 March 1964*

The Alaska earthquake of 28 March 1964 with magnitude $M_{\rm w} = 9.2$ produced a catastrophic trans-Pacific tsunami, the second strongest in the 20th century (after the 1960 Chilean tsunami). The maximum water rise was 20 m in the source area, while numerous subaerial and submarine landslides, initiated by the earthquake generated local tsunamis, had runups of up to 70 m (Lander 1996). The earthquake, which was the greatest instrumentally recorded earthquake in the North Pacific Ocean and one of the most powerful earthquakes ever recorded, occurred in the region of Prince William Sound, leading to the widely used name of the "Prince William Sound Earthquake" (Spaeth and Berkman 1967; Johnson et al. 1996). Due to the date (28 March 1964), this earthquake and associated tsunami are also called the "Good Friday Earthquake and Tsunami". The event produced the strongest tsunami response ever observed on the coast of British Columbia. Because the 1964 Alaska tsunami was the highest and the most important tsunami to impact the coast of British Columbia, we give more attention to this event than to any of the earlier tsunamis.

The 1964 earthquake occurred within the Alaska-Aleutian megathrust zone, where the Pacific Plate subducts under the North American Plate (Fig. 15). This zone has the greatest potential to generate destructive tsunamis and is one of the most seismically active fault zones in the North Pacific [cf. The Great Alaska Earthquake of 1964 (1972)]. The earthquake rupture extended for 600–800 km from Prince William Sound along the Kenai Peninsula to Kodiak Island and included a 200–250 km wide zone (Fig. 15) with an average fault slip of more than 10 m (Suito and Freimueller 2009). The 1964 megathrust Alaska earthquake caused the most destructive tsunami in Alaskan history and, further south, strongly impacted the west coasts of the USA and Canada (Johnson et al. 1996; Myers and Baptista 2001; Suleimani et al. 2013). In addition to the major tectonically-generated tsunami, more than 20 local tsunamis were generated by landslides in coastal areas of Alaska. Of 132 fatalities associated with the 1964 earthquake, 122 were caused by the tsunami (Lander 1996). The tsunami spread over the entire Pacific Ocean and was recorded by a great number of coastal tide gauges (Spaeth and Berkman 1967), including those located on the coast of British Columbia (Wigen and White 1964; herein, WW64 in the following text).

The tsunami swept southward from the source area in Prince William Sound along the British Columbia coast, causing about $10 million in damage (1964-dollar values). If such an earthquake and tsunami were to occur today, the cost of the damage would be a factor of 10–100 times greater (Rabinovich et al. 2018). The 1964 tsunami was the strongest tsunami ever observed on the Pacific coast of Canada. All tsunamis detected on the BC coast before or after this event, including the 1960 Chile and 2011 Tohoku tsunamis, were considerably smaller (Fig. 2). The tsunami struck the coast of British Columbia near the time of high tide (Figs. 17 and 18). The cumulative effect of the high tide (which was from 1.5 m on the south coast to almost 3.5 m on the north coast) and the tsunami was one of the main reasons for the devastating impact of this event on the BC coast. The earliest recorded arrival was at 05:32 UTC at Tasu Sound on the west coast of Moresby Island, the Queen Charlotte Islands[4] (Fig. 16), the tsunami wave arrived at this station only 1 h 56 min after the main shock (Table 5). The highest wave reported in Canada was in Shields Bay on the west coast of Graham Island (Fig. 16), where the crest wave was ~5.2 m above the spring high water (9.8 m above tidal datum). A logging camp in this remote area of the coast was damaged by the wave. The most severe damage from the event occurred at the twin towns of Alberni and Port Alberni, with maximum tsunami height at Port Alberni reaching almost 8 m (White 1966; Stephenson et al. 2007). The waves at these two towns, located at the head of long and narrow Alberni Inlet, were strongly amplified as they moved up Barkley Sound and into the inlet. The wave heights at Port Alberni were approximately 2.5 times higher than at Tofino and Ucluelet on the open coast of Vancouver Island (Fig. 16) (Clague 2001; Stephenson et al. 2007). As originally assumed by Murty and Boilard (1970), and confirmed by Henry and Murty (1995) and Fine et al. (2008), strong amplification of tsunami waves in Alberni Inlet was caused by resonant matching of the dominant period of incoming waves with the fundamental (Helmholtz) eigen period (~110 min) of this inlet. The major destruction that occurred at Alberni and Port Alberni has been described in a number of reports, papers and books, including WW64, White (1966), Clague (2001) and Stephenson et al. (2007).

Figure 13 ▶
Frequency-time plots (*f–t* diagrams) for the 22 May 1960 Great Chile tsunami records at nine stations located on the northern mainland coast of British Columbia and on the outer coast of Vancouver Island. The solid vertical red line labelled "E" denotes the time of the earthquake; the dashed white line indicates the tsunami arrival

Tsunami waves also created severe damage at Prince Rupert. The details of the tsunami effects in Prince Rupert are described by local newspapers, in particular by the Prince Rupert Daily News (1964). The vessel *Yaloa* sank at her Digby Island moorings, 8–9 million feet of logs were cut loose and the float and pierhead at Metlakatla were torn out. Also, there was extensive breakwater damage at Fairway Bay and the submarine cable providing telephone service and the main power supply to Digby Island was put out of action, causing Digby Island airport and homes to go on emergency power (Stephenson et al. 2007; Rabinovich et al. 2018).

In addition to Prince Rupert on the mainland, several sites along the outer coast of Vancouver Island were also affected by the 1964 tsunami (Fig. 16). Tsunami waves sank a fishing boat and damaged log booms in Ucluelet, damaged wharf facilities and the municipal water pipeline in Tofino, destroyed houses in a village in Hot Springs Cove and swept buildings off their foundations in Zeballos. The waves did considerable damage to buildings and a radio-telephone communication system in the small logging community of Amai on the northwest coast of Vancouver Island (Stephenson et al. 2007).

[4] Renamed *Haida Gwaii* in 2010.

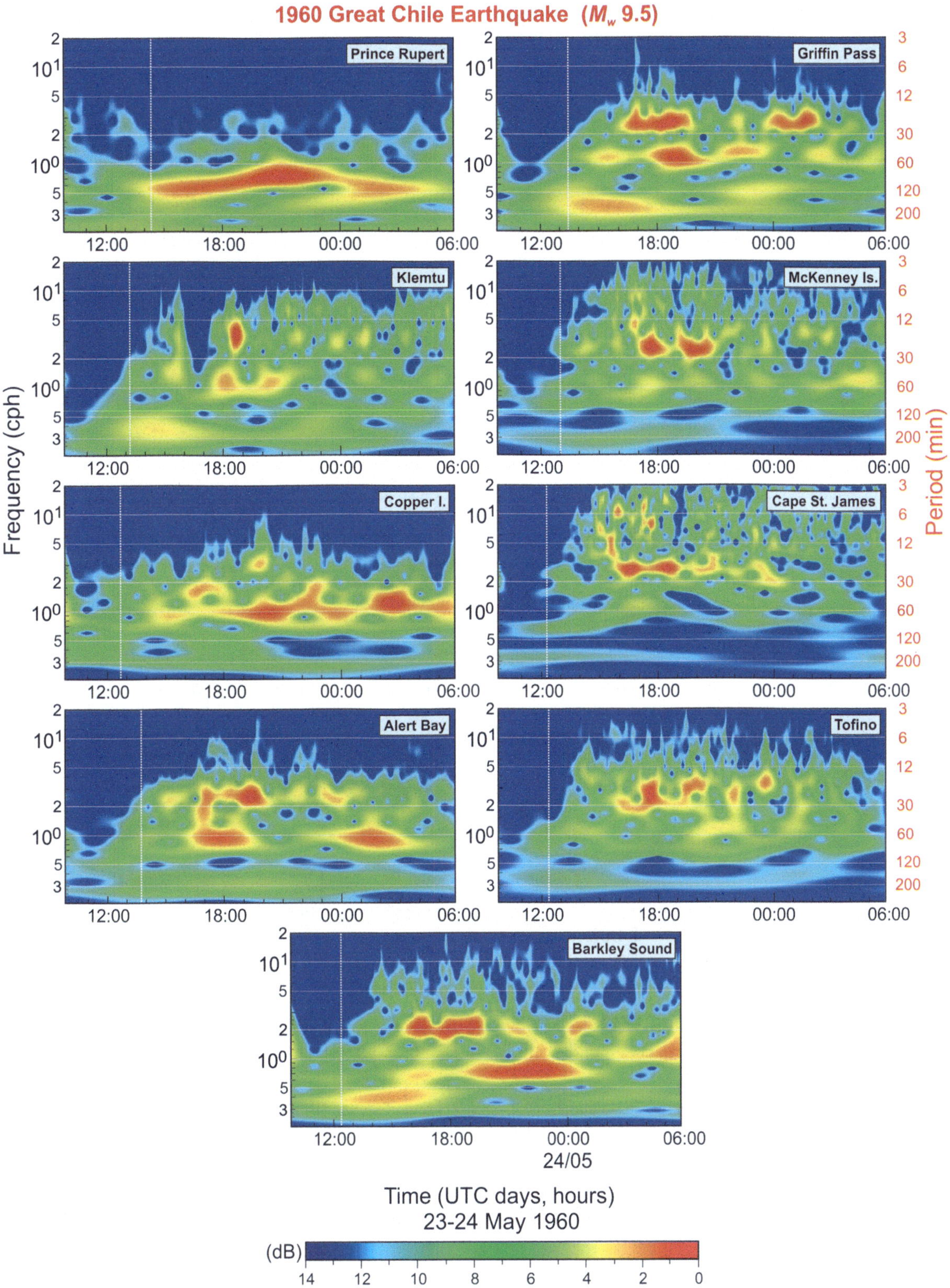
1960 Great Chile Earthquake (M_w 9.5)
Prince Rupert
Griffin Pass
Klemtu
McKenney Is.
Copper I.
Cape St. James
Alert Bay
Tofino
Barkley Sound
Frequency (cph)
Period (min)
2
10^1
5
2
10^0
5
3
3
6
12
30
60
120
200
12:00
18:00
00:00
06:00
24/05
Time (UTC days, hours)
23-24 May 1960
(dB)
14
12
10
8
6
4
2
0

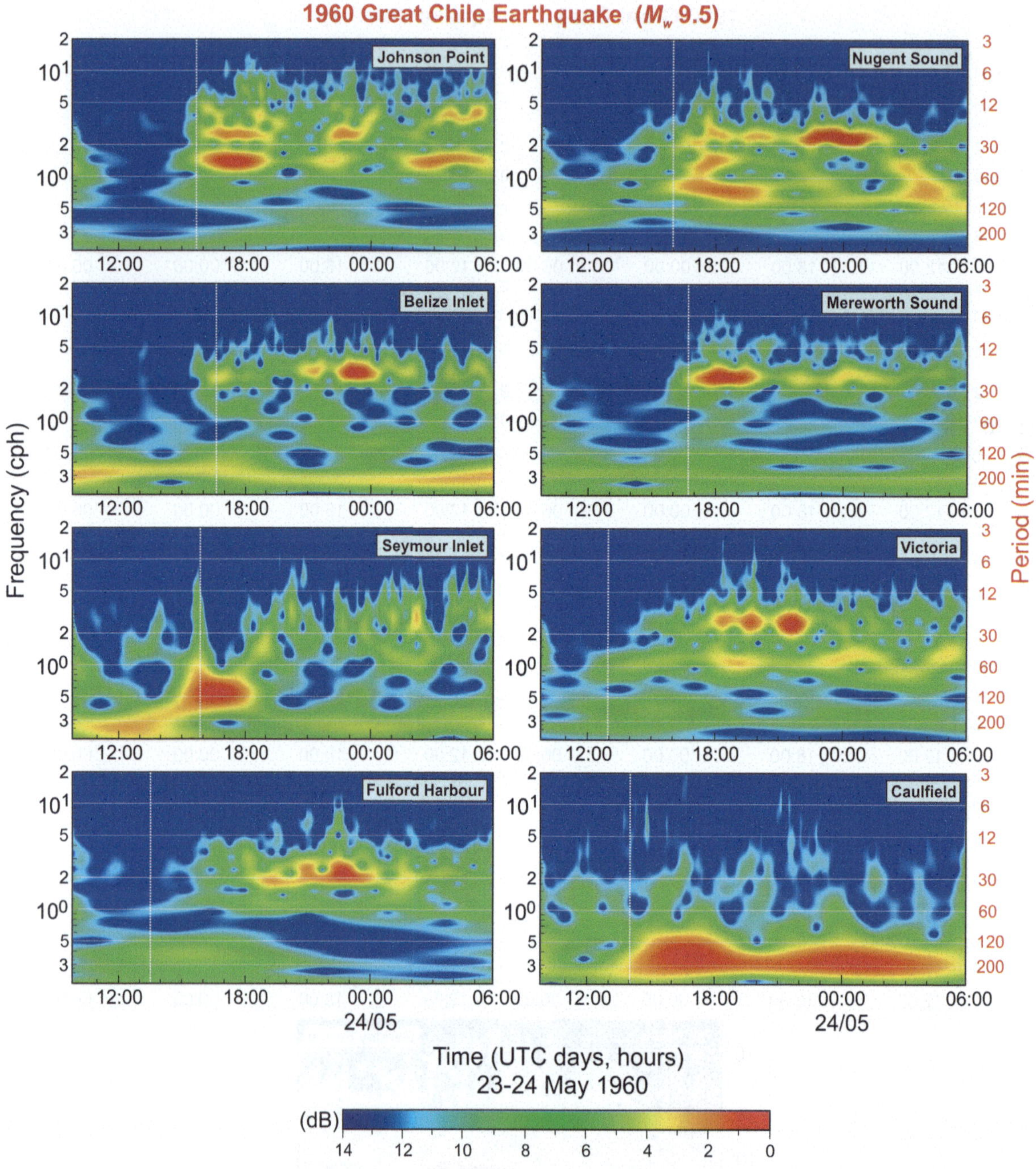

Figure 14
The same as in Fig. 13 but for eight stations located in Juan de Fuca Strait, the Strait of Georgia and in inlets and channels of the west mainland coast

A remarkable feature of the 1964 tsunami on the coast of British Columbia was that *all sea level records* located by WW64, contained the tsunami signal. According to WW64, the tsunami was recorded by twelve CHS tide gauges on the BC coast, consisting of nine permanent and three temporary gauges (Fig. 16). Six of the permanent stations were the same as the ones that recorded the 1960

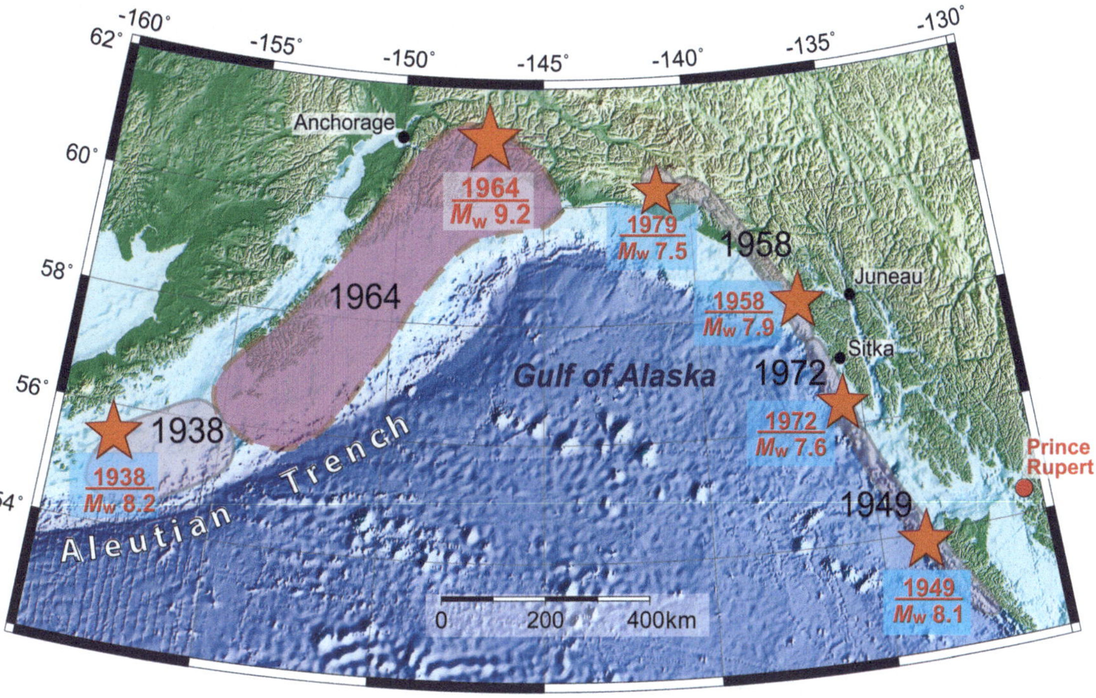

Figure 15

Map of south-central and southeastern Alaska with the rupture zones of six major historical earthquakes (shaded); the rupture zone of the 1964 Alaska earthquake is contoured by a thick brown dashed line. Red stars indicate epicentres of the earthquakes; the sizes of the stars are proportional to the earthquake magnitudes (modified from Suleimani et al. 2013). Tide gauges at Sitka, Juneau and Anchorage are indicated by solid black circles; the red circle denotes the tide gauge at Prince Rupert

Chile tsunami, namely, *Prince Rupert, Alert Bay, Tofino, Victoria, Fulford Harbour* and *Point Atkinson* (formerly, Caulfeild). The three additional permanent stations were *Bella Bella* on the mainland coast of central British Columbia, *Port Alberni* at the head of Alberni Inlet on the Pacific coast of Vancouver Island and *Vancouver* in the Strait of Georgia. The three temporary stations were *Tasu Sound* on the outer coast of Moresby Island, *Ballenas Islands* in the Strait of Georgia, and *Port Moody* located on the southeastern side of Burrard Inlet (Fig. 16). Tasu and Ballenas Islands were instrumented with pressure gauges; all the other stations had standard float-actuated gauges deployed in stilling wells (WW64).

An additional 1964 tsunami record was provided by Crown Zellerbabach Ltd. for Ocean Falls on the central BC mainland coast (Fig. 16). This gauge was located in the power house about 300 m from the river mouth. As low tides were not measured at the site, parts of the record were chopped off (Fig. 17). WW64 was also able to obtain five river gauge records from the Fraser River delta region provided by the Water Resources Branch, Department of Northern Affairs and National Resources: *North Arm, Steveston, New Westminster, Deas Island and Pitt Lake* (see the insert in Fig. 16).

Taken together, WW64 collected and inspected 18 records of the 1964 Alaska tsunami on the coast of British Columbia. They reviewed, slightly revised and enlarged these records and included all of them into their report, except three records: Ballenas Islands, Port Moody and Deas Island "*since they differ only slightly from Point Atkinson, Vancouver and Steveston, respectively*". In the analyses that follow, we used not the original records (which were unavailable) but the 15 records presented in the WW64 report. These records were digitised and examined in the same manner as the records for the four other great events. Figures 17 and 18 show 14 outer and inner coast records with tides: tsunamis

Figure 16
Location of tide gauges on the coast of British Columbia that recorded the 1964 Alaska (Prince William Sound) tsunami and certain villages affected by tsunami waves; "HSC" = Hot Springs Cove. The inset shows "inner" stations situated in Juan de Fuca Strait (Victoria), the Strait of Georgia (Fulford Harbour) and eight stations in the area of metro Vancouver. The Queen Charlotte Islands are now officially known as "Haida Gwaii"

superimposed on high tides significantly strengthen the destructive effect, promoting extensive inundation of lowland areas, in particular, at the head of Alberni Inlet and the adjacent delta of the Somass River. Also, Fig. 17 reveals the low-tide gaps in the Ocean Falls record that required careful processing.

WW64 indicated that all records had been adjusted to Pacific Standard Time (PST = GMT − 8 h); however, an accurate prediction of the tides for the Ocean Falls record shows that the time was in PST +1 h (i.e., in GMT—7 h). This means that all previous timing estimates for this station (WW64; Stephenson et al. 2007) had an error

Table 5

Parameters of the Alaska (Prince William Sound) tsunami of 28 March 1964 recorded by CHS analogue tide gauges on the coast of British Columbia (Main shock, $M_w = 9.2$ at 03:36 UTC)

Station	First wave			Maximum wave			Visible period (min)
	Arrival time (UTC)	Travel time (hh:mm)	Amplitude (cm) and sign	Amplitude (cm)	Time (UTC) of max amplitude	Height (cm)	
Prince Rupert	06:30 (06:52)	2:54	+ 62	132	09:05	271 (271)	110
Tasu Sound	05:32 (05:33)	1:56	+ 96	96	05:52	191 (192)	62
Ocean Falls	06:46 (08:00[a])	3:10	+ 154	154	07:28	376 (381)	32
Bella Bella	06:49 (06:53)	3:13	+ 92	92	07:21	190 (192)	55
Alert Bay	07:23 (07:39)	3:47	+ 120	120	08:52	222 (174)	50
Tofino	07:01 (07:00)	3:25	+ 99	99	07:15	237 (247)	25
Port Alberni	07:50 (08:00)	4:14	+ 300 (?)	420 (?)	11:25 (?)	– (~ 770)	110
Victoria	08:08 (08:02)	4:32	+ 70	70	08:20	147 (146)	25, 90
Fulford Harbour	08:35 (08:35)	4:59	+ 24	42	12:41	83 (61)	30, 110
Pt. Atkinson	09:14 (09:07)	5:38	+ 11	12	09:45	28 (24)	120
Ballenas Is.	– (09:14)	– (5:38)	–	–	–	–	–
Vancouver	09:23 (09:20)	5:47	+ 6	11	09:52	22 (18)	120
Port Moody	– (10:53[b])	– (7:17[b])	–	–	–	–	–
Steveston	09:36 (09:45)	6:00	+7	13	13:48	32 (–)	120
Deas Island	– (10:00)	– (6:24)	–	–	–	–	–
North Arm	09:49 (10:15)	6:13	+ 13	22	14:11	50 (–)	120
New Westminster	10:29 (10:30)	6:53	+ 10	16	12:38	34 (–)	120
Pitt Lake	11:49 (12:00)	8:13	+ 1	5	13:33	9 (–)	120

Parameters of the tsunami waves were estimated using digitised analogue tide gauge records that had been re-sampled to 1-min

Non-CHS tide gauge stations are indicated by italics

Tsunami arrival times and maximum wave heights estimated by WW64 are given in brackets

The question mark ("?") indicates uncertain values at Port Alberni due to damage to the tide gauge during the event

[a]As found from tidal predictions, the indicated time in the Ocean Falls record was not in the Pacific Standard Time (PST), as was assumed by WW64, but in PST + 1 h. Thus, the tsunami arrival estimated by the previous authors should be 1 h earlier

[b]Probably erroneous values; the tsunami arrival times at Vancouver and Port Moody have to be very close

of one hour. Tsunami oscillations are evident and prominent at all stations on the outer coast (Fig. 17), as well as at Victoria and Fulford Harbour (Fig. 18). However, at six stations located near Vancouver and the Fraser River delta, the tsunami signal is more difficult to distinguish because the oscillations are relatively weak, have quite long periods and are strongly masked by tides (Fig. 18).

The one tsunami record not presented in Figs. 17 and 18 is that for Port Alberni. This record (Fig. 19), which had the highest tsunami waves, ever instrumentally measured tsunami waves for the coast of Canada, needs a more detailed discussion. The first tsunami wave arrived at Port Alberni at 07:50 UTC on 28 March (4 h 14 min after the main shock) and damaged the tide gauge, putting it out of use for about 1.5 h (WW64). The gauge then recorded a portion of the second wave between 09:17 and 09:30 UTC and a small portion of the third wave before stopping for several hours. According to WW64, the gauge returned to normal operation at 16:08 UTC (Fig. 19). The first three waves propagated up the Somass River and were recorded by the Water Resources Branch water gauge located ~ 5.5 km up-river from Alberni. Although WW64 did not provide a copy of the water gauge record, they indicated the arrival times and heights of the first three wave crests:

1st wave: 24 cm at ~10:16 UTC;
2nd wave: 80 cm at ~11:53 UTC;
3rd wave: 30 cm at ~13:30 UTC.

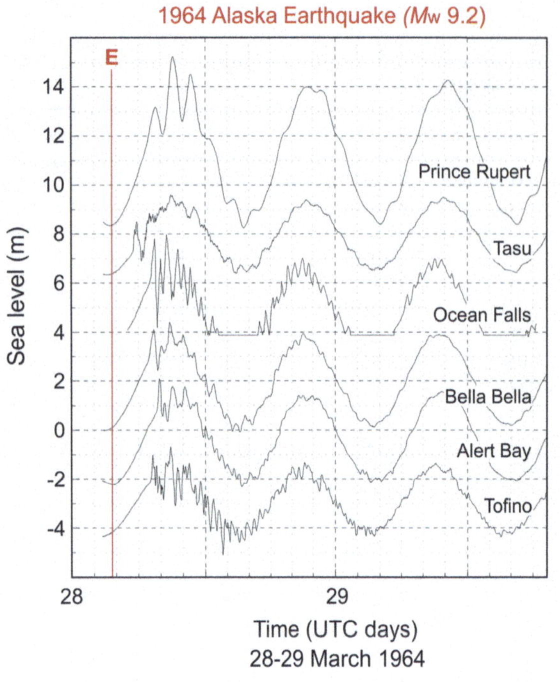

Figure 17
The 28 March 1964 Alaska tsunami, including tides, as recorded at six stations located on the northern mainland coast of British Columbia and on the outer coast of Vancouver Island. For display purposes, the records have been vertically shifted relative to each other. The solid vertical red line labelled "E" denotes the time of the earthquake

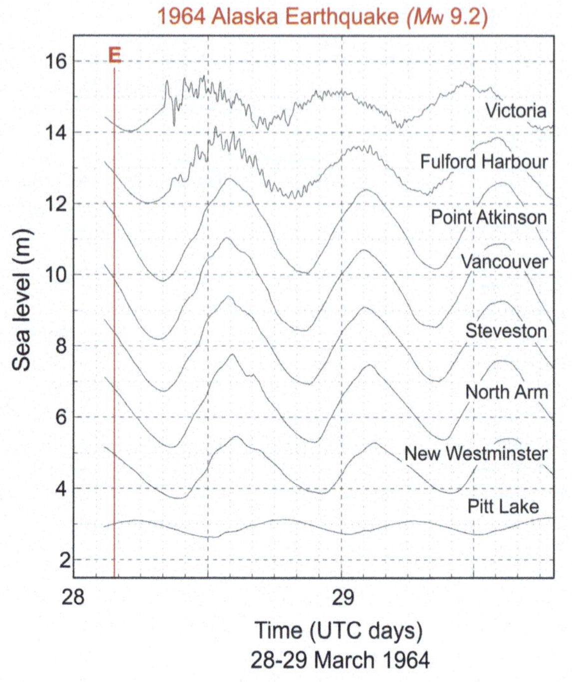

Figure 18
The same as in Fig. 17 but for eight stations located in Juan de Fuca Strait (Victoria), in the Strait of Georgia (Fulford Harbour and Point Atkinson) and in the area of Vancouver and the Fraser River delta. The solid vertical red line labelled "E" denotes the time of the earthquake

The time differences between the 1st, 2nd and 3rd wave crests were the same, 97 min. WW64 assumed that the same height and time intervals took place at Port Alberni as at the Somass River gauge. In retrospect, we suspect that more precise estimates could have been obtained from the de-tided records, but this had not been done at the time and now the Somass River record of the 1964 tsunami is unavailable. Therefore, all existing estimates of the 1964 Alaska tsunami parameters from this region are derived from the WW64 report, which, in addition to the Somass River gauge, used water marks on Port Alberni buildings and eyewitness evidence at the time of the event.

As indicated by the partly restored Port Alberni tide gauge record (Fig. 19), the maximum absolute water level (tsunami + high tide) was associated with the second tsunami wave crest that reached about 27 feet (8.2 m) above chart datum (Fig. 19). However, the highest trough-to-crest height of the tsunami wave alone (without inclusion of the tide) of ~7.7 m was due to the 3rd wave. Fine et al. (2008) digitised the continuous part of the Port Alberni record (after 16:08 UTC) and, from spectral analysis of this record, found a prominent spectral peak with a period of 110 min.[5] The computed period of the fundamental (Helmholtz) mode of Alberni Inlet was ~112 min, in close agreement with the observations. These results confirm the resonant nature of the 1964 tsunami oscillations in Alberni Inlet proposed earlier by Murty and Boilard (1970) and Henry and Murty (1995), who tried to explain the strong amplification of tsunami waves at Port Alberni.

The fourteen tide gauge records of the 1964 tsunami, shown in Figs. 17 and 18, were de-tided (Figs. 20 and 21) and used to estimate the principal

[5] Fine et al. (2008) suggested that the peak corresponds to the period 107 min; however, additional analysis of the spectrum corrected this value slightly to 110 min (see the inset in Fig. 19).

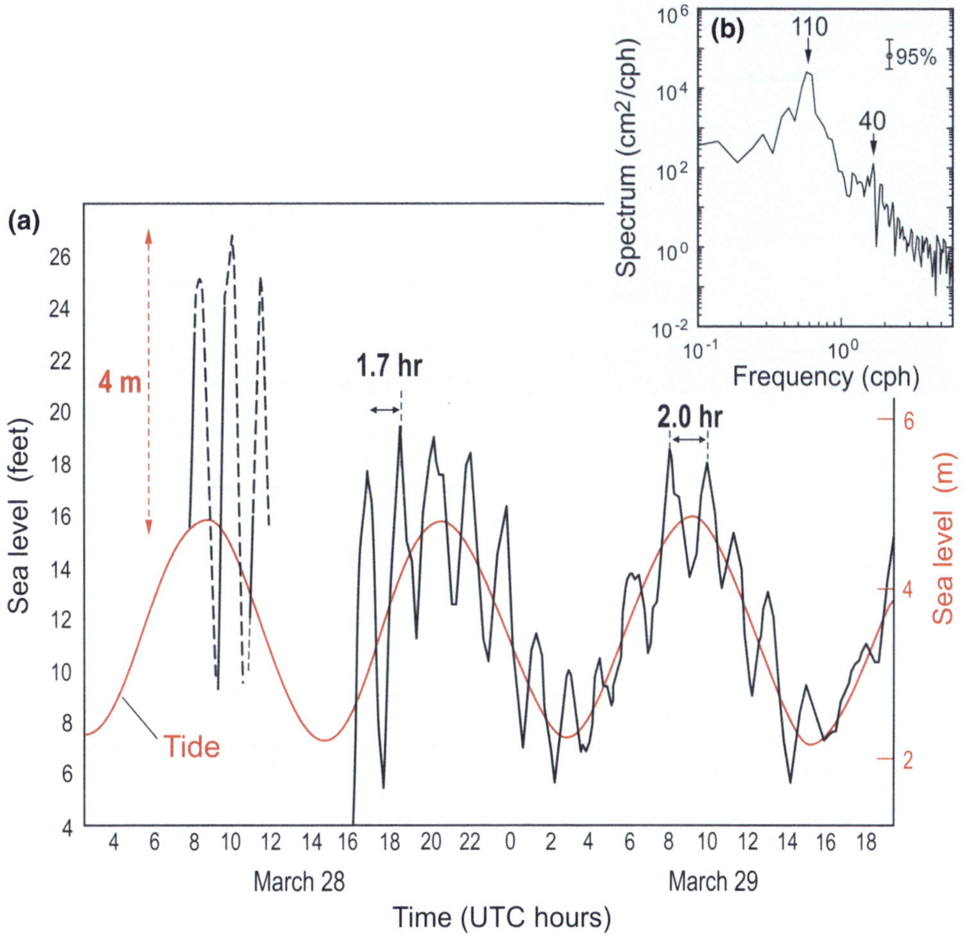

Figure 19
The 1964 Alaska tsunami as recorded at Port Alberni, British Columbia (modified from Fine et al. 2018a, b). **a** A restored copy of the Port Alberni record (from Wigen and White 1964). The predicted tide is shown by the red line. The dashed lines indicate the record segments restored by Wigen and White (1964) based on land marks and the Somass River water gauge observations. The tide gauge stopped working on March 28 after 12:00 UTC, but resumed operation four hours later. The highest recorded sea level, about 4 m above high tide at 10:00 UTC, was due to the second incoming tsunami wave, while the corresponding maximum (peak-to-trough) wave height in Port Alberni was substantially higher than 6 m and occurred during the third, or later wave. The actual sequence numbers of the observed waves and maximum wave height of the largest wave are uncertain because of the 4-h gap in the data. However, based on beach marks, the maximum height was roughly 7.7 m. Oscillations in Alberni Inlet persisted for at least 48 h, with wave periods from 1.7 to 2 h. **b** The spectrum of the tsunami record at Port Alberni shown in **a** for the time period between 16:00 of 28 March and 18:00 UTC of 29 March. Periods (in min) of the two main spectral peaks are indicated

tsunami parameters (Table 5). The 1964 Alaska tsunami was a unique event in that the maximum recorded trough-to-crest wave heights at eight stations were >1 m and at five stations >2 m (Table 5). According to the results in WW64 and those from the numerical modelling of Fine et al. (2018a, b), wave heights along the entire outer coast of British Columbia were close to, or higher, than 2 m. As indicated above, tsunami waves first arrived at Tasu Sound and then to the other five "northern" stations (Fig. 20). According to our estimates, the frontal wave of the 1964 tsunami arrived at Prince Rupert at 06:30 UTC, 2 h and 54 min after the main shock (previous estimates by WW64 showed an arrival time of 06:52). This wave had an amplitude of 62 cm (43 cm according to WW64); the second wave was the highest (Fig. 20), with a wave height of 271 cm (the same as was estimated by WW64). These

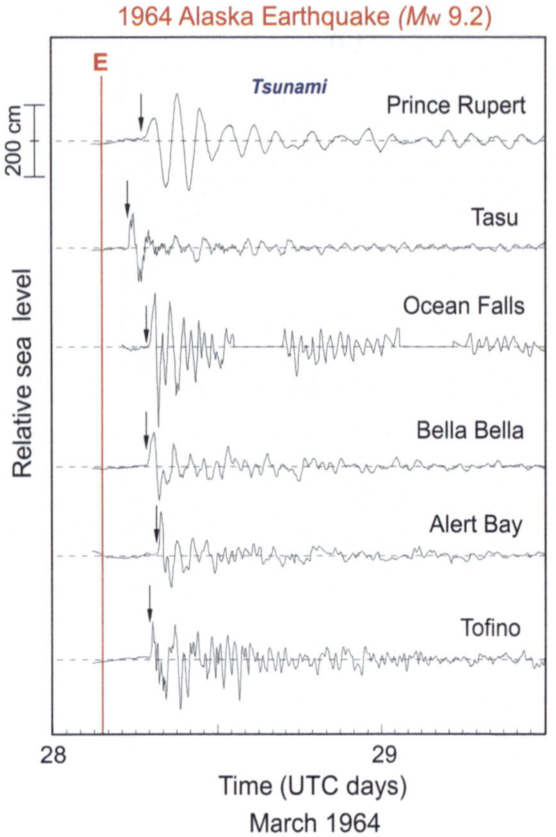

Figure 20
De-tided but unfiltered records of the 1964 tsunami from six "northern" stations on the coast of British Columbia that recorded the 1964 Alaska (Prince William) tsunami. The solid vertical red line labelled "E" denotes the time of the earthquake; the arrows indicate the tsunami arrival

estimates are in good agreement with the numerical results of Fine et al. (2018a). The dominant period of the observed waves at Prince Rupert was about 110 min, which is approximately the same as during the 1952, 1957 and 1960 tsunamis.

As it rounded the Queen Charlotte Islands, the 1964 tsunami entered Queen Charlotte Sound, arriving at Ocean Falls at 6:46 UTC and then, 3 min later, at Bella Bella. These two stations are located on the central mainland BC coast, relatively close to each other (Fig. 16). The shape and general character of the first few oscillations (which were the largest), look similar at these stations (Fig. 20), however the maximum recorded tsunami wave height at Ocean Falls (3.76 m) was almost two times larger than at Bella Bella (1.90 m). The difference demonstrates the strong influence of local topography on the observed tsunami heights. After entering Queen Charlotte Sound, the tsunami travelled southward along the northern coast of Vancouver Island and at 07:23 UTC (3 h 47 min after the earthquake) reached Alert Bay, where it produced a maximum wave height of 2.22 m (Table 5).

The main branch of the 1964 Alaska tsunami propagated along the west coast of Vancouver Island, arriving at Tofino at 7:01 UTC; 49 min later, the tsunami inundated Alberni and Port Alberni. Had the Tsunami Warning System been in operation in 1964, this time difference could have provided an effective early warning. The maximum tsunami wave height of 7.7 m at Port Alberni was $\sim$ 3.2 times higher than the corresponding height of 2.37 m at Tofino, demonstrating once again the strong tsunami amplification from resonance within Alberni Inlet. It is also important to note that at most stations along the outer coast—Tasu Sound, Ocean Falls, Bella Bella (Fig. 20), and Victoria (Fig. 21a)—the first wave was the strongest, whereas at Port Alberni, it was the third wave: Due to the resonance, the fundamental inlet mode consequently accumulated energy and amplified.

As it propagated southward along the outer coast of Vancouver Island, part of the energy of the 1964 tsunami entered Juan de Fuca Strait. The first wave reached Victoria at 8:08 UTC and then entered the Strait of Georgia, arriving at Fulford Harbour at 8:35 UTC. As the tsunami propagated from the ocean inward along the straits, it began to gradually attenuate, with maximum wave heights falling from 2.37 m at Tofino, to 1.47 m at Victoria and 0.83 m at Fulford Harbour (Table 5, Fig. 21a). In Victoria Harbour, the leading wave was the highest. The wave arrived as a pronounced wave crest of 70 cm, followed by an abrupt trough of 77 cm; the period of this wave was about 100 min. Although the same wave is evident in the Fulford Harbour record, the highest wave was the third wave, not the first, apparently as a result of local harbour resonant effects. In contrast to other major tsunamis (1946, 1952, 1957, 1960, 2010 and 2011), the 1964 tsunami at Victoria and Fulford, as well as at other stations of the British Columbia coast (Figs. 20 and 21), decayed relatively quickly (Rabinovich et al. 2018). The reason for this fast decay appears to be the closeness of the 1964 source region to the observation area. As shown by Rabinovich et al. (2011) for the global 2004 Sumatra tsunami, near-field tsunami oscillations decay much faster than more distant far-field oscillations. The prominent feature of this tsunami is the close similarity of the first 2–3 oscillations along the entire outer coast (Fig. 20) and at Victoria and Fulford Harbour (Fig. 21a). The observed tsunami oscillations at Victoria were in good agreement with the results of a high-resolution numerical model of the 1964 tsunami for Victoria Harbour, constructed by Fine et al. (2018b).

After it entered the Strait of Georgia, the 1964 tsunami reached Point Atkinson (formerly, Caulfeild) at 9:14 UTC on 28 March and then propagated into the numerous narrow inlets on the mainland coast and into the Fraser River (see the inset in Fig. 16 for station locations). Although the tsunami oscillations are barely recognisable in the original tidal records for these areas (Fig. 18), the waves are clearly evident in the residuals (de-tided) (Fig. 21b). The character of these oscillations is significantly different from those observed in other regions of the BC coast in that the wave oscillations at all six stations were very similar and much more regular than at other stations. Short-period oscillations were not observed at these stations, while long-period oscillations were very consistent among the records, with a predominant period $T \sim 2$ h. The maximum tsunami wave heights at all stations (ignoring Pitt Lake) ranged from 22 cm at Vancouver to 50 cm at the North Arm in the Fraser River (Table 5).

A unique feature of the 1964 event was tsunami oscillations recorded at Pitt Lake. Pitt Lake has a tidal range of ~1 m and is one of the few tidal lakes on the BC coast. The water gauge is located at the southern tip of the lake, about 18 km from the confluence of the Pitt and Fraser rivers, which, in turn, is about 46 km from the Strait of Georgia. To reach the Pitt Lake gauge from the strait, the 1964 tsunami had to propagate ~ 64 km upstream. The first wave arrived at the lake at 11:49 UTC, 2 h 35 min later than at Point Atkinson and 1 h 20 min later than at New Westminster (Table 5; the station locations are shown in the inset of Fig. 16). Despite the small wave height of 9 cm, the tsunami is clearly detectable in the Pitt Lake record (Fig. 21b).

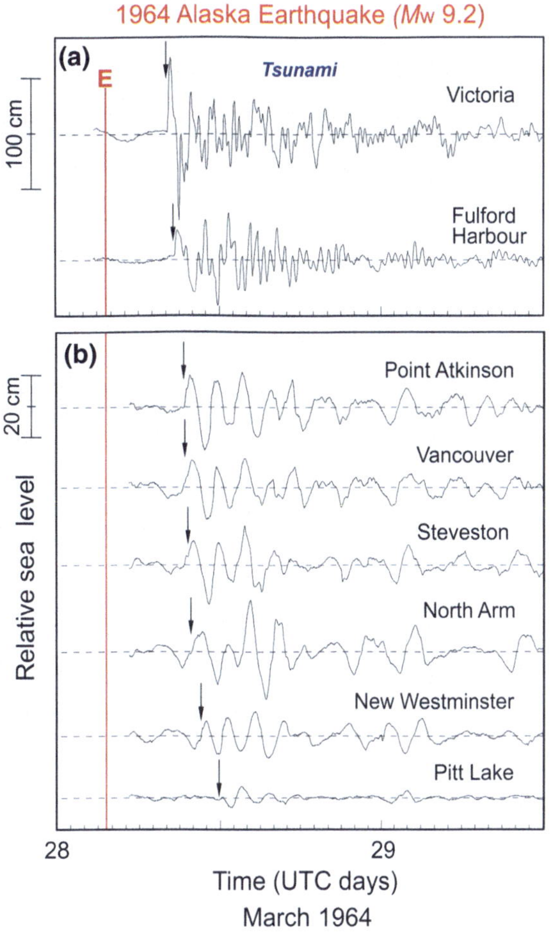

Figure 21

The same as in Fig. 20, but for the "southern" stations located in **a** Juan de Fuca Strait (Victoria) and the Strait of Georgia (Fulford Harbour); and **b** six stations in the area of Vancouver and the Fraser River delta

The *f–t* diagrams in Figs. 22 and 23 provide details of the wave structure and time evolution of the 14 records of the 1964 tsunami shown in Figs. 20 and 21. These diagrams are substantially different from those of the four other tsunamis discussed above: As indicated in *f–t* diagrams in Figs. 4, 6, 8, 13 and 14, there is considerable *individuality* among the various stations but a strong *similarity* at a given station for the different events. In contrast, the 1964 *f–t* diagrams for all BC stations were alike: the tsunami energy was mainly at low frequencies of ~ 0.4–0.8 cph (periods of 2.5–1.2 h) and decayed abruptly within the first 5–6 h. Higher-frequency, shorter duration oscillations with $f \sim 2$ cph

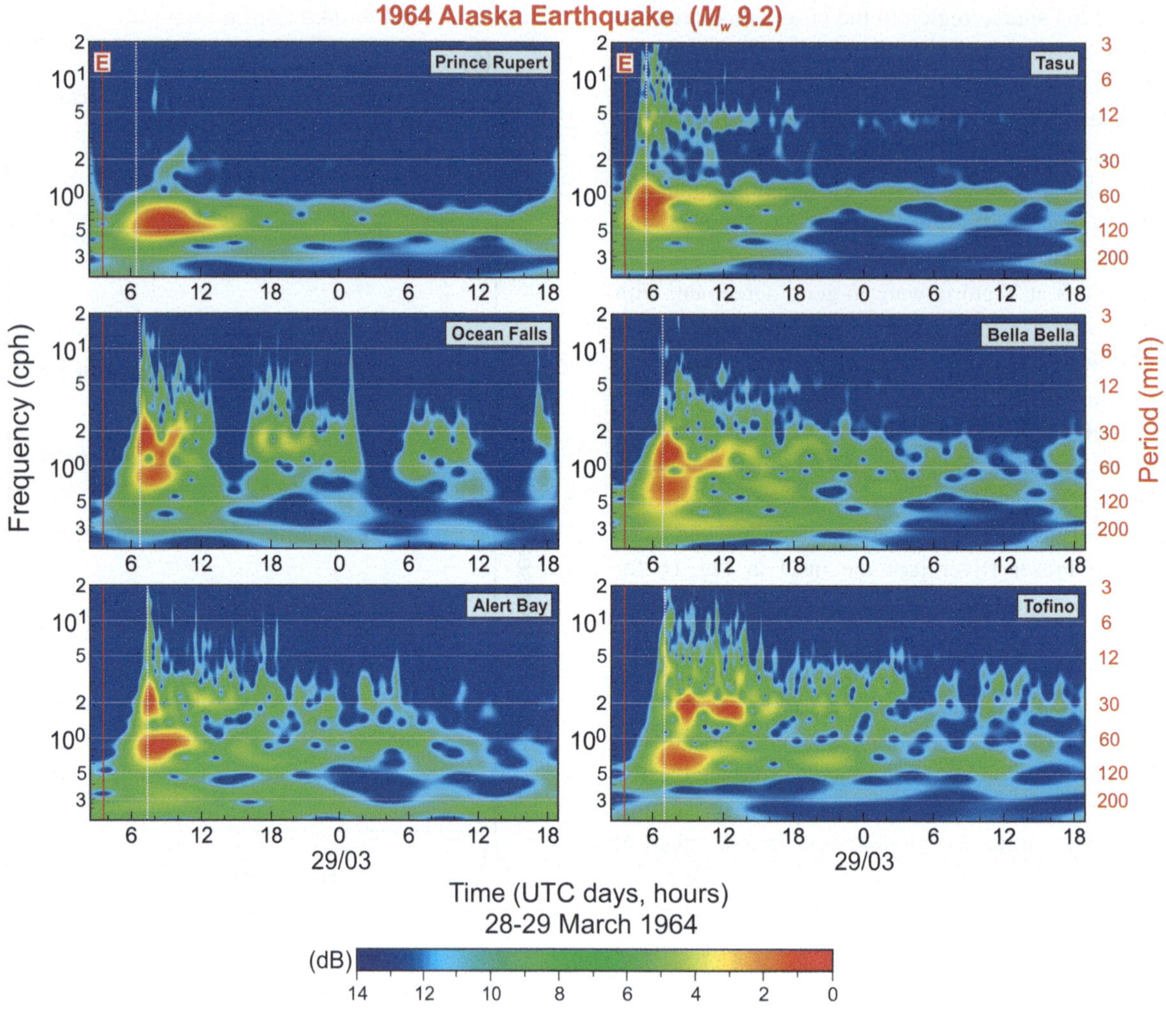

Figure 22

Frequency-time plots (*f–t* diagrams) for the 1964 Alaska tsunami recorded at six "northern" stations on the coast of British Columbia. The solid vertical red line labelled "E" denotes the time of the earthquake; the dashed white line indicates the tsunami arrival

($T \sim 30$ min) are evident at Ocean Falls, Bella Bella, Alert Bay, Tofino, Victoria and Fulford Harbour. At the two northern stations (Prince Rupert and Tasu Sound), as well as at the six stations located in the Fraser River delta region, high-frequency oscillations were totally absent, with the dominant wave period typically around 2 h. Based on these observations, it appears that, in contrast to previous events, the properties of the 1964 oscillations were strongly determined by the *source*, while for the other events, the influence of the regional/local *topography* prevailed. We believe that this is because the 1964 tsunami source was located much closer to the BC coast than the 1946, 1952, 1957 or 1960 sources.

Although the characteristics of the source region have a marked effect on the incoming waves from the 1964 tsunami on the BC coast, the local topography also plays a major role in determining their properties. The extreme waves and pronounced damage in Alberni/Port Alberni were caused by the close match between the dominant period of the arriving waves (~ 2 h according to the *f–t* diagrams shown in Figs. 22 and 23) and the fundamental period of Alberni Inlet. Witness reports and water gauge measurements in several rivers (Wigen and White 1964; White 1966) indicate significant amplification of tsunami waves entering other inlets and fjords of the BC coast (Fig. 24). Also, certain straits and

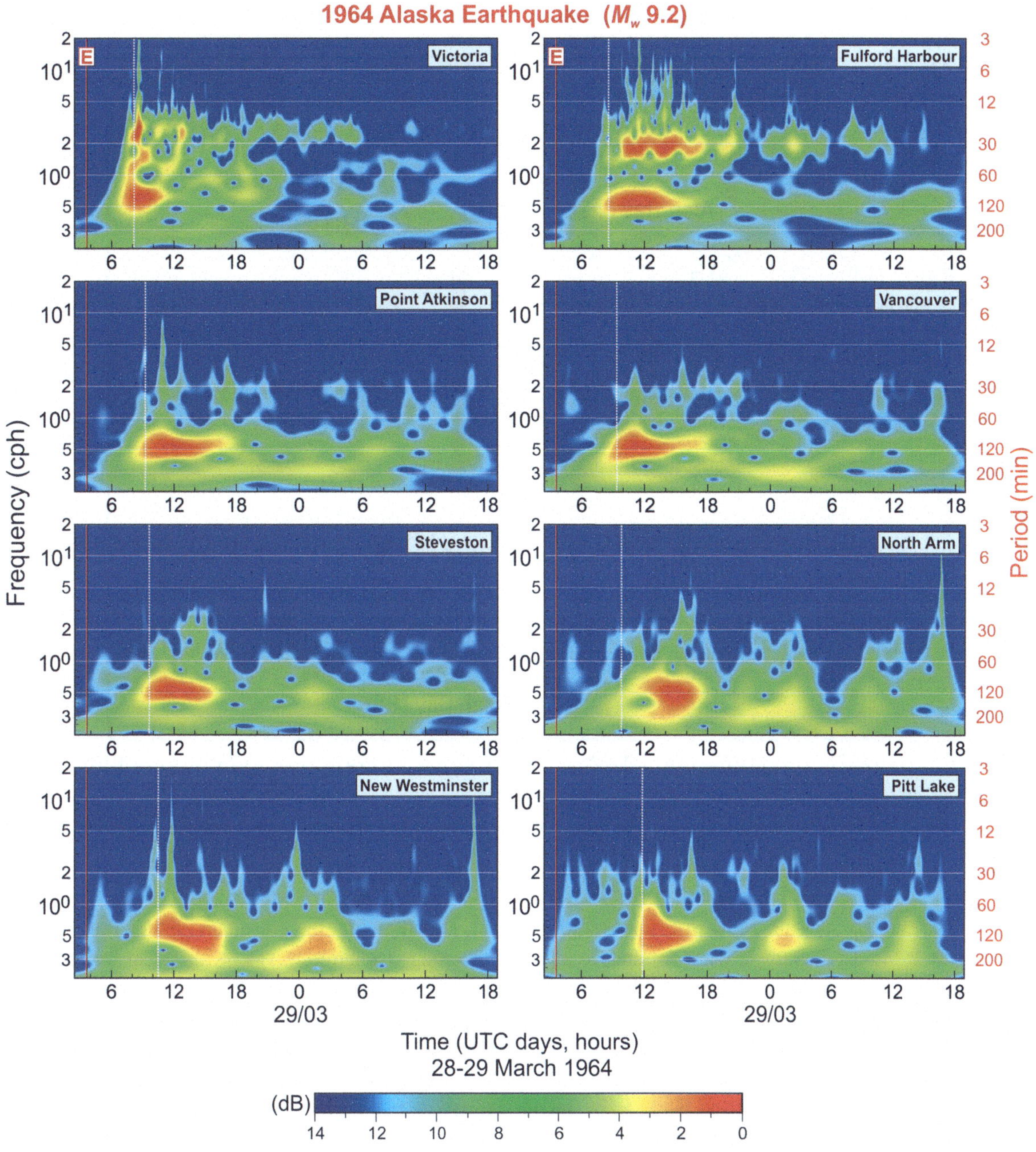

Figure 23

The same as in Fig. 22, but for eight stations located in Juan de Fuca Strait, the Strait of Georgia and the Fraser River delta region (see inset in Fig. 16 for station locations)

channels act as a low-pass filter that effectively suppresses high-frequency (HF) oscillations, while allowing low-frequency (LF) waves to pass through. In particular, Dixon Entrance filters out waves with periods shorter than 1–1.5 h, which is why only long-period waves were recorded at this station during the 1964 event, as well as during the 1952, 1957 and 1960 tsunamis. It appears that Tasu Sound plays a similar role by preventing penetration of the HF "tail" of the 1964 tsunami (Fig. 22). Another region that is "blocked" from HF waves is metro Vancouver and the Fraser River delta. This blockage is the

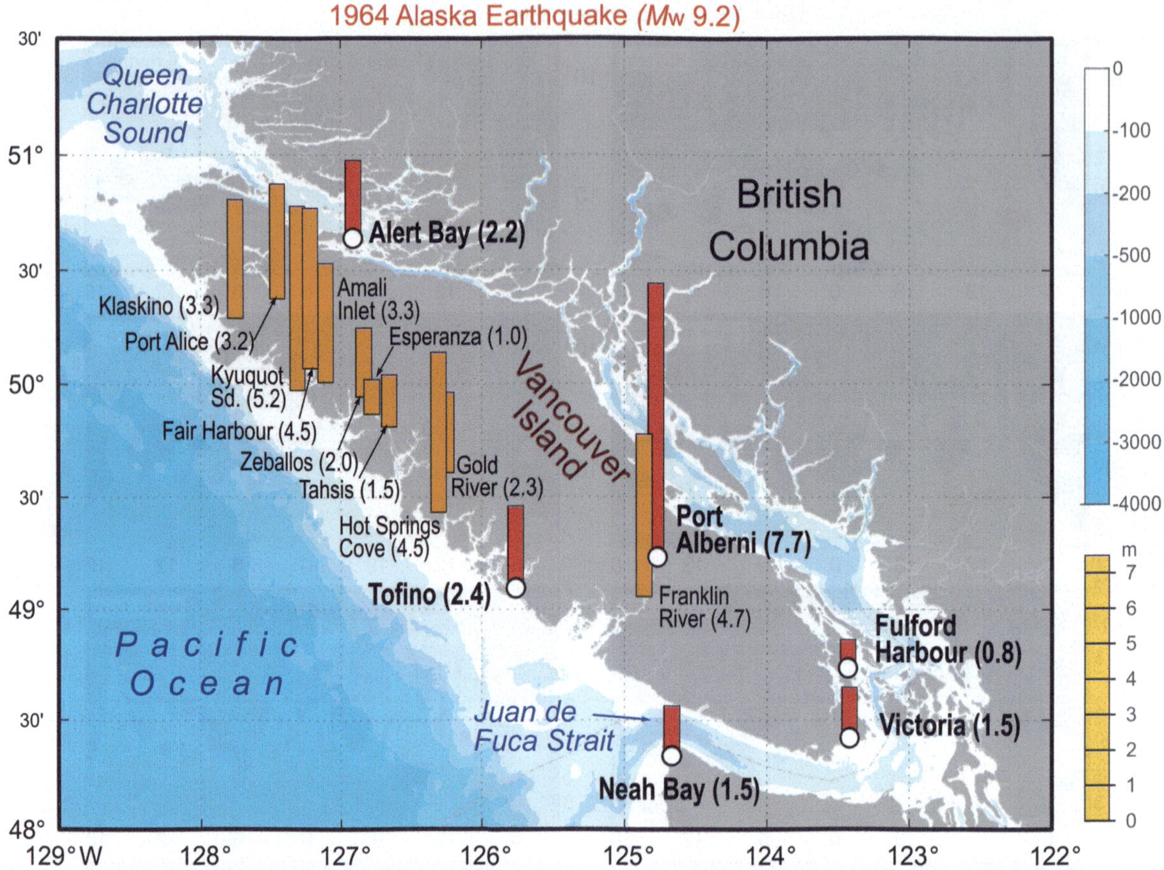

Figure 24
Map of Vancouver Island showing maximum tsunami runup along the coast. The white circles denote tide gauges that were working during the 1964 Alaska event. The red bars indicate maximum wave heights estimated from the tide gauge data (Table 5); the orange bars are runup heights collected by Wilson and Tørum (1972) from Wigen and White (1964), White (1966), witness reports and in-river gauges

reason why the six *f-t* diagrams for this region (Fig. 23) look almost identical, with dominance of long-period motions and the absence of HF oscillations (periods shorter than 1 h).

Wilson and Tørum (1972) summarised available tide gauge/river water gauge measurements and witness reports for the 1964 Alaska tsunami along the coast of Vancouver Island. We re-examined this information and expanded the findings based on our analysis of digitised versions of all existing tsunami records. This enabled us to construct a map of observed runup heights for the island (Fig. 24). In addition to providing a general overview of the 1964 tsunami waves on the coast of Vancouver Island, we are able to show that large wave heights of more than 3–5 m were typical along the entire oceanic coast of the island. These strong oscillations were mainly associated with the resonant interaction of the incoming tsunami waves with local topographic features of the region. The results provide emergency response managers with insight into runup heights to be expected from future tsunamis originating from a $M_w \sim 9.2$ Alaska subduction zone failure in the northeast Pacific.

4. Discussion

One important question that arises from our analysis of the coastal tide gauge records of the five great tsunamis of the 20th century is: *How can we use the information from one "reference" station for real-time tsunami warning and forecasting of tsunami heights and arrival times at other sites?* Our analysis of these five events enables us to examine this problem for three selected stations: Tofino, Victoria and Port Alberni.

The Tsunami Warning Station at Tofino is the most important tsunami monitoring station on the west coast of Canada and one of the most important on the Pacific coast of North America. The station has been in operation for 114 years and has detected 75 tsunamis (Fig. 2), forming one of the most complete compilations of instrumentally recorded tsunamis in the World Ocean. For the Canadian Tsunami Warning System, Tofino can be considered the "*bell-weather station*". Thus, we can use this station as the "*reference (warning) station*" and correlate tsunami parameters at TWS Tofino with those measured at Victoria and Port Alberni.

To complete this study, we have used additional data from two recent great tsunamis: The tsunamis of 27 February 2010 Chile (Maule) and 11 March 2011 Tohoku (East Japan). Both tsunamis were recorded along the entire coast of British Columbia (cf. Rabinovich et al. 2013, 2018) and are the strongest tsunamis ever recorded by high-precision digital tide gauges on the coast of Canada. Estimated parameters of these two tsunamis for Tofino, Port Alberni and Victoria are presented in Table 6. The epicentres of all seven great tsunamis are indicated in Fig. 1.

All seven great tsunamis considered in this study were recorded at both Tofino and Victoria. Therefore, we can use the resulting parameters to estimate empirical relationships between observed maximum tsunami wave heights and arrival times for the two stations. As shown in Fig. 25, the mean ratio of maximum tsunami wave heights at Victoria versus those at Tofino, $H_{\text{Vict}}/H_{\text{Tof}} \approx 0.57 \pm 0.03$ ($N = 7$), where N is the number of the events, is well constrained. Similarly, the mean square correlation coefficient, $R^2 = 0.89$, among events is very high (Fig. 25a). As a consequence, the maximum tsunami wave height expected at Victoria will almost invariably be roughly half that for Tofino. The situation with arrival times is more complicated. In this case, the time differences (specifically, the lead time at Tofino relative to Victoria) depends on the source location. The sources of the seven great tsunamis were located in three distinct regions: (1) Aleutian-Alaska; (2) Kamchatka-Japan; and (3) Chile (Fig. 1). Relative to the coast of Vancouver Island, these source regions were located to the (1) northwest (2) west and (3) south. Accordingly, tsunami waves arrived at Tofino and Victoria from three different directions. In Fig. 25b, the three regions are indicated by source-region specific symbols: Maximum time differences arise when the tsunami waves arrive from the northwest, while minimum time differences occur

Table 6

Parameters of the Chile (Maule) tsunami of 27 February 2010 and the Tohoku (East Japan) tsunami of 11 March 2011 at three CHS stations on the coast of Vancouver Island

Station	First wave			Maximum wave			Visible period (min)
	Arrival time (UTC)	Travel time (hh:mm)	Amplitude (cm) and sign	Amplitude (cm)	Time (UTC) of max amplitude	Height (cm)	
27 February 2010 Chile $M_w = 8.8$ at 06:34 UTC[b]							
Tofino	23:15	16:41	+ 20.3	20.3	23:42	36.6	20–30
Port Alberni	23:49	17:15	+ 31.6	45.0	05:12[a]	97.8	100
Victoria	23:48	17:14	+ 16.2	16.2	00:34[a]	23.2	25
11 March 2011 Tohoku ($M_w = 9.0$ at 05:46 UTC)							
Tofino	15:23	9:37	+ 51.3	56.3	21:31	124.5	25
Port Alberni	16:05	10:19	+ 64.3	71.5	17:50	135.7	85
Victoria	16:17	10:31	+ 20.0	26.4	22:24	52.0	25

[a]28 February 2010

[b]In their estimation of parameters for the 2010 tsunami on the coast of British Columbia, Rabinovich et al. (2013) erroneously considered the frontal trough ("negative phase") as the tsunami arrival. Later, Watada et al. (2014) demonstrated that this negative phase is mainly related to the elasticity of the Earth (see also Eblé et al. 2015). These parameters for the 2010 tsunami waves for Tofino, Port Alberni and Victoria have been corrected for this study

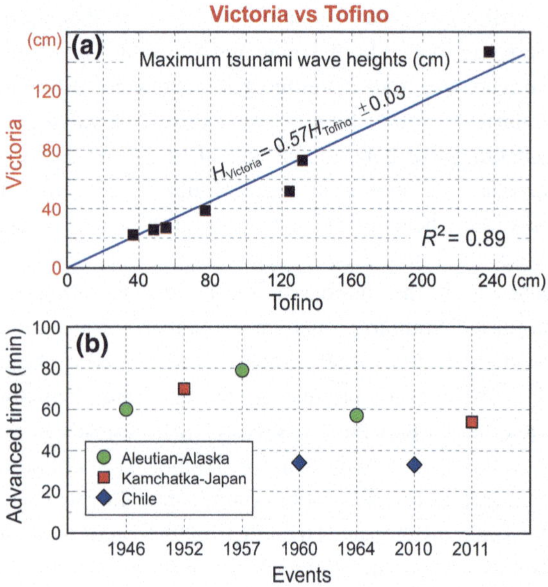

Figure 25

a An empirical relationship between the maximum tsunami wave heights at Victoria vs Tofino for the seven great events (1946, 1952, 1957, 1960, 1964, 2010 and 2011). **b** The differences in tsunami arrival times ("lead time") between Tofino and Victoria for the three groups of source regions: Aleutian-Alaska (1946, 1957 and 1964); Kamchatka-Japan (1952 and 2011); and Chile (1960 and 2011)

when the waves arrive from the south. The mean values for the three source regions are: (1) 69 min; (2) 62 min; and (3) 34 min. These estimates show that TWS Tofino can be effectively used to forecast tsunami waves at Victoria. For tsunami sources located near the coast of South America, the lead time will be a little more than 30 min, while for sources located near Alaska, the Aleutian Islands, the Kamchatka, Kuril Islands and Japan, it will be ~60–70 min.

An attempt to generate similar statistics using tsunami values for Port Alberni was only marginally successful. There were only three great tsunamis recorded at Port Alberni (1964, 2010 and 2011) and these led to markedly different height ratios, H_{PA}/H_{Tof}, for Port Alberni versus Tofino. Specifically, 3.2 (1964), 2.6 (2010) and 1.1 (2011), making it impossible to provide definite conclusions regarding the expected tsunami heights at Port Alberni based solely on the information from Tofino. However, the estimates of arrival time differences for the three events are much steadier: 49, 34 and 42 min, respectively. This means that for "southern sources" (Chile), the lead time of Tofino over Port Alberni is approximately the same as for Victoria, while for "northwestern" and "western" sources, it is ~ 40–50 min.

5. Conclusions

The five great earthquakes examined in this study—1946 Aleutian Islands (M_w 8.6), 1952 Kamchatka (M_w 9.0), 1957 Andreanof Islands (M_w 8.6), 1960 Chile (M_w 9.5), and 1964 Alaska (M_w 9.2) earthquakes—generated destructive trans-Pacific tsunamis that were recorded by analogue tide gauges along the coast of British Columbia. Analysis of these analogue records is important because the five tsunamis, together with the recent 2010 Chile and 2011 Tohoku tsunamis, were the strongest events ever instrumentally recorded on the west coast of Canada. Moreover, one them—the 1964 Alaska tsunami—was the most destructive tsunami in Canadian history, causing about $10 million in damage (in 1964 dollars). If such a tsunami were to occur today, the cost of the damage would be much greater. The maximum tsunami wave heights ever recorded on the BC coast are associated with this particular event.

Altogether, we found and processed 2 records for 1946, 5 for 1952, 6 for 1957, 17 for 1960 and 15 for 1964. All five tsunamis were recorded at Tofino and Victoria, and four of the tsunamis (except 1946) were also measured at Prince Rupert and Alert Bay. The results of these analyses are essential for addressing three major problems:

(1) *Cataloguing of tsunami events.*
Based on the new data and tsunami wave estimates, we were able to correct earlier information on the five events for the BC coast presented in the catalogue of Stephenson et al. (2007).

(2) *Estimation of the potential tsunami risk to the BC coast.*
Precise values of maximum tsunami heights and amplitudes are critical for long-term tsunami forecasting. This information has already been used to evaluate possible tsunami wave heights at

Prince Rupert and Victoria (Rabinovich et al. 2018) and can be used for other sites on the BC coast. These estimates also help identify the probable sources of tsunami waves that are a major threat for this coast. Based on our analysis of the five tsunamis, and taking into account the 2010 and 2011 events, we conclude that the Alaska region, and major earthquakes in this region similar to the 1964 event, have highest potential as a source region for major tsunamis on the coast of British Columbia.

(3) *Verification and calibration of numerical tsunami models.*
The research has enabled us to provide model verification data for reliable modelling of maximum tsunami wave heights and associated currents for specific sites along the coast of British Columbia. In particular, the 1964 tsunami records at Prince Rupert, Victoria and Fulford Harbour were used to verify the 1964 numerical model of Fine et al. (2018a, b).

Each of the five examined great tsunami events has its own individuality and specific properties. At the same time, the waves have many similar features. The five tsunamis had different amplitudes on the BC coast and produced quite different wave heights. However, the general character of the recorded oscillations, their dominant frequencies and evolution in time are mostly alike. It is well known that the main spectral peaks for different tsunami events at the same site are similar, while the spectral peaks from the same tsunami event, but at different sites, are substantially different (cf. Honda et al. 1908; Rabinovich 1997; see also Fig. 8 in Zaytsev et al. 2017). The results of the present study demonstrate that not only spectra, but also *f–t* diagrams (wavelets), for the same site are similar for different events. This means that tsunami waves for different events not only oscillate at the same frequencies (likely as a result of frequency-specific resonance) but also evolve and decay in comparable ways. It is evident that this similarity is related to the local/regional topography of the sites and that, for far-field tsunamis (like those in 1946, 1952, 1957 and 1960), the topographic admittance function (frequency response) determines the character of the observed waves. Rabinovich and Stephenson (2004) used long time series of background oscillations to estimate the frequency response of the primary BC tide gauge stations. Not unexpectedly, the peaks of these functions for the respective sites match the high-energy frequency bands at the corresponding *f–t* plots for these four tsunamis. One of the important properties of the four earliest (1946–1960) tsunamis examined in this study, is the long ringing duration of 3–5 days and the slow energy decay. For some sites, it was quite common to find maximum tsunami wave heights occurring several hours after the first wave arrival.

The 1964 Alaska tsunami for the BC coast was distinct in several ways. Specifically, this was the strongest tsunami ever recorded on this coast, with waves at eight stations reaching maximum trough-to-crest wave heights that were higher than 1 m. This included Port Alberni (7.7 m), Ocean Falls (3.76 m), Prince Rupert (2.71 m), Tofino (2.37 m) and Alert Bay (2.22 m). At all of these stations, the maximum wave occurred within the first three waves. Wave heights then rapidly attenuated. In comparison with other events, the "ringing" of the 1964 tsunami was much shorter (< 1.5 days) and the energy decay much faster. While the wave spectra for four of the other tsunamis differed among stations, with each having its own spectral "individuality" (as evident in their *f–t* diagrams), corresponding diagrams for the 1964 tsunami records were quite similar. It appears that the differences in the *f–t* diagrams for individual stations for the earlier (1946–1960) tsunamis are determined by the differences in topographic response, while the similarity of the 1964 diagrams is determined by the source characteristics. The distinct properties of the 1964 tsunami waves are explained by the proximity of the 1964 source area to the BC coast. For this coast, the 1964 tsunami can be considered as a "near-field" event; at stations in the Pacific Ocean located at greater distances from the source, the ringing was much longer, the energy decay much slower and the spectral structure markedly different for different stations (cf. Spaeth and Berkman 1967).

An intriguing feature of the five tsunamis is their deep penetration inside the narrow channels, fjords and inlets of the BC coast. For example, the 1952 Kamchatka tsunami was recorded at Kitimat at the head of Kitimat Arm, which itself is a continuation of the long and narrow Douglas Channel. The wave propagated $\sim$ 80 km from Hecate Strait to the

Kitimat tide gauge. The 1957 Andreanof Islands tsunami was measured at Bella Coola, located at the head of North Bentinck Arm, the remotest part of Burke Channel, a distance of ~ 110 from the open ocean (Queen Charlotte Sound). The 1960 Chile tsunami was recorded at 17 stations on the BC coast; five of these stations were located well within the Seymour—Belize Inlet System on the central mainland coast of British Columbia. To reach these stations, the tsunami had to go through Nakawakto Rapids, a narrow constriction having the fastest tidal currents in the world ocean (up to 17 knots). The 1960 tsunami also propagated over 100 km through Juan de Fuca Strait and a series of narrow channels in the Gulf Islands and San Juan Islands into the Strait of Georgia.

The 1964 Alaska tsunami was recorded at 18 tide gauge stations. More than half of these stations were located in the Strait of Georgia and the Fraser River. The tsunami waves propagated far upstream in the river and were recorded at Pitt Lake, which is connected with the Strait of Georgia by a 64-km route through the Fraser and Pitt rivers. It took about 2.5 h for the tsunami wave to travel this route, all the time the attenuating waves maintaining their shape and spectral structure. The recorded waves at this station, as well as at other stations in the region of metro Vancouver and the Fraser River delta, had an almost monochromatic character with one strongly dominant period of about two hours.

With respect to tsunami wave penetration into fjords and inlets, we emphasise that, although the channels hinder propagation of the waves, they can also lead to strong resonant amplification when the dominant frequency of the incoming waves coincides with the eigen (natural) frequency of the interconnected basins. The most noteworthy example of this kind for the BC coast, is the resonant amplification of the 1964 tsunami waves in Alberni Inlet, which resulted in severe damage to the towns of Alberni and Port Alberni.

Acknowledgements

Most of the analogue records digitised and processed in our study were originally collected and examined by Sydney Wigen. We thank Drs. Mohammad Heidarzadeh (Brunel University, London, UK) and Slava Gusiakov (Institute of Computational Mathematics and Mathematical Geophysics, Novosibirsk, Russia) for their positive criticism and valuable comments. We would like to acknowledge the help of Dr. Elena Suleimani from the University of Alaska, Fairbanks for providing us with essential information on the 1964 Alaska earthquake and Dr. Isaac Fine of the Institute of Ocean Sciences, Sidney, BC, Canada, for useful discussion and advice. The work for AR was partially supported by the Russian State Assignment of IO RAS # 0149-2019-0005.

References

Abe, K. (1973). Tsunami and mechanism of great earthquakes. *Physics of Earth and Planetary Interiors,7*, 143–153.

Abe, K. (1979). Size of great earthquakes of 1873–1974 inferred from tsunami data. *Journal of Geophysical. Research,84*(B4), 1561–1568.

Berkman, S. C., & Symons, J. M. (1960). *The tsunami of May 22, 1960 as recorded at tide gauge stations* (p. 79). Washington, D.C.: US Department of Commerce, Coast and Geodetic Survey.

Clague, J. J. (2001). Tsunamis. In G. R. Brooks (Ed.), *A Synthesis of Geological Hazards in Canada* (vol. 548, pp. 27–42). Ottawa: Geol. Surv. Canada Bull.

Clague, J. J., Bobrowsky, P. T., & Hutchinson, I. (2000). A review of geological records of large tsunamis at Vancouver Island, British Columbia, and implications for hazard. *Quaternary Science Reviews,19*, 849–863.

Clague, J. J., Munro, A., & Murty, T. S. (2003). Tsunami hazard and risk in Canada. *Natural Hazards,28*(2–3), 407–434.

Eblé, M. C., Mungov, G., & Rabinovich, A. B. (2015). On the leading negative phase of major 2010–2014 tsunamis. *Pure and Applied Geophysics,172*(12), 3493–3508. https://doi.org/10.1007/s00024-01.

Fine, I. V., Cherniawsky, J. Y., Rabinovich, A. B., & Stephenson, F. E. (2008). Numerical modeling and observations of tsunami waves in Alberni Inlet and Barkley Sound British Columbia. *Pure and Applied Geophysics,165*(11/12), 2019–2044.

Fine, I.V., Thomson, R.E., Lupton, L.M., & Mundschutz, S. (2018a). Numerical modelling of an Alaska 1964-type tsunami at the Canadian Coast Guard Base in Seal Cove, British Columbia. Canadian Technical Report on Hydrography and Ocean Sciences Fs97-18/321E-PDF; 978-0-660-24928-5.

Fine, I.V., Thomson, R.E., Lupton, L.M., & Mundschutz, S. (2018b). Numerical modelling of an Alaska 1964-type tsunami at the Canadian Coast Guard Base in Victoria, British Columbia.

Canadian Technical Report on Hydrography and Ocean Sciences Fs97-18/323E-PDF; 978-0-660-25253-7.

Fuchs, V. (1982). *Of Ice and Men: The Story of the British Antarctic Survey 1943–1973* (p. 383). Avon: Antony Nelson Ltd.

Green, C. K. (1946). Seismic sea wave of April 1, 1946 as recorded on tide gages. *Transactions of the American Geophysical Union,27*, 490–500.

Gusiakov, V. K. (2014). Strongest tsunamis in the World Ocean and the problem of marine coastal security. *Izvestiya, Atmospheric and Oceanic Physics,50*(5), 435–444.

Henry, R. F., & Murty, T. S. (1995). Tsunami amplification due to resonance in Alberni Inlet: Normal modes. In Y. Tsushiya & N. Shuto (Eds.), *Tsunami: Progress in Prediction, Disaster Prevention and Warning* (pp. 117–128). Dordrecht: Kluwer.

Honda, K., Terada, T., Yoshida, Y., & Isitani, D. (1908), An investigation on the secondary undulations of oceanic tides. Journal of the College of Science, Imperial University of Tokyo, vol 24, p 1–110.

Johnson, J. M., Tanioka, Y., Ruff, L. J., Satake, K., Kanamori, H., & Sykes, L. R. (1994). The 1957 Great Aleutian earthquake. *Pure and Applied Geophysics,142*(1), 3–28.

Johnson, J. M., Satake, K., Holdahl, S. R., & Sauber, J. (1996). The 1964 Prince William Sound earthquake—Joint inversion of tsunami waveforms and geodetic data. *Journal of Geophysical. Research,101*(B1), 523–532.

Kanamori, H. (1972). Mechanism of tsunami earthquakes. *Physics of Earth and Planetary Interiors,6*(5), 346–359.

Kong, L. S. L., Dunbar, P. K., & Arcos, N. (Eds.). (2015). *Pacific Tsunami Warning System: A Half-Century of Protecting the Pacific, 1965–2015* (p. 188). Honolulu: ITIC/NOAA.

Kowalik, Z., Knight, W., Logan, T., & Whitmore, P. (2007). The tsunami of 26 December 2004: Numerical modeling and energy considerations. *Pure and Applied Geophysics,164*, 379–393.

Lander, J. F. (1996). *Tsunamis Affecting Alaska, 1737–1996* (p. 195). Boulder: USDC/NOAA.

Lander, J. F., & Lockridge, P. A. (1989). *United States Tsunamis (Including United States Possessions), 1690–1988* (p. 265). Boulder: National Geophysical Data Center, USDC/NOAA.

Leonard, L. J., Rogers, G. C., & Mazotti, S. (2014). Tsunami hazard assessment of Canada. *Natural Hazards,70*, 237–274.

Miller, G. R., Munk, W. H., & Snodgrass, F. E. (1962). Long-period waves over California's borderland. Part II, Tsunamis. *Journal of Marine Research,20*(1), 31–41.

Murty, T. S. (1992). Tsunami threat to the British Columbia coast. *Geotechnique and Natural Hazards* (pp. 81–89). Vancouver: BiTech. Publ.

Murty, T. S., & Boilard, L. (1970). The tsunami in Alberni Inlet caused by the Alaska earthquake of March, 1964. In W.M. Adams (Ed.), *Tsunami in the Pacific Ocean* (pp. 165–187). Honolulu: East West Center Press.

Myers, E. P., & Baptista, A. M. (2001). Analysis of factors influencing simulations of the 1993 Hokkaido Nansei-Oki and 1964 Alaska tsunamis. *Natural Hazards,23*, 1–28.

Okal, E., Synolakis, C., Fryer, G., et al. (2002). A field survey of the 1946 Aleutian tsunami in the far field. *Seismological Research Letters,73*, 490–502.

Prince Rupert Daily News (1964), Newspaper articles for March 30—April 7, 1964.

Rabinovich, A. B. (1997). Spectral analysis of tsunami waves: Separation of source and topography effects. *Journal of Geophysical Research,102*(C6), 12663–12676.

Rabinovich, A. B., & Stephenson, F. E. (2004). Longwave measurements for the coast of British Columbia and improvements to the tsunami warning capability. *Natural Hazards,32*(3), 313–343.

Rabinovich, A. B., Thomson, R. E., & Stephenson, F. E. (2006). The Sumatra tsunami of 26 December 2004 as observed in the North Pacific and North Atlantic oceans. *Surveys in Geophysics,27*, 647–677.

Rabinovich, A. B., & Thomson, R. E. (2007). The 26 December 2004 Sumatra tsunami: Analysis of tide gauge data from the World Ocean. Part 1: Indian Ocean and South Africa. *Pure and Applied Geophysics,164*(2/3), 261–308.

Rabinovich, A. B., Candella, R., & Thomson, R. E. (2011). Energy decay of the 2004 Sumatra tsunami in the world ocean. *Pure and Applied Geophysics,168*(11), 1919–1950. https://doi.org/10.1007/s00024-01-0279-1.

Rabinovich, A.B., Thomson, R.E. & Fine, I.V. (2013), The 2010 Chilean tsunami off the west coast of Canada and the northwest coast of the United States. *Pure and Applied Geophysics, 170*, 1529–1565. https://doi.org/10.1007/s00024-012-0541-1.

Rabinovich, A. B., Thomson, R. E., Lupton, L. M., & Mundschutz, S. (2018). Historical Tsunamis at the Seal Cove and Victoria Canadian Coast Guard Stations, British Columbia. *Canadian Technical Report of Hydrography and Ocean Sciences,325*, 44.

Salsman, G. G. (1959). *The tsunami of March 9, 1957, as recorded at tide stations* (6th ed., p. 18). Washington: US Department of Commerce, Coast and Geodetic Survey, Techn. Bull.

Satake, K., & Atwater, B. (2007). Long-term perspectives on giant earthquakes and tsunamis at subduction zones. *Annual Review of Earth and Planetary Sciences,35*, 349–374.

Sheppard, F. P., MacDonald, G. A., & Cox, D. C. (1950). The tsunami of April 1, 1946. *Bulletin of the Scripps Institution of Oceanography,* University of California, La Jolla,*5*, 391–528.

Spaeth, M. G., & Berkman, S. C. (1967). *The Tsunami of March 28, 1964, as Recorded at Tide Stations*, ESSA Technical Report C&GS 33 (p. 86). Md: Rockville.

Stephenson, F.E., Rabinovich, A.B., Solovieva, O.N., Kulikov, E.A. & Yakovenko, O.I. (2007), *Catalogue of Tsunamis, British Columbia, Canada: 1700–2007*. Preprint. Moscow: P.P. Shirshov Inst. Oceanology, p. 134.

Suito, H., & Freymueller, J. T. (2009). A viscoelastic and afterslip postseismic deformation model for the 1964 Alaska earthquake. *Journal of Geophysical Research,114*, B11404. https://doi.org/10.1029/2008JB005954.

Suleimani, E.N., Nicolsky, D.J., & Koehler, R.D., (2013), *Tsunami Inundation Maps of Sitka, Alaska*, Report of Investigations 2013-3, State of Alaska, Department of Natural Resources, Division of Geological and Geophysical Surveys, Fairbanks, AK, p. 76.

Takahashi, R., & Hatori, T. (1961). A Summary Report on the Chilean Tsunami of May 1960. *Report on the Chilean Tsunami* (pp. 23–34). Tokyo: Comm. Field Investigation on Chilean Tsunami.

The Great Alaska Earthquake of 1964 (1972), Committee on the Alaska Earthquake of the Division of Earth Sciences National Research Council, National Academy of Sciences, Washington, D.C., p. 596.

Thomson, R. E. (1981), *Oceanography of the British Columbia Coast*. Can. Special Pub. Fish. Aquat. Sci., 56. Ottawa, p. 291.

Thomson, R. E., & Emery, W. J. (2014). *Data Analysis Methods in Physical Oceanography* (3rd ed., p. 716). New York: Elsevier Science.

Titov, V. V., Rabinovich, A. B., Mofjeld, H., Thomson, R. E., & González, F. I. (2005). The global reach of the 26 December 2004 Sumatra tsunami. *Science,309*, 2045–2048.

Van Dorn, W. G. (1984). Some tsunami characteristics deducible from tide records. *Journal of Physical Oceanography,14*, 353–363.

Watada, S., Ksumoto, S., & Satake, K. (2014). Traveltime delay and initial phase reversal of distant tsunamis coupled with the self-gravitating elastic Earth. *Journal Geophysical Research, Solid Earth,119*, 4287–4310. https://doi.org/10.1002/2013JB010841.

Weeks, S., & Studds, R.F.A. (1953), The *Tsunami of November 4, 1952 as Recorded at the Tide Gauges*. Spec. Publ. #300, U.S. Department of Commerce, Coast and Geodetic Survey, Washington, D.C., p. 62.

White, W.R.H. (1966), The Alaska Earthquake—Its Effect in Canada, *Canadian Geographic Journal*, 210–219.

Wigen, S. O. (1960). *Tsunami of May 22, 1960*. West Coast of Canada: Unpublished Report, Canadian Hydrographic Service, Sidney, BC.

Wigen, S. O. (1983). Historical studies of tsunamis at Tofino, Canada. In K. Iida & T. Iwasaki (Eds.), *Tsunamis—Their Science and Engineering* (pp. 105–119). Tokyo, Japan: Terra Sci. Publ. Comp.

Wigen, S. O., & White, W. R. H. (1964). *Tsunami of March 27–29, 1964, West Coast of Canada* (p. 12). Victoria, BC, Canada: Dept. Mines Techn. Surv.

Wilson, B. W., & Tørum, A. (1972). Runup heights of the major tsunami on North American coasts. *The Great Alaska Earthquake of 1964: Oceanography and Coastal Engineering* (pp. 158–180). Washington, D.C.: National Academy of Sciences.

Zaytsev, O., Rabinovich, A.B., and Thomson, R.E. (2017), The 2011 Tohoku tsunami on the coast of Mexico: A case study. *Pure and Applied Geophysics, 174* (8), 2961-2986. https://doi.org/10.1007/s00024-017-1593-z.

(Received November 26, 2018, revised February 5, 2019, accepted February 11, 2019, Published online March 1, 2019)

Pure Appl. Geophys. 176 (2019), 2925–2949

https://doi.org/10.1007/s00024-019-02151-1

Pure and Applied Geophysics

The 20th July 2017 Bodrum–Kos Tsunami Field Survey

Gozde Guney Dogan,[1] Alessandro Annunziato,[2] Gerassimos A. Papadopoulos,[3] Hasan Gokhan Guler,[1] Ahmet Cevdet Yalciner,[1] Tarık Eray Cakir,[4] Ceren Ozer Sozdinler,[5,6] Ergin Ulutas,[7] Taro Arikawa,[8] Mehmet Lutfi Suzen,[9] Isikhan Guler,[1] Pamela Probst,[2] Utku Kânoğlu,[10] and Costas Synolakis[11,12]

Abstract—The July 20, 2017 Bodrum–Kos Earthquake caused tsunami wave motions and damage in the south of Bodrum Peninsula, Turkey, and on Kos Island, Greece. Immediately after the earthquake, we conducted several post-tsunami field surveys including interviews in coastal zones impacted by the tsunami, i.e., the coastlines of Bodrum Peninsula, Karaada Islet and Akyaka Town in Gökova Bay, Turkey, and eastern Kos Island, Greece. We present observations and measurements to document the variation of the tsunami effects along the coast. The largest tsunami runup was about 1.9 m and observed at the mouth of a small dry streambed at Gumbet Bay, Bodrum. No significant water motions were reported at the northern and western coasts of Bodrum Peninsula. The tsunami runup distribution along the coast of eastern Kos was overall regular, with runup not exceeding 1 m except in the Port of Kos where a 1.5 m tsunami runup was measured.

Key words: Bodrum, Kos, earthquake, tsunami, post-tsunami field survey, tsunami effects, tsunami runup.

1. Introduction

The Aegean Sea is one of the most active and complicated parts of the western Alpine-Himalayan belt (e.g., McKenzie 1972) and suffers from large earthquakes and volcanic eruptions, some of which have caused destructive tsunamis (Galanopoulos 1960; Ambraseys 1962; Kuran and Yalciner 1993; Minoura et al. 2000; Ambraseys and Synolakis 2010; Papadopoulos et al. 2012, 2014; England et al. 2015; Howell et al. 2015; Papadopoulos 2016). In the South Aegean several tsunamis were reported in the last century or so (Ambraseys 1962; Papadopoulos and Chalkis 1984; Papadopoulos et al. 2007; Altinok et al. 2011), with the most powerful being the Amorgos tsunami of 9 July 1956 triggered by a Mw = 7.4 earthquake (e.g. Galanopoulos 1957; Papadopoulos and Pavlides 1992; Yalciner et al. 1995; Cakir and Yalciner 2002; Beisel et al. 2009; Okal et al. 2009).

On July 20, 2017 22:31 UTC, a strong (Mw = 6.6) shallow earthquake ruptured the area between Gökova Bay, between the Bodrum Peninsula, southwest of Turkey, and Kos Island, Greece (Fig. 1a). The earthquake triggered a tsunami, which caused strong agitations and damage in south coast of Bodrum Peninsula and in Kos. The source characteristics of the earthquake have been discussed in several studies such as Ganas et al. (2017), Saltogianni et al. (2017), Batıgün et al. (2018), Karasözen et al. (2018),

[1] Ocean Engineering Research Center, Department of Civil Engineering, Middle East Technical University, Ankara, Turkey. E-mail: gguneydogan@gmail.com
[2] Joint Research Centre, European Commission, Ispra Site, Italy.
[3] Institute of Geodynamics, National Observatory of Athens, Athens, Greece.
[4] Turkish Chamber of Civil Engineers, Bodrum Branch, Turkey.
[5] Department of Geophysics, Kandilli Observatory and Earthquake Research Institute, Bogazici University, Istanbul, Turkey.
[6] Institute of Education, Research and Regional Cooperation for Crisis Management Shikoku, Kagawa University, Takamatsu, Japan.
[7] Department of Geophysical Engineering, Kocaeli University, Kocaeli, Turkey.
[8] Department of Civil and Environmental Engineering, Chuo University, Hachioji, Japan.
[9] Department of Geological Engineering, Middle East Technical University, Ankara, Turkey.
[10] Department of Aerospace Engineering, Middle East Technical University, Ankara, Turkey.
[11] USC, Los Angeles, USA.
[12] Technical University of Crete, Chania, Greece.

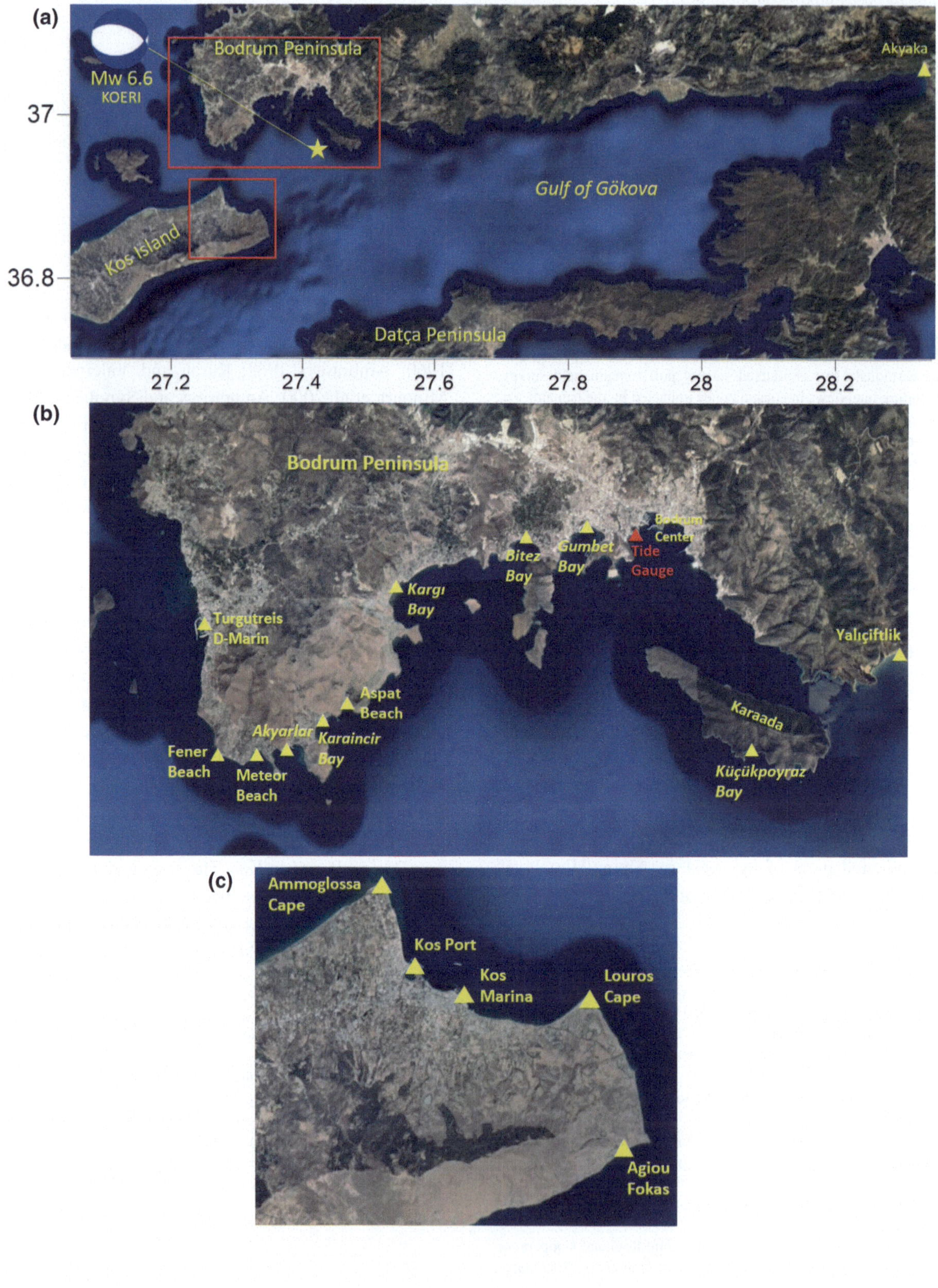
(a)
Mw 6.6
KOERI
Bodrum Peninsula
Akyaka
37
36.8
Gulf of Gökova
Kos Island
Datça Peninsula
27.2
27.4
27.6
27.8
28
28.2
(b)
Bodrum Peninsula
Bodrum Center
Tide Gauge
Bitez Bay
Gumbet Bay
Kargı Bay
Turgutreis D-Marin
Yalıçiftlik
Aspat Beach
Akyarlar
Karaincir Bay
Fener Beach
Meteor Beach
Karaada
Küçükpoyraz Bay
(c)
Ammoglossa Cape
Kos Port
Kos Marina
Louros Cape
Agiou Fokas

◀Figure 1
Map showing the study area (red rectangles and Akyaka) **a** Gokova Gulf including earthquake epicenter (yellow star) and the focal mechanism solution reported by the Kandilli Observatory and Earthquake Research Institute (KOERI) (2017), **b** southern coast of Bodrum Peninsula and Karaada showing the field survey locales, **c** northeastern Kos Island showing the field survey locales. Images are taken from Google Earth©

Kiratzi and Koskosidi (2018), Konca et al. (2018), Ocakoğlu et al. (2018), Öztürk and Şahin (2018), and Tiryakioğlu et al. (2018). In addition, the numerical modeling of the tsunami is presented in Heidarzadeh et al. (2017), Cordrie et al. (2018), Dimova and Raykova (2018), and Macías et al. (2018). After this event, the challenges and requirements for near-field tsunami warning in the Aegean Sea and Eastern Mediterranean are also discussed in Necmioglu et al. (2018).

In this paper, we present the results of the 2017 post-tsunami field surveys following the instructions and recommendations of Synolakis and Okal (2005) and the Post-Tsunami Field Survey Guide (UNESCO IOC 2014). The surveys were conducted including the southern, western and northern coastlines of Bodrum Peninsula, Karaada Islet and Akyaka Town in Turkey, and eastern Kos Island in Greece (Fig. 1). Preliminary results of the post-tsunami field surveys were presented in Yalciner et al. (2017), documenting once again that the tsunamis are a real threat in the Mediterranean. It is also important to note that, since the runup heights are relatively low at some locations (0.3–0.5 m), Bodrum–Kos area is under a micro-tidal regime as the tidal range does not exceed 0.3 m (on spring tides), and the principal tidal constituents decrease southwards in the Aegean Sea (Tsimplis 1994; Alpar et al. 2000; Vousdoukas et al. 2009; Pavlopoulos 2010; Poulos et al. 2012). Alpar et al. (2000) gives the mean spring tidal range as small as 0.11 m at Bodrum. Also, the locals and eyewitnesses reported that the sea was calm at the time of the event as it was summertime and storm waves are observed in winter months, from December to February, in Bodrum region (Derebay 2007; Ergin et al. 2008). The wave climate in the eastern coast of Kos Island is also similar as it faces the southern coast of Bodrum Peninsula.

2. Brief Information on the Earthquake

The source of the earthquake is located east-northeast offshore Kos Island (Fig. 1), just southwest of the small island of Karaada and the northwest side of the gulf of Gökova (e.g., Karasözen et al. 2018; Tiryakioğlu et al. 2018). The Gulf of Gökova is a narrow gulf of the Aegean Sea, between the Bodrum Peninsula to the north and Datça Peninsula to the south. Tectonically, this part of western Anatolia has undergone rapid extension (Taymaz et al. 1991), and currently it is characterized by north–south regional extensional tectonic regimes (Kurt et al. 1999; Yılmaz et al. 2000; Gürer et al. 2013; Tur et al. 2015). The Gökova graben structure, where the earthquake occurred, has evolved with east–west faulting related to the north–south extension and counterclockwise rotation towards the Hellenic Arc of the Aegean–Anatolian plate block (e.g., Ocakoğlu, et al. 2018; Tur et al. 2015; Yolsal-Çevikbilen and Taymaz 2012; Yolsal-Çevikbilen et al. 2014). The seismic activity is mostly accumulated along the northern branch of Gökova Fault Zone, and the Gökova Gulf has experienced hazardous earthquakes in history and especially in the twentieth century (Gürer et al. 2013). In 1493, the town of Bodrum, and also Kos Isl., were totally destroyed (Luttrell 1999). Earthquake fault plane solutions show that normal mechanisms with a strike-slip component have been observed on the E–W-oriented graben and normal fault systems in the Gulf of Gökova, which supports that the extension is in a north–south direction (Yolsal and Taymaz 2005, 2007, 2010). The extensional character of this region has been confirmed by several studies investigating the fault mechanisms by using GPS data (Aktug et al. 2009, 2010; Dogru et al. 2014; McClusky et al. 2000; Ozener et al. 2013; Reilinger et al. 2006; Tiryakioglu et al. 2013; Yolsal and Taymaz 2010). Tur et al. (2015), also suggests that Gökova Graben is not a local tectonic feature, but is part of a large regional lazy-S-shaped structure extending from the Gulf of Gökova westward to the Island of Karpathos north of the Cretan Trough, mentioned as the Gökova–Nisyros–Karpathos Graben in their study.

Focal mechanism solutions and slip distribution studies reveal that the 20 July 2017 earthquake was associated with normal faulting, in accordance with the tectonically extensional regime, inferred for this region [Heidarzadeh et al. 2017; Kandilli Observatory and Earthquake Research Institute (KOERI), 2017; Saltogianni et al. 2017; Ganas et al. 2017; USGS 2017; Tiryakioğlu et al. 2018; Karasözen et al. 2018]. Many aftershocks recorded after the mainshock were distributed along an east–west trending rupture area and likely occurred due to tensional stress change caused by the coseismic slip. Sözbilir et al. (2017) examined the earthquakes of $Mw \geq 5$ occurring in the last century in Gökova Fault Zone and found that there is an earthquake migration based on the energy transfer from east to west in the fault segments within the fault zone. The migration to the west-southwest has been revealed by earthquake swarms, which mostly occurred during 2004 and 2005, and the recent seismic activity in Gökova region can be characterized by these earthquake swarms (Kalafat and Horasan 2012; Sözbilir et al. 2017).

Despite these analyses, the dip polarity of the fault is still under discussion. Saltogianni et al. (2017) and Tiryakioğlu et al. (2018) inferred the dip polarity as south dipping while Karasözen et al. (2018) and Ganas et al. (2017) preferred a north dipping fault. On the other hand, Kiratzi and Koskosidi (2018) attempted to resolve dip polarity and dip angle from the slip model, and damage patterns, by using strong ground motion records. Their test inversion results for the dip polarity showed that a south dipping fault is acceptable for steep angles ($\sim$ 50°), and north dipping is for a smaller dip angle ($\sim$ 35°). The results of the tsunami field survey of this study might help constrain the dipping polarity of the earthquake, but this is beyond our scope.

According to the intensity maps produced by the Earthquake Loss Estimation Routine (ELER) in KOERI, the earthquake intensity in Bodrum was of VII degree, indicating that the perceived shaking was powerful and the damage was moderate (KOERI, 2017). On the other hand, the earthquake intensity in Kos Island, Bodrum Peninsula and north of Datça Peninsula was VI, representing strong perceived shaking and potential light damage. However, in various spots of Kos city the intensity reached up to VIII.

3. *Observations on the Tsunami*

3.1. *Tide Gauge Data*

The tsunami was recorded by a shore-based tide gauge installed by KOERI inside the Bodrum Municipality Marina (27.4167°E–37.0333°N) (Fig. 1b). Shore-based tide gauge stations record tsunami together with sea tides. The raw record of the Bodrum tide gauge is given in Fig. 2a. Several postprocessing steps can be applied to the data prior to the estimate of arrival times and height of tsunami waves, in order to remove the non-tsunami signals from the data (e.g., Yolsal-Çevikbilen et al. 2018). Here, a polynomial of degree 11 is used in order to fit the polynomial functions of sea tides. Then, sea tides are subtracted from the original time series of the record (Fig. 2b). The analysis revealed that the wave attacking Bodrum Peninsula was largely a leading depression N-wave (Tadepalli and Synolakis 1994) with 0.18 m trough and 0.11 m crest, hence the trough-to-crest height of 0.29 m (Fig. 2b). The tide gauge was calibrated using a temporary installation of an Inexpensive Device for Sea Level (IDSL) at a structure of the Bodrum Voyage Hotel (27.4170°E–37.0275°N) to verify that the readings of the Bodrum tide gauge were consistent with the measurements of the IDSL. The IDSL was left to record for more than 29 h. The comparison between the two instruments gives consistent results (Fig. 2c). The small difference might be due to the different installation location. On the other hand, a closer installation was found impossible in the time frame of the field survey. The calibration indicated that the tide gauge records are reliable (Yalciner et al. 2017).

3.2. *Observations along Bodrum Peninsula, Karada Islet and Akyaka Town*

The first field survey was performed on July 22–23, 2017 along the southern coasts of Bodrum Peninsula in between 37.0016°N–27.2575°E (Turgutreis Marina) and 36.9928°N–27.5279°E (Yaliciftlik region). The second survey was held on July 28–31, 2017 and added more observation points to the survey area, as well as the regions of Karaada Islet and Akyaka Town. Figure 3 shows the spatial distribution

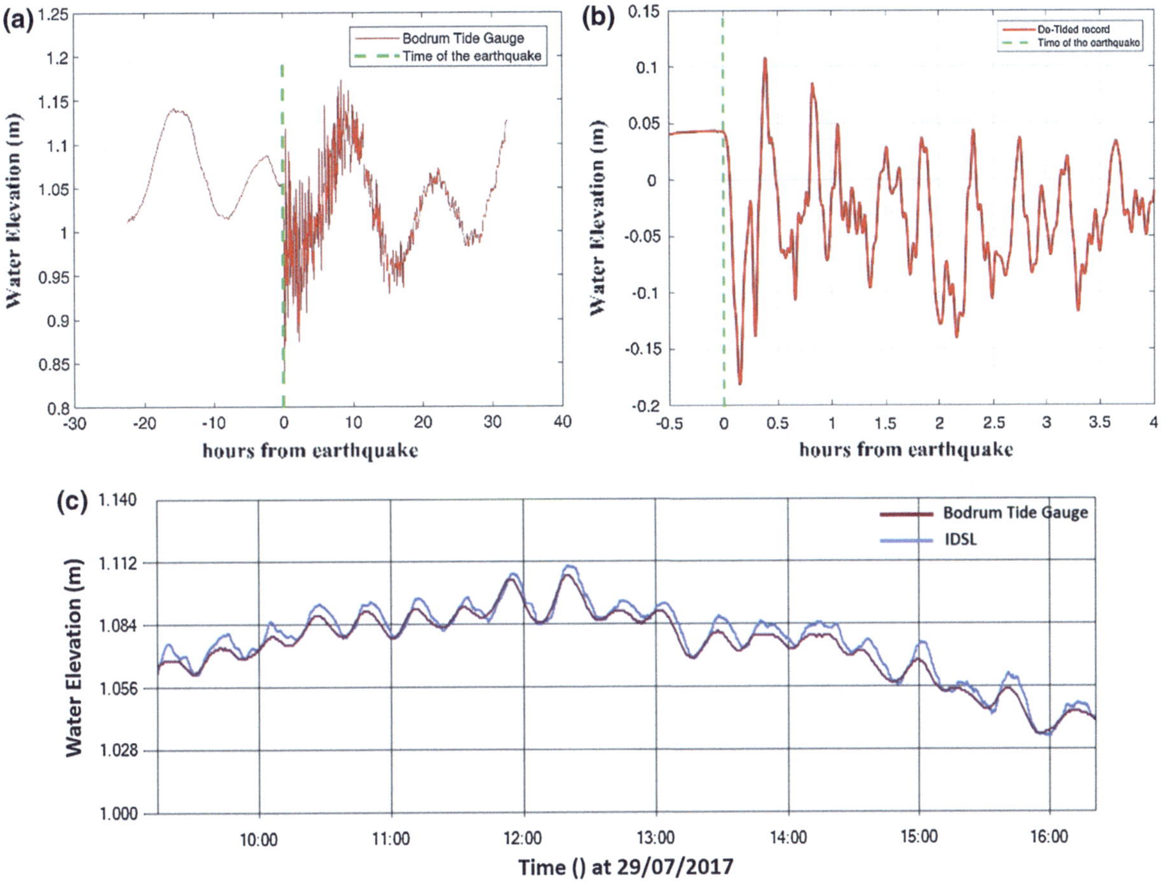

Figure 2
a The raw record of the Bodrum tide gauge, **b** de-tided and filtered waveform and **c** comparison of the Bodrum tide gauge and IDSL buoy recordings

of the measurements along southern coasts of Bodrum Peninsula and Karaada Islet, while Table 1 provides details of these measurements.

3.2.1 Gumbet Bay

Gumbet Bay was the worst hit area in Bodrum Peninsula (Fig. 1b). Five minutes after the earthquake, the shoreline initially receded consistent with the tide gauge recording, then 12–13 min later, the first wave arrived, as inferred from a security camera video of a hotel located near the eastern side of Gumbet Bay. Waves reached the entrance door of the hotel about 13 min after the earthquake. The tsunami inundation distance was about 60 m, and the inundation depth reached up to 0.5–0.6 m near this hotel.

At the eastern side of Gumbet Bay, small boats were berthed at a relatively shallow region. The tsunami dragged all boats together and moved them to the same inland location (approximate coordinates: 37.0281°N–27.4050°E). More than 30 boats were damaged, and more than 10 boats sunk (Fig. 4). The boat captains reported that there was a very strong swirling water motion at three locations in the bay. The runup height measured in this region was above 1.7 m.

The inundation distance near the municipality cafe located in northeast of Gumbet Bay (Point 8-Table 1) was reported as about 20 m with a 0.6 m inundation depth. Inundation distance was reported as 100 m near Cesar Beach Hotel (Point 22-Table 1). Eyewitnesses at this area of Gumbet Bay reported successive waves and water level changes until the

Figure 3
Representative runup distributions according to survey locations along the southern coast of Bodrum Peninsula and Karaada Islet

sunrise, i.e., about 3 h after the earthquake. People found dead fish mostly in this area and all along the coastline of the bay.

The waves penetrated through a dry streambed located in the northeastern part of Gumbet Bay, used as a road and parking lot. The width of the streambed is 3.3 m (5.7 m with side walls) and the flow accelerated. The waves dragged 12 cars parked near the shoreline along the concrete bed of the dry stream (Fig. 5). Here, the inundation distance and tsunami runup height were measured as 100 m and 1.9 m, respectively. The maximum inundation depth in the streambed was around 0.85 m (37.0314°N–27.4067°E), while the cars and the concrete planters in front of these cars were collected around the same location (near 37.0311°N–27.4069°E) along the streambed near the estuary. The inundation depth reached ~ 1 m in front of the headwall (37.0314°N–27.4068°E), along the streambed.

At the western part of Gumbet Bay, 1.15 m tsunami runup height was measured at Nagi Beach (Point 21-Table 1), with an inundation distance of 35 m. Figure 6 illustrates the tsunami motions inside Gumbet Bay according to the eyewitness accounts and measurements in this region.

3.2.2 *Turgutreis*

Minor impact of the tsunami was observed in Turgutreis (Fig. 1b), located at the western coast of Bodrum Peninsula. The manager of a local marina (D-Marin, Point 1-Table 1) reported that strong currents were observed along the circulation channels on the marina breakwater. On the other hand, no tsunami runup was reported. According to a fisherman at the Turgutreis Fishery Harbor (Point 11-Table 1), the water level increased approximately 60 cm in front of the berthing place along a vertical

Table 1

Details of measurements along Bodrum Peninsula, Karaada and Akyaka

Point #	Survey #	Place	Region	Latitude (N)	Longitude (E)	Tsunami runup (m)
1	1	Turgutreis D-Marin	Turgutreis	37.0016	27.2575	NMWO
2	1	Olivia Restaurant	Fener Beach, Akyarlar	36.9648	27.2646	NMWO
3	1	Kilavuz Motel Beach	Akyarlar Bay	36.9674	27.2909	1.2
4	1	Balmahmut's Place	Karaincir	36.9730	27.3005	1.4
5	1	Aspat Beach	Aspat Bay	36.9796	27.3124	1.0
6	1	Camel Beach	Kargi Bay	37.0136	27.3306	0.5
7	1	Bitez Beach	Bitez Bay	37.0256	27.3836	0.5
8	1	Municipality Café (MC) and River Stream (RS)	Gumbet Bay	37.0309	27.4053	1.9
9	1	Bodrum Marina	Bodrum Center	37.0358	27.4248	1.0–1.2
10	1	Yali Café	Yaliciftlik	36.9928	27.5279	NMWO
11	2	Turgut Reis Fishery Harbor	Turgutreis	37.0060	27.2564	0.6
12	2	Small Beach Hotel	Turgutreis	36.9821	27.2568	NMWO
13	2	Meteor Beach	Dogu Beach	36.9662	27.2788	0.9
14	2	Xanadu Island Hotel	Akyarlar Peninsula Tip	36.9607	27.2862	0.5
15	2	La Brezza Beach	Western Karaincir Bay	36.9699	27.3002	1.6
16	2	Meteor–2 Beach	Eastern Karaincir Bay	36.9735	27.3009	1.4
17	2	Tek Sitesi	Eastern Karaincir Bay	36.9758	27.3041	0.8–1.0
18	2	Bitez Mor Beach	Western Bitez Bay	37.0238	27.3724	–
19	2	Western Bitez Beach	Western Bitez Bay	37.0269	27.3729	0.9
20	2	Eastern Bitez Beach	Eastern Bitez Bay	37.0248	27.3845	0.5
21	2	Nagi Beach	Gumbet Bay	37.0315	27.4036	1.15
22	2	Cesar Beach Hotel	Gumbet Bay	37.0310	27.4074	–
23	2	Voyage Hotel Bodrum	Bodrum Marina Region	37.0275	27.4170	1.0
24	2	Kucukpoyraz	Karaada (Black Island)	36.9663	27.4614	1.5–2.0
25	3	Akyaka Beach and River Branch	Akyaka	37.0508	28.3244	0.6

NMWO no major wave oscillation

wall. Here, about 1 m subsidence was observed. However, the beach-bar workers at the Small Beach Hotel (Point 12-Table 1) reported that no water anomalies at the water level.

3.2.3 Akyarlar

Fener Beach, Meteor Beach and Akyarlar Bay were surveyed (Fig. 7) in Akyarlar region (Fig. 1b). Fener Beach (Point 2-Table 1) is located at the corner of western and southern coasts of the Bodrum Peninsula and was the most probable boundary for significant tsunami impacts (approximately 36.9648°N–27.2646°E). Eyewitnesses at Meteor Beach observed an initial 2–2.5 m withdrawal and 1-2 m inundation distance. Tsunami runup measurement in this bay was approximately 0.9 m (Point 13-Table 1).

In Akyarlar Bay shown in Fig. 7, there is a small peninsula at the western end where Xanadu Island Hotel is located (Point 14-Table 1). Here, there was no observed water-level change, according to the hotel manager and several beach workers. However, seaweed residuals were observed 3 m onshore of the beach, at an elevation of 50 cm, and might have been dragged by the tsunami. The small fishery port in this bay was relatively safe, considering the source of the earthquake. According to information from fishermen, the sea level first decreased and then rose, which is interpreted as again a leading depression N-wave (Tadepalli and Synolakis 1994). They observed strong currents at the entrance of the port, described as "the strong flow of a stream."

An approximately 40-cm tsunami overflow over the breakwater was observed for a short period. Summerhouse residents in Akyarlar Bay reported an approximately 10-m inundation distance, corresponding to 1.2 m tsunami runup height, near Kılavuz Beach Motel (Point 3-Table 1); the inundation depth

Figure 4
View of damaged boats and shoreline of eastern side of Gumbet Bay

was approximated as 50 cm by the eyewitnesses. At the eastern side of Akyarlar Bay, the area up to Akyarlar Motel (36.9678°N–27.2946°E) was affected by the tsunami and the water level increased by about 40 cm, in front of the vertical wall at the shoreline.

3.2.4 Karaincir Bay

Karaincir Bay is adjacent to the eastern side of Akyarlar Bay (Fig. 1b). Considering the tsunami source, Karaincir Bay was not as protected as Akyarlar Bay. At the western part of Karaincir Bay, inundation distance was measured as 125 m at La Brezza Beach (Point 15-Table 1) with 1.6 m tsunami runup and 40 cm inundation depth.

Here, the water level first drewdown to 1.2 m, corresponding to 20 m shoreline recession near Balmahmut's Place, then rose resulting in 1.4 m tsunami runup (Point 4-Table 1). The inundation distance was reported as 60 m here. There is a road and a streambed near each other at the eastern side of Balmahmut's Place, located around 36.9736°N–27.3007°E (Fig. 8). In the 2.5 m wide streambed, the tsunami runup and the inundation depth were measured as 1.4 m and 0.8 m, respectively, 100 m away from the shoreline by following the tsunami traces on the sidewall (Fig. 8). From the traces on the walls, we infer at least three waves reached at the Karaincir locality. The tsunami inundation distance was approximately 250 m along the streambed. On the other hand, a 1.5 m tsunami runup and 0.8 m inundation depth were measured along the road, respectively, 100 m away from the shoreline, from traces on the sidewall of the road. Eyewitnesses reported several cars dragged by the tsunami.

Another survey point in Karaincir Bay is the Meteor-2 Beach (Point 16-Table 1), located at the eastern part of the bay. Beach workers reported that the sea receded about 70 m from the shoreline and inundated 30 m. The tsunami runup height was measured as 1.4 m with an inundation depth of 0.6 m.

Figure 5
View of the stream bed that the tsunami inundated (37.0298°N–27.4078°E)

In Karaincir region, there is a small bay called TEK Sitesi Bay (Fig. 8). Here, (Point 17-Table 1), the wave receded 60 m corresponding to a 1.5 m drop in water level. Tsunami runup height was measured as 1 m.

3.2.5 Aspat Beach

In Aspat Beach (Fig. 1) (Point 5-Table 1), there is a stream bed called Azmak. Along this stream, there are side walls up to 0.75 m above the still water level. The tsunami inundation distance was approximately measured as 60 m. Rising seawater in the stream flooded the beach area and caused minor damage to the beach-bar area. Tsunami runup was measured as 1 m. A pier suffered minor damage as well at its connection to the coast due to the earthquake or tsunami.

3.2.6 Kargı Bay (Camel Beach)

Kargı Bay (Camel Beach) was another area affected by the earthquake and tsunami (Fig. 1) (Point 6-Table 1). According to interviews with restaurant

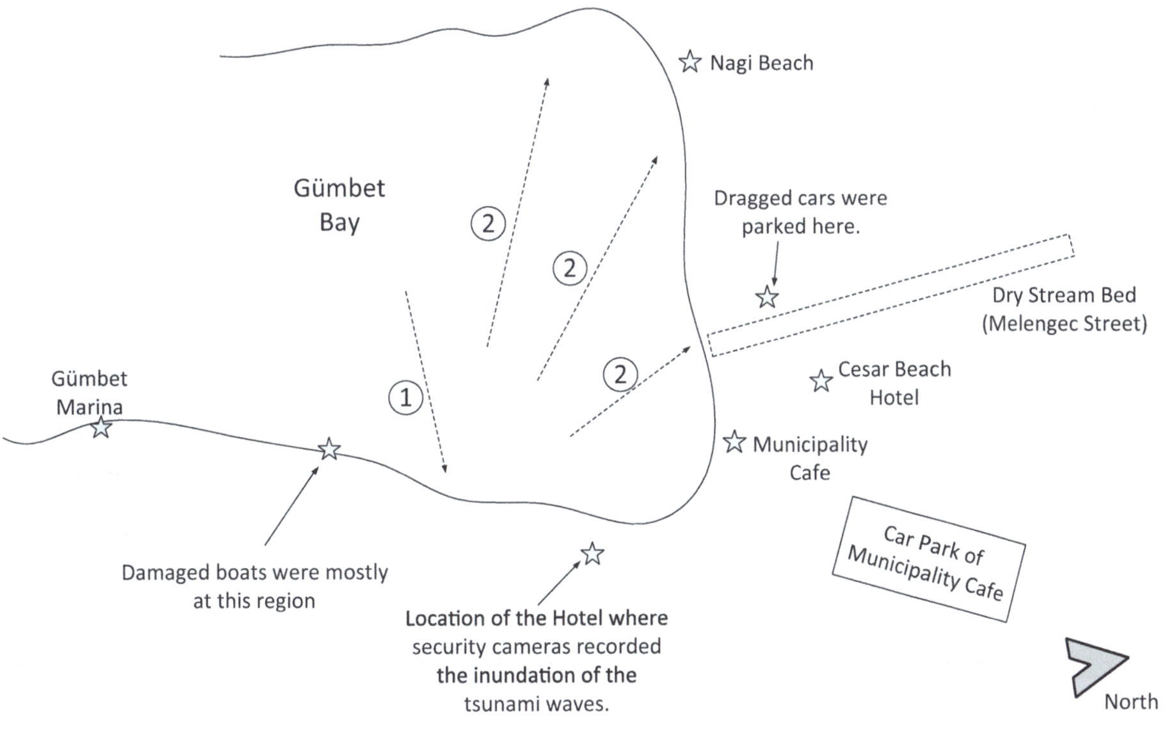

Figure 6
Sketch of the motion of the tsunami at the eastern end of Gumbet Bay (numbers show approximate direction of motion)

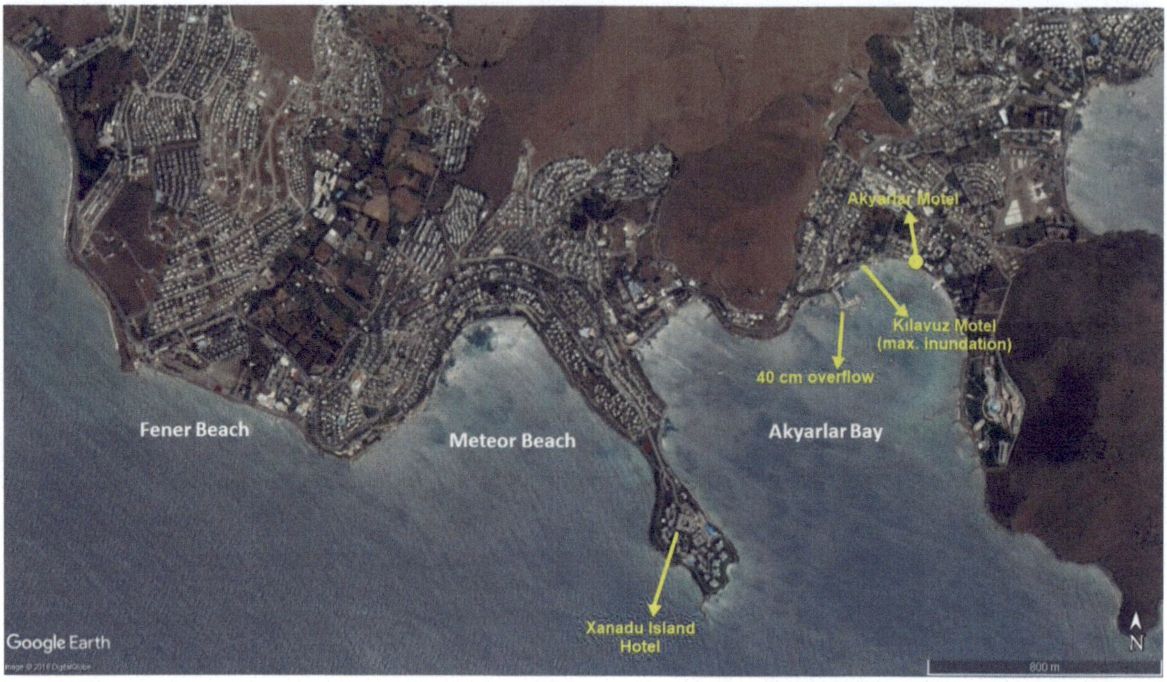

Figure 7
Surveyed areas and inspected localities in Akyarlar region

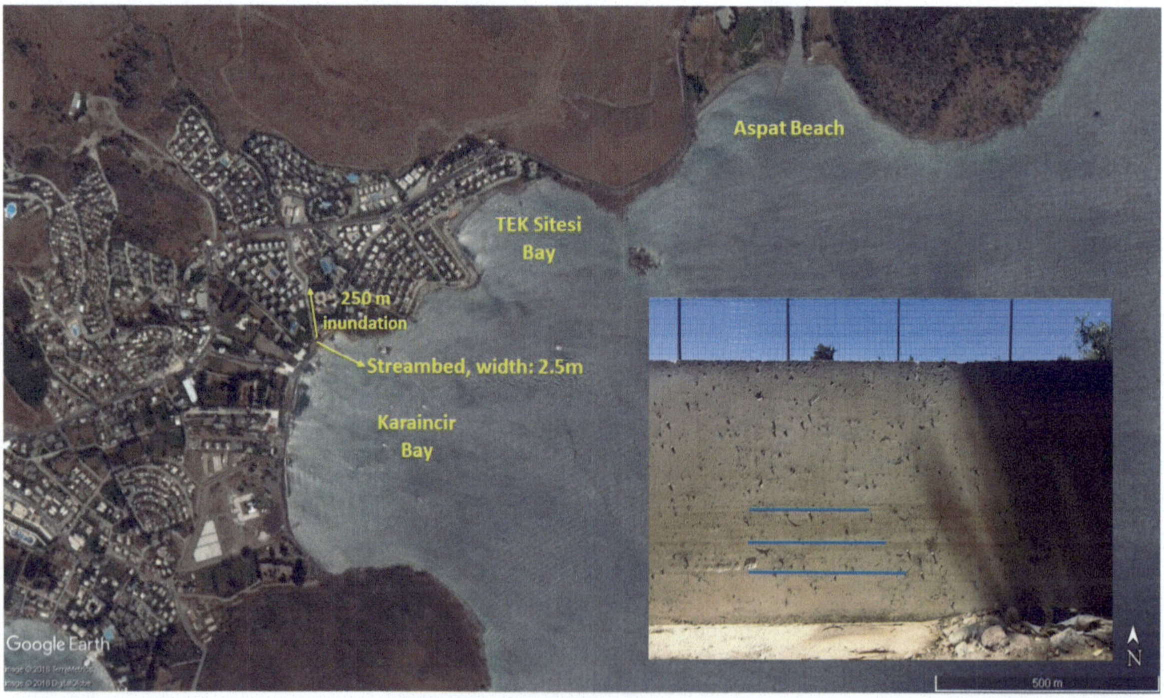

Figure 8
Tsunami inundation along the roads in Karaincir Bay and the sidewall of the streambed

workers and beach goers, the first wave arrived approximately 10 min after the earthquake, while the second reached the region after approximately 80 min. According to the observations, the motion was a leading depression N-wave. The tsunami runup height was approximately 40–50 cm, whereas draw-down was reported as 120 cm. The inundation distance was measured as 7–8 m.

3.2.7 Bitez Bay

At the western end of Bitez Bay (Point 18-Table 1), at Mor Beach, the shoreline withdrew 7–8 m and about 15 min after the earthquake. The inundation distance was approximately 20 m. Along this streambed (Point 19-Table 1), the inundation distance was 250 m. Elsewhere, in the vicinity of the streambed, the maximum inundation distance was reported as 30 m, whereas the inundation depth and the tsunami runup was measured as 0.5 m and 0.9 m, respectively. At the eastern side of Bitez Bay (Point 7-Table 1), the first sea motion was reported as a leading depression N-wave of up to 3 m, after 10–15 min of earthquake, and the inundation distance was reported as approximately 10 m. Close to the berthing place at the eastern part of Bitez Bay, the inundation distance was reported as 10 m (Point 20-Table 1). There was no record of boat damage in Bitez Bay. The inundation distances in Bitez is summarized in Fig. 9.

3.2.8 Bodrum Bay (Centrum of Bodrum Peninsula)

The staff of Voyage Hotel Bodrum Beach (Point 23-Table 1) located at the southwest of the entrance of Bodrum Marina reported that the sea receded 30 m corresponding to a depth of around 1.7 m. The inundation distance at this location was around 20 m, and the tsunami runup was measured as 1 m. An eyewitness who was on his boat at Bodrum Marina (Point 9-Table 1) also reported 1.2 m of tsunami runup. In the marina, he observed swirling water with 4–5 cycles of substantial vertical motions.

3.2.9 Yalıçiftlik

Yalıçiftlik Bay (Point 10-Table 1) very likely was the most probable eastern boundary for significant

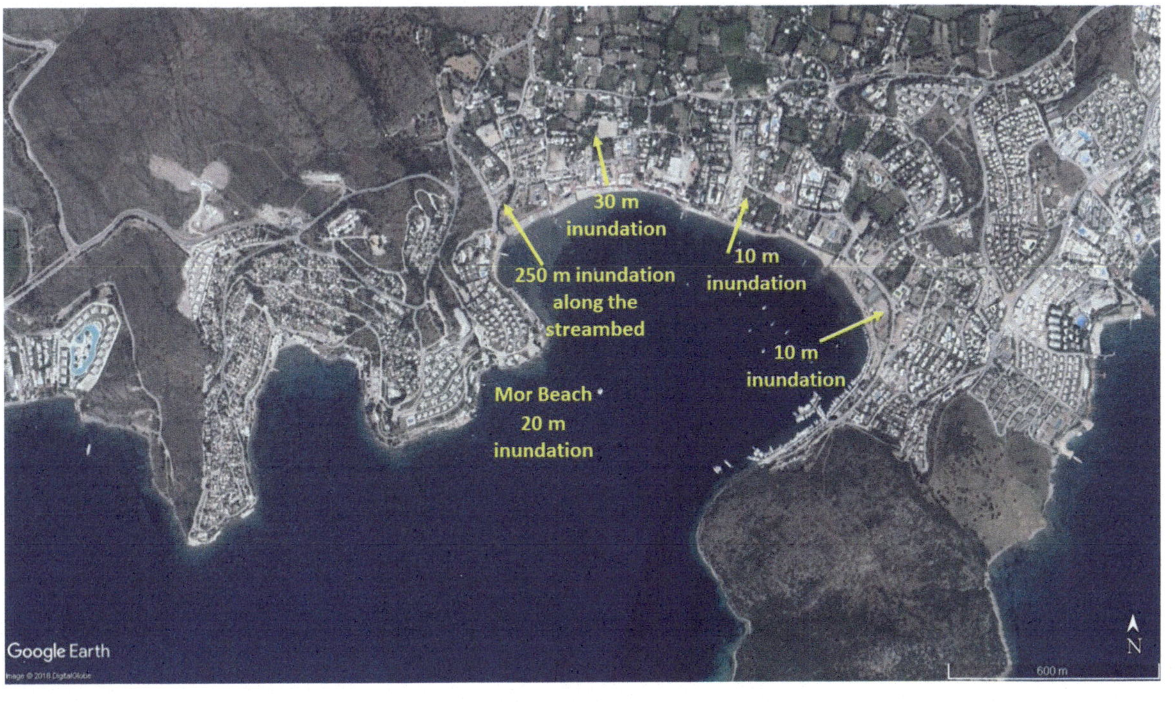

Figure 9
Tsunami inundation in Bitez Bay

tsunami inundation at the eastern end of the southern coastline of Bodrum Peninsula. No major damage or tsunami inundation were reported here where the sea bottom slope is relatively steep in this region. The only damage observed along the coastline was toppling of boats hauled out at the dockyard likely due to the earthquake shaking.

3.2.10 Karaada (Karaada Islet)

Karaada Islet is located in between Bodrum and Kos Island; it is approximately 7 km long along northwest-southeast direction and 1.2 km wide along southwest-northeast direction. There are two small gravel beaches at Küçükpoyraz Bay (Point 24-Table 1) at the southern coastline of the island, facing the epicenter of the earthquake. In this region, small gravel deposits filling the gaps between the rock surfaces were found. It is assumed that these deposits were carried by the tsunami, thus we inferred 1.5–2.0 m of tsunami runup height.

3.2.11 Akyaka

Akyaka (Point 25-Table 1) is located at the eastern tip of Gökova Gulf, about 80 km away from the earthquake epicenter (Fig. 1a). The survey team interviewed the cafe workers along Akyaka Beach and Fisheries cooperative. The eyewitnesses reported the wave receding around 35 min after the earthquake, in Akyaka withdrawing up to about 50 m distance. The tsunami inundation distance was about 18 m. The staff of fisheries cooperative reported 160 m inundation distance along the small stream. The measured runup height here indicated about 60 cm. No flooding was observed on the banks of this stream. The tsunami inundation in Akyaka is summarized in Fig. 10.

3.2.12 Datça

Datça region (Fig. 1a), located at the southern part of Gökova Gulf, was not surveyed by the field survey team. However, reliable information was obtained from Datça Branch of Turkish Chamber of Civil

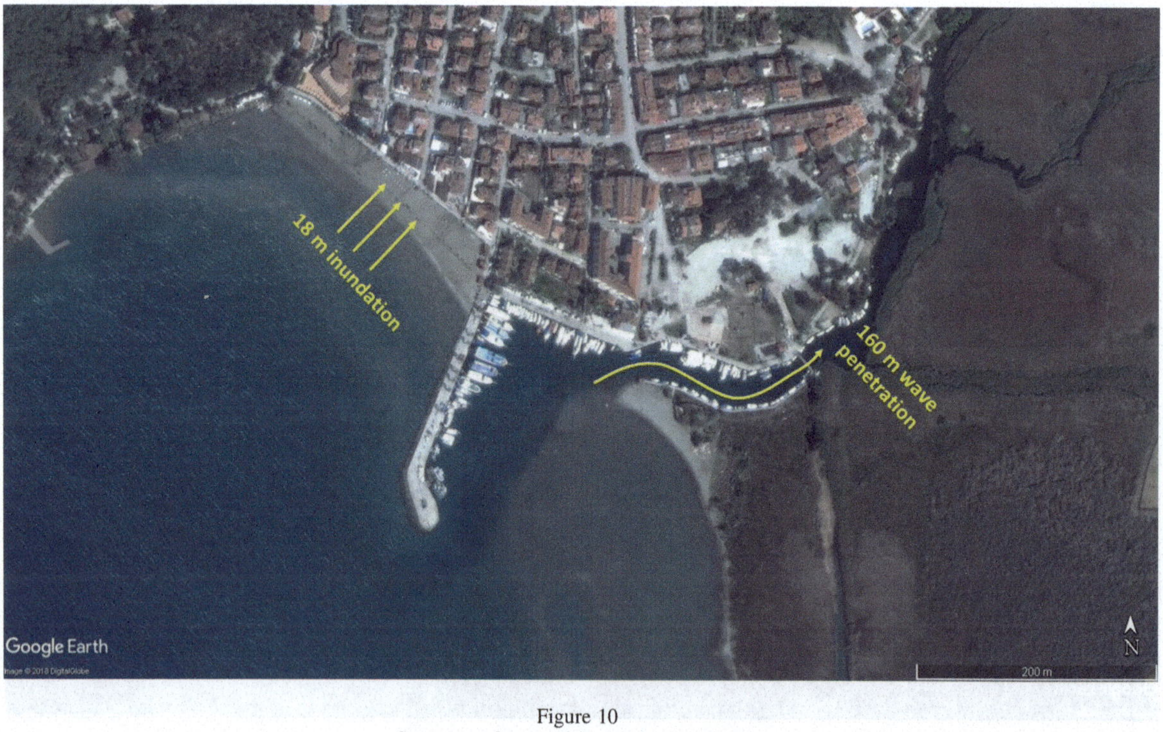

Figure 10
Summary of tsunami inundation in Akyaka

Figure 11
Results of Kos tsunami survey

Engineers (President, Mr. Levent Ozbek, personal communication) and an experienced fisherman (Mr. Ozkan Yigiterhan, personal communication). At Karakoy Kormen Marina (36.7714°N–27.6177°E), located at the north of Datça Peninsula, rise and drop in water level was reported as about 1.2 m and 1 m, respectively, at the vertical quay wall. Furthermore, very strong currents were observed starting 10 min after the earthquake, lasting until the morning inside and outside of a fishery harbor at Palamutbuku (36.6693°N–27.5033°E). One of the boats in the fishery harbor sank, due to the strong motions inside the harbor.

3.3. Observations along Kos Island

The team surveyed the area from 36.8903°N–27.2073°E to 36.8577°N–27.3498°E on 1–3 August 2017 (Fig. 1c). Most of the locations were identified by looking for tsunami traces, mainly in algae or other form of debris. In addition, the survey team interviewed eyewitnesses and collected photos and videos.

The tsunami runup height distribution along the coast of Kos was in general uniform, i.e., quite different from that in Bodrum area, where bays with high runup height were followed by bays with negligible runup height. In general, the tsunami runup in the Kos coastal zone reached no more than ~ 1 m. In two measurement spots it likely reached 1.4 m and 1.5 m, but only locally. However, the Amphora-shaped Kos Port is an exemption since the tsunami wave clearly amplified within the basin and reached up to 1.5 m on the quay with maximum inland penetration of ~ 60 m. A possible reason could be the resonance effect in the port, driven by the unique form of the bay with a narrow converging entrance. A summary of the results of the Kos Island field survey is summarized in Fig. 11. The survey in Kos Island is divided into five parts, and the

Table 2

Details of measurements along Kos Island

Point #	Place	Region	Latitude (N)	Longitude (E)	Tsunami runup (m)
1	Byron Apartments	Northeast Kos	36.8903	27.2073	0.05
2	La Branda Acqua Marine Resort	Northeast Kos	36.8965	27.2383	0.1
3	Diamond Hotel	Northeast Kos	36.9043	27.2581	0.5
4	Along the shoreline	Northeast Kos	36.9122	27.2725	1.5
5	Along the shoreline	Northeast Kos	36.9137	27.2752	1.4
6	Along the shoreline	Northeast Kos	36.9143	27.2757	0.7
7	Ammoglossa Cape	Ammoglossa Cape	36.9157	27.2798	0.6
8	East Kos before Kos Port	East Kos before Kos Port	36.9149	27.2811	0.6
9	Atlantis Hotel	East Kos before Kos Port	36.9106	27.2828	0.6
10	East Kos before Kos Port	East Kos before Kos Port	36.9056	27.2851	0.3
11	East Kos before Kos Port	East Kos before Kos Port	36.9008	27.2857	1.2
12	The Port of Kos	Kos city	36.8959	27.2875	1.5
13	Hotel Aktis	Between Kos Port and Louros Cape	36.8932	27.2941	0.70
14	Kos Marina	Between Kos Port and Louros Cape	36.8928	27.3010	0.80
15	4 Roses Cafe	East Kos after Port	36.8877	27.3052	0.7
16	Discharge channel	East Kos after Port	36.8852	27.3192	0.8
17	Along the shoreline	East Kos after Port	36.8846	27.3256	0.8
18	Psalidi beach	East Kos after Port	36.8872	27.3324	0.6
19	Louros Cape	Louros Cape	36.8898	27.3387	0.5
20	Big Blue Surf Center	Southeast Kos	36.8820	27.3464	0.8
21	Watersport Flyboard	Southeast Kos	36.8710	27.3522	0.7
22	Agiou Fokas	Southeast Kos	36.8577	27.3498	0

The coordinates indicated are the average of the various points (3–5) that compose the "location". In the text, instead, the real coordinates of each image are indicated. For this reason, they may not be identical. When only one point was used, they coincide

Figure 12
Images taken on **a** Kos Island (36.9137°N–27.2752°E) showing tsunami deposits (The beach is very narrow, the distance from the sea is less than 5 m.), **b** Kos Port (36.8975°N–27.2879°E)

observations are given in the following sections and in Fig. 11. Table 2 provides observation locations.

3.3.1 North Coast (between 36.8902°N–27.2073°E and 36.9157°N–27.2798°E)

Along the north coast of Kos Island (Fig. 11), the tsunami wave runup height ranged from 0.1 m up to 0.4 m. However, in two observation points the runup measurements were of 1.0–1.4 m and 1.3–1.5 m. Figure 12a shows tsunami deposits (Point 3-Table 2), where one witness reported that the beach was inundated by 16 m and at least 0.5 m of runup height.

3.3.2 North Coast (between 36.9157°N–27.2798°E and Kos Port)

In this region, most of the measured tsunami runup heights are 0.5–0.6 m, with only one exception where a runup height of 1.2 m was reported. No tsunami

Figure 13
Ioanidi road 36.8945°N–27.2884°E at Kos port flooded with sea water up to the pharmacy sign

trace was found, but an eyewitness reported a substantial water withdrawal and then water return indicating a leading depression N-wave. The runup height here was estimated as 0.3 m (Point -Table 2).

3.3.3 The Kos Port

The Port of Kos is an Amphora-shaped basin with narrow entrance at the north (Fig. 1c). Observations were collected from four points of the Kos Port area, as shown in Fig. 11. The observed water traces given in Fig. 12b, show an estimated tsunami runup height of 1.4 m. Figure 13 shows Ioanidi Road at the Port, which was flooded with seawater up to the "Pharmacy" sign. The maximum runup was about 1.5 m. According to an eyewitness, the wave arrived 30 min after the event. Figure 14 shows a boat resting on the pavement near the observation point, which is at the east side of Kos Port and locally called "Bridge."

A marina official arrived near the "Bridge" point of the Port, approximately 1–1.5 h after the earthquake. The area was already wet. Bikes and motorbikes had fallen and drifted away. Cars crashed and drifted (Figs. 15, 16). At that time, he described that the sea level was approximately 1.5–2.0 m below its usual level. In about 1–2 min, the sea level increased quickly by approximately 0.5 m above the dock level, and inundated about 25–30 m inland, transporting motorbikes. After that, the sea water retreated, and remained at its normal level. As the dock is at about 1 m from the sea level, the estimated runup height here was 1.5 m.

In Kos port, various videos were collected from security cameras. One proved extremely useful, since

Figure 14
Boat resting on the pavement after its removal inland near the observation point 36.8948°N–27.2896°E, east side of Kos Port, locally called "Bridge" (photo credit Reuters)

Figure 15
Bikes, motorbikes and other objects overthrown and drifted by the tsunami in Port of Kos, near the "Bridge" point (photo credit ANA-MPE)

it allowed identifying the time of the earthquake and the times of tsunami wave arrivals in the south part of the port. The particular security video was kindly provided by the travel agency "Blue Star Ferries." The camera was shooting exactly to the southern tip of the port and was installed at a distance of 40 m from the shoreline.

Figure 16
People trying to take away cars that crashed due to tsunami attack in Port of Kos (photo credit ANA-MPE)

We analyzed the video images and were able to reconstruct the timeline of the wave motion after carefully measuring the field distances and dimensions of characteristic structures, such as a plant box, seen in the video. Four snapshots from the video record are given in Fig. 17, showing a) the time of first wave flooding the road, but not overpassing the sidewalks elevation, b) the time of minimum height of the water in the port (the boat is down and no fender is visible), c) the time of maximum height (second wave), the white car in the image starts floating and changing position and d) the time of water disappear and start of people passing. Figure 18 represents the absolute height of the water, considering the location of the plant box (at 1 m elevation from sea level) and the height of the plant box itself (49 cm). From the video analysis we concluded that a first strong wave arrived 12 min from the earthquake origin time with runup height of 1 m, while the strongest wave arrived 21 min after the earthquake with runup height of 1.49 m.

3.3.4 *Between Kos port and Louros Cape*

The impact along the north coast of Kos, between the port and Louros Cape (Fig. 11), is rather uniform with a decreasing trend towards Louros Cape. The manager of the Hotel Akti Kos (Point 13-Table 2), reported (Prof. E. Lekkas, personal communication) that the tsunami runup height was approximately 0.7 m. Further, in Kos Marina (Point 14-Table 2), eyewitnesses reported that initially the water level decreased by approximately 1 m about 15 min after the earthquake. Then, it increased about 0.8 m above the dock level and inundated up to 10 m inland. A second wave arrived about 25–30 min after the earthquake, and inundated approximately 10 m inland. The sea level oscillation happened about three times in about 1-h time intervals.

Another observation point was along Georgiou Papandreou Avenue, which is close to Point (14) in Table 2. The inundation distance at that location was about 33 m inside the stream and the maximum runup height was observed as between 0.5 and 0.7 m (Fig. 19).

In another small stream, located at 36.8844°N–27.3252°E and corresponding to Point (15) in Table 2, we observed several deposits indicating a possible tsunami inundation distance of 39 m; at that location, the estimated runup was estimated at 0.7 m (top-right picture in Fig. 11). A couple, residents in a house exactly next to the stream (Point 15-Table 2), verified that the tsunami inundation distance was about 15 m in their house location.

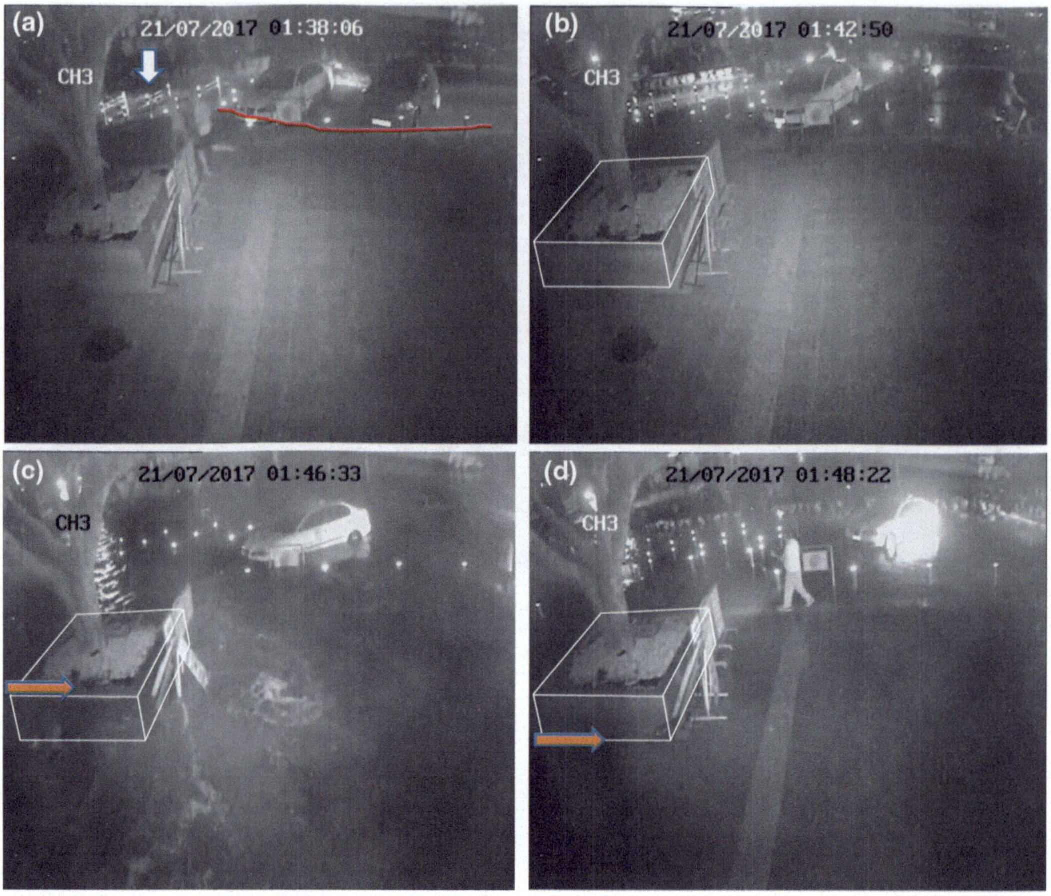

Figure 17
Video snapshots showing **a** first wave flooding the road, **b** minimum height of the water in the port, **c** maximum height of the water in the port, **d** water disappear

Louros Cape was an interesting observation point (Fig. 1c). Here, there is a tall antenna that has been installed on a large concrete foundation. From Google Earth position and pictures (Fig. 20a, b), it is clear that its base was standing about 10 m inland in December 2011. On the 2nd of August 2017, the base practically was situated at the shoreline, as can be seen in Fig. 20c. This may be due either to strong beach erosion owing to the 2017 tsunami, or to gradual beach erosion, or to coastal subsidence due to the 2017 earthquake.

Two scuba divers have reported to the mayor of Kos that the sea floor around Cape Louros had deepened significantly after the earthquake, suggesting that local subsidence occurred. This was consistent with eyewitness accounts that a small water pool was caused by the earthquake about 30 m inland. This observation may imply co-seismic subsidence.

Subsidence evidence comes also from the marshy field situated next to the lake. In this field, the survey team observed ground failures typical for soil liquefaction due to earthquake shaking.

3.3.5 Between Louros Cape and Agios Focas

Figure 21 is a photo taken at Point (21) in Table 2. It was provided by an eyewitness identifying himself as Mike and shows the water receding during one of the multiple waves arrivals. However, the eyewitness mentioned that this was not the first wave, but he could not indicate the exact time of the image. The

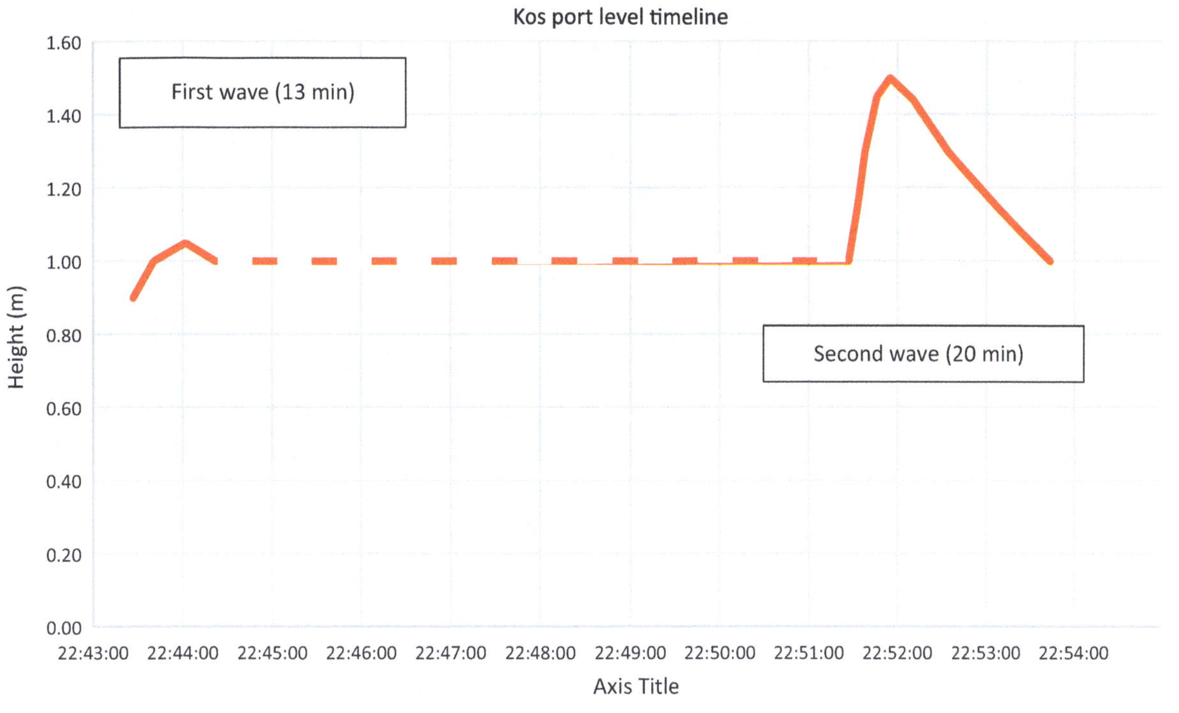

Figure 18
Kos Port level timeline according to Kos Port video analysis

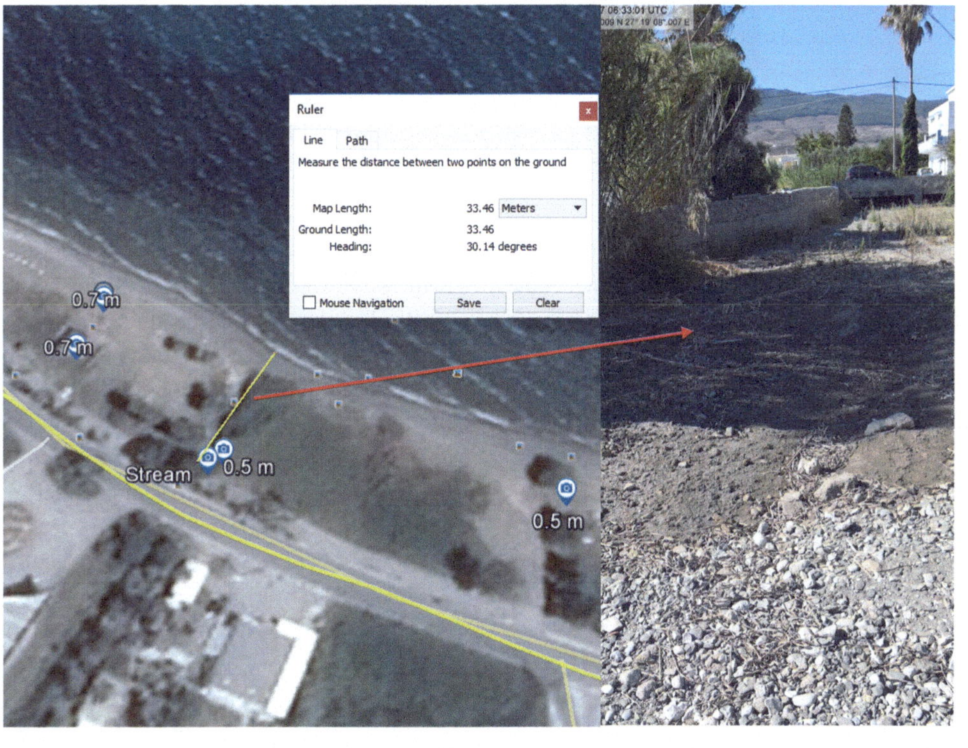

Figure 19
Water penetration along Georgiou Papandreou Avenue

Figure 20

a Foundation of the antenna at the shoreline from Google Earth View, **b** showing the antenna foundation (photo courtesy by Sofia Karagianni, uploaded December 2011, http://www.panoramio.com/user/6522978?with_photo_id=63318452), **c** figures showing the current condition of the antenna

shoreline advanced from its normal position (Fig. 21 at the bottom). From white sandbags, one may estimate the inundation distance as approximately 7–8 m.

According to the survey, the tsunami impact appears to have terminated around the Point (22) in Table 2 (36.8577°N–27.3498°E). However, no algae deposits or other tsunami evidence was documented here. Eyewitnesses of a local café bar reported no tsunami wave there.

4. Discussion and Conclusions

Tsunamis in the eastern Mediterranean have been documented from instrumental records, historical sources, archaeological evidence and geological signatures. However, large tsunami events appear only rarely. For this reason, the study of smaller events is of particular importance for understanding the tsunami generation mechanisms and the assessment of future tsunami impact. In addition, smaller events are useful for testing the tsunami early warning systems (e.g. Kânoglu et al. 2015; Heidarzadeh et al. 2017; Necmioglu et al. 2018).

In this sense, the Bodrum–Kos seismic tsunami of 20 July 2017 is a good recent example for further developing the tsunami science and engineering in the Mediterranean. It was a good opportunity to remind the general public about the tsunami threat in the Mediterranean Sea. The results of questionnaires focused on the human response against the tsunami

Figure 21
The water receding during one of the multiple waves arrivals (at 36.8869°N–27.3071°E), Kos Island

conducted during the field surveys at the southern and northern coastlines of Bodrum Peninsula showed that tsunami awareness is extremely low in this region (Yalciner et al. 2017; Arikawa et al. 2017). Heidarzadeh et al. (2017) has highlighted the need for promoting education and preparedness programs for the people at risk in this region.

Immediately after the event, we organized several field surveys to collect data regarding the hydrodynamics and impact of the tsunami along the coastal zones of Bodrum Peninsula, SW Turkey, and Kos Island, Greece. The data collected are based on field observations, pictures, video records and interviews with eyewitnesses. This type of macroscopic evidence provides a useful data set to better understand the 2017 tsunami since it was recorded by only one tide-gauge in Bodrum.

In southern Bodrum Peninsula, Gumbet Bay was the worst hit by the tsunami. The inundation distance reached up to 60 m, while the maximum tsunami runup and inundation depth reached as 1.9 m and 1 m, respectively. Higher tsunami inundation distances were observed in several streambeds located in Gumbet, Karaincir, and Bitez Bays. In a streambed, eyewitnesses reported 280 m inundation distance. In Gumbet Bay, significant tsunami inundation, strong whirlpools, and currents dragged small crafts, and damaged about 30 boats and causing the sinking of another 10, while in Bitez Bay there was no boat damage reported.

It appears that Fener Beach was the most likely boundary of significant inundation at the western end of the southern coast of Bodrum Peninsula, whereas Yaliciftlik Bay is the most probable boundary at the eastern end. It is very likely that Karaada Islet, situated between Bodrum and Kos, acted as a tsunami breakwater and protected Bodrum town. The maximum runup at Küçükpoyraz Bay at the south of Karaada Islet ranged high as 1.5–2.0 m. Perhaps higher runup occurred at some other small narrow bays at the south of Karaada Islet where, unfortunately, the survey could not be performed at those critical bays facing to the earthquake source. Minor tsunami impacts were observed in Akyaka and Datça regions.

In Kos, the tsunami arrived and inundated only in the eastern part of the island. In general, in the northwestern, eastern and southeastern coasts of Kos the tsunami runup did not exceed 1.0 m with two local exceptions. The first is in the northwest where a maximum runup of 1.4–1.5 m likely occurred, but no damage was reported. The second is the Amphora-shaped bay of the Kos Port with a narrow entrance facing towards the earthquake source. In that bay, the maximum runup was accurately measured from video records and found to be 1.5 m. The maximum water penetration inland in the Kos Port was about 65 m. From video records it appears that the first wave of 1 m runup arrived 12 min after the earthquake occurrence while the main wave arrived at 21 min after the earthquake occurrence with runup 1.5 m. The tsunami arrival caused some damage along the Kos Port coastal zone, such as drifting away of various motor bikes and cars, as well as light damage to some small size vessels. Future numerical simulation studies that would be in agreement with the results of the field survey presented here may help in better understanding of the faulting process in the earthquake source and the mechanism of tsunami generation.

Acknowledgements

Bodrum and Datca branches of Turkish Chamber of Civil Engineers (TCCE) and Bodrum Municipality are acknowledged for their close cooperation and support in logistics and data collection. This study was partly supported by EC Project Assessment, Strategy and Risk Reduction for Tsunamis in Europe (ASTARTE)—FP7-ENV2013 6.4-3, Grant 603839, EC DG ECHO funded ECHO/SUB/2015/718568/PREV26—"Probabilistic Tsunami hazard maps for the NEAM Region" (TSUMAPS-NEAM), Japan–Turkey Joint Research Project by JICA on earthquakes and tsunamis in Marmara Region by MarDim SATREPS, Erasmus + Program EC Turkish National Agency funded Contract number 2015-3-TR01-KA205-024506, Training Of Youth for Preparedness Against Marine Induced Hazards (TROYO), TUBITAK 113M556, TUBITAK 108Y227 Projects and UDAP-Ç-12-14 project granted by Disaster Emergency Management Presidency of Turkey (AFAD) which provided invaluable support to gain experience and develop knowledge for investigating tsunamis. The municipality of Kos that allowed to obtain the Security Cameras images from the Blue Star Agency of Kos. Satellite images are taken from Google Earth 7.1.8.3036, the image on 1/17/2017, at 12: 38:00 am, server kh.google.com. METU Faculty of Engineering, DOLFEN Engineering and Consultancy Company, Oya Butik Hotel in Bodrum, KOERI-Regional Earthquake-Tsunami Monitoring Center (RETMC), Department of Earthquake Engineering from KOERI, Underwater Research Society Turkey, Philip England, Chiara Projetti, Ocal Necmioglu, Naeimeh Sharghivand, Emrecan Isik, Duygu Tufekci, Ghazal Khodkar, Can Goztepe, Bora Yalciner, Andrey Zaytsev, Marinos Charalampakis, Şükrü Ersoy, and Tuncay Taymaz are also acknowledged.

References

Aktug, B., Kaypak, B., & Çelik, R. N. (2010). Source parameters of 03 February 2002 Çay Earthquake, Mw6.6 and aftershocks from GPS Data, Southwestern Turkey. *Journal of Seismology, 14,* 445–456.

Aktug, B., Nocquet, J. M., Cingoz, A., Parsons, B., Erkan, Y., England, P. C., et al. (2009). Deformation of western Turkey from a combination of permanent and campaign GPS data: Limits to block-like behavior. *Journal of Geophysical Research, 114,* B10404.

Alpar, B., Dogan, E., Yuce, H., & Altiok, H. (2000). Sea level changes along the Turkish coasts of the Black Sea, the Aegean Sea and the Eastern Mediterranean. *Mediterranean Marine Science, 1*(1), 141–156.

Altinok, Y., Alpar, B., Özer, N., & Aykurt, H. (2011). Revision of the tsunami catalogue affecting Turkish coasts and surrounding regions. *Natural Hazards and Earth System Sciences, 11*(2), 273–291.

Ambraseys, N. N. (1962). Data for the investigation of the seismic sea-waves in the Eastern Mediterranean. *Bulletin of the Seismological Society of America, 52*(4), 895–913.

Ambraseys, N., & Synolakis, C. (2010). Tsunami catalogs for the Eastern Mediterranean, revisited. *Journal of Earthquake Engineering, 14*(3), 309–330.

Arikawa, T., Guler, H. G., & Yalciner, A. C. (2017). Comparison of human response against earthquake and tsunami. In *American Geophysical Union Fall Meeting 2017*, December 11–15, 2017, New Orleans, USA.

Batıgün, E., Yolsal-Çevikbilen, S., & Taymaz, T. (2018). Source Characteristics of 2017 Ayvacik, Lesvos, and Bodrum–Kos Earthquakes Obtained from Regional Moment Tensor Inversion. EGU2018-761, Geophysical Research Abstracts, Vol. 20, Poster, 08–13 April 2018, Vienna, Austria.

Beisel, S., Chubarov, L., Didenkulova, I., Kit, E., Levin, A., Pelinovsky, E., et al. (2009). The 1956 Greek tsunami recorded at Yafo, Israel, and its numerical modeling. *Journal of Geophysical Research: Oceans, 114,* C09002. https://doi.org/10.1029/2008JC005262.

Cakir, T. E., & Yalciner, A. C. (2002). Effects of 1956 South Aegean tsunami to Bodrum Peninsula. In *Turkish chamber of civil engineers 4th national coastal engineering symposium*, Oct 24–26, 2002, Antalya. (**in Turkish**).

Cordrie, L., Gailler, A., Heinrich, P., Hébert, H., & Schindelé, F. (2018). Tsunami modeling of the 20th July 2017 Mw 6.6 earthquake in the Aegean Sea. In *EGU General Assembly Conference Abstracts* (Vol. 20, p. 15114).

Derebay, S. K. (2007). Wind and swell wave climate for Turkish coast of the Aegean and Mediterranean Sea (Doctoral dissertation, Master thesis at the Science in Civil Engineering Department, Middle East Technical University, Turkey).

Dimova, L., & Raykova, R. (2018) Numerical simulations of the earthquake-induced tsunami of July 20, 2017 (Mw = 6.6) in Bodrum–Kos, Aegean Sea. Review of the Bulgarian Geological Society, 79(1).

Dogru, A., Gorgun, E., Ozener, H., & Aktug, B. (2014). Geodetic and seismological investigation of crustal deformation near Izmir (Western Anatolia). *Journal of Asian Earth Sciences, 82,* 21–31.

England, P., Howell, A., Jackson, J., & Synolakis, C. (2015). Palaeotsunamis and tsunami hazards in the Eastern

Mediterranean. *Philosophical Transactions of the Royal Society A, 373*(2053), 20140374.

Ergin, A., Yalciner, A.C., Guler, I., Baykal, C., Esen, M., Ozyurt, G., Arikan, G.F. (2008). Bodrum Gumbet Yacht Marina Project Research Report, Middle East Technical University, Ocean Engineering Research Center, Ankara, Turkey, p. 2–9. **(in Turkish)**

Galanopoulos, A. G. (1957). The seismic sea-wave of 9 July 1956. *Praktika Academy Athens, 32,* 90–101. **(in Greek with Engl. abstr.)**.

Galanopoulos, A. G. (1960). Tsunamis observed on the coasts of Greece from antiquity to present time. *Annals of Geophysics, 13*(3–4), 369–386.

Ganas, A., Elias, P., Valkaniotis, S., Briole, P., Kapetanidis, I., Kassaras, I., Barberopoulou, A., Argyrakis, P., Chouliaras, G., & Moschou, A. (2017). Co-seismic deformation and preliminary fault model of the July 20, 2017 M6.6 Kos earthquake. In *Aegean Sea Report, EMSC,* 30 July 2017, Athens, Greece.

Gürer, Ö. F., Sangu, E., Özburan, M., Gürbüz, A., & Sarıca-Filoreau, N. (2013). Complex basin evolution in the Gökova Gulf region: Implications on the Late Cenozoic tectonics of SW Turkey. *International Journal of Earth Sciences*. https://doi.org/10.1007/s00531-013-0909-1.

Heidarzadeh, M., Necmioglu, O., Ishibe, T., & Yalciner, A. C. (2017). Bodrum–Kos (Turkey-Greece) Mw 6.6 earthquake and tsunami of 20 July 2017: A test for the Mediterranean tsunami warning system. *Geoscience Letters, 4,* 31. https://doi.org/10.1186/s40562-017-0097-0.

Howell, A., Jackson, J., England, P., Higham, T., & Synolakis, C. (2015). Late Holocene uplift of Rhodes, Greece: Evidence for a large tsunamigenic earthquake and the implications for the tectonics of the eastern Hellenic Trench System. *Geophysical Journal International, 203*(1), 459–474.

Kalafat, D., & Horasan, G. (2012). A seismological view to Gkova region at southwestern Turkey. *International Journal of Physical Sciences, 7*(30), 5143–5153.

Kânoğlu, U., Titov, V., Bernard, E., & Synolakis, C. (2015). Tsunamis: Bridging science, engineering and society. *Philosophical Transactions of the Royal Society A, 373*(2053), 20140369.

Karasözen, E., Nissen, E., Büyükakpınar, P., Cambaz, M. D., Kahraman, M., Kalkan Ertan, E., et al. (2018). The 2017 July 20 Mw 6.6 Bodrum–Kos earthquake illuminates active faulting in the Gulf of Gökova, SW Turkey. *Geophysical Journal International, 214*(1), 185–199.

Kiratzi, A., & Koskosidi, A. (2018). Constrains on the near source motions of the Kos-Bodrum 20 July 2017 Mw 6.6 earthquake. In *16th European conference of earthquake engineering*, 18–21 June 2018, Thessaloniki, Greece.

KOERI, Kandilli Observatory and Earthquake Research Institute (2017) *Report of 21 July 2017 Gokova Bay - Mediterranean Earthquake distributed by KOERI Regional Earthquake and Tsunami monitoring Center (in Turkish).* http://www.koeri.boun.edu.tr/sismo/2/wp-content/uploads/2017/07/Gokova.Korfezi.v4.pdf.

Konca, A. O., Cetin, S., Ozdemir, A., Isik, S. E., Dogan, U., Karabulut, H., Ergintav, S., Funning, G., Floyd, M., & Reilinger, R. (2018). The geometry and coseismic slip of 2017 Mw6.6 Bodrum–Kos earthquake inferred from geodetic, seismic data and aftershocks. In: *EGU General Assembly Conference Abstracts* (Vol. 20, p. 7442).

Kuran, U., & Yalciner, A. C. (1993). Crack propagations, earthquakes and tsunamis in the Vicinity of Anatolia. In Stefano Tinti (Ed.), *Tsunamis in the World.*, Advances in natural and technological hazards research New York: Springer.

Kurt, H., Demirbağ, E., & Kuşçu, İ. (1999). Investigation of the submarine active tectonism in the Gökova gulf, southwest Anatolia–southeast Aegean Sea, by multi-channel seismic reflection data. *Tectonophysics, 305*(4), 477–496. https://doi.org/10.1016/S0040-1951(99)00037-2.

Luttrell, A. (1999). Earthquakes in the Dodecanese: 1303–1513. In E. Zachariadou (Ed.), *Natural disasters in the Ottoman Empire* (pp. 145–151). Rethymnon: Crete University Press.

Macías, J., Annunziato, A., Papadopolous, G., Ortega, S., González-Vida, J. M., & Agalos, A. (2018). Numerical simulation of Aegean Sea summer 2017 tsunami events. In: EGU General Assembly Conference Abstracts (Vol. 20, p. 8191).

McClusky, S., Balassanian, S., Barka, A., Demir, C., Ergintav, S., Georgiev, I., et al. (2000). Global Positioning System constraints on plate kinematics and dynamics in the eastern Mediterranean and Caucasus. *Journal of Geophysical Research-Solid Earth, 105,* 5695–5719. https://doi.org/10.1029/1999jb900351.

McKenzie, D. (1972). Active tectonics of the Mediterranean region. *Geophysical Journal of the Royal Astronomical Society, 30,* 109–185.

Minoura, K., Imamura, F., Kuran, U., Papadopoulos, G., Takahashi, T., & Yalciner, A. C. (2000). Discovery of Minoan Tsunami deposits. *Geology, 28*(1), 59–62.

Necmioglu, O., Özer Sözdinler, C., Cevdet Yalçiner, A., Heidarzadeh, M., & Ishibe, T. (2018). Challenges and requirements for near-field tsunami warning in the Aegean Sea and Eastern Mediterranean. In *EGU General Assembly Conference Abstracts* (Vol. 20, p. 12273).

Ocakoğlu, N., Nomikou, P., Isçan, Y., Loreto, M. F., & Lampridou, D. (2018). Evidence of extensional and strike-slip deformation in the offshore Gökova-Kos area affected by the July 2017 Mw6.6 Bodrum–Kos earthquake, eastern Aegean Sea. *Geo-Marine Letters, 14,* 15. https://doi.org/10.1007/s00367-017-0532-4.

Okal, E. A., Synolakis, C. E., Uslu, B., Kalligeris, N., & Voukouvalas, E. (2009). The 1956 earthquake and tsunami in Amorgos, Greece. *Geophysical Journal International, 178*(3), 1533–1554.

Ozener, H., Dogru, A., & Acar, M. (2013). Determination of the displacements along the Tuzla fault (Aegean region-Turkey): Preliminary results from GPS and precise leveling techniques. *Journal of Geodynamics, 67,* 13–20.

Öztürk, S., & Şahin, Ş. (2018). A statistical space-time-magnitude analysis on the aftershocks occurrence of the July 21th, 2017 MW = 6.5 Bodrum–Kos, Turkey, earthquake. *Journal of Asian Earth Sciences, 172,* 443–457.

Papadopoulos, G. A. (2016). Historical and geological evidence of tsunamis in Europe and the Mediterranean. In *Tsunamis in the European-Mediterranean Region by G. A. Papadopoulos*, Hardcover. Elsevier. ISBN: 9870124202245.

Papadopoulos, G. A., & Chalkis, B. J. (1984). Tsunamis observed in Greece and the surrounding area from antiquity up to the present times. *Marine Geology, 56*(1–4), 309–317.

Papadopoulos, G. A., Daskalaki, E., Fokaefs, A., & Giraleas, N. (2007). Tsunami hazard in the eastern Mediterranean: Strong earthquakes and tsunamis in the east hellenic arc and trench system. *Natural Hazards & Earth System Science, 7,* 57–64.

Papadopoulos, G. A., Gràcia, E., Urgeles, R., Sallares, V., De Martini, P. M., Pantosti, D., et al. (2014). Historical and pre-historical tsunamis in the Mediterranean and its connected seas: Geological signatures, generation mechanisms and coastal impacts. *Marine Geology, 15,* 14. https://doi.org/10.1016/j.margeo.2014.04.014.

Papadopoulos, G. A., Minoura, K., Imamura, F., Kuran, U., Yalciner, A. C., Fokaefs, A., et al. (2012). Geological evidence of tsunamis and earthquakes at the eastern hellenic arc: Correlation with historical seismicity in the eastern Mediterranean Sea. *Research in Geophysics, 2,* e12.

Papadopoulos, G. A., & Pavlides, S. B. (1992). The large 1956 earthquake in the South Aegean: Macroseismic field configuration, faulting and neotectonics of Amorgos Island. *Earth and Planetary Science Letters, 113,* 383–396.

Pavlopoulos, K. (2010). Relative sea level fluctuations in Aegean coastal areas from middle to late Holocene. *Geodinamica Acta, 23*(5–6), 225–232.

Poulos, S. E., Plomaritis, T. A., Ghionis, G., Collins, M. B., & Angelopoulos, C. (2012). The role of coastal morphology in influencing sea level variations induced by meteorological forcing in microtidal waters: Examples from the Island of Crete (Aegean Sea, Greece). *Journal of Coastal Research, 29*(2), 272–282.

Reilinger, R., McClusky, S., Vernant, P., Lawrence, S., Ergintav, S., Cakmak, R., et al. (2006). GPS constraints on continental deformation in the Africa–Arabia–Eurasia continental collision zone and implications for the dynamics of plate interactions. *Journal of Geophysical Research, 111,* B05411.

Saltogianni, V., Taymaz, T., Yolsal-Çevikbilen, S., Eken, T., Gianniou, M., Öcalan, T., Pytharouli, S., & Stiros, S. (2017). Fault-model of the 2017. Kos-Bodrum (east Aegean Sea) mw 6.6 earthquake from inversion of seismological and GPS data—Preliminary report. Department of Civil Engineering, University of Patras, Greece. https://www.emsc.eu/Files/event/606346/2017-SALTOGIANNI-TAYMAZ-ETALKOS-BODRUM-EARTHQUAKE-PRELIMINARY-REPORT.pdf.

Sözbilir, H., Uzel, B., Sümer, Ö., Eski, S., Softa, M., Tepe, Ç. (2017). 21 Temmuz 2017 Gökova Körfezi Depremleri ve Bölgenin Depremselliği Raporu. (in Turkish), Dokuz Eylül University Earthquake Research and Application Center Active Fault Research Group, İzmir, Turkey.

Synolakis, C. E., & Okal, E. A. (2005). 1992–2002: Perspective on a decade of post-tsunami surveys. In K. Satake (Ed.), *Tsunamis: Case studies and recent developments* (pp. 1–30)., Advances in natural and technological sciences series New York: Springer.

Tadepalli, S., & Synolakis, C. E. (1994). The run-up of N-waves on sloping beaches. *Proceedings of the Royal Society of London A, 445*(1923), 99–112.

Taymaz, T., Jackson, J., & McKenzie, D. (1991). Active tectonics of the north and central Aegean Sea. *Geophysical Journal International, 106*(2), 433–490.

Tiryakioğlu, I., Aktuğ, B., Yiğit, C. Ö., Yavaşoğlu, H. H., Sözbilir, H., Özkaymak, Ç., et al. (2018). Slip distribution and source parameters of the 20 July 2017 Bodrum–Kos earthquake (Mw6.6) from GPS observations. *Geodinamica Acta, 30*(1), 1–14. https://doi.org/10.1080/09853111.2017.1408264.

Tiryakioglu, I., Floyd, M., Erdogan, S., Gulal, E., Ergintav, S., McClusky, S., et al. (2013). GPS constraints on active deformation in the Isparta Angle region of SW Turkey. *Geophysical Journal International, 195,* 1455–1463.

Tsimplis, M. N. (1994). Tidal oscillations in the Aegean and Ionian Seas. *Estuarine, Coastal and Shelf Science, 3,* 201–208.

Tur, H., Yaltırak, C., Elitez, İ., & Sarıkavak, K. T. (2015). Pliocene–Quaternary tectonic evolution of the Gulf of Gökova, southwest Turkey. *Tectonophysics, 638,* 158–176.

UNESCO IOC (2014). The UNESCO Intergovernmental Oceanographic Commission International Tsunami Survey Team (ITST) Post-Tsunami Field Survey Guide 2nd Edition. http://itic.ioc-unesco.org/images/stories/itst_tsunami_survey/survey_documents/field_survey_guide/ITST_FieldSurveyGuide_229456E.pdf. Accessed 30 March 2018.

United States Geological Survey (USGS) (2017). Earthquake hazards program. Global CMT Project Moment Tensor Solution. https://earthquake.usgs.gov/earthquakes/eventpage/us20009ynd#moment-tensor. Accessed 2017.

Vousdoukas, M. I., Velegrakis, A. F., Dimou, K., Zervakis, V., & Conley, D. C. (2009). Wave run-up observations in microtidal, sediment-starved pocket beaches of the Eastern Mediterranean. *Journal of Marine Systems, 78,* S37–S47.

Yalciner, A. C., Annunziato, A., Papadopoulos, G., Dogan, G. G., Guler, H. G., Cakir, T. E., Sozdinler, C. O., Ulutas, E., Arikawa, T., Suzen, L., Kanoglu, U., Guler, I., Probst, P., & Synolakis, C. (2017). The 20th July 2017 (22:31 UTC) Bodrum/Kos earthquake and tsunami; post tsunami field survey report. http://users.metu.edu.tr/yalciner/july-21-2017-tsunami-report/Report-Field-Survey-of-July-20-2017-Bodrum-Kos-Tsunami.pdf. Accessed 30 March 2018.

Yalciner, A. C., Kuran, U., Akyarli, A., & Imamura, F. (1995). An investigation of the propagation of tsunamis in the Aegean Sea by mathematical modeling. In Y. Tsuchiya & N. Shuto (Eds.), *Tsunami: Progress in prediction, disaster prevention and warning*. New York: Springer.

Yılmaz, Y., Genç, S.C., Gürer, O.F., Bozcu, M., Yılmaz, K., Karacık, Z., Altunkaynak, Ş. Elmas A. (2000). When did the western Anatolian grabens begin to develop? In: Bozkurt, E., Winchester, J. A., Piper, J. D. A. (Eds.), Geological Society, London, Special Publications (Vol. 173, pp. 353–384).

Yolsal, S., & Taymaz, T. (2005). Potential source regions of earthquakes and tsunamis along the hellenic and cyprus arcs, eastern Mediterranean. In T. Taymaz (Ed.), *International symposium on the geodynamics of eastern Mediterranean: Active tectonics of the Aegean region, 15–18, June 2005* (p. 240). Istanbul: Kadir Has University.

Yolsal, S., & Taymaz, T. (2007). Source mechanism and rupture histories of the recent Gulf of Gökova and Sığacık Bay earthquakes. Geophysical Research Abstracts, 9, EGU2007-A-01776.

Yolsal, S., & Taymaz, T. (2010). Gökova Körfezi depremlerinin kaynak parametreleri ve Rodos-Dalaman bölgesinde tsunami riski. *İTÜ dergisi/d, Mühendislik, 9–3,* 53–65.

Yolsal-Çevikbilen, S., & Taymaz, T. (2012). Earthquake source parameters along the Hellenic subduction zone and numerical simulations of historical tsunamis in the eastern Mediterranean. *Tectonophysics, 536–537,* 61–100. https://doi.org/10.1016/j.tecto.2012.02.019.

Yolsal-Çevikbilen, S., Taymaz, T., & Helvacı, C. (2014). Earthquake mechanisms in the gulfs of Gökova, Sığacık, Kuşadası, and the Simav region (western Turkey): Neotectonics, seismotectonics and geodynamic implications. *Tectonophysics, 635,* 100–124. https://doi.org/10.1016/j.tecto.2014.05.001.

Yolsal-Çevikbilen, S., Ulutaş, E., & Taymaz, T. (2018). Source models of the 2012 Haida Gwaii (Canada) and 2015 Illapel (Chile) earthquakes and numerical simulations of related tsunamis. *Pure and Applied Geophysics, 15,* 14. https://doi.org/10.1007/s00024-018-1996-5.

(Received October 10, 2018, revised February 26, 2019, accepted March 2, 2019, Published online March 11, 2019)

Pure Appl. Geophys. 176 (2019), 2951–2967

https://doi.org/10.1007/s00024-019-02161-z

Pure and Applied Geophysics

Revealing the Source of the 27 August 1810 Loreto, Baja California, Tsunami from Historical Evidence and Numerical Modelling

María Teresa Ramírez-Herrera,[1] Néstor Corona,[2] and Rocío Castillo-Aja[3,4]

Abstract—Historical documents revealed that on 27 August 1810 an earthquake affected the Loreto town and the surrounding region, along 200 km of the coast of Baja California, Mexico, and damage was considerable due to strong ground motions. However, the 1810 Loreto earthquake, unlike other earthquakes of the nineteenth and twentieth centuries, produced an extensive tsunami flooding on the eastern coast of Baja California, from Loreto to La Paz. We report here the historical account of this event and analyse the source of the related tsunami. Based on the historical record and the tectonic setting, we infer that this earthquake was produced along a transform fault, right-lateral strike slip fault, Farallon fault, on the Baja California Gulf. However, the associated tsunami was disproportionally large if compared with the earthquake intensity (IX) and approximate magnitude (7.4 Mw), fault style, and the relatively late tsunami arrival, i.e. an hour after the earthquake, suggesting that a submarine landslide, triggered by this earthquake, was the suspect source for the Loreto tsunami. We present here historical data and a numerical model of the Loreto 1810 tsunami to elucidate the source, using both the 1810 earthquake and a submarine landslide, and confirm that a submarine landslide is most likely the tsunami-triggering mechanism that followed the 1810 earthquake.

Key words: Tsunami, submarine landslide, earthquake, strike-slip fault, Loreto, Baja California.

Electronic supplementary material The online version of this article (https://doi.org/10.1007/s00024-019-02161-z) contains supplementary material, which is available to authorized users.

[1] Laboratorio de Tsunamis y Paleosismología, Instituto de Geografía, Universidad Nacional Autónoma de México, Mexico City, Mexico. E-mail: tramirez@igg.unam.mx; maria_teresa_ramirez@yahoo.com

[2] Centro de Estudios en Geografía Humana, El Colegio de Michoacán A.C., La Piedad, Michoacán, Mexico.

[3] Universidad Nacional Autónoma de México, Mexico City, Mexico.

[4] *Present Address*: Departamento de Geografía, CUCSH-Universidad de Guadalajara, Guadalajara, Jalisco, Mexico.

1. Introduction

Tsunami hazard caused by landslide-generated tsunamis has been underestimated. Even more underestimated is the tsunami hazard related to strike-slip faults, because they consist mostly of horizontal movement and relatively modest vertical motion. However, ground shaking and modest uplift produced by strike-slip earthquakes can generate submarine and coastal ground failure capable of generating catastrophic tsunamis (Tinti et al. 2006; Hornbach et al. 2010; Fritz et al. 2013). The most recent 28 September 2018 Sulawesi strike-slip earthquake produced a devastating tsunami in a confined bay. Several hypotheses for the Sulawesi tsunami source have been proposed and are still being debated by the scientific community, such as the liquefaction-induced collapse of coastal land at several sites (Sassa and Takagawa 2019), submarine landslides (Dongeren et al. 2018; Heirarzadeh et al. 2019), massflows, volcano slope collapse, and submarine step-over faults.

Here we present and analyse data from historical accounts held in archives, and through numerical modelling we document submarine landslide as the cause of the Loreto tsunami of 1810. Based on the historical record and the tectonic setting, we infer that this earthquake was produced along a transform fault, a right-lateral strike slip fault, the Farallon fault, on the Baja California Gulf. However, the associated tsunami was disproportionally large if compared with the earthquake intensity (IX) and approximate magnitude (7.4 Mw; Suter 2018), fault style, and the relatively late tsunami arrival, i.e. an hour after the earthquake, suggesting that a submarine landslide, triggered by this earthquake, was the suspect source for the Loreto tsunami. We used multiple signatures and criteria to determine the tsunamigenesis caused by slope failure, including (1) documented large run-up values close to the source and focusing effects (Assier-Rzadkieaicz et al. 2000; Ward 2001; Okal and Synolakis 2003, 2004; Satake and Tanioka 2003; Harbitz et al. 2006; Fritz et al. 2007); (2) limited far-field effects when compared with earthquake tsunamis (Satake and Tanioka 2003; Okal and Synolakis 2004) due to the dipole nature of landslide tsunamis that decays faster in the far field (Satake and Tanioka 2003; Mohammed and Fritz 2012); (3) propagation tends to produce radial directivity patterns (Ben-Menahem and Rosenman 1972a; Iwasaki et al. 1997; Okal 2000; Okal et al. 2003); and (4) landslide tsunamis can propagate and reach coastlines in a very short time because the source is often located in continental margins at shallow waters (Zaniboni et al. 2014), for instance the 1994 Skagway, Alaska earthquake and the 2010, Haiti earthquake reached the coast in 1 min (Kulikov et al. 1996; Synolakis et al. 2002; Fritz et al. 2013). Lastly, our analysis using numerical modelling of both scenarios, i.e. earthquake and submarine landslide sources, suggests that a submarine landslide is most likely the tsunami-triggering mechanism that followed the 1810 earthquake.

1.1. Seismic and Tectonic Setting

The Baja California peninsula is located between latitude 32°N and 23°N. The peninsula was separated from mainland Mexico, the North American Plate, during the Cenozoic when opening of the Gulf of California occurred about 12 million years ago (Stock and Hodges 1989; Atwater and Stock 1998). The Gulf of California is characterized by relative Pacific-North American plate motion forming an active oblique-divergent plate boundary, i.e. active transtensional rifting (Stock and Hodges 1989; Sedlock 2003) (Fig. 1). The different fault systems that accommodate deformation comprise mainly right-lateral strike slip faults, short spreading centres, and an array of normal faults remain active in the southwestern gulf margin of the Baja California peninsula and offshore in the southern sector of the gulf (Fletcher and Munguía 2000; Munguia et al. 2006; Plattner et al. 2007; Busch et al. 2011; Coyan et al. 2013; Umhoefer et al. 2014). Most recent studies show a significant number of focal mechanisms indicative of right-lateral strike-slip faulting related to the transform fault system along the Gulf of California rift, while the southwestern portion of the gulf focal mechanism indicates normal faulting (Fletcher and Munguía 2000; Munguia et al. 2006).

The study area is located in the southern part of the Baja California peninsula (Mexico) bordering the Gulf of California, at 26°N and 24.3°N, between the bays of Loreto and La Paz (Fig. 1). Loreto is located on the coastal plain of the western part of the Gulf, where active normal faults displace Quaternary deposits (Umhoefer et al. 2014). At the plate boundary, in the Gulf of California, a 125-km-long transform fault extends east of Loreto, the Farallon transform fault, and this is where seismic activity is relatively high and large earthquakes indicate right-lateral strike slip motion (Goff et al. 1987; Sumy et al. 2013; Castro et al. 2016, 2017). Most recent analysis of historical earthquakes suggest that the Loreto 1810 earthquake was related to the Farallon transform fault (Suter 2018).

2. Methods

2.1. Historical Analysis

We used primary sources to find new data for earthquakes with tsunamigenic evidence. With this purpose, we searched in repositories available online and historical archives. To confirm data sources (authorship) and the reliability of time and place of reported events, we reviewed thoroughly historical documents applying hermeneutic and philological text analysis to interpret and put in context the historical facts (Guidoboni and Ebel 2009). To validate the location of cited sites, we used historical and modern cartography.

2.2. Numerical Modelling

2.2.1 Data

We used the topo-bathymetric data obtained from the global multi-resolution topography model (Ryan et al. 2009) and the topographic model from the National Institute of Statics and Geography (INEGI

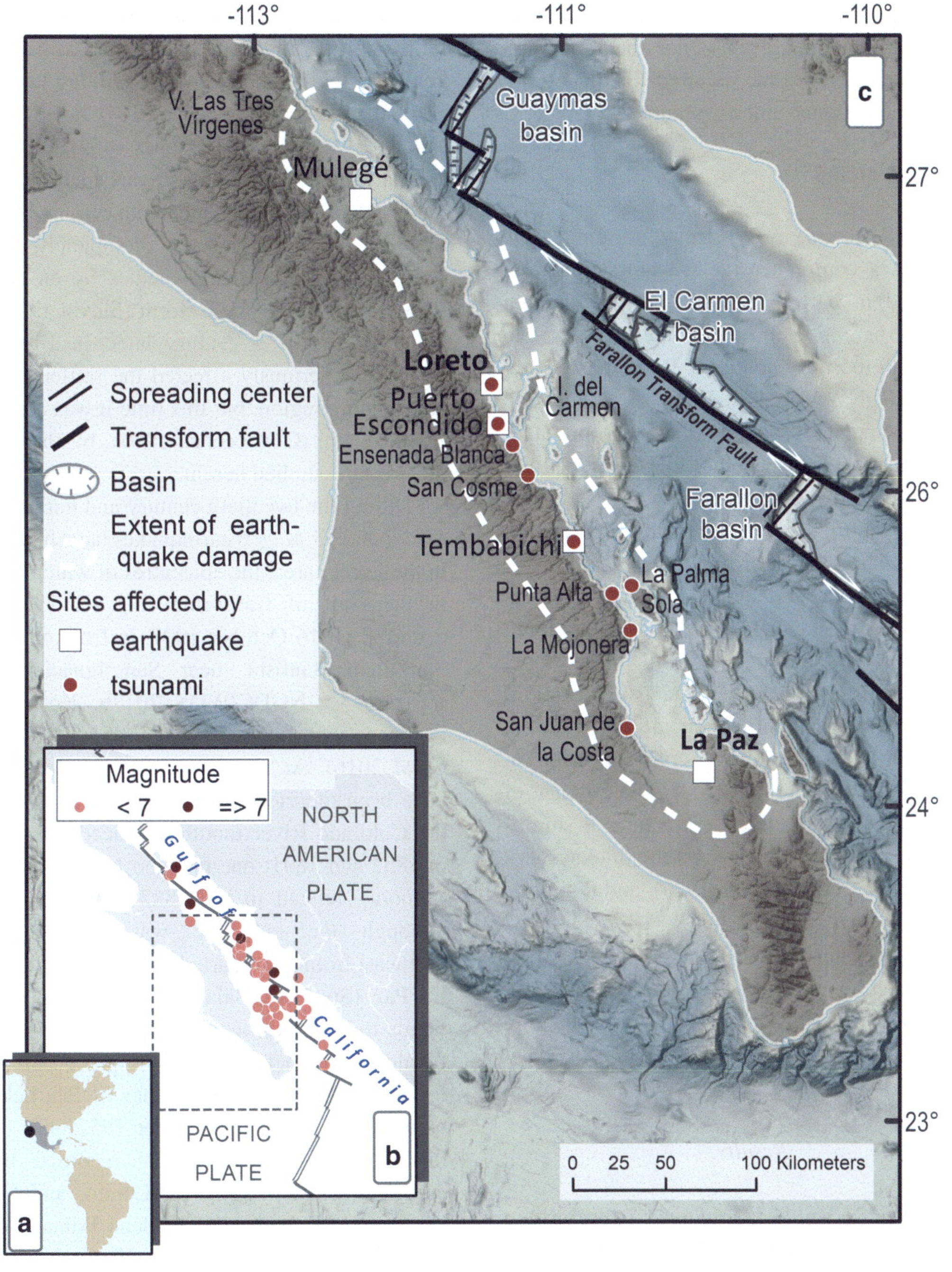

Figure 1

Seismotectonic setting of the 1810 Loreto earthquake and tsunami. The white dashed line indicates sites and area damaged by earthquake and tsunami reported by Felipe Goycoechea (1813). **a** Context location; **b** Gulf of California. Pink and red dots: epicentres; dashed line: study area shown in **c**. **c** Study area and tectonic context; white squares and red dots indicate sites reported in historical archives with earthquake and tsunami effects, accordingly

2013). These data were merged to create a detailed topographic model depicting coastal landforms. Data were converted to contour models to be edited and merged; later, a raster model of 80 × 80 m-pixel was recalculated (Fig. 2). This pixel size shows coastal features that might influence the tsunami inundation at local scale. This procedure does not affect the quality of the bathymetric chart nor the modelling procedure, but only the time required for modelling; i.e. more computational time is required.

2.2.2 Tsunami Modelling

The GEOWAVE code was selected to model the tsunami. It is a comprehensive tsunami simulation model formed in part by combining the Tsunami Open and Progressive Initial Conditions System (TOPICS) with the fully nonlinear Boussinesq water wave model FUNWAVE [see Watts et al. (2003)]. GEOWAVE simulates tsunami generation, propagation, and inundation using the fourth-order fully nonlinear and fully dispersive Boussinesq wave model with a multiple wave dissipation mechanism, wave breaking, and land overflow.

Input data for the earthquake source is listed in Table 1, and for the landslide, assumed as a slump; data (longitude, latitude, length, width, mean slope, orientation, depth, and distance of landslide motion) were obtained from geomorphological interpretation. The bulk density was taken as constant value of 1.906 kg/m^3 (Løvholt et al. 2015). In the absence of marine geology data, the thickness values of slumps were inferred by taking into account the thickness-to-length ratio (Prior and Coleman 1979), and the limits of the ratio of curvature (R), as $R = (b^2/8T) + T/2$ (Turner and Schuster 1996). Tsunami models were run for 60 min of modulation time, and 72 h of computational processing using a PC with an intel Core i7 processor at 3.0 GHz and 32 GB of RAM.

3. Results

Data presented here was obtained from written reports by the governor of Baja California, Felipe Goycoechea, in 1810 (Goycoechea 1810a, b), preserved in the Archivo General de la Nación (AGN) de México, in which he describes the occurrence of an earthquake and tsunami that affected the region between Loreto and La Paz. [Please see Supplementary material in this paper and in Suter (2018), for detailed historical data on the 1810 earthquake.]. The port of La Paz by the mid-1800s housed a population of 400; it included the buildings of the garrison and the mission with a surrounding wall and dyke and with a well in the centre. In 1813, Loreto and its outlying farms housed 824 inhabitants (Goycoechea 1813); however, this had fallen to 493 by 1857 (Lassépas 1859). Crosby (2007) estimates that the sea has advanced by at least 100 yards (91 m) since the eighteenth century. It is known that in 1717, Loreto was completely destroyed, although it is not clear whether by an earthquake (Clavijero 1937) or by a storm (Slevin 1922; Salazar Rovirosa 1980). A century later, in 1810, another earthquake seriously affected the settlement and the surrounding region, but this time it was accompanied by a tsunami (Goycoechea 1810a, b); the account of this event is studied herein.

During the twentieth century and until today, four $M = 7$ and 33 $M \geq 6$ earthquakes have been recorded in the Loreto area, the epicentres of which are located in the Gulf of California (Fig. 1). The strongest occurred on 16 October 1907, in front of the Bay of San Juan Bautista, near San Ignacio (Figueroa 1970, 1972; NGDC/WDS 2017b, Kostoglodov & Pacheco 1999; Lopez-Pineda 2005, GEM 2013; ISC-GEM 2016). At least four historical local tsunamis have been reported in the Gulf of California: two in the Colorado River mouth produced by earthquakes in 1852 and 1891; one in Cabo Haro, near Guaymas (Sonora), related to the 1887 Bavispe Earthquake, although the epicentre was located $\sim$ 350 km northeast from the reported tsunami; and, one more in La Paz (Southern Baja California) in 1894 which seems to not have been produced by an earthquake. Ewald (1997) mentions an earthquake as responsible for flooding of the salt marshes of Isla del Carmen between 1941 and 1942. After 1952 until today, 17 far-field tsunamis have been recorded by mareographs located in La Paz, Loreto, and Cabo San Lucas, for which the 1960 Chilean Tsunami produced the maximum runup value (0.73 m) [National Geophysical Data Center/World Data Service (NGDC/WDS 2017a)].

3.1. Historical Data Regarding Effects on Loreto by the 1810 Earthquake and Tsunami

At 05:20 h (local time) on 27 August 1810, an earthquake was accompanied by a tsunami that was reported by the Governor of Baja California in two letters sent to the Viceroy on separate dates: 2 September and 13 October. The letters are held in the General Archive of the Nation (AGN) in two different collections: Instituciones Coloniales/Indiferente Virreinal/4096 Exp. 041; and Instituciones Coloniales/Operaciones de Guerra/Vol. 750. Exp. 9 (52–54).

The lack of response by the authorities can be explained by the historical context of the time: the insurgency leading to the War of Independence was barely within 3 weeks of its onset. The letters also reported the presence of the brigantine *Activo*, under the command of pilot José María de Narváez, anchored in Puerto Escondido. It was perhaps by this vessel that correspondence was carried to the mainland. Unfortunately, upon its arrival at San Blas in November when the insurgents were taking the port, it was compelled to confront them rather than deal with any other matter.

The eyewitness and author of the reports was Don Felipe Goycoechea (1747–1814), who was promoted to commandant and official in charge of the army, the California garrisons at Santa Bárbara, California; in 1805, he was appointed Governor of Antigua (Baja) California, and was transferred to Loreto where he took up office in 1806 and would perform his duties until his death in 1814. The situation that the Governor described in his letters was desperate: *I am at present under a tree because my house has*

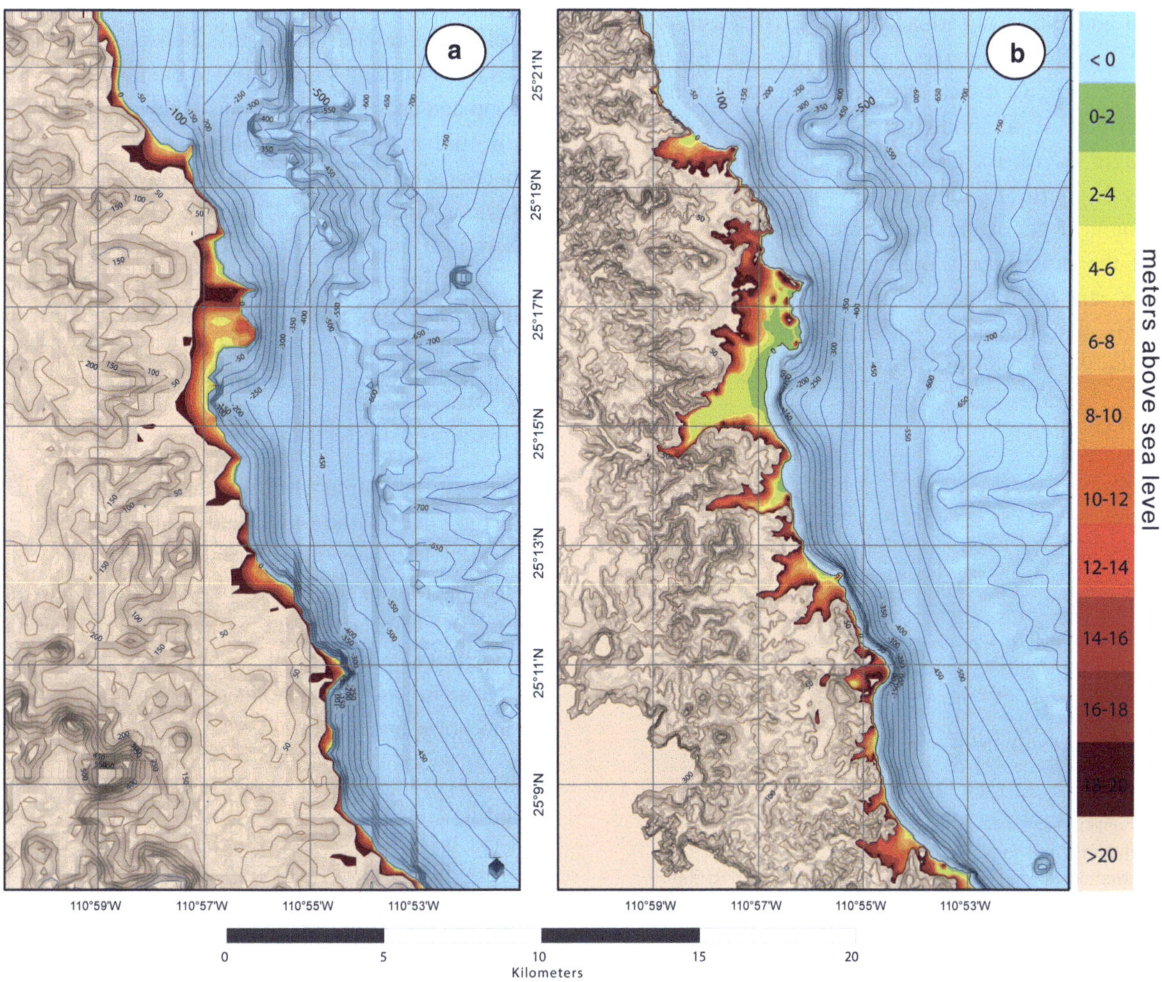

Figure 2
Comparison between topo-bathymetric chart obtained from GEBCO (**a**); Merged model with finest topographic model (**b**). It shows relevant changes on coastal land morphology, crucial for accurate tsunami modelling results

been destroyed and he even suggests that the garrison be moved elsewhere in view of the *advantages of relocating this community to the Mulegé Mission* (13 October 1810); he seeks the help of the Viceroy for the inhabitants and for the reconstruction of the town. His requests were not granted, since during the years of the insurgency *the San Blas naval station remains at the barest minimum, and fails to fulfil its duty to bring to Alta California the salaries and reports of the garrisons* (Fuster Ruiz 1997).

A common practice in all the garrisons of the Californias was the report in the form of the Memoranda of the Garrison (*Memorias del Presidio*), which had to be sent to Mexico City in vessels that were sent to collect them. That was the case of the brigantine *Activo*, which set sail from San Blas for Alta and Baja California in June 1810 (Fuster Ruiz 1997) and which *carries the reports of this garrison*; from 23 August it was anchored at Puerto Escondido, described as a good anchorage for ships of any draught (Cárdenas de la Peña 1969). According to Fray Norberto de Santiago (de Santiago 1813), "on account of the extenuating circumstances, the report that the procurator should submit annually has not been sent for the past three years", which helps to clarify the conditions that existed on the peninsula during the early years of the insurgency.

Table 1

Summary of tsunami modelling parameters using an earthquake source

Parameter	Value
Fault centroid longitude	−110.287652
Fault centroid latitude	25.940738
X	571,327
Y	2,869,315
Orientation of strike from north	128
Fault dip angle	79
Slip rake angle	168
Earthquake magnitude	7.4
Maximum fault slip	3
Centroid depth	10,000
Typical fault length	125,000
Typical fault width	20,000
Shear modulus	4.00E + 10
Radius for slip to drop 50%	125,000
Depth for slip to drop to 1%	150
Typical water depth	1950

The first letter, dated 2 September 1810, records the hour and date of the earthquake—05:20 on 27 August 1810—as well as its extent:

> "According to the dispatch brought to me by the soldier who came with the mail from 40 leagues [1 royal league = 6.687 km; 1 marine league = 5.573 km (Dirección General del Instituto Geográfico y Estadístico 1886; Peñalver and de Olive 1869)] away to the north, and the commandant of the southern part of Baja California, this earthquake was widespread."

and damage to buildings:

> "... an earthquake that has destroyed its main buildings, for example the Governor's residence, the two royal arsenals ... the guardhouse and lockup, forge and lieutenant's house ... and various other buildings belonging to the families of the troop and crew, which, because they are small and low, did not suffer so much damage. However, there was no tragedy since everybody was sleeping outdoors on account of the excessive heat that there is at this time of year. The church shifted on its foundations and its walls split into several parts; the main arch of the presbytery has come adrift on one side of the keystone and is threatening collapse"

and the physical effects:

> "The surface-ruptures of this earthquake occurred on the edge of the beach and some 150 varas (1 vara = 83.59 cm; 3 varas = 2.5 m) away just a little in front of the Governor's residence, leaving the earth displaced and with several holes that opened and gushed water. The whole expanse of land as far as the port of La Paz split off with a loud noise and rumbling"

as well as a description of the tsunami that accompanied it:

- In Loreto:

"... after an hour the sea became rough and then broke its confines with such force that we initially thought we would be submerged, with no chance of escape by retreat to higher ground were it not for the fact that the violent waves were directed towards the island that faces this garrison and is called del Carmen".

- In Puerto Escondido:

"… and although [the brigantine *Activo*] had been moored securely with four anchors, these were not sufficient to hold it because the swirls and strong eddies of the sea water caused it to make more than forty turns submerging and twisting all the anchor cables so that, had the fury of the waters lasted a little longer, the anchors would have finished up in the air and the ship would have stranded, and, in brief, it was some eight hours in continuous movement"

The second letter, dated 13 October 1810, reported that the earthquakes (aftershocks?) that started on 27 August had lasted until 30 September, but with less intensity than the first, and that this had exacerbated the damage to the buildings. It records how the earthquakes left:

"…this displaced earth, cracked and hollow underneath as has been seen in various houses where, when the families were going about their daily tasks, their feet have become buried; when the people probed these cavities with stakes more than three vara long, they have not reached the bottom"

1 vara = 83.59 cm, 3 varas = 2.5 m

It mentions the fear of the inhabitants on finding themselves at the edge of the beach, since

"…they saw the vast movements that the sea made and they subsequently learned that between this garrison and the port of La Paz in an inlet that they call de Tembabichi (alternatively Tembabiche) the sea went out more than a league; it destroyed the piers and left for the vessels a good anchorage, and at other sites made less damage".

3.2. Numerical Modelling

Considering that (1) the Gulf of California, in the Baja California sector, is dominated by transform and onshore normal faults near La Paz bay, and apparently older, perhaps Miocene in age, offshore normal faults in the same bay (Umhoefer et al. 2014), thus the main earthquake source in this region was more likely associated to a strike-slip fault offshore Loreto, and could not produce a tsunami of such characteristics as the ones described for the 1810 Loreto tsunami; (2) The historical description of the tsunami suggests localized large waves and approximately "40 leagues (ca. 5 km) inundation distance", which does not agree with the 27 August 1810 earthquake as the direct source mechanism. Therefore, we hypothesize that the 1810 earthquake was not capable of producing the accompanied tsunami, but rather it triggered a tsunamigenic submarine landslide. To prove our hypothesis, we first modelled a tsunami using the 1810 earthquake as the source, and, secondly, we analysed the seafloor morphology and modelled the 1810 tsunami to determine the potential submarine landslide source for the 1810 Loreto tsunami.

We merged the topo-bathymetric model from the Global Multi-Resolution Topography model with ~ 200-m resolution (Ryan et al. 2009) and a digital elevation model (DEM) with 15-m resolution (INEGI 2013; Fig. 2). Both models were merged and a topo-bathymetric model with 80-m pixels was recalculated. This resolution preserves the topographic elements of the coast at an acceptable level, and is the limit in which the proportion of the bathymetric model is not changed. This pixel size was also selected since GEOWAVE software does not accept ASCII files larger than 1400 × 1800 pixels. The topo-bathymetric model with pixels of 80 m had a pixel dimension of 1238 rows × 1771 columns.

3.2.1 Tsunami Numerical Modelling for an Earthquake Source Scenario

Our earthquake source scenario was based on the Gulf of California's 125-km-long transform fault that extends east of Loreto, the Farallon transform fault, with a right-lateral strike slip motion (Goff et al. 1987; Sumy et al. 2013; Castro et al. 2016, 2017). We also agree with the recent results on historical earthquakes that suggest that the Loreto 1810 earthquake was related to the Farallon transform fault (Suter 2018). For the tsunami model with the 1810 earthquake source, we used the parameters described in Table 1, and also validated the fault parameters with those derived from instrumentally medium-sized earthquakes recorded on the Farallon fault (Rodríguez-Lozoya et al. 2008). Our tsunami model shows that with the 1810 earthquake source, the maximum tsunami amplitudes are of only 1.5 m without significant inundation distances (Fig. 3). We thus can exclude the 27 August 1810 earthquake as the tsunami source because it do not agree with those characteristics described in historical records.

3.2.2 Tsunami Numerical Modelling for a Submarine Landslide Scenario

Our modelling results show that model "b" (Fig. 2) allows gauging of detailed topographic features, such as narrow valleys (rias coast), that directly influence the tsunami inundation behaviour. For instance, waves of 8 m do not produce any inundation in model "a", while in model "b", they can reach almost 3 km inland. Bathymetric data remains identical and it was not modified in "a" and "b" models.

The analysis of seafloor morphology revealed at least 22 probable features related with submarine landslides, presumably slumps (Fig. 4). We ran pre-modelling scenarios for each one of them and considered a potential tsunami source to select the most likely source of the 27 August 1810 tsunami. Our criteria for excluding the unlikely tsunami sources were the maximum wave amplitude and horizontal inundation produced on Tembabichi, because historical accounts indicate that "the tsunami flooded 5 km inland in Tembabichi". Therefore, the most probable source, a submarine landslide, that caused the 1810 tsunami is located about 70 km northeast and offshore of Tembabichi, shown in Fig. 4a in a black square, and it is marked in the detailed bathymetry in Fig. 4b. This submarine landslide is most likely a slump 12,000 m long, 21,000 m wide and about 600 m thick, and the average depth is 1450 m below m.a.s.l. (Fig. 4b). The head of the slump faces east.

We ran pre-modelling scenarios to exclude those submarine landslides that were not a source of the 1810 Loreto tsunami. Submarine landslides that were excluded as possible tsunami sources of the 1810 event showed none to small sea deformation, and the distance to the shoreline of the affected area was relatively far (Table 2).

Data extracted from modelling included the maximum tsunami amplitude (MTA), the horizontal inundation (HI), and maximum flow directions (Fig. 5). MTA on shoreline reached ca. 20 m in the shallowest bays, such as the one located northwest of La Paz bay in an inhabited shore. The model indicates three main zones that concentrate the maximum amplitudes, where MTAs are concentrating in a central zone, and decaying gradually (Fig. 5a). From north to south, from Puerto Escondido to Palma Sola, MTAs reached almost 20 m at Ensenada Blanca and San Cosme. Horizontal inundations (HI) though are not accordant with maximum wave amplitudes; for instance, in Tembabichi, HI value is less than 3 km inland, with a MTA of less than 15 m; while at Puerto Escondido, the HI is close to 0 m. The mountain and cliff coastal morphology conditioned here the tsunami behaviour reflected in our model.

The second region, from Punta Alta to El Primer Bosque characterized by high coastal relief, shows MTA reaching 15 m , and the maximum HI is less than 500 m . The third region, from San Juan de la Costa to near La Paz bay, MTA reached 25 m at only one site, but the mean values range around 10 m. The maximum HI was concentrated around La Paz bay gradually descending to the north (Fig. 5a, b).

The predominant flow direction shows several zones where seawater swirls occurred (Fig. 5c). The continuous turbulence produced by the reflectance effect during the tsunami propagation phase generated them in the spaces between islands, and island-peninsula sectors. Figure 5c shows some locations around Isla del Carmen, where seawater swirls were formed close to Juncalillo and Puerto Escondido. The historical accounts refer to this kind of swirls in the sector shown in Fig. 5c.

4. Discussion and Conclusions

4.1. Historical Data Interpretation

We interpreted the historical data documented above and summarized it as follows:

- Exact date and time of the main earthquake: 27 August 1810; 05:20 h (local time).
- Extent of and damage caused by the earthquake (Fig. 1):
  - All buildings of stone and mortar were destroyed and the church was badly damaged in Loreto.
  - Cracks formed at the edge of the beach and up to 150 m inland, and more than 2.5 m deep, along Loreto shore.
  - Landslides occurred along approximately 230 km, between Loreto and La Paz, probably along Sierra de La Giganta.
  - The 1810 earthquake was felt from Santa Gertrudis and San Ignacio to La Paz, along approximately 193 km, with the greatest damage being in the Loreto community.
  - Probable liquefaction effects and sand volcanoes associated to the main event.
  - Smaller aftershocks continued for 35 days.
  - A tsunami occurred 1 h after the earthquake
  - The tsunami reached up to 5 km inland of Tembabichi harbour; the surge and eddies lasted 8 h in Puerto Escondido; the surge was in the direction of Isla del Carmen.
  - Erosion of sand banks at the entrance of Tembabichi, which deepened the anchorage.
  - The explanation for the absence of human casualties was that: "… everybody was sleeping outdoors on account of the excessive heat present at this time of year" at the moment of the earthquake; also, the houses that were "small and low did not suffer so much damage". Goycoechea even suggests in his second letter that "there is no need to construct expensive masonry such as buildings made of mortar and stone, because it is apparent that it was those that suffered; those constructed of adobe are sufficient and cost less".

Based on the damage and associated effects (liquefaction, landslides, and the description of a tsunami), we estimate an earthquake intensity of VIII–IX level in the Modified Mercalli Scale, an approximate peak ground acceleration (PGA) of 100–250 cm/s^2, and an equivalent magnitude $M \geq 6$ to ≤ 7 (Richter scale). Our results and interpretations agrees with the intensity of IX and Mw 7.4 magnitude proposed by Suter (2018). This approximate PGA (100–250 cm/s^2, or 0.10 g to 0.25 g) is capable of triggering landslides since the minimum limit of PGA than can trigger a landslide is 0.05–0.07 g (Yu and Xuebao 2010). Also, an earthquake magnitude > 4.0 is the smallest magnitude known to produce landslides (Keefer 1984). However, for small-magnitude earthquakes, the main provoked events would mainly be represented by rock falls, rock slides, soil falls, disrupted soil slides, soil slumps, and soil block slides. Thus, the 1810 M 7.4 earthquake was capable of producing even larger landslides, and because the earthquake fault is located offshore, it could generate a large submarine landslide.

The associated tsunami occurred 1 h after the earthquake, reaching up to 5 km inland of Tembabichi harbour; the surge and eddies lasted 8 h in Puerto Escondido; and the surge was in the direction of Isla del Carmen. Thus, based on the time of the tsunami arrival, the tsunami's relatively narrow extension along the coast, and the tsunami's inundation distance, we proposed that the 27 August 1810 earthquake most probably occurred on a strike-slip fault, most likely the Farallon transform fault, as suggested by Suter (2018), producing mainly horizontal motion and relatively modest uplift, which was not the source of the Loreto tsunami, but rather the earthquake-triggered ground motion that is destabilizing and produces a tsunamigenic submarine landslide.

4.2. Numerical Modelling

To prove our hypothesis of a tsunami generated by a submarine landslide, we first ran a tsunami model using an earthquake source scenario for the 1810 tsunami; however, our model results show that the 27 August 1810 earthquake was not capable of producing a tsunami with those characteristics described in the historical records. Thus, we focus on those results from our tsunami model using a submarine landslide source. Because detailed descriptions of the tsunami behaviour along the coast are scarce, since the coast was barely populated at the time, we focused on specific sites along the continental shoreline and the islands. Nevertheless, a

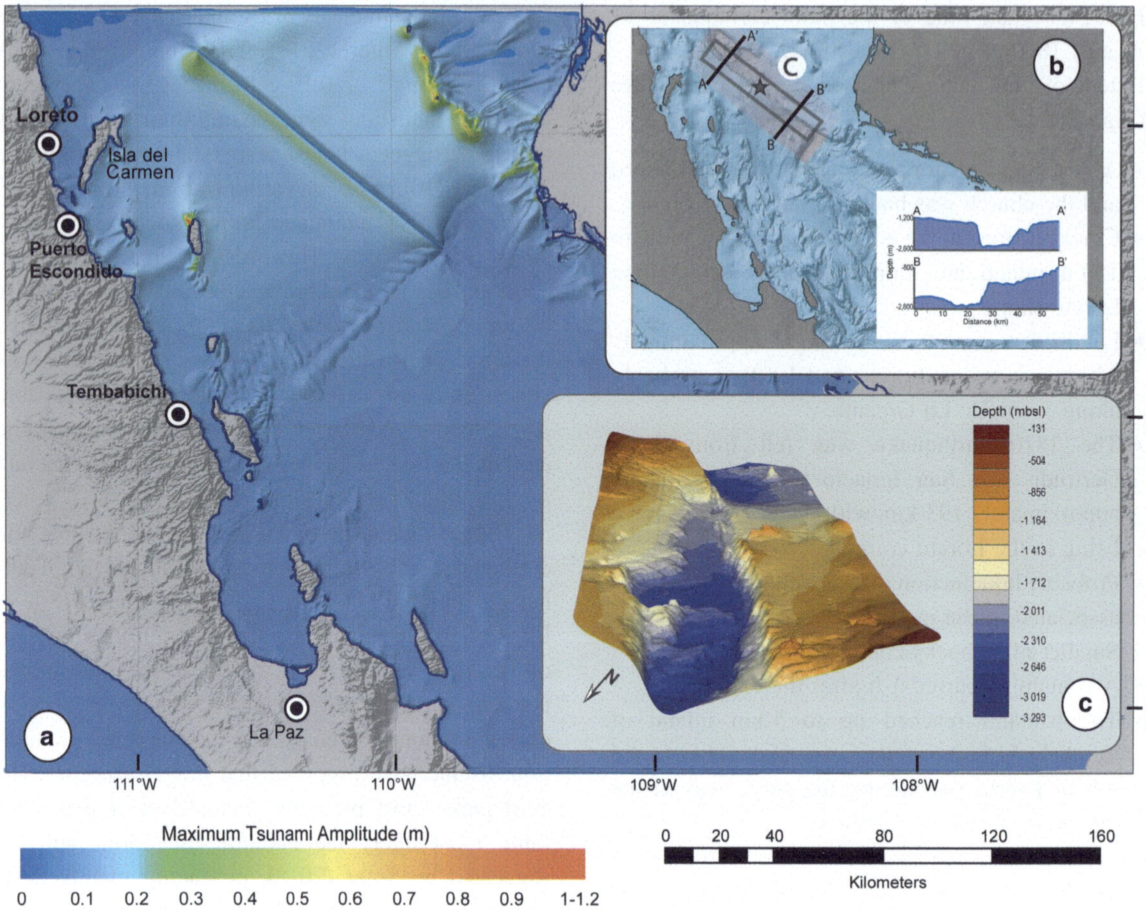

Figure 3

Tsunami model using an earthquake source scenario. **a** Tsunami model showing the Farallon transform fault as the earthquake source and results of maximum tsunami amplitude of 1–1.2 m. **b** Sea floor of the Gulf of California across Loreto; grey rectangle—area of the Farallon transform fault; A–A′ and B–B′ cross-profiles showing fault scarps facing NE across the north part of the fault, and facing SW on the southern part. **c** 3D model of the sea floor along the Farallon fault

regional tsunami behaviour was reconstructed with mathematical modelling. Regarding local effects, the Tembabichi site stands as the location where the maximum tsunami horizontal inundation was observed. This can be explained by the local coastal morphology at Tembabichi bay characterized by rias, very narrow valleys, which have produced similar effects along other coasts, i.e. increasing the inundation distances in narrow valleys due to a channelling effect, such as during the 2011 Tohoku tsunami (e.g. Ramírez-Herrera and Navarrete Pacheco 2012; Abe and Hori 2016). Even though the MTA reached less than 15 m, due to the amplification effect by adjacent mountains, the horizontal inundation reached near 2.5 km (Fig. 5 shows general tsunami behaviour and Fig. 6 shows detailed effects in Tembabichi). The historical accounts also described these effects. Almost all the rias coasts showed the same wave amplification effect.

Our data agree with the following criteria proposed to determine the tsunamigenesis by a submarine landslide:

1. Documented large run-up values close to the source and focusing effects (Assier-Rzadkieaicz et al. 2000; Ward 2001; Okal et al. 2003; Satake and Tanioka 2003; Okal and Synolakis 2004; Harbitz et al. 2006; Fritz et al. 2007) described in

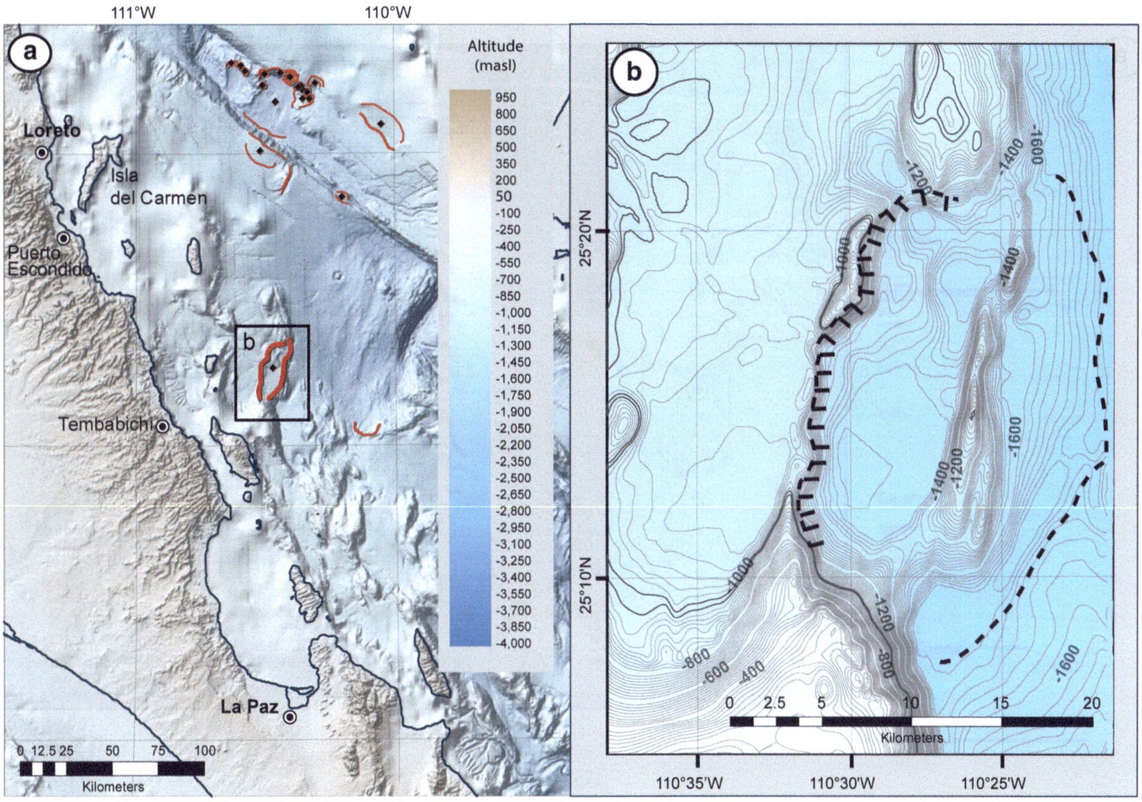

Figure 4

Morphological characterization of the Gulf of Baja California. **a** Topo-bathymetric model (from Ryan et al. 2009); land—brown scale; bathymetry—blue scale; diamonds—landslide centroid locations; red lines—morpholineaments related to submarine landslides (headscarps, slump toes, and other characteristic features associated to submarine landslides). **b** Location and details of the slump; dashed barbed lines—head scarp, dashed line—probable slump toe

Table 2

Summary of tsunami modelling parameters using a submarine landslide source

Parameters	Values
SMF longitude (m)	553,275
SMF latitude (m	2,795,028
Orientation from north (°)	255
Average depth (m)	1450
Average slope (°)	9
Length (m)	12,000
Maximum thickness (m)	600
Maximum width (m)	21,000
Tsunami cutoff width (m)	25,000
Slump motion distance (m)	6000
Bulk density (kg/m^3)	1900

SMF slope mass failure

the literature were documented for the Loreto tsunami and confirmed by modelling.

2. We found no documentation of this local tsunami on other localities other than the ones mentioned in previous sections. Limited far-field effects have been documented when compared with earthquake tsunamis (Satake and Tanioka 2003; Okal and Synolakis 2004; Mohammed and Fritz 2012; McFall and Fritz 2016) and explained by the dipole nature of landslide tsunamis that decays faster in the far field (Satake and Tanioka 2003).
3. The Loreto tsunami model shows that propagation tends to produce radial directivity patterns as described in the literature (Ben-Menahem and Rosenman 1972b; Iwasaki et al. 1997; Okal 2000; Okal et al. 2003); it is likely that the Sonora coast, located east of the detected tsunamigenic

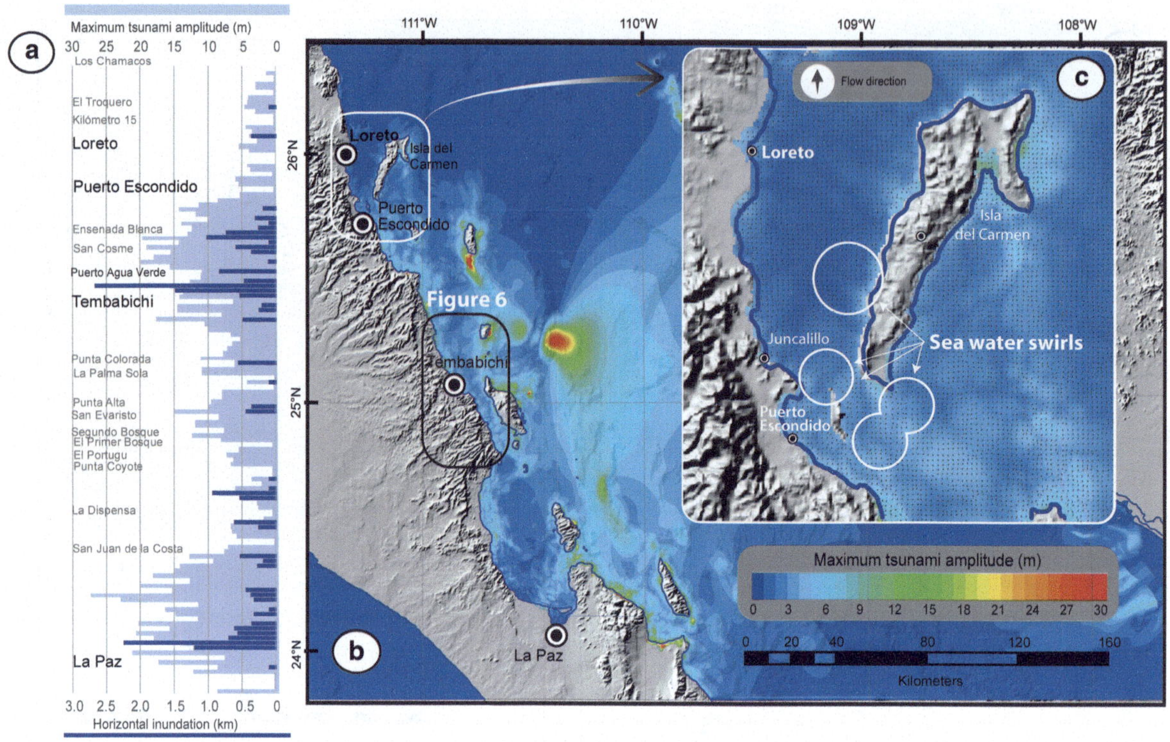

Figure 5

Tsunami model using a submarine landslide source. **a** Maximum tsunami amplitude and maximum horizontal inundation charts. Values where extracted from modelling. **b** Maximum tsunami amplitude resulted from the suspected submarine landslide. **c** Predominant tsunami flow direction pattern around Loreto, Puerto Escondido and Isla del Carmen. White circles show seawater swirls resulted from the reflection during the tsunami propagation time. For sites location, see Fig. 1

submarine landslide was also affected by this tsunami; however, due to the scarce population of this coast at the time, no documentation of this event is available.

4. Because the submarine landslide is located farther away from the coast, the time lapse between the earthquake/submarine landslide and the tsunami was longer than when the source is located in continental margins at shallow waters, as shown in other cases (Kulikov et al. 1996; Synolakis et al. 2002; Fritz et al. 2013). We tested the tsunami wave speeds considering the depth at which the Farallon transform fault is located, as the suspected source of the 1810 Loreto earthquake (Suter 2018); and the time lapse between the fault and the Loreto coast is about 24 min (Fig. 7). We tested the tsunami arrival time by modelling the tsunami travel time (TTT) from the proposed submarine landslide. The TTT model shows the first time the tsunami reaches each point on the surface, and does not necessarily correspond to the waves of greater amplitude. When using the submarine landslide located south of Loreto as the tsunami source, the modelled time lapse (the TTT) indicates that the tsunami arrives from two fronts to Loreto, north and south, and the TTT ranges from 30 to 50 min, respectively (Fig. 7). The reason for this difference in time lapse is explained by the TTT and arrival of the first waves coming from the north to Loreto at about 30–35 min, while later tsunami waves arrive from the south taking 50 min time from the source. This is caused by the trajectory of the tsunami to the north of Loreto where fewer topographic barriers are faced during the tsunami propagation process; meanwhile, at the southern front, there is a tsunami arrival delay because Isla del Carmen hinders the tsunami wave free flow, and after

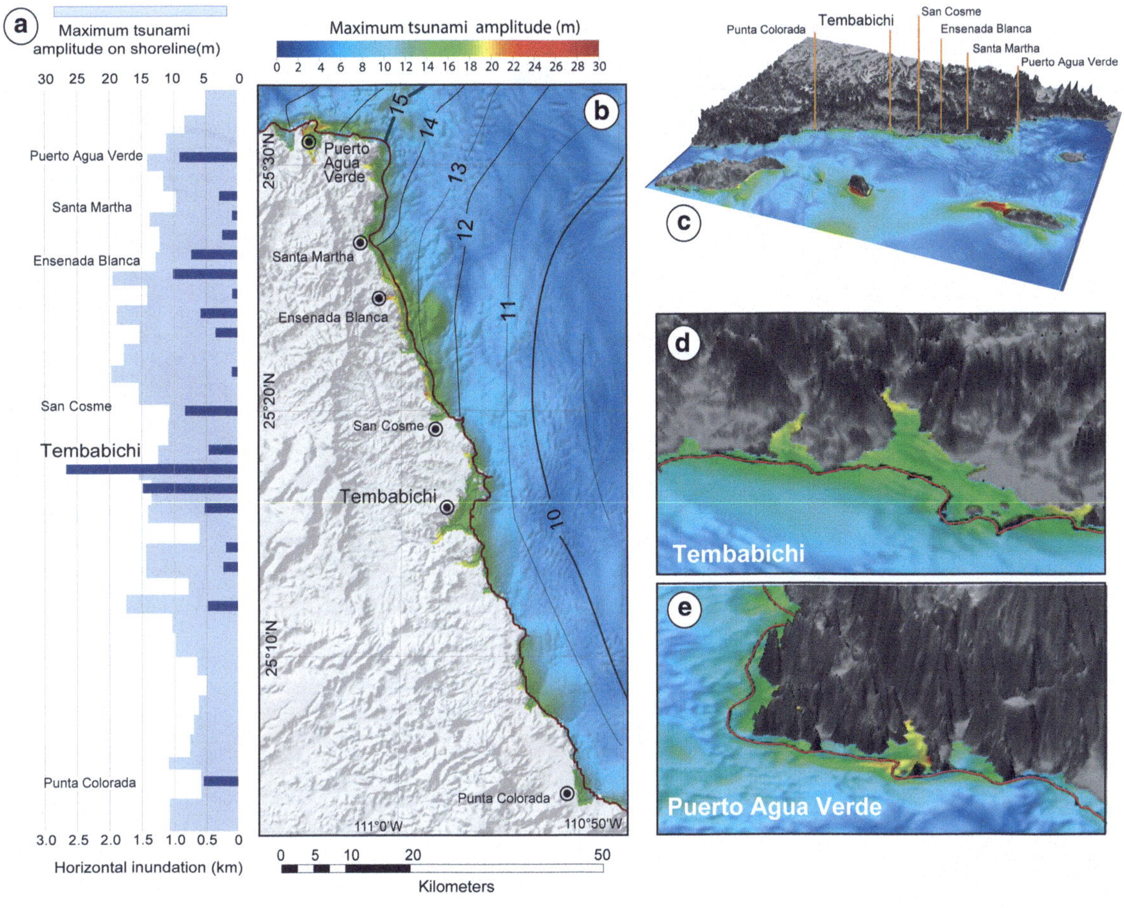

Figure 6

Detailed model of the Tembabichi zone. **a** Maximum tsunami amplitude on shore and horizontal inundation. **b** Detailed model of maximum tsunami amplitude at Tembabichi, red line—coast, black lines with numbers—isochrones showing tsunami arrival time at the coast. **c** 3D model showing amplification effects due to rias morphology. **d** Tsunami flooding at Tembabichi bay and rias morphology, red line—coast. **e** Tsunami flooding at Puerto Agua Verde, red line—coast

overcoming this topographic obstacle, off the coast of Puerto Escondido, the tsunami continues to Loreto, where it arrives as a second train of waves with arrival times of 50 min (Fig. 7). The latest is explained by the wave behaviour when the tsunami waves find physical barriers, in this case, several islands, along its path. Our tsunami modelling wave arrival is also more accordant with the historical description.

We thus demonstrated here that the source of the Loreto 1810 tsunami was a submarine landslide triggered by a Mw 7.4 earthquake on the Farallon strike-slip fault. The Loreto tsunami emphasizes the need to consider landslide tsunamis generated by earthquakes triggered by strike-slip faults in tsunami hazard assessments, using historical data and numerical modelling in areas with relatively high earthquake activity.

Acknowledgements

Ramírez-Herrera acknowledges grant numbers PAPIIT-109117 and CONACYT-SEP-284365. Castillo-Aja acknowledges the postgraduate scholarship by CONACYT and Posgrado de Geografía UNAM. We thank Ann Grant for the English proofreading and editing of this manuscript.

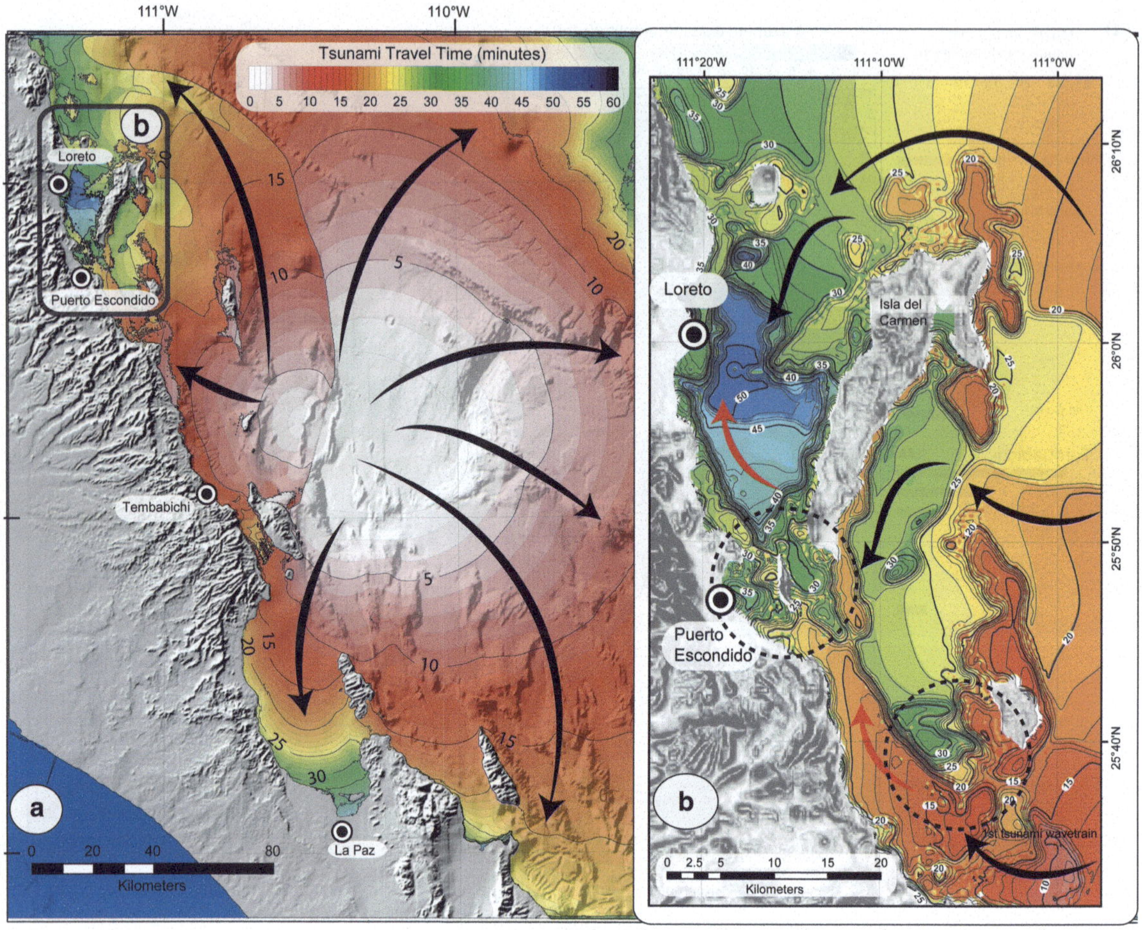

Figure 7

Tsunami travel time (TTT) model. **a** Regional model of TTTs. **b** Detail of the wave pattern propagation and refraction attenuation effects in the Loreto and Puerto Escondido area. The colour scale shows the TTT from the source to each location in the study area. Dark lines with numbers—isochrones with 5-min interval; light grey lines—secondary isochrones at 1-min intervals. Black arrows show the direction of tsunami propagation; red arrows—direction of tsunami propagation in areas where there is a delay due to effects of attenuation, tsunami wave braking and refraction due to physical barriers, i.e. islands off the coast. Dashed circles show areas where the wave train experiences refraction effects and loss of velocity in the propagation, observed in the TTT

References

Abe, T., & Hori, K. (2016). Coastal geomorphology and tsunami disaster by the 2011 off the Pacific coast of Tohoku. In S. Haruyama & T. Sugai (Eds.), *Natural disaster and coastal geomorphology*. Cham: Springer. https://doi.org/10.1007/978-3-319-33814-9_3.

Assier-Rzadkieaicz, S., Heinrich, P., Sabatier, P. C., Savoye, B., & Bourillet, J. F. (2000). Numerical modelling of a landslide-generated tsunami: The 1979 Nice event. *Pure and Applied Geophysics*. https://doi.org/10.1007/PL00001057.

Atwater, T., & Stock, J. (1998). Pacific-North America plate tectonics of the Neogene southwestern United States: An update. *International Geology Review*. https://doi.org/10.1080/00206819809465216.

Ben-Menahem, A., & Rosenman, M. (1972a). Amplitude patterns of tsunami waves from submarine earthquakes. *Journal of Geophysical Research, 77*(17), 32.

Ben-Menahem, A., & Rosenman, M. (1972b). Amplitude patterns of tsunami waves from submarine earthquakes. *Journal of Geophysical Research*. https://doi.org/10.1029/JB077i017p03097.

Busch, M. M., Arrowsmith, J. R., Umhoefer, P. J., Coyan, J. A., Maloney, S. J., & Gutiérrez, G. M. (2011). Geometry and evolution of rift-margin, normal-fault-bounded basins from gravity and geology, La Paz-Los Cabos region, Baja California Sur, Mexico. *Lithosphere*. https://doi.org/10.1130/L113.1.

Cárdenas de la Peña, E. (1969). *Visión y presencia de Baja California*. México: Secretaría de Marina.

Castro, R. R., Stock, J. M., Hauksson, E., & Clayton, R. W. (2016). Source functions and path effects from earthquakes in the Farallon transform fault region, Gulf of California, Mexico that Occurred on October 2013. *Pure and Applied Geophysics*,. https://doi.org/10.1007/s00024-016-1346-4.

Castro, R. R., Stock, J. M., Hauksson, E., & Clayton, R. W. (2017). Active tectonics in the Gulf of California and seismicity (M > 3.0) for the period 2002–2014. *Tectonophysics*. https://doi.org/10.1016/j.tecto.2017.02.015.

Clavijero, F. J. (1937). The history of (Lower) California. In S. E. Lake, & A. A. Gray (Eds.), Palo Alto: Standford University Press.

Coyan, M. M., Arrowsmith, J. R., Umhoefer, P., Coyan, J., Kent, G., Driscoll, N., et al. (2013). Geometry and Quaternary slip behavior of the San Juan de los Planes and Saltito fault zones, Baja California Sur, Mexico: Characterization of rift-margin normal faults. *Geosphere*. https://doi.org/10.1130/GES00806.1.

Crosby, H. W. (2007). Loreto, capital of Antigua California: The first century. In P. Ganster, O. Arizpe, & A. Ivanova (Eds.), *Loreto: The future of the first capital of the Californias* (pp. 77–102). San Diego: San Diego State University Press, Institute for Regional Studies of the Californias.

de Santiago, N. (1813). *Carta al Exelentísimo Señor Virrey y Capitán General de la Nueva España. 20 de febrero 1813*. México: Archivo General de la Nación (AGN), Instituciones Coloniales/Indiferente Virreinal/Cajas 2000-2999/Caja 2356.

Dirección General del Instituto Geográfico y Estadístico. (1886). *Equivalencias entre las Pesas y Medidas usadas antiguamente en las Provincias de España y las legales del sistema métrico-decimal*. Madrid: Publicadas de Real Orden. Madrid Imp. de la Dirección General del Instituto Geográfico y Estadístico.

Dongeren A. V., Vatvani D., & Ormondt M. V. (2018). Simulation of the 2018 tsunami along the coastal areas in the Palu Bay. In *Abstract. American Geophysical Union Fall Meeting 2018, Washington DC, USA*.

Ewald, U. (1997). *La industria salinera de México, 1560–1994*. México: Fondo de Cultura Económica.

Figueroa, A. J. (1970). *Catálogo de Sismos Ocurridos en la República Mexicana*. Mexico: UNAM.

Figueroa, A. J. (1972). Sismicidad en Baja California. *Geofísica Internacional, 12*(3), 213–225.

Fletcher, J. M., & Munguía, L. (2000). Active continental rifting in southern Baja California, Mexico: Implication for plate motion partitioning and the transition to seafloor spreading in the Gulf of California. *Tectonics, 19*(6), 1107–1123.

Fritz, H. M., Hillaire, J. V., Molière, E., Wei, Y., & Mohammed, F. (2013). Twin tsunamis triggered by the 12 January 2010 Haiti earthquake. *Pure and Applied Geophysics*. https://doi.org/10.1007/s00024-012-0479-3.

Fritz, H. M., Kongko, W., Moore, A., McAdoo, B., Goff, J., Harbitz, C., et al. (2007). Extreme runup from the 17 July 2006 Java tsunami. *Geophysical Research Letters*. https://doi.org/10.1029/2007GL029404.

Fuster Ruiz, F. (1997). *El final del descubrimiento de América: California, Canada y Alaska (1765–1882); aportación documental del Archivo General de la Marina*. Murcia: Univ., Servicio de Publicaciones.

Global Earthquake Model (GEM). (2013). GEM global historical earthquake catalogue GHCE v1.0. https://www.emidius.eu/GEH/download/GEM-GHEC-v1.kmz. Accessed 25 June 2017.

Goff, J. A., Bergman, E., & Solomon, S. C. (1987). Earthquake source mechanisms and transform fault tectonics in the Gulf of California. *Journal of Geophysical Research, 92*(B10), 10485–10510.

Goycoechea, F. (1810a). Carta al Exelentísimo Señor Virrey de Nueva España. 2 de septiembre de 1810. In *Archivo General de la Nación (AGN), México. Instituciones Coloniales/Indiferente Virreinal/4096 Exp.041*.

Goycoechea, F. (1810b). Carta al Muy Poderoso Señor. 13 de octubre de 1810. In *Archivo General de la Nación (AGN), México. Operaciones de Guerra/Vol 750. Exp. 9* (pp. 52–54).

Goycoechea, F. (1813). Estadoque manifiesta el número de vasallos de ambos sexos, de indios y gente de razón que tiene el Rey V.S. (Q.D.G.) en esta provincia, y por notas el estado en que se hallan las Misiones que Administran los Religiosos de Santo Domingo. In *Archivo General de la Nación (AGN), México. Instituciones Coloniales/Indiferente virreinal/Caja 2356/Exp. 034 (Californias)*.

Guidoboni, E., & Ebel, J. E. (2009). *Earthquakes and tsunamis in the past: A guide to techniques in historical seismology*. Cambridge: Cambridge University Press.

Harbitz, C. B., Lovholt, F., Pedersen, G., & Masson, D. G. (2006). Mechanisms of tsunami generation by submarine landslides; a short review. *Norsk Geologisk Tidsskrift, 86*(3), 255–264.

Heirarzadeh, M., Muhari, A., & Wijanarto, A. B. (2019). Insights on the source of the 28 September 2018 Sulawesi Tsunami, Indonesia base on spectral analyses and numerical simulations. *Pure and Applied Geophysics, 176*, 25–43. https://doi.org/10.1007/s00024-018-20165-9.

Hornbach, M. J., Braudy, N., Briggs, R. W., Cormier, M.-H., Davis, M. B., Diebold, J. B., et al. (2010). High tsunami frequency as a result of combined strike-slip faulting and coastal landslides. *Nature Geoscience*. https://doi.org/10.1038/ngeo975.

INEGI. (2013). *Contínuo de Elevaciones Mexicano 3.0 (CEM 3.0)*. México: Instituto Nacional de Estadística y Geografía. http://www.inegi.org.mx/geo/contenidos/datosrelieve/continental/descarga.aspx. Accessed 25 June 2018.

ISC-GEM Earthquake Catalogue (2016). ISC-GEM Global Instrumental Earthquake Catalogue Version 3.0. International Seismological Centre. https://doi.org/10.31905/D808B825.

Iwasaki, S. I., Eguchi, T., Fujinawa, Y., Fujita, I., Watanabe, I., Fukuyama, E., Fujiwara, H., Hishiki, K. (1997). Precise tsunami observation system in deep ocean by an ocean bottom cable network for the prediction of earthquakes and tsunamis. In: Perspectives on tsunami hazard reduction; observations, theory and planning (pp. 47–66). Kluwer Academic Publishers.

Keefer, D. K. (1984). Landslides caused by earthquakes. *Geological Society of America Bulletin, 95*, 406–421. https://doi.org/10.1130/0016-7606(1984).

Kostoglodov, V., & Pacheco, J. F. (1999). *100 años de Sismicidad en México*. Instituto de Geofísica, UNAM. http://usuarios.geofisica.unam.mx/vladimir/sismos/100a%F1os.html. Accessed 20 June 2018.

Kulikov, E. A., Thomson, R. E., & Bornhold, B. D. (1996). The landslide tsunami of November 3, 1994, Skagway Harbor, Alaska. *Journal of Geophysical Research, 101*(C3), 6609–6615.

Lassépas, U. (1859). *De la colonizacion de la Baja California y decreto de 10 de marzo de 1857,/por el ciudadano Ulisses*

Urbano Lassépas. Primer memorial. V. García Torres, México. https://hdl.handle.net/2027/uc1.31822043015254, Accessed 26 June 2018.

Lopez-Pineda, L. (2005). Source characteristics of the mw 6.2 loreto earthquake of 12 march 2003 that occurred in a transform fault in the middle of the gulf of California, Mexico. *Bulletin of the Seismological Society of America, 95*(2), 419–430. https://doi.org/10.1785/0120030227.

Løvholt, F., Pedersen, G., Harbitz, C. B., Glimsdal, S., & Kim, J. (2015). On the characteristics of landslide tsunamis. *Philosophical Transactions of the Royal Society A: Mathematical, Physical and Engineering Sciences*. https://doi.org/10.1098/rsta.2014.0376.

McFall, B. C., & Fritz, H. M. (2016). Physical modelling of tsunamis generated by 3D deformable granular landslides on planar and conical island slopes. *Proceedings of the Royal Society A, 472*(2188), 20160052. https://doi.org/10.1098/rspa.2016.0052.

Mohammed, F., & Fritz, H. M. (2012). Physical modeling of tsunamis generated by three-dimensional deformable granular landslides: Landslide generated tsunamis. *Journal of Geophysical Research: Oceans*. https://doi.org/10.1029/2011JC007850.

Munguia, L., González, M., Mayer, S., & Aguirre, A. (2006). Seismicity and state of stress in the La Paz-Los Cabos Region, Baja California Sur, Mexico. *Bulletin of the Seismological Society of America*. https://doi.org/10.1785/0120050114.

National Geophysical Data Center/World Data Service (NGDC/WDS). (2017a). *NCEI/WDS Global Historical Tsunami Database*. NOAA National Centers for Environmental Information. https://doi.org/10.7289/v5pn93h7.

National Geophysical Data Center/World Data Service (NGDC/WDS). (2017b). NCEI/WDS Global Significant Earthquake Database. *NOAA National Centers for Environmental Information., 2017*. https://doi.org/10.7289/V5TD9V7K.

Okal, E. A. (2000). T waves from the 1998 Sandaun, PNG sequence: Definitive timing of the slump. In *2000 Western Pacific Geophysics Meeting*. AGU.

Okal, E. A., Plafker, G., Synolakis, C., & Borrero, J. C. (2003). Near-field survey of the 1946 Aleutian Tsunami on Unimak and Sanak Islands. *Bulletin of the Seismological Society of America*. https://doi.org/10.1785/0120020198.

Okal, E. A., & Synolakis, C. E. (2003). A Theoretical comparison of tsunamis from dislocations and landslides. *Pure and Applied Geophysics*. https://doi.org/10.1007/s00024-003-2425-x.

Okal, E. A., & Synolakis, C. E. (2004). Source discriminants for near-field tsunamis: Near-field tsunamis. *Geophysical Journal International*. https://doi.org/10.1111/j.1365-246X.2004.02347.x.

Peñalver, J., & de Olive, P. M. (1869). *Nuevo diccionario de la lengua castellana: que comprende la última edición del de la Academia Española con un suplemento que contiene el diccionario de la rima y el de sinónimos por una sociedad literaria. 1225, iii, 273 p.* Paris: Rosa y Bouret.

Plattner, C., Malservisi, R., Dixon, T. H., LaFemina, P., Sella, G. F., Fletcher, J., et al. (2007). New constraints on relative motion between the Pacific Plate and Baja California microplate (Mexico) from GPS measurements. *Geophysical Journal International*. https://doi.org/10.1111/j.1365-246X.2007.03494.x.

Prior, D. B., & Coleman, J. M. (1979). Submarine landslides—geometry and nomenclature. *Zeitschrift für Geomorphologie, 23*(4), 415–426.

Ramírez-Herrera, M. T., & Navarrete Pacheco, J. A. (2012). Satellite data for a rapid assessment of tsunami inundation areas after the 2011 Tohoku earthquake. *Pure and Applied Geophysics, 170,* 1067–1080. https://doi.org/10.1007/s00024-012-0537-x.

Rodríguez-Lozoya, H. E., Quintanar, L., Ortega, R., Rebollar, C. J., & Yagi, Y. (2008). Rupture process of four mediumsized earthquakes that occurred in the Gulf of California. *Journal of Geophysical Research, 113,* B10301. https://doi.org/10.1029/2007JB005323.

Ryan, W. B. F., Carbotte, S. M., Coplan, J., O'Hara, S., Melkonian, A., Arko, R., et al. (2009). Global Multi-Resolution Topography synthesis. *Geochemistry, Geophysics, Geosystems, 10,* Q03014. https://doi.org/10.1029/2008GC002332.

Salazar Rovirosa, A. (1980). *Historia del Estado de Baja California: de 1500 a 1980* (3rd ed.). Guadalajara: Ediciones Económicas.

Sassa, S., & Takagawa, T. (2019). Liquefied gravity flow-induced tsunami: First evidence and comparison from the 2018 Indonesia Sulawesi earthquake and tsunami disasters. *Landslides, 16,* 195. https://doi.org/10.1007/s10346-018-1114-x.

Satake, K., & Tanioka, Y. (2003). The July 1998 Papua New Guinea Earthquake: Mechanism and quantification of unusual tsunami generation. *Pure and Applied Geophysics,*. https://doi.org/10.1007/s00024-003-2421-1.

Sedlock, R. L. (2003). Geology and tectonics of the Baja California peninsula and adjacent areas. In S. E. Johnson, S. R. Paterson, J. M. Fletcher, G. H. Girty, D. L. Kimburgh, & A. Martín-Barajas (Eds.), *Tectonic evolution of northwestern México and the southwestern USA* (pp. 1–42). Boulder: Geological Society of America.

Slevin, J. R. (1922). General Account of Expedition of the California Academy of Sciences to the Gulf of California in 1921. In *Proceedings of the California Academy of Sciences, XII*.

Stock, J. M., & Hodges, K. V. (1989). Pre-Pliocene extension around the Gulf of California and the transfer of Baja California to the Pacific plate. *Tectonics, 8*(1), 99–115.

Sumy, D. F., Gaherty, J. B., Kim, W., Diehl, T., & Collins, J. A. (2013). The Mechanisms of Earthquakes and Faulting in the Southern Gulf of California. *Bulletin of the Seismological Society of America*. https://doi.org/10.1785/0120120080.

Suter, M. (2018). The historical seismicity of the Loreto Region, Baja California Peninsula, Mexico (1684–1878). *Seismological Research Letters*. https://doi.org/10.1785/0220170204.

Synolakis, C. E., Yalciner, A., Borrero, J. C., & Plafker, G. (2002). Modeling of the November 3, 1994 Skagway, Alaska tsunami. In *Solutions to coastal disasters'02 Conference Proceedings* (pp. 915–927). University of Southern California.

Tinti, S., Armigliato, A., Manucci, A., Pagnoni, G., Zaniboni, F., Yalçiner, A. C., et al. (2006). The generating mechanisms of the August 17, 1999 İzmit bay (Turkey) tsunami: Regional (tectonic) and local (mass instabilities) causes. *Marine Geology*. https://doi.org/10.1016/j.margeo.2005.09.010.

Turner, A. K., & Schuster, R. L. (1996). *Landslides: investigation and mitigation: Transportation research board special report 247* (p. 673). Washington, DC: National Research Council.

Umhoefer, P. J., Maloney, S. J., Buchanan, B., Arrowsmith, J. R., Martinez-Gutiérrez, G., Kent, G., et al. (2014). Late Quaternary faulting history of the Carrizal and related faults, La Paz region, Baja California Sur, Mexico. *Geosphere*. https://doi.org/10.1130/GES00924.1.

Ward, S. N. (2001). Landslide tsunami. *Journal of Geophysical Research: Solid Earth, 106*(6), 11201–11215.

Watts, P., Grilli, S. T., Kirby, J. T., Fryer, G. J., & Tappin, D. R. (2003). Landslide tsunami case studies using a Boussinesq model and a fully nonlinear tsunami generation model. *Natural Hazards and Earth System Science*. https://doi.org/10.5194/nhess-3-391-2003.

Yu, Y., & Xuebao, G. (2010). Research on relationship between landslides and peak ground accelerations produced by Wenchuan earthquake. *Chinese Journal of Rock Mechanics and Engineering, 29*(1), 82–89.

Zaniboni, F., Armigliato, A., Pagnoni, G., & Tinti, S. (2014). Continental margins as a source of tsunami hazard: The 1977 Gioia Tauro (Italy) landslide–tsunami investigated through numerical modeling. *Marine Geology*. https://doi.org/10.1016/j.margeo.2014.08.011.

(Received September 10, 2018, revised February 23, 2019, accepted March 12, 2019, Published online March 27, 2019)

Pure Appl. Geophys. 176 (2019), 2969–2983

https://doi.org/10.1007/s00024-018-2078-4

Pure and Applied Geophysics

Multi-Data-Type Source Estimation for the 1992 Flores Earthquake and Tsunami

IGNATIUS RYAN PRANANTYO[1] and PHIL R. CUMMINS[1]

Abstract—We revisit the source of the 1992 Flores earthquake and tsunami using finite-fault inversion. We simultaneously invert teleseismic body and surface waves together with coseismic uplift/subsidence datasets. We then verify the inverted source against tsunami run-up heights along the northern coast of Flores Island and the only tide gauge recording of the tsunami. Our preferred source model provides a good fit to all the datasets, whereas previous models only explained a subset of the available data. We show that the fault geometry implies segmentation of the back-arc thrust system in the eastern Sunda Arc.

Key words: Flores Island tsunami 1992, finite-fault source inversion, tsunami hazard, seismic hazard.

1. Introduction

A devastating earthquake with moment magnitude (Mw) of 7.8 followed by a tsunami severely impacted the northern coast of Flores Island, Nusa Tenggara Timur, Indonesia on 12 December 1992. It killed more than 2000 people and left thousands injured, with more than 25,000 buildings and much infrastructure destroyed (World Bank 1993; Yeh et al. 1993; Tsuji et al. 1995).

To understand why this event had such a major impact, we need to know those details of its source characteristics that determine the generation of tsunami and strong ground motion, viz. which fault ruptured, how large the slip was, and how it was distributed over the fault surface. Several studies have attempted to constrain these details using different datasets: Beckers and Lay (1995) used seismic waveforms, Hidayat et al. (1995) and Imamura and Kikuchi (1994) used a combination of seismic and tsunami data, and Griffin et al. (2015) used a combination of seismic and uplift data. However, it can be argued that none of these models fit all the data well.

In this paper, the source of the Flores 1992 earthquake and tsunami is revisited by conducting finite-fault source inversion using teleseismic waveforms and coastal uplift data. The estimated source is then adjusted systematically to fit tsunami run-up along the northern coast of Flores Island and a tide gauge waveform at Palopo, Sulawesi. First, the tectonic setting of the Flores region is discussed. Second, a brief explanation is given of the methodologies used, followed by the results of the inversion. Lastly, further investigations that could be done and the implications for the tsunami and seismic hazard associated with the Flores Back-Arc Thrust are discussed.

2. Tectonic Framework

Eastern Indonesia encompasses a complex tectonic environment, involving the convergence of four major and at least four minor tectonic plates (Fig. 1). Its southern margin is dominated by convergence of the Australian Plate with the Sunda Block and other minor plates comprising Nusa Tenggara (aka the Lesser Sunda Islands) and the Banda Island Arc. Flores Island lies in the Nusa Tenggara region, where the style of plate convergence transitions from ocean–continent subduction in the Java Trench to arc-continent collision near Timor.

The plate boundary along Indonesia's southern margin is often depicted as a contiguous subduction zone running offshore Sumatra and Java and continuing along the Timor Trough and its extension along

Electronic supplementary material The online version of this article (https://doi.org/10.1007/s00024-018-2078-4) contains supplementary material, which is available to authorized users.

[1] Research School of Earth Sciences, The Australian National University, Canberra, Australia.
E-mail: ryan.pranantyo@anu.edu.au; phil.cummins@anu.edu.au

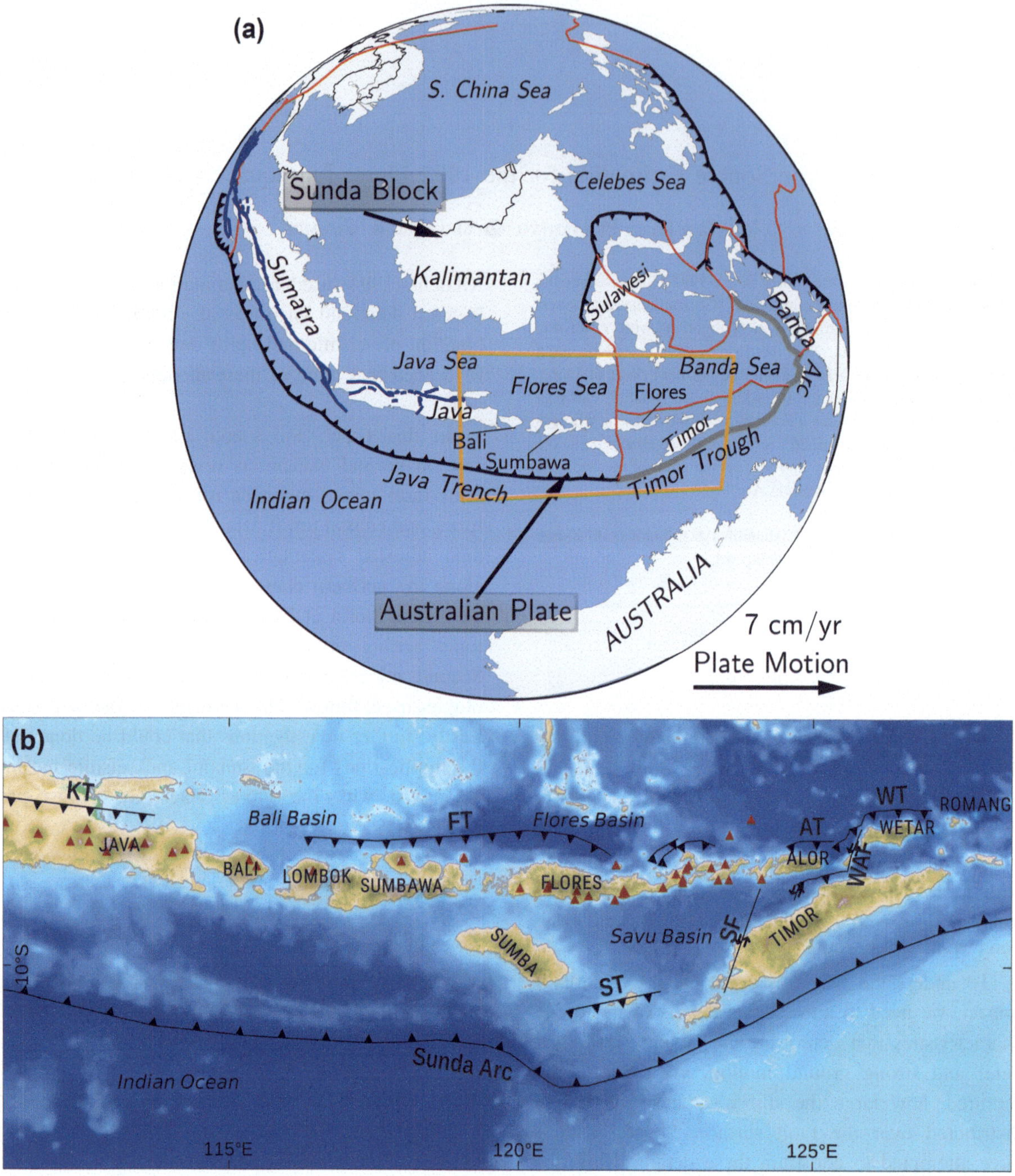

Figure 1

a Summary of the tectonics of Indonesia, with major plates and plate boundaries from Bird (2003), with the Nusa Tenggara region indicated; **b** inset indicated in **a**, illustrating details of the Flores Back-Arc Thrust zone. The faults are compiled from Hamilton (1979), Silver et al. (1983, 1986), Koulali et al. (2016), and Breen et al. (1989). *KT* Kendeng Thrust, *FT* Flores Thrust, *AT* Alor Thrust, *WT* Wetar Thrust, *WAF* Wetar-Atauro Fault, *SF* Semau Fault, *ST* Savu Thrust the Banda Arc (the heavy gray curve in Fig. 1). Although there is a well-developed Benioff zone along the southern Banda Arc (Sandiford 2008), recent interpretations of tomography and the plate tectonic evolution of the region suggest that active subduction ceased along the Banda Arc 4 Myr ago (Spakman and Hall 2010), while the Timor Trough has been recognized as a foreland basin of the Banda orogeny rather than a subduction trench (Audley-Charles 2011). Plate convergence east of the Java Trench, to the extent that it is not accommodated by the Timor Trough or Banda Arc, must be accommodated elsewhere.

An alternative for accommodating Australian–Sunda Plate convergence is the Flores Back-Arc Thrust, first identified in marine seismic and seismicity studies (Hamilton 1979; Silver et al. 1983, 1986; McCaffrey and Nabalek 1984). These surveys showed clear evidence of back-arc thrusting along two distinct segments, viz. the 450-km-long Flores Thrust north of Sumbawa and western Flores, and the 350-km-long Wetar Thrust north of Timor (see Fig. 1b). It was thought that, elsewhere, convergence may be accommodated by strike-slip faults that cut the arc at angles oblique to the convergence.

The Flores Back-Arc Thrust is depicted in the global plate model of Bird (2003) (see Fig. 1a) as a continuous back-arc thrust stretching from western Flores to the Tanimbar Islands in the eastern Banda Arc. Koulali et al. (2016) (see also Nugroho et al. 2009) used Global Positioning System (GPS) measurements of present-day crustal movement to suggest that a continuous zone of back-arc thrusting extends all the way from the Kendeng Thrust zone in eastern Java, through the Flores Thrust zone, to the Wetar Thrust zone in the east. According to Koulali et al. (2016), convergence between the Australian Plate and Indonesia is progressively transferred from over 90 % accommodation on the Java Trench west of Lombok, to over 90 % accommodation on a back-arc thrust system along the southern margin of the Flores Sea east of Timor. Near Flores, the accommodation of convergence is roughly equally split between the Timor Trough and the Flores Back-Arc Thrust, with 33 and 26 mm/year convergence, respectively. This implies that a slip deficit of 10 m could accumulate on the back-arc thrust north of Flores every 400 years.

The detailed geometry of the Flores Back-Arc Thrust, including its segmentation and orientation of faulting, is important for understanding the earthquake and tsunami hazard of the Nusa Tenggara region. Does earthquake rupture occur only offshore, so that tsunami generation is enhanced but the seismic hazard onshore is reduced? Or does earthquake rupture extend onshore, so that both tsunami and ground shaking are major perils? Over 20 shallow ($\leq$ 40 km depth) earthquakes of magnitude 6.5 or larger have been recorded on the Flores Back-Arc, and many of these have caused damage through ground shaking. Although a few historical events causing major tsunamis have been attributed to the Flores Back-Arc Thrust (Griffin et al. 2018, in review), the only instrumentally recorded event causing a major tsunami is the Mw 7.9, 1992 Flores earthquake (Tsuji et al. 1995).

3. Data for the 1992 Flores Earthquake

Because it occurred prior to the post-2004 expansion of broadband seismographic networks worldwide, the number of stations that recorded seismic waveforms of the 1992 Flores earthquake is fewer than are typically available for more modern earthquakes. Similarly, there is only one tide gauge that recorded a waveform of the tsunami generated by the earthquake. On the other hand, because it was the target of one of the world's first detailed post-tsunami surveys by an international team of experts (Tsuji et al. 1995), considerable data exist in the from of tsunami run-up and coastal uplift and subsidence.

3.1. Teleseismic Body and Surface Waveforms

Seismic waveform data from 26 stations of the Global Seismographic Network (GSN) were retrieved from the Incorporated Research Institutions for Seismology's (IRIS) Data Management Center (Fig. 2). Only stations at distance between 30° and 90° were used, to reduce the influence of either upper

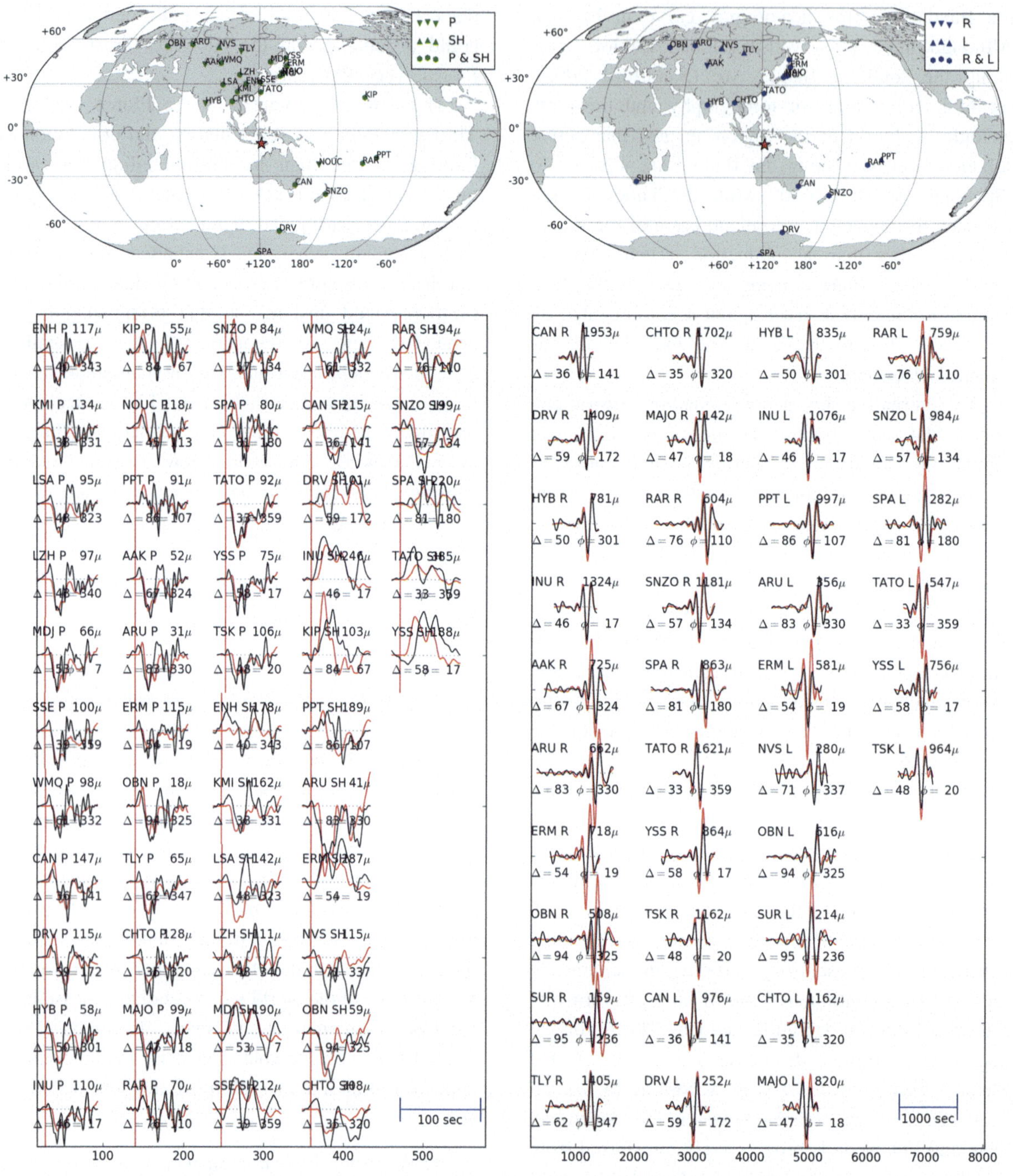

Figure 2

Teleseismic data coverage. Black and red curves show the original and inverted waveform fit for each station, respectively. The station and wave phase are indicated at the top left of each waveform, and the maximum peak value at the top right; Δ and ϕ are the distance and azimuth from the epicenter, respectively

mantle or core–mantle boundary structure on the modeled waveforms. Waveforms were corrected for instrument response, and bandpass Butterworth filters with passbands of 0.02–0.0125 and 0.004–0.5 Hz were applied to surface and body wave data, respectively.

3.2. Coseismic Uplift/Subsidence

The post-tsunami survey of Tsuji et al. (1995) documented extensive coastal uplift on the northern side of Flores Island due to the 1992 earthquake. The maximum uplift and subsidence was 1.1 and −1.6 m, respectively (see Fig. 3). The distance over which the maximum uplift shifted to the maximum subsidence was about 15 km, so the point where the "hinge line" (i.e., between uplift and subsidence where there is no change in elevation) crosses the coast is very well constrained. In total, there are eight locations where coastal uplift data collected by Tsuji et al. (1995) are used in the source inversion. Tsuji et al. (1995) note that the 1.6 m subsidence measurement at Kolisia may be biased by liquefaction, while the 0.75 m subsidence at Babi Island was reported by a local and may therefore be less accurate. However, these measurements both fit in well with the overall spatial pattern of the other measurements made by the team, so we include them in the inversion.

3.3. Tsunami Run-Up Heights

Tsunami run-up heights of up to 5.2 m were measured near the epicenter along Maumere Bay, with the highest measurements on its western side and gradually decreasing run-up towards the east (Fig. 4). Due to its circular shape, Babi Island appears to have experienced the remarkable phenomenon of edge waves that flowed around either side of the island and constructively interfered at its far side, resulting in particularly strong tsunami inundation that killed 700, about 25 % of Babi Island's population (Liu et al. 1995). Finally, extremely high tsunamis were observed around Hading Bay, with tsunami run-up of up to 26.2 m measured at Riangkroko. Yeh et al. (1993), Hidayat et al. (1995), and Tsuji et al. (1995) suggested that these high run-ups were due to landslides, which were observed at several locations along Hading Bay. All tsunami run-up data used here to verify the inverted source model are taken from Tsuji et al. (1995).

3.4. Tsunami Waveforms

Gonzalez et al. (1993) reported that only one tsunami waveform for this event is available, from the Palopo tide gauge at Bone Bay in Sulawesi, Indonesia, just over 600 km NNE from the earthquake epicenter. The tsunami arrived 104 min after the earthquake, with a predominant wave period of 69 min. The maximum and minimum peak amplitude recorded are 27 and −47 cm, respectively. Unfortunately, it is difficult to retrieve the original data, so the tsunami waveform is digitized from Hidayat et al.

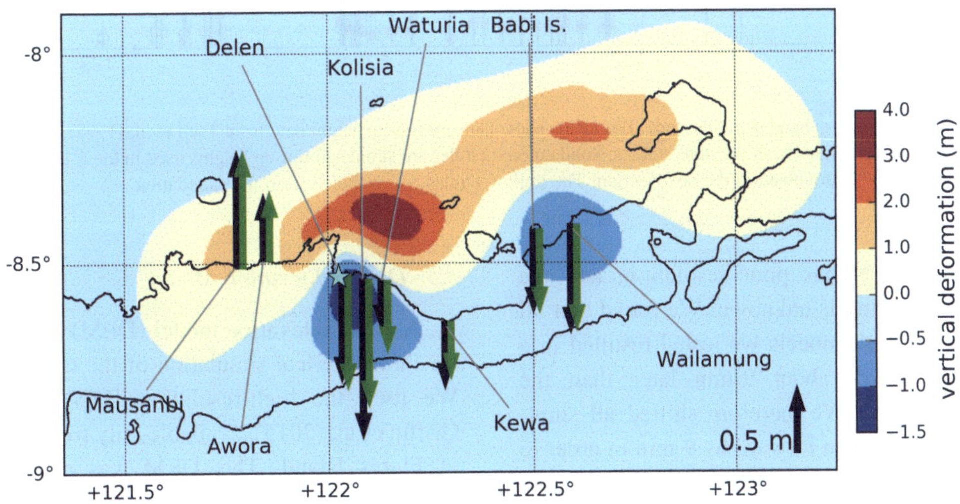

Figure 3
Coastal uplift/subsidence. Black arrows are coastal uplift observations from Tsuji et al. (1995); green arrows show the uplift simulated from the preferred slip model

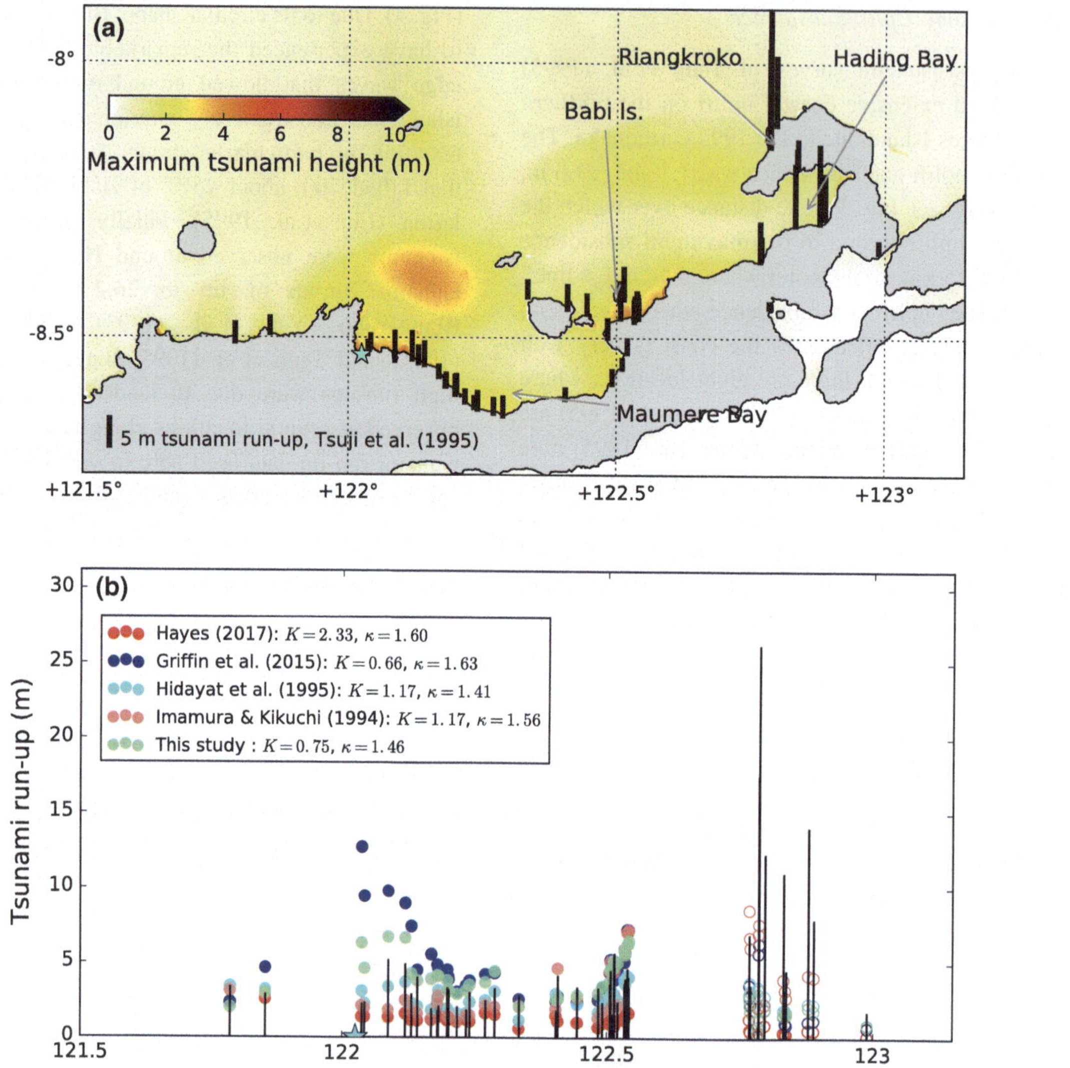

Figure 4

a Maximum simulated tsunami heights in this study. Bar charts show tsunami run-up height data from Tsuji et al. (1995). **b** Comparison with simulated tsunami run-up heights from previous studies. Solid colored circles are simulated run-up heights used in the K and κ calculation, as discussed in the next section. The hollow circles are excluded from the calculation

(1995). However, it has poor resolution, and the accuracy of its timing is unknown. We found that all of the plausible fault models we tested resulted in a tsunami that arrives about 9 min later than the observed waveform. We therefore shifted all simulated tsunami waveform forward by 9 min in order to align the peaks in observed and modeled waveforms (Fig. 5), and ascribe this to a timing error at the Palopo tide gauge.

3.5. *Digital Elevation Model*

A digital elevation model (DEM) is used as input for the numerical simulation of the tsunami (Fig. 6). We used the high-resolution DEM developed by Griffin et al. (2015), available only for the north coast of Flores Island. This DEM was extended to the Palopo tide gauge using a combination of data from the General Bathymetric Chart of the Oceans (GEBCO), the 90-m Shuttle Radar Topography Mission (SRTM-90), and a 90-m commercial

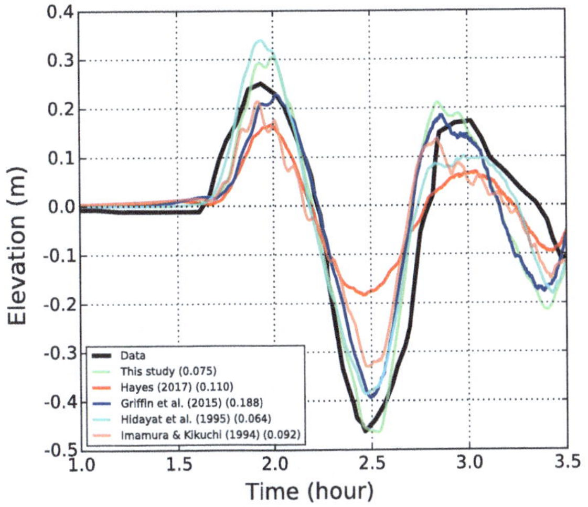

Figure 5
Tsunami waveforms at Palopo tide gauge of all sources. Tsunami waveform data are digitized from Hidayat et al. (1995). Simulated waveform is shifted forward by 9 min to match the maximum peak arrival time (discussed in Sect. 6). Numbers on the legend show the RMS error value for each model

bathymetry dataset provided by TCarta Marine for depths shallower than 100 m (see Griffin et al. 2015).

4. Methodologies

4.1. Finite-Fault Source Inversion

Trifunac (1974) and then Olson and Apsel (1982) introduced the finite-fault source inversion (FFI) technique, utilizing near-field seismic waveform datasets to invert for subsurface slip of an earthquake of a known discrete fault model. A stable inverted slip is obtained by applying a constrained linear least-squares method, and the basic methodology is similar in many subsequent studies. Satake (1987) proposed the use of tsunami waveform data for FFI, and Satake (1993) extended FFI by jointly inverting tsunami and geodetic data. After these early studies, FFI utilizing two or more datasets have been commonly done (Johnson et al. 1996; Konca et al. 2007; Baba et al. 2009; Tanioka et al. 1995; Gusman and Tanioka 2014; Hill et al. 2012; Benavente and Cummins 2013).

We utilize teleseismic waveform ($\mathbf{d}_s$) and coseismic uplift ($\mathbf{d}_u$) data to invert for a fault slip model ($\mathbf{m}$). Since two data types are used, proper weighting coefficients (γ_s, γ_u) are needed to balance the different datasets. To obtain a stable inverse solution, we also need to regularize using a physically reasonable spatial smoothing operator. We apply a Laplacian smoothing matrix ($\mathbf{S}$) multiplied by a smoothing coefficient (β) to minimize the second spatial derivative of the slip. Our inversion problem is written as Eq. (1), where $\mathbf{I}$ is an identity matrix, N_s and N_u are the number of seismic waveform and uplift data points, respectively, and $\mathbf{G}_s$ and $\mathbf{G}_u$ are subfault "Green's function" matrices, the columns of which predict the data that would be observed if unit slip occurred on a single subfault. $\mathbf{G}_s$ includes rows for body and for surface waves, which are numerically calculated using Kikuchi and Kanamori (1991, 1982, 1986) and Kanamori and Stewart (1976) [including phase velocity corrections for three-dimensional (3D) structure from Ekström et al. (1997)], respectively. The elements of $\mathbf{G}_u$ are calculated using the Okada (1985) formulae.

$$\begin{pmatrix} \gamma_s \mathbf{I}_{N_s} & 0 & 0 \\ 0 & \gamma_u \mathbf{I}_{N_u} & 0 \\ 0 & 0 & \beta \mathbf{I}_{N_m} \end{pmatrix} \begin{pmatrix} \mathbf{G}_s & 0 & 0 \\ 0 & \mathbf{G}_u & 0 \\ 0 & 0 & \mathbf{S} \end{pmatrix} \mathbf{m} = \begin{pmatrix} \gamma_s \mathbf{d}_s \\ \gamma_u \mathbf{d}_u \\ 0 \end{pmatrix} \tag{1}$$

A systematic series of inversions are conducted to reproduce the best-fit source model, with trial-and-error adjustment of the weighting parameters γ_s, γ_u, and β. The earthquake epicenter is also adjusted by 15 km in all directions to account for potential error in epicenter location. This is followed by analyzing the fault plane orientation (strike and dip angles as well as epicenter depth). Lastly, rupture propagation of the earthquake source is investigated. In addition, tsunami propagation and inundation modeling is considered to analyze the run-up heights along the northern coast of Flores Island and tsunami waveform at Palopo tide gauge.

4.2. Tsunami Modeling

The JAGURS code is utilized for tsunami propagation and inundation simulation (Baba et al. 2015). The code numerically solves the nonlinear Bossinesq

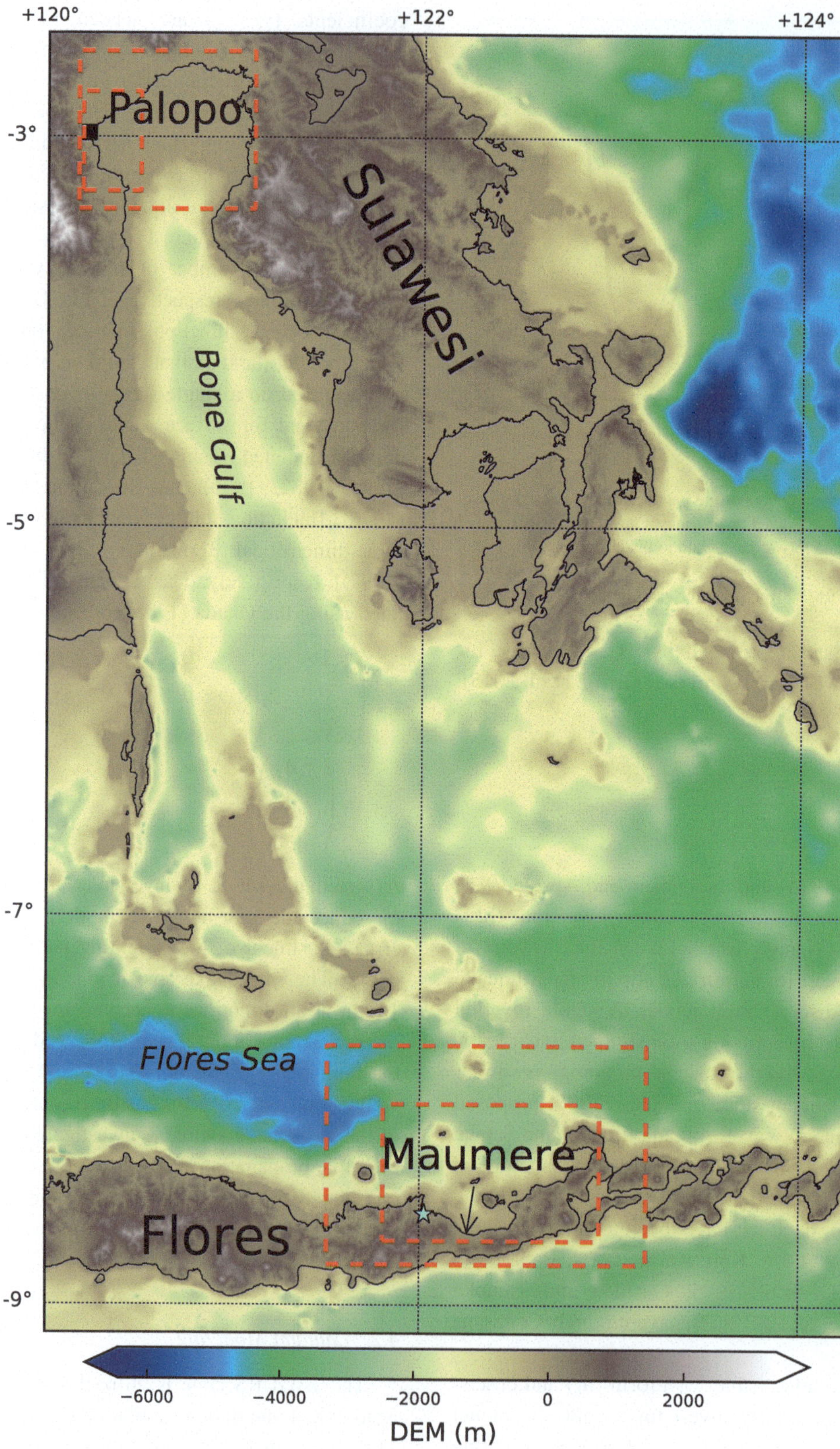
+120°
+122°
+124°
-3°
-5°
-7°
-9°
Palopo
Sulawesi
Bone Gulf
Flores Sea
Maumere
Flores
−6000
−4000
−2000
0
2000
DEM (m)

◀Figure 6
Elevation model for tsunami modeling. Red boxes are the nested grids. Resolutions of the DEM (from the finest) are 25 m, 75 m, and 225 m, respectively

equations in spherical coordinates. An initial sea surface elevation as the input is assumed to be equal to the seafloor deformation due to fault rupture calculated using the Okada (1985) formula.

Tsunami simulation is conducted on a nested grid domain model. The coarsest and finest grid resolution is 225 m and 25 m, respectively (Fig. 6). Each simulation is run for 4 h with a 0.15 s time step to satisfy the Courant stability condition.

We validate simulated tsunami run-up heights using the Aida numbers of K and κ (Eq. 2) (Aida 1978; Nakamura 2009; Gusman et al. 2014). K represents the mean ratio between data measurements (Obs_i) and simulation (Sim_i), whereas κ represents the standard deviation. In addition, we calculate the root-mean-square (RMS) misfit of observed and calculated tsunami waveforms at the Palopo tide gauge. We consider K of 1 ± 0.4 and a smaller κ as well as a small RMS value as indicating an acceptable source model.

$$\begin{aligned} \log K &= \frac{1}{N}\sum_{i=1}^{N} \log K_i \\ \kappa &= \sqrt{\frac{1}{N}\sum_{i=1}^{N} \left((\log K_i)^2 - (\log K)^2\right)} \\ K_i &= \frac{\mathrm{Obs}_i}{\mathrm{Sim}_i}. \end{aligned} \tag{2}$$

5. Results and Discussion

5.1. Original Fault Plane (ORI)

As a starting model, we use a fault plane with 150 km along-strike length and 60 km down-dip width, discretized into 30 and 12 rectangular subfaults in these respective directions with dimension of 5 km × 5 km. The strike and dip angles as well as hypocenter depth are set at 70°, 28°, and 16 km, respectively, taken from the seismic waveform analysis of Beckers and Lay (1995). The epicenter is taken at 121.902° E and 8.475° S (IRIS-Wilber 2018). These parameters are referred to below as the ORI fault plane model.

5.2. Weighting and Smoothing

There are only eight coastal uplift data available. In contrast, the body and surface waves have about 3900 and 3200 data points (i.e., $N_S = 3900 + 3200$ and $N_u = 8$ in Eq. 1). Therefore, while γ_s for the seismic waveforms (both body and surface waves) is fixed at 1.0, γ_u for the coastal uplift needs to be adjusted to ensure that the misfits in the two datasets are reasonably balanced.

Figure S1 shows how the misfit of the different datasets varies with γ_u and β. As might be expected, misfits for all datasets are low when β is small, $\beta = 0.01$, and increase with increasing β. The misfits for the seismic waveforms are low and those for uplift data are high when $\gamma_u = 1$, while the opposite is true for $\gamma_u = 500$. Some of the corresponding slip models are depicted in Fig. S2, which shows that slip models with $\beta = 0.01$ have very high spatial variability, while some of the models obtained using $\beta = 1.0$ have slip that is almost uniform over the rupture plane. Furthermore, the misfit to the uplift data becomes unacceptably high for $\gamma_u \leq 20$, while Fig. S1 shows that the misfit of the seismic waveform data begins to increase rapidly at around $\gamma_u = 40$.

Thus, visual inspection of the variation of both data misfit and model roughness with γ_u and β in Figs. S1 and S2 suggests that acceptable ranges for γ_u and β are 20–40 and 0.3–1.0, respectively. Since $\beta = 0.3$ is the minimal smoothing that gives rises to models with a reasonable level of spatial variability, and $\gamma = 20$ and 40 bound the region for which the misfits to both the uplift and seismic waveform data are acceptable, these values are taken for further analysis. The models with the ORI fault orientation and $\gamma_u = 20$ and $\gamma_u = 40$ are denoted as ORI20 and ORI40, respectively.

5.3. Variations in Hypocenter and Fault Plane Orientation

From their body wave analysis, Beckers and Lay (1995) suggested best-fitting strike and dip angles of

70° ± 30° and 28° ± 10°, respectively. Although the distribution of aftershocks is in rough agreement with this fault plane orientation (see Fig. 7), the uncertainties on the estimated parameters are large. Furthermore, although the various published hypocentral solutions are in good agreement and have small (5–15 km) formal uncertainties, the lack of near-field data suggests a potential for bias in the hypocenter estimate. It therefore seems prudent to investigate the sensitivity of our source model to variations in the hypocenter as well as to azimuth and dip angles.

Figure S3 shows the variance reductions that result from moving the epicenter 15 km from the original epicenter in different directions, with $\gamma_u = 20$ (S20, SE20, SW20, etc.) and $\gamma_u = 40$ (S40, SE40, SW40, etc.). Again, applying a larger γ_u produces a much better fit to the coastal uplift data, with S40 being the model that best fits the uplift data (Fig. S4). Furthermore, by validating the inverted slip model against tsunami run-up data, we note that simulated tsunami run-up heights along the northern coast of Flores become smaller when moving the epicenter to the north (Fig. S5). Then by giving more weight to the coastal uplift data, it produces a better fit to the tsunami amplitude at the Palopo tide gauge (Fig. S6). Because $\gamma_u = 40$ improves the misfit to both uplift and tsunami data, this value is chosen for further analysis. Moving the epicenter to the southeast also gives a better fit to the tsunami data. Therefore, SE40 is chosen as the preferred model.

The misfits to the data are also sensitive to the depth and fault plane orientation, as indicated in

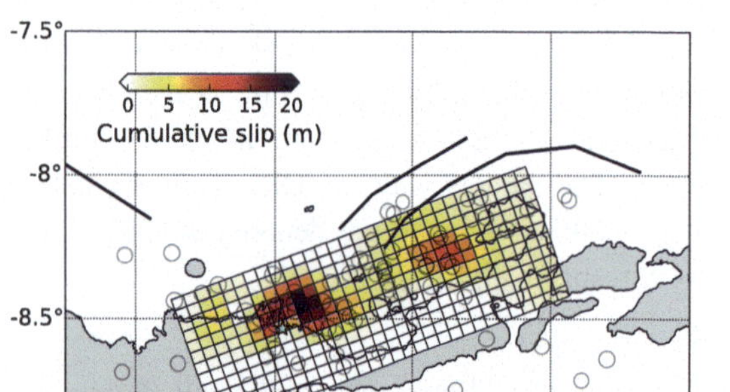

Figure 7
Final inverted slip model for Flores 1992 earthquake

Fig. S7. A fault plane with 10 km hypocentral depth and strike greater than 80° produces smaller variance for all the datasets. However, a shallower depth tends to overfit the uplift data (Fig. S9). A larger strike angle also results in simulated tsunami waveforms at the Palopo tide gauge with amplitude much smaller than is observed (Fig. S11). This waveform amplitude is reduced as well by using dip angles other than 28°, but the misfit is slightly improved by a deeper hypocenter depth. However, a 16 km depth is least preferred by the Rayleigh waves (Fig. S7). The simulated tsunami run-up heights are less sensitive to the fault plane orientation (Fig. S10). Therefore, SE40 is still regarded as the best fault plane model.

5.4. *Rupture Propagation*

Compared with other datasets, the P-waveforms are more sensitive to the rupture propagation variables, viz. the maximum rupture velocity V_r, half-duration of subfault rise times t_h, and number of time windows (NTW) (see Fig. S12). With NTW ≤ 5, the smallest variance is obtained with $V_r = 2.0\,\mathrm{km/s}$. At larger NTW, $V_r = 2.5\,\mathrm{km/s}$ is preferred. However, the other datasets prefer a slower V_r. Furthermore, P-waveforms are fit better using $t_h = 2.0\,\mathrm{s}$. On the other hand, other datasets produce small variance with a longer t_h. In general, a large NTW is preferred by all datasets.

Changing the rupture variables—NTW, t_h, and V_r—does not significantly change the misfit of model SE40 to the coastal uplift data (Fig. S13). It also has only a small effect on the tsunami run-up heights (Fig. S14). However, it has a significant influence on the tsunami amplitude at the Palopo tide gauge (Fig. S15). By decreasing V_r and increasing NTW, the tsunami waveform amplitude is increased. From the P-wave variances, NTW = 9 and $V_r = 2.5$ km/s produce the smallest variance. However, the NTW = 7 with $V_r = 2.5$ km/s model results in only slightly higher variance than the best-fit one. Even though this model slightly overestimates the maximum amplitude at the Palopo tide gauge, it generates a good fit to the minimum amplitude. In addition, to keep the rupture model as simple as possible while still fitting the data, NTW = 7 is selected as the preferred rupture propagation model (Table 1).

Table 1

Final inversion parameters used

Parameter	Flores 1992
Strike	70°
Dip	28°
Hypocenter	122.025° E,8.532° S, and 16 km
Fault dimension	150 km × 60 km
Number of subfaults	32 × 12
NTW	7
t_h	2.0
V_{rmax}	2.5 km/s
β	0.3
γ_s	1
γ_u	40

6. Comparison with Previous Models

Our preferred rupture model (SE40) for the 1992 Flores earthquake has two patches of high slip resulting from the inversion. The highest slip patch is located near the hypocenter. The second patch is located about 70 km northeast from the first one. This result is similar in some respects to those of Imamura and Kikuchi (1994), Beckers and Lay (1995), Hidayat et al. (1995), and Griffin et al. (2015). Imamura and Kikuchi (1994) and Hidayat et al. (1995) used simpler fault models, although they estimated a higher slip on the northeast part from the epicenter. Most previous studies also have a similar fault plane orientation in the ENE direction to fit the aftershock distribution, with the exception of the Hayes (2017) model. The latter causes the Hayes (2017) model to have three high-slip regions, but the maximum slip is much smaller than in other studies. We consider these models in more detail below.

The Hayes (2017) model was constructed by inverting seismic waveform data, so it fits these observations very well. However, it grossly under-predicts both the tsunami run-up and tide gauge waveform data (Figs. 4, 5, respectively), as well as the uplift data (Fig. 3).

The models of Imamura and Kikuchi (1994) and Hidayat et al. (1995) were constructed to fit both seismic waveform and tsunami run-up data [and, in the case of Hidayat et al. (1995), the Palopo tide gauge waveform]. Their fit to the seismic waveforms is similar to that of this study and Hayes (2017), and they also fit the tsunami run-up data better (17 % underprediction) than the model obtained here (30 % overprediction). However, their fits to the coastal uplift data are very poor (Fig. 3). Although they provide better fits to the Palopo tide gauge waveform than Hayes (2017), they do not fit these data substantially better than the model obtained here.

Of the previously published models for the 1992 earthquake, the model of Griffin et al. (2015) is most similar to our preferred model. Griffin et al. (2015) optimized the misfit of their model to the seismic waveform and coastal uplift data, and also compared its predictions with the tsunami run-up data. Although their model indeed provides excellent fit to the seismic waveform and coastal uplift data, its fit to the tsunami run-up data as well as the Palopo tide gauge waveform are considerably poorer than those of our preferred model (Figs. 4, 5, respectively). We suspect that the Griffin et al. (2015) model overfits the coastal uplift data, which Tsuji et al. (1995) note are subject to some uncertainty (Fig. 8).

In summary, we believe that our preferred model (SE40) provides a better fit to all of the datasets, even though its fit to, e.g., the tsunami run-up data is marginally poorer than some of the previous models. Moreover, even though the Hidayat et al. (1995) model gives the smallest RMS for the tsunami waveform fit at the Palopo tide gauge (Fig. 5), the RMS of our preferred model is only slightly greater and is much less than those of the other three models (Fig. 5). We also attempt to match the macroseismic observations of Tsuji et al. (1995) at Maumere and Ende, where Modified Mercalli Intensity (MMI) of 9–10 and 8–9, respectively, was estimated. Comparison with the MMI modeled using the fault modeling of Pagani et al. (2014) showed that these were matched only by the SE40 model, with the Griffin et al. (2015) model slightly underpredicting the MMI at Maumere but the other models giving values almost a unit of intensity too low at both locations (see Table 2).

7. Further Investigation

7.1. Tsunami Landslide Model

Because our preferred source model is unable to explain the extreme run-up heights observed around

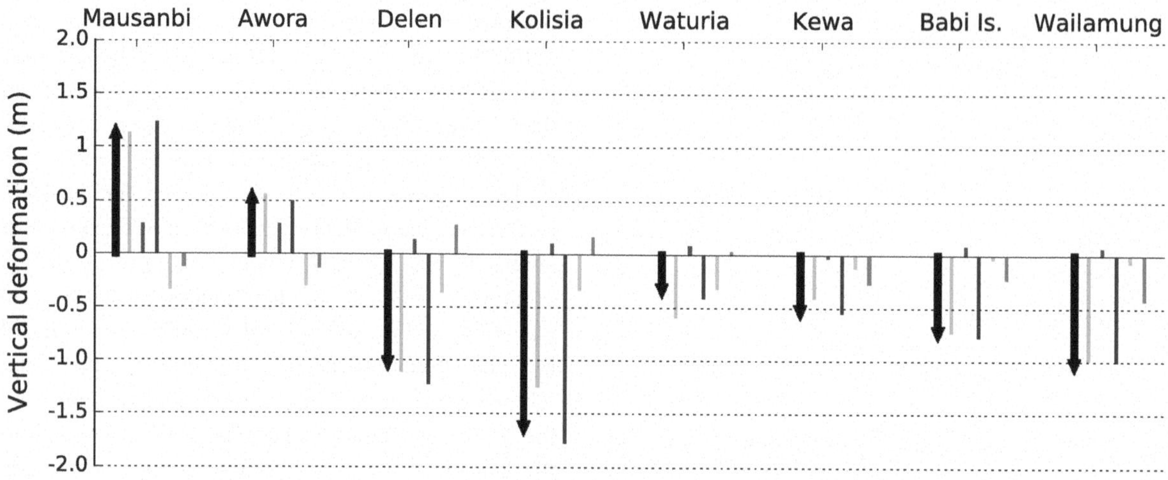

Figure 8
Vertical deformation comparison from all studies. Black = data, green = this study, red = Hayes (2017), blue = Griffin et al. (2015), cyan = Hidayat et al. (1995), salmon = Imamura and Kikuchi (1994)

Table 2

Modified Mercalli Intensity (MMI) calculated for the different published source models at Maumere and Ende, where Tsuji et al. (1995) estimated MMI of 9–10 and 8–9, respectively

Source model	Maumere	Ende
Hidayat et al. (1995)	7.8	8.3
Imamura and Kikuchi (1994)	8.3	7.1
Hayes (2017)	7.1	6.8
Griffin et al. (2015)	8.7	8.2
This study (SE40)	9.1	8.0

Hading Bay, we agree with previous studies in ascribing these to coastal landslides (Yeh et al. 1993; Tsuji et al. 1995; Hidayat et al. 1995; Imamura and Kikuchi 1994). Even though the locations of suspected coastal landslides are reported, it is difficult to build a tsunami landslide model without more detailed information.

We generate models for four landslides based on the limited information available in Yeh et al. (1993). We can produce very high tsunami run-up around Hading Bay (Fig. 9a). However, it is not an ideal model. Initial sea surface elevation is estimated using Watts et al. (2005) by assuming a Gaussian shape model with a slump mechanism (Fig. 9b). The equations estimate the maximum acceleration and duration of the mass movement as well as the maximum initial sea surface elevation. But the equations do not consider the details of the dynamic process, which is crucial in a tsunami landslide model. However, these four landslides have very little effect on the Palopo tide gauge waveform, and we believe this is unlikely to change with more careful consideration of landslide physics. Therefore, modeling of the Palopo tide gauge waveform using a fault source alone, as considered above, seems justified.

7.2. *Segmentation of the Flores Back-Arc Thrust*

Back-arc thrusting north of the eastern Sunda Arc is portrayed in the literature as southward-dipping thrusting that emerges at the surface in the Flores Sea. Beginning north of Bali in the west, its surface trace follows the bathymetric low that extends eastward to the western end of Flores, after which it appears only intermittently before reappearing north of the Island of Wetar in the east (see Fig. 1b, and McCaffrey and Nabalek 1984; Silver et al. 1986). Every published reference to the 1992 Flores earthquake we are aware of ascribes it to the Flores Back-Arc Thrust, but none have explained how the western end of the fault rupture extending beneath Flores can be connected to the surface expression of the Flores Back-Arc Thrust off the northern shore of the island's western end.

There seems to us no question that the 1992 Flores earthquake ruptured a major fault inclined at

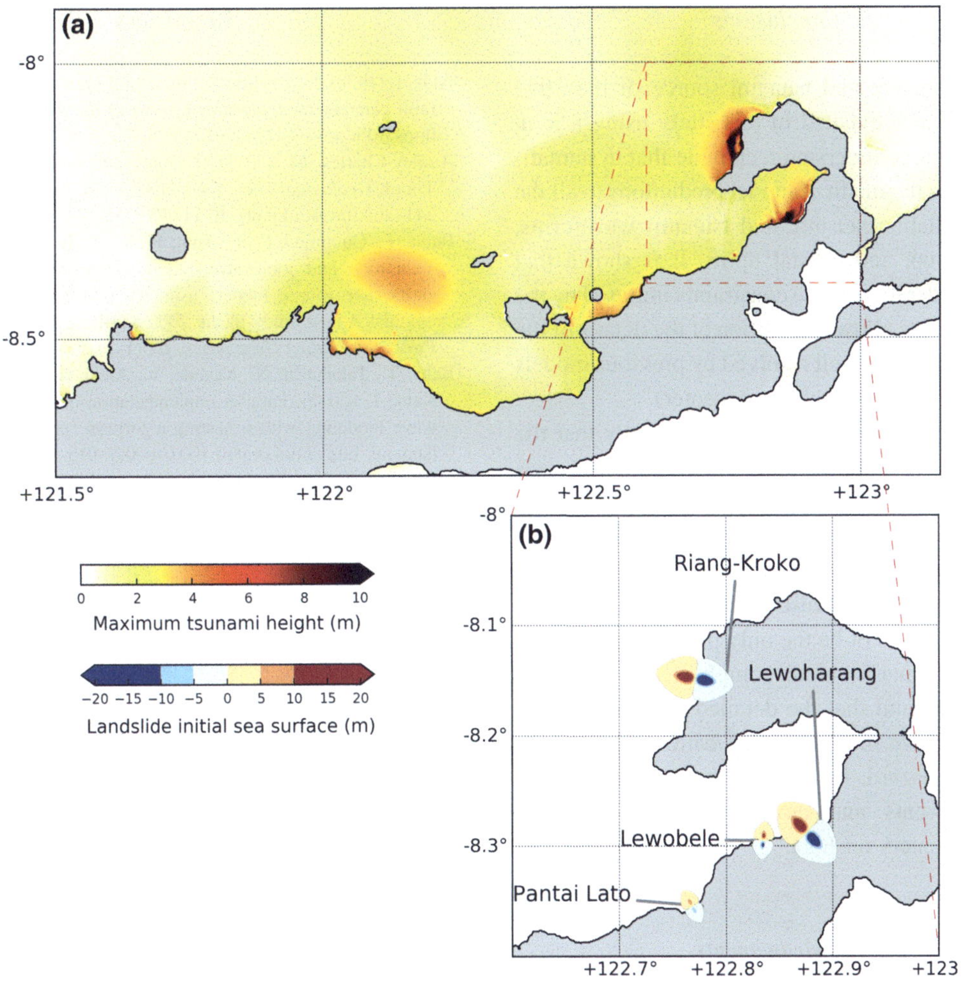

Figure 9
a Maximum tsunami heights when adding landslide sources, **b** arbitrary landslide initial sea surface elevation

about 70° azimuth, extending onshore at its western end but offshore at its eastern. Beckers and Lay (1995) first showed that the seismic data are best fit by a strike of 70°, that the aftershock zone is aligned ENE extending from onshore in the west to offshore in the east, and that the point where the "hinge line" (i.e., boundary between coseismic uplift ad subsidence) crosses the coast is consistent with this geometry. Finally, a substantial portion of the fault's eastern end must extend offshore in order to generate a large tsunami.

Whatever the relationship of the earthquake fault's onshore western end to the eastern end of the Flores Back-Arc Thrust offshore, it seems clear that the 1992 Flores earthquake ruptured a distinct segment of the back-arc thrust system. This implies a limit to the maximum magnitude of thrust earthquakes in Flores, and suggests that the back-arc thrust system may be more complicated than previously thought. We note that the 2004 Mw 7.5 Alor earthquake had a strike of 67°, implying it may have ruptured a separate, en echelon segment of the thrust system to the east. If such ENE-oriented segments exist elsewhere along the back-arc thrust system, the implications for hazard are important. As with the 1992 earthquake, the western end may extend onshore, where it can threaten population centers with strong ground shaking.

8. Conclusions

The earthquake and tsunami source of the 1992 Flores event are revisited in this study using a joint finite-fault source inversion technique that systematically explores the misfit of model predictions to all the datasets available: seismic and tsunami waveforms, tsunami run-up, and coastal uplift. It is shown that careful consideration of the constraints imposed by the multiple datasets allows resolution of the details of the source that were not well resolved by previous models (i.e., the larger slip near the hypocenter).

The study also establishes unequivocally that the 1992 Flores earthquake ruptured a fault that was inclined to the ENE, deviating from the overall EW trend of back-arc thrusting in the eastern Sunda Arc. This implies that the event ruptured a distinct segment of the back-arc thrust system. We suggest that this segment may not be the only part of the back-arc thrust system that is more complicated than previously thought, and that the detailed geometry of the system may have important implications for seismic and tsunami hazard, by limiting the maximum magnitude of events and directing rupture close to population centers onshore.

Acknowledgements

This work was partly supported by the Government of Australia's Australian Awards Scholarship program. The high-resolution digital elevation model around Flores Island was provided by the Australia–Indonesia Facility for Disaster Reduction (AIFDR), which was supported by the Australian Department of Foreign Affairs and Trade's Australian Aid program. The tsunami numerical simulation was undertaken with the assistance of resources and services from the National Computational Infrastructure (NCI), which is supported by the Australian Government. We also thank two reviewers, who provided rigorous comments on the paper.

References

Aida, I. (1978). Reliability of a tsunami source model derived from fault parameters. *Journal of Physics of the Earth*, *26*(1), 57–73. https://doi.org/10.4294/jpe1952.26.57.

Audley-Charles, M. (2011). Tectonic post-collision processes in Timor. *Geological Society London Special Publications*, *355*(1), 241–266. https://doi.org/10.1144/SP355.12.

Baba, T., Cummins, P. R., Thio, H. K., & Tsushima, H. (2009). Validation and joint inversion of teleseismic waveforms for earthquake source models using deep ocean bottom pressure records: A case study of the 2006 Kuril megathrust earthquake. *Pure and Applied Geophysics*, *166*(1–2), 55–76.

Baba, T., Takahashi, N., Kaneda, Y., Ando, K., Matsuoka, D., & Kato, T. (2015). Parallel implementation of dispersive tsunami wave modeling with a nesting algorithm for the 2011 Tohoku tsunami. https://doi.org/10.1007/s00024-015-1049-2

Beckers, J., & Lay, T. (1995). Very broadband seismic analysis of the 1992 Flores, Indonesia, earthquake (Mw = 7.9). *Journal of Geophysical Research Solid Earth*, *100*(B9), 18179–18193. https://doi.org/10.1029/95JB01689.

Benavente, R., & Cummins, P. R. (2013). Simple and reliable finite fault solutions for large earthquakes using the w-phase: The Maule (Mw= 8.8) and Tohoku (Mw= 9.0) earthquakes. *Geophysical Research Letters*, *40*(14), 3591–3595.

Bird, P. (2003). An updated digital model of plate boundaries. *Geochemistry Geophysics Geosystems*, *4*, 3. https://doi.org/10.1029/2001GC000252.

Breen, N. A., Silver, E. A., & Roof, S. (1989). The Wetar Back Arc Thrust Belt, eastern Indonesia: The effect of accretion against an irregularly shaped arc. *Tectonics*, *8*(1), 85–98. https://doi.org/10.1029/TC008i001p00085.

Ekström, G., Tromp, J., & Larson, E. W. (1997). Measurements and global models of surface wave propagation. *Journal of Geophysical Research Solid Earth*, *102*(B4), 8137–8157.

Gonzalez, F., Sutisna, S., Hadi, P., Bernard, E., & Winarso, P.(1993). Some observations related to the Flores Island earthquake and tsunami. In: TSUNAMI, International Tsunami Symposium, Wakayama, Japan

Griffin, J., Latief, H., Kongko, W., Harig, S., Horspool, N., Hanung, R., et al. (2015). An evaluation of onshore digital elevation models for modelling tsunami inundation zones. *Frontiers in Earth Science*, *3*, 32. https://doi.org/10.3389/feart.2015.00032.

Gusman, A., Tanioka, Y., MacInnes, B., & Tsushima, H. (2014). A methodology for near field tsunami inundation forecasting: Application to the 2011 Tohoku tsunami. *Journal of Geophysical Research Solid Earth*, *119*(11), 8186–8206.

Gusman, A. R., & Tanioka, Y. (2014). W-phase inversion and tsunami inundation modeling for tsunami early warning: Case study for the 2011 Tohoku event. *Pure and Applied Geophysics*, *171*(7), 1409–1422.

Hamilton, W. (1979). *Tectonics of the Indonesian region* (4th ed.). Washington: United States Government Printing Office.

Hayes, G. (2017). Preliminary finite fault results for the Dec 12, 1992 Mw 7.7 -8.5100,121.8900 earthquake (version 1). https://earthquake.usgs.gov/earthquakes/eventpage/usp0005j5a

Hidayat, D., Barker, J., & Satake, K. (1995). Modeling the seismic source and tsunami generation of the December 12, 1992 Flores Island, Indonesia, earthquake. *Pure and Applied Geophysics*, *144*(3–4), 537–554. https://doi.org/10.1007/BF00874382.

Hill, E. M., Borrero, J. C., Huang, Z., Qiu, Q., Banerjee, P., Natawidjaja, D. H., et al. (2012). The (2010) Mw 7.8 Mentawai earthquake: Very shallow source of a rare tsunami earthquake determined from tsunami field survey and near-field GPS data. *Journal of Geophysical Research Solid Earth, 117*, 6.

Imamura, F., & Kikuchi, M. (1994). Moment release of the 1992 Flores Island earthquake inferred from tsunami and teleseismic data. *Science Tsunami Hazards, 12*, 67–76.

IRIS-Wilber (2018) Iris-wilber 3 dataset. www.iris.edu

Johnson, J. M., Satake, K., Holdahl, S. R., & Sauber, J. (1996). The 1964 Prince William Sound earthquake: Joint inversion of tsunami and geodetic data. *Journal of Geophysical Research Solid Earth, 101*(B1), 523–532.

Kanamori, H., & Stewart, G. S. (1976). Mode of the strain release along the Gibbs fracture zone, Mid-Atlantic ridge. *Physics of the Earth and Planetary Interiors, 11*(4), 312–332.

Kikuchi, M., & Kanamori, H. (1982). Inversion of complex body waves. *Bulletin of the Seismological Society of America, 72*(2), 491–506.

Kikuchi, M., & Kanamori, H. (1986). Inversion of complex body waves—II. *Physics of the Earth and Planetary Interiors, 43*(3), 205–222.

Kikuchi, M., & Kanamori, H. (1991). Inversion of complex body waves—III. *Bulletin of the Seismological Society of America, 81*(6), 2335–2350.

Konca, A. O., Hjorleifsdottir, V., Song, T. R. A., Avouac, J. P., Helmberger, D. V., Ji, C., et al. (2007). Rupture kinematics of the 2005 Mw 8.6 Nias–Simeulue earthquake from the joint inversion of seismic and geodetic data. *Bulletin of the Seismological Society of America, 97*(1A), S307–S322.

Koulali, A., Susilo, S., McClusky, S., Meilano, I., Cummins, P., Tregoning, P., et al. (2016). Crustal strain partitioning and the associated earthquake hazard in the Eastern Sunda-Banda Arc. *Geophysical Research Letters, 43*, 1943–1949. https://doi.org/10.1002/2016GL067941.

Liu, P. L. F., Cho, Y. S., Briggs, M. J., Kanoglu, U., & Synolakis, C. E. (1995). Runup of solitary waves on a circular island. *Journal of Fluid Mechanics, 302*, 259–285.

McCaffrey, R., & Nabalek, J. (1984). The geometry of back arc thrusting along the eastern Sunda Arc, Indonesia: Constraints from earthquake and gravity data. *Journal of Geophysical Research, 89*(B7), 6171–6179.

Nakamura, M. (2009). Fault model of the 1771 Yaeyama earthquake along the Ryukyu trench estimated from the devastating tsunami. *Geophysical Research Letters, 36*, 19. https://doi.org/10.1029/2009GL039730.

Nugroho, H., Harris, R., Lestariya, A. W., & Maruf, B. (2009). Plate boundary reorganization in the active Banda Arc-continent collision: Insights from new GPS measurements. *Tectonophysics, 479*(1–2), 52–65.

Okada, Y. (1985). Surface deformation due to shear. *Bulletin of the Seismological Society of America, 75*(4), 1135–1154.

Olson, A. H., & Apsel, R. J. (1982). Finite faults and inverse theory with applications to the 1979 Imperial Valley earthquake. *Bulletin of the Seismological Society of America, 72*(6A), 1969–2001.

Pagani, M., Monelli, D., Weatherill, G., Danciu, L., Crowley, H., Silva, V., et al. (2014). Openquake engine: An open hazard (and risk) software for the global earthquake model. *Seismological Research Letters, 85*(3), 692–702.

Sandiford, M. (2008). Seismic moment release during slab rupture beneath the Banda Sea. *Geophysical Journal International, 174*(2), 659–671.

Satake, K. (1987). Inversion of tsunami waveforms for the estimation of a fault heterogeneity: Method and numerical experiments. *Journal of Physics of the Earth, 35*(3), 241–254.

Satake, K. (1993). Depth distribution of coseismic slip along the Nankai trough, Japan, from joint inversion of geodetic and tsunami data. *Journal of Geophysical Research Solid Earth, 98*(B3), 4553–4565.

Silver, E. A., Breen, N. A., & Prasetyo, H. (1986). Multibeam study of the Flores Back-Arc thrust belt, Indonesia. *Journal of Geophysical Research, 91*(B3), 3489–3500.

Silver, E. A., Reed, D., & McCaffrey, R. (1983). Back arc thrusting in the Eastern Sunda arc, Indonesia: A consequence of arc-continent collision. *Journal of Geophysical Research, 88*(B9), 7429–7488.

Spakman, W., & Hall, R. (2010). Surface deformation and slab-mantle interaction during Banda arc subduction rollback. *Nature Geoscience, 3*, 562–566. https://doi.org/10.1038/NGEO917.

Tanioka, Y., Satake, K., & Ruff, L. (1995). Total analysis of the 1993 Hokkaido Nansei-Oki earthquake using seismic wave, tsunami, and geodetic data. *Geophysical Research Letters, 22*(1), 9–12.

Trifunac, M. (1974). A three-dimensional dislocation model for the San Fernando, California, earthquake of February 9, 1971. *Bulletin of the Seismological Society of America, 64*(1), 149–172.

Tsuji, Y., Matsutomi, H., Imamura, F., Takeo, M., Kawata, Y., Matsuyama, M., et al. (1995). Damage to coastal villages due to the 1992 Flores Island earthquake tsunami. *Pure and Applied Geophysics, 144*(3–4), 481–524. https://doi.org/10.1007/BF00874380.

Watts, P., Grilli, S., Tappin, D. R., & Fryer, G. J. (2005). Tsunami generation by submarine mass failure II: Predictive equations and case studies. *Journal of Waterway Port Coastal and Ocean Engineering, 131*, 298–310. https://doi.org/10.1061/ASCE)0733-950X(2005)131:6(298).

World Bank. (1993). *Indonesia Flores earthquake reconstruction project*. The World Bank: Report

Yeh, H., Imamura, F., Synolakis, C., Tsuji, Y., Liu, P., & Shi, S. (1993). The Flores Island tsunamis. *EOS Transactions American Geophysical Union, 74*(33), 369.

(Received October 15, 2018, revised December 6, 2018, accepted December 14, 2018, Published online January 3, 2019)

Pure Appl. Geophys. 176 (2019), 2985–2993
https://doi.org/10.1007/s00024-019-02100-y

Pure and Applied Geophysics

The 2015 Illapel Tsunami Source Recovery by Inversion of DART Tsunami Waveforms Using the R-Solution Method

Tatyana A. Voronina,[1] Vladislav V. Voronin,[2] and Vladimir A. Cheverda[3]

Abstract—An approach to recovering an initial tsunami waveform with its application to the 16 September 2015 Illapel tsunami is proposed. The approach is based on inverting the remote measurements of water-level data from the deep-ocean DART buoys without a priori information on a source except for the common information about its spatial localization. The ill-posed inverse problem in question is regularized by means of a least-squares inversion using the truncated singular value decomposition method and the *r*-solution method. The method proposed suppresses the instability of the numerical solution of the ill-posed problem under consideration. The computational algorithm allows one to find the way to improve the inversion by selecting the most informative set of available observation stations. The method proposed was successfully applied to the 16 September 2015 Illapel tsunami.

Key words: Tsunami numerical simulation, tsunami waveform inversion, ill-posed inverse problem, singular value decomposition, *r*-solution.

1. Introduction

According to the United States Geological Survey, the 2015 Illapel earthquake occurred on 16 September, at 22:54:32 (UT), at 31.573°S, 72.674°W and at 22.4-km depth. This earthquake (*Mw* 8.3, Global CMT) generated local and trans-Pacific tsunamis that caused economic losses, fatalities and devastation along the Pacific coast of Chile. The Peru–Chile subduction zone has been historically characterized by high seismic activities due to the collision of the two plates along tectonic boundaries: the Nazca plate subducts beneath the South American plate. In the last decade, there were numerous earthquakes ($Mw > 8$) located in this region and there is a high probability of new similar events in the same area. Thus, the common objective of all studies is to understand the characteristics of the tsunamis triggered along the Peru–Chile Trench. One of such characteristic features of these earthquakes and tsunamis is that their sources are in proximity to the coast which, in turn, causes an essential decrease in the time of identifying the regions at risk and of taking decisions on evacuation. There are two important aspects of assessing the tsunami hazard in the coastal areas: the initial waves generated in the source area and their subsequent propagation. This study deals with the first of these challenges.

The sources of the earthquake and of the associated tsunami of the 16 September 2015 Illapel Chile event have been studied by many researches from various viewpoints using different observation data (Heidarzadeh et al. 2016; Li et al. 2016; Ruiz et al. 2016; Tilmann et al. 2016; Melgar et al. 2016; Lee et al. 2016; Calisto et al. 2016; Fuentes et al. 2016; Tang et al. 2016). An overview of these studies on tsunami source and slip distribution was presented by Satake and Heidarzadeh (2017). As a result, they have summarized the following: the largest slip is located near 31°S, 72.31°W and it is displaced, approximately, 70 km to the NW of the epicenter; the slip amount varies from 5 to 16 m in different studies. According to this review, the tsunami source zone is centered around the 31°S. This zone is assumed to be limited by a travel time curve (Heidarzadeh et al. 2016) and is extending west to 73°W, while the main

[1] Institute of Computational Mathematics and Mathematical Geophysics of Siberian Branch of Russian Academy of Science, Pr. Ac. Lavrentieva, 6, 630090 Novosibirsk, Russia. E-mail: tanvor@bk.ru
[2] Novosibirsk State University, Novosibirsk, Russia.
[3] A.A.Trofimuk Institute of Petroleum Geology and Geophysics of Siberian Branch of Russian Academy of Science, Novosibirsk, Russia.

tsunami source extends to the trench axis. Observed data don't give a bound of the tsunami zone on the eastern edge.

In the present paper, the problem of inferring the space distribution of the sea levels in the tsunami source area (denoted below as the *tsunami source*) is treated as an ill-posed problem of the hydrodynamic inversion with tsunami sea-level records. Unavoidable noise of real data contributes to the instability of the numerical solution of the inverse ill-posed problem and, thereby, it creates a serious obstacle in applying mathematical methods for solving these problems in any real application. In the approach proposed, the regularization is performed by means of the truncated singular value decomposition (SVD) that was implemented by using the *r*-solution method (Cheverda and Kostin 2010). The *r*-solution obtained is a projection of the exact solution onto a subspace that is the linear hull of the first right singular vectors of the direct problem operator. This subspace is chosen by analyzing the properties of the singular spectrum of the direct problem operator, which, in turn, is determined by the observation system and bathymetry.

The first results of the numerical simulation have shown that the approach proposed reduces the numerical instability. The application of this approach to the tsunami waveforms inversion was described in detail (Voronin et al. 2015; Voronina and Romanenko 2016; Voronina 2016).

It should be noted the tsunami waveform inversion has an advantage in recovering a tsunami source, as compared with the seismic waveform inversion, because the tsunami wave propagation can be more accurately simulated than seismic waves.

Although considerable attention has been given to developing the inversion methods to infer the initial tsunami waveform (Satake 1989; Piatanesi et al. 2001; Pires and Miranda 2001; Titov et al. 2005; Percival et al. 2011; Satake et al. 2013; Tinti and Tonini 2013; Mulia and Asano 2016; Romano et al. 2016), a lesser number of studies has been devoted to revealing the influence of such characteristics of observation systems as the number and the spatial distribution of the recording devices. In order to correlate these notions, a series of the numerical experiments based on the approach proposed have been carried out.

The results of the numerical implementation of this approach using real data of the 2015 Illapel Chile tsunami are presented below.

2. The Inversion Method

In the present paper, the tsunami wave propagation is considered within the scope of the linear shallow-water theory. The wave run-up processes of the shore structures are not taken into account. A function of the water surface elevation relative to the mean sea level is considered to be the solution of the linear shallow-water equations:

$$\begin{array}{c} \eta_t + g\nabla\cdot(h\mathbf{V}) = 0 \\ \mathbf{V}_t + g\nabla\eta = 0 \end{array}, \qquad (1)$$

with the following initial conditions:

$$\eta|_{t=0} = \varphi(x,y), \qquad \mathbf{V}|_{t=0} = 0; \qquad (2)$$

and the boundary conditions:

$$\mathbf{V}\cdot\mathbf{n} = 0 \quad \text{on the solid boundary}; \qquad (3)$$

$$-c\mathbf{V}\cdot\mathbf{n} - \eta_{tt} + \left.\frac{c^2\partial^2\eta}{2\partial\tau^2}\right|_{\Gamma} = 0 \text{ on the open boundary}. \qquad (4)$$

In the above equations, the vector $\mathbf{V} = (v_x, v_y)$ is the horizontal fluid velocity vector, $h(x, y)$ is the water depth, g is the gravity acceleration, $c(x,y) = \sqrt{gh(x,y)}$ is the wave phase velocity, $\mathbf{n}$ and τ are the unit vectors, respectively, outward directed, normal and tangential to the boundary; $\varphi(x,y)$ is the initial water elevation defined in the tsunami source area $\Omega : \{(x,y) \in [0,l_1] \times [0,l_2]\}$.

The inverse problem in question is to obtain the unknown initial water displacement $\varphi(x,y)$ as data output, while the observed tsunami waveforms form the vector η_0 as data input. This vector is assumed to be known on a set of the points $M = \{(x_i,y_i)\}, i = 1,\ldots,P$ (below called *receivers*) and at the time instants t_j, $j = 1,\ldots,N_t$:

$$\eta_0 = \left(\eta_{11},\ldots,\eta_{1N_t},\eta_{21},\eta_{2N_t},\ldots,\eta_{P1},\ldots,\eta_{PN_t}\right)^{\mathrm{T}}, \qquad (5)$$

$$\eta_{ij} = \eta(x_i, y_i, t_j),\ i = 1, 2, \ldots P;\ j = 1, 2, \ldots, N_t.$$

In the present paper, to solve problems (1)–(5) we use an inversion method, already described in its fundamentals by Voronina (2012), Voronin et al. (2015) and Voronina and Romanenko (2016).

It is common practice to present an unknown function as a series of certain functions with unknown coefficients. For example, there is a number of inversion approaches, following Satake (1989), presenting the unknown function in terms of the Green function. In this paper, an unknown initial tsunami waveform is represented as a part of the spatial harmonics series $\{\varphi_{mn}(x, y)\}$ in the source area Ω:

$$\varphi(x, y) = \sum_{m=1}^{I} \sum_{n=1}^{J} c_{mn} \varphi_{mn}, \tag{6}$$

with $\varphi_{mn} = \sin\frac{m\pi}{l_1}x \cdot \sin\frac{n\pi}{l_2}y$ and unknown coefficients $\mathbf{c} = \{c_{mn}\}$. Then the vector $\eta_0 = \{\eta_i\}$ containing the observed tsunami waveforms according to the ratio (5) is expressed as follows:

$$\eta_0 = \mathbf{Ac}, \tag{7}$$

where $\mathbf{A}$ is a matrix whose columns consist of computed waveforms for every spatial harmonic $\varphi_{mn}(x, y)$ which is used as the initial condition for the direct problems (1)–(4). Now, the dimensions of the solution and the data space are equal to: $dim(sol) = K = I \times J$; $dim(data) = L = P \times N_t$. Then the matrix $\mathbf{A}$ is $K \times L$ matrix, $L \gg K$. To calculate the matrix $\mathbf{A}$, problems (1)–(4) were approximated by means of a finite-difference approach. It should be noted that the calculation of the matrix $\mathbf{A}$ is the most time-consuming part of the algorithm.

After applying the singular value decomposition (SVD) of the matrix $\mathbf{A}$, the r-solution of Eq. (7) and the r-solution of the inverse problem in question are represented in the forms, respectively,

$$\mathbf{c}^{[r]} = \sum_{k=1}^{r} \alpha_k \mathbf{e}_k,$$
$$\varphi^{[r]}(x, y) = \sum_{j=1}^{r} \alpha_j \sum_{m=1}^{I} \sum_{n=1}^{J} \beta_{mn}^{j} \varphi_{mn}(x, y), \tag{8}$$

where $\alpha_k = \frac{(\eta_0, \mathbf{l}_k)}{s_k}$, $\{s_k\}$, $k = 1, \ldots, K$ are the singular values of the matrix $\mathbf{A}$, $\{\mathbf{e}_j = (\beta_{11}^{j}, \beta_{12}^{j} \ldots, \beta_{IJ}^{j})^T\}$, $\{\mathbf{l}_j\}$, $j = 1, \ldots, L$ are the right and the left singular vectors of this matrix. The ill-posedness of the problem in question results in a rapid decrease in the singular values of the matrix as their numbers increase. Setting a certain conditioning number (below denoted as $cond\mathbf{A}^{[r]}$) of the matrix obtained after truncation denoted as $\mathbf{A}^{[r]}$, one can determine the number r from the condition:

$$r = \max\{k : s_k/s_1 \geq 1/cond\mathbf{A}^{[r]}\}. \tag{9}$$

The truncated series in (8) is stable for any fixed parameter r with respect to perturbations of the right-hand side. It is clear that the value r depends on the rate of decreasing a singular spectrum of the matrix $\mathbf{A}$, which is tightly bounded with the parameters of an observation system, and the noisiness of data. The location of tsunami waveform recorders affects the choice of number r by such a way: the better configuration of an observation system is, the longer a weakly decreasing part of the singular spectrum. The number r is taken to be much smaller than the minimal dimension of the matrix. It is natural that as r increases, the information about the r-solution obtained also increases while its stability decreases. By changing r, one can control the solution instability.

Comparison of the singular spectra of the matrices $\mathbf{A}$ calculated by using the different sets of receivers makes it possible to highlight the *most informative* part of the observation system that provides the best recovery result.

3. Application to the Real Event

In this paper, the approach proposed was applied to infer the tsunami source of the 2015 Illapel tsunami by the inversion of the tsunami waveforms recorded only in the deep ocean by the Deep-ocean Assessment and Reporting of Tsunamis (DART) buoys nearest to the epicenter of the earthquake. It is assumed that the tsunami wave excitation is caused by an instantaneous abrupt rising and sinking of the seafloor at the initial time instant from the state of rest with transferring of the formed shape at the sea surface. The DART data are provided by the National Oceanic and Atmospheric Administration (NOAA) of the USA (http://www.ndbc.noaa.gov/dart.shtml).

Figure 1 shows the simulation domain (top) and the source area Ω (bottom). As was mentioned above, the wave run-up is not considered in this study. Therefore, the coast line was set when the depth was equal to 50 m with the total reflection conditions. The simulation domain is the water part of the rectangle $\Pi = \{(x; y) : 102°W \leq x \leq 70°W; 12°N \leq y \leq 38°S\}$. It is covered with a 1-min rectangular grid of 1921 × 3001 points to follow GEBCO bathymetry (http://www.gebco.net/). The source area is assumed to be

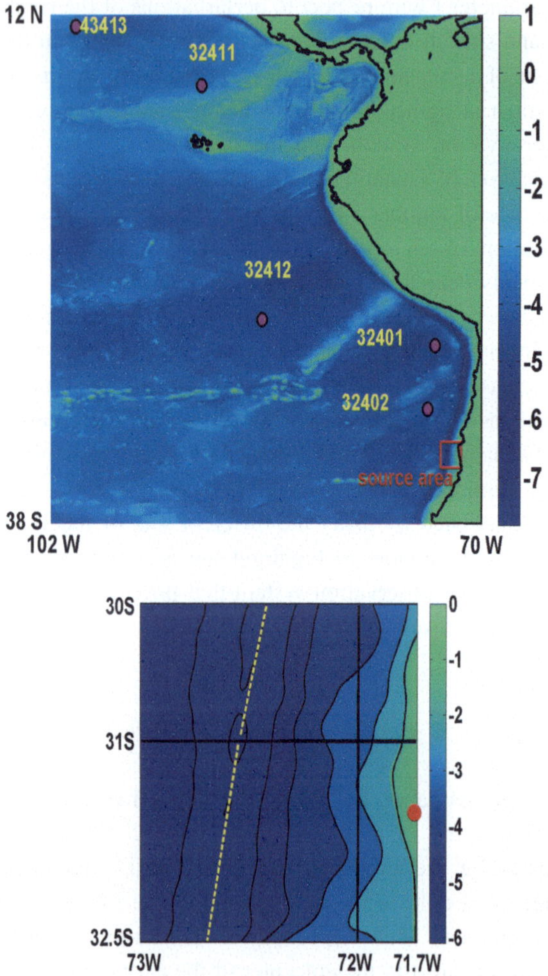

Figure 1
The calculation domain (top): the magenta circles indicate the localization of DART stations with their ID (yellow); the red rectangle marks the target domain; the green domain corresponds to the dry land; the black line corresponds to zero depth scaled in kilometers. The target domain (bottom): solid black contours show the depth levels with a 1-km interval; the yellow dashed line corresponds to the trench axis; the red circle denotes the earthquake epicenter

$\Omega = \{71.5°W \leq x \leq 73°W; 30°S \leq y \leq 32.5°S\}$. It is covered with a grid of 77 × 151 points. The observation system consists of five deep-water DART buoys numbered counter-clock-wise according to Fig. 1 (top) and marked as magenta circles: 1—32402; 2—32401; 3—32412; 4—32411; 5—43413.

The original code is used to numerically simulate the tsunami propagation. It is a FORTRAN code based on a finite-difference algorithm which uses an explicit–implicit difference scheme constructed with a four-point stencil on a uniform rectangular staggered grid (Marchuk et al. 1983). The scheme has a second order of spatial approximation and a first order w.r. to time. The time step equals 4 s in all experiments and the total number of steps is equal to 9001 for the matrix calculation. The program was executed on the HP DL980 G7 server, using 2 of 10-core Intel Xeon E7-4870 processors (20 cores in total). In this programming environment, calculating one column of the matrix **A** requires, approximately, 47 s.

Because the truncation of the SVD decomposition is the key ingredient of the method, we would like to know how it affects the solution. Since we represent the unknown source function as a series of the spatial harmonics according to formula (6), first of all, we have investigated the dependence between the quality of the reconstruction of each harmonic and the value of r. We have carried out a lot of numerical experiments with reconstructing the spatial harmonics that vary in their frequencies.

As a simple example we represent here the reconstruction of the first harmonic $\varphi_{11}(x, y) = \sin\frac{\pi}{l_1}x \cdot \sin\frac{\pi}{l_2}y$ as a source, with the extreme values $\varphi_{11\,\max} = 1.0$m, $\varphi_{11\,\min} = 0.0$m and using the real Chile offshore bathymetry. The synthetic waveforms were calculated at the location of the real DART buoys by solving the direct problems (1)–(4). The initial data $\eta_0(x,\ y,\ t)$ were calculated by means of Eq. (7).

We use the following notation for calculating variants: v$XY\ldots Z$, where numbers $X, Y, \ldots, Z$ after the letter "v" are the indices of the DART buoys whose records were used in the inversion [Fig. 1 (top) shows the DART buoys indices]. Figure 2 shows four examples of reconstructing the first harmonic with the following configurations: $I = 8$; $J = 8$; $N_t = 1201$; $L = P \times 1201$. Here, P is the number of receivers

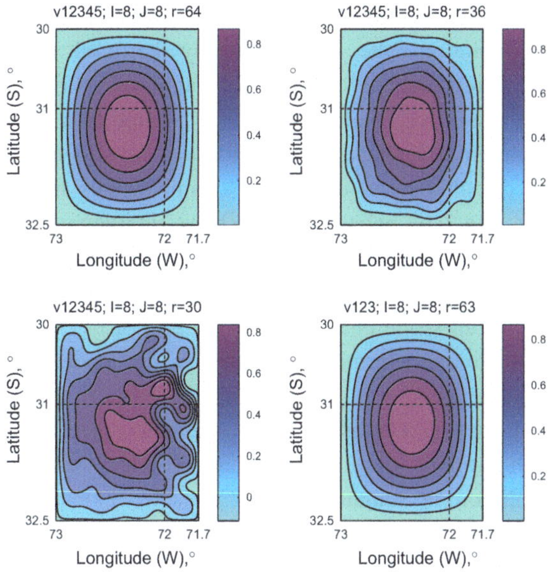

Figure 2
The reconstruction results for the first harmonic. Configuration v12345: (top left) $\varphi_{max} = 1.0$m, $\varphi_{min} = 0.0$m, $r = 64$; (top right): $\varphi_{max} = 1.03$m, $\varphi_{min} = 0.0$m, $r = 36$; (bottom left): $\varphi_{max} = 0.9867$m, $\varphi_{min} = -0.121$m, $r = 30$. Configuration v123 (bottom right): $\varphi_{max} = 1.0$m, $\varphi_{min} = 0$m, $r = 63$

Table 1

The influence of the configuration of observation systems and number ***r*** *on the accuracy of the first harmonic reconstruction*

N	Configuration	P	r	φ_{max}	φ_{min}
1	v12345	5	64	1.00	0.0
2	v12345	5	36	1.03	0.0
3	v12345	5	30	0.9867	– 0.121
4	v123	3	63	1.0	0.0

used in the reconstruction; φ_{max}, φ_{min} are the extreme values of the function recovered (scaled in meters). Table 1 illustrates how the configuration of the observation system in truncated SVD affects the accuracy of the first harmonic reconstruction.

It should be noted that noise-free synthetic data allow us to use a full set of right singular vectors. The cases under consideration differ in the number of synthetic waveforms that were used in the inversion and in the dimension of the subspace where the approximate solution has been projected. In Fig. 2 (top left, top right and bottom left), one can notice that the solution distortion grows when the subspace dimension decreases. In Fig. 2 (bottom left), where $r = 30$, the solution has already been strongly distorted. However, even in this case, the synthetic and recovered marigrams are in good agreement with a difference of about 2% in amplitude.

Comparing the top-left and bottom-right picture of Fig. 2 , we notice that when the full space of the right singular vectors is used, the reconstruction results are equally good for configurations v12345 and v123. Moreover, in the case of v123, there is a good matching between the synthetic and the recovered marigrams, not only in the 1st, 2nd and 3rd receivers (whose marigrams were used in the inversion), but also in the 4th and 5th receivers. This fact is greatly important for applications since once having a reliable receiver configuration for the inversion, one can compute marigrams at the points different from those used in the inversion. In other words, the absence of the waveforms from the 4th and 5th receivers has only insignificant influence on the result of the inversion. This will also be true for the subsequent numerical experiments with real data.

In similar experiments for other harmonics with noise-free data, it was found that each harmonic is reconstructed almost exactly (99.8%) in the space based on a full set of the right singular vectors. However, when a truncated subspace is used for reconstruction, numerical artifacts can appear including a noticeable distortion of the amplitude and a series of "additional" harmonics (mostly high-frequency) like in Fig. 2 (bottom left) for the first harmonic. The higher the frequency is, the larger are the artifacts.

Thus, there should be a sort of trade-off for the dimension of the subspace. On the one hand, in order to have better approximation properties, one would want to increase the number of harmonics. On the other hand, inevitable truncation of the SVD brings about the numerical artifacts. Also, the time required for calculating the matrix **A** depends on the number of harmonics used. Hence, the problem is to decrease the number of harmonics used in the inversion without distorting the form of the source, for the fixed r.

It is obvious that for each specific bathymetry, for each observation system and for each source area, the optimal number of harmonics will be different. Below we present the analysis for the considered setup. First of all, we need to know how the right singular vectors are represented by the spatial harmonics.

Figure 3 shows 64 right singular vectors decomposed to the spatial harmonics. For each right singular vector, we present its coefficients on the basis of spatial harmonics. The harmonics are numbered in the following order:

$$\varphi_k = \varphi_{mn} = \sin\frac{m\pi}{l_1}x \cdot \sin\frac{n\pi}{l_2}y,$$
$$k = n + (m-1)\times 15;\; m = 1,\ldots,15;\; n = 1,\ldots,15,$$

where "m" is the indices of harmonic in latitude and "n" is the indices of harmonic in longitude.

Based on the analysis of the singular values of the matrix $\mathbf{A}^{[r]}$, we know that a reasonable choice of r is $r \leq 40$. This is confirmed by Fig. 3 since we notice that 40 first right singular vectors contain the components corresponding to both the low- and high-frequency harmonics. Using singular vectors with large indices would result in a "dispersion" of the form of the source into numerous high-frequency sea surface oscillations. On the contrary, when we use a small number of the first singular vectors to represent our solution, the high-frequency harmonics are poorly represented for this set of singular vectors, and as a result, we obtain a too "rude" source form with low extreme values. There is a worse coincidence in the calculated and observed tsunami signals in this case, too. The behavior of the right singular vectors presented in Fig. 3 w.r. to the spatial harmonics remains structurally the same even if the dimension of the full space changes, say, to $225 = 15 \cdot 15$. The structure is determined mostly by the underlying bathymetry.

It may also be useful to estimate the projection of each harmonic onto the truncated subspace where the r-solution is sought. To this end, we extend the number of spatial harmonics to $225 = 15 \cdot 15$ and compute the cosine of the angle between each harmonic φ_{mn} and the r solution subspace ($r = 37$) . The results are plotted in Fig. 4. Indices on the curves denote the indices of harmonics in latitude (m). Indices of harmonics in longitude (n) are marked on the horizontal axis. As we can notice, the harmonics with a higher frequency (more than 8) in latitude have smaller angles and low-frequency (in latitude) harmonics have a sharp decrease in the cosine values when the longitude index is higher than 8. The smaller values of cosine mean that the corresponding harmonics are poorly represented within the truncated

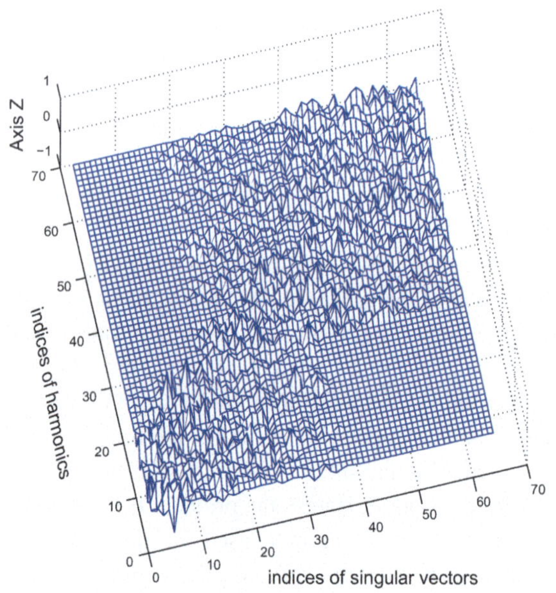

Figure 3
The right singular vectors in terms of spatial harmonics. On the axis X, indices of the right singular vectors; on the axis Y, indices of harmonics $\{\varphi_k\} = \{\varphi_{mn}\}$, $k = n + (m-1) \times J$, $m = 1,\ldots,I$ (in longitude), $n = 1,\ldots,J$ (in latitude); on the axis Z, the values of the components of singular vectors

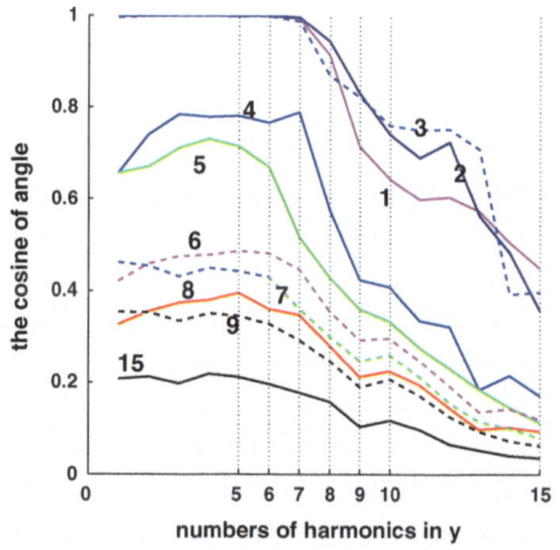

Figure 4
On the vertical axis—the cosine of the angle between each harmonic φ_{mn} and a subspace of the quasi-solution; on the horizontal axis—indices of harmonics in longitude corresponded to variable y (n); indices on the curves denote indices of harmonics in latitude (m). Plots computed for $r = 37$

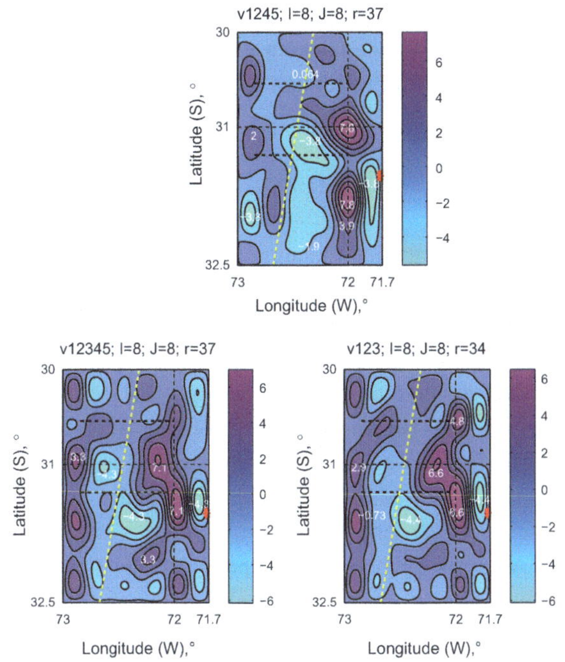

Figure 5
The recovered tsunami sources in the target domain inverted with parameters: $I = 8; J = 8;$ $cond\mathbf{A}^{[r]} = 100$; variants of calculations: v1245 (top); v12345 (bottom left) and v123 (bottom right). The black dotted line indicates the inferred tsunami source area; the yellow dashed line indicates the axis of the trench; the red circle indicates the earthquake epicenter

subspace and thus their contribution in the resulting r-solution would be small, whereas numerical artifacts would be large. Hence, we can conclude that in our case the optimal number of spatial harmonics is close to $64 = 8 \cdot 8$.

Figure 5 shows the tsunami sources recovered with the following variants of observation configurations: v1245 (top) , v12345 (bottom left) and v123 (bottom right) with the parameters: $I = 8,\ J = 8,\ N_t = 1201,\ L = P \times 1201,\ cond\mathbf{A}^{[r]} = 100\}$ in all the cases. Table 2 illustrates how the configuration of the observation system in the truncated SVD affects the accuracy of the tsunami source reconstruction.

Our results of numerical experiments do not contradict the ones obtained by other researchers from seismic data have been mentioned in an overview (Satake and Heidarzadeh 2017). We concluded that the tsunami source area is located, approximately, at 70 km to the NW of the epicenter. In Fig. 5, the black dotted lines indicate the inferred tsunami source area according to Heidarzadeh et al.

Table 2

The influence of the configuration of the observation systems and number $\boldsymbol{r}$ *on the accuracy of a tsunami source reconstruction*

N	Configuration	P	r	$\varphi_{\max}$	$\varphi_{\min}$
1	v1245	4	37	9.67	− 5.7
2	v12345	5	37	9.01	− 6.3
3	v123	3	34	7.4	− 6.2

(2016). The main tsunami uplift is placed near the point (31°S, 72.31°W). However, the amplitudes of the displacement are slightly different from the ones obtained in other studies. In this study, the extreme values vary from 9 m in the positive to − 6.3 m in the negative displacement. It can be assumed that some uplifts (especially, oscillatory), located to the west of the trench, are numerical artifacts and associated with the truncation of SVD. However, these artifacts did not prevent us from attaining good agreement between the observed and computed marigrams, which is one of our main objectives.

We have conducted numerous inversion experiments for the 2015 Chile tsunami with different sets of receivers, and each time the presence of the waveform recorded by the third receiver (DART 32412) significantly improved the result of the inversion. According to Fig. 1 (top), this receiver is located on a more western direction relative to the source area as compared with the first (DART 32402) and second (DART 32401) receivers. The fourth receiver (DART 32411) and the fifth receiver (DART 43413) fix only weak signals and, therefore, their influence on the inversion result is also insignificant. The paramount importance of the western direction for the inversion quality strongly agrees with the energy propagation of the tsunami wave in the considered topography (according to Fig. 5 in Tang et al. 2016). Developing this idea, it seems reasonable to think about the western direction as the "most informative direction" for the tsunami receiver locations in such regions. Moreover, the above facts are in good agreement with the relation between the direction of the largest variability of the form of the tsunami source and the "most informative" location of recorders revealed in Voronin et al. (2015). In the case in question, the western direction is perpendicular to the axis of the Atamakan Trench near Illapel, and, thus, it matches with the largest variability direction of the form of a feasible

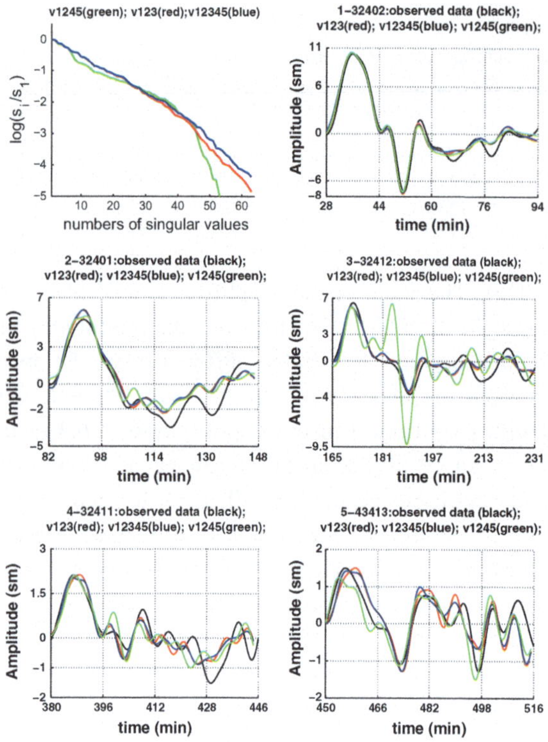

Figure 6
Plots of singular values in the common logarithmic scale of matrix **A** with respect to their indices (upper left) for the following variants: v1245 (green); v12345 (blue); v123 (red). The five other pictures demonstrate the comparison of the observed tsunami waveforms (the black line) with the calculated ones in cases: v12345 (the blue line), v123 (the red line) and v1245 (the green line). The parameters used: $I = 8$; $J = 8$; $cond\mathbf{A}^{[r]} = 100$. The labels on the horizontal axis indicate the time (in minutes) after the earthquake origin time

source. To justify the choice of r and, also, to compare the quality of the observation systems w.r. to the singular spectrum, we present the plots of singular values in log-scale [Fig. 6 (top left)]. The singular spectrum of the matrix $\mathbf{A}^{[r]}$ calculated with observation systems both containing the third receiver (the red and the blue curves) and the non-containing one (the green curve) are plotted in Fig. 6 (top left) [configurations: v12345, v123 and v1245, respectively]. Notice that the green curve corresponding to the observation system without the third receiver has a sharper decrease in singular values thus resulting in a lesser stability in the solution.

Particular attention should be given to the fact that the waveform in the third receiver was not sufficiently restored in the configuration v1245. Meanwhile in case v123 (the observation system consists of the first three receivers), the tsunami waveforms are successfully restored in the 4th and 5th receivers.

4. Conclusion

We have applied the approach based on the SVD and r-solution methods to recovering the initial water elevation in the source area for the Illapel tsunami of 16 September, 2015.

The method proposed suppresses the instability due to the ill-posedness of the problem. The characteristic features of the r-solution obtained are largely determined by the underlying bathymetry and the observation system.

Analysis of the singular spectra of the matrix **A** allows one to compare the efficiency of different observation systems with respect to the result of the destined inversion and, consequently, makes it possible to improve the quality of inversion by selecting the most informative set among the available observation stations.

Based on the numerical experiments, we conclude that the method presented provides a good matching in the observed and calculated tsunami waveforms. Moreover, the algorithm provides a possibility to simultaneously restore the tsunami source and calculate tsunami waveforms even at the points where there are no observational data.

The method seems to be attractive from the computational point of view since the main efforts are required for calculating the matrix **A** which depends on the bathymetry, the temporal extent of the signal, the observation system and the number of harmonics used to represent an unknown source. If an observation system is fixed and tsunami-prone areas are detected, one can compute the matrix only once as a preliminary stage of solving the desired problem. The inversion procedure takes only a few seconds.

The complicated shape of a source requires a larger number of harmonics in the source representation, but this could cause increase of both the time of calculating the matrix **A** as well as appearing the addition oscillation in the solution. The study of singular vectors allows one to obtain the desired balance with respect to the number of harmonics used.

Acknowledgements

The authors thank Artem Loskutov, Ph.D., from the Institute of Marine Geology and Geophysics Far Eastern Branch Russian Academy of Science, for his assistance in the preparation of data, and Aleksei Romanenko, Ph.D., from Novosibirsk State University for his assistance in arranging calculations. The research proposed was made under the State Budget Program no. 0315-2016-0005.

Publisher's Note Springer Nature remains neutral with regard to jurisdictional claims in published maps and institutional affiliations.

REFERENCES

Calisto, I., Miller, M., & Constanzo, I. (2016). Comparison between tsunami signals generated by different source models and the observed data of the Illapel 2015 earthquake. *Pure and Applied Geophysics*, *173*, 1051–1061.

Cheverda, V. A., & Kostin, V. I. (2010). r-Pseudoinverse for compact operator. *Siberian Electronic Mathematical Reports*, *7*, 258–282.

Fuentes, M., Riquelme, S., Hayes, G., Medina, M., Melgar, D., & Vargas, G., et al. (2016). A study of the 2015 Mw 8.3 Illapel earthquake and tsunami: Numerical and analytical approaches. *Pure and Applied Geophysics*, *173*, 1847–1858.

Heidarzadeh, M., Murotani, S., Satake, K., Ishibe, S., & Gusman, A. R. (2016). Source model of the 16 September 2015 Illapel, Chile, Mw 8.4 earthquake based on teleseismic and tsunami data. *Geophysical Research Letters*, *43*, 643–650.

Lee, S.-J., Yeh, T.-Y., Lin, T.-C., Lin, Y.-Y., Song, T.-R. A., & Huang, B.-S. (2016). Two-stage composite megathrust rupture of the 2015 Mw 8.4 Illapel, Chile, earthquake identified by spectral-element inversion of teleseismic waves. *Geophysical Research Letters*, *43*, 4979–4985.

Li, L., Lay, T., Cheung, K. F., & Ye, L. (2016). Joint modeling of teleseismic and tsunami wave observations to constrain the 16 September 2015 Illapel, Chile, Mw 8.3 earthquake rupture process. *Geophysical Research Letters*, *43*, 4303–4312.

Marchuk, A. G., Chubarov, L. B., & Shokin, Y. I. (1983). *The numerical simulation of the tsunami wave* (p. 176). Novosibirsk: Nauka. (in Russian).

Melgar, D., Fan, W., Riquelme, S., Geng, J., Liang, C., & Fuentes, M. (2016). Slip segmentation and slow rupture to the trench during the 2015, Mw 8.3 Illapel, Chile earthquake. *Geophysical Research Letters*, *43*, 961–966.

Mulia, I. E., & Asano, T. (2016). Initial tsunami source estimation by inversion with an intelligent selection of model parameters and time delays. *Journal of Geophysical Research: Oceans*, *121*, 441–456.

Percival, D. B., Denbo, D. W., Eble, M. C., Gica, E., Mofjeld, H. O., Spillane, M. C., et al. (2011). Extraction of tsunami source coefficients via inversion of DARTr buoy data. *Natural Hazards*, *58*(1), 567–590.

Piatanesi, A., Tinti, S., & Pagnoni, G. (2001). Tsunami waveform inversion by numerical finite-elements Green's functions. *Natural Hazards and Earth System Sciences*, *1*, 187–194.

Pires, C., & Miranda, P. M. A. (2001). Tsunami waveform inversion by adjoint methods. *Journal of Geophysical Research: Oceans*, *106*, 19773–19796.

Romano, F., Piatanesi, A., Lorito, S., Tolomei, C., Atzori, S., & Murphy, S. (2016). Optimal time alignment of tide-gauge tsunami waveforms in nonlinear inversions: Application to the 2015 Illapel (Chile) earthquake. *Geophysical Research Letters*, *43*, 11226–11235.

Ruiz, S., Klein, E., del Campo, F., Rivera, E., Poli, P., Metois, M., et al. (2016). The seismic sequence of the16 September 2015 Mw 8.3 Illapel, Chile, earthquake. *Seismological Research Letters*, *87*, 789–799.

Satake, K. (1989). Inversion of tsunami waveforms for the estimation of heterogeneous fault motion of large submarine earthquakes: The 1968 Tokachi-oki and the 1983 Japan sea earthquake. *Journal of Geophysical Research: Oceans*, *94*, 5627–5636.

Satake, K., Fujii, Y., Harada, T., & Namegaya, Y. (2013). Time and space distribution of coseismic slip of the 2011 Tohoku earthquake as inferred from tsunami waveform data. *Bulletin of the Seismological Society of America*, *103*(2B), 1473–1492.

Satake, K., & Heidarzadeh, M. A. (2017). Review of source models of the 2015 Illapel, Chile earthquake and insights from tsunami data. *Pure and Applied Geophysics*, *174*(1), 1–9.

Tang, L., Titov, V. V., Moore, C., & Wei, Y. (2016). Real-time assessment of the 16 September 2015 Chile tsunami and implications for near-field forecast. *Pure and Applied Geophysics*, *173*, 369–387.

Tinti, S., & Tonini, R. (2013). The UBO-TSUFD tsunami inundation model: Validation and application to a tsunami case study focused on the city of Catania. *Italy Natural Hazards and Earth System Sciences*, *13*, 1795–1816.

Titov, V. V., Gonzalez, F. I., Bernard, E. N., Eble, M. C., Mofjeld, H. O., Newman, J. C., et al. (2005). Real-time tsunami forecasting: Challenges and solutions. *Natural Hazards*, *35*(1), 41–58.

Tilmann, F., Zhang, Y., Moreno, M., Saul, J., Eckelmann, F., & Palo, M. (2016). The 2015 Illapel earthquake, central Chile: A type case for a characteristic earthquake? *Geophysical Research Letters*, *43*, 574–583.

Voronina, T. A. (2012). Reconstruction of initial tsunami waveforms by a truncated SVD method. *Journal of Inverse and Ill-Posed Problems*, *19*, 615–629.

Voronina, T. A. (2016). Recovering a tsunami source and designing an observational system based on an r-solution method. *Numerical Analysis and Applications*, *9*(4), 267–276.

Voronina, T. A., & Romanenko, A. A. (2016). The new method of tsunami source reconstruction with r-solution inversion method. *Pure and Applied Geophysics*, *173*(12), 4089–4099.

Voronin, V. V., Voronina, T. A., & Tcheverda, V. A. (2015). Inversion method for initial tsunami waveform reconstruction. *Natural Hazards and Earth System Sciences*, *15*, 1251–1263.

(Received July 6, 2018, revised January 3, 2019, accepted January 11, 2019, Published online January 29, 2019)

Pure Appl. Geophys. 176 (2019), 2995–3033

https://doi.org/10.1007/s00024-018-1996-5

Pure and Applied Geophysics

Source Models of the 2012 Haida Gwaii (Canada) and 2015 Illapel (Chile) Earthquakes and Numerical Simulations of Related Tsunamis

SEDA YOLSAL-ÇEVIKBILEN,[1] ERGIN ULUTAŞ,[2] and TUNCAY TAYMAZ[1]

Abstract—Seismological observations provide essential input parameters for numerical tsunami simulations. Here, we present source mechanism parameters, finite-fault source rupture models and numerical tsunami simulation results for the destructive October 28, 2012 Haida Gwaii-Canada (M_w 7.7) and September 16, 2015 Illapel-Chile (M_w 8.3) earthquakes and resulting tsunamis. These two earthquakes were controlled by active tectonic features along the subduction zones that had developed in response to the convergent movements of lithospheric plates. The faulting geometry (strike, dip, and rake angles), focal depth, fault dimensions, average and maximum slip values on the fault planes and seismic moments of the earthquakes are estimated by analyzing teleseismic long-period P- and SH-waves and broadband P-waveforms and using waveform inversion and hybrid back-projection methods. The obtained slip models of the earthquakes reveal heterogeneous slip distributions on fault planes with long source durations ($\sim$ 80 s and 150 s) and low stress drop values (10–15 bars). Numerical simulations of tsunami wave propagation are further performed using the uniform and non-uniform slip models and nonlinear long-wave equations in spherical coordinates. The shape and arrival times of leading tsunami waves are adequately constrained particularly with the heterogeneous slip distribution models. The general characteristics of synthetic tsunami waveforms (e.g., amplitude, shape, arrival time) calculated using the non-uniform slip model, are more consistent with the observed tsunami records than those of a uniform slip model. It is further seen that simulation results using preliminary and fast slip models for both earthquakes give only approximate early tsunami estimates; tsunami wave heights and arrival times to the coasts are mostly not well simulated. The results indicate that tsunami simulations based on finite-fault source slip models likely contribute to the determination of tsunamigenic coastal regions by revealing locations, arrival times, amplitudes, and directions of tsunami waves within a close approximation to observed records off-shore and far from the source region. They provide sufficient information to facilitate tsunami warning and mitigation challenges after the destructive earthquakes. We further suggest that joint inversions of GPS, tsunami, teleseismic and strong ground motion records and higher resolution bathymetry data are needed in order to obtain better correlations between observed and synthetic tsunami data, especially for the later arriving waves.

Key words: Earthquake source parameters, finite-fault slip distribution, hybrid back-projection, numerical tsunami simulation.

1. Introduction

Earthquake-induced tsunamis, which typically have wavelengths of several hundreds of kilometers, are long and rapidly moving ocean waves caused by a dislocation of the ocean floor. It is known that a large amount of water mass above the slip area can be vertically displaced when an earthquake occurs beneath the sea. The computation of major vertical dislocations on the ocean floor is needed for estimating the initial height of tsunami waves in numerical wave simulations. It is expected that the vertical displacement on the fault plane of the earthquake mimics the ocean bottom displacement, causing the generation of initial tsunami waves. The accurate prediction of maximum tsunami wave heights and arrival times to the coastal areas mainly rely on the robust estimation of initial wave height. Seismological observations from earthquake source studies and finite-fault slip distribution modeling methods can be used to derive the displacement on the fault plane of an earthquake. Most numerical tsunami models are based on shallow water equations derived from the principles of conservation of mass and conservation of momentum. In the theory of shallow water waves, the vertical acceleration of water particles is ignored compared to the gravitational acceleration (Kajiura 1963). The pressure is assumed to be hydrostatic (Imamura et al. 1988), and

[1] Department of Geophysical Engineering, The Faculty of Mines, İstanbul Technical University, Maslak, 34469 Istanbul, Turkey. E-mail: yolsalse@itu.edu.tr; taymaz@itu.edu.tr

[2] Department of Geophysical Engineering, Faculty of Engineering, Kocaeli University, 41380 Kocaeli, Turkey. E-mail: ergin@kocaeli.edu.tr

the vertical motion of water particles has no effect on the pressure distribution. The finite difference, finite element or finite volume methods are widely used to compute the shallow water equations in the context of tsunami modeling (e.g., Shuto and Goto 1978; Imamura 1996; Titov and González 1997; Franchello 2010; Piatanesi et al. 2001; Yalçıner et al. 2004; George and LeVeque 2006; Lorito et al. 2008; Yolsal-Çevikbilen and Taymaz 2012). The slip/rupture distribution pattern throughout the faulting area is critical in calculating the vertical seafloor deformation in response to an earthquake (Geist and Dmowska 1999; Geist 2002). A simplified elastic dislocation model constrained by a uniformly distributed slip is commonly used in numerical tsunami simulations. However, seismic, geodetic, and joint inversion methods indicate that the earthquakes rupture heterogeneously over the entire faulting area (Satake 1987; Ammon et al. 2011; Yagi et al. 2012a, b; Gica et al. 2015 among many others). Following earthquakes, within a short period of time, preliminary slip distribution models are reported by several institutions (i.e., USGS-United States Geological Survey), but they are rapid and operational solutions. Fine et al. (2015) observed that although the amplitudes and periods of synthetic tsunami waves obtained from numerical simulations based on USGS preliminary slip model of the 2012 Haida Gwaii earthquake are consistent with the observed data, the arrival times deviate considerably from the observed times, suggesting that the actual source location differs from the initial location provided by USGS (2012a). Calisto et al. (2016) also suggested that preliminary source models produce usable initial tsunami estimates; however, they mostly underestimate the actual wave heights along the coasts. Similarly, Fuentes et al. (2016) showed that rapid estimations of slip models cannot reproduce the observed run-up values; the most accurate match to run-up observations comes from the joint inversions of multiple types of data, such as tide gauges, InSAR and GPS, providing improved azimuthal coverage.

The main aim of this study is to utilize detailed seismological models to obtain reasonable tsunami simulation results. For this purpose, we analyzed two large ($M_w \geq 7.5$) subduction zone earthquakes with thrust faulting mechanisms: the 2012 Haida Gwaii (British Columbia, Canada) and 2015 Illapel (Chile) earthquakes (Fig. 1a, b). Source mechanism parameters and finite-fault slip distribution models of those earthquakes are estimated by using conventional teleseismic P- and SH-waveform inversion and hybrid back-projection techniques and by inverting accurately selected waveforms, which show good azimuthal coverages on focal spheres to improve the resolution of results. Then, we numerically simulated propagation of tsunami waves to estimate the arrival times and maximum heights of tsunami waves along the coasts. The synthetic waveforms are compared with those coming from the observed tide gauge and Deep-ocean Assessment and Reporting of Tsunami (DART) records. This study also provides new findings for comparisons of tsunami simulations based on uniform and non-uniform earthquake slip models. In addition, we showed the simulation results obtained based on fast and preliminary slip models provided by USGS (2012b, 2015b) to check the effect of slip distribution models on tsunami calculations. Schematic diagrams of possible structures in the study areas are shown in Fig. 1a, b.

1.1. The October 28, 2012 M_w 7.7 Haida Gwaii (British Columbia, Canada) Earthquake

A major earthquake occurred on October 28, 2012 at 03:04:06.30 UTC in western British Columbia, Canada. This earthquake, with a moment magnitude of 7.7 and a shallow focal depth of 21 km, is the second largest earthquake ever instrumentally recorded in Canada. The epicenter (Lat. 52.622°N and Lon. 132.103°W) is located just west of Moresby Island, the southern end of Haida Gwaii (Fig. 1a; Kao et al. 2015). Despite its large magnitude, the observed damage was limited due to the long distances (more than 60–80 km) between the earthquake epicenter and the nearest communities (Bird and Lamontagne 2015). The tsunami spawned by this sizable earthquake prompted a local tsunami warning in the region. The impact of the tsunami was directly measured from watermarks on trees, buildings and walls, and was also recorded at several DART stations and numerous tide gauges on the coasts of the Pacific Ocean (NGDC-NOAA 2012). Tsunami run-up heights exceeding 13 m in a small inlet on the

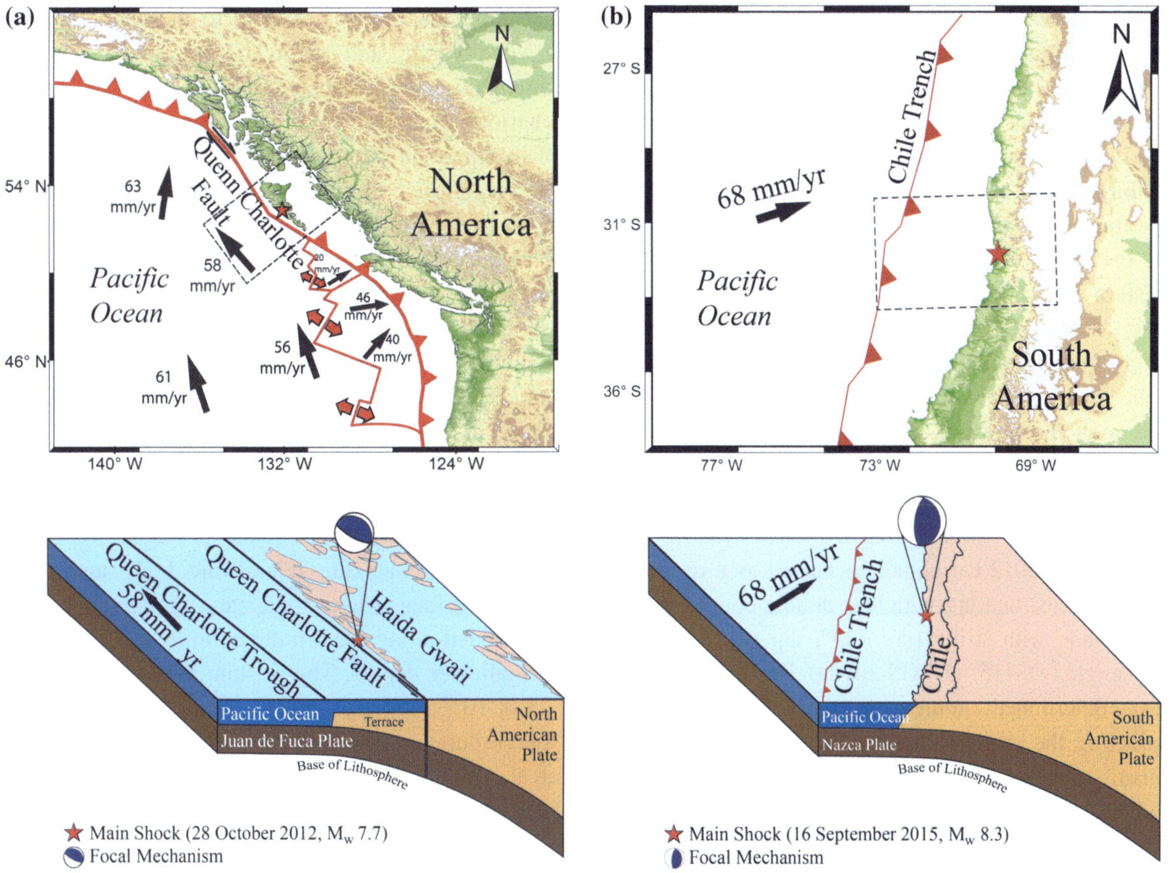

Figure 1
Tectonic sketch map of active boundaries along **a** the Pacific margins of the North American Plate (near Canada), and **b** the South American Plate (near Chile) superimposed on GEBCO30 bathymetry with a schematic diagram of possible structures across the subduction zones associated with these plates (compiled from Hyndman and Ellis 1981; Bostwick 1984; Angermann et al. 1999; Vigny et al. 2009). Source mechanism solutions of the October 28, 2012 Haida Gwaii-Canada (M_w 7.7) and the September 16, 2015 Illapel-Chile (M_w 8.3) earthquakes, determined in this study, are also presented

west coast of Haida Gwaii were reported by Leonard and Bednarski (2014, 2015). The largest tide gauge recorded peak-to-trough wave heights after this earthquake were 1.52 m at Maui, Hawaii area, and 0.52 m at Henslung Cove, Langara Island. Approximately 40 min after the main shock, the tsunami wave was first recorded at a tsunameter/DART 46419 with a 7 cm maximum height (NGDC-NOAA 2012). Although the recorded waves were relatively small, it is important to simulate the waves at the coastal sites near the source and in the far field. The proposed source models could be useful for understanding the initializations and propagation of tsunamis in the region.

The origin of earthquakes in coastal British Columbia are mainly associated with the subduction of the Pacific and Juan de Fuca Plates beneath the northerly portion of the western side of the North American Plate (Fig. 1a). In this region, where there have been several large earthquakes in the past (Rogers 1983), the Pacific Plate moves approximately north at a rate of 58 mm/year (Riddihough 1977; Hyndman and Ellis 1981; Lay et al. 2013; Hyndman 2015). In 1949, the epicenter of the largest instrumentally recorded earthquake in Canada (M_s 8.1) was located at the middle of the Queen Charlotte Fault. It had a mainly strike-slip transform fault motion, so the rupture displacement produced only a small tsunami

(Rogers 1983; Bostwick 1984; Cassidy et al. 2014). In addition, a moderate M_w = 6.1 thrust earthquake that took place near the southern portion of the Queen Charlotte Fault in 2001 resulted in a tsunami with maximum recorded amplitudes of up to 23 cm (Rabinovich et al. 2008). The Queen Charlotte Fault is a transform fault that marks the boundary between the Pacific Plate and North American Plate along the western edge of central British Columbia and southeast Alaska. Current motion along the fault, as determined from global plate motions, is primarily right-lateral strike-slip (Minster and Jordan 1978). Although the previous tectonic studies suggest that plate motions in this region are primarily related to strike-slip faulting parallel to the plate boundary (Hyndman and Ellis 1981; Bostwick 1984), the October 28, 2012 earthquake occurred as a result of oblique thrust faulting with best double-couple of the fault plane with strike (ϕ): 323°, dip (δ): 25°, and rake angle (λ): 115° (USGS 2012a). Kao et al. (2015) also determined source mechanism parameters reflecting the shallow dipping thrust faulting mechanism (strike: 311°, dip: 17°, rake: 121°, h: 21 km, and M_0: 1.526 $\times$ 10^{20} Nm) from the moment tensor inversion. The origin of the faulting type could be deduced from the convergence associated with underthrusting that formed a trough or trench off the west coast of Haida Gwaii region (Riddihough 1977). The convergence could stem from the oblique subduction of the Pacific plate or compressive deformation of the study region, or a combination of both (Perez and Jacob 1980). The angle of the oblique interaction between Pacific and North American plates is small at the northern end of the Queen Charlotte Fault, but increases towards the southern end (Bostwick 1984). Hyndman and Ellis (1981) proposed a model that the main Queen Charlotte strike-slip fault may cut what was formerly North American lithosphere, so the terrace could be underlain by continental crust. According to this model, the component of convergence is implemented either by periodic unlocking and down-dip motion on the upper portion of the thrust, or by compressive deformation and shortening of the whole oceanic lithosphere under the terrace and the terrace sediments exhibit uplift. Wang et al. (2015) reported that the 2012 Haida Gwaii earthquake is an example of a megathrust earthquake related to the subduction of a very young and warm oceanic plate and the kinematic change of plate motions across the Queen Charlotte Fault coincides with a down-dip transition of the thermally controlled seismogenic behavior of the megathrust. Gosselin et al. (2015) presented the first evidence for the extension of subducting oceanic plate beneath the entirety of Moresby Island from the receiver function study. They obtained a positive velocity contrast at about 18–26 km depth indicating a shallow continental Moho, and a low shear-wave velocity zone about 12–17 km thick along the direction of plate convergence interpreted as subducting oceanic material.

The preliminary finite-fault slip model for the 2012 Haida Gwaii earthquake provided by USGS (2012b) by analyzing teleseismic body and long-period surface waves indicates a rupture mainly propagated along the dip direction of the fault plane at shallower depths. The seismic moment release of 7.3 $\times$ 10^{20} Nm, maximum slip of about 6 m and source duration of about 95 s are also reported. Lay et al. (2013) obtained the best double couple with strike 317.1°, dip 18.5° and rake 103.3° angles with a centroid depth of 9.5 km, maximum slip of 7.7 m, average slip of 3.3 m and seismic moment of 6.9 $\times$ 10^{20} Nm for this earthquake by inverting long-period W-phase signals. They showed a rupture expanded from the hypocenter at about 2.3 km/s along a fault length of about 150 km with an up-dip directed slip from the 15.9 km deep hypocenter below the western part of the region.

1.2. The September 16, 2015 M_w 8.3 Illapel (Chile) Earthquake

The September 16, 2015 earthquake occurred at 22:54:32 UTC near the coastal area of north central Chile (Lat. 31.573°S, Lon. 71.674°W) with a moment magnitude (M_w) of 8.3 and a focal depth of 22.4 km. The epicenter of the main shock was 46 km west of Illapel city (USGS 2015a; Fig. 1b). Many aftershocks with a maximum magnitude of M_w 7.0 followed this earthquake. It generated tsunami waves that were responsible for local damage observed in several areas such as Coquimbo, La Serena, Tongoy, Concon, and Illapel in Chile along the Pacific region. The

Pacific Tsunami Warning Center issued a Tsunami Warning Message for a large part of the entire Pacific coasts, and then "Red Alert" was issued for the coastal regions of Chile (JRC 2015). The earthquake triggered a significant tsunami immediately after the mainshock, with tsunami waves arriving in a short time to the coastal areas of Chile (Bahlburg et al. 2017).

The epicenter of the 2015 Illapel (Chile) earthquake is located near a major convergent zone, where the oceanic Nazca Plate subducts beneath the South American Plate (Fig. 1b) with a velocity of about 70 mm/year (Angermann et al. 1999; De Mets et al. 2010). The Peru–Chile subduction zone hosted several great tsunamigenic earthquakes, including the November 11, 1922 (M_s 8.3), the April 6, 1943 (M_s 7.9) Illapel (central Chile), the May 22, 1960 (M 9.5) Valdivia (southern Chile), the February 27, 2010 (M_w 8.8) Maule (Chile) and the April 01, 2014 (M_w 8.2) Iquique (northern Chile) earthquakes (i.e., Araujo and Suárez 1994; Beck et al. 1998; Lemoine et al. 2002). The initial dip angle of the zone is reported to be 16°± 2° with the dip angle increasing down to 80 km depth (Tichelaar and Ruff 1991). Beck et al. (1998) suggested that most of the destructive earthquakes (e.g., 1922 and 1943 events) in the region are related to underthrusting of the Nazca Plate beneath the South American Plate along the central Chile subduction zone, and they have mostly shallow focal depths. Source characteristics of the 2015 Illapel (Chile) earthquake have been analyzed by several authors (e.g., Heidarzadeh et al. 2015; Aranguiz et al. 2016; Calisto et al. 2016; Fuentes et al. 2016; Melgar et al. 2016; Okuwaki et al. 2016; Omira et al. 2016; Ruiz et al. 2016; Tilmann et al. 2016; Ye et al. 2016; Yin et al. 2016; Chunyan et al. 2017; Klein et al. 2017; Satake and Heidarzadeh 2017). USGS (2015b) reported a preliminary slip model that shows a rupture concentrated in the shallower part of the subduction interface, with a seismic moment release of 3.2×10^{21} Nm and a maximum slip of about 8 m. Heidarzadeh et al. (2015) showed a large-slip area located about 70 km to the northwest of the earthquake epicenter, with a size of 80 km (along strike) × 100 km (along dip) by inversion of teleseismic P-waveforms. They also found a seismic moment of 4.42×10^{21} Nm, an average slip of 5.0 m, moment magnitude (M_w) of 8.4, and a focal depth of 12–33 km for this earthquake. However, Melgar et al. (2016) determined two large slip patches on the fault plane with ~ 10 m peak slip at ~ 30 km and 15 km depths, respectively. Okuwaki et al. (2016) found that a maximum slip of 10.4 m in the shallow part of the seismic source region caused the devastating tsunami waves. Aranguiz et al. (2016) calculated run-up values exceeding 6 m at several locations with the maximum run-up of 10.8 m at Totoral (Chile), a small bay in the tsunami source area. They observed an asymmetric variation of tsunami wave heights to the north and south of the earthquake epicenter and considerably varied run-up values and phase of the maximum waves along the coasts. Additionally, Chunyan et al. (2017) proposed that a co-seismic slip mainly concentrated in a shallow area to the northwest of the earthquake source through the modelling of InSAR data. They calculated a rupture length of 340 km along the strike with a maximum slip of about 8.16 m.

2. *Data and Methods*

2.1. *Seismic Source Inversion*

In this study, we initially performed double-couple point-source inversions using teleseismic (epicentral distances between 30° and 90°) long-period P- and SH-waves and broadband P-waveforms and the moment tensor algorithm (MT5) provided by McCaffrey et al. (1991) and Zwick et al. (1994). Earthquake data are recorded by the International Federation of Digital Seismograph Networks (FDSN) and the Global Digital Seismograph Network (GDSN) stations (Fig. 2) and archived and distributed by IRIS Consortium (*Incorporated Research Institutions for Seismology*). The MT5 algorithm inverts P- and SH-waveforms to find strike, dip and rake angles, focal depth, seismic moment and source time function of an earthquake. In this study, the P-wave first motion polarities recorded by teleseismic and near field stations is further used to confirm the nodal planes on focal spheres. Uncertainties are estimated using the inversion routine to carry out a variety of experiments (Taymaz and Price 1992), and the error

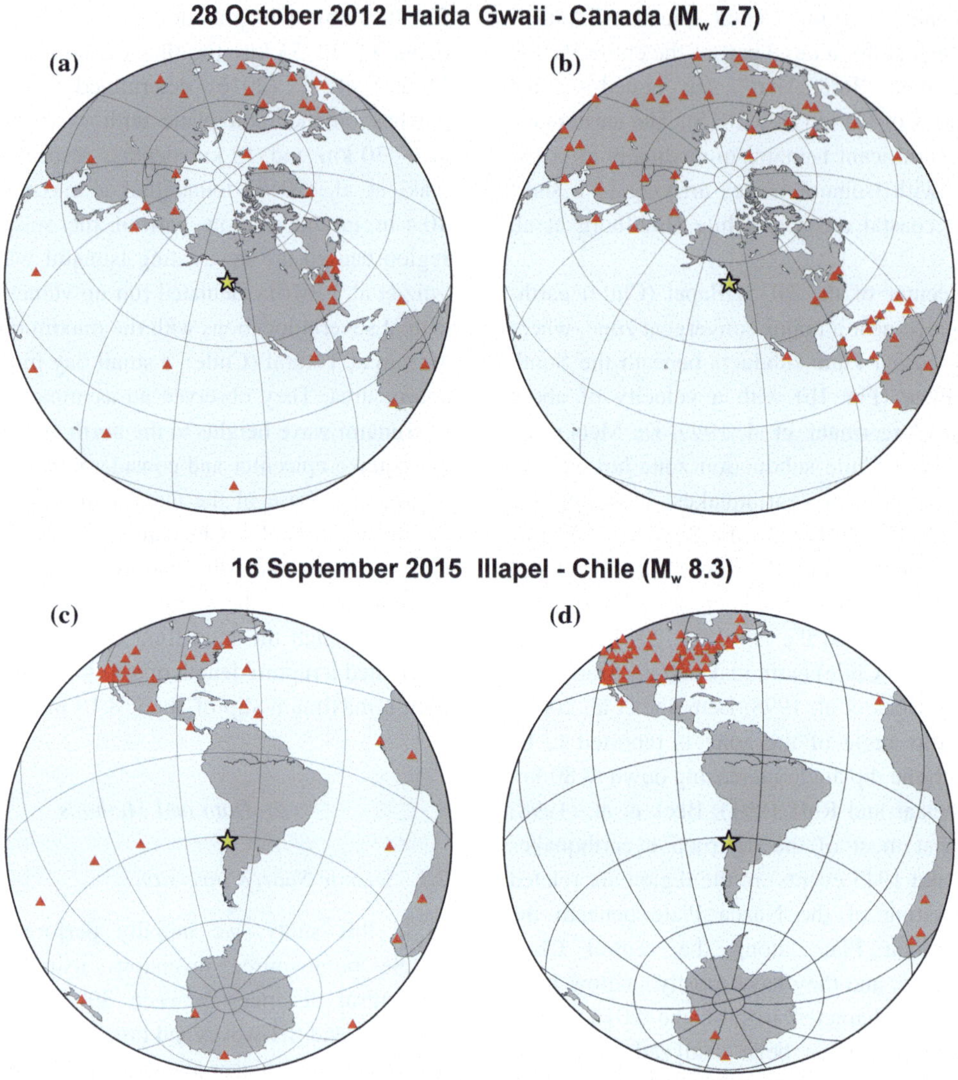

Figure 2
Distribution of teleseismic stations (red triangles) used in **a**, **c** teleseismic P- and SH-waves and **b**, **d** finite-fault slip inversions for the 2012 Haida Gwaii (Canada) and 2015 Illapel (Chile) earthquakes. The yellow stars indicate the earthquake epicenters

limits of strike, dip and rake angles are found as ± 5° and of focal depths as ± 2 km. The accurate and stable focal depth of earthquakes can be specifically constrained by the MT5 algorithm, and then compared to focal depths reported by the fast and automatic moment tensor solutions, such as Harvard-CMT and USGS-NEIC. These catalogues use long-period and mantle waves for obtaining source parameters, and restrict the crust centroid depths to be at least 15 km to minimize the instabilities in the moment tensor components (Arvidsson and Ekström 1998). Those authors stated that long period and intermediate period seismograms are used in two frequency bands in the standard CMT analysis. For moderate earthquakes ($M_w \leq 6.0$), complete body waves arriving before the fundamental-mode surface wave are inverted with a low-pass cut-off period of 45 s; however, the longer-period surface waves ($T > 135$ s) are used in analysis for large earthquakes ($M_w > 6.0$). Furthermore, in the modified CMT analysis, the first arriving surface waves (e.g., G1, R1) are filtered with a low-pass cutoff period (40–60 s) before inversions. These waves have not been used in previous CMT analysis due to their substantial travel-time delays and their sensitivity to the highly heterogeneous structure of the lithosphere and crust. The waveform types (surface wave or body wave) are quite important for inversions, since the frequency content, ray path corrections and attenuation parameters affect the source and focal depth estimations. This is the main reason for the different source mechanism parameters reported by different moment tensor catalogues. Therefore, the re-assessments of source mechanism solutions with carefully selected good-quality data, obtained from azimuthally distributed seismic stations, and a different methodology based on appropriate seismic phases are mostly needed.

In addition, the high-frequency hybrid back (HBP) projection method of Yagi and Fukahata (2011) and Yagi et al. (2012a, b) is used for estimating the spatio-temporal distribution of the seismic energy release of earthquakes on a fault plane which was divided into $M \times N$ sub-faults with length dx and width dy. The dimensions of each sub-fault are selected after several numerical inversion trials for constructing the fault plane. It is also assumed that faulting occurs on a single fault plane, and the rake angle remains free to change during the rupture. This method employs teleseismic broadband P-waveforms to integrate the direct P-phase with reflected phases from structural discontinuities near the source. The 1-D radial Earth model (*PREM*—Preliminary Reference Earth Model; http://www.iris.edu/dms/products/emc/models/PREM.htm) established by Dziewonski and Anderson (1981) is used in slip inversions. In future source studies, enhanced velocity-depth (Earth) models that can be resolved after multi-scale full waveform modeling (Fichtner et al. 2013) are important. We determine the source rupture parameters such as faulting area, maximum and average slip values, seismic moment, stress drop of the earthquakes, and also estimate how the rupture propagates on the fault plane. In this study, we choose the HBP method (Yagi et al. 2012a; Okuwaki et al. 2014) for obtaining finite-fault slip distributions since it ensures better resolution at depth by using the differences of travel times and phase amplitudes between the P-phase and depth phases (pP and sP), extracted by cross correlation of observed waveforms and the theoretically calculated Green's functions. Yagi et al. (2012a) suggested that this algorithm provides higher resolution images of seismic energy release than the conventional back-projection methods that produced more blurred images; hence it gives the more robust distribution of the high-frequency radiation with the detailed motion of rupture front and frequency dependence. It is known that the high-frequency radiation reflects the complex rupture process (e.g., Spudich and Frazer 1984). The HBP method also differs from waveform inversion in that it uses a deconvolution process to mitigate the effect of subsurface structure. Another advantage of the HBP method is that it allows estimation of a seismic energy distribution without the non-negative and smooth constraints that apply to waveform inversions. By using the cross-correlation of observed waveforms and Green's functions, the HBP method can mitigate the effects of subsurface structure and thus model the distinct rupture episodes of megathrust earthquakes (Yagi and Fukahata 2011; Yagi et al. 2012a). Before slip inversions, the observed teleseismic P-waveforms (vertical components) are aligned on the basis of their first arrival time, and then converted into displacement waveforms (Yagi et al. 2012a, b). Average slip on the fault plane and stress drop ($\Delta\sigma$) values of earthquakes are estimated using the equations of Aki (1972) and Kanamori (1994).

$$\text{Seismic Moment } (M_o) = \mu \times S \times D$$

$$\Delta\sigma = \frac{7 \times \pi^{3/2}}{16} \times \frac{Mo}{S^{3/2}},$$

where μ (crustal rigidity) = 3×10^{11} dyne/cm^2; S (faulting area) = $L \times W$ (L = fault length; W = fault width); D = average slip, $\Delta\sigma$ = stress drop.

The other details can be found in the relevant articles of Taymaz et al. (1990), Yagi and Kikuchi (2000), Fukahata et al. (2003), Yagi et al. (2012a, b), Yolsal-Çevikbilen and Taymaz (2012), Okuwaki et al. (2014, 2016) and Saltogianni et al. (2016).

2.2. *Numerical Tsunami Simulations*

Numerical tsunami simulations are performed using the SWAN code developed by Mader (1988), which solves incompressible non-linear long wave

Table 1

Source mechanism parameters of the October 28, 2012 Haida Gwaii (Canada) and September 16, 2015 Illapel (Chile) earthquakes obtained from teleseismic P- and SH-waveform modeling and finite-fault slip distribution inversions

Earthquake parameters	October 28, 2012 Haida Gwaii (Canada) earthquake	September 16, 2015 Illapel (Chile) earthquake
Strike/dip/rake (°)	**NP1: 326°/25°/100°** NP2: 135°/65°/85°	**NP1: 350°/25°/80°** NP2: 181°/65°/95°
Focal depth (km)	20	26
Seismic moment (Nm) (P-SH/finite-fault)	$4.33 \times 10^{20}/4.08 \times 10^{20}$	$7.89 \times 10^{20}/4.06 \times 10^{21}$
M_w (P-SH/finite-fault)	7.7/7.7	7.9/8.3
Fault length (km)	160	350
Fault width (km)	60	100
D_{max} (m)	4.30	6.5
D_{av} (m)	1.41	3.86
$\Delta\sigma$ (bar)	11	15.5

The fault plane (NP1) is shown in bold

NP nodal plane, M_w moment magnitude, D_{max} maximum slip, D_{av} average slip, $\Delta\sigma$ stress drop

(shallow water) equations of the fluid flow with a finite difference scheme in time and space including Coriolis and frictional effects. The shallow water equations proposed by Mader (1988) are implemented into the SWAN code (Annunziato 2007). The static vertical displacement initiated by an earthquake source is computed from the planar fault model of Okada (1985). This model defines the kinematic seafloor deformation with time varying sub-fault contributions, and provides seafloor displacements as an uplift and subsidence due to a specific finite-fault slip inversion.

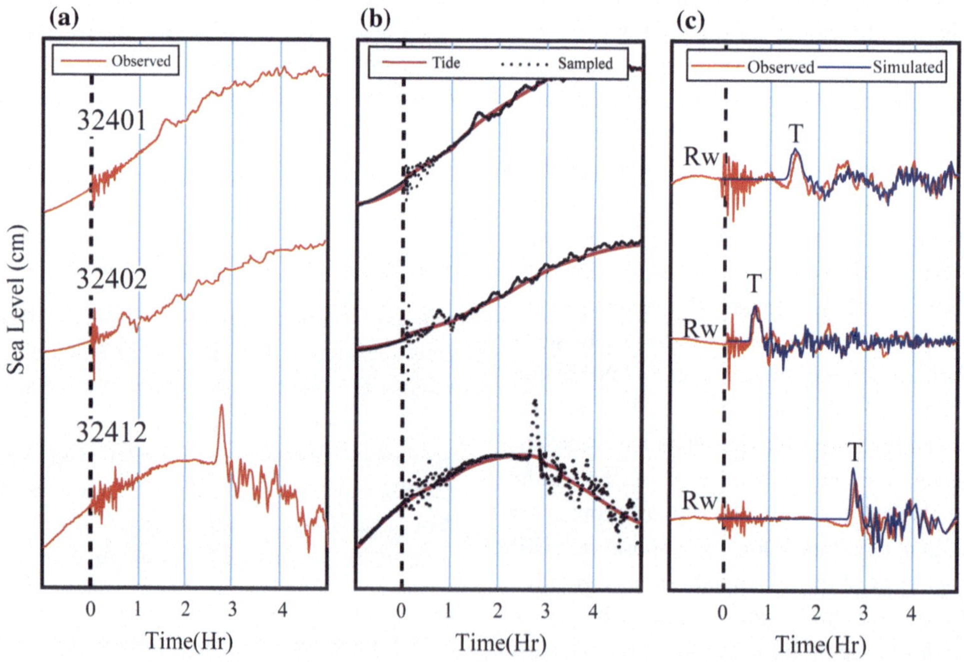

Figure 3
Example tsunami signals recorded by three DART stations in the SE Pacific for the September 16, 2015 Illapel, Chile (M_w 8.3) earthquake demonstrating tsunami simulation steps of fitting and filtering. **a** Observed water surface elevations at the three DART stations; **b** fitting by polynomial functions on the records; **c** comparison of observed (red) and simulated (blue) tsunami waves. Rw and T stand for the earthquake Rayleigh waves and the tsunami arrival time, respectively

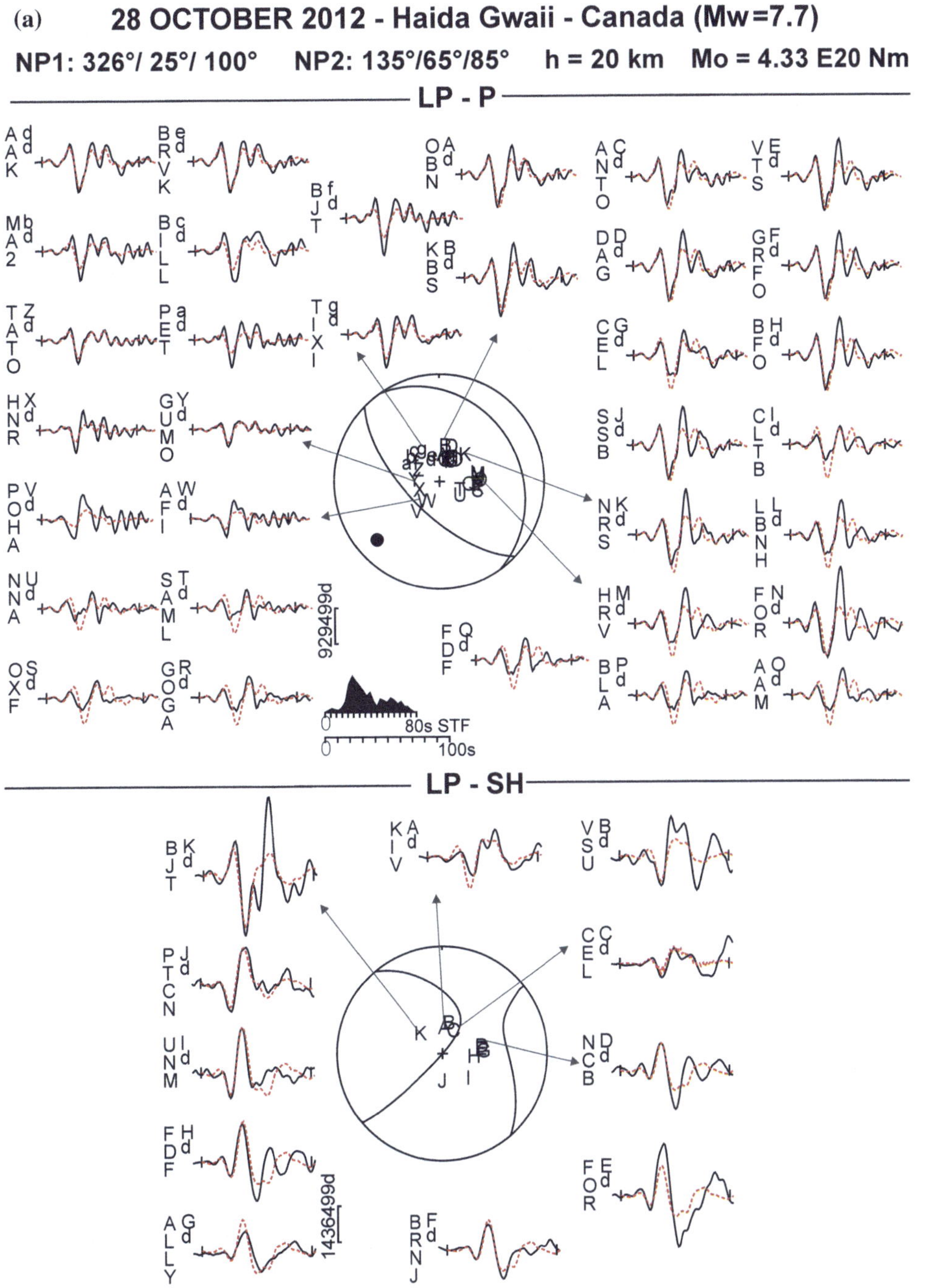

Figure 4

a The best-fitting minimum misfit solution of the October 28, 2012 Haida Gwaii (Canada) earthquake. Comparison of the long-period (LP) observed (solid black lines) and synthetic (dashed red lines) waveforms (33 P- and 11 SH-waves). Strike, dip, and rake angles of the first and second nodal planes (NP), focal depth (*h*), and seismic moment (M_o) of the earthquake are indicated in the header. The source time function (STF) and the time scale used for the waveforms are given below the P-focal sphere; P- and *T*-axes are denoted by the solid and open circles, respectively. International station codes are indicated by letters; **b** distributions of the P-wave first motion polarities of the October 28, 2012 Haida Gwaii (Canada) earthquake. The filled and open circles indicate the up and down P-wave first motions, respectively. The station codes and epicentral distances (Δ) are shown near the waveforms

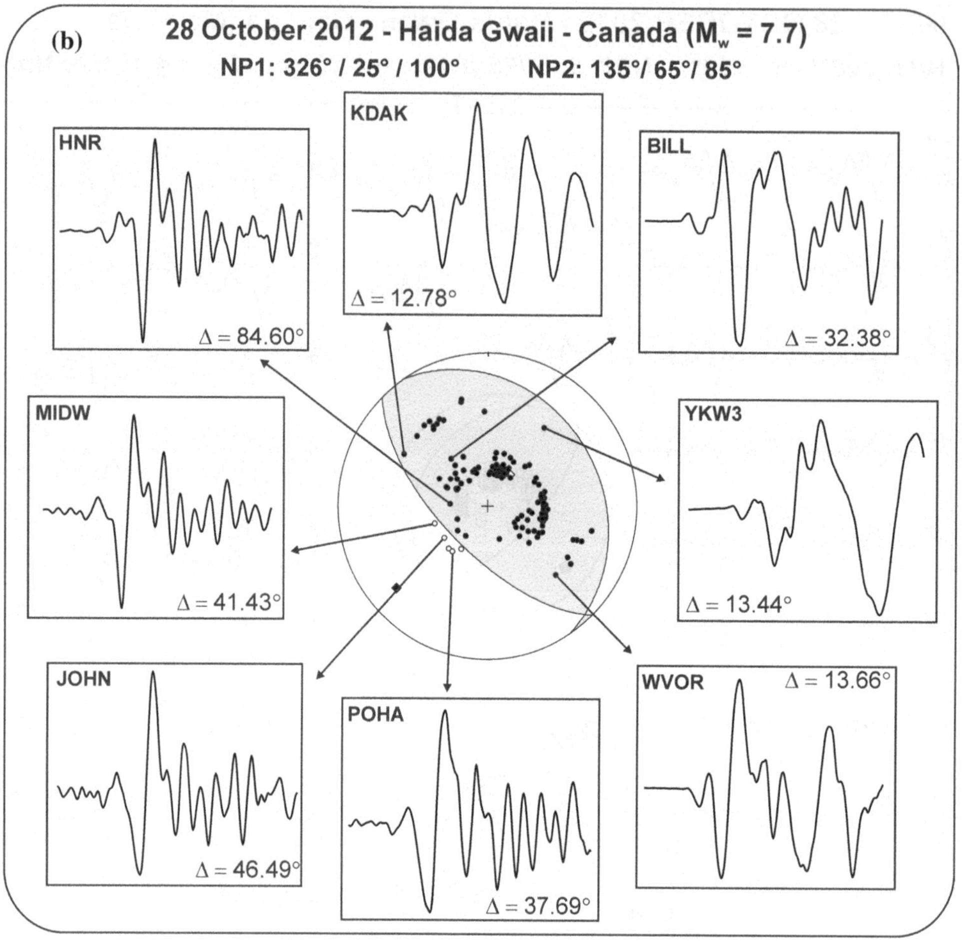

Figure 4
continued

In this study, we performed numerical tsunami calculations using uniform and non-uniform finite-fault slip distribution models of earthquakes. For the uniform slip model case, only the strike, dip and rake angles of the fault plane, the length and width of the faulting area, focal depth of the earthquake and average slip on the fault plane are used in simulations (Table 1), whereas for the non-uniform slip model case, all those parameters for each sub-fault are used to calculate the height and propagation of tsunami waves. The geographical locations of each sub-fault and varying slip values throughout the faulting area are also defined. We used General Bathymetry Chart of the Ocean (GEBCO–BODC; https://www.bodc.ac.uk/data/online_delivery/gebco/) with 30 arc-sec grid bathymetry data for computation domains. Time steps are adjusted for satisfying the Courant–Friedrichs–Lewy (CFL) stability criterion. It is assumed the Manning's roughness coefficient is 0.025 for shallow water areas or natural beach (Kotani et al. 1998; Kaiser et al. 2011). In simulations, computation domains are set in two parts as regions of A and B. To calculate tsunami wave heights, bathymetry data of 2-arcmin and 0.25-arcmin interval grids are used in the computation domains of A and B, respectively. The tsunami propagation times in the A and B domains are selected as 4.0 h and 2.0 h to enable adequate comparison between the simulated and observed wave heights in the open ocean and nearest coastal areas, respectively. Tsunami Analysis Tool (TAT) developed by Annunziato (2007) is further used to visualize travel times and

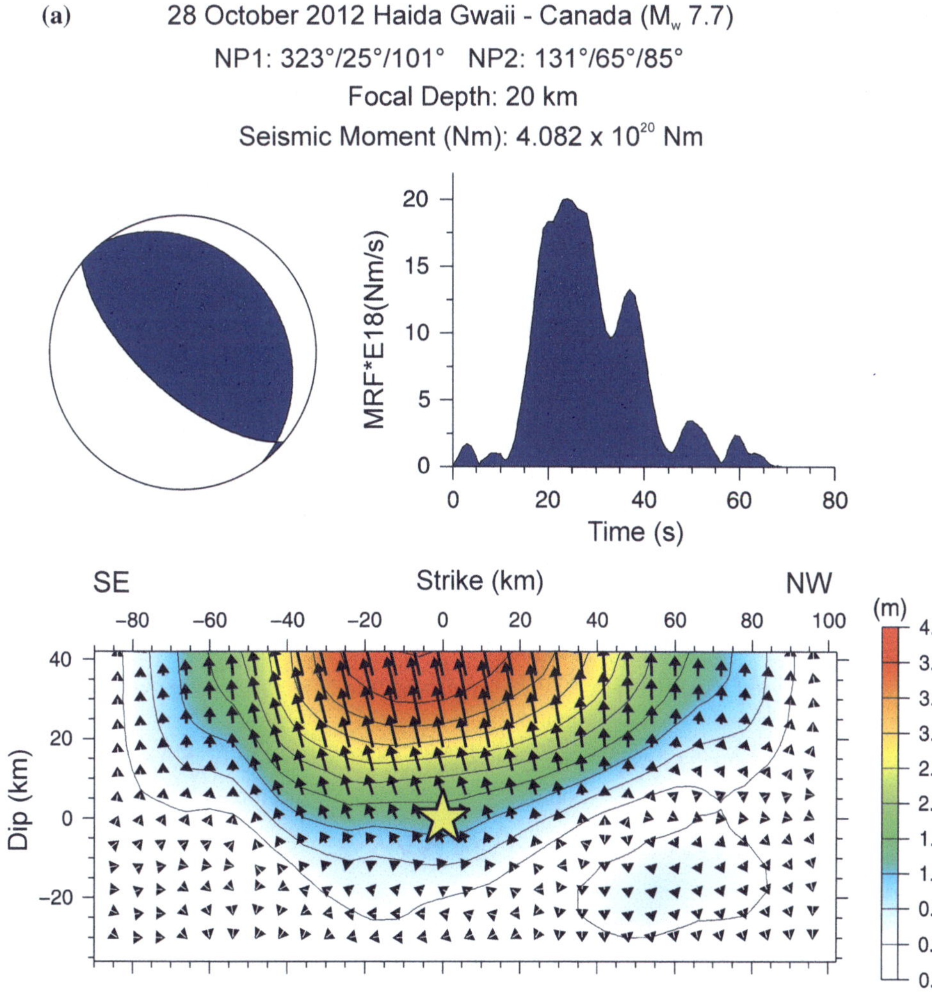

Figure 5

a Focal mechanism, co-seismic slip distribution and total moment rate function of the October 28, 2012 Haida Gwaii (Canada) earthquake. Slip distribution and slip vectors are shown on the fault plane. The slip model is estimated on the first nodal plane (NP1). The yellow star denotes the location of the initial break situated at a focal depth of about 20 km. The colored scale given at the right of the slip model shows the displacement in meters; **b** comparison of 56 observed (black) and synthetic (red) teleseismic broadband P-waveforms used in the finite-fault slip inversion of the October 28, 2012 Haida Gwaii (Canada) earthquake. Station code, amplitude, station azimuth and epicentral distances are given above the waveforms

maximum heights of tsunami waves at the locations of deep-water DART buoys and tide gauges. It allows comparison of the calculated and measured waveforms available from the Intergovernmental Oceanographic Commission (IOC) and the National Oceanic and Atmospheric Administration (NOAA) web sources (Annunziato 2007; Ulutaş 2011, 2013; Ulutaş et al. 2012). The observed data were recorded at shore-based tide gauges (PSMSL 2012, 2015; SLSMF 2012, 2015) and DART buoys (González et al. 1998, 2005; NDBC 2012, 2015; Santos et al. 2016). DART stations consist of a surface buoy and a seafloor bottom pressure recorder (BPR) package that detects pressure changes caused by tsunamis (Table 3). Tide gauges (http://www.ioc-sealevelmonitoring.org), installed by various operators, are located along the coastal areas of the Pacific Ocean (Table 4). In addition to those stations, SHOA (the Hydrographic and Oceanographic Service of the Chilean Navy) tide gauges situated close to the

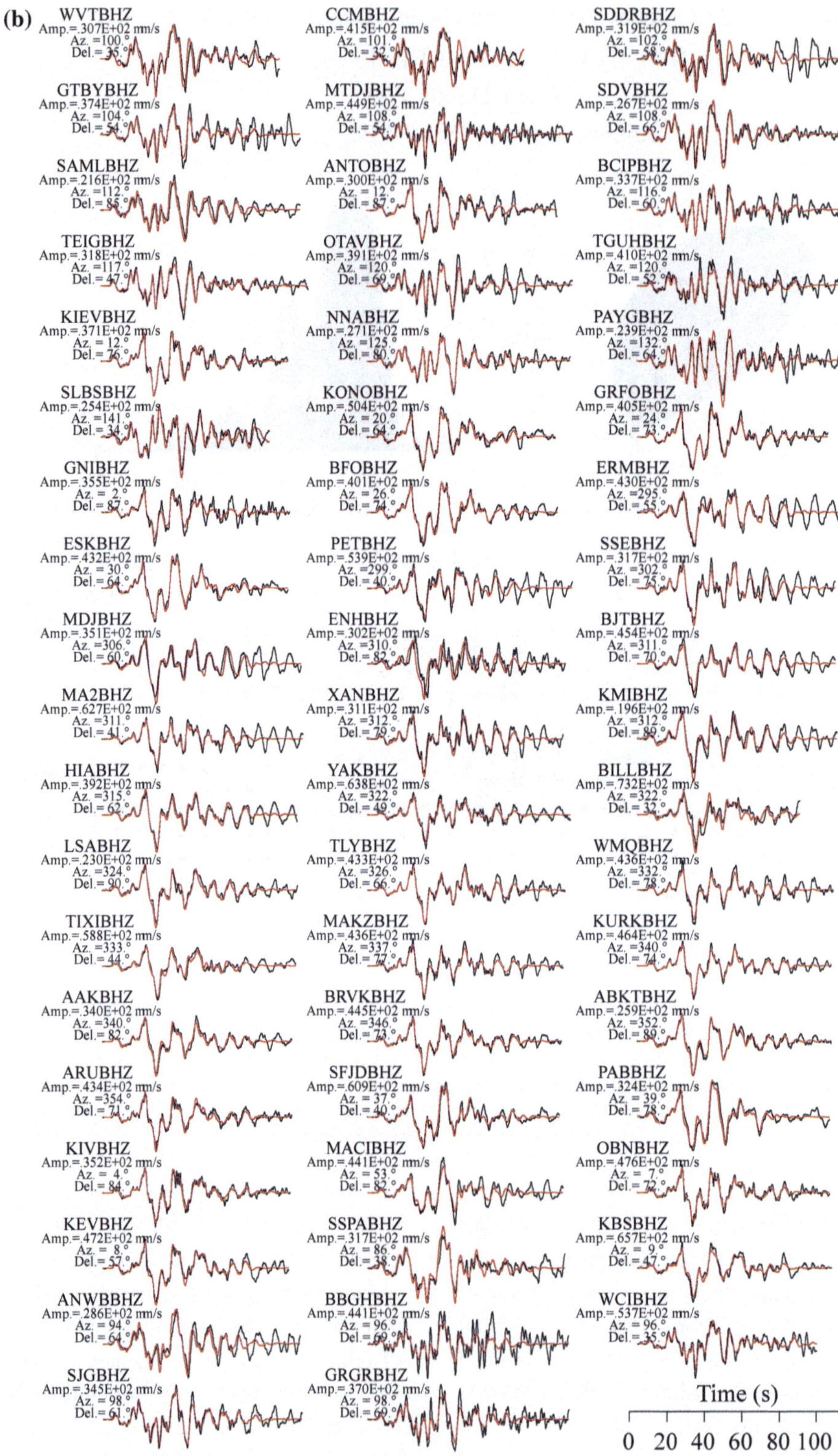

Figure 5
continued

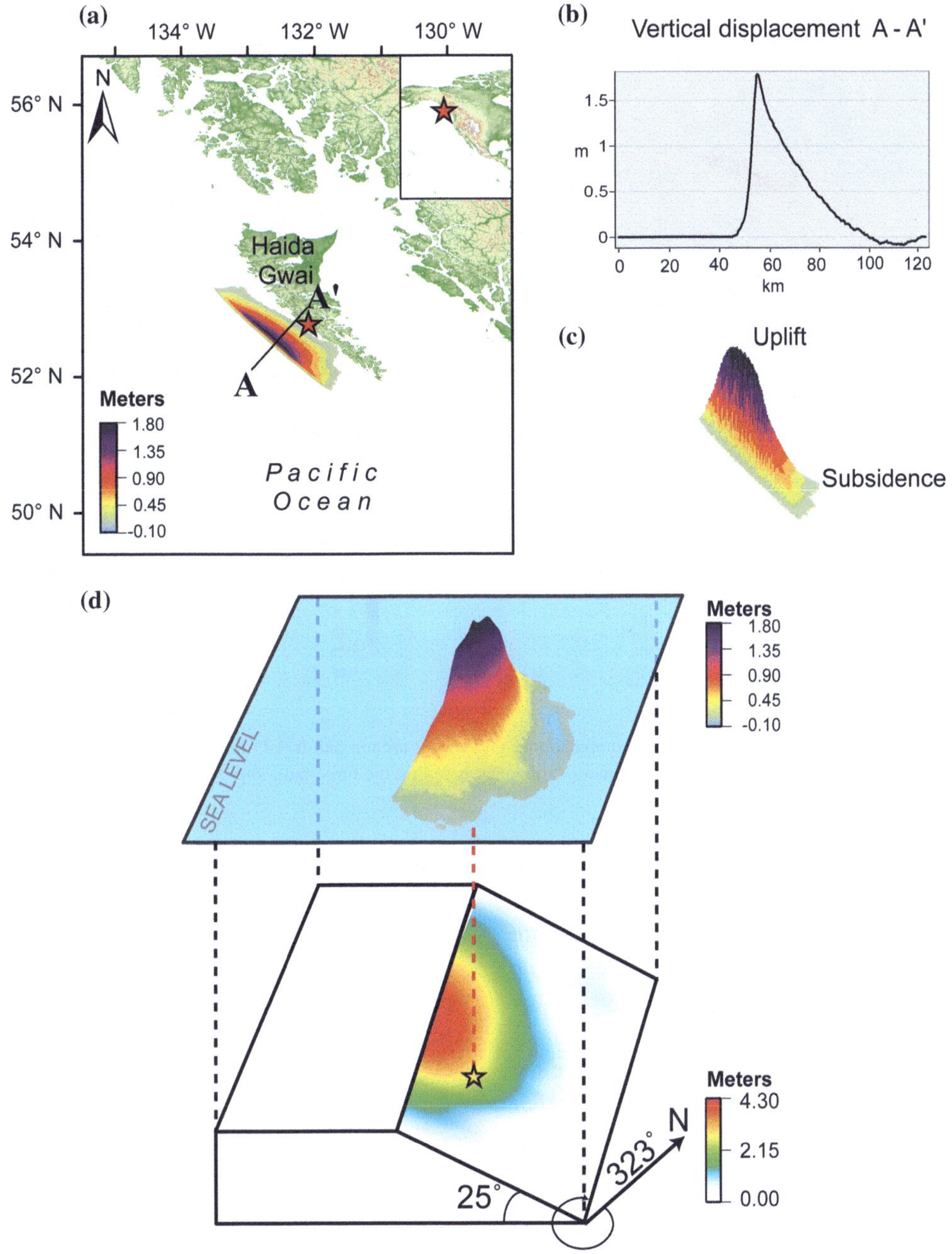

Figure 6

a Initial surface elevations from the finite-fault slip distribution of the October 28, 2012 Haida Gwaii earthquake. The red star indicates the earthquake epicenter; **b** cross-section of the surface displacement along *A–A′* based on the calculated seafloor deformation; **c** a 3D view of the calculated deformation; **d** a sketch illustration of the earthquake fault geometry and initial sea level elevations. The slip distribution is from the finite-fault model and the initial sea surface height is calculated from the dislocation model of Okada (1985)

coastal cities of Chile are provided for the 2015 Illapel tsunami.

The sea level measurements from DART systems and coastal tide gauges ensure us detailed information that improves our understanding of the influence of bathymetric features and coastal areas on tsunami simulations. However, bottom pressure sensors of DART stations record tsunami waves together

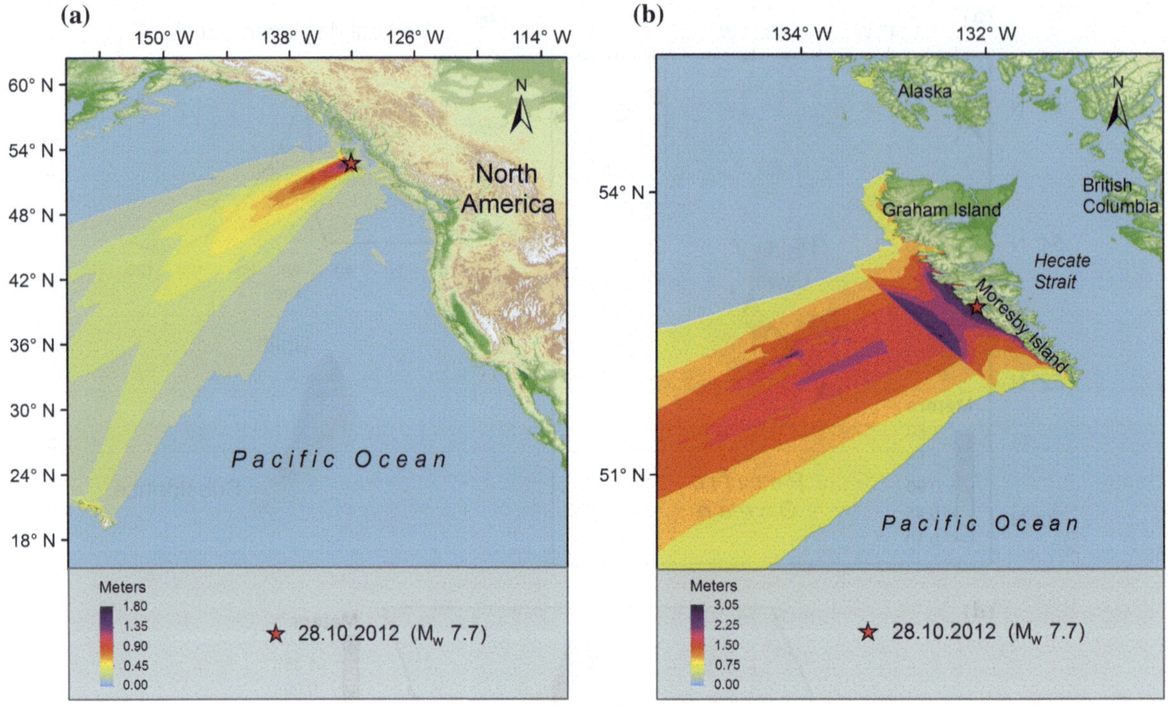

Figure 7
Computed maximum tsunami wave heights with a computational grid of **a** 2 arcmin and **b** 0.25 arcmin from numerical simulation of the October 28, 2012 Haida Gwaii (Canada) tsunami earthquake based on the finite-fault slip distribution earthquake model

with ocean tides and earthquake-generated seismic waves, such as surface Rayleigh waves that can be clearly seen on all deep-ocean bottom pressure records (Filloux 1982, 1983; Rabinovich and Eblé 2015). Hence, we applied several pre-processing steps to the data prior to the estimate of arrival times and height of tsunami waves, in order to remove the non-tsunami signals from the data (Fig. 3). For example, the astronomical tides are predicted by fitting the polynomial functions, and are then subtracted from the original series. A polynomial of degree 15 is used, for which we applied the function "polyfit" in the Matlab program (MathWorks 2012) for fitting. A sampling interval of 1 min is used for both DART and tide gauge records. The DART records are averaged internally to a default sampling interval Δt = 15 min in the absence of tsunami. The records are then switched from Δt = 15 min mode to the Δt = 15 s mode for several minutes, and then they are transmitted Δt = 1 min data until the end of the event (Rabinovich et al. 2013). The huge artificial jumps at the place of the Rayleigh waves on DART records are also filtered by taking into account the "event mode" DART data. For high-pass filtering, the Butterworth digital filter from the Matlab signal processing toolbox is employed for which a cut-off frequency of 0.00015–0.00025 Hz was chosen (MathWorks 2012).

3. *Results*

Source mechanism parameters of 2012 Haida Gwaii-Canada (M_w 7.7) and 2015 Illapel-Chile (M_w 8.3) earthquakes used in this study are summarized in Table 1. The interpretation of numerical tsunami wave simulation results of both earthquakes has been done by comparing the arrival times and maximum heights of synthetic tsunami waves along the coast and those at observed tide gauges and DART systems. Furthermore, we compared our simulation results with those from preliminary fault slip models

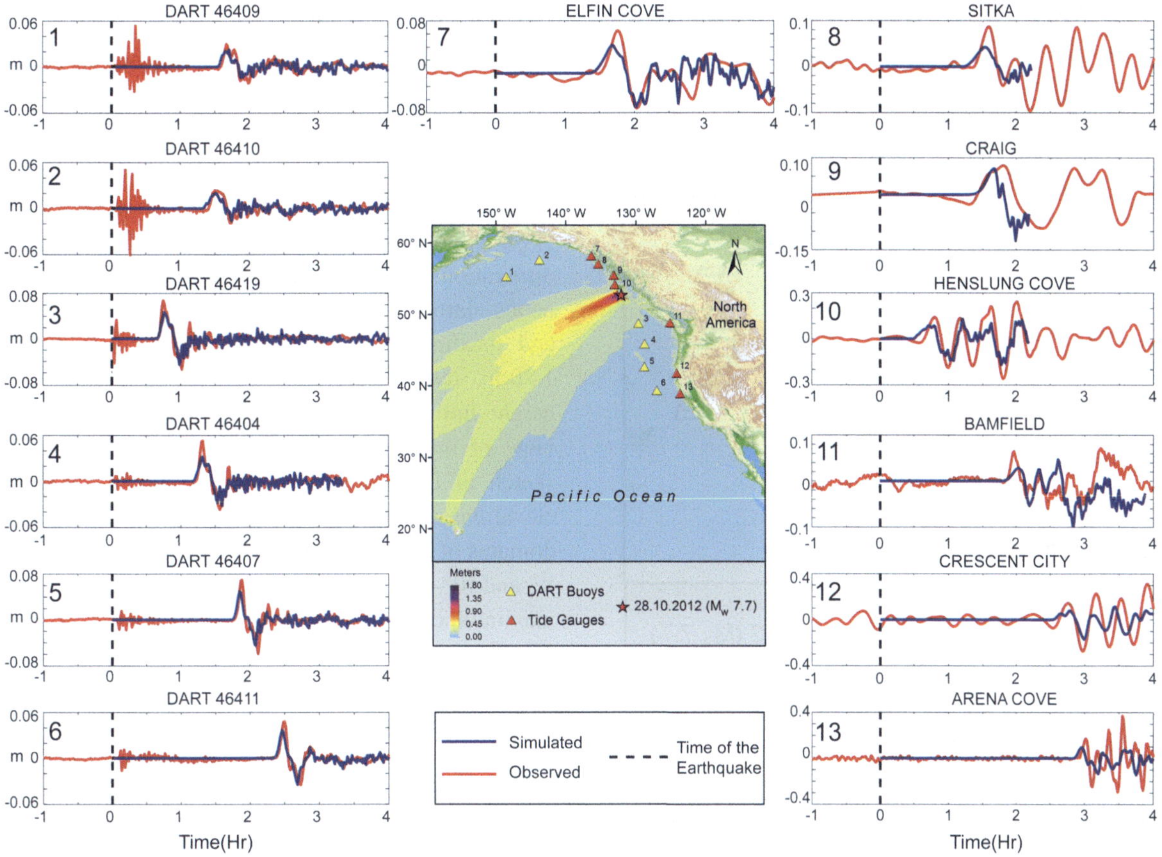

Figure 8

Computed maximum tsunami wave heights (inner map) for the October 28, 2012 Haida Gwaii earthquake with a computational grid of 2 arcmin. based on the finite-fault slip distribution model, and comparison of the observed and calculated (synthetic) tsunami waveforms of DART and tide-gauge records

reported by the USGS (2012b, 2015b) in the "Appendix".

3.1. Source Mechanism and Slip Distribution Inversion of the 2012 Haida Gwaii Earthquake

The best fitting waveform point-source solution (strike: 326°, dip: 25°, rake: 100°, focal depth (*h*): 20 km and seismic moment (M_o): 4.33 × 10^{20} Nm; Fig. 4a and Table 1) exhibits a thrust faulting mechanism with a small strike-slip component for the 2012 earthquake. We inverted long period 33 P- and 11 SH-waveforms (Fig. 2a). The two nodal planes dip to the northeast and southwest, respectively, and have a small amount of oblique slip. The P-wave first motion polarities recorded by near-field (e.g., KDAK (Δ = 12.78°); YKW3 (Δ = 13.44°)) and teleseismic stations (e.g., JOHN (Δ = 46.49°); POHA (Δ = 37.69°)) are rather coherent with the point-source solution within ± 5–10° (Fig. 4b).

The finite-fault slip model is further obtained from the inversion of 56 teleseismic broadband P-waveforms (Figs. 2b, 5a, b). The NW–SE directed nodal plane with a dip angle of 25°, consistent with tectonic characteristics of the region, is selected as the fault plane. It is divided into 31 × 13 sub-faults with dimensions of 6 × 6 km^2 , which has been selected by experimenting within inversions among various grid-sizes and fault dimensions. The source-time (slip-rate) function of each sub-fault is expanded in a series of 39 overlapping triangle functions each with a rise time (τ) of 1.0 s. The preferred model has

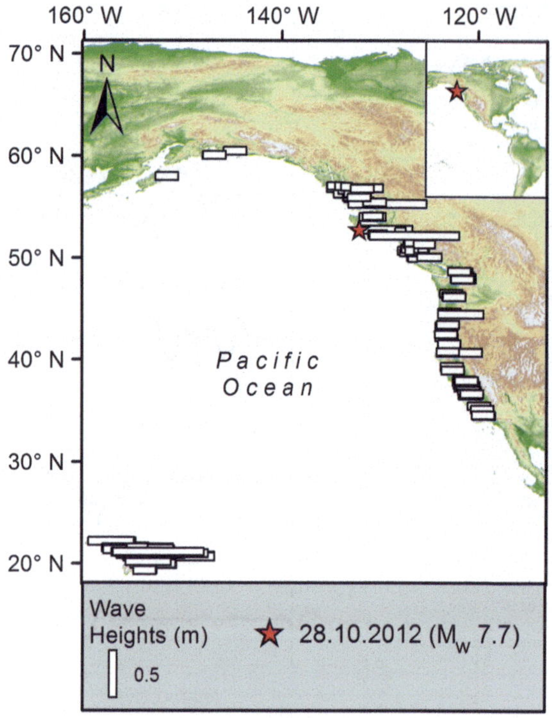

Figure 9
Estimated maximum tsunami wave heights with bar chart visualization for several locations along the coasts of Alaska, British Columbia and the Hawaiian Islands. The horizontal white bars indicate the zero-to-crest tsunami wave heights larger than 0.25 m

most of the fault slip close to the epicenter well above the hypocenter, as expected from the peak in the moment rate function after about 20 s and lasted until about 50 s after the origin time of the earthquake. It shows that rupture propagated along the dip direction of the fault plane. It has also a second area of smaller than 1 m slip, which is less well-constrained. The rake angle is estimated to be 101°. The total seismic moment (M_0) and maximum displacement are found to be 4.082×10^{20} Nm, and about 4.3 m, respectively (Fig. 5a). The source time function reveals that the earthquake duration was about 70 s, but most of the moment release occurred in the first 40–50 s. The faulting area (160 km fault length × 60 km fault width) is estimated to be ~ 9600 km^2 from the slip model (Fig. 5a). The stress drop and average slip value on the fault plane are calculated as to be 11 bars and 1.41 m, respectively (Table 1). The moment magnitude (M_W) of the 2012 Haida Gwaii earthquake is calculated to be 7.7 using both seismic moment values and the Kanamori (1977) equation.

3.2. Numerical Tsunami Simulation of the 2012 Haida Gwaii Earthquake

The co-seismic vertical displacement of the sea bottom is identified by considering the slip distribution model in which slip is spatially variable at all resolvable scales. To achieve this, we used the Okada (1985) model that assumes the superposition of planar dislocations (i.e., finite faults) embedded in homogeneous elastic half-spaces. The maximum subsidence and maximum uplift are found as − 0.10 m and 1.80 m, respectively, for the initial tsunami wave height. Initial surface elevations, cross-section and 3D view of the calculated seafloor deformation and a sketch of fault geometry with the initial sea level elevations are displayed in Fig. 6. The computation domains of A and B cover 15°–62° N and 112°–158° W, and 50°–55° N and 130°–136° W, respectively. The maximum calculated wave heights are estimated to be 1.80 m and 3.05 m for the computational grids of 2.0 arcmin and 0.25 arcmin, respectively (Fig. 7).

We particularly focus on the near field wave characteristics of the 2012 Haida Gwaii tsunami along the North American west coast. The records from six DART stations (46404, 46407, 46409, 46410, 46411 and 46419) and seven tide gauges (Elfin Cove, Sitka, Craig, Henslung Cove, Bamfield, Crescent City and Arena Cove) are simulated (Tables 3, 4). The earthquake magnitude (M_w 7.7) was relatively large and maximum wave amplitudes computed in the source area were about 3 m (Fig. 7b). However, simulated tsunami waves along the main coast of British Columbia were rather small, in agreement with actual observations (Fig. 8). Fine et al. (2015) suggested that those small recorded tsunami amplitudes might be explained by the radiation direction of the main flux tsunami energy southwestward from Haida Gwaii (Fig. 7), and the shadowing effect of local topography. Overall, the arrival times and amplitudes of simulated leading waves are found to be consistent with those recorded at DART stations, while they are partially underestimated for tide gauges (Fig. 8). Actual tsunami waves might have been amplified by submarine canyons, channels and coastal embayments, which are not well resolved by the 30 arcs bathymetry data (see Lay et al. 2013). Numerically modeled tsunami waves also do not match well the

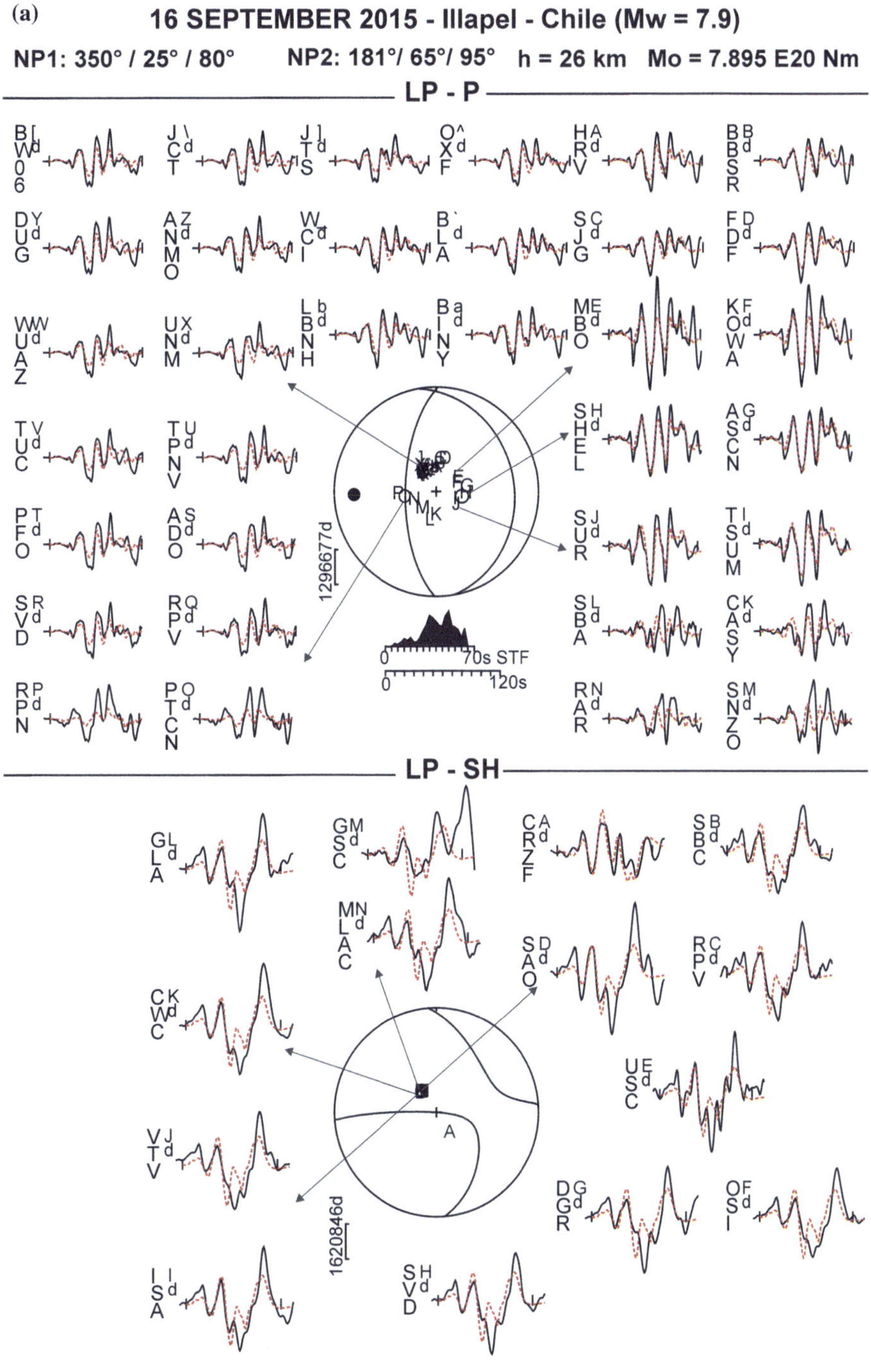

Figure 10

a The best-fitting minimum misfit solution of the September 16, 2015 Illapel (Chile) earthquake. Comparison of long-period 34 P- and 14 SH-observed waveforms (solid black lines) and synthetic waveforms (dashed red lines) is shown. The figure caption is as in Fig. 4a; **b** distributions of the P-wave first motion polarities of the September 16, 2015 Illapel (Chile) earthquake. The figure caption is as in Fig. 4b

waveforms of later arriving waves recorded at tide gauges. The calculated maximum wave heights (zero-to-crest) larger than 0.25 m along the coast of North America and at the Hawaiian Islands are shown in Fig. 9. The modeled maximum uplift generates an initial wave of 1.80 m height off Moresby Island. The

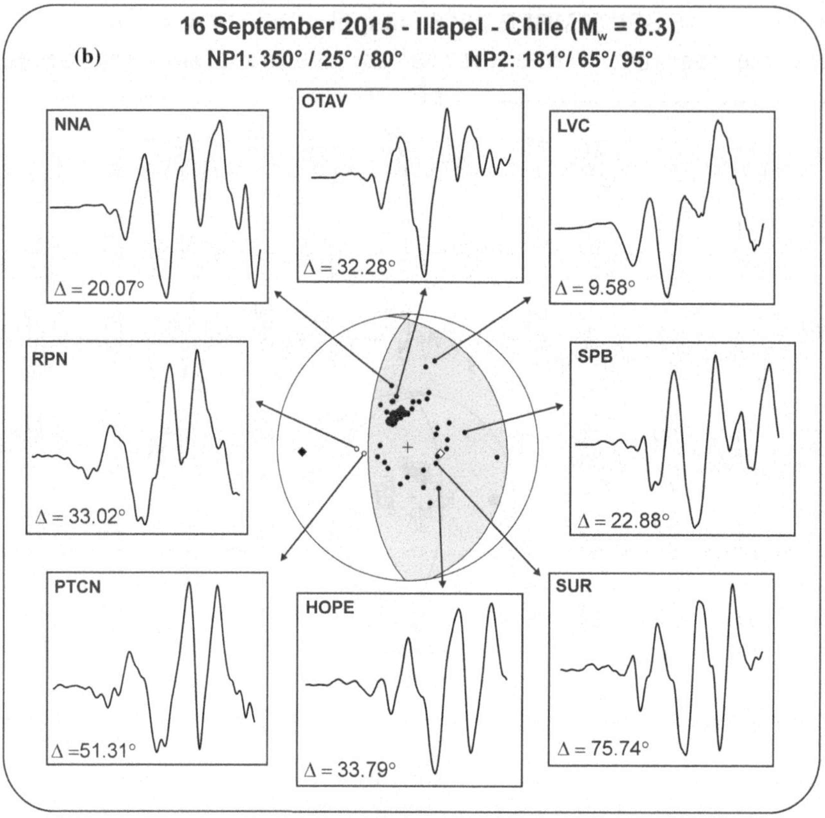

Figure 10
continued

highest tsunami waves calculated near Moresby Island, compared to adjacent northerly Graham Island of Haida Gwaii, might have been due to its location where is rather close to the initial tsunami wave and the narrow embayments of the island. Alaska, Vancouver Island and the west coast of the USA, which are more than 500 km to the north and south, are in a direction of expected low estimated tsunami wave heights for thrusting orthogonal to the Haida Gwaii margin. Those locations were not exposed to the main focus of tsunami energy. However, the tsunami energy was focused on the Hawaiian Islands more than 4000 km away to the southwest (see Fig. 9).

3.3. Source Mechanism and Slip Distribution Inversion of the 2015 Illapel (Chile) Earthquake

The 2015 Illapel (Chile) earthquake occurred as a result of a thrust fault on the interface between the Nazca and South American plates in Central Chile. The approximate parameters of the NW–SE directed fault plane are: strike = 350°, dip = 25°, rake = 80°. In inversion, long period 34 P- and 14 SH-waveforms are used (Figs. 2c, 10a). The seismic moment and focal depth of this earthquake are determined to be 7.895×10^{20} Nm and 26 km, respectively (Table 1). First motion polarities of P-waveforms recorded by near-field and teleseismic stations confirm the focal mechanism solution of this earthquake (Fig. 10b).

The detailed slip distribution along the east dipping plane is determined using 60 broadband teleseismic P-waveforms (Figs. 2d, 11a, b). The nodal plane that has a low dip angle (25°) is assumed to be the fault plane, and it is divided into 40 × 12 sub-faults with dimensions of 10 × 10 km^2. The source-time (slip-rate) function of each sub-fault is expanded in a series of 39 overlapping triangle functions each with a rise time of 2.5 s. The preferred model revealed slip distributions mainly along the dip direction of the fault plane with an area of maximum

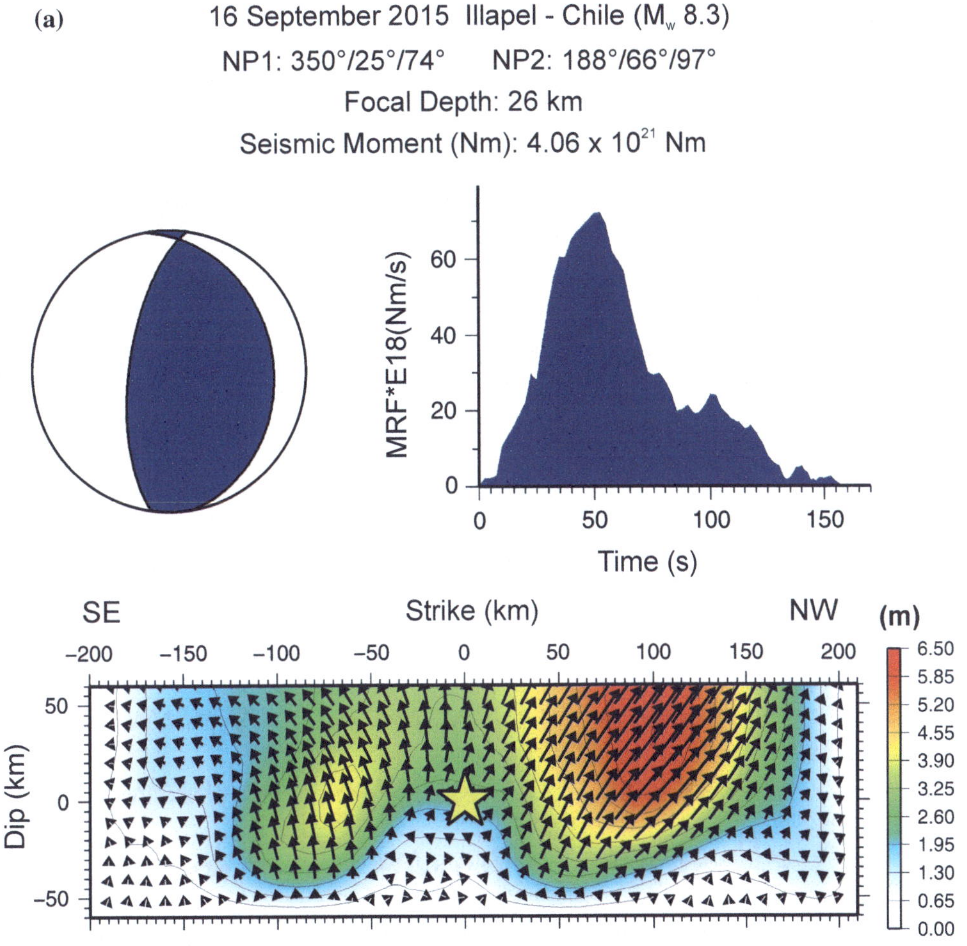

Figure 11

a Focal mechanism, co-seismic slip distribution and total moment rate function of the September 16, 2015 Illapel (Chile) earthquake. The yellow star denotes the location of the initial break situated at a focal depth of about 26 km. The figure caption is as in Fig. 5a; **b** comparison of 60 observed (black) and synthetic (red) teleseismic broadband P-waveforms used in finite-fault slip inversion of the September 16, 2015 Illapel (Chile) earthquake. Station code, amplitude, station azimuth and epicentral distances are given above the waveforms

slip (6.50 m) at the NW of the centroid (Fig. 11a). It shows two slip patches on the fault plane, the large-slip area is located ~ 75 to 100 km to the northwest of the epicenter at shallow depth. The faulting area (fault length = 350 km, fault width = 100 km) is determined to be ~ 35,000 km^2. By using the seismic moment value (M_o = 4.06 × 10^{21} Nm) and the size of the faulting area, we calculated the stress drop and average slip to be 15.5 bars and 3.86 m, respectively. The amount of maximum slip (6.5 m) and the source duration (~ 150 s) of this earthquake are found similar to slip models described above. Our model has relatively larger fault dimensions displaying slip distributions in a larger area than most of the other models. The seismic moment of the earthquake obtained from point-source waveform inversion (M_o = 7.895 × 10^{20} Nm) is smaller than the value of finite-fault slip inversion (M_o = 4.06 × 10^{21} Nm). The reason is that we cannot model full waveforms in P- and SH-waveform modeling, while we thoroughly modelled 150 s long P-waveforms in slip inversion (Table 1).

3.4. *Numerical Tsunami Simulation of the 2015 Chile Earthquake*

The co-seismic vertical seafloor displacement was calculated by the superposition of 40 × 12 sub-faults

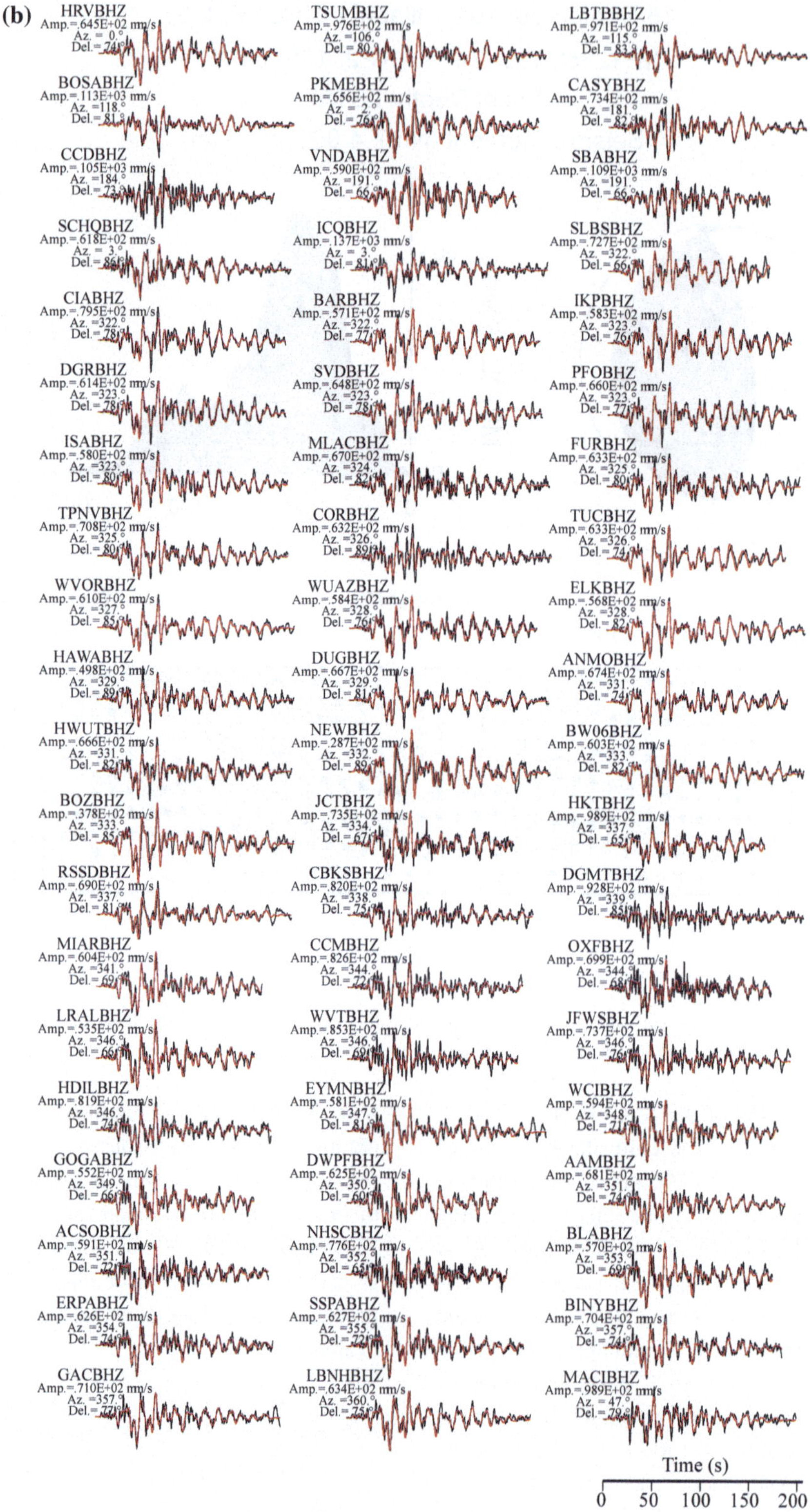

Figure 11
continued

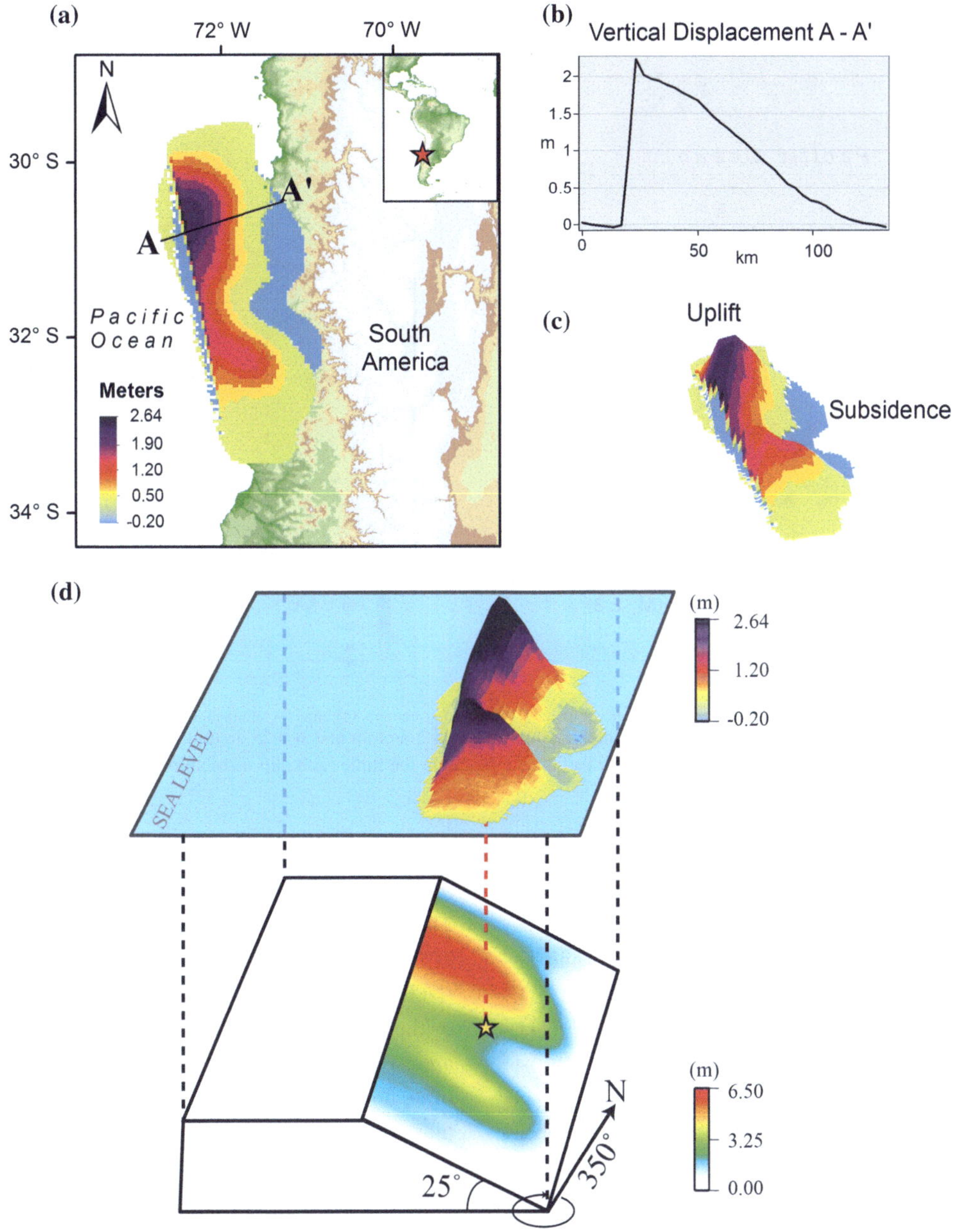

Figure 12

a Initial surface elevations from the finite-fault slip distribution model of the September 16, 2015 Illapel (Chile) earthquake. The red star indicates the earthquake epicenter; **(b)** cross-section of the surface displacement along *A–A′* based on the calculated seafloor deformation; **c** a 3D view of the calculated deformation; **d** a sketch illustration of the earthquake fault geometry and initial sea level elevations. The slip distribution is from the finite-fault model and the initial sea surface height is calculated from the dislocation model of Okada (1985)

with dimensions of 10 × 10 km^2, using the Okada (1985) model. The maximum subsidence and the maximum uplift for the initial wave height was estimated to be − 0.20 m and 2.64 m, respectively (Fig. 12). The initial surface elevations, cross-section and 3D view of the calculated seafloor deformation and a sketch of the fault geometry are displayed in Fig. 12. The computation domains of A and B span areas of 4°–45°S and 69°–111°W, and of 28°–35°S and 69°–77°W, respectively. The maximum

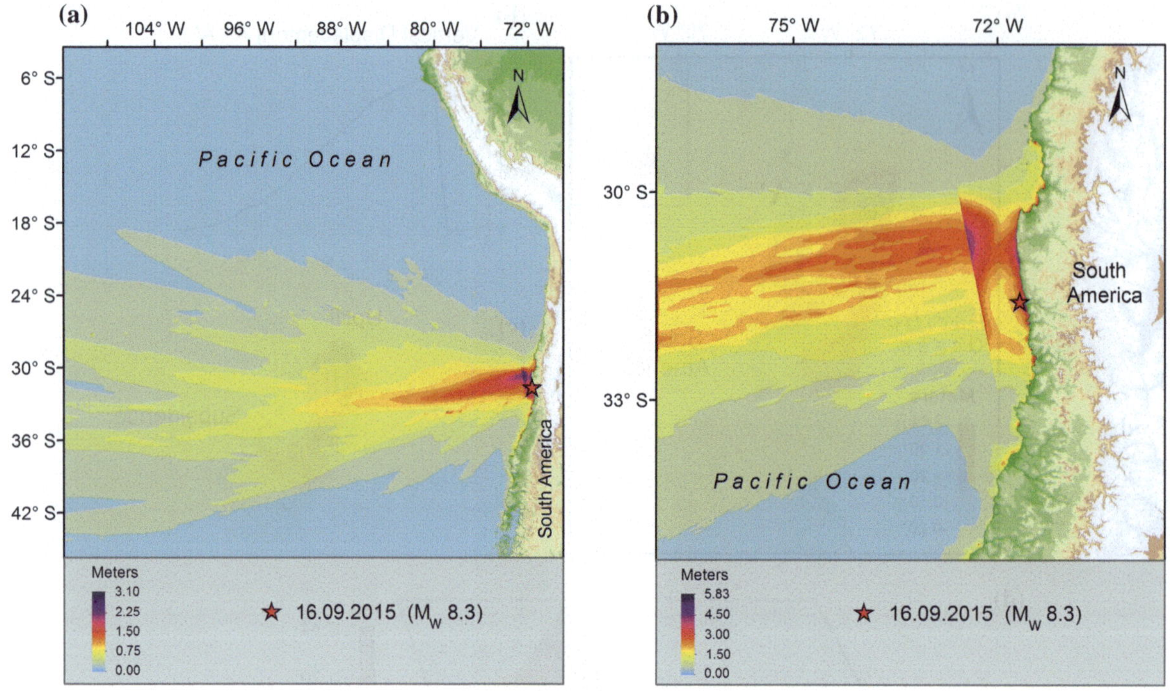

Figure 13
Computed maximum tsunami wave heights with a computational grid of **a** 2 arcmin and **b** 0.25 arcmin from numerical simulation of the September 16, 2015 Illapel (Chile) tsunami earthquake based on the finite-fault slip distribution earthquake model

calculated tsunami wave heights were found to be 3.10 m for the computational grid of 2 arcmin and 5.83 m for 0.25-arcmin grids (Fig. 13).

The waveforms from three DART stations and 13 tide gauges are used in simulations (Tables 3, 4). Similar to the 2012 Haida Gwaii earthquake (M_w 7.7), the simulated leading tsunami waves seem to almost coincide with the observed waveforms except at Valparaiso (no 8 in Fig. 14) in terms of shape, arrival times and wave amplitudes. However, simulated later arriving waves at tide gauges are in poorer agreement with the observed ones. At the Valparaiso tide gauge located in Valparaiso Bay, synthetic tsunami waves are underestimated for the entire simulation period. The shape of first tsunami peaks for this station are also not modeled well. Specific limitations of tide gauge data, such as complexity of near-shore bathymetry, bay resonances, perturbations in waveforms, which are outside of scope of this study, could be responsible for this underestimation of tsunami wave heights at Valparaiso (Fig. 14). However, our study focuses mainly on the tsunami source and its effects on tsunami wave propagation, and hence we examined specifically the shape of the first wave in actual tsunami records. The reported largest tsunami wave generated by the 2015 Illapel (Chile) earthquake, with a 4.75 m height (NGDC-NOAA 2015), was at Coquimbo, which is one of the nearest tide gauges to the earthquake epicenter area. For this station, we adequately modeled arrival time, amplitude and periods of tsunami waves, particularly for the leading wave, however, the amplitudes of following tsunami waves are, in general, underestimated. Maximum calculated tsunami wave heights (zero-to-crest) larger than 0.50 m, with bar chart visualization, for several populated locations along the Chilean coast are presented in Fig. 15. At locations farther north and south from the earthquake epicenter, we observed a significant decrease in tsunami wave heights, not exceeding 1 m. However, the calculated tsunami heights are remarkably larger at the stations located closer to the earthquake epicenter and rupture patch (see Fig. 15).

In addition, we simulated tsunami waves of the 2015 Illapel (Chile) earthquake using a uniform slip

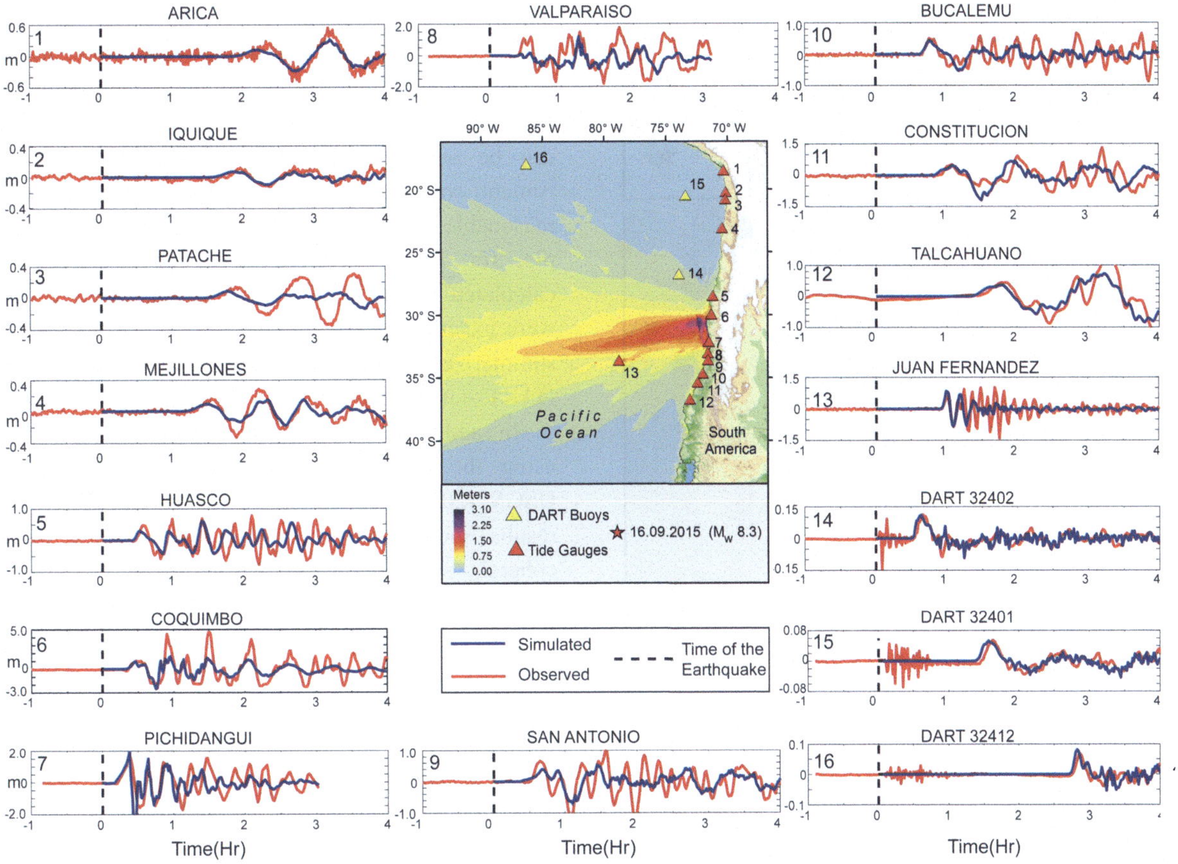

Figure 14
Computed maximum tsunami wave heights (inner map) for the September 16, 2015 Illapel (Chile) earthquake with a computational grid of 2 arcmin based on the finite-fault slip distribution model, and comparison of the observed and calculated (synthetic) tsunami waveforms elevations of DART and tide-gauge records

model to compare the effects of different earthquake models on tsunami simulations. As in many tsunami papers, one can simply calculate the height of the initial tsunami wave using earthquake source parameters estimated by empirical scaling relationships (e.g., Wells and Coppersmith 1994; Henry and Das 2001; Mai and Beroza 2000; Papazachos et al., 2004; Leonard 2010) based on the moment magnitude (M_w) of earthquakes (see Yolsal et al. 2007; Yolsal and Taymaz 2010; Løvholt et al. 2012; Yolsal-Çevikbilen and Taymaz 2012; Ulutaş 2013; Goda et al. 2016). In order to check the consistency with the empirical values and our source parameters obtained from teleseismic waveform inversions, we calculated the probable fault width, fault length, and average slip of the 2015 M_w 8.3 Chile earthquake using the Wells and Coppersmith (1994) equations that are valid for major earthquakes. We observed that rupture dimensions for the uniform slip model of the 2015 Chile earthquake (i.e., fault length = 350 km, fault width = 100 km), are rather compatible with the empirical values (fault length = 250 km; fault width = 62 km). Using the uniform slip model (Fig. 16), we calculated the maximum subsidence and maximum uplift of the initial wave height as to be − 0.37 m and 1.55 m, respectively. For comparison, these values are found as − 0.20 m and 2.64 m for the non-uniform slip model (see Fig. 12). To test the performance of selected slip models on tsunami wave propagation and simulated waveforms (Fig. 17), we utilized the tide gauge and DART records for stations located near and far from the earthquake epicenter.

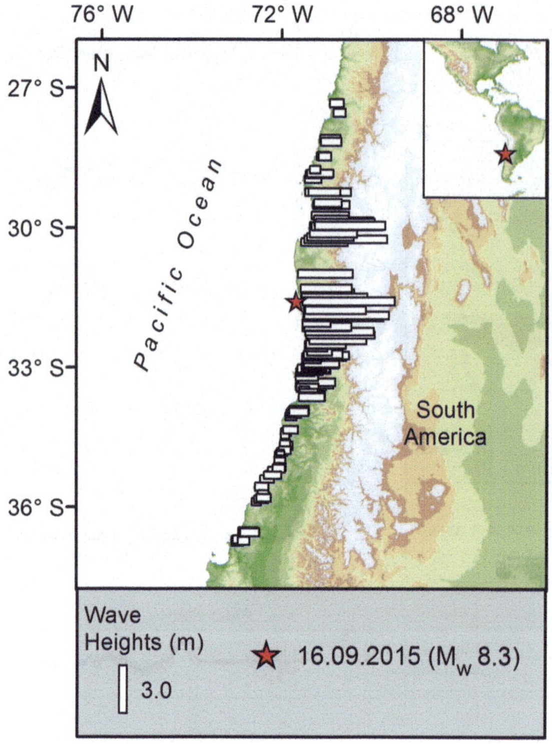

Figure 15
Estimated maximum tsunami wave heights with bar chart visualization for several locations along the coasts of Chile. The horizontal white bars indicate the zero-to-crest tsunami wave heights larger than 0.50 m

It is clearly observed that the general characteristics of synthetic tsunami waveforms (i.e., amplitude, shape, arrival time, etc.) calculated using the non-uniform (finite-fault) slip model, are more consistent with the observed tsunami records than those of the uniform slip model (Figs. 17, 18a, b). For example, the tsunami data recorded at Huasco, Pichidangui, Constitución, Talcahuano and DART 32401 stations are not properly simulated using the uniform slip model; the calculated amplitudes and arrival times of tsunami waveforms are not accurate enough. The largest mismatch in arrival time of the leading wave is observed for the record of Constitución. It is seen that the observed arrival time of the first tsunami wave is about 50 min after the main earthquake shock; however, for the uniform slip model, the computed leading wave arrives at the station in about 46 min. In contrast, the differences between the observed and estimated times are much smaller for the tsunami simulations based on the non-uniform slip model. The time misfits calculated for the near-field tide gauge stations might be a result of the lack of high-resolution bathymetry data used in simulations (Fig. 18b; Table 2). On the other hand, it can be seen that the estimated arrival times of tsunami waves and first wave maxima values are in reasonable agreement for DART stations. For example, for DART stations 32402 and 32412, the calculated arrival times of the first tsunami waves were exactly the same as for the observations. The summarized results for selected tide gauge and DART stations are shown in Table 2.

In addition, we simulated tsunami waveforms using the preliminary finite-fault slip models of USGS (2012b, 2015b) for both earthquakes. The results show that our slip models provide more coherent simulated tsunami waveforms than those obtained using preliminary, fast and operational models (see Figs. 19, 20, 21, 22). As previously discussed by several authors (e.g., Fine et al. 2015; Calisto et al. 2016; Fuentes et al. 2016), these operative slip source models can only give preliminary tsunami estimates, and not precise tsunami wave heights and arrival times.

4. *Discussion and Conclusions*

In this study, we propose new finite-fault slip models for the October 28, 2012 Haida Gwaii (British Columbia, Canada; M_w 7.7) and the September 16, 2015 Illapel (Chile; M_w 8.3) earthquakes occurred at two different subduction zones by carefully analyzing earthquake waveforms. Tsunami generation, wave propagation, and near-field effects of tsunamis have been numerically simulated based on the estimated earthquake source parameters (Table 1).

The slip models of the 2012 Haida Gwaii and 2015 Illapel earthquakes reveal heterogeneous slip distributions on fault planes with long source durations ($\sim$ 80 s and 150 s) and low stress drop values ($\Delta\sigma$ = 10–15 bars). The 2012 earthquake generated maximum slip ($\sim$ 4.30 m) near the western coast of Moresby Island where maximum calculated tsunami wave heights were about 3.05 m (Fig. 7). Similarly, related to the maximum slip ($\sim$ 6.5 m) observed at the NW part of the fault plane of the 2015 Illapel

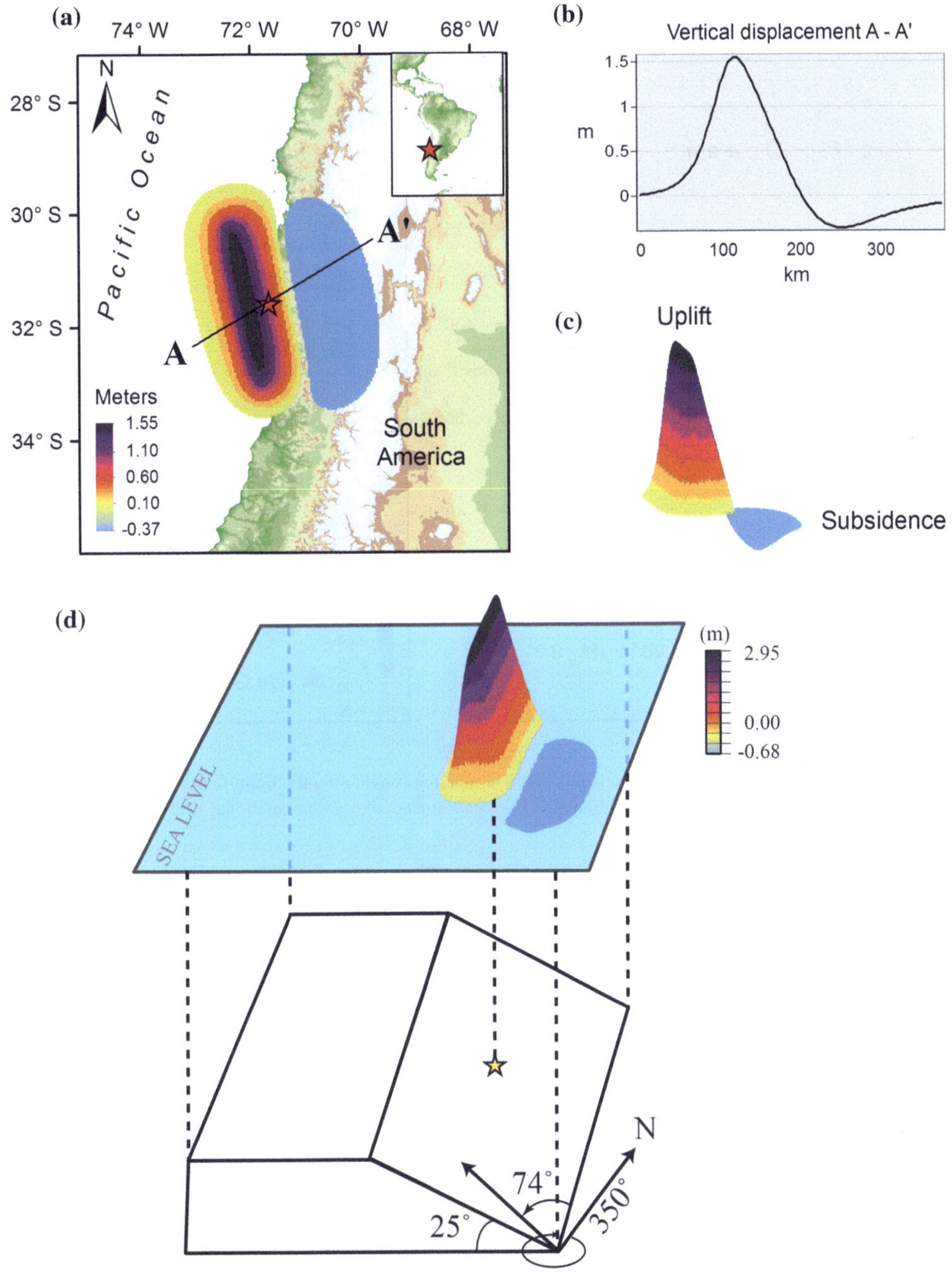

Figure 16
a Initial surface elevations from the uniform slip distribution model of the September 16, 2015 Illapel (Chile) earthquake. The red star indicates the earthquake epicenter; **b** cross-section of the surface displacement along *A–A′* based on the calculated seafloor deformation; **c** a 3D view of the calculated deformation; **d** a sketch illustration of the earthquake fault geometry, and initial sea level elevations

earthquake, maximum 5.83 m amplitude tsunami waves were computed for this location (Fig. 13). It is also seen that tsunami waves generated by the 2012 Haida Gwaii (M_w 7.7) earthquake propagate over a larger area despite its relatively small slip value (4.3 m) in comparison to the slip of 6.5 m of the 2015

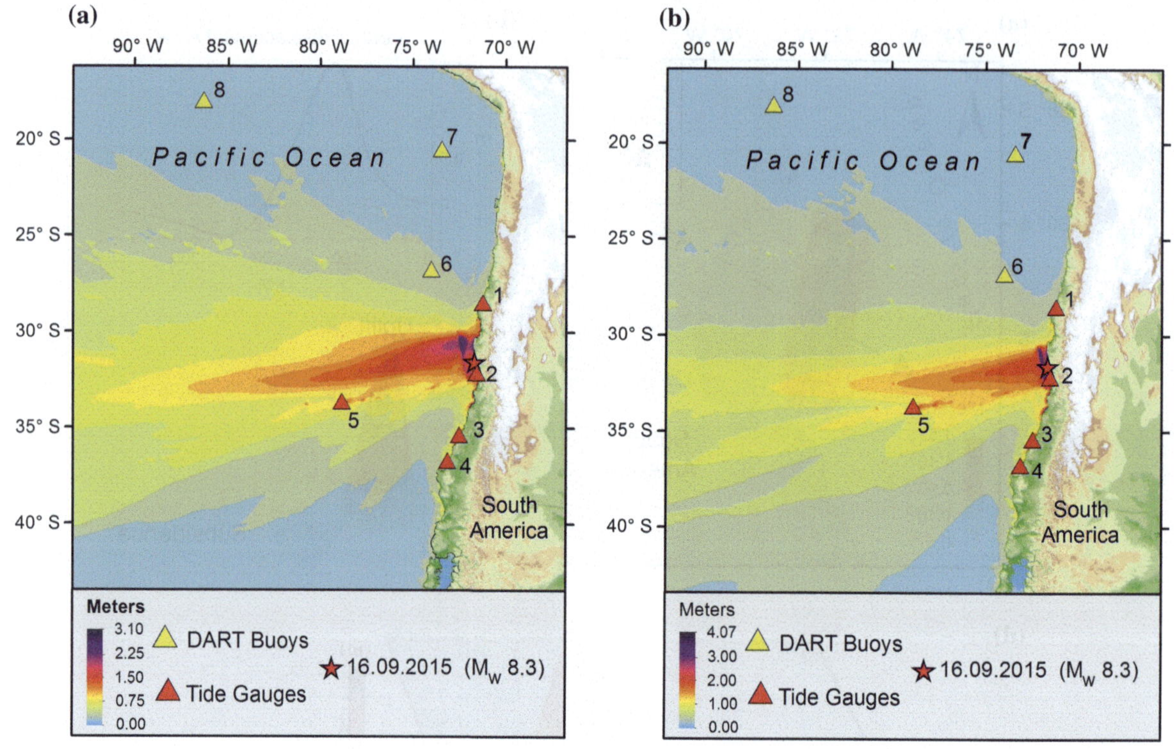

Figure 17
Computed maximum tsunami wave heights from numerical tsunami simulations based on **a** the finite-fault and **b** the uniform slip distribution models of the September 16, 2015 Illapel (Chile) earthquake with a computational grid of 2 arcmin

Illapel (Chile) earthquake. For example, at the Hawaiian Islands, located far from the epicenter, it produced tsunami waves of about 1 m (Fig. 9). However, tsunami waves of the 2015 Illapel (Chile) earthquake (M_w 8.3) only affected the local area by producing high amplitude ($\sim$ 4 m) tsunami waves near the epicenter; the waves were rather small in far-field regions. This might be related to the fact that some part of the vertical co-seismic displacements of the 2015 Illapel (Chile) earthquake were distributed on land, while almost the entire vertical co-seismic displacements of the 2012 Haida Gwaii earthquake were situated on the ocean floor. It caused only a small water displacement, which limited far-field wave propagation. Similarly, Omira et al. (2016) emphasized that the loss of the initial tsunami potential energy in inland deformation (more than 5%) caused a remarkable local earthquake impact and reduced far-field effects of the 2015 Chile tsunami. Based on the finite-fault slip distribution model of the 2015 Illapel-Chile earthquake, we can conclude that it involved multiple ruptures of sub-faults and various crustal displacements both on the seafloor and onshore. However, wide leading uplift occurred close to the coast, while most of the other uplift parts occurred inland, according to our initial height model calculated by assuming superposition of sub-faults. This could be also the reason why the highest simulated waves were next to the wide leading uplift area along the coast of Coquimbo (Fig. 14). Further, we observed that most of the tsunami energy propagated perpendicular to the strike of the fault plane, which is consistent with the theory of directivity (Okal 1988). For example, related to the 2012 Haida Gwaii earthquake, high amplitude tsunami waves were computed to be at the Hawaiian Islands, far from the earthquake epicenter (Fig. 9), since they are positioned on the path of the main tsunami energy flux (Fig. 7). This result is in a good accordance with the instrumentally recorded data. Similarly, after the

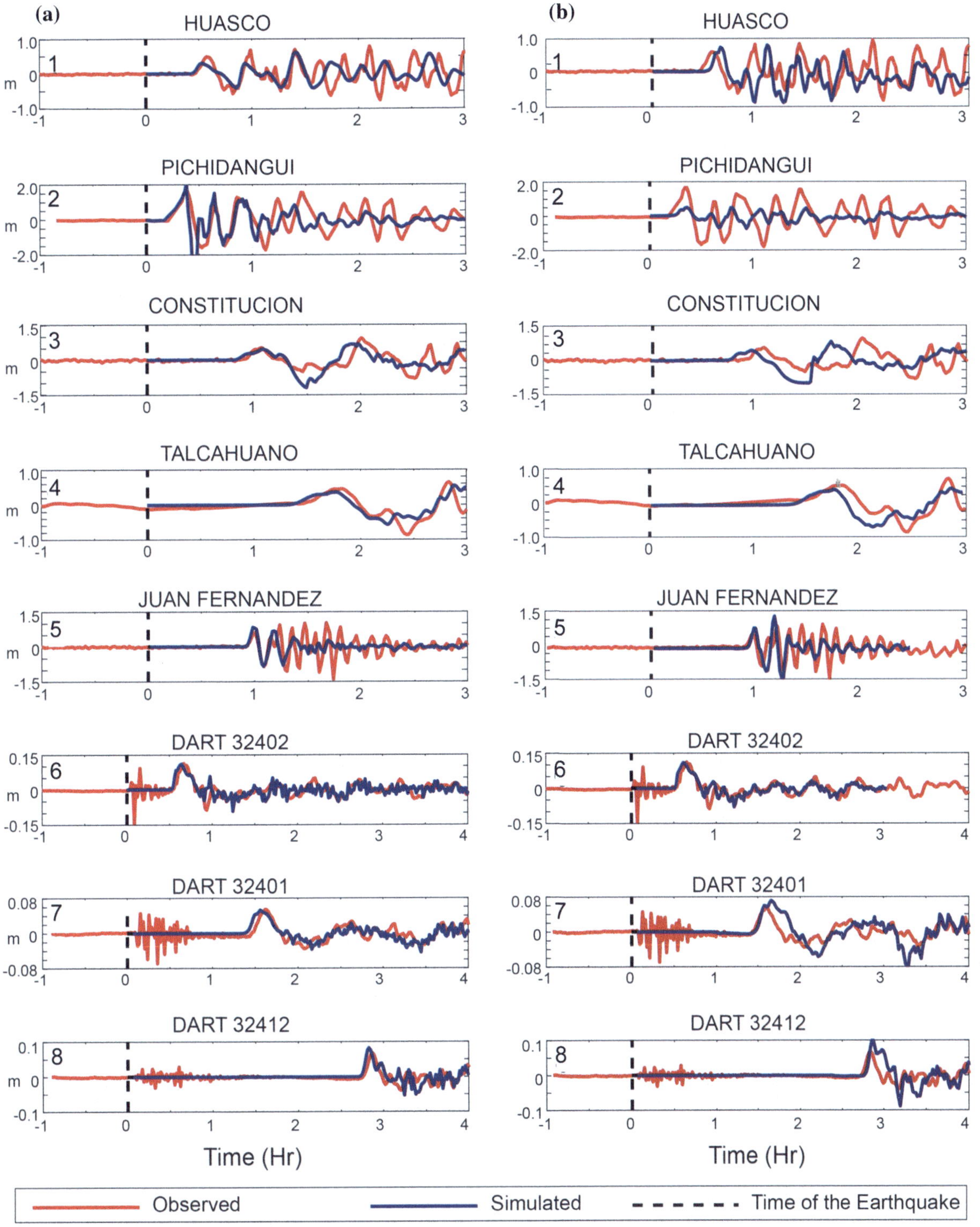

Figure 18
Comparison of the observed tsunami wave elevations with the calculated (synthetic) ones from numerical tsunami simulations based on **a** the finite-fault (non-uniform) and **b** the uniform slip distribution models of the 2015 Illapel (Chile) earthquake. The numbers correspond to the locations of DART and tide gauge stations shown in Fig. 17

Table 2

Comparison of the observed tsunami wave characteristics recorded at tide gauges and DART stations with the calculated (synthetic) ones for the 16 September 2015 Illapel (Chile) earthquake obtained from simulations based on uniform and finite fault (non-uniform) slip models

#	Station	Observed values for tide gauges and DART systems		Calculated values from the non-uniform slip model		Calculated values from the uniform slip model	
		Travel time (min)	First wave maximum (m)	Travel time (min)	First wave maximum (m)	Travel time (min)	First wave maximum (m)
1	Huasco	26.166	0.499	25.602	0.287	30.234	0.662
2	Pichidangui	12.168	1.769	12.105	1.996	16.032	0.425
3	Constitucion	50.316	0.505	50.406	0.470	46.116	0.888
4	Talcahuano	88.620	0.400	83.340	0.378	80.232	0.360
5	Juan Fernandez	55.365	0.575	55.368	0.811	54.373	1.450
6	DART 32402	31.602	0.113	31.602	0.112	31.680	0.110
7	DART 32401	84.462	0.055	82.620	0.053	84.600	0.071
8	DART 32412	163.620	0.068	163.62	0.082	166.620	0.104

Travel time from the earthquake in minutes and the first wave maximum values in meters are given for the observed and calculated tsunami waves. The numbers correspond to the locations of DART and tide gauge stations shown in Fig. 18

2015 Illapel earthquake, most of the tsunami wave energy was directed towards the Coquimbo area, located at the north of the epicenter, where our source model indicates the largest slip. The highest tsunami waves were observed along the coast of Coquimbo, Tongoy and Concón, where prominent flooding was reported. The tsunami energy was also directed towards the Juan Fernández Islands that are located perpendicular to the fault plane orientation. The altered far-field wave characteristics in the coastal areas of the Juan Fernández Islands might be from the existence of the Juan Fernández Ridge and the O'Higgins Seamount (Fig. 14).

Our results imply that we adequately simulated the leading waves of the observed tsunamis (Figs. 8, 14), while later arriving waves are not clearly modeled. It is known that the leading waves are less influenced from the local geomorphology such as shelf and bay resonance (Calisto et al. 2016; Yamazaki and Cheung 2011). A tsunami source mainly contributes to the shape of the leading wavelength through the calculation of vertical sea floor displacement and the initial tsunami wave field rather than the later arriving waves. The obtained results reveal that our finite-fault slip distribution models are sufficiently good to constrain particularly the shape and arrival times of the first tsunami waves (see Table 2). Following waves, particularly on tide gauge records (see Valparaiso tide gauge in Fig. 14) might be modeled by using higher resolution (i.e., 0.05 arcmin) bathymetry and topography data especially for the nearest coastal areas, since they significantly affect the near-field tsunami wave characteristics (Mofjeld et al. 2001; Yolsal et al. 2007; Yolsal and Taymaz 2010; Gusman et al. 2014). Geist (2002) already reported that the local conditions (shallow bathymetry, coastal topography, local resonance, trapped modes of edge waves, and run-up dynamics), where the tide gauge is installed can affect their response. In contrast to tide gauge records, DART records are not affected by coastal effects; hence they provide more reliable information about the slip distribution and are more appropriate for tsunami models. The differences between the observed DART records and synthetic waveforms might be due to the source errors in slip models (Calisto et al. 2016). Gusman et al. (2016), among others, also pointed out the importance of dispersion effects on tsunami wave propagation. They found that the dispersion effects reduce the maximum wave amplitudes by up to 20% compared to the conventional linear long-wave propagation models. From this point of view, our results are reasonably coherent with the observed shape and first peak amplitudes in DART records, revealing the reliability of our slip models (Figs. 8, 14). In addition, tsunami simulation results acquired from uniform and non-uniform slip models indicate that the heights of the leading waves and the arrival times were not coherently estimated when uniform source models were used (see Fig. 18a, b for details). Average slip values and the approximate dimensions of the faulting area determined by using the aftershock area for many cases do not give accurate tsunami predictions. In order to adequately estimate the fit between the observed and synthetic tsunami data, the non-uniform (finite fault) slip models including variable depth, slip, rake angle, rupture time and seismic moment for multiple sub-faults on the fault plane are more appropriate than the uniform slip models. The differences between observed and simulated waveforms along the coasts are evident in simulation results. Nevertheless, the use of uniform slip model approaches in simulations has principal advantages, like their simplicity and necessity of limited source parameters resulting in fast tsunami observations (see also Geist and Dmowska 1999; Ulutaş 2013; MacInnes et al. 2013).

Overall, this study enhances our understanding of the potential tsunami risk associated with similar magnitude earthquakes on the Cascadia and Peru–Chile subduction zones along populated coastal areas. As suggested by Calisto et al. (2016), although tsunami simulations based on teleseismic finite-fault slip models do not produce perfect results, the arrival times and amplitudes of tsunami waves arriving at the coast provide sufficient information to facilitate tsunami warning and risk mitigation challenges. The results can be used for evaluating tsunami wave characteristics of potential future shallow tsunamigenic earthquakes ($M_w > 7.5$) near the Haida Gwaii and Chile regions. We further emphasize that joint inversions of GPS, tsunami, teleseismic and strong ground motion records and higher resolution bathymetry data are needed to obtain better correlations between observed and synthetic tsunami data, especially for the later arriving waves.

Acknowledgements

This study is supported by the Scientific and Technological Research Council of Turkey (TÜBİTAK, Project No: ÇAYDAG-114Y066). We would like to thank the İstanbul Technical University Research Fund (İTÜ-BAP) and Alexander von Humboldt-Stiftung (AvH) for their financial support. Generic Mapping Tools (GMT) (Wessel and Smith 1998) and SAC2000 (Goldstein et al. 2003) software packages were used to prepare figures, and to process conventional earthquake data, respectively. Waveform data are recorded by the Global Digital Seismograph Network (GDSN) and by the International Federation of Digital Seismograph Networks (FDSN) stations, and archived and distributed by the Incorporated Research Institutions for Seismology-Data Management Center (IRIS-DMC). Earthquake hypocenters are taken from the USGS-NEIC earthquake catalogue. Tsunami waves and their interaction with various topographies were numerically modeled using the SWAN code (Mader 1988). ArcMap tools (ESRI 2010) were used to prepare initial wave height and tsunami simulation figures. We would like to thank Alessandro Annunziato for sharing the Tsunami Analysis Tool (TAT) (Annunziato 2007). We appreciate much judicial evaluations of Alexander B. Rabinovich, Fred Stephenson, anonymous referees and beneficial discussions with Tuna Eken to improve the manuscript.

Appendix

In the Appendix, we presented tsunami simulation results based on preliminary finite-fault slip distribution models reported by the USGS (2012b, 2015b) for the October 28, 2012 Haida Gwaii (Canada) and the September 16, 2015 Illapel (Chile) earthquakes. We further compared the calculated tsunami wave elevations from numerical tsunami simulations based on finite-fault slip distribution models of earthquakes obtained by (a) this study and (b) USGS (2012b, 2015b) with the observed data.

We also provided two tables showing the general information on DART and tide gauge stations. This section consists of four Appendix figures and two tables with captions (see Figs. 19, 20, 21, 22; Tables 3, 4).

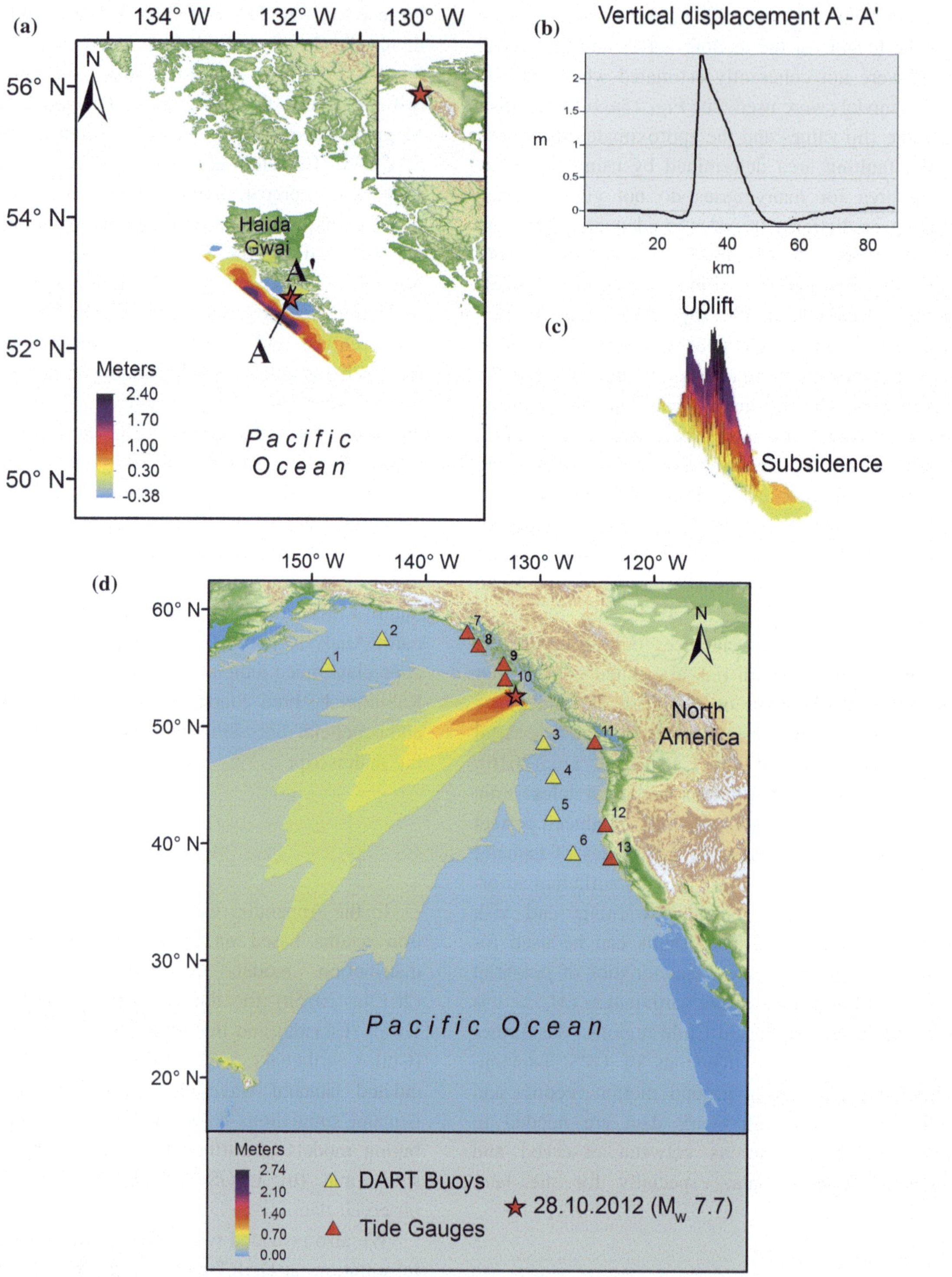

Figure 19

a Initial surface elevations from the finite-fault slip distribution model of the October 28, 2012 Haida Gwaii (Canada) earthquake reported by USGS (2012b). The red star indicates the earthquake epicenter; **b** cross-section of the surface displacement along *A*–*A′* based on the calculated seafloor deformation; **c** a 3D view of the calculated deformation; **d** computed maximum tsunami wave heights for the October 28, 2012 Haida Gwaii (Canada) earthquake with a computational grid of 2 arcmin. The triangles with numbers show the locations of DART and tide gauge stations

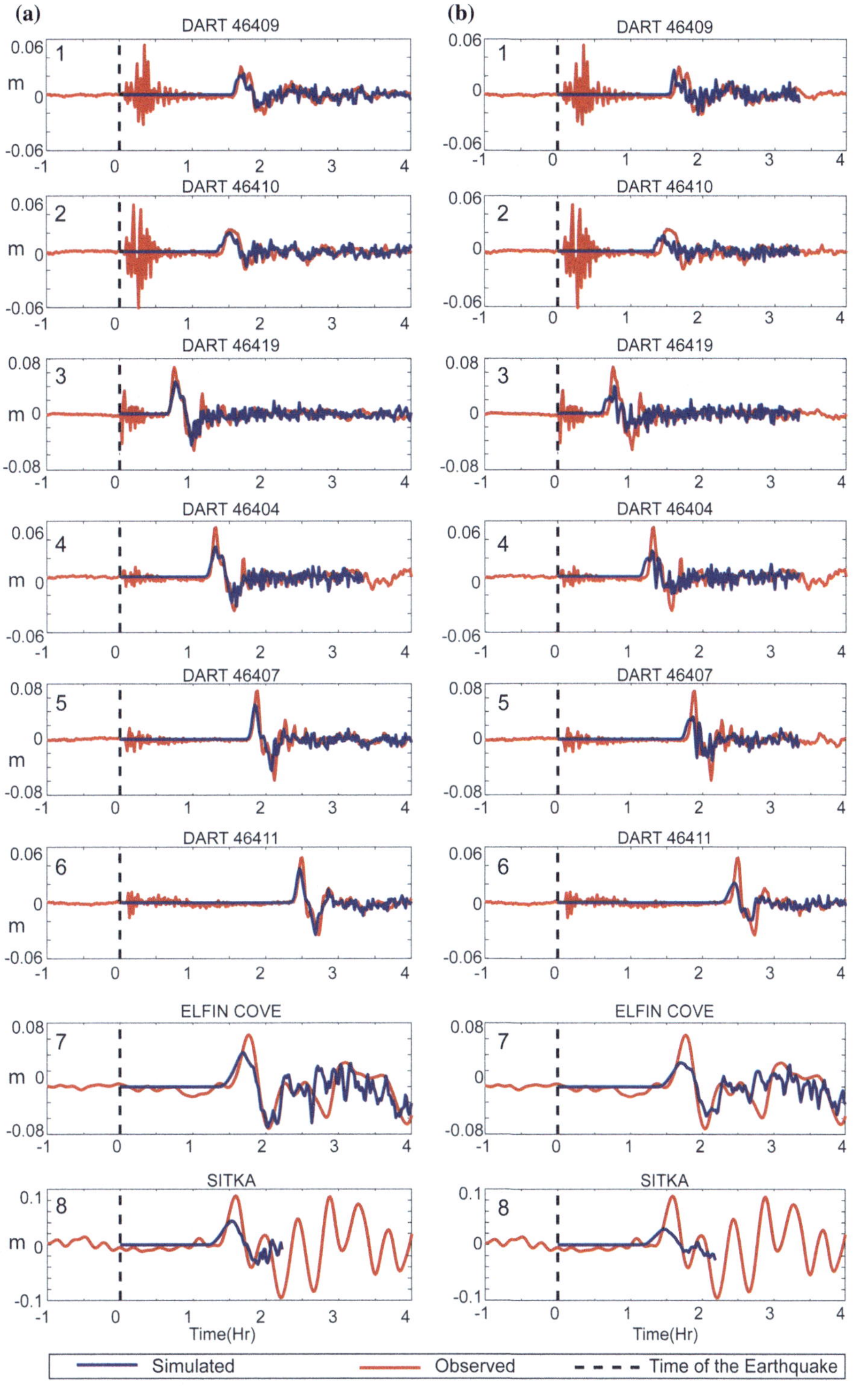

Figure 20
Comparison of the observed tsunami wave elevations with the calculated (synthetic) ones from numerical tsunami simulations based on the finite-fault slip distribution models of the 2012 Haida Gwaii (Canada) earthquake determined by **a** this study and **b** USGS (2012b). The numbers correspond to the locations of DART and tide gauge stations shown in Fig. 19

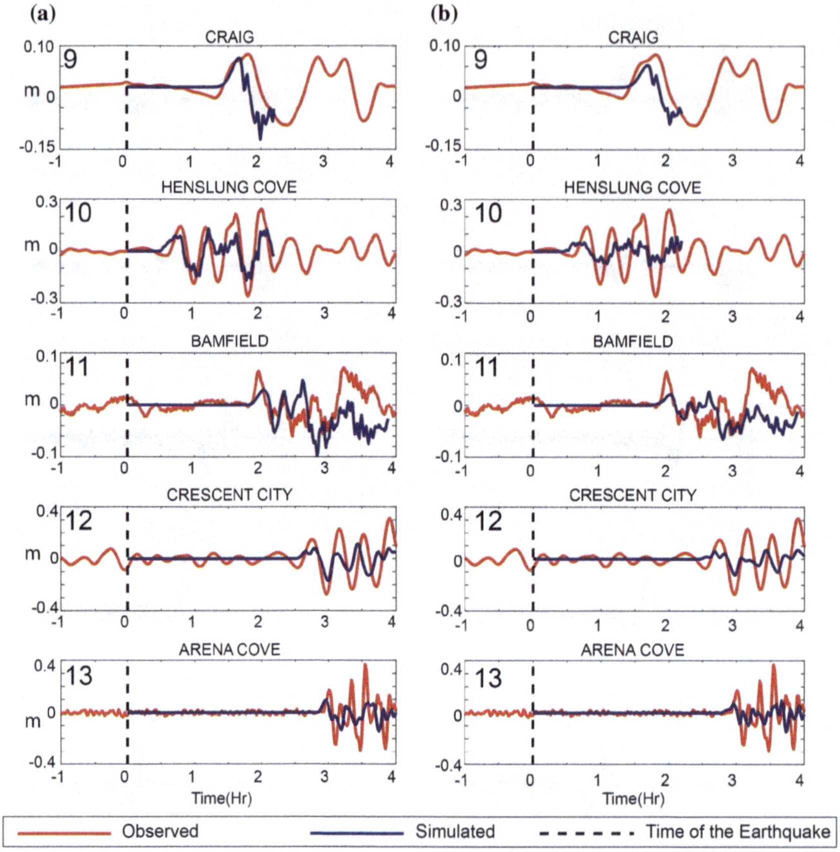

Figure 20
continued

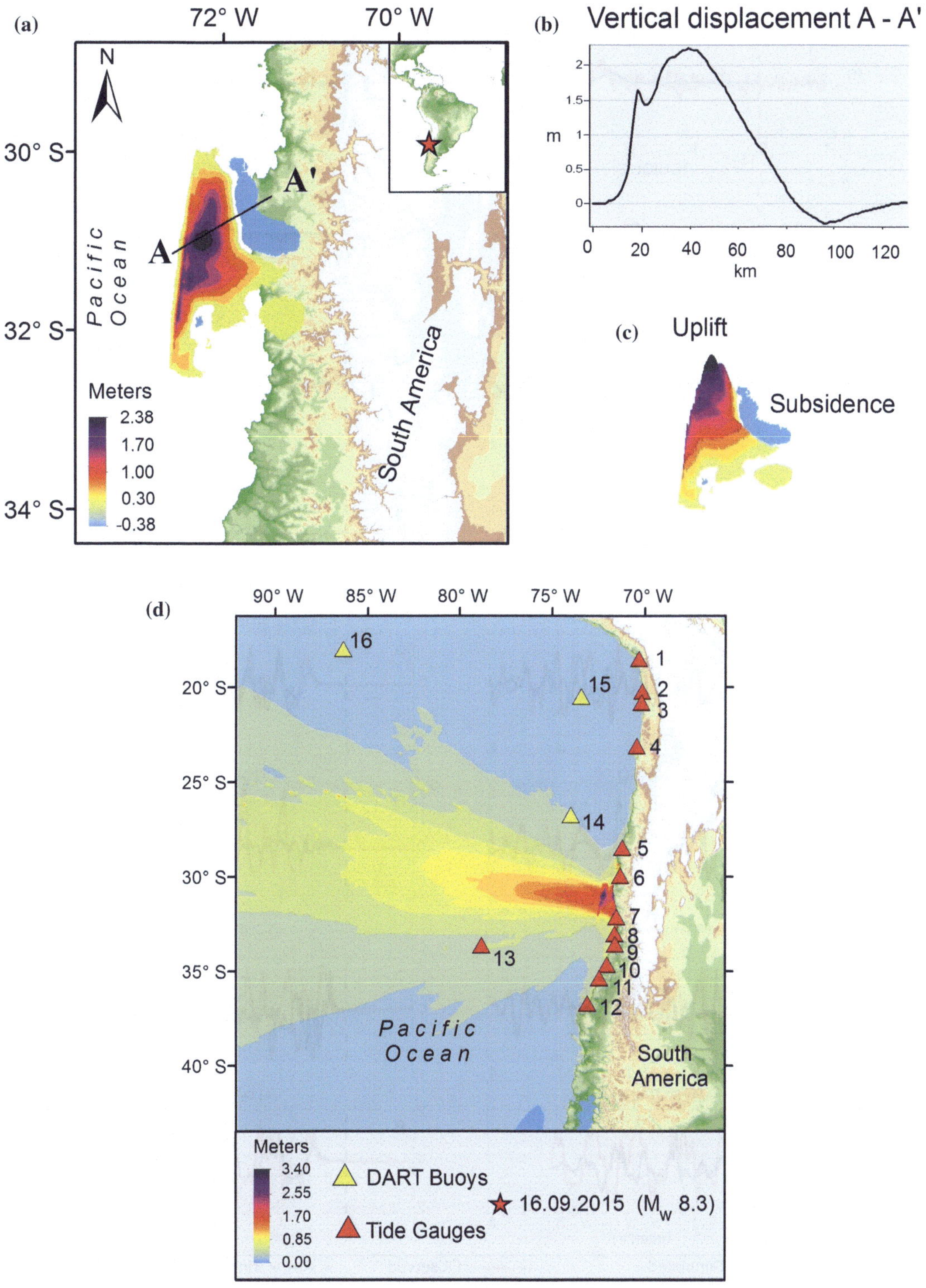

Figure 21

a Initial surface elevations from the finite-fault slip distribution model of the September 16, 2015 Illapel (Chile) earthquake reported by the USGS (2015b). The red star indicates the earthquake epicenter; **b** cross-section of the surface displacement along *A–A′* based on the calculated seafloor deformation; **c** a 3D view of the calculated deformation; **d** computed maximum tsunami wave heights for the September 16, 2015 Illapel (Chile) earthquake with a computational grid of 2 arcmin. The triangles with numbers show the locations of DART and tide gauge stations

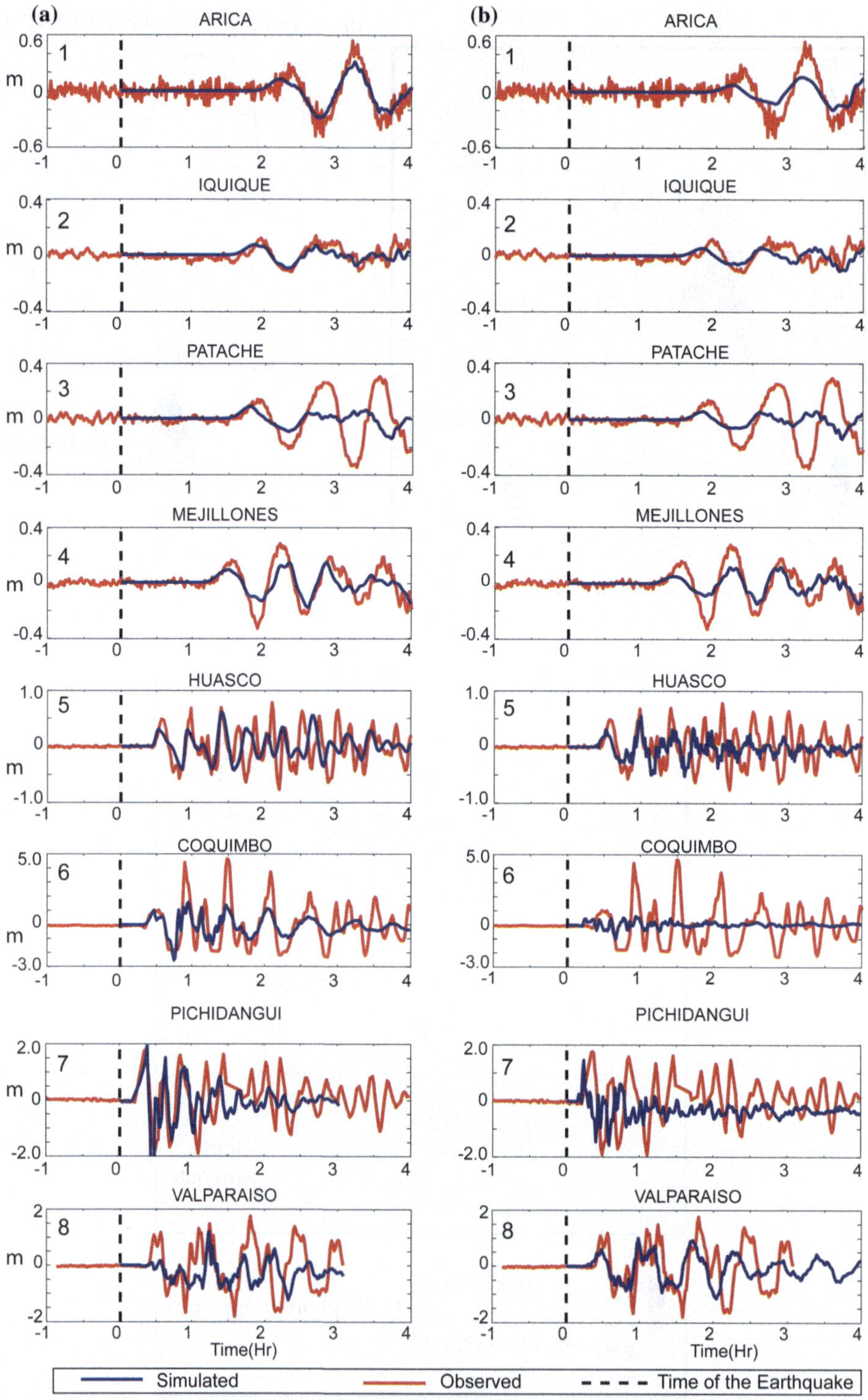

Figure 22
Comparison of the observed tsunami wave elevations with the calculated (synthetic) ones from numerical tsunami simulations based on finite-fault slip distribution models of the 2015 Illapel (Chile) earthquake determined by **a** this study and **b** USGS (2015b). The numbers correspond to the locations of DART and tide gauge stations shown in Fig. 21

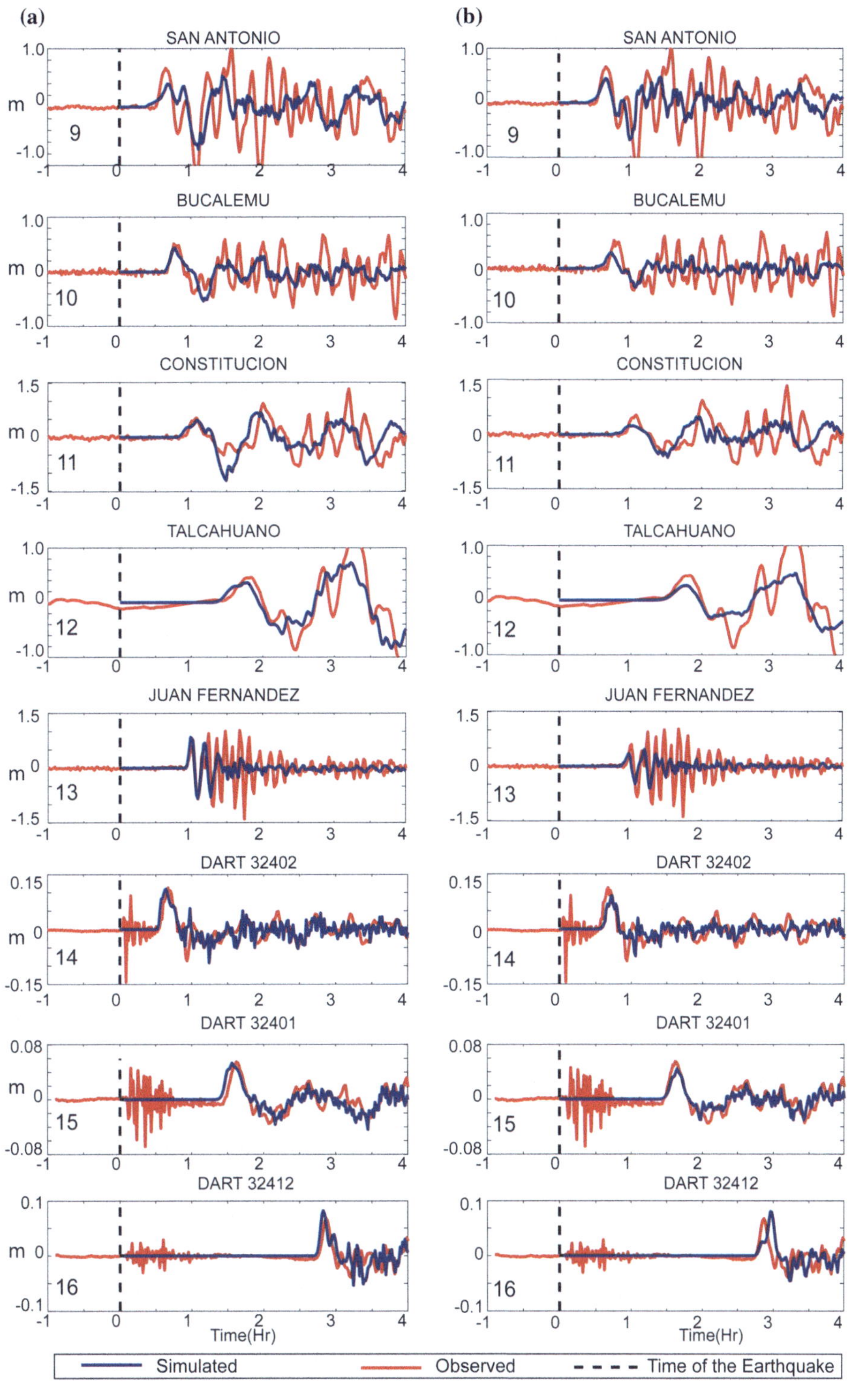

Figure 22
continued

Table 3

List of DART buoys used in this study (NDBC 2012, 2015)

Station	Lat. (°N)	Lon. (°E)	Water depth (m)
46404	45.855	− 128.775	2793
46407	42.664	− 128.807	3322
46409	55.300	− 148.515	4200
46410	57.635	− 143.786	3783
46411	39.349	− 127.021	4259
46419	48.766	− 129.633	2775
32412	− 17.980	− 86.330	4387
32401	− 20.473	− 73.429	4797
32402	− 26.743	− 73.983	4070

Table 4

List of tide gauge stations used in this study (SLSMF 2012, 2015; PSMSL 2012, 2015)

Station	Country	Lat. (°N)	Lon. (°E)	Connection
Elfin Cove	USA	58.195000	− 136.346667	Web
Sitka	USA	57.051667	− 135.341667	Web
Craig	USA	55.488333	− 133.141667	Web
Henslung Cove	Canada	54.183330	− 133.000000	Ftp
Bamfield	Canada	48.840000	− 125.140000	Ftp
Crescent City	USA	41.745000	− 124.183333	Web
Arena Cove	USA	38.913333	− 123.708333	Web
Coquimbo	Chile	− 29.95010	− 71.33530	Web
Valparaiso	Chile	− 33.02730	− 71.62590	Web
Talcahuano	Chile	− 36.70090	− 73.10630	Web
Pichidangui	Chile	− 32.13560	− 71.52930	Web
Huasco	Chile	− 28.46080	− 71.22380	Web
Arica	Chile	− 18.47580	− 70.32323	Web
Iquique	Chile	− 20.20458	− 70.14783	Web
Mejillones	Chile	− 23.09771	− 70.45066	Web
Patache	Chile	20.80321	− 70.19803	Web
Quintero	Chile	− 32.77549	− 71.52543	Web
Juan Fernandez	Chile	− 33.63602	− 78.82988	Web
San Antonio	Chile	− 33.58163	− 71.61819	Web
Bucalemu	Chile	− 34.63938	− 72.04624	Web
Constitucion	Chile	− 35.35606	− 72.45798	Web

References

Aki, K. (1972). Earthquake mechanism. *Tectonophysics, 13*(1–4), 423–446.

Ammon, C. J., Lay, T., Kanamori, H., & Cleveland, M. (2011). A rupture model of the 2011 off the Pacific coast of Tohoku-Oki earthquake. *Earth Planets Space, 63,* 693–696.

Angermann, D., Klotz, J., & Reigber, C. (1999). Space-geodetic estimation of the Nazca-South America Euler vector. *Earth and Planetary Science Letters, 171,* 329–334.

Annunziato, A. (2007). The Tsunami assesment modelling system by the joint research center. *Science of Tsunami Hazards, 26,* 70–92.

Aranguiz, R., González, G., González, J., et al. (2016). The 16 September 2015 Chile tsunami from the post-tsunami survey and numerical modeling perspectives. *Pure and Applied Geophysics, 173,* 333–348.

Araujo, M., & Suárez, G. (1994). Geometry and state of stress of the subducted Nazca Plate beneath central Chile and Argentina: Evidence from teleseismic data. *Geophysical Journal International, 116,* 283–303.

Arvidsson, R., & Ekström, G. (1998). Global CMT analysis of moderate earthquakes, $M_w \geq 4.5$, using intermediate-period surface waves. *Bulletin of the Seismological Society of America, 88*(4), 1003–1013.

Bahlburg, H., Nentwig, V., & Kreutzer, M. (2017). The September 16, 2015 Illapel tsunami, Chile—Sedimentology of tsunami deposits at the beaches of La Serena and Coquimbo. *Marine Geology*. https://doi.org/10.1016/j.margeo.2016.12.011.

Beck, S., Barrientos, S., Kausel, E., & Reyes, M. (1998). Source characteristics of historic earthquakes along the central Chile subduction zone. *Journal of South American Earth Sciences, 11,* 115–129.

Bird, A. L., & Lamontagne, M. (2015). Impacts of the October 2012 magnitude 7.8 earthquake near Haida Gwaii, Canada. 105. *Bulletin of the Seismological Society of America, 105*(2B), 1178–1192.

Bostwick, T.K. (1984). *Re-examination of the August 22, 1949 Queen Charlotte Earthquake*. Ph.D. Thesis, The Faculty of Graduate Studies Department of Geophysics and Astronomy, The University of British Columbia, Canada, p. 115.

Calisto, I., Miller, M., & Constanzo, I. (2016). Comparison between tsunami signals generated by different source models and the observed data of the Illapel 2015 Earthquake. *Pure and Applied Geophysics, 173,* 1051–1061.

Cassidy, J. F., Rogers, G. C., & Hydnman, R. D. (2014). An overview of the 28 October 2012 M_w 7.7 earthquake in Haida Gwaii, Canada: A tsunamigenic thrust event along a predominantly strike-slip margin. *Pure and Applied Geophysics, 171,* 3457–3465.

Chunyan, Q., Ronghu, Z., XinJian, S., et al. (2017). Coseismic and post-seismic deformation fields mapped using satelliteradar interferometry and fault slip inversion of the 2015 M_w 8.3 Illapel earthquake. *Chile. Journal of Geodynamics, 104,* 36–48.

De Mets, C., Gordon, R. G., & Argus, D. F. (2010). Geologically current plate motion. *Geophysical Journal International, 181,* 1–80.

Dziewonski, A. M., & Anderson, D. L. (1981). Preliminary reference Earth model. *Physics of the Earth and Planetary Interiors, 25,* 297–356.

ESRI. (2010). *ArcGIS desktop: Release 10*. Redlands, CA: Environmental Systems Research Institute.

Fichtner, A., Trampert, J., Cupillard, P., Saygin, E., Taymaz, T., Capdeville, Y., et al. (2013). Multiscale full waveform inversion. *Geophysical Journal International, 194*(1), 534–556.

Filloux, J. H. (1982). Tsunami recorded on the open ocean floor. *Geophysical Research Letters, 9*(1), 25–28.

Filloux, J. H. (1983). Pressure fluctuations on the open ocean floor off the Gulf of California: Tides, earthquakes, tsunamis. *Journal of Physical Oceanography, 13*(5), 783–796.

Fine, I. V., Cherniawsky, J. Y., Thomson, R. E., Rabinovich, A. B., & Krassovski, M. V. (2015). Observations and numerical modeling of the 2012 Haida Gwaii tsunami off the coast of British Columbia. *Pure and Applied Geophysics, 172,* 699–718.

Franchello, G. (2010). Shoreline tracking and implicit source terms for a well-balanced inundation model. *International Journal of Numerical Methods Fluids, 63,* 1123–1146.

Fuentes, M., Riquelme, S., Hayes, G., et al. (2016). A Study of the 2015 M_w 8.3 Illapel earthquake and tsunami: Numerical and analytical approaches. *Pure and Applied Geophysics, 173,* 1847–1858.

Fukahata, Y., Yagi, Y., & Matsuura, M. (2003). Waveform inversion for seismic source processes using ABIC with two sorts of prior constraints: Comparison between proper and improper formulations. *Geophysical Research Letters*. https://doi.org/10.1029/2002gl016293.

Geist, E. L. (2002). Complex earthquake rupture and local tsunamis. *Journal of Geophysical Research, 107,* B5. https://doi.org/10.1029/2000JB000139.

Geist, E. L., & Dmowska, R. (1999). Local tsunamis and distributed slip at the source. *Pure and Applied Geophysics, 154,* 485–512.

George, D. L., & LeVeque, R. J. (2006). Finite volume methods and adaptive refinement for global tsunami propagation and indundation. *Science of Tsunami Hazards, 24*(5), 319–328.

Gica, E., Titov, V. V., Moore, C., & Wei, Y. (2015). Tsunami simulation using sources inferred from various measurement data: Implictions for the model forecast. *Pure and Applied Geophysics, 172,* 773–789.

Goda, K., Yasuda, T., Mori, N., & Maruyama, T. (2016). New scaling relationships of earthquake source parameters for stochastic tsunami simulation. *Coastal Engineering Journal*. https://doi.org/10.1142/S0578563416500108.

Goldstein, P., Dodge, D., Firpo, M., & Minner, L. (2003). SAC2000: Signal processing and analysis tools for seismologists and engineers. In W. H. K. Lee, H. Kanamori, P. C. Jennings, & C. Kisslinger (Eds.), *Contribution to "The IASPEI International Handbook of Earthquake and Engineering Seismology"*. London: Academic.

González, F. I., Bernard, E. N., Meinig, C., Eblé, M. C., Mofjeld, H. O., & Stalin, S. (2005). The NTHMP tsunameter network. *Natural Hazards, 35,* 25–39.

González, F.I., Milburn, H.M., Bernard, E.N., & Newman, J.C. (1998). Deep-ocean Assessment and Reporting of Tsunamis (DART®): Brief overview and status report. In *Proceedings of the International Workshop on Tsunami Disaster Mitigation, 19–22 January 1998*, Tokyo, Japan.

Gosselin, J. M., Cassidy, J. F., & Dosso, S. E. (2015). Shear-wave velocity structure in the vicinity of the 2012 M_w 7.8 Haida Gwaii earthquake from receiver function inversion. *Bulletin of the Seismological Society of America, 105*(2B), 1106–1113.

Gusman, A. R., Mulia, I. E., Satake, K., Watada, S., Heidarzadeh, M., & Sheehan, A. F. (2016). Estimate of tsunami source using optimized unit sources and including dispersion effects during tsunami propagation: The 2012 Haida Gwaii earthquake. *Geophysical Research Letters, 43,* 9819–9828.

Gusman, A. R., Tanioka, Y., MacInnes, B. T., & Tsushima, H. (2014). A methodology for near-field tsunami inundation forecasting: Application to the 2011 Tohoku tsunami. *Journal of Geophysical Research Solid Earth, 119,* 8186–8206.

Heidarzadeh, M., Murotani, S., Satake, K., Ishibe, T., & Gusman, A. R. (2015). Source model of the 16 September 2015 Illapel, Chile, M_w 8.4 earthquake based on teleseismic and tsunami data. *Geophysical Research Letters, 43,* 643–650.

Henry, C., & Das, S. (2001). Aftershock zones of large shallow earthquakes: Fault dimensions, aftershock area expansion and scaling relations. *Geophysical Journal International, 147,* 272–293.

Hyndman, R. D. (2015). Tectonics and structure of the Queen Charlotte Fault Zone, Haida Gwaii, and large thrust earthquakes. *Bulletin of the Seismological Society of America*. https://doi.org/10.1785/012014018.

Hyndman, R. D., & Ellis, R. M. (1981). Queen Charlotte Fault Zone: Microearthquakes from a temporary array of land stations and ocean bottom seismographs. *Canadian Journal of Earth Sciences, 18,* 776–788.

Imamura, F. (1996). Review of tsunami simulation with a finite-difference method. In H. Yeh, P. Liu, & C. Synolakis (Eds.), *Long-Wave Runup Models* (pp. 25–42). Singapore: World Scientific.

Imamura, F., Shuto, N., & Goto, C. (1988). Numerical simulations of the transoceanic propagation of tsunamis. 6th Congress APD-IAHR, pp. 265–272.

JRC. (2015). *M_w 8.3 Chile—Earthquake and tsunami*. Ispra: European Commision Joint Research Centre.

Kaiser, G., Scheele, L., Kortenhaus, A., Løvholt, F., Romer, H., & Leschka, S. (2011). The influence of land cover roughness on the results of high resolution tsunami inundation modeling. *Natural Hazards and Earth System Sciences, 11,* 2521–2540.

Kajiura, K. (1963). The leading wave of a tsunami. *Bulletin of the Earthquake Research Institute, 41,* 535–571.

Kanamori, H. (1977). The energy release in great earthquakes. *Journal of Geophysical Research, 82,* 2981–2986.

Kanamori, H. (1994). Mechanics of earthquakes. *Annual Review of Earth and Planetary Sciences, 22,* 207–237.

Kao, H., Shan, S.-J., & Farahbod, A. M. (2015). Source characteristics of the 2012 Haida Gwaii earthquake sequence. *Bulletin of the Seismological Society of America, 105*(2B), 1206–1218.

Klein, E., Vigny, C., Fleitout, L., et al. (2017). A comprehensive analysis of the Illapel 2015 M_w 8.3 earthquake from GPS and InSAR data. *Earth and Planetary Science Letters, 469,* 123–134.

Kotani, M., Imamura, F., & Shuto, N. (1998). Tsunami run-up simulation and damage estimation by using geographical information system. *Proceedings of Coastal Engineering JSCE, 45,* 356–360. **(in Japanese)**.

Lay, T., Ye, L., Kanamori, H., Yamazaki, Y., Cheung, K. F., Kwong, K., et al. (2013). The October 28, 2012 M_w 7.8 Haida Gwaii underthrusting earthquake and tsunami: Slip partitioning along the Queen Charlotte Fault transpressional plate boundary. *Earth and Planetary Science Letters, 375,* 57–70.

Lemoine, A., Madariaga, R., & Campos, J. (2002). Slab-pull and slab-push earthquakes in the Mexican, Chilean and Peruvian subduction zones. *Physics of the Earth and Planetary Interiors, 132,* 157–175.

Leonard, M. (2010). Earthquake fault scaling: Self-consistent relating of rupture length, width, average displacement, and moment release. *Bulletin of the Seismological Society of America, 100,* 1971–1988.

Leonard, L. J., & Bednarski, J. M. (2014). Field survey following the 28 October 2012 Haida Gwaii tsunami. *Pure and Applied Geophysics, 171*(12), 3467–3482.

Leonard, L. J., & Bednarski, J. M. (2015). The preservation potential of coastal coseismic and tsunami evidence observed following the 2012 M_w 7.8 Haida Gwaii thrust earthquake. *Bulletin of the Seismological Society of America, 105*(2B), 1280–1289.

Lorito, S., Romano, F., Piatanesi, A., & Boschi, E. (2008). Source process of the September 12, 2007, M_w 8.4 southern Sumatra earthquake from tsunami tide gauge record inversion. *Geopysical Research Letters, 35,* L02310. https://doi.org/10.1029/2007gl032661.

Løvholt, F., Pedersen, G., Bazin, S., Kuhn, D., Bredesen, R. E., & Harbitz, C. (2012). Stochastic analysis of tsunami runup due to heterogeneous coseismic slip and dispersion. *Journal of Geophysical Research, 117,* C03047.

MacInnes, B. T., Gusman, A. R., Leveque, R. J., & Tanioka, Y. (2013). Comparison of earthquake source models for the 2011 Tohoku Event using tsunami simulations and near-field observations. *Bulletin of the Seismological Society of America, 103*(2B), 1256–1274.

Mader, C. (1988). *Numerical modeling of water waves* (p. 206). Berkeley, California: University of California Press.

Mai, P. M., & Beroza, G. C. (2000). Source scaling properties from finite-fault-rupture models. *Bulletin of the Seismological Society of America, 90,* 604–615.

MathWorks (2012). Signal processing toolbox, Butterworth filter design: User's Guide (R2012a), https://www.mathworks.com/help/signal/ref/butter.html. Accessed Nov 2017.

McCaffrey, R., Zwick, P., & Abers, G. A. (1991). SYN4 program. *IASPEI Software Library, 3,* 81–166.

Melgar, D., Fan, W., Riquelme, S., et al. (2016). Slip segmentation and slow rupture to the trench during the 2015, M_w 8.3 Illapel, Chile earthquake. *Geophysical Research Letters, 43,* 961–966.

Minster, J. B., & Jordan, T. H. (1978). Present day plate motions. *Journal of Geophysical Research, 83,* 5331–5354.

Mofjeld, H. O., Titov, V. V., González, F. I., & Newman, J. C. (2001). Tsunami scattering provinces in the Pacific Ocean. *Geophysical Research Letters, 28*(2), 35–337.

NDBC (2012, 2015). National Data Buoy Center. http://www.ndbc.noaa.gov/. Accessed Sept 2015.

NGDC-NOAA (2012, 2015). National Geophysical Data Center. http://www.ngdc.noaa.gov/hazard/recenttsunamis.shtml. Accessed Sept 2015.

Okada, Y. (1985). Surface deformation due to shear and tensile faults in a half-space. *Bulletin of the Seismological Society of America, 75,* 1135–1154.

Okal, E. A. (1988). Seismic parameters controlling far-field tsunami amplitudes: A review. *Natural Hazards, 1,* 67–96.

Okuwaki, R., Yagi, Y., Aranguiz, R., González, J., & González, G. (2016). Rupture process during the 2015 Illapel, Chile earthquake: Zigzag-along-dip rupture episodes. *Pure and Applied Geophysics, 173,* 1011–1020.

Okuwaki, R., Yagi, Y., & Hirano, S. (2014). Relationship between high frequency radiation and asperity ruptures, revealed by hybrid backprojection with a non-planar fault model. *Scientific Reports, 4*(1), 7120. https://doi.org/10.1038/srep07120.

Omira, R., Baptista, M. A., & Lisboa, F. (2016). Tsunami characteristics along the Peru-Chile Trench: Analysis of the 2015 M_w 8.3 Illapel, the 2014 M_w 8.2 Iquique and the 2010 M_w 8.8 Maule tsunamis in the near-field. *Pure and Applied Geophysics, 173,* 1063–1077.

Papazachos, B. C., Scordilis, E. M., Panagiotopoulos, D. G., Papazachos, C. B., & Karakaisis, G. F. (2004). Global relations between seismic fault parameters and moment magnitude of earthquakes. *Bulletin of the Geological Society of Greece, 36,* 1482–1489.

Perez, O. J., & Jacob, K. H. (1980). Tectonic model and seismic potential of the eastern Gulf of Alaska and Yakataga seismic gap. *Journal of Geophysical Research, 85,* 7132–7150.

Piatanesi, A., Tinti, S., & Pagnoni, G. (2001). Tsunami waveform inversion by numerical finite-elements Green's functions. *Natural Hazards Earth System Sciences, 1,* 187–194.

PSMSL (2012, 2015). Permanent service for mean sea level. Obtaining tide gauge. http://www.psmsl.org/data/obtaining/. Accessed Sept 2015.

Rabinovich, A. B., Candella, R. N., & Thomson, R. E. (2013). The open ocean energy decay of three recent trans-Pacific tsunamis. *Geophysical Research Letters, 40,* 3157–3162. https://doi.org/10.1002/grl.50625.

Rabinovich, A. B., & Eblé, M. C. (2015). Deep-ocean measurements of tsunami waves. *Pure and Applied Geophysics, 172,* 3281–3312.

Rabinovich, A. B., Thomson, R. E., Titov, V. V., Stephenson, F. E., & Rogers, G. C. (2008). Locally generated tsunamis recorded on the coast of British Columbia. *Atmosphere–Ocean, 46*(3), 343–360.

Riddihough, P. (1977). A model for recent plate interactions off Canada's west coast. *Canadian Journal of Earth Sciences, 14,* 384–396.

Rogers, G. (1983). *Seismotectonics of British Columbia.* Ph.D. thesis, University of British Columbia.

Ruiz, S., Klein, E., del Campo, F., et al. (2016). The seismic sequence of the 16 September 2015 M_w 8.3 Illapel, Chile earthquake. *Seismological Research Letters.* https://doi.org/10.1785/0220150281.

Saltogianni, V., Taymaz, T., Yolsal-Çevikbilen, S., Eken, T., Moschas, F., & Stiros, S. (2016). Fault model for the 2015 Leucas (Aegean Arc) earthquake: analysis based on seismological and geodetic observations. *Bulletin of the Seismological Society of America, 107*(1), 433–444.

Santos, A., Tavares, A. O., & Queiros, M. (2016). Numerical modelling and evacuation strategies for tsunami awareness: Lessons from the 2012 Haida Gwaii Tsunami. *Geomatics, Natural Hazards and Risk, 7*(4), 1442–1459.

Satake, K. (1987). Inversion of tsunami waveforms for the estimation of a fault heterogeneity: Method and numerical experiments. *Journal of Physics of the Earth, 35*(3), 241–254.

Satake, K., & Heidarzadeh, M. (2017). A review of source models of the 2015 Illapel, Chile earthquake and insights from tsunami data. *Pure and Applied Geophysics, 174,* 1–9.

Shuto, N., & Goto, C. (1978). Numerical simulation of tsunami run-up. *Coastal Engineering in Japan, 21,* 13–20.

SLSMF (2012, 2015). Sea Level Station Monitoring Facility. Station Lists. http://www.ioc-sealevelmonitoring.org/list.php. Accessed Sept 2015.

Spudich, P., & Frazer, L. (1984). Use of ray theory to calculate high-frequency radiation from earthquake sources having spatially variable rupture velocity and stress drop. *Bulletin of the Seismological Society of America, 74*(6), 2061–2082.

Taymaz, T., Jackson, J., & Westaway, R. (1990). Earthquake mechanisms in the Hellenic Trench near Crete. *Geophysical Journal International, 102,* 695–731.

Taymaz, T., & Price, S. (1992). The 1971 May 12 Burdur earthquake sequence, SW Turkey: A synthesis of seismological and geological observations. *Geophysical Journal International, 108,* 589–603.

Tichelaar, B. W., & Ruff, L. J. (1991). Seismic coupling along the Chilean subduction zone. *Journal of Geophysical Research, 96,* 11997–12022.

Tilmann, F., Zhang, Y., Moreno, M., et al. (2016). The 2015 Illapel earthquake, central Chile: A type case for a characteristic earthquake? *Geophysical Research Letters, 43,* 574–583.

Titov, V. V., & González, F. I. (1997). *Implementation and testing of the method splitting tsunami (MOST) model*. NOAA Technical Memorandum ERL PMEL-112.

Ulutaş, E. (2011). Tsunami simulation of the October 25, 2010, South Pagai Island, Sumatra earthquake. *International Journal of Physical Sciences, 6,* 459–475.

Ulutaş, E. (2013). Comparison of the seafloor displacement from uniform and non-uniform slip models on tsunami simulation of the 2011 Tohoku-Oki earthquake. *Journal of Asian Earth Sciences, 62,* 568–585.

Ulutaş, E., Inan, A., & Annunziato, A. (2012). Web-based tsunami early warning system: A case study of the 2010 Kepulaunan Mentawai earthquake and tsunami. *Natural Hazards Earth System Sciences, 12,* 1855–1871.

USGS (2012a). United States Geological Survey, Earthquake Hazards Program, Origin of the October 28, 2012 earthquake. https://earthquake.usgs.gov/earthquakes/eventpage/usp000juhz#origin. Accessed Nov 2012.

USGS (2012b). United States Geological Survey, Earthquake Hazards Program, Preliminary Finite Fault Results for the October 28, 2012 Haida Gwaii earthquake. https://earthquake.usgs.gov/earthquakes/eventpage/usp000juhz#finite-fault. Accessed Nov 2012.

USGS (2015a). United States Geological Survey, Earthquake Hazards Program, Origin of September 16, 2015 Chile earthquake. https://earthquake.usgs.gov/earthquakes/eventpage/us20003k7a#origin. Accessed Nov 2015.

USGS (2015b). United States Geological Survey, Earthquake Hazards Program, Preliminary Finite Fault Results for the Sep 16, 2015 M_w 8.2 earthquake. https://earthquake.usgs.gov/earthquakes/eventpage/us20003k7a#finite-fault. Accessed Nov 2015.

Vigny, C., Rudloff, A., Ruegg, J.-C., Madariaga, R., Campos, J., & Alvarez, M. (2009). Upper plate deformation measured by GPS in the Coquimbo Gap, Chile. *Physics of the Earth and Planetary Interiors, 175*(1–2), 86–95.

Wang, K., He, J., Schulzeck, F., Hyndman, R. D., & Riedel, M. (2015). Thermal condition of the 27 October 2012 M_w 7.8 Haida Gwaii subduction earthquake at the obliquely convergent Queen Charlotte margin. *Bulletin of the Seismological Society of America, 105*(2B), 1290–1300.

Wells, D. L., & Coppersmith, K. J. (1994). New empirical relationships among magnitude, rupture length, rupture width, rupture area, and surface displacement. *Bulletin of the Seismological Society of America, 84*(4), 974–1002.

Wessel, P., & Smith, W. H. F. (1998). New, improved version of the generic mapping tools released. *EOS Transactions American Geophysical Union, 79,* 579.

Yagi, Y., & Fukahata, Y. (2011). Introduction of uncertainty of Green's function into waveform inversion for seismic source processes. *Geophysical Journal International, 186,* 711–720.

Yagi, Y., & Kikuchi, M. (2000). Source rupture process of the Kocaeli, Turkey, earthquake of August 17, 1999, obtained by joint inversion of near-field data and teleseismic data. *Geophysical Research Letters, 27,* 1969–1972.

Yagi, Y., Nakao, A., & Kasahara, A. (2012a). Smooth and rapid slip near the Japan Trench during the 2011 Tohoku-oki earthquake revealed by a hybrid back-projection method. *Earth and Planetary Science Letters, 355–356,* 94–101.

Yagi, Y., Nishimura, N., & Kasahara, A. (2012b). Source process of the 12 May 2008 Wenchuan, China, earthquake determined by waveform inversion of teleseismic body waves with a data covariance matrix. *Earth Planets Space, 64,* e13–e16.

Yalçıner, A. C., Pelinovsky, E., Talipova, T., Kurkin, A., Kozelkov, A., & Zaitsev, A. (2004). Tsunamis in the Black Sea: Comparison of the historical, instrumental, and numerical data. *Journal of Geophysical Research Oceans, 109,* C12023. https://doi.org/10.1029/2003JC002113.

Yamazaki, Y., & Cheung, K. F. (2011). Shelf resonance and impact of near-field tsunami generated by the 2010 Chile earthquake. *Geophysical Research Letters, 38,* L12605. https://doi.org/10.1029/2011GL047508.

Ye, L., Lay, T., Kanamori, H., & Koper, K. (2016). Rapidly estimated seismic source parameters for the 16 September 2015 Illapel, Chile M_w 8.3 earthquake. *Pure and Applied Geophysics, 173,* 321–332.

Yin, J., Yang, H., Yao, H., & Weng, H. (2016). Coseismic radiation and stress drop during the 2015 M_w 8.3 Illapel, Chile megathrust earthquake. *Geophysical Research Letters*. https://doi.org/10.1002/2015GL067381.

Yolsal, S., & Taymaz, T. (2010). Sensitivity analysis on relations between earthquake source rupture parameters and tsunami waves: Case studies in the Eastern Mediterranean region. *Turkish Journal of Earth Sciences, 19*(3), 313–349.

Yolsal, S., Taymaz, T., & Yalçıner, A. C. (2007). Understanding tsunamis, potential source regions and tsunami prone mechanisms in the Eastern Mediterranean. *The Geodynamics of the Aegean and Anatolia, Special Publication Geological Society, London, Special Publications, 291,* 201–230.

Yolsal-Çevikbilen, S., & Taymaz, T. (2012). Earthquake source parameters along the Hellenic subduction zone and numerical simulations of historical tsunamis in the Eastern Mediterranean. *Tectonophysics, 536–537,* 61–100.

Zwick, P., McCaffrey, R., & Abers, G. A. (1994). MT5 program, IASPEI Software Library 4.

(Received November 23, 2017, revised September 7, 2018, accepted September 12, 2018, Published online September 24, 2018)

Pure Appl. Geophys. 176 (2019), 3035–3057

https://doi.org/10.1007/s00024-019-02123-5

Pure and Applied Geophysics

Numerical Modeling of the June 17, 2017 Landslide and Tsunami Events in Karrat Fjord, West Greenland

Alexandre Paris,[1,3] Emile A. Okal,[2] Cyrielle Guérin,[3] Philippe Heinrich,[3] François Schindelé,[3] and Hélène Hébert[3]

Abstract—On June 17 2017, the western coast of Greenland was the site of a tsunami which flooded several villages, killing 4 people and destroying 11 houses in the village of Nuugaatsiaq. This tsunami was triggered by a subaerial landslide which occurred in a fjord 32 km ENE of Nuugaatsiaq. This paper presents the numerical modeling of this landslide of $\sim$ 50 million m^3 and of the tsunami propagation from its source to Nuugaatsiaq. The landslide is considered as a granular flow under gravity forces and the water waves generated are related to the displacement of the sea bottom. The results obtained are similar in amplitude to our inferences from videos, *i.e.*, three water waves between 1 and 1.5 m arriving at Nuugaatsiaq with a period of roughly 3 min, and are also in general agreement with the amplitude (1 m) resulting from deconvolution of oscillations recorded on a horizontal seismogram operating at Nuugaatsiaq (NUUG). According to the field survey performed by Fritz et al. (EGU General Assembly Conference Abstracts, Vol. 20 of EGU General Assembly Conference Abstracts, p 18345, 2018a) on July 2017, a second mass next to the landslide is threatening Karrat Fjord. A sensitivity study is realized on its volume, with 2, 7, 14 and 38 million m^3 reaching the sea. The shape of the water waves is found to be independent of volume, and linearity is observed between the volume and the water wave heights. Finally, the orientation of the slide does not seem to influence either the period or the shape of the generated water waves.

Key words: Tsunami, landslide, Greenland, simulation.

1. Introduction

Tsunamis generated by landslides, either subaerial or submarine, can be as dangerous for coastal populations as earthquake-generated tsunamis (Thomson et al. 2001; Synolakis et al. 2002). Because landslides are often (but not always) triggered by major earthquakes, and because, especially in the marine environment, they can remain conspicuously hidden to direct observation, it can occasionally be difficult to determine the exact generation process of major tsunamis (Geist 2000; Okal and Synolakis 2001; Synolakis et al. 2002). On the other hand, the different physical nature of the source has allowed the definition of robust discriminants (Okal and Synolakis 2003). A review of problems associated with a simulation of landslide tsunamis can be found for example in Fine et al. (2003).

Subaerial landslides, which initiate above sea level and penetrate the water column, pose additional challenges. Their aerial components lend themselves to direct observation, which can provide important constraints on the initial dynamics of the sliding process. However, their evolution into the water column, and in particular the degree of cohesion of the material, is often poorly documented. The landmark subaerial slide remains to this day the Lituya Bay event of 10 July 1958, which was triggered by a major strike-slip earthquake on the Fairweather fault in the panhandle of Alaska (Miller 1960) and resulted in a record-breaking 525-m run-up on the opposite side of Gilbert Inlet, a geometry repeated on a smaller scale during the 2007 Aysén earthquake in Southern Chile (Sepúlveda and Serey 2009). By contrast, "orphan" subaerial landslides have taken place in the absence of any detectable seismic trigger, a typical example being the 1999 event at Fatu Hiva, Marquesas (Okal et al. 2002; Hébert et al. 2002). The 2015 Taan Fjord landslide at Icy Bay, Alaska featured a geometry of sliding strikingly similar to that of the 1958 Lituya

[1] Laboratoire de Géologie, Ecole Normale Supérieure, CNRS UMR8538, PSL Research University, Paris, France. E-mail: alexandre.antoine.paris@gmail.com
[2] Department of Earth and Planetary Sciences, Northwestern University, Evanston, IL, USA.
[3] CEA, DAM, DIF, 91297 Arpajon Cedex, France.

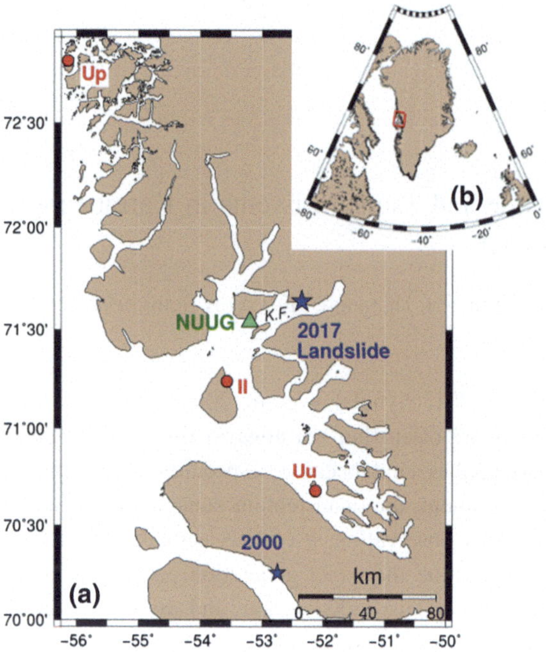

Figure 1
a Close-up Mercator projection of the western coast of Greenland, showing location of the 2017 landslide (star) in Karrat Fjord (K.F.), and of the village of Nuugaatsiaq (triangle). The red dots identify other locations where the tsunami was observed (*Il* Illorsuit, *Up* Upernavik, *Uu* Uummannaq). The location of the 2000 slide is also shown. **b** Map of Greenland showing boundaries of the close-up map (**a**) in thick red

Bay event, albeit on a smaller scale reaching "only" 193 m of run-up, and with no detectable seismic trigger (Higman et al. 2018). Subaerial landslides have been studied in the laboratory and successfully modeled analytically, notably by Fritz (2002), Weiss et al. (2009) and Viroulet et al. (2013).

This paper deals with a subaerial landslide that occurred on June 17, 2017, on the Northern slope of the Karrat Fjord along the western coast of Greenland (see Fig. 1 for context and location). It was followed by a tsunami which was observed in several villages up to a distance of 160 km. In Nuugaatsiaq (pop. ~ 80), 11 houses were damaged or washed away, 4 persons were reported killed or missing, and 9 injured (Clinton et al. 2017). Several videos,[1] posted on the Internet after the event, document water flooding houses in Nuugaatsiaq. Thanks to the ruler tool and the location of houses at different times before and after the event on GoogleEarth, the maximum inundation length is estimated at 150 m and maximum run-up heights at about 10 m in Nuugaatsiaq.

The 2017 landslide was identified the next day by an aerial survey of the Arktisk Kommando of the Danish Defence,[2] about 32 km ENE of Nuugaatsiaq, upstream Karrat Fjord. The failed mass consists of a portion of bluff about 1000 m wide and 500 m tall (see red ellipse on Fig. 2); its scar is well identified on photographs from the aerial survey, and from before-and-after satellite imagery. We note (see blue ellipse on Fig. 2) that a second mass seems about to fail, immediately to the West of the 2017 slide.

As discussed below, three previous landslide-generated tsunamis had been documented in Greenland in 1954, 2000 and 2012, with the 2000 event similar in nature to the 2017 one, making this kind of event a recurring hazard in the region.

The 2017 landslide in Karrat Fjord was detected as a seismic event worldwide (up to distances of 95°); however, the NEIC of the USGS did not locate the event, but simply used the location reported from satellite imagery (71.640°N; 52.344°W), for which they obtained an origin time of 23:39:12 GMT. The event was given magnitudes M_L = 3.3, m_b = 3.6 and M_s = 4.0 by the International Data Center of the CTBTO, and M_s = 4.2 by the NEIC. The growth of magnitude with period is typical of a non dislocative event, whose source is expected to have a longer duration than a genuine seismic source of comparable size, and as a result to be deficient in high frequencies (Okal 2003).

In addition to teleseismic recordings, the 2017 Karrat Fjord event was observed in Nuugaatsiaq (NUUG) on a 3-component Streckeisen STS-2 seismometer operated by the Danish Geological Survey. As will be detailed in Sect. 3.3, a remarkable aspect of the resulting waveforms is that they include a quantifiable recording of the tsunami, a priceless occurrence in the absence of a tidal gauge station.

[1] To see the videos, click on the links below:
https://www.youtube.com/watch?v=jBmkT5y52ng
https://www.youtube.com/watch?v=LzSUDBbSsPI
https://www.youtube.com/watch?v=amWshLXe74s
https://www.youtube.com/watch?v=tWvYFMo2LsQ.

[2] Photos of the aerial survey are visible here: https://goo.gl/XRJomU.

Figure 2
Aerial photograph of the 2017 slide (red ellipse) and the western potential landslide (blue ellipse)

The purpose of this paper is to provide a numerical modeling of the landslide and of the tsunami, as well as a quantitative interpretation of the seismic record at NUUG. We model both the Karrat slide and its tsunami using the numerical code Avalanche (Heinrich et al. 2001a), developed at the Commissariat à l'énergie atomique et aux énergies alternatives, and widely applied in previous landslide studies, both submarine (Rodriguez et al. 2013; Poupardin et al. 2017) and subaerial (Heinrich et al. 2001b). Assuming the collapse of a 53 million m^3 landslide, we simulate water waves of 1–1.5 m amplitude, at 8 m water depth, reaching the village after 8 min.

In addition, we simulate the potential failure of the "second", precarious mass identified to the West of the 2017 slide (see blue ellipse on Fig. 2), using several volume scenarios, the purpose being to try to identify the potential threat for the village of Nuugaatsiaq.

2. Previous Similar Events

Three other tsunamis are documented for Greenland in the NOAA/PMEL database in 1954, 2000 and 2012. However, the 1954 event is a rogue wave whose origin is undetermined (ICAO 1955), and that in 2012 is due to calving of a fjord glacier into the ocean (NOAA 2018). By contrast, the November 21, 2000 tsunami, investigated in detail by Pedersen et al. (2002) and Dahl-Jensen et al. (2004), is directly comparable to the 2017 event. It consisted of a large volume of 90 million m^3 of rock, sliding at Paatut, only 155 km South of Karrat Fjord, between altitudes of 1400 and 1000 m, of which 30 million m^3 sunk into the ocean. The resulting tsunami reached a height of 50 m at the source, and inundated 250 m for a run-up of 28 m at the village of Qullissat which, fortunately, had been abandoned 28 years earlier. Dahl-Jensen et al. (2004) also document another tsunamigenic landslide at Qullissat in 1952, which killed 1 person.

Landslides having generated tsunamis of comparable characteristics have been described worldwide and can be triggered by several mechanisms. Paramount among them are local earthquakes, in a context reminiscent of the record 1958 event at Lituya Bay (Miller 1960): for example, the 2007 landslides in Aysen, Southern Chile (Naranjo et al. 2009), totalling $\sim$ 20 million m^3 in volume, generated a tsunami which ran up $\sim$ 30 m at the source and 14 m on the opposite side of the fjord, following a strike-slip earthquake of magnitude $m_b = 6.1$, only 30 km away. Volcanic eruptions can result in the build-up of unstable or precarious structures, failing through large scale submarine or aerial landslides, as happened on December 30, 2002, at Stromboli, Italy, with two slides separated by 7 min, one submarine (20 million m^3), the other aerial with a volume of 4–9 million m^3 (Tinti et al. 2006). The aerial one was more similar to the 2017 Karrat case, as it had an altitude of 550 m, and generated waves 12 m high at a village located 2 km away.

In many cases, the triggering mechanism is of such small magnitude as to remain unnoticed. That would be the case, for example, of the 1999 Fatu Hiva landslide in the Marquesas Islands (4 million m^3 of brecciated volcanic material), which flooded the nearby village of Omoa (Hébert et al. 2002; Okal et al. 2002), or of the large, tragic landslides in Norwegian fjords described by Hermanns et al. (2006) at Tafjord (1934) and Loen (1905 and 1936), the latter involving the closed Lovatnet lake. The Tafjord failure (1.5 million m^3) took place $\sim$ 800 m above sea level and produced a tsunami which ran up 62 m and killed 41 people; the 1905 Loen failure was smaller (0.4 million m^3) but its tsunami ran up 41 m, and killed 61 people, while the 1936 slide involved 1 million m^3, a 74-m runup and 73 fatalities. The 2007 landslide into Chehalis Lake, B.C. involved 3 million m^3 of rock and produced a run-up of 38 m on the opposite side of the lake (Wang et al. 2015), fortunately without loss of life. A particularly tragic case of a landslide-generated tsunami in a closed lake occurred in 1963 in the Vajont reservoir in Northern Italy, where a 200-m tall splash overtopped the dam and wiped out the village of Longarone in its lee, killing upwards of 1900 people (Ward and Day 2011). A possible, if not probable, repeat of this situation at Sarez Lake, Tajikistan could reach even more catastrophic dimensions, as it would probably destroy the precarious natural dam holding the lake, itself the result of the blockage of the valley by a much larger, 2.4-km^3 landslide during the earthquake of 18 February 1911 (Schuster and Alford 2004; Ambraseys and Bilham 2012).

With the exception of the latter, the selection of events listed above are generally comparable to the 2017 Karrat Fjord landslide. In the context of global warming, and of the melting of ice caps, it is expected that the weakening of permafrost will lead to an increase in catastrophic landslides in that part of the world (Haeberli and Gruber 2009; Huggel et al. 2012).

3. *Observations of the 17 June 2017 Event*

3.1. *Numerical Reconstruction of the Landslide*

In order to perform a numerical modeling of the landslide, two digital surface models (DSM) of the area were compiled before and after the event. The "before" DSM was realized from Spot6 stereoscopic images, acquired on July 22, 2013 at 1.5-m spatial resolution. The "after" DSM was obtained from Pleiades tri-stereoscopic images, acquired right after the event, on July 08, 2017, at 0.7-m spatial resolution, with estimated 1 m in vertical precision (Guérin et al. 2014).

The DSMs are calculated through an automatic pipeline which performs the data co-registration and the DSM generation, as described in Guérin et al. (2014). The data co-registration is realized as a preprocessing step in order to ensure that the image (and hence the DSM) are finely registered between them. This operation consists in a bundle-block adjustment of all the available data which can be performed without any Ground Control Points (GCP), as none are available over the area of interest. The complete methodology is presented in Guérin (2017) and is based on a tie-points detection according to a pairwise image correlation followed by an iterative refinement of the image acquisition models provided as Rational Polynomial Coefficients. Tie-points detection remains a sensitive part of the method, especially when dealing with areas featuring steep slopes. In order to get accurate tie-points, the detection is then performed on the images after their orthorectification with the most accurate digital terrain model (DTM) available (Guérin 2017).

Once the registration is performed, the DSMs are generated at each date according to the methodology described in Guérin et al. (2014) and based on so-called ground space image matching, performed with

Figure 3
Difference between the "before" and "after" digital surface models (DSM), superimposed on the "after" DSM, computed from the pleiades tri-stereoscopic images from 2017. The variations of the difference elevation values (in m) are represented with the color bar

the open source software MICMAC, developed by the French National Geographic Institute (IGN) (Pierrot-Deseilligny and Paparoditis 2006). This methodology allows the calculation of an elevation value for each point on a grid defined on the ground space with both planimetric and altimetric steps. In practice, for each ground point of the grid, the image coordinates are obtained according to refined acquisition models of the images that link the ground position of the point and its coordinates on the images. A correlation score is then computed between pixel windows selected over each image and at each altitude of the grid. The final altitude value is chosen considering the correlation score and a regularization term (Guérin et al. 2014). Figure 4 represents the DSM obtained from the pleiades tri-stereoscopic images acquired after the event. For this study, both DSM were generated with a 2 m planimetric step and 1 m altimetric step.

Finally, one computes the difference between the DSMs, at the same resolution and hence perfectly comparable. Figure 3 represents the resulting differential DSM over the collapsed area, superimposed on the "after event" DSM. We use it to infer that the collapsed area reached a height of 240 m and a total volume of 48 million m^3. However, part of this volume did not spill into the ocean, as confirmed by the presence of important subaerial debris on the mountain side (Fig. 2); we estimate a volume of 45 million m^3 for the effective spill into the ocean. The inferred width of the slide is 1000 m, and its length 500 m (Fig. 4).

These numbers are in general agreement with Gauthier et al. (2018), who describe the Karrat landslide as a rock avalanche of 58 million m^3, of which only 45 million m^3 reached the water, the slide being initially 950 m wide and 800 m long, at altitudes between 800 and 1200 m. The origin of the 10% discrepancy in volume may be rooted in the decimation of the high-resolution satellite images (sampled at 1.5 and 0.7 m) when building the simulation grid (sampled at 25 m).

In a recent contribution, Chao et al. (2018) have used regional recordings of the Karrat Fjord event across Greenland to invert the source characteristics, using Ekström and Stark's (2013) methodology. They

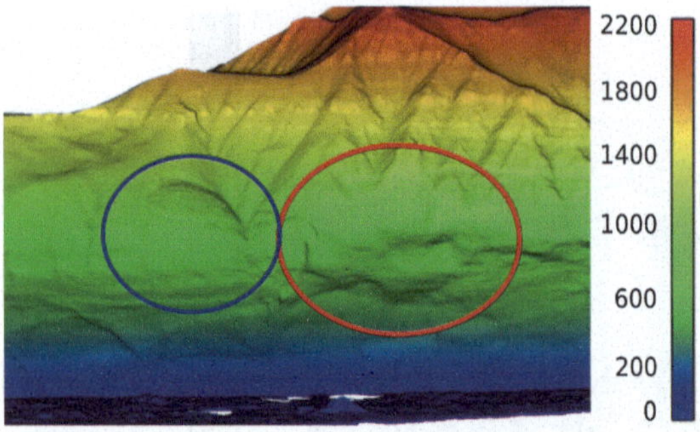

Figure 4
Digital surface model computed from the pleiades tri-stereoscopic images from 2017 with a 2 m planimetric step and a 1 m altimetric step. The color bar corresponds to the elevation values (in m). The red ellipse identifies the collapsed area; the blue ellipse identifies the potential West slide

obtain a significantly larger volume penetrating the sea of about 75 million m^3.

3.2. Damage

The tsunami reached and flooded the two main villages in the area, first Nuugaatsiaq, 32 km WSW of the landslide, and later Illorsuit, an additional 35 km to the SSW. At Nuugaatsiaq, 4 people were killed and 11 houses were completely washed away by the water waves. Figure 5 shows examples of damage by the tsunami at Nuugaatsiaq. Much smaller impact and flood were observed at Illorsuit.

GoogleEarth images taken before and after the event clearly show holes where houses once stood (Fig. 6), which document scouring of their foundations, possibly during an ebbing phase. Several photos and videos taken during the event are available and allow a further evaluation of the damage caused by the tsunami. On some videos,[3] recording starts before any visually detectable wave arrival, which suggests a first wave with either a negative amplitude (leading depression, withdrawal), or a positive one of very small amplitude.

One of the videos[4] shows the water waves stopping before the cemetery at Nuugaatsiaq. As an effect of perspective, the exact line of inundation is difficult to evaluate. According to GoogleEarth, the distance between the coast and the cemetery is about 200 m, so we can conclude that the inundation distance was about 150 m at this location, with an estimated run-up of about 10 m.

3.3. Analysis of the Seismic Signal

Figure 7 shows a record of the event at the seismic station NUUG (71.538°N, 53.200°W), located at Nuugaatsiaq, 32 km from the landslide, and part of the Danish Seismological Network. Note that the horizontal components were, at the time, rotated 79° clockwise from their standard orientation (Clinton, pers. comm., 2018). In addition to classical P and S phases (interpreted here as crustal P_g and S_g), a long-period oscillation with a period of about 3 min is present starting at 23:47 GMT, *i.e.*, 8 min after the body waves. This waveform is much too late to be interpreted as a traditional surface wave from the same source as the body waves (as it would have reached NUUG in at most 15 s); we also note that it is not observed at others stations in Greenland and across Baffin Bay in Canada. Accordingly, we tentatively interpret it as a seismic record of the tsunami.

We note the remarkable similarity between the waveforms obtained at NUUG and those recorded at Panarea Island during the tsunamigenic landslides of 30 December 2002 at Stromboli Volcano, Italy (La Rocca et al. 2004). In particular, a spectrogram analysis of the S11°E seismogram (Fig. 8) shows that the long-period oscillation is peaked between 6 and 8 mHz, within the range of frequencies (6–15 mHz) quoted by La Rocca et al. (2004).

However, we observe on Figs. 7 and 9 that the principal component of ground motion is horizontal and polarized in the direction $N20°$W which is close to 90° away from the azimuth from NUUG to the source ($\beta = 69°$), as was the case at Panerea. Rather, the observed polarization of the signals at NUUG is essentially across the width of the narrow Karrat channel offshore of Nuugaatsiaq (Fig. 9), and as such reminiscent of the geometry of seiching reported in the Panama Canal by McNamara et al. (2011), where the seismic signals recorded by a local seismometer were polarized perpendicular to the axis of the canal. This observation indicates that part of the wave

[3] This video for example: https://www.youtube.com/watch?v=amWshLXe74s.

[4] Around 0:46 on this video: https://www.youtube.com/watch?v=jBmkT5y52ng.

Figure 5
a, **b** Examples of damage from 2017 at Nuugaatsiaq. **c** Screenshot from a YouTube video (https://www.youtube.com/watch?v=jBmkT5y52ng) showing the tsunami flooding Nuugaatsiaq. The blue house is being washed away. Rocks in the foreground delimit the cemetery of Nuugaatsiaq. The running man (yellow arrow) escaped safely

activity in the channel, as recorded on the NUUG seismometer, may involve seiching of the channel.

In the absence of a tidal gauge at Nuugaatsiaq, we next attempt to quantify the seismic record in order to obtain an estimate of the amplitude of the tsunami in Karrat Fjord. For this purpose, we recall that seismic recording of tsunamis was documented in the far field by Yuan et al. (2005) and Hanson and Bowman (2005) at stations located on islands in the Indian Ocean during the 2004 Sumatra tsunami. Using Ward's (1980) representation of tsunamis as a special branch of the Earth's normal modes, Okal (2007) later showed that such recordings could be successfully interpreted and quantified by assuming that the horizontal seismometer is simply deployed on the ocean floor in the absence of the island, and responds to the combination $u_x^{app.}$ of the horizontal displacement of the solid Earth, of the tilt induced on the ocean floor by the passage of the tsunami wave, and of a change of gravitational potential accompanying the tsunami, as detailed by Gilbert (1980), who had been motivated by the need to apply small corrections (of at most 10%) to the precise quantification of the conventional spheroidal modes of the Earth. By contrast, Okal (2007) showed that these "corrections" could reach several orders of magnitudes in the case of tsunami modes, and defined a "Gilbert Response Function", $GRF(\omega)$, that could be used in the frequency domain to restore the spectral amplitude of the vertical sea-surface motion of the tsunami, $\eta(\omega)$, from that of the apparent ground motion recorded by the instrument, $u_x^{app.}(\omega)$:

$$\eta(\omega) = \frac{u_x^{app.}(\omega)}{GRF(\omega)} \tag{1}$$

with $GRF(\omega)$ readily computed from the various components of the tsunami eigenfunction at the ocean floor:

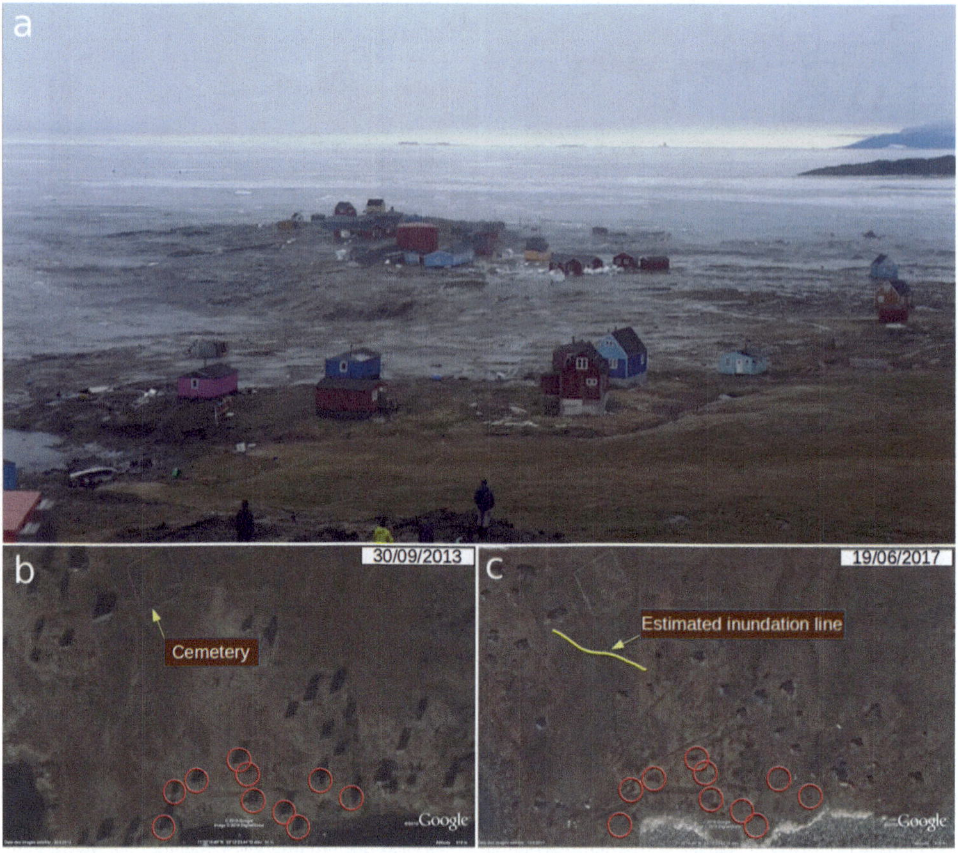

Figure 6
a Looking south, view of Nuugaatsiaq in the aftermath of the tsunami, identifying the inundated zone. Photo from: https://goo.gl/2CYybt. **b**, **c** GoogleEarth images before (30/09/2013) and after (19/06/2017) the event. The yellow line represents the estimated inundation line of the tsunami. Red circles surround the houses washed away by the water waves

$$GRF(\omega) = ly_3 - \frac{l}{r\omega^2}(gy_1 - y_5) \quad (2)$$

in the notation of Saito (1967) and with all details given in Okal (2007).

Despite the extreme simplifying assumptions of this model (ignoring the island or receiving shore), Okal (2007) showed in particular that a deconvolution of a seismic recording using (1) compared favorably with time series obtained on the high seas by DART buoys, and could actually be used quantitatively to estimate an acceptable value of the seismic moment of the parent earthquake.

In the present case, and notwithstanding the added complexity of the probable seiching in Karrat Channel, we adapt the formalism to the case of propagation in very shallow waters, ranging from a probable 400 m in the fjord to $\sim$ 100 m in the channel facing the village of Nuugaatsiaq. In this context, a remarkable aspect of the function $GRF(\omega)$ is that it is essentially independent of water depth H. This property had been verified numerically in the range $H = 4$–5 km and a simple justification provided, in the Appendix to Okal (2007). On Fig. 10 we extend this investigation by computing systematically values of $GRF(\omega)$ for ocean models with depths varying between 100 m and 4 km; we verify that this function varies only minimally with H, and can be approximated by regressing it logarithmically as

$$\log_{10} GRF(\omega) = -2 \log_{10} f - 1.317 \quad (3)$$

where f is the frequency in mHz and the slope has been forced to the exact value − 2, on the basis of the approximation suggested in Equation (A.3) of Okal (2007); we further verify that the constant (− 1.317)

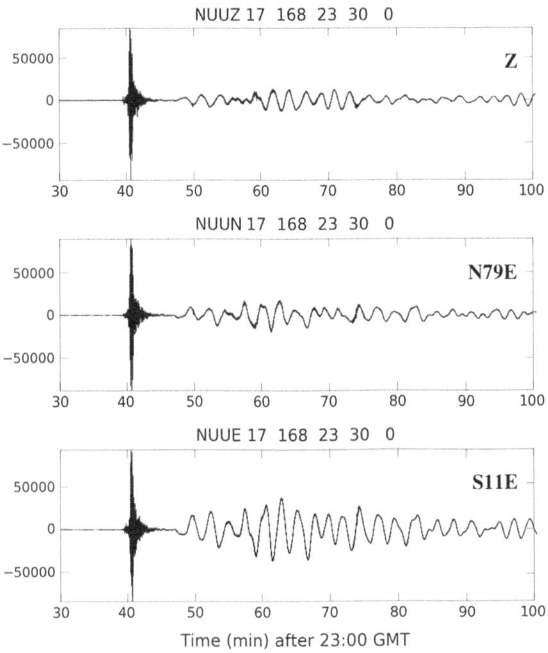

Figure 7
Three-component record of the 2017 Karrat landslide and tsunami at Nuugaatsiaq (NUUG), 32 km away. These are raw seismograms, expressed in digital units, and uncorrected for instrument response (common to the three components). Note that the horizontal components were misoriented, and therefore recorded motion along the N79°E and S11°E directions, respectively (Clinton, pers. comm., 2018). The tsunami is recorded as the long-period oscillation starting at 23:47 GMT; note that it is maximum on the S11°E component

is in excellent agreement with the value predicted by that equation, assuming a rigidity $\mu = 4 \times 10^{11}$ dyn/cm^2, intermediate between crustal and mantle values.

Equation (3) expresses the response to a tsunami of the ocean-solid Earth system in terms of an apparent horizontal displacement of the ocean floor. It is then possible to combine it with the instrument, in our case the STS-2 operated at NUUG. The latter has a broadband *velocity* response, essentially flat below 100 s, before falling like ω^2 at longer periods. This means that the resulting combination has a *displacement* response peaked around 100 s (10 mHz), and falling like ω and ω^{-1}, respectively on each side; in other words, an STS-2 recording a tsunami acts like a moderate band-pass filter centered around 10 mHz.

In this context, we present on Fig. 11 the result of the deconvolution of the S11°E component of the NUUG record of the Karrat Fjord event. This figure is conceptually similar to Figure 10 of Okal (2007), except for adjusted bandwidth parameters. Frame (a) reproduces the raw seismogram in Fig. 7. Frame (b) shows the result of deconvolving the instrument response in the frequency band 1.7–17 mHz, and thus represents the apparent horizontal motion of the ocean floor, $u_x^{app.}(t)$; note that this particular time series has no direct mechanical interpretation. Finally, Frame (c) is obtained by deconvolving the Gilbert Response Function and is thus representative of the particle motion $\eta(t)$ at the surface of the ocean. We recall that Okal (2007) was able to compare favorably the amplitude of the similarly deconvolved record of the 2004 Sumatra tsunami at Amsterdam Island with that of the direct detection of the tsunami by the JASON altimeter satellite (Scharroo et al. 2005), therefore validating the deconvolution procedure. In the present situation, a major unknown regarding the propagation of the tsunami remains the poorly charted and *a priori* variable depth H of the water column, but as we have seen, the function $GRF(\omega)$ is independent of H, which further justifies our procedure. Finally, we have verified that the spectrogram of the resulting time series $\eta(t)$ does not differ significantly from Fig. 8, with maximum spectral amplitudes in the 6–8 mHz range.

We conclude that the record of the tsunami by seismic station NUUG suggests an offshore zero-to-peak amplitude of $\sim$ 1.9 m. However, we note that this maximum is reached only 15 min after the first arrival (around 00:02 GMT on the 18th), probably under the influence of seiching. During the first 10 min, corresponding to the time window of our simulations, the maximum wave amplitude is only $\sim$1 m.

4. Methods

4.1. Landslide Model

For the purpose of simulating the tsunami generated by the Karrat landslide, we use the parameters determined in Sect. 3.1 as a source condition. Some authors (Løvholt et al. 2008; Abadie et al. 2012) generally use a 3D model for such landslides, with full 3D Navier-Stokes equations applied to

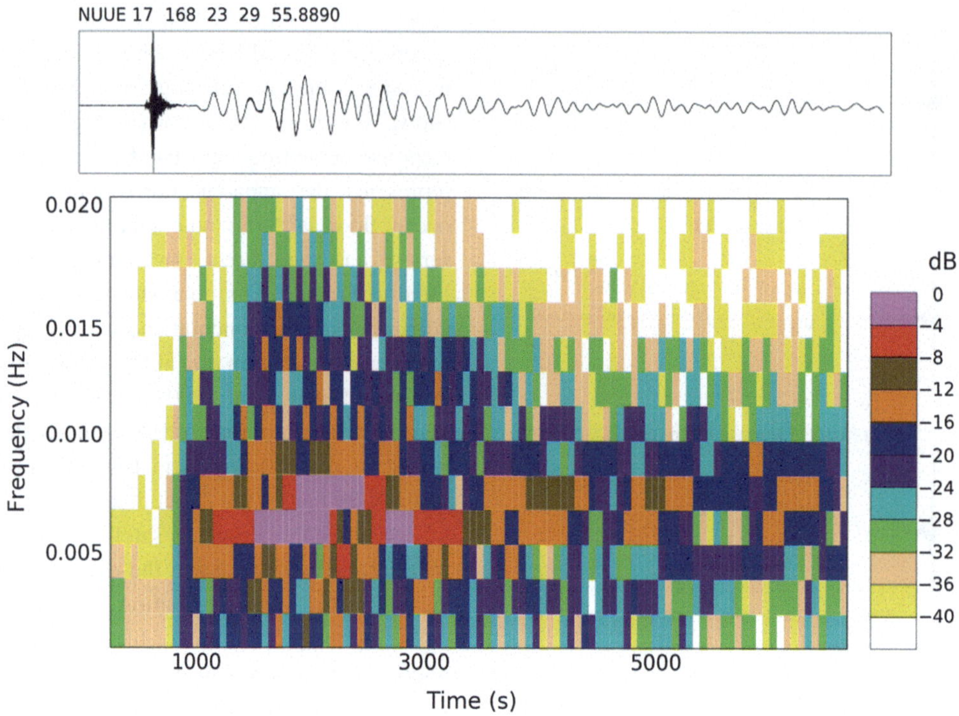

Figure 8
Spectrogram of the raw S11°E seismogram at NUUG. A standard Fourier transform is performed in a 600-s long window moving across the seismogram in increments of 60 s, and the resulting spectral amplitude in each time and frequency pixel is color coded according to the logarithmic scale at right. Note the dominant frequency component between 6 and 8 mHz

multimaterial flow. Here, and following our previous studies, we use a simpler, depth-averaged model to simulate both the landslide and the propagation of water waves (Assier-Rzadkiewicz et al. 2000; Labbé et al. 2012).

Examination of photos of the scar suggests that the landslide can be considered as a homogeneous, incompressible fluid-like flow of granular material following a Coulomb-type friction law. We further assume that the entire mass suddenly fails in one block after losing its equilibrium, and that it has no initial velocity.

For simplicity, basal friction is modeled in this study by a Coulomb-type friction law with a constant friction angle. This hypothesis may show limitations, since the friction angle depends on the velocity, as shown by laboratory experiments on granular flows but it should also be valid in the case of a rough bed with high inclination angles (Pouliquen 1999).

Following the one-phase grain-flow model of Savage and Hutter (1989) and taking into account Coulomb basal friction and gravity, we model the slide by solving the equations of conservation of mass and momentum in a (x', y') coordinate system linked to the topography :

$$\frac{\partial h_s}{\partial t} + \frac{\partial}{\partial x'}(h_s u_s) + \frac{\partial}{\partial x'}(h_s v_s) = 0; \quad (4)$$

$$\frac{\partial}{\partial t}(h_s u_s) + \frac{\partial}{\partial x'}(h_s u_s^2) + \frac{\partial}{\partial y'}(h_s u_s v_s) = -\frac{1}{2}\kappa\frac{\partial}{\partial x'}(g h_s^2 \cos\theta) + \kappa g h_s \sin\theta_{x'} + F_{x'}; \quad (5)$$

$$\frac{\partial}{\partial t}(h_s v_s) + \frac{\partial}{\partial x'}(h_s v_s u_s) + \frac{\partial}{\partial y'}(h_s v_s^2) = -\frac{1}{2}\kappa\frac{\partial}{\partial y'}(g h_s^2 \cos\theta) + \kappa g h_s \sin\theta_{y'} + F_{y'} \quad (6)$$

where h_s is the slide's thickness in a direction perpendicular to the slope, $\mathbf{u} = (u_s, v_s)$ the depth-averaged velocity vector parallel to the bed, $\kappa = 1 - \rho_w/\rho_s$ where ρ_w and ρ_s are the water and rock densities with a ratio $\rho_s/\rho_w = 1.7$ (for the subaerial part of the slide, κ is identically equal to 1), $\theta(x, y)$

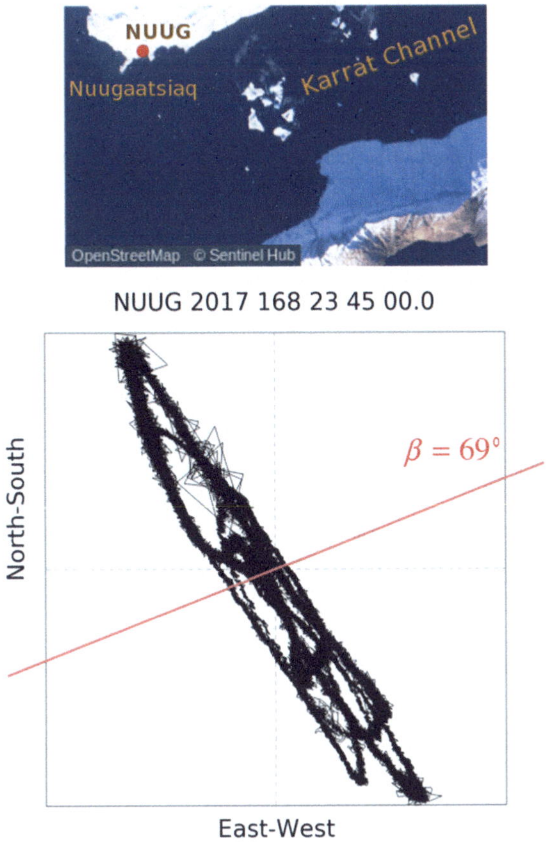

Figure 9
Two-dimensional plot of the horizontal components of the seismogram at NUUG, after counterclockwise rotation of 79° to correct for instrument misorientation (Clinton, pers. comm., 2018). The time window starts at 23:45 GMT and lasts 55 min, thus excluding the seismic waves. Note that the motion is not polarized in the direction of the back-azimuth to the epicenter ($\beta = 69°$), shown as the red line, but rather across the Karrat Channel (Sentinel inset at upper), whose width is $\sim$ 3.5 km in front of Nuugaatsiaq. On the scale of the inset, the red dot corresponds to the location of station NUUG

the local steepest slope angle, $\theta_{x'}$ and $\theta_{y'}$ the slope angles along the x' and y' axes respectively, and $\mathbf{F} = -\kappa gh\cos(\theta)\tan(\phi)\mathbf{u}/\|\mathbf{u}\|$ where ϕ is the friction angle. Curvature terms representing the effects of coordinate transformations (Savage and Hutter 1991) are considered as second-order terms in this paper. A sketch of the situation is visible on Fig. 12.

These equations are solved by the code Avalanche, that simulates both landslides and generated tsunamis. While it has been used mainly with sources fully contained under water, e.g., Papua New Guinea (Heinrich et al. 2001a) and the 1979 landslide at

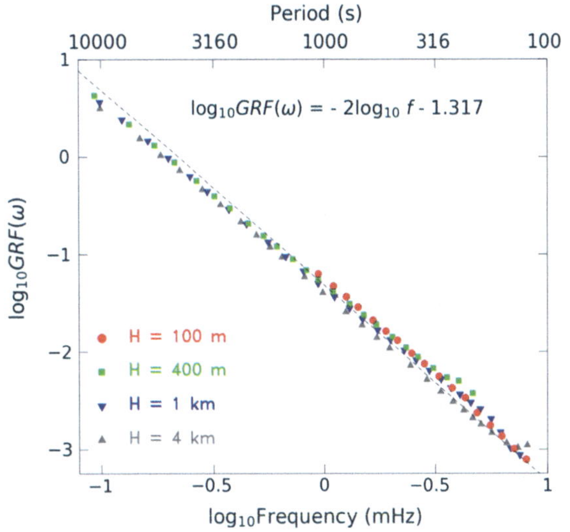

Figure 10
"Gilbert Response Function" $GRF(\omega)$ defined by Okal (2007), computed theoretically as a function of frequency for oceanic models of various depths. This figure is a generalization of Figure 10 of Okal (2007), and uses a logarithmic scale for frequency, to emphasize the power law behavior of $GRF(\omega)$; note that it is essentially independent of the water depth H. The dashed line is the regression (3) used in the deconvolution

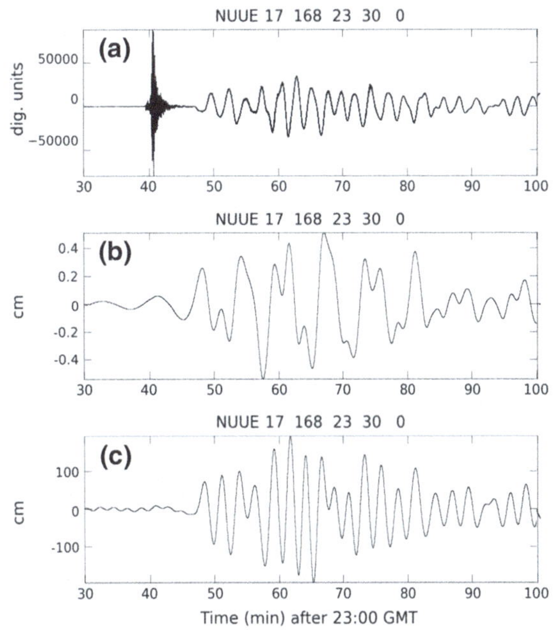

Figure 11
Deconvolution of the sea-surface tsunami amplitude $\eta(t)$ from the seismic record at NUUG. **a** Raw S11°E seismogram (see Fig. 7). **b** Apparent horizontal motion recorded by the instrument, after deconvolution of the instrument response. **c** Reconstructed time series $\eta(t)$ after deconvolution of the Gilbert Response Function

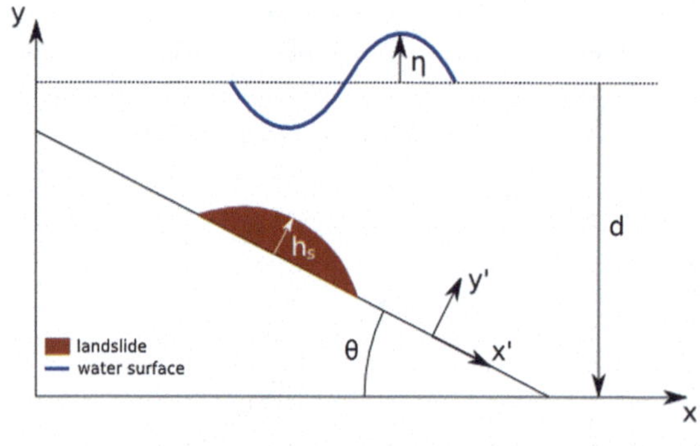

Figure 12
Sketch of our landslide-generated tsunami model. Along the x′ and y′ axes, h_s is the slide's thickness and along the x and y axes, η is the water surface elevation, d the unperturbed depth and θ the steepest slope angle

Mururoa (Poupardin et al. 2017), it was developed with the capacity to handle hybrid sources involving both subaerial debris flow and underwater sliding, as validated by the successful modeling of the 1997 Montserrat tsunami (Heinrich et al. 2001b) and the 2003 Montagne Pelée collapse event (Le Friant et al. 2003), this approach being similar to Weiss et al.'s (2009) simulation of the Lituya Bay event.

The shape of the bathymetry is typical of a fjord, with an abrupt slope (ranging from 40° to 60°) along the coasts. Due to this topography, ϕ is fixed at 50°. While this value may appear high, it results in a physically acceptable speed of the landslide: for a smaller friction angle of 30° we would obtain unrealistic aerial speeds of more than 200 m/s; with $\phi = 50°$, we obtain a maximum aerial speed of 85 m/s, which is still very high: while a maximum speed of 100 m/s has been suggested for aerial landslides (e.g., Satake et al. 2002), this figure was proposed based on the value of 70 m/s, documented by filming the avalanche during the 1980 eruption of Mount Saint Helens (Voight 1981), which involved a powerful atmospheric explosion as the triggering mechanism. In the present case, we obtain a runout length of $\sim$ 5000 m and a vertical fall height of $\sim$ 1900 m.

4.2. Tsunami Model

Our strategy for the tsunami propagation is to start with the Saint–Venant equations during 80 s to simulate the tsunami generation in shallow water, then to continue with the Boussinesq model in order to take into account any possible dispersive effects. We use the code Avalanche to solve the Saint–Venant equations, which are Navier–Stokes equations integrated over depth:

$$\frac{\partial \eta}{\partial t} + \frac{\partial (hu)}{\partial x} + \frac{\partial (hv)}{\partial y} = -\frac{\partial d}{\partial t}; \tag{7}$$

$$\frac{\partial u}{\partial t} + u\frac{\partial u}{\partial x} + v\frac{\partial u}{\partial y} = -g\frac{\partial \eta}{\partial x} + F_x; \tag{8}$$

$$\frac{\partial v}{\partial t} + u\frac{\partial v}{\partial x} + v\frac{\partial v}{\partial y} = -g\frac{\partial \eta}{\partial y} + F_y \tag{9}$$

where η is the surface elevation, $h = \eta + d$ the water column height where d is the unperturbed depth, u and v the depth-averaged velocities along the x and y axes and F_x and F_y the friction and Coriolis forces along the x and y axes, which remain negligible on the scale of the present experiment.

The sea-bottom deformation $\partial d/\partial t$ in Eq. (7) is computed as a forcing term:

$$\frac{\partial d}{\partial t} = \frac{1}{\cos\theta}\frac{\partial h_s}{\partial t} \tag{10}$$

Avalanche can also use a Boussinesq model, following Løvholt et al. (2008):

$$\frac{\partial \eta}{\partial t} + \frac{\partial (hu)}{\partial x} + \frac{\partial (hv)}{\partial y} = -\frac{\partial d}{\partial t}; \tag{11}$$

$$\begin{aligned}\frac{\partial u}{\partial t} + u\frac{\partial u}{\partial x} + v\frac{\partial u}{\partial y} &= -g\frac{\partial \eta}{\partial x} + \frac{d^3}{3}\left[\frac{\partial^2 u_t}{\partial x^2} + \frac{\partial^2 v_t}{\partial x \partial y}\right] \\ &+ \frac{d}{2}\left[u_t\frac{\partial^2 d}{\partial x^2} + u_t\frac{\partial^2 d}{\partial x \partial y}\right] + d\frac{\partial d}{\partial x}\frac{\partial u_t}{\partial x} \\ &+ \frac{d}{2}\left[\frac{\partial d}{\partial x}\frac{\partial v_t}{\partial y} + \frac{\partial d}{\partial y}\frac{\partial v_t}{\partial x}\right] + F_x;\end{aligned} \tag{12}$$

$$\begin{aligned}\frac{\partial v}{\partial t} + u\frac{\partial v}{\partial x} + v\frac{\partial v}{\partial y} &= -g\frac{\partial \eta}{\partial y} + \frac{d^3}{3}\left[\frac{\partial^2 v_t}{\partial y^2} + \frac{\partial^2 u_t}{\partial x \partial y}\right] \\ &+ \frac{d}{2}\left[v_t\frac{\partial^2 d}{\partial y^2} + v_t\frac{\partial^2 d}{\partial x \partial y}\right] + d\frac{\partial d}{\partial y}\frac{\partial v_t}{\partial y} \\ &+ \frac{d}{2}\left[\frac{\partial d}{\partial x}\frac{\partial u_t}{\partial y} + \frac{\partial d}{\partial y}\frac{\partial u_t}{\partial x}\right] + F_y\end{aligned} \tag{13}$$

where u_t and v_t the time-derivatives of u and v.

Both the landslide and Saint–Venant equations, which are very similar, are solved by Godunov's finite-volume scheme, extended to second order by a Van Leer scheme (Heinrich and Piatanesi 2000; Labbé et al. 2012).

The lone bathymetric dataset available in the region, the International Bathymetric Chart of the Artic Ocean (IBCAO) (Jakobsson et al. 2012), only features a 500-m resolution; we had to interpolate it to a sampling of 25 m, in order to project the slide over more than a single pixel. After cropping, we obtain a bathymetric map of a 25 $\times$ 60 km^2 area, covering the slide's location and the village of Nuugaatsiaq located 32 km SSW from the source.

However, we found that the maximum depth proposed in Karrat Fjord (430 m) is clearly too shallow, as it results in propagation times to Nuugaatsiaq of 14 min, incompatible with the difference in time of only 8 min documented on the seismic record at NUUG between the seismic and tsunami waves. For this reason, we later increased all depths in the fjord by an admittedly arbitrary factor of 2.

5. Results and Discussion

5.1. Results for the 2017 Event

The results are analyzed at a number of virtual gauges (Fig. 13) located in front of the landslide (Gauge 1), then along the path of the tsunami (Gauges 2 and 3), and finally in front of the village of Nuugaatsiaq (Gauge 4).

Our first simulations, using the bathymetry of Model IBCAO (Jakobsson et al. 2012) with a maximum depth of 430 m in Karrat Fjord, predict a travel time of $\sim$ 12 min to Nuugaatsiaq, about 4 min longer than observed on the seismic recording at NUUG (Fig. 14). We first note that the misfit in travel is much longer than the initial duration of the wave at Gauge 1 (Fig. 15); thus it cannot be a source effect, and is clearly due to propagation, from Gauge 1 to Gauge 4. In addition, Chao et al. (2018) have similarly noticed that the tsunami is observed on the NUUG seismometers about 5 min earlier than simulated using GEBCO bathymetry. They explain this discrepancy by interpreting the seismic signal as generated by the impact of the tsunami on a coastline located about 15 km upstream Karrat Fjord. Based on our observation that onland seismic recordings of tsunami falter at distances of a fraction of wavelength (Okal 2007), and on the horizontal polarization of the recordings (Fig. 9), we prefer to invoke an inaccurate bathymetry in Karrat Fjord.

In this context, we decided to increase the water depth by a factor of two across the board, with a new maximum value of 860 m. Under the Saint–Venant approximation, this admittedly arbitrary correction increases the tsunami velocity (and reduces the travel

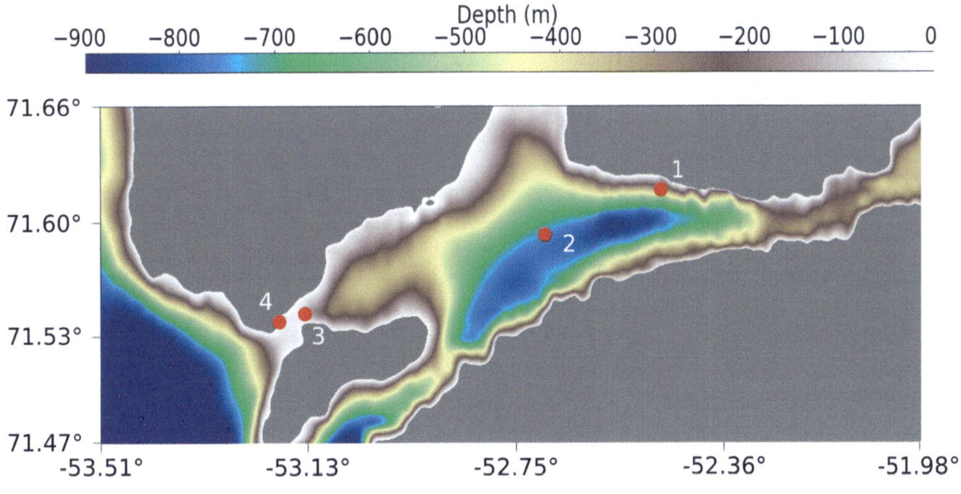

Figure 13
Bathymetric grid in Karrat Fjord and gauge locations (red dots) along the path of the tsunami. The tsunami is initialized at Gauge 1, and progresses to Gauges 2, 3 and finally 4 in front of Nuugaatsiaq

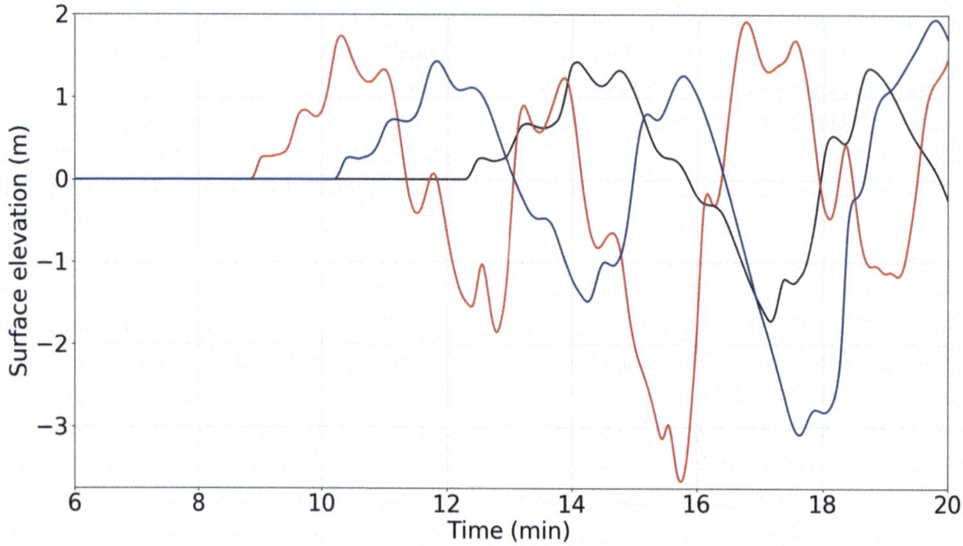

Figure 14
Surface elevation at Gauge 4 (in front of Nuugaatsiaq) for the initial bathymetry (black), increased by a factor 1.5 (blue) and by a factor 2 (red), simulated by the Saint–Venant model

time) by a factor of $\sqrt{2}$ (Fig. 14); in the Boussinesq model used herein, where the relationship between velocity and square root of depth is not linear, the effect is only slightly smaller, and in both models the leading wave now reaches the village in $\sim$ 8 min, in much better agreement with the observed arrival time.

While our correction of the fjord bathymetry may appear drastic, we stress that a similar situation was documented recently in Palu Bay, Sulawesi; following the earthquake and tsunami of 28 September 2018, it became clear that the bathymetry available from global models such as IBCAO or GEBCO grossly underestimated (by a factor as large as 4) the

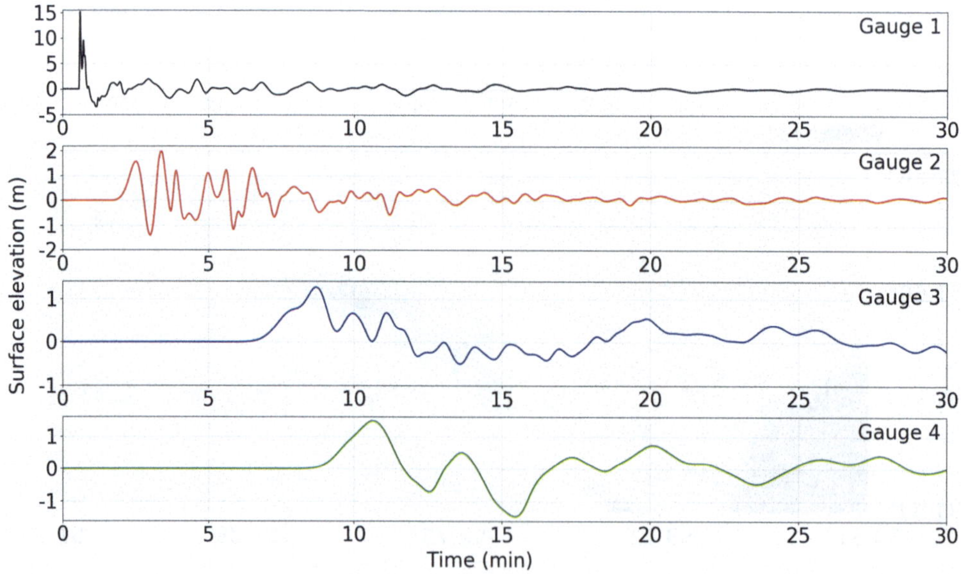

Figure 15
Surface elevations for Gauges 1, 2, 3 and 4, obtained with the Boussinesq model

water depth in the bay (Fritz et al. 2018b). We must conclude that such models cannot pretend to give a reliable small-scale bathymetry, especially in profoundly indented bays such as Karrat Fjord, where the proposed bathymetry is most likely inaccurate.

Under our model of deepened bathymetry, and as shown on Fig. 15, water waves reach 15.5 m at the source (Gauge 1), 2 m at the maximum depth (Gauge 2), 1.3 m before Nuugaatsiaq (Gauge 3) and 1.5 m in front of Nuugaatsiaq (Gauge 4). These figures are not directly comparable to the results of Fritz et al.'s (2018a) field survey, who reported run-up reaching 90 m in the source area, and 50 m on the opposite side of the 6-km wide fjord. We stress that our simulations do not include run-up computations, and that our virtual gauges are located in deep water (340 and 750 m, respectively for Gauges 1 and 2).

We obtain a wave-train at Gauge 4 (in front of Nuugaatsiaq) similar to the signal recorded at the seismic station (Fig. 16), featuring three waves with a period of $\sim$ 3 min (see Gauge 4 on Fig. 15).

Finally, we note on Fig. 16 that while the amplitude of the simulated wave is on the same order (1.5 m) as that deconvolved in a comparable time window from the S11°E component of the NUUG seismometer (1.0 m), the simulated wave features significantly lower frequencies (4.6–5.4 mHz vs. 6.2 mHz), presumably as a result of seiching in the channel, which may also explain the longer duration (and later higher amplitudes) of the deconvolved time series. Nevertheless, the acceptable agreement between them serves as an *a posteriori* validation of the deconvolution procedure in Sect. 3.3.

Although we used the Boussinesq model to simulate the tsunami, results show (Fig. 17) that three water waves are still observed within the first 12 min when using the Saint–Venant equations. This suggests that the generation of these three water waves depends on the bathymetry's shape and is not a result of dispersive effects, but more probably of reflection on the coasts.

This is further confirmed through the use of Glimsdal et al.'s (2013) dimensionless parameter $\tau = 6ht/gT^3$, where h is water depth, t travel time and T dominant period. In the present case, the packet of three waves corresponds to $\tau \approx 0.03$, which is clearly below the value of 0.5 given by the authors as a threshold for significant dispersion effects (following Shuto 1991).

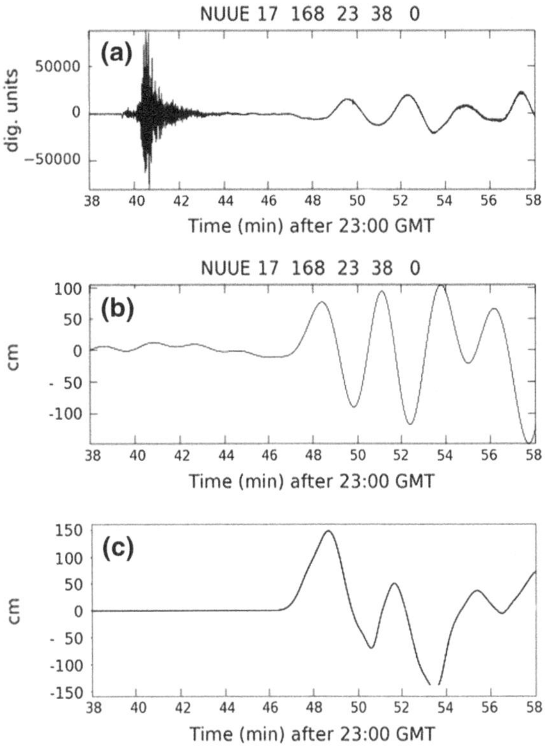

Figure 16
Comparison of recorded and simulated tsunami waveshapes. **a**, **b** Respectively the raw seismogram and the deconvolved tsunami amplitude η for a time window starting at 23:38 GMT (72 s before the initiation of the slide), and lasting 20 min; they are close-ups of Frames (**a**, **c**) of Fig. 11. **c** Simulated tsunami wave at Gauge 4 (in front of Nuugaatsiaq); this is simply a 20-min window of frame **b** of Fig. 15. Note the good agreement of amplitudes between (**b**) and (**c**), but the lower-frequency character of the simulated wave

On Fig. 20, maximum elevations through the entire simulation show that the water level reaches up to 40 m at the source, and we then observe a decrease of the water heights due to propagation in deeper water. Finally the water waves reach between 1 and 1.5 m at Nuugaatsiaq.

While we eventually obtained a good agreement between our simulations and the seismic recording at NUUG, we note that we had to significantly increase (by a factor of 2) the field of depths available from published bathymetric charts, confirming if need be, the critical role of an accurate bathymetry in allowing realistic simulations of the propagation of tsunamis in

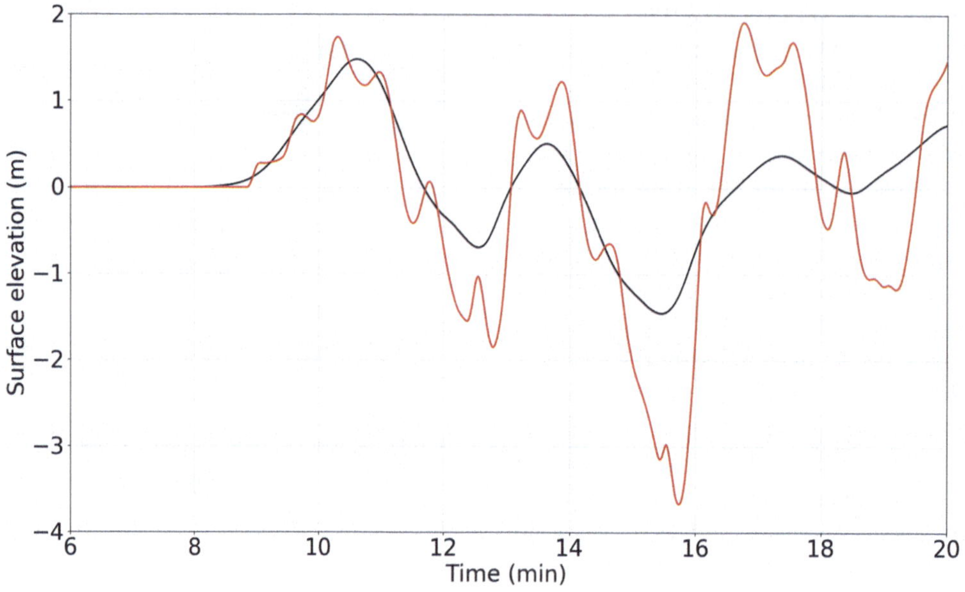

Figure 17
Surface elevation at Gauge 4 simulated by the Boussinesq model (black curve) and the Saint–Venant model (red curve)

complex, poorly chartered marine environments. For the sake of simplicity, we used an across-the-board, constant, and indeed arbitrary, factor of 2 which provided an adequate gain of about 4 min in travel-time. In particular, we did not try more sophisticated, laterally variable, corrections to the available

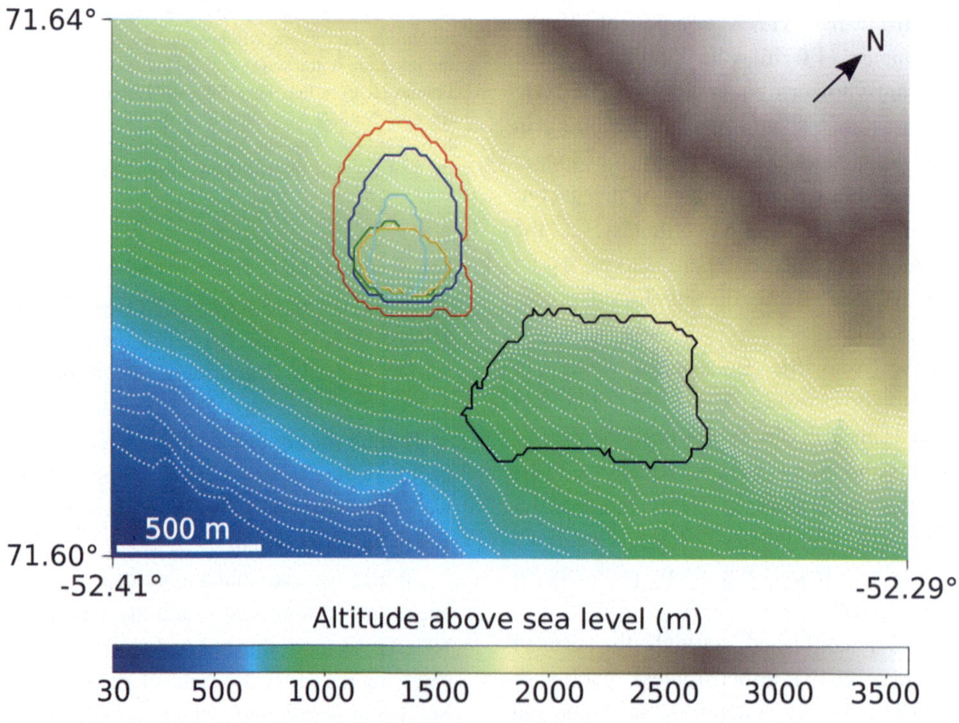

Figure 18
Footprints of the various landslides, the 2017 event (black) and the western hypothetic volumes of 38 million m^3 (red), 14 million m^3 (blue), 7 million m^3 (green), 2 million m^3 "E–W" (orange) and 2 million m^3 "N–S" (cyan)

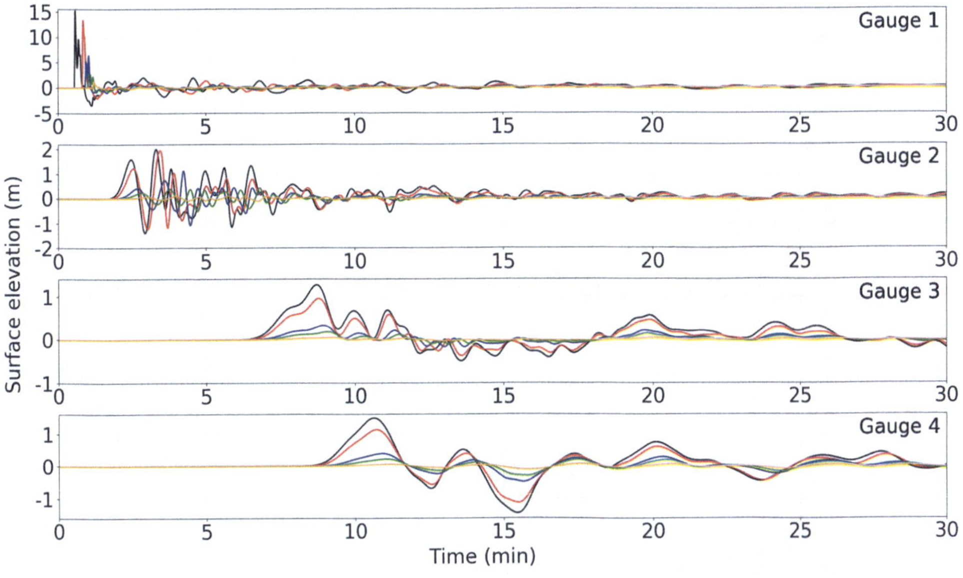

Figure 19
Surface elevations for different volumes of landslide on Gauges 1, 2, 3 and 4, for the 2017 event of 53 million m^3 (black) and for the potential events of 38 million m^3 (red), 14 million m^3 (blue), 7 million m^3 (green) and 2 million m^3 (orange)

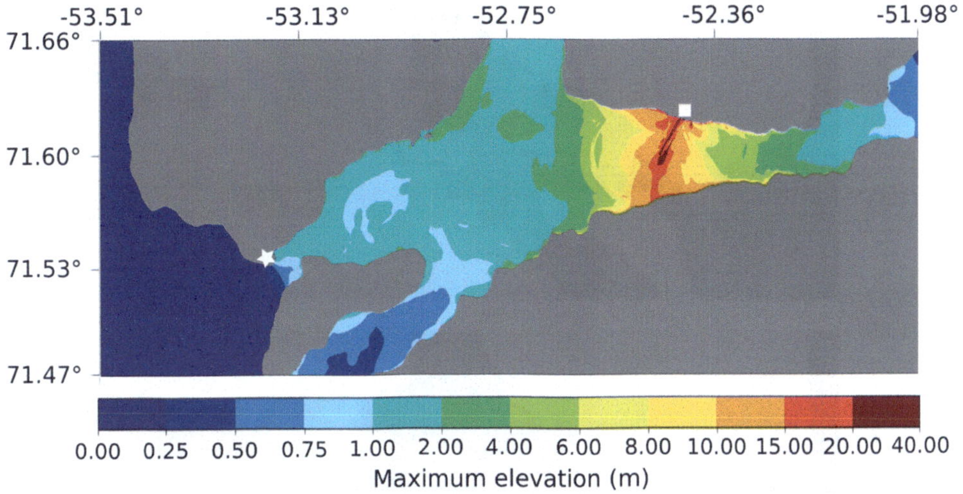

Figure 20
Maximal surface elevations simulated between the source (white square) and the village of Nuugaatsiaq (white star) for the 2017 event of 53 million m^3

bathymetry, which would have been futile, given its present quality, which the original misfit of travel times documents as grossly erroneous. In view of the potential for failure of the Western slide, and in the context of the general regional hazard expressed by the previous tsunamis of 1952, 2000, and to a lesser extent, 1954 and 2012 (see Sect. 2 for details), a high-quality bathymetric survey would be desirable in Karrat Fjord, and in the other fjords with populated settlements on their shores.

5.2. The Next, Potential, Event: Sensitivity Studies on Its Volume and Shape

We now address the question of the "next" slide, *i.e.*, the potential failure of the material immediately

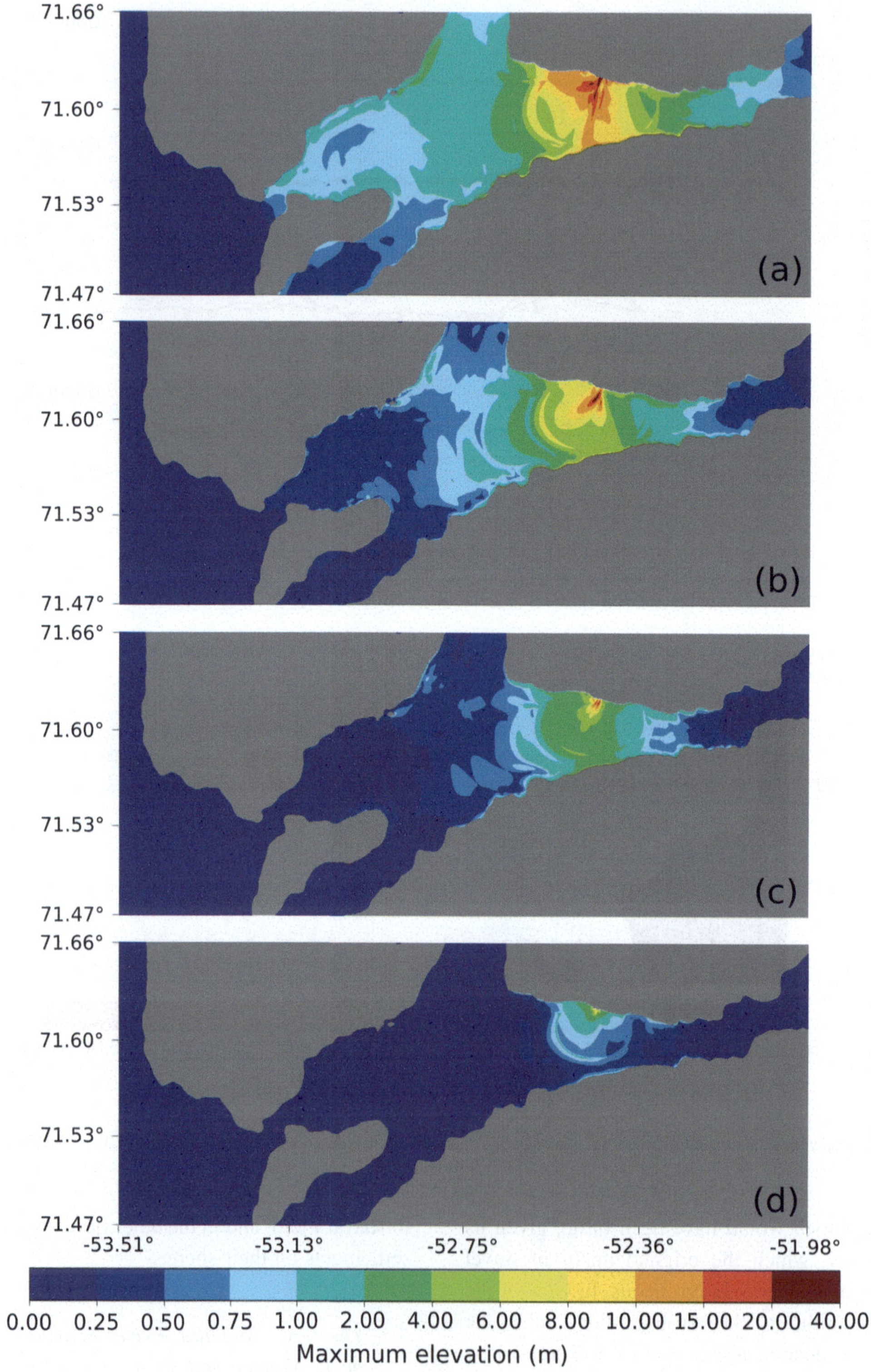

Figure 21
Maximal surface elevations simulated between the source and the village of Nuugaatsiaq for different volumes of the western landslide, **a** 38 million m^3; **b** 14 million m^3; **c** 7 million m^3; **d** 2 million m^3

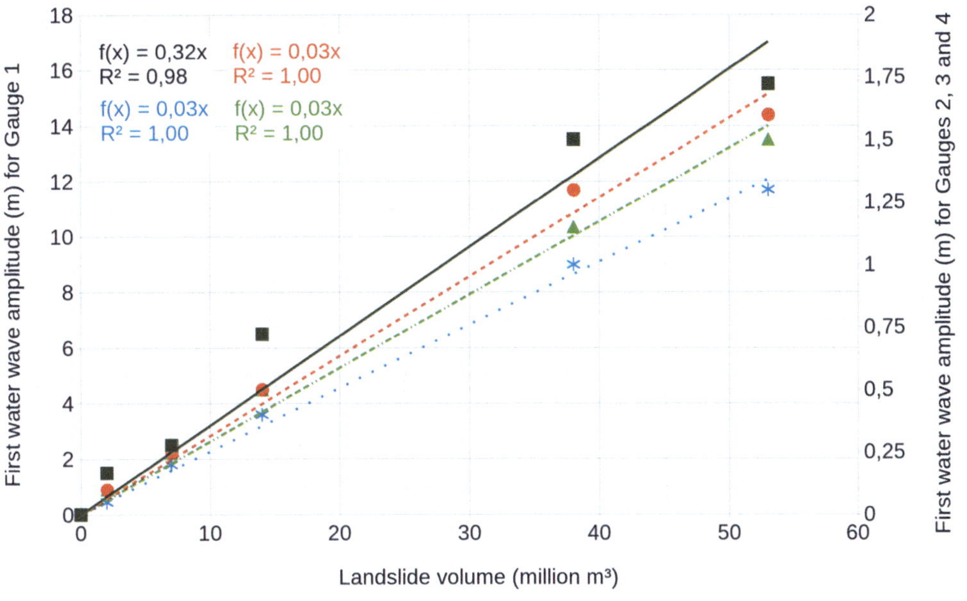

Figure 22
Amplitude (in m) of the first generated water wave *vs.* the landslide volume (in million m^3), for Gauges 1 (black squares, solid line, left vertical scale), 2 (red circles, dashed line, right vertical scale), 3 (blue asterisks, dotted line, right vertical scale) and 4 (green triangles, dash-dotted line, right vertical scale). Gauges 2, 3 and 4 are referred to the right scale while Gauge 1 is referred to the left scale

to the West of the 2017 slide (Fig. 2). Unlike for the latter, we can only speculate as to its volume, shape and of course timing. In order to investigate the impact of that potential event, and in particular, the influence of its volume, we consider four landslide geometries of variable volume, ranging from 2 million m^3 (hereafter Model "E–W") to 7, 14 and 38 million m^3, respectively (Fig. 18; Table 1), and also compare our results with those of the 2017 slide (53 million m^3).

The range of volumes proposed here (2–38 million m^3) remains small, as compared to that studied by Scheidegger (1973), and thus the range of friction angles derived from the application of his power law (with an exponent of $-1/6$) between volume and friction coefficient, is only a few degrees (from 15 to 23). A similar range of friction angles was studied in Mergili et al. (2017), showing some difference in runout between 15° and 23° , for a rock avalanche of 6.4 million m^3. Nevertheless, we have argued in Sect. 4.1 that such angles would lead to unrealistic velocities upon reaching the water. In this context, we keep a friction angle of 50° for the present calculations. The landslide rheology is a major source of uncertainties when considering the landslide dynamics. However, its effect is of the second order when considering the water waves amplitudes.

Simulated time series are compiled on Fig. 19; note that for the various scenarios of the Western slide, Gauge 1 is moved 1 km to the West, without significant difference in water depth (Fig. 20). In addition, Fig. 21 maps maximum wave elevations for the various western slides.

Figure 19 documents a good linearity between the volume of the landslides and the simulated water heights, as evidenced directly in front of the slide (Gauge 1), as well as farther along the path of the tsunami (Gauges 2, 3, and 4). For example, the largest case simulated (38 million m^3) from the

Table 1

Summary of dimensions and volumes of the western landslide

Volume (million m^3)	Width × length × thickness (m, max)	Altitude (m, center)
38	400 × 600 × 200	1500
14	300 × 500 × 146	1550
7	250 × 200 × 171	1500
2 "E–W"	250 × 200 × 110	1500
2 "N–S"	125 × 400 × 103	1500

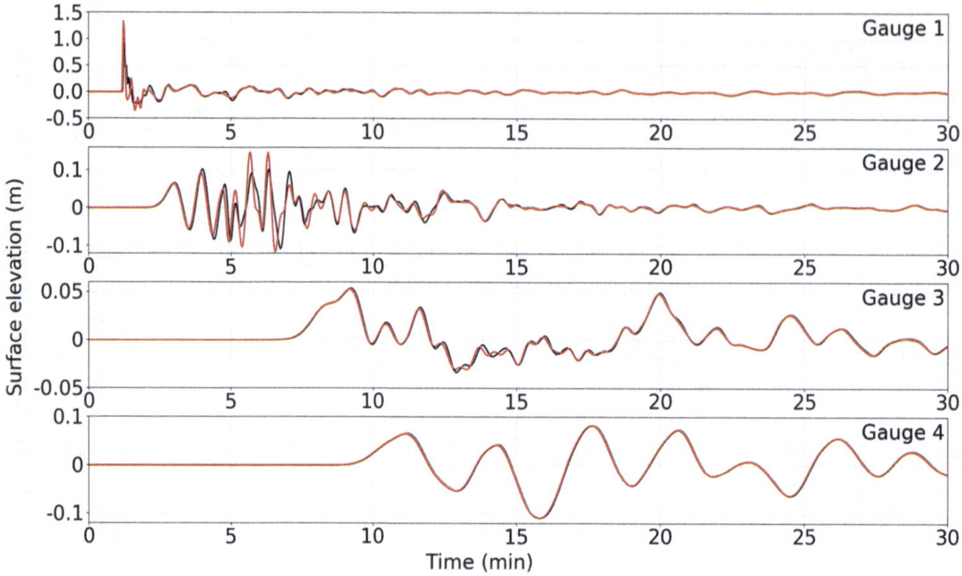

Figure 23
Surface elevations comparison for the 2 million m^3 "E–W" (black) and "N–S" (red), at Gauges 1, 2, 3 and 4

western landslide is smaller than the 2017 slide by a factor of 1.5, and the resulting water waves by a comparable factor (1.3). At the next lower level (14 million m^3), the water height remains 40 cm, which still presents a risk for the population. Dividing this volume by two, we obtain a water height of 20 cm, which is considered the limit of hazard for flooding by a tsunami. This linear relationship between the first generated water wave amplitude and the landslide volume is confirmed on Fig. 22 by excellent regression coefficients [R^2= 0.99 for Gauge 1, black: R^2 = 1.00 with two significant digits for Gauges 2, 3, 4 (red, blue, green)]. We note that the least linear relationship is for Gauge 1, located in the area where we use the Saint–Venant equations, so it appears that the linearity is most valid for the Boussinesq model.

Finally, we explore the possible influence of the shape of the landslide on the characteristics of the tsunami, such as its amplitude and dominant period. For this purpose, we consider a landslide (Model "N–S") with the same volume (2 million m^3) as our smallest "E–W" scenario (*d*), but with a narrower and taller footprint (shown in cyan on Fig. 18, and detailed in Table 1). Figure 23 presents the surface elevation comparison at the different gauges for the "E–W" and "N–S" 2 million m^3 western landslides and shows no significant difference either in amplitude or dominant period between the two cases, allowing us to conclude that the total volume of the slide is the primary factor controlling the characteristics of the wave.

6. Conclusion

Simulation of the tsunami generated by the landslide of June 17, 2017 in Karrat Fjord, Greenland satisfactorily matches both inferred wave activity at Nuugaatsiaq, 32 km WSW of the source, and the early portion of the signal recorded at the seismic station NUUG. Using a slide volume of $\sim$ 50 million m^3, we model three water waves with periods of $\sim$ 3 min, reaching the village of Nuugaatsiaq after 8 min. While these results are satisfying, they require doubling the depths documented by available bathymetry; this effect cannot be simply an artifact of our forced interpolation (from a 500 to a 25-m resolution) of the IBCAO dataset, and thus attests to the poor quality of the latter.

Our sensitivity study on different scenarios for the potential landslide, immediately to the West of the 2017 scar, supports a general linearity between their volumes and the heights of resulting tsunami waves. Using a threshold of 30 cm (at Gauge 4, considered as a "last", initially wet grid point) for a potentially hazardous tsunami at Nuugaatsiaq, we estimate that it corresponds to a 7 million m^3 slide, considerably smaller than the more realistic volume of 38 million m^3 regarded as precarious to the West of the 2017 event, a scenario which would lead to a level flooding and destruction comparable to the 2017 event.

Acknowledgements

We thank John Clinton, Director of Seismic Networks, ETH, Zürich, for access to the NUUG seismograms, and for critical metadata concerning their misorientation. Some figures were produced using the GMT software (Wessel and Smith 1991). This work was supported by the LRC Yves Rocard (Laboratoire de Recherche Conventionné CEA-ENS). The paper was improved by the constructive comments of David Tappin and a second, anonymous, reviewer.

REFERENCES

Abadie, S. M., Harris, J. C., Grilli, S. T., & Fabre, R. (2012). Numerical modeling of tsunami waves generated by the flank collapse of the Cumbre Vieja Volcano (La Palma, Canary Islands): Tsunami source and near field effects. *Journal of Geophysical Research Oceans, 117*(5), 1–26.

Ambraseys, N., & Bilham, R. (2012). The Sarez–Pamir earthquake and landslide of 18 February 1911. *Seismological Research Letters, 83*(2), 294–314.

Assier-Rzadkiewicz, S., Heinrich, P., Sabatier, P. C., Savoye, B., & Bourillet, J. F. (2000). Numerical modelling of a landslide-generated Tsunami: The 1979 nice event. *Pure and Applied Geophysics, 157*(10), 1707–1727.

Chao, W.-A., Wu, T.-R., Ma, K.-F., Kuo, Y.-T., Wu, Y.-M., Zhao, L., et al. (2018). The large greenland landslide of 2017: Was a Tsunami warning possible? *Seismological Research Letters, 89*(4), 1335–1344.

Clinton, J., Larsen, T., Dahl-Jensen, T., Voss, P., & Nettles, M. (2017). Special event: Nuugaatsiaq Greenland landslide and tsunami. *Incorporated Research Institutions for Seismology Washington DC* . https://ds.iris.edu/ds/nodes/dmc/specialevents/2017/06/22/nuugaatsiaq-greenland-landslide-and-tsunami/.

Dahl-Jensen, T., Larsen, L. M., Pedersen, S. A. S., Pedersen, J., Jepsen, H. F., Pedersen, G., et al. (2004). Landslide and Tsunami 21 November 2000 in Paatuut, West Greenland. *Natural Hazards, 31*(1), 277–287.

Ekström, G., & Stark, C. P. (2013). Simple scaling of catastrophic landslide dynamics. *Science, 339*(6126), 1416–1419.

Fine, I. V., Rabinovich, A. B., Thomson, R. E., & Kulikov, E. A. (2003). *Numerical modeling of Tsunami generation by submarine and subaerial landslides* (pp. 69–88). Dordrecht: Springer.

Fritz, H. M. (2002). *Initial phase of landslide generated impulse waves.* PhD Thesis, ETH Zurich.

Fritz, H. M., Giachetti, T., Anderson, S., & Gauthier, D. (2018a). Field survey of the 17 June 2017 landslide generated Tsunami in Karrat Fjord, Greenland. In *EGU General Assembly Conference Abstracts*, Vol. 20 of *EGU General Assembly Conference Abstracts*, p. 18345.

Fritz, H. M., Synolakis, C., Kalligeris, N., Skanavis, V., Santoso, F., Rizal, M., et al. (2018b). Field survey of the 28 September 2018 Sulawesi tsunami. *Eos Transactions American Geophysical Union, 99*, 53. **(NH22B-04, [abstract])**.

Gauthier, D., Anderson, S. A., Fritz, H. M., & Giachetti, T. (2018). Karrat Fjord (Greenland) tsunamigenic landslide of 17 June 2017: Initial 3D observations. *Landslides, 15*(2), 327–332.

Geist, E. L. (2000). Origin of the 17 July 1998 Papua New Guinea Tsunami: Earthquake or landslide. *Seismological Research Letters, 71*(3), 344–351.

Gilbert, J. (1980). An introduction to low-frequency seismology. In A. Dziewopnski & E. Boschi (Eds.), *Proceedings of the International School of Physics "Enrico Fermi'* (Vol. 78, pp. 41–81). Amsterdam: North Holland.

Glimsdal, S., Pedersen, G. K., Harbitz, C. B., & Løvholt, F. (2013). Dispersion of tsunamis: Does it really matter? *Natural Hazards and Earth System Sciences, 13*, 1507–1526.

Guérin, C. (2017). Effect of the DTM quality on the bundle block adjustment and orthorectification process without GCP: Example on a steep area. In *Proceedings of 2017 IEEE international geoscience remote sensing symposium (IGARSS). IEEE*, pp. 1067–1070.

Guérin, C., Binet, R., & Pierrot-Deseilligny, M. (2014). Automatic detection of elevation changes by differential DSM analysis: Application to urban areas. *IEEE Journal of Selected Topics in Applied Earth Observations and Remote Sensing, 7*(10), 4020–4037.

Haeberli, W., & Gruber, S. (2009). *Global warming and mountain permafrost* (pp. 205–218). Berlin: Springer.

Hanson, J. A., & Bowman, J. R. (2005). Dispersive and reflected tsunami signals from the 2004 Indian Ocean tsunami observed on hydrophones and seismic stations. *Geophysical Research Letters, 32*, 17.

Hébert, H., Piatanesi, A., Heinrich, P., & Schindelé, F. (2002). Numerical modeling of the September 13, 1999 landslide and tsunami on Fatu Hiva Island (French Polynesia). *Geophysical Research Letters, 29*(10), 10–13.

Heinrich, P., Boudon, G., Komorowski, J. C., Sparks, R. S. J., Herd, R., & Voight, B. (2001b). Numerical simulation of the December 1997 Debris Avalanche in Montserrat, Lesser Antilles. *Geophysical Research Letters, 28*(13), 2529–2532.

Heinrich, P., & Piatanesi, A. (2000). Near-field modeling of the July 17, 1998 tsunami in Papua New Guinea. *Geophysical Research Letters, 27*(19), 3037–3040.

Heinrich, P., Piatanesi, A., & Hébert, H. (2001a). Numerical modelling of tsunami generation and propagation from submarine slumps: The 1998 Papua New Guinea event. *Geophysical Journal International, 145*(1), 97–111.

Hermanns, R. L., Blikra, L. H., Naumann, M., Nilsen, B., Panthi, K. K., Stromeyer, D., et al. (2006). Examples of multiple rock-slope collapses from Köfels (Ötz valley, Austria) and western Norway. *Engineering Geology, 83*(1–3), 94–108.

Higman, B., Shugar, D. H., Stark, C. P., Ekström, G., Koppes, M. N., Lynett, P., et al. (2018). The 2015 landslide and tsunami in Taan Fiord, Alaska. *Scientific Reports, 8*(1), 12993.

Huggel, C., Clague, J. J., & Korup, O. (2012). Is climate change responsible for changing landslide activity in high mountains? *Earth Surface Processes and Landforms, 37*(1), 77–91.

ICAO. (1955). International Civil Aviation (ICAO) Meteorological Stations in Greenland. *ICAO Bulletin, 10*(7), 7–11.

Jakobsson, M., Mayer, L., Coakley, B., Dowdeswell, J. A., Forbes, S., Fridman, B., et al. (2012). The international bathymetric chart of the Arctic Ocean (IBCAO) Version 3.0. *Geophysical Research Letters, 39*, L12609.

La Rocca, M., Galluzzo, D., Saccorotti, G., Tinti, S., Cimini, G. B., & Del Pezzo, E. (2004). Seismic signals associated with landslides and with a tsunami at Stromboli volcano, Italy. *Bulletin of the Seismological Society of America, 94*(5), 1850–1867.

Labbé, M., Donnadieu, C., Daubord, C., & Hébert, H. (2012). Refined numerical modeling of the 1979 tsunami in Nice (French Riviera): Comparison with coastal data. *Journal of Geophysical Research Earth Surface, 117*, F1.

Le Friant, A., Heinrich, P., Deplus, C., & Boudon, G. (2003). Numerical simulation of the last flank-collapse event of Montagne Pelée, Martinique, Lesser Antilles. *Geophysical Research Letters, 30*, 2.

Løvholt, F., Pedersen, G., & Gisler, G. (2008). Oceanic propagation of a potential tsunami from the La Palma Island. *Journal of Geophysical Research Oceans, 113*(9), 1–21.

McNamara, D., Ringler, A., Hutt, C., & Gee, L. (2011). Seismically observed seiching in the Panama Canal. *Journal of Geophysical Research Solid Earth, 116*, B4.

Mergili, M., Fischer, J.-T., Krenn, J., & Pudasaini, S. P. (2017). r.avaflow v1, and advances open-source computational framework for the propagation and interaction of two-phase mass flows. *Geoscientific Model Development, 10*(2), 553–569.

Miller, D. J. (1960). Giant waves in Lituya Bay, Alaska. *US Geological Survey Professional Paper, 354-C*.

Naranjo, J. A., Arenas, M., Clavero, J., & Muñoz, O. (2009). Mass movement-induced tsunamis: Main effects during the Patagonian Fjordland seismic crisis in Aisén (45° 25'S), Chile. *Andean Geology, 36*, 1.

NOAA. (2018). *National Geophysical Data Center/ World Data Service: NCEI/WDS Global Historical Tsunami Database. NOAA National Centers for Environmental Information.* https://data.nodc.noaa.gov/cgi-bin/iso?id=gov.noaa.ngdc.mgg.hazards:G02151

Okal, E. A. (2003). *T* waves from the 1998 Papua New Guinea earthquake and its aftershocks: Timing the tsunamigenic slump. *Pure and Applied Geophysics, 160*, 1843–1863.

Okal, E. A. (2007). Seismic records of the 2004 Sumatra and other tsunamis: A quantitative study. *Pure and Applied Geophysics, 164*, 325–353.

Okal, E. A., Fryer, G. J., Borrero, J. C., & Ruscher, C. (2002). The landslide and local tsunami of 13 September 1999 on Fatu Hiva (Marquesas islands; French Polynesia). *Bulletin de la Société géologique de France, 173*(4), 359–367.

Okal, E. A., & Synolakis, C. E. (2001). Comment on "Origin of the 17 July 1998 Papua New Guinea tsunami: Earthquake or landslide?" by EL Geist. *Seismological Research Letters, 72*(3), 363–366.

Okal, E. A., & Synolakis, C. E. (2003). A theoretical comparison of tsunamis from dislocations and landslides. *Pure and Applied Geophysics, 160*(10–11), 2177–2188.

Pedersen, S. A. S., Larsen, L. M., Dahl-jensen, T., Jepsen, H. F., Krarup, G., Nielsen, T., et al. (2002). Tsunami-generating rock fall and landslide on the south coast of Nuussuaq, central West Greenland. *Geology of Greenland Survey Bulletin, 191*, 73–83.

Pierrot-Deseilligny, M., Paparoditis, N. (2006). A multiresolution and optimization-based image matching approach: An application to surface reconstruction from SPOT5-HRS stereo imagery. In *IAPRS vol XXXVI-1/W41 in ISPRS Workshop On Topographic Mapping From Space (With Special Emphasis on Small Satellites, Ankara, Turquie).*

Pouliquen, O. (1999). Scaling laws in granular flows down rough inclined planes. *Physics of Fluids, 11*(3), 542–548.

Poupardin, A., Heinrich, P., Frère, A., Imbert, D., Hébert, H., & Flouzat, M. (2017). The 1979 submarine landslide-generated Tsunami in Mururoa, French Polynesia. *Pure and Applied Geophysics, 174*, 3293–3311.

Rodriguez, M., Chamot-Rooke, N., Hébert, H., Fournier, M., & Huchon, P. (2013). Owen Ridge deep-water submarine landslides: Implications for tsunami hazard along the Oman coast. *Natural Hazards and Earth System Science, 13*, 417–424.

Saito, M. (1967). Excitation of free oscillations and surface waves by a point source in a vertically heterogeneous earth. *Journal of Geophysical Research, 72*(14), 3689–3699.

Satake, K., Smith, J., & Shinozaki, K. (2002). *Three-dimensional reconstruction and tsunami model of the Nuuanu and Wailau giant landslides* (pp. 333–346). American Geophysical Union Geophysical Monograph Series: Hawaii.

Savage, S. B., & Hutter, K. (1989). The motion of a finite mass of granular material down a rough incline. *Journal of Fluid Mechanics, 199*, 177–215.

Savage, S. B., & Hutter, K. (1991). The dynamics of avalanches of granular materials from initiation to runout. Part I: Analysis. *Acta Mechanica, 86*(1), 201–223.

Scharroo, R., Smith, W., Titov, V., & Arcas, D. (2005). Observing the Indian Ocean tsunami with satellite altimetry. *Geophysical Research Abstracts, 7*, 230.

Scheidegger, A. E. (1973). On the prediction of the reach and velocity of catastrophic landslides. *Rock Mechanics and Rock Engineering, 5*(4), 231–236.

Schuster, R. L., & Alford, D. (2004). Usoi landslide dam and lake sarez, Pamir mountains, Tajikistan. *Environmental and Engineering Geoscience, 10*(2), 151–168.

Sepúlveda, S. A., & Serey, A. (2009). Tsunamigenic, earthquake-triggered rock slope failures during the April 21, 2007 Aisén earthquake, southern Chile (45.5 S). *Andean Geology, 36*, 1.

Shuto, N. (1991). Numerical simulation of tsunamis—its present and near future. *Natural Hazards, 4*, 171–191.

Synolakis, C. E., Bardet, J.-P., Borrero, J. C., Davies, H. L., Okal, E. A., Silver, E. A., Sweet, S., & Tappin, D. R. (2002). The slump origin of the 1998 Papua New Guinea tsunami. In *Proceedings of the Royal Society of London, Series A: Mathematical, Physical and Engineering Sciences, Vol. 458, The Royal Society*, pp. 763–789.

Thomson, R. E., Rabinovich, A. B., Kulikov, E. A., Fine, I. V., & Bornhold, B. D. (2001). *On Numerical simulation of the landslide-generated Tsunami of November 3, 1994 in Skagway Harbor* (pp. 243–282). Dordrecht: Springer.

Tinti, S., Pagnoni, G., & Zaniboni, F. (2006). The landslides and tsunamis of the 30th of December 2002 in Stromboli analysed through numerical simulations. *Bulletin of Volcanology, 68*(5), 462–479.

Viroulet, S., Cébron, D., Kimmoun, O., & Kharif, C. (2013). Shallow water waves generated by subaerial solid landslides. *Geophysical Journal International, 193*(2), 747–762.

Voight, B. (1981). The 1980 eruptions of Mount St. Helens, Washington. Time scale for the first moments of the May 18 eruption. *US Geological Survey Professional Paper, 1250*, 69–86.

Wang, J., Ward, S. N., & Xiao, L. (2015). Numerical simulation of the December 4, 2007 landslide-generated tsunami in Chehalis Lake, Canada. *Geophysical Journal International, 201*(1), 372–376.

Ward, S. N. (1980). Relationships of tsunami generation and an earthquake source. *Journal of Physics of the Earth, 28*(5), 441–474.

Ward, S. N., & Day, S. (2011). The 1963 landslide and flood at Vajont Reservoir Italy. A tsunami ball simulation. *Italian Journal of Geosciences, 130*(1), 16–26.

Weiss, R., Fritz, H. M., & Wünnemann, K. (2009). Hybrid modeling of the mega-tsunami runup in Lituya Bay after half a century. *Geophysical Research Letters, 36*, 9.

Wessel, P., & Smith, W. H. F. (1991). Free software helps map and display data. *Eos Transactions American Geophysical Union, 72*(41), 441–446.

Yuan, X., Kind, R., & Pedersen, H. A. (2005). Seismic monitoring of the Indian Ocean tsunami. *Geophysical Research Letters, 32*, 15.

(Received September 13, 2018, revised January 29, 2019, accepted January 31, 2019, Published online February 11, 2019)

Pure Appl. Geophys. 176 (2019), 3059–3098

https://doi.org/10.1007/s00024-018-1978-7

Pure and Applied Geophysics

Landslide Tsunami Hazard Along the Upper US East Coast: Effects of Slide Deformation, Bottom Friction, and Frequency Dispersion

Lauren Schambach,[1] Stephan T. Grilli,[1] James T. Kirby,[2] and Fengyan Shi[2]

Abstract—Numerical simulations of Submarine Mass Failures (SMFs) are performed along the upper US East Coast to assess the effect of slide deformation on predicted tsunami hazard. Tsunami generation is simulated using the three-dimensional non-hydrostatic model NHWAVE. For rigid slumps, the geometry and law of motion are specified as bottom boundary conditions. Deforming slide motion is modeled using a depth-integrated bottom layer of dense Newtonian fluid, fully coupled to the overlying fluid motion. Once the SMFs are no-longer tsunamigenic, tsunami propagation simulations are performed using the Boussinesq wave model FUNWAVE-TVD, using nested grids of increasingly fine resolution towards shore and employing a one-way coupling methodology. Probable maximum tsunamis are simulated for Currituck SMF proxies sited in four areas of the shelf break slope that have enough sediment accumulation to cause large failures. Deforming slides have a slightly larger initial acceleration, but still generate a smaller tsunami than rigid slumps due to their spreading and thinning out during motion, which gradually makes them less tsunamigenic. Comparing the maximum envelope of surface elevations along a 5 m isobath, consistent with earlier work, the bathymetry of the wide shelf is found to strongly control the spatial distribution of tsunami inundation. Overall, tsunamis caused by rigid slumps are worst case scenarios, providing up to 50% more inundation than for deforming slides having a moderate level of viscosity set in the upper range of debris flows. Tsunamis from both types of SMFs are able to cause water withdrawal to the 5 m isobath or deeper. Bottom friction effects are assessed by performing some of the simulations using two different Manning coefficients, one 50% larger than the other. With increased bottom friction, the largest tsunami inundations at the coast are reduced by up to 15%. Selected simulations are rerun by turning off dispersion in the model, which leads to moderate changes in maximum surface elevations nearshore (− 10 to + 5% changes), but to more significant effects in the far field (− 40 to 80% changes). Onshore, dispersion causes the appearance of short period undular bores that eventually break nearshore without significantly affecting inundation at the coast. However, these bores increase wave-induced maximum flow velocity and impulse forces, the latter by up to 40%, which may affect the design of coastal structures.

Electronic supplementary material The online version of this article (https://doi.org/10.1007/s00024-018-1978-7) contains supplementary material, which is available to authorized users.

[1] Department of Ocean Engineering, University of Rhode Island, Narragansett, RI 02882, USA. E-mail: lschambach@uri.edu; grilli@uri.edu

[2] Center for Applied Coastal Research, University of Delaware, Newark, DE 19716, USA. E-mail: kirby@udel.edu; fyshi@udel.edu

1. Introduction

This study is part of tsunami hazard assessment work carried out by the authors since 2010 to develop tsunami inundation maps for the US East Coast (USEC), under the auspices of the US National Tsunami Hazard Mitigation Program (NTHMP). While Probabilistic Tsunami Hazard Analyses (PTHAs) have been considered or performed in other areas of the US and the world (e.g., González et al. 2009; Heidarzadeh and Kijko 2011; Horspool et al. 2014; Geist and Lynett 2014; Grezio et al. 2017), the paucity of paleo-tsunami information makes it difficult at present to conduct such an analysis for the USEC. This absence of a coastal sedimentary record of the past tsunamis may be attributed to glaciers and frequent large tropical cyclones, which have caused significant erosion and geomorphic changes nearshore and onshore.

Consequently, in earlier NTHMP-USEC work, tsunami generation, propagation, and coastal impact were modeled for a series of extreme sources in the Atlantic Ocean basin causing Probable Maximum Tsunamis (PMTs), which were selected based on a variety of studies that can be found in references (Grilli et al. 2010, 2017a; Tehranirad et al. 2015). Hence, the resulting NTHMP inundation maps provide envelopes of maximum inundation from the combined coastal impact of these PMTs, without consideration of return periods (ECMAP 2017). In the earlier work, the generation and propagation of each PMT were simulated using three-dimensional (3D) non-hydrostatic (Ma et al. 2012) and two-dimensional (2D) Boussinesq (Shi et al. 2012; Kirby et al. 2013) wave models, respectively. A series of nested spherical and Cartesian grids were used, with increasingly fine resolution towards the coast (down to 10–30 m meshes) and commensurately accurate bathymetric and topographic data.

This earlier work, as well as other studies, have shown that coastal hazard along the upper USEC is dominated by near-field tsunamis that could potentially be generated by large submarine mass failures (SMFs) (ten Brink et al. 2008, 2009a, b, 2014; Grilli et al. 2009, 2015, 2017a, b). The moderate seismicity typical of the region would not be expected to cause significant near-field co-seismic tsunamis, but could trigger large SMFs where large amounts of sediment has accumulated over steep enough slopes, on or near the continental shelf break. This is particularly the case off of major estuaries such as Chesapeake Bay or the Hudson River (Fig. 1). In fact, the largest earthquake ever measured in the USEC area, with a Mw 7.2 magnitude in the the upper Laurentian Fan, was responsible for triggering the 1929 landslide and tsunami off of the Grand Banks (Piper et al. 1999; Fine et al. 2005; Løvholt et al. 2018). The large tsunami generated by this 100 km^3 landslide caused a maximum runup of 13 m, 28 fatalities, and widespread destruction of coastal communities in Newfoundland, Canada. The SMF turned into a turbidity current that reached speeds of 17–28 m/s, breaking 12 underwater communication cables along its way [more details about this important historical case study can be found in Løvholt et al. (2018)]. Confirming that the 1929 landslide was not an isolated event, Chaytor et al. (2009) and ten Brink et al. (2014) mapped numerous paleo-SMFs on the US Atlantic continental shelf and margin, with the largest one being the Currituck slide complex (CSC) (Locat et al. 2009), off of Chesapeake Bay (Fig. 1). The CSC has been extensively studied from geological and slide triggering points of views (e.g., Locat et al. 2009, and references herein). Tsunami generation from a reconstituted Currituck SMF was first modeled by Geist et al. (2009), using a simplified SMF tsunami generation model, and then by Grilli et al. (2015) using a more complete model. Both modeling studies showed that, if it had failed in present days, this paleo-SMF might have generated a destructive tsunami for the USEC, from the Carolinas to Massachusetts. Recent field work (Hill et al. 2017) has dated this old slide complex to 16–50 Ka, but Chaytor (personal communication, 2017) indicates that its age is likely to be on the younger end of this range. Considering when this failure occurred, the sea level would have been much lower than in present days.

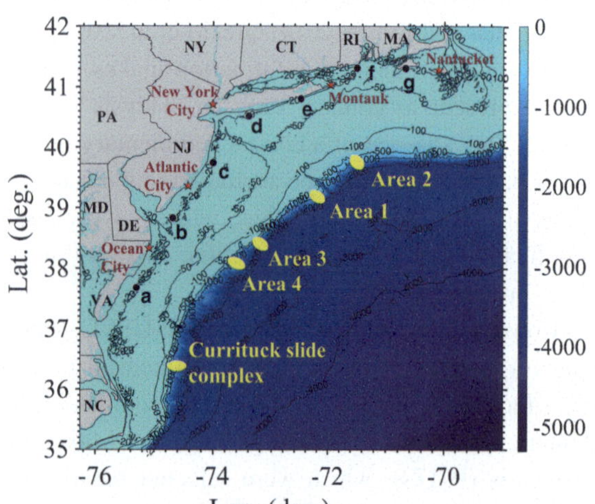

Figure 1
Geography of study area [with marked state limits and names, and names of a few cities (red stars)]. Areas 1–4, identified by Grilli et al. (2015) as having high potential for large tsunamigenic SMFs, and the location of the historical Currituck slide complex are marked by yellow ellipses. Numerical gage stations a–g (Table 3) are marked by black bullets. Color scale and contours show bathymetry in meters

When considering coastal hazard assessment for SMF tsunamis, many associated sources of uncertainty should be considered, related to SMF triggering, source characteristics, rheology/deformation, law of motion, epistemic uncertainties in the model, and others. With the goal to evaluate the first-order SMF tsunami hazard along the USEC, Grilli et al. (2009) attempted to account for many of these sources of uncertainty by performing Monte Carlo Simulations (MCS) of slope instabilities triggered by seismicity and causing tsunami generation and coastal impact. This study, which involved many assumptions and simplified models, was initially carried out from southern New Jersey to Cape Cod, MA, but was later extended to southern Florida (Krauss 2011). Two main failure mechanisms were considered, translational slides or slumps, both assumed to be rigid (i.e., solid block motion), in non-cohesive or cohesive sediment, respectively. Earlier surveys of the USEC margin by Booth et al. (2003) identified 43% of rotational failures (slumps) versus 57% of translational failures (slides). In cohesive sediment, the former can be considered as undertaking a mostly rigid motion; hence, slumps, although less frequent, are a realistic failure mode on the USEC. MCS results provided 100 and 500 year return period SMF tsunami runups along the entire USEC, as well as the respective percentages of slides and slumps, which were found to be consistent with field surveys Booth et al. (2003). MCS results predicted a 500 year maximum runup of 5–6 m north of Virginia and a significantly reduced runup (1–2 m) south of it. This study, however, did not consider whether there was sufficient sediment accumulation for large SMFs to occur off the coast, where large runup was predicted. To this effect, Eggeling (2012) performed geophysical and geotechnical analyses of seafloor data collected in regions facing coastal areas with the largest runup, which led to selecting 4 areas for further consideration and modeling (Areas 1–4 in Fig. 1). These areas typically have a large bottom slope and a sediment thickness sufficient to make a large tsunamigenic SMF possible, given sufficient seismicity. The likelihood for failure was confirmed by more detailed slope stability analyses performed using SLIDE (2017). These yielded low factors of safety of bottom slopes in these areas in the presence of the maximum expected seismicity obtained using the USGS 2002 Interactive Deaggregation Tool (USGS 2002). Details can be found in Eggeling (2012), but overall, the mean factors of safety for transects in Areas 1–4 were found to be as low as 0.82–0.92 (with standard deviation of 0.11–0.12), yielding probabilities of failure of 73–88%.

In light of these results, and considering the many uncertainties on SMF mechanisms and the scarcity of geotechnical field data on the upper USEC shelf, the NTHMP modeling committee decided that the inundation mapping work for the USEC would be performed on the basis of SMF PMTs. Accordingly, Grilli et al. (2015) simulated tsunami generation from large SMFs sited in Areas 1–4 (Fig. 1), which were all parameterized based on the characteristics of the largest known historical failure in the region, the CSC, as a single large failure with a volume in the 128–165 km^3 range estimated for the entire CSC by Locat et al. (2009).

For this reason, they were referred to as "Currituck SMF proxy" sources. In addition, to model worst case scenarios, each SMF was assumed to fail as a rigid slump, which, based on earlier work, was expected to maximize tsunami generation and coastal impact (Grilli and Watts 2005). Tsunami generation was computed using the 3D non-hydrostatic model NHWAVE (Ma et al. 2012) in which both the slump Gaussian-like geometry and rigid block law of motion were specified as bottom boundary conditions (details of this method, which is used in this paper to simulate rigid slumps, can be found in "Appendix"). The maximum SMF velocity used in the law of motion was that estimated by Locat et al. (2009) at $\simeq 35$ m/s, using a Bingham model. Once the slump stopped moving and tsunami generation was complete, simulations of tsunami propagation and coastal impact were continued using the 2D Boussinesq model FUNWAVE-TVD (Shi et al. 2012) (referred to hereafter as FUNWAVE), in a series of nested grids. Results of this work were used to develop the NTHMP tsunami inundation maps currently released for the upper USEC region (ECMAP 2017).

To confirm that rigid slumps indeed lead to worst case scenarios for tsunami coastal hazard along the USEC, Grilli et al. (2017b) investigated effects of SMF deformation on tsunami generation. They used the recent two-layer version of NHWAVE by Kirby et al. (2016), in which deforming slides are simply modeled as a dense Newtonian fluid, in a depth-integrated lower layer two-way coupled to the upper water flow simulated with the standard multi-σ-layer NHWAVE model. Earlier work based on simulations with the Bingham model BING concluded that it may not be necessary for proper tsunami generation to use a complex deforming slide rheology (Watts and Grilli 2003). Other work pointed out the importance of SMF mass, initial acceleration, and submergence for first-order tsunami generation, whereas SMF deformation should be second order (Grilli and Watts 2005). This justifies why more complex deformation models were not investigated here. Note, it is not the purpose of this paper to provide an exhaustive review of the various models and modeling methodologies that have been used to simulate underwater landslides and landslide tsunami generation, nor of historical landslide tsunami case studies. Fairly comprehensive reviews of such works to date can be found in Tappin et al. (2014), Kirby et al. (2016), and Grilli et al. (2015, 2017b). A broader scope recent review can be found in Yavari et al. (2016). Note that Løvholt et al. (2017) recently considered another mode of failure for giant slides, retrogressive failure, in which the landslide fails as a continuous sequence of blocks. They found that this failure mode could cause less tsunami generation than anticipated based on the slide size.

Grilli et al. (2017b) first revisited the simulation of the historical Currituck SMF with this model, to estimate values of slide viscosity ν_s and bottom friction Manning coefficient n (between slide and substrate) yielding a slide center of mass motion $S(t)$ similar to that used by Grilli et al. (2015) to simulate rigid slumps. This ensured that maximum velocity during motion and runout S_f after a time of motion t_f were similar to those of the slumps and thus consistent with values derived from field observations by Locat et al. (2009). These simulations resulted in a fairly large viscosity, $\nu_s = 0.5$ m^2/s, which was then used to simulate a Currituck SMF proxy in Area 1 (Hudson River canyon) as a deforming slide. For this case, maximum surface elevations simulated nearshore over the 5 m isobath were found to be smaller at most locations than those of the rigid slump, but nearly identical at some locations. Despite the fairly high viscosity, there was significant spreading of the deforming slide during its motion that led to a reduction of the acceleration during part of its initial motion and related maximum tsunami elevations. In addition, because the deforming slide flowed on its own (rather than having its motion prescribed), it followed the steepest bottom slope. Hence, the generated tsunami was more asymmetrical than for the slump, which affected both where the maximum tsunami impact occurred along the coast and the level of hazard. This preliminary study led Grilli et al. (2017b) to conclude that considering SMFs all failed as rigid slumps [as in Grilli et al. (2015)] was likely too conservative to estimate the coastal tsunami hazard along the USEC. Hence, modeling SMF tsunami generation and coastal hazard, with even a moderate level of deformation (i.e., here a high viscosity), should be more realistic, and also better cover the range of failures observed in the field (Booth et al. 2003; ten Brink et al. 2014).

Another aspect of the SMF tsunami hazard assessment relates to using proper physics in the wave generation and propagation models. In particular, one should assess whether using models that include frequency dispersion or non-hydrostatic pressure effects is important for assessing tsunami coastal hazard. Earlier work addressed the importance of dispersion during the generation phase of SMF tsunamis when vertical seafloor accelerations as well as 3D effects may be significant (Grilli and Watts 1999, 2005; Grilli et al. 2002). This showed that the important parameter to consider is $\mu = kd$, where d denotes the water depth above the SMF center of mass initial location and $k = 2\pi/\lambda_0$, where λ_0 is the dominant wavelength of generated tsunami waves. The latter can be approximated as $\lambda_0 \simeq 2b$, with b the SMF downslope length. Thus, according to linear long-wave theory, initial SMF tsunami waves will be dispersive when $\mu > \pi/10$ or $d > b/10$. For such a case, for instance, Ma et al. (2012) simulated Enet and Grilli's laboratory experiments for a rigid quasi-Gaussian shape slide moving down a slope using their 3D non-hydrostatic model NHWAVE, and showed a good agreement with time series of surface elevations measured at several wave gages. Re-running simulations with the non-hydrostatic pressure terms turned off in the model led to large errors.

For the very large SMFs considered here, however, it will be shown that $d < b/10$ and, hence, the initially generated tsunami waves are not expected to be dispersive. During their subsequent propagation onshore and offshore, however, tsunami wave trains will experience increasing dispersive effects, the larger the magnitude of $\mu = kh$, here based on ocean depth h, with $k = 2\pi/\lambda$ and λ the local tsunami wavelength. The cumulative effects of dispersion will gradually become larger, the longer the propagation distance L. Assuming regular waves propagating over constant depth, Glimsdal et al. (2013) quantified dispersive effects by way of a "dispersion time" $\tau = 6h^2L/\lambda^3$ and showed that these typically become significant when $\tau > 0.1$. For landslide tsunamis, they pointed out as discussed above that the initial wavelength λ_0 is more significant to dispersion than depth or propagation distance [see also Løvholt et al. (2015) for additional considerations and discussions]. Dispersive effects also affect long waves propagating shoreward, into shallow water, causing the eventual occurrence of undular bores around steep crests (e.g., Madsen et al. 2008; Glimsdal et al. 2013), which may locally nearly double maximum tsunami elevation and hence affect the coastal tsunami hazard. Glimsdal et al. (2013) indicated that undular bores may break before reaching the shore, as they are made of short and steep waves; hence, their effect on maximum elevation is more uncertain. However, breaking waves may increase flow velocity and impulse force. To be able to resolve such short waves in numerical simulations, tsunami models require fine resolution nearshore grids. With a 32 m resolution grid, Grilli et al. (2015) were barely able to simulate the initial stages of the appearance of undular bores in SMF tsunamis propagating into Chesapeake Bay. Using one-dimensional simulations, Grilli et al. (2013a) showed that to fully resolve such undular bores at this location, the resolution needed to be about 5 m.

Earlier work has shown a significant effect of bottom friction on tsunami propagation. Not only is tsunami inundation affected by land cover roughness (Jakeman et al. 2010; Kaiser et al. 2011), but for wide continental shelves, tsunami elevations can significantly decrease due to bottom friction (Geist et al. 2009; Tehranirad et al. 2015; Grilli et al. 2015). Indeed, based on linear long-wave theory, the depth-averaged tsunami-induced velocity $U \propto h^{-3/4}$, for a depth h. Hence, energy dissipation due to bottom friction $\mathcal{D} \propto C_d U^3$. Grilli et al. (2015) performed tsunami propagation simulations with FUNWAVE using a constant bottom friction coefficient $C_d = 0.0025$ (Shi et al. 2012); hence, bottom friction dissipation was $\mathcal{D} \propto C_d h^{-9/4} \propto h^{-2.25}$. In more recent nearshore tsunami simulation work with FUNWAVE (Shelby et al. 2016), a more realistic depth-dependent friction coefficient was used based on Manning's formula, $C_d = g n^2/h^{1/3}$, function of the Manning n parameter; this yields larger values of the bottom friction dissipation, $\mathcal{D} \propto h^{-31/12} \propto h^{-2.58}$, i.e., energy dissipation more than quadratically increases as depth decreases when long waves propagate onshore into shallower water.

In this paper, the landslide tsunami hazard is more realistically assessed along the upper USEC by considering effects of slide deformation on tsunami generation. Similar to Grilli et al. (2017b), tsunami generation and propagation are modeled for Currituck SMF proxies sited in Areas 1–4 (Fig. 1) off of the upper USEC. Tsunami generation is compared for SMFs failing as deforming slides or rigid slumps, with the latter being re-simulated to compare results for both types of SMFs in the same higher resolution coastal grids used here. Bottom friction is modeled using $n = 0.025$ s^2/m$^{2/3}$ throughout FUNWAVE simulations, which corresponds to a coarse sand substrate. However, since sandy and gravel/boulder substrates can be found in various areas of the USEC a limited comparison is performed for tsunami inundation computed with $n = 0.025$ or 0.0375 (which reflects the rougher substrate). This is done in one of the most impacted areas of the USEC, in northern New Jersey and western Long Island, NY. Finally, by re-running selected simulations without dispersion, the effects of dispersion are evaluated on tsunami propagation (both far- and near-field) and coastal hazard (maximum inundation height, velocity, and impulse force). Tsunami simulations for the four considered areas allow assessing the tsunami hazard from extreme SMFs along the USEC from Virginia to Cape Cod, MA. Results are provided as instantaneous surface elevations maps, time series of surface elevation at numerical wave gages, envelopes of tsunami surface elevations for single or the combined SMFs, and percent differences between envelopes of results without and with dispersion.

Section 2 summarizes the modeling methodology. Section 3 presents simulations of SMF tsunami generation with NHWAVE and Sect. 4 of SMF tsunami propagation and coastal impact with FUNWAVE. Section 5 is devoted to evaluating the importance of dispersive effects on the tsunami hazard and is followed by a discussion in Sect. 6, and conclusions in Sect. 7.

2. Modeling Methodology

2.1. Numerical Models

Tsunami generation by SMFs is modeled with the 3D non-hydrostatic model NHWAVE. For rigid SMFs, the geometry, velocity, and acceleration are specified as bottom boundary conditions, following the methodology of Ma et al. (2012) and Grilli et al. (2015) [see also, Grilli and Watts (2005), Watts et al. (2005) and Enet and Grilli (2007); a summary of this methodology is given in "Appendix"]. For deforming slides, the two-layer version of NHWAVE is used with the lower layer representing the SMF as a dense Newtonian fluid and the upper-layer water domain being divided into multiple σ layers (Kirby et al. 2016; Grilli et al. 2017b). In the bottom layer, equations of mass and momentum conservation are depth-integrated, similar to long-wave models. Both volumetric (i.e., viscous) and bottom friction (i.e., basal shear) dissipation terms are included in the equations. This is different from the standard depth-integrated models, which typically only include a single bottom friction term. Modeling experiments for slides made of glass beads, Grilli et al. (2017b) observed that results were not very sensitive to slide viscosity, perhaps due to this parameterization of dissipation. Hence, they simulated a deforming Currituck SMF proxy in Area 1 using a fairly high viscosity $v_s = 0.5$ m^2/s, which is in the range of suggested values for debris flows ($v_s = 0.2$–0.6 m^2/s), and compared results obtained for various substrate friction coefficients, $n = 0.05, 0.10,$ or 0.15 s^2/m$^{2/3}$ to those of the rigid slump (Grilli et al. 2015). They found that $n = 0.1$ led to approximately the same maximum runout of the slide center of mass at time $t_f = 11.9$ min as for rigid slumps, and a similar velocity of their center of mass during the first 10 min of each SMF motion. [Note that Grilli et al. (2017b) used this v_s value in their simulations, but wrongly listed the dynamic viscosity in their paper as $\mu_s =$ 500 kg/(m s), while it should have been 950 kg/(m s), based on bulk density.]

As in earlier work, the initial geometry of both rigid and deforming SMFs is modeled as a sediment mound of quasi-Gaussian cross sections with an elliptical footprint over the slope (see, Enet and Grilli (2007); Grilli et al. (2015), and Eqs. 1–3 in "Appendix", for details). In all cases, NHWAVE uses a horizontal Cartesian grid of resolution $(\Delta x, \Delta y)$ and a boundary fitted σ-coordinate grid with N_σ layers in the vertical direction.

Once the tsunami is fully generated, simulations of wave propagation are pursued in the 2D fully nonlinear and dispersive Boussinesq model FUNWAVE (this will be detailed below). As in earlier work, these simulations are performed by one-way coupling in a series of nested grids of increasingly fine resolution in shallower water towards the coast [see, e.g., Grilli et al. (2013b, 2015, 2017a, b), Tappin et al. (2014) and Shelby et al. (2016), for details and examples of this approach]. When simulating tsunami propagation in regional grids of small geographic extent, the Cartesian coordinate version of FUNWAVE is used (Shi et al. 2012), whereas when simulating propagation in a grid at the scale of the northern Atlantic ocean, the spherical version of the model is used (Kirby et al. 2013).

NHWAVE and FUNWAVE are non-hydrostatic, i.e., dispersive, wave models, in which dispersion can be turned off as an option. As discussed in the introduction, earlier studies have shown the importance of including frequency dispersion effects to model SMF tsunami generation and propagation, particularly when the initial SMF footprint is small compared to depth (e.g., Ma et al. 2012). In this case, the generated tsunami waves have a relatively small wavelength compared to depth (e.g., Grilli and Watts 1999, 2005; Glimsdal et al. 2013). Moreover, when simulating the long distance (far-field) propagation of any tsunami [i.e., for a large "dispersion time" τ in the sense of Glimsdal et al. (2013)], constructive–destructive wave–wave interactions will be affected by dispersion, potentially leading to significant differences in wave height and phase as compared to non-dispersive simulations (Kirby et al. 2013). Finally, when tsunamis propagate onshore in shallow water, dispersion leads to the development of undular bores near their wave crests. This was observed for many tsunamis: (1) experimentally (e.g., Matsuyama et al. 2007 and references listed in Grilli et al. 2017b; (2) in the field (e.g., Madsen et al. 2008; Tappin et al. 2014); or (3) numerically (e.g., Løvholt et al. 2008; Madsen et al. 2008; Abadie et al. 2012; Glimsdal et al. 2013; Grilli et al. 2013a, 2015; Tehranirad et al. 2015).

2.2. Methodology and Computational Grids

For rigid slumps, which are rotational, tsunami generation is completed when they stop moving upon reaching their maximum runout S_f at time t_f. By contrast, deforming slides keep "flowing" down the continental slope for a much longer time. However, they may no longer be tsunamigenic when becoming thin and/or reaching deep enough water. Here, no measurable wave generation was observed in NHWAVE for $t \geq 20$ min in all deforming slides simulations; hence, this time was selected to initialize FUNWAVE. This model, however, assumes a steady bottom but, at $t \geq 20$ min, deforming slides are still slightly moving and thus imposing a time-dependent lower boundary condition to the upper water layer in NHWAVE. In their simulations of a deforming slide in Area 1, Grilli et al. (2017b) observed that this boundary condition mismatch caused the generation of spurious (rebound) waves in FUNWAVE, which they eliminated by filtering out velocities over the slide location before initializing simulations. Because at this time, the slide was far offshore and no-longer tsunamigenic, and this filtering did not affect simulations of the tsunami onshore propagation (as was verified by performing a longer simulation with NHWAVE). A similar method is applied here, in combination with a larger grid (details of this procedure are given later).

Tsunami generation from each SMF sited in Areas 1–4 is computed with NHWAVE, using four distinct Cartesian grids with $\Delta x = \Delta y = 500$ m horizontal resolution and $N_\sigma = 5$ σ-layers centered around each SMF initial location (x_0, y_0) (Fig. 2a; Table 1). Each grid covers a 500 by 500 km footprint (hence, it has 1000 by 1000 cells; see grid SW corner coordinates in Table 1) and is nested within FUNWAVE's $\Delta x = \Delta y = 500$ m Cartesian grid G0 (Fig. 2). This first-level (or base) grid, in which simulations of tsunami propagation are continued, covers an 800 km by 800 km area (hence, it has 1600 by 1600 cells; see details in Table 2). To eliminate reflection along offshore boundaries, 125 km-thick sponge layers are specified on both the eastern and southern boundaries of grid

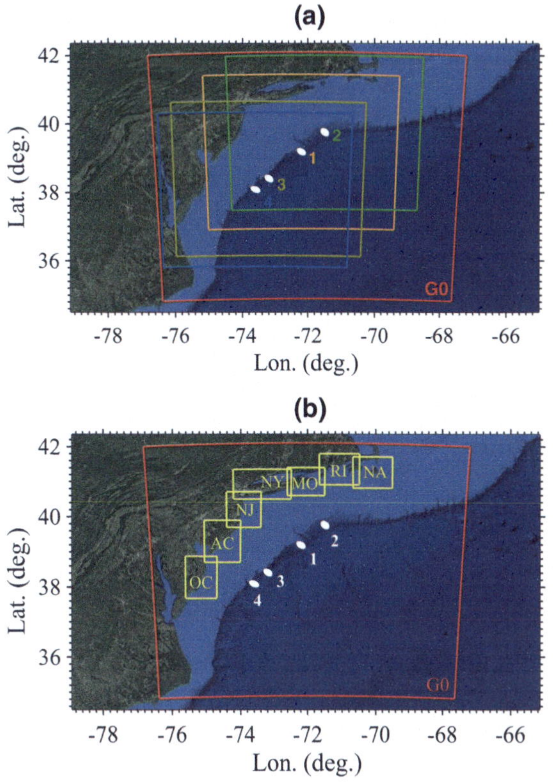

Figure 2
Horizontal footprints of computational grids, with (red) FUNWAVE $\Delta x = \Delta y = 500$ m resolution Cartesian grid G0 encompassing the SMFs and their grids, and **a** 4 color coded $\Delta x = \Delta y = 500$ m horizontal resolution nested Cartesian NHWAVE grids (with $N_\sigma = 5$ vertical σ-layers), used to simulate motion and tsunami generation for SMFs sited in Areas 1–4 (numbered ellipses; Table 1) and **b** 7 labeled $\Delta x = \Delta y = 120$ m resolution Cartesian FUNWAVE grids (OC, AC, NJ, NY, MO, RI, and NA from southwest to northeast; Table 2)

G0. At $t = 20$ min, the free surface elevation, bottom bathymetry, and horizontal velocities (u, v) [at the required $0.531h$ depth to ensure extended dispersion properties in FUNWAVE; see Shi et al. (2012)] computed in NHWAVE are interpolated onto G0 and used as initial conditions to continue simulations for each SMF. For the onshore tsunami propagation, simulations are then continued with FUNWAVE by one-way coupling in seven nested Cartesian grids of $\Delta x = \Delta y = 120$ m resolution (Fig. 2b; Table 2). To study dispersive effects nearshore, additional finer nested grids are used in FUNWAVE, with 30 m (NJ30, OC30) and 7.5 m (NJ7.5) resolutions (Table 2; see details later). Likewise, to study dispersive effects on far-field tsunami propagation, simulations are also

Table 1

Parameters of NHWAVE grids ($\Delta x = \Delta y = 500$ m resolution; $N_\sigma = 5$ σ-layers; 1000 by 1000 cells), and locations (center) of SMFs of width $b = 30$ km, width $w = 20$ km and thickness $T = 0.75$ km, in Areas 1–4 (Figs. 1, 2a)

Grids and SMFs	Area 1	Area 2	Area 3	Area 4
SW corner (Lon., Lat.)	(− 75.00, 36.90)	(− 74.32, 37.47)	(− 75.97, 36.12)	(− 76.37, 35.80)
Center (x_0, y_0)	(− 72.19, 39.19)	(− 71.49, 39.76)	(− 73.19, 38.41)	(− 73.6, 38.09)
Azimuth θ (CW from N)	136°	153°	140°	126°

For rigid slumps, the average local slope is assumed to be $\alpha = 4°$ (Fig. 3)

Table 2

Parameters of FUNWAVE computational grids (Figs. 2, 14a, 16a, 21a)

Grid name	Cells (M_x, N_y)	Resol.	SW corner (Lat., Lon.)
Base grid (G0)	(1600, 1600)	500 m	(34.82N, 76.37W)
Ocean City (OC)	(672, 1104)	120 m	(37.68N, 75.60W)
Atlantic City (AC)	(752, 1104)	120 m	(38.70N, 75.04W)
New Jersey (NJ)	(720, 936)	120 m	(39.70N, 74.40W)
New York (NY)	(1212, 792)	120 m	(40.49N, 74.21W)
Montauk (MO)	(780, 744)	120 m	(40.60N, 72.60W)
Rhode Island (RI)	(816, 828)	120 m	(40.90N, 71.65W)
Nantucket (NA)	(828, 832)	120 m	(40.81N, 70.67W)
North Atlantic (ATL)	(5236, 3136)	1 min	(5.903N, 88.154W)
Seaside heights (NJ30)	(984, 984)	30 m	(39.92N, 74.17W)
Ocean City (OC30)	(984, 984)	30 m	(38.26N, 75.23W)
Seaside heights (NJ7.5)	(1200,1200)	7.5 m	(39.91N, 74.13W)

continued with FUNWAVE in the 1 arc-min (about 1800 m) resolution spherical grid ATL, which covers the North Atlantic ocean; this large grid has 5236 by 3136 cells in the longitudinal and latitudinal directions, respectively (Table 2), and uses 150 km-thick sponge layers along the outer boundaries. Simulations in the ATL grid are also initialized using NHWAVE results computed at $t = 20$ min (details of this grid are given later as well).

The bathymetry used in the 500 and 120 m resolution grids (Table 2) was interpolated from the 3-arc-s ($\simeq$ 90 m) resolution Northeast Atlantic and Southeast Atlantic US Coastal Relief Models (NGDC 1999a, b), wherever available (mostly up to the shelf break); otherwise, data were obtained from the 1 arc-min resolution ETOPO-1 Global Relief Model (Amante and Eakins 2009). The bathymetry of the 30 m NJ30 and 7.5 m NJ7.5 resolution grids, centered on Seaside Heights, NJ, was interpolated from the 1/9 arc-s ($\simeq$ 3 m) post Hurricane Sandy survey data (NCEI 2014) and that of the 30 m OC30 grid, centered on Ocean City, MD, from Grothe et al. (2010) (Table 2). For each Cartesian grid, the bathymetry/topography was constructed using a Mercator transverse projection centered on each grid center. The ETOPO1 bathymetry in the study area is plotted in Fig. 1, as both a color scale and contour lines. North of North Carolina, it shows the typical pattern of a wide shallow USEC shelf (depth $h \leq 100$ m), a steep shelf break slope, and a deep abyssal plain ($h = 2000$–5000 m). Similar to the Currituck slide complex, the four study Areas 1–4 are all located on the shelf break slope.

3. SMF Tsunami Generation with NHWAVE

3.1. SMF Geometry, Rheology, and Kinematics

Tsunami generation from Currituck SMF proxies sited in Areas 1–4 off of the USEC (Fig. 1) is modeled with NHWAVE, assuming each SMF fails

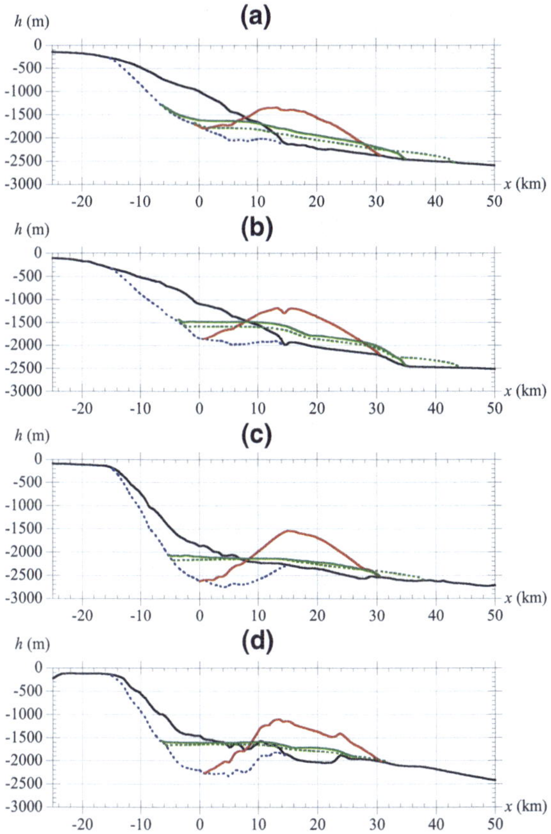

Figure 3
Vertical bathymetric transects in each NHWAVE grid (Fig. 2a), through Areas 1–4 (plots **a–d**) SMF centers (x_0, y_0), in azimuthal direction θ (Table 1): (solid black) current bathymetry; (dash blue) initial SMF profile; (solid red) final slump profile after displacement (runout) $S_f = 15.8$ km at $t_f = 715$ s (11.9 min); (solid green/dash green) deforming slide profiles at $t = 715$ and 1200 s (20 min) [note, slides continue moving beyond this time but are no longer significantly tsunamigenic.] The x-axis measures distances from each SMF center; vertical exaggeration is 25 times

either as a rigid slump [similar to Grilli et al. (2015)] or as a deforming slide [similar to that considered in Area 1 by Grilli et al. (2017b)]. Consistent with earlier work (Grilli et al. 2015), each SMF is assumed to have a quasi-Gaussian geometry with an initial elliptical footprint of downslope length $b = 30$ km, width $w = 20$ km, and thickness $T = 0.75$ km (Enet and Grilli 2007; Grilli et al. 2015, 2017b). [These values were inferred by Grilli et al. (2015) from field data of the CSC, for which the total material in the two main failure lobes identified by Locat et al. (2009) were lumped into a single initial failure originating in the headwall scar located up the slope.] Using a shape coefficient $\epsilon = 0.717$, this yields a SMF volume $V_s = 158$ km^3, which is in the range of the total volume estimated for the entire Currituck slide complex (see details in "Appendix"). The initial SMF geometry is specified below current seafloor in the model (Fig. 3), with the elliptical footprint centered at longitude–latitude (x_0, y_0) and its major axis oriented downslope in an initial azimuthal direction of motion θ (Table 1). Figure 3 shows vertical cross sections in direction θ through each SMF sited in Areas 1–4 (both seafloor bathymetry and initial/later SMF profiles). The SMFs are all initially located on the continental shelf slope, with the water depth above their initial center of mass location varying between $d = 1000$ and 1800 m. Hence, $d < b/10 = 3$ km and the dominant SMF tsunami waves that will be generated are expected to be in the long-wave regime. The initial elliptical footprints of each SMF are plotted at scale in Fig. 4; the rigid slumps are all specified to move in the direction of the major axis of each ellipse.

As in earlier work (Grilli and Watts 2005; Grilli et al. 2015, 2017b), for rigid slumps, the bottom boundary conditions are specified analytically as a function of SMF geometry and other parameters, assuming a pendulum-like motion parallel to slope, $S(t)$. [Note, it is not the scope of this paper to revisit this earlier work on solid block motion, whose details can be found in reference; a summary is provided in "Appendix" (see Eq. 4).] In this motion, maximum runout S_f and duration t_f are computed as a function of slump parameters: beside geometry, these include bulk sediment density $\rho_s = 1900$ kg/m^3, water density $\rho_w = 1025$ kg/m^3, average local slope α, radius of (pendulum) motion R, and (hydrodynamic) added mass coefficient C_M. In the transects of Fig. 3, the average seafloor slopes are 3.2°, 3.0°, 4.0°, and 3.1°, in Areas 1–4 SMFs, respectively. However, considering that there are steeper parts in each transect, $\alpha = 4°$ is used for each case, which also yields identical maximum runout and time of motion in each case. As in Grilli et al. (2015, 2017b), $C_M = 1$ is used and the radius is estimated as, $R \simeq b^2/(8T) = 150$ km. With these parameters, Eqs. (4)–(8) in "Appendix" yield $S_f = 15.8$ km and $t_f = 715$ s (11.9 min). Figure 3 shows the final profile of each rigid

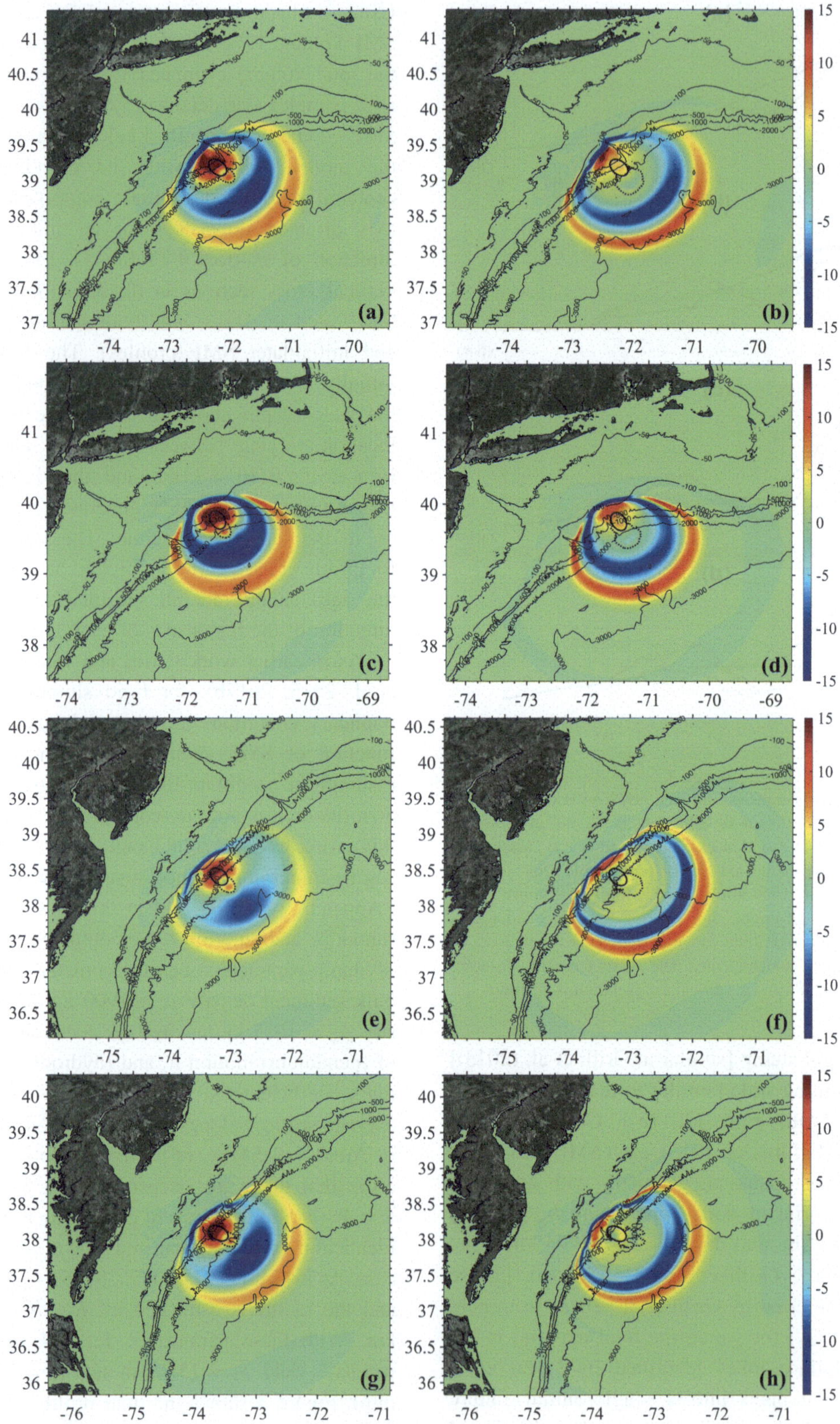
(a)
(b)
(c)
(d)
(e)
(f)
(g)
(h)
15
10
5
0
-5
-10
-15

◀Figure 4
Free surface elevations (color scale in m) simulated at $t = 800$ s (13.3 min) with NHWAVE, in 500 m grids (Fig. 2a; Table 1), for Currituck SMF proxies sited in Areas 1–4 (from top to bottom; Fig. 1), modeled as: (**a**, **c**, **e**, **g**) rigid slumps (similar to Grilli et al. 2015); or (**b**, **d**, **f**, **h**) deforming slides (with $\nu_s = 0.5$ m/s^2, $n = 0.1$). Black ellipses mark: (solid) initial footprints of each SMF (Table 1); (dashed) SMF footprints at 715 s for the slumps and 20 min for the slides. All SMFs have a volume $V_s = 158$ km^3, density $\rho_s = 1900$ kg/m^3, and similar runout S_f at $t_f = 715$ s (when the slumps stop moving)

slump in Areas 1–4 after they have moved a distance S_f down the slope (for $t \geq t_f$).

The motion of deforming slides is modeled with the two-layer NHWAVE, which computes it together with tsunami generation based on the same initial geometry and bulk sediment density as for slumps and on a specified SMF viscosity (assuming a dense Newtonian fluid) and substrate-to-slide friction. Following the above discussions, the kinematic viscosity is set to $\nu_s = 0.5$ m^2/s and slide-to-substrate Manning friction coefficient to $n = 0.1$ s^2/m$^{2/3}$, as in earlier work by Grilli et al. (2017b). This will make deforming slide kinematics consistent with that of rigid slumps, allowing to compare tsunami generation and coastal impact for both SMF types. Figure 5 shows the kinematics of each SMF center of mass, i.e., motion $S(t)$, velocity $U(t)$, and acceleration $A(t)$, computed during NHWAVE simulations, for rigid slumps and deforming slides. For the former, equations of the specified kinematics are listed in "Appendix"; for the latter, kinematics is derived from calculating each deforming slide center of mass location as a function of time in NHWAVE results.

Figure 3 shows vertical cross sections through SMFs computed in Areas 1–4, in their initial direction of motion θ. Profiles are shown at time $t_f = 715$ s when rigid slumps stop moving, and at $t = 1200$ s (20 min) when NHWAVE's solution is moved into FUNWAVE as an initial condition. As expected, the deforming slides' center of mass locations at $t = t_f$ appears to be similar to those of the rigid slumps. However, while rigid slumps have preserved their geometry and thickness all the way to the end of their motion, the deforming slides have flowed both downslope and sideways, leading to a significant reduction in thickness. For $t > t_f$, the deforming slides continue flowing down the slope, some by a considerable distance, but due to both their reduced thickness and the large depth, they are no longer significantly tsunamigenic. Figure 4 confirms this behavior by showing, for Areas 1–4, the final footprints of the rigid slumps at 715 s and those of the deforming slides at 20 min. These different behaviors during motion are further detailed in

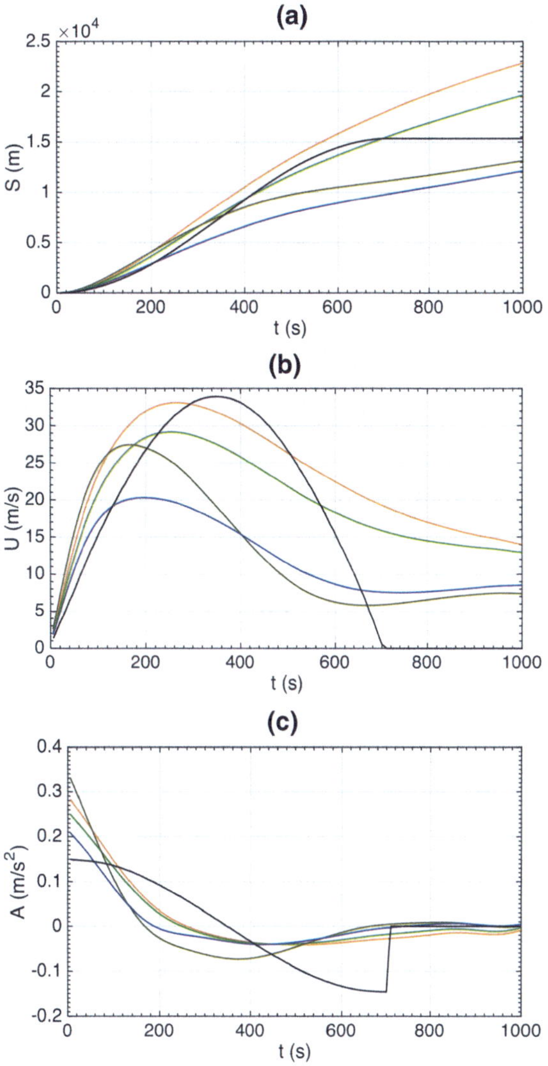

Figure 5
Kinematics of Currituck SMF proxies sited in Areas 1–4 (Fig. 1), during tsunami generation. SMF center of mass: **a** motion; **b** velocity; and **c** acceleration, specified for rigid slumps (black) (based on Eqs. 4–8 in "Appendix"), and computed for deforming slides (for $\nu_s = 0.5$ m^2/s and $n = 0.1$), in Area: (orange) 1; (green) 2; (brown) 3; (blue) 4

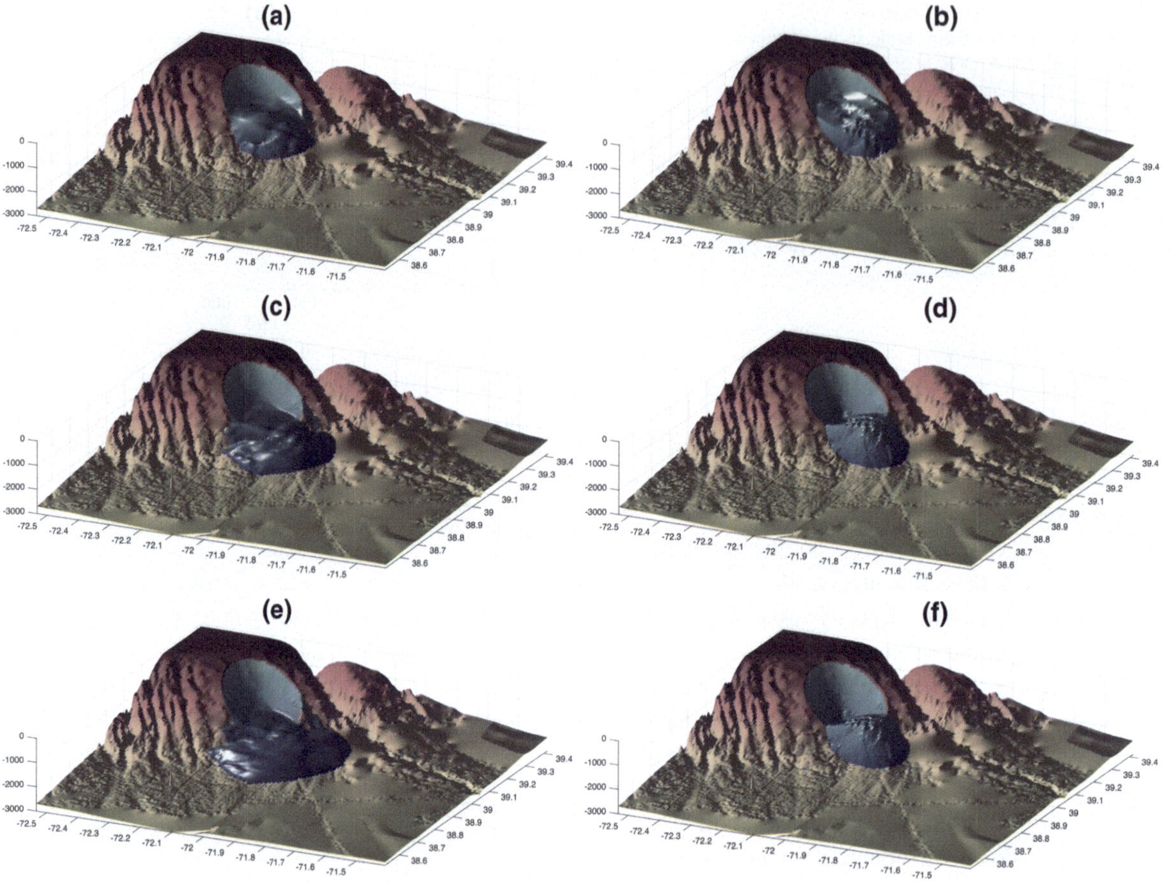

Figure 6
Instantaneous 3D geometry (grey volume) of Currituck SMF proxy in Area 1 (Fig. 1), simulated with NHWAVE in a 500 m resolution grids with $N_\sigma = 5$ vertical layers (Fig. 2a), modeled as a: (**a**, **c**, **e**) deforming slide; or (**b**, **d**, **f**) rigid slump, at $t =$ (**a**, **b**) 5, (**c**, **d**) 10, and (**e**, **f**) 15 min. The vertical axis denotes depth in meter. Note, the rigid slump reaches its maximum runout at $t = 11.9$ min (hence this is the situation for **f**), while the slide keeps moving beyond $t = 15$ min (see Fig. 3)

Fig. 6, which show the instantaneous 3D geometry at $t = 5$, 10, and 15 min, of the Currituck SMF proxy modeled in Area 1 with NHWAVE, either as a deforming slide or a rigid slump. The SMFs both start moving from the same initial elliptical footprint and geometry below seafloor, on the West side of the Hudson canyon apron (visible on the figures), but as time increases the deforming slide continuously flows down the steepest slope, while the rigid slump undertakes its forced pendulum motion in direction θ. In the case considered here, it is apparent that the deforming slide sediment flows asymmetrically, more towards the west side, whereas the rigid slump stays symmetrical with respect to its axis of motion. This will lead to different alongshore tsunami generation in each case (see below). Finally, as the slump stops moving at $t_f = 11.9$ min, there is not much additional motion between $t = 10$ and 15 min visible in Fig. 6d, f.

The SMF kinematics displayed in Fig. 5 are consistent with these observations. The theoretical laws of motion of rigid slumps are plotted as a reference. These are identical for Areas 1–4, since slumps are all similarly parameterized to reach a maximum runout and stop moving at $t_f = 715$ s. They reach a maximum velocity $U_{\max} = 34.7$ m/s at mid-course and have a maximum acceleration $A_0 = 0.153$ m/s^2 at $t = 0$, with an identical negative value at the end of their rigid motion. By contrast, the center of mass kinematics of deforming slides computed during NHWAVE simulations is site specific to each area and thus different for each slide. Figure 5c shows that, up to $t = 60$–100 s depending on the considered area, their acceleration is larger than that of rigid slumps. In Areas 1 and 2, whose seafloor features a simple fairly convex continental shelf slope (Fig. 3a, b), deforming slides reach a maximum velocity during motion and runout at t_f similar to those of rigid slumps (Fig. 5a, b). This is expected from their parameterization, consistent with the work of Grilli et al. (2017b). In Areas 3 and 4, however, whose shelf slope is concave, despite a motion similar to the other two areas for $t < 350$ s, the slides slow down quicker and reach a smaller runout at time t_f. This is clearly seen in the transects of Fig. 3c, d, where the slide material ends up filling the bottom of the initial cavity, which is below the deeper part of the shelf slope; hence, only a smaller part of each slide continues moving more slowly down the slope. It will be shown that this pattern of velocity and acceleration of deforming slides will lead to the generation of smaller onshore propagating tsunami waves, but larger offshore propagating tsunami waves compared to those of rigid slumps.

3.2. Tsunami Generation

Figure 4 shows free surface elevations simulated with NHWAVE at $t = 800$ s (13.3 min) for Currituck SMF proxies of Areas 1–4 (Fig. 1), whose motion is modeled as: (left panels) rigid slumps, or (right panels) deforming slides. For each area, computations are performed in individual grids with 500 m horizontal resolution and $N_\sigma = 5$ vertical layers (Fig. 2a; Table 1). By design, rigid slumps stop moving at $t_f = 715$ s after covering a downslope runout distance $S_f = 15.8$ km (Fig. 3), and at this time, deforming slides have approximately reached the same runout (see Fig. 5). As a result of their similar kinematics, waves generated at $t = 800$ s by deforming slides have a similar horizontal pattern as those generated by rigid slumps. However, they have different elevations due to the lateral spreading and thickness reduction of deforming slides late in their motion (Figs. 3, 6). In all cases, onshore propagating tsunamis feature a 10–15 m-deep leading depression wave followed by an equally large crest; these leading waves are larger for the rigid slumps. Deforming slides generate more asymmetric waves (alongshore) than rigid slumps, as a result of the effects of bathymetry on their motion. In all cases in Fig. 4, the offshore propagating tsunamis feature concentric (cylindrical) waves, with a 5–10 m leading crest followed by a deeper trough. The size of this leading wave is function of the SMF initial acceleration, which is slightly larger for deforming slides (Fig. 5c); hence, the leading crest elevation is larger for deforming slides than for rigid slumps.

Although the rigid slumps stop moving and no longer generate waves at $t_f = 715$ s, simulations were performed with NHWAVE for both types of SMFs up to $t = 1200$ s (20 min), before continuing simulations in FUNWAVE, to make sure 3D effects had become negligible in the generated wave trains. At this time, Fig. 3 shows that deforming slides are quite thin (i.e., have a large length-to-thickness ratios) and have reached a large depth ($>$ 1500 m); hence, these are no longer significantly tsunamigenic. As indicated before, flow velocities are filtered out of NHWAVE results above each slide location to prevent triggering numerical instabilities when initializing simulations in FUNWAVE, due to a mismatch in bottom boundary condition.

4. Onshore Tsunami Propagation and Coastal Impact

Results of simulations with FUNWAVE of the onshore propagation and coastal inundation of tsunamis generated by rigid slumps and deforming slides sited in Areas 1–4 are detailed in the following. Simulations initialized at $t = 20$ min based on NHWAVE results are first performed in the 500 m resolution grid G0; then, these are continued by one-way coupling in 120 m resolution nearshore nested grids (Fig. 2b; Table 2). Results of additional simulations in higher resolution grids of the deforming slide in Area 1 are presented in the next Section, to analyze dispersive effects during nearshore propagation. Specifically, we use: (1) two 30 m resolution grids nested within their respective 120 m grids, NJ30 off of Seaside Heights, NJ and OC30 off Ocean City, MD and then (2) one 7.5 m resolution grid, NJ7.5,

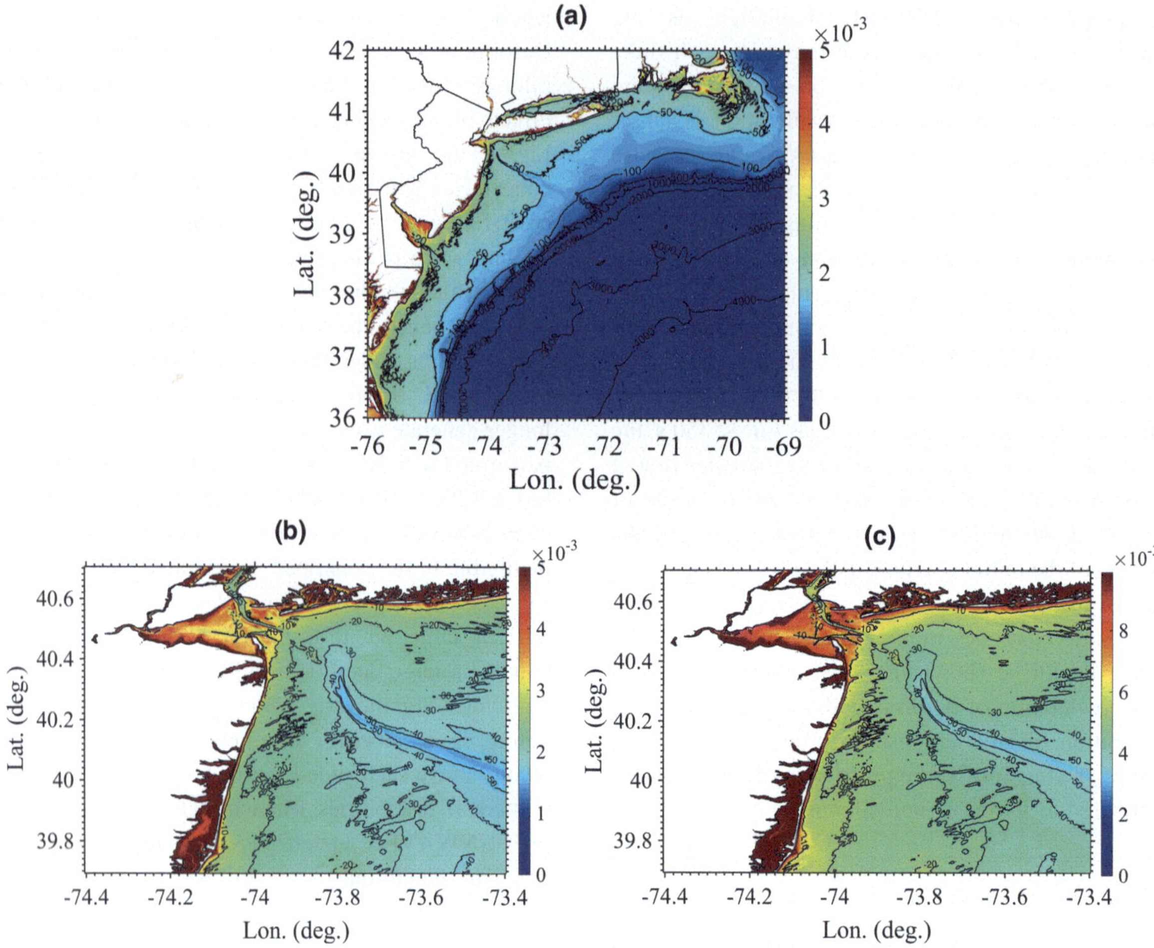

Figure 7
Spatial variation of friction coefficient $C_d = g\,n^2/h^{1/3}$ (color scale) used in FUNWAVE simulations, based on local depth h (contour lines in meter) using a Manning coefficient $n =$ **a** 0.025 over grid G0 (Fig. 2); **b**, **c** 0.025 and 0.0375 $s^2/m^{2/3}$ over grid NJ, respectively (Fig. 2b; Table 2). Note, a minimum depth of 0.1 m is assumed in the C_d calculations

nested within NJ30 (Table 2). These results will also allow assessing the convergence with grid resolution of the computed maximum surface elevations near the coastline.

Bottom friction was neglected in NHWAVE simulations of tsunami generation, except for slide-to-substrate friction, which was modeled with a Manning $n = 0.1$ in the bottom layer of the model. This was acceptable since wave generation took place over short times in fairly large depth. In the FUNWAVE simulations of the onshore tsunami propagation over the wide USEC shelf, bottom friction effects, which are more significant, are specified as a function of a Manning coefficient. In the absence of site specific data, a coarse sand substrate was assumed in most simulations, for which $n = 0.025$ $s^2/m^{2/3}$. However, the seafloor of the upper USEC shelf, whose geology is post-glacial, features many areas covered with gravels and boulders (McMaster 1960),

Figure 8 ▶
Instantaneous free surface elevations (color scale in m) simulated with NHWAVE and FUNWAVE in grid G0 (zoom-in of Fig. 2) for the deforming Currituck SMF proxy in Area 1 (Figs. 1, 4b) at $t =$ **a** 10, **b** 30, **c** 50, **d** 70, **e** 90, **f** 110, **g** 130, and **h** 150 min. Black bullets mark locations of wave gage stations **a**–**g** (Fig. 1; Table 3). Black contours mark depth in meter

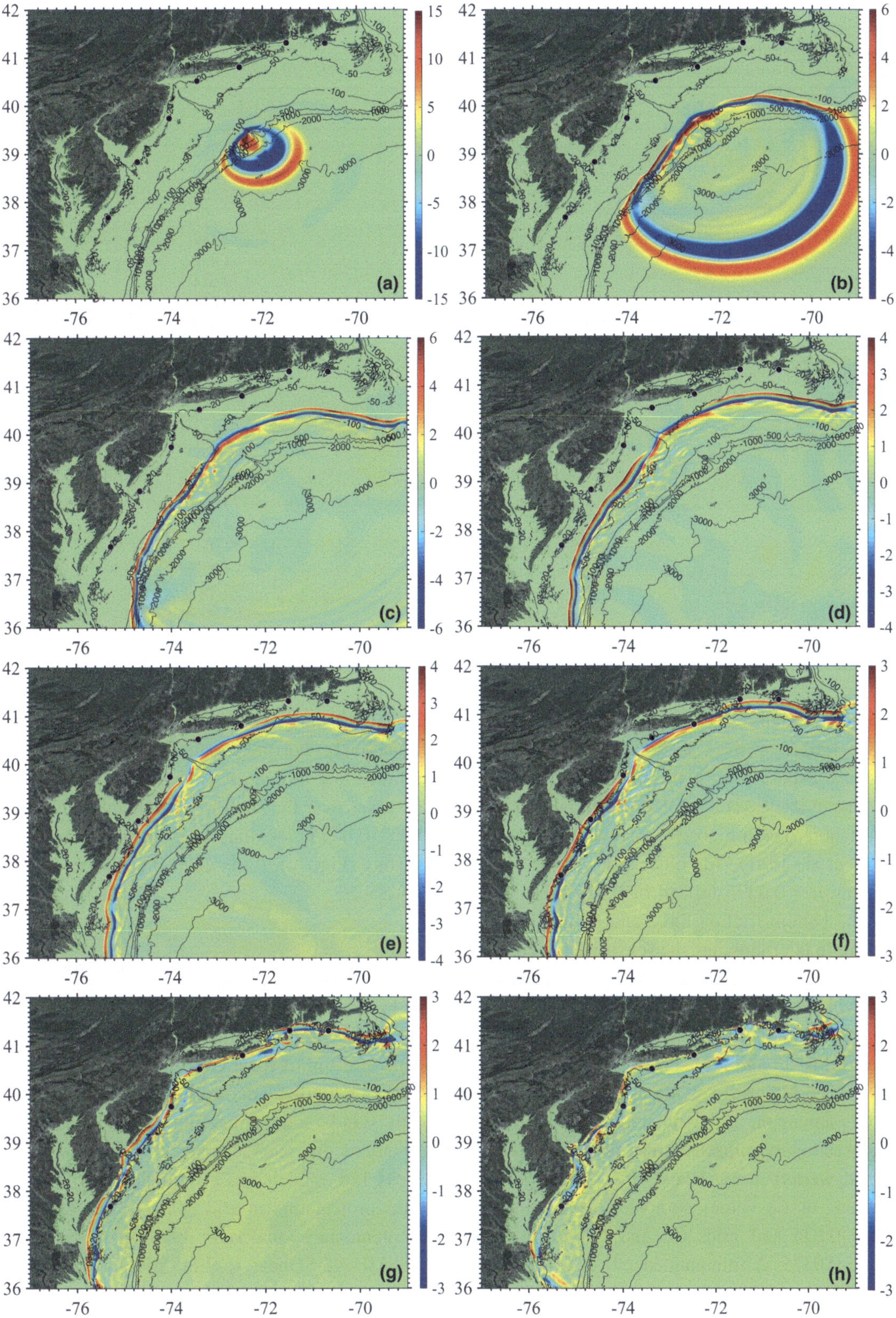
(a)
(b)
(c)
(d)
(e)
(f)
(g)
(h)

Table 3

Name, location, and depth of deep and shallow water wave gage stations (Figs. 1, 14a, 16a), where time series of surface elevations (Figs. 9, 15, 18) are computed in 500 m G0, 1 min ATL, 30 m NJ30, and 30 m OCE30 grids (Figs. 2, 14a, 16a, 21a)

Number	Label	N Lat. (deg.)	Lon. (deg.)	Depth (m)
a		37.678	− 75.303	20
b		38.832	− 74.682	20
c		39.744	− 74.000	20
d		40.522	− 73.391	20
e		40.807	− 72.489	20
f		41.316	− 71.492	20
g		41.313	− 70.674	20
1	DART 44401	37.592	− 50.030	5494
2	DART 44402	39.298	− 70.659	2649
3	DART 41424	33.000	− 72.660	5250
4	DART 41420	23.508	− 67.287	5743
5	DART 41421	23.380	− 64.112	5845
6		51.812	− 13.077	985
7		45.163	− 4.652	4556
8		38.675	− 10.181	3349
9		29.713	− 19.715	4676
10		39.856	− 74.069	15
11		39.919	− 74.062	15
12		39.990	− 74.051	15
13		40.037	− 74.037	15
14		38.2910	− 75.0960	10
15		38.3565	− 75.0648	10
16		38.4208	− 75.0469	10
17		38.4784	− 75.0452	10

for which the Manning coefficient would be larger, up to $n = 0.0375$ $s^2/m^{2/3}$ or so. In other shallow water areas exposed to high wave energy from storms, fine particles could be preferentially transported offshore, leaving coarser sediment nearshore (Twichell et al. 1981). All other parameters being equal, using $n = 0.0375$ instead of 0.025 more than doubles the bottom friction force and related energy dissipation. In the following, the sensitivity of coastal inundation to the magnitude of bottom friction (in relation with changes in substrate) is assessed by repeating simulations of the deforming slide sited in Area 1 in grids G0 and NJ using $n = 0.0375$ $s^2/m^{2/3}$ (Table 2). The NJ grid was selected for this sensitivity study among the 120 m resolution grids, since it encompasses highly impacted areas in northern New Jersey and western Long Island (Fig. 2b). Figure 7 shows maps of C_d values computed over grid G0 using $n = 0.025$ and grids G0 and NJ, using $n =$ 0.025 and 0.0375. A minimum depth of 0.1 m was specified in the model, which based on Manning's formula limits the value of C_d to $C_d^{max} = 21.14n^2$, i.e., 0.013 and 0.030 for $n = 0.025$ and 0.0375, respectively. For the smaller n value (Fig. 7a, b), C_d is fairly low in deeper water (less than 0.0015), but increases over the shelf to more than 0.002 (reaching 0.003–0.005 in less than 10 m depth); using the larger n value (Fig. 7c), C_d is 2.25 times larger than in the first case.

4.1. *Instantaneous Propagation and Time Series*

Figure 8 shows a sequence of instantaneous surface elevations computed at $t = 10$–150 min (by steps of 20 min) for a Currituck SMF proxy sited in Area 1, assuming a deforming slide. [Four video files, with side-by-side animations of simulated surface elevations for each Currituck SMF proxy sited in Areas 1–4, modeled as a rigid slump or a deforming slide, propagating in grid G0, are provided as online supplementary material: EtaSMF-A1-SDnear.mp4, EtaSMF-A2-SDnear, EtaSMF-A3-SDnear, and EtaSMF-A4-SDnear (color scale in m).] Tsunami generation is modeled with NHWAVE (upper left

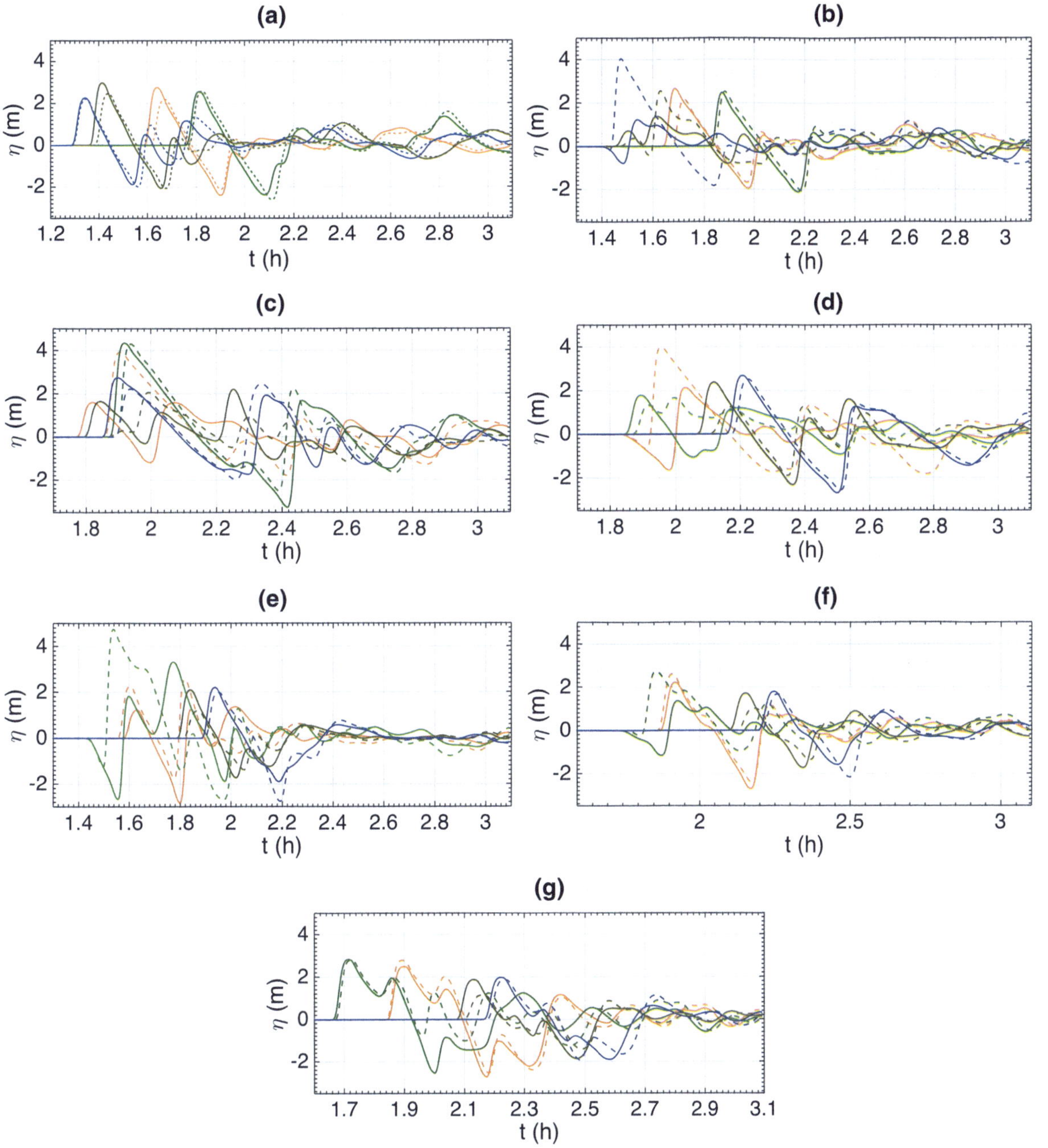

Figure 9
Time series of free surface elevation simulated with NHWAVE and FUNWAVE in grid G0 (Fig. 2) at: **a–g** shallow water wave gage stations **a–g** (Fig. 1; Table 3), as: (solid lines) rigid slumps; or (dash lines) deforming, Currituck SMF proxies, sited in Areas (Fig. 1): (orange) 1; (green) 2; (brown) 3; and (blue) 4

panel) and the subsequent wave propagation in grid G0 is modeled with FUNWAVE (other panels). As time increases, the initially onshore propagating tsunami waves continue their propagation towards the nearest shores. At the same time, the northern and southern parts of the offshore propagating waves gradually refract over the shelf break bathymetry and eventually orient themselves parallel to local

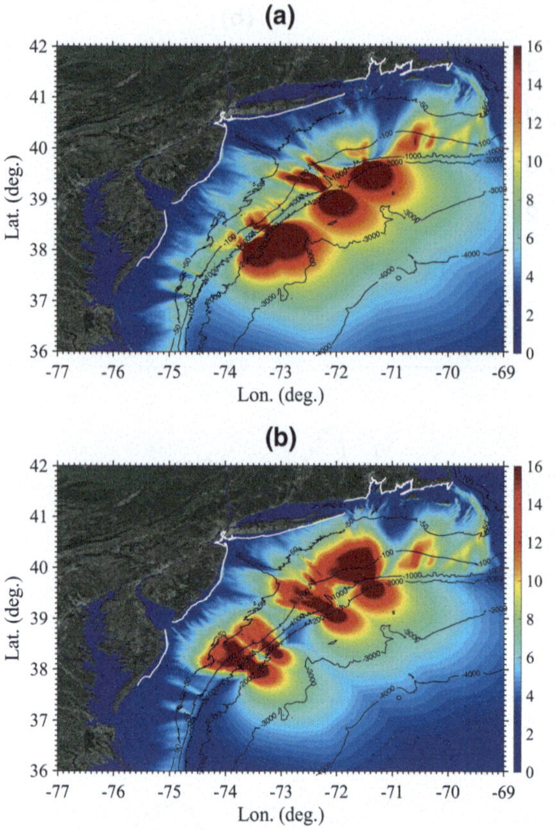

Figure 10
Combined envelope of maximum surface elevations (color scale in m) computed with FUNWAVE in 500 m grid G0 or 120 m nested grids wherever available (Fig. 2b; Table 2), for the four SMFs sited in Areas 1–4 (Fig. 1): **a** deforming slides; or **b** rigid slumps. The white lines mark the 5 m isobath along which maximum and minimum wave heights are computed (see, Fig. 12). Black contours mark depth in meter

isobaths, propagating onshore towards the upper and lower parts of the coastline. Although there is some decrease in elevation up and down the coast due to energy spreading, the onshore propagating waves remain quite large (several meters) over the entire study area. At 110 min, Fig. 8f shows a nearly continuous elevation or depression wave, spanning from south to north and propagating towards the coast. This wave approximately reaches the 20 m isobath at the same time at most locations (the 7 black bullets denote wave gage stations a–g on the 20 m isobath; Fig. 1; Table 3). This implies that the SMF tsunami generated in the Hudson river canyon in Area 1 ends up affecting the coast over the entire study area. This observation also applies to tsunamis generated by SMFs sited in other Areas 2–4 (as can be seen in the videos provided as supplementary material).

Note that in FUNWAVE simulations of all SMF cases, it was observed that outbound waves were not fully absorbed by the 125 km wide sponge layers specified along the eastern and southern boundaries of grid G0, which caused some reflection. Faced with the same problem, Grilli et al. (2017b) had filtered out the outbound wave surface elevations and horizontal velocities, before continuing simulations in a finer resolution nested grid. Such filtering, however, affects the ability of outbound waves to properly refract up and down the shelf break and contribute to the coastal tsunami hazard. Here, to prevent affecting the tsunami impact up and down the coast, the footprint of grid G0 was extended by 400 km in both the east and south directions and simulations of each tsunami were pursued up to $t = 50$ min in this larger grid (stage of the third panel in Fig. 8). At this time, the outbound waves, which had propagated beyond the offshore boundary of grid G0, were simply truncated and simulations restarted with these waves in grid G0, up to $t = 187$ min (3h7').

Figure 9 shows time series of surface elevations computed with FUNWAVE at wave gage Stations a–g in grid G0 (Fig. 1) for tsunamis generated by the 8 Currituck SMF proxies sited in Areas 1–4 (both slumps and deforming slides). Note that these stations are all located at a 20 m depth on the boundary of one of the finer nested grids, but results shown in Fig. 9 were computed in the coarse 500 m resolution grid G0. Hence, higher frequency waves that might be generated are not resolved here. In a following section, similar results computed at shallow water wave gage stations located in finer resolution nested grids will be shown and discussed. Consistent with the instantaneous surface elevations shown in Fig. 8, depending on the SMF case and considered station, incoming tsunami waves feature either a leading depression or elevation wave, which in most cases has a N-wave shape. For the deforming slide in Area 1 (Fig. 8), for instance, the tsunami arrives at all stations as an elevation wave, except at station d, which is located most directly onshore of the source. Earlier work has shown that such long N-wave-

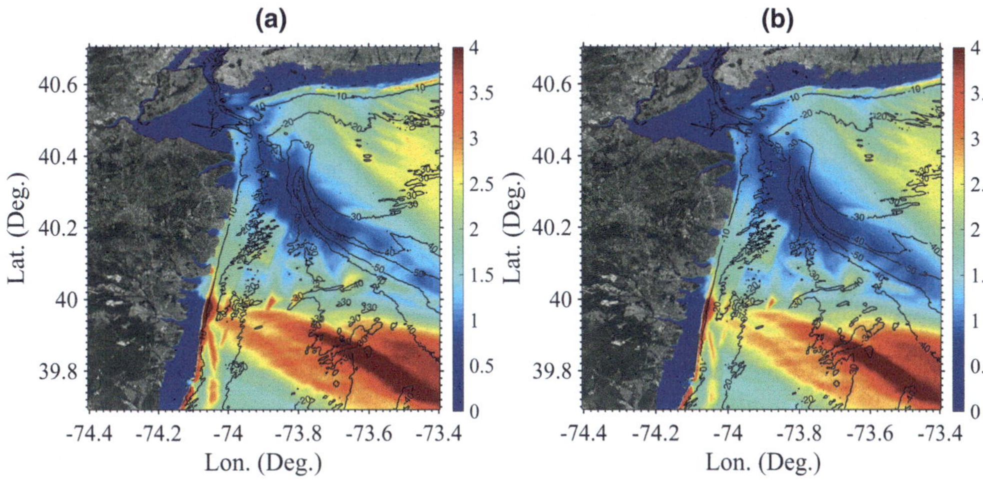

Figure 11
Envelopes of maximum surface elevations (color scale in m) computed with FUNWAVE in 120 m grid NJ (Fig. 2; Table 2), for the deforming Currituck SMF proxy sited in Area 1 (Fig. 1), with a Manning $n =$ **a** 0.025; or **b** 0.0375 $s^2/m^{2/3}$ (Fig. 7b, c). Black contours mark depth in meter

shaped waves typically develop undular bores in shallow water (a.k.a. dispersive shock waves), made of shorter oscillations occurring at wave crests and/or behind wave troughs (Matsuyama et al. 2007; Madsen et al. 2008; Grilli et al. 2013a, 2015). Here, grid G0 is too coarse to resolve undular bores, but these will appear in results of finer resolution grids presented later. Finally, depending on the considered SMF and station, the period of the leading wave varies between 6 and 29 min, i.e., from quite short to much longer, and these waves are followed by a train of multiple waves, which can be shorter or longer with many shorter waves riding on top of them.

4.2. *Maximum and Minimum Surface Elevations, and Bottom Friction Effects*

Figure 10a, b shows the combined envelope of maximum surface elevations computed for the 4 deforming and 4 rigid slump Currituck SMF proxies sited in Areas 1–4, respectively. These figures are based on results of simulations in grid G0 and in the 7 nested 120 m grids, whichever is available, using a Manning friction coefficient $n = 0.025$. While offshore surface elevations significantly differ for the two different SMF rheologies, particularly over their generation areas (in accordance with Fig. 4), patterns and directionality of maximum nearshore surface elevations are similar for both SMF types. This results from the strong control of the shelf bathymetry on the focusing and defocusing of long tsunami waves, which start significantly refracting in fairly deep water (Tehranirad et al. 2015). This bathymetric control effect is more apparent in Fig. 11a, which shows the envelope of maximum surface elevation computed in the 120 m resolution grid NJ with $n = 0.025$ (Fig. 2b; Table 2), for the deforming Currituck SMF proxy sited in Area 1 (Fig. 1), which is the same case as in Fig. 8. The Hudson River canyon bathymetry causes incident tsunami waves, which are nearly parallel to the local isobath ahead of the canyon in 100 m depth (Fig. 8, second panel), to gradually refract away from the canyon and the entrance of New York Harbor. As a result, waves are focused towards both the northern New Jersey and western Long Island areas. In particular, Fig. 11a shows maximum surface elevations of more than 4 and 2.5 m in the highly developed barrier beaches of Seaside Height, NJ and Long Beach, NY, respectively, with the lagoons being fully flooded behind these. This defocusing-focusing phenomenon was already pointed out in earlier work performed in this area by Grilli et al. (2010, 2015) and Shelby et al. (2016).

Bottom friction effects on tsunami coastal impact are assessed next by comparing results of simulations of the deforming SMF in Area 1 using a larger Manning friction coefficient $n = 0.0375$ to the above results using $n = 0.025$ (Fig. 7b, c). Figure 11a, b shows the respective envelopes of maximum surface elevation computed in the 120 m resolution grid NJ. While the patterns of wave focusing–defocusing caused by refraction are not significantly affected when increasing n there is an overall reduction in maximum nearshore surface elevation, particularly along the barrier beaches skirting the coastline. Figure 12 better illustrates focusing/defocusing and bottom friction effects on the coastal tsunami hazard by plotting maximum surface elevations computed along the 5 m isobath (marked as a white line in Fig. 10) in the seven 120 m resolution grids for the two SMF types, as a function of the curvilinear distance s measured along the isobath (from Chincoteague Island, VA, south of Ocean City, MD to Cape Cod, MA).

Figure 12a first shows results for the Currituck SMF proxy sited in Area 1 simulated in grid NJ: (1) rigid slump with $n = 0.025$ and (2) deforming slide with either $n = 0.025$ (Fig. 11a) or $n = 0.0375$ (Fig. 11b). As mentioned before, the pattern of higher and lower maximum surface elevations is nearly identical for the three cases. However, as already noted by Grilli et al. (2017b), given the same friction coefficient $n = 0.025$ and a single high-resolution grid, the maximum elevations caused by the deforming slide are nearly half those of the rigid slump, and there is another small reduction in elevation when using the larger friction coefficient in the former case. Figure 12b shows similar results, but for the combined envelope of the rigid slumps or deforming slides sited in Areas 1–4, simulated in the seven 120 m resolution nested grids (as in Fig. 10). Here, the picture appears more complex: due to wave refraction and wave-guiding effects over the complex geometry of the shelf break, as discussed above, all SMF sources can generate tsunamis that affect the entire upper USEC in some manner. Due to the larger onshore propagating waves, they generate, rigid slumps cause the largest inundation at all sites, with a 11.5 m maximum around the eastern end of long Island near Montauk, NY (Fig. 1). [This maximum impact occurs onshore of a marked ridge in the shelf bathymetry that causes high wave focusing (see, e.g., Fig. 1)]. However, because deforming slides generate larger offshore propagating waves, which upon refracting on the shelf break can affect distant areas, the reduction in coastal impact shown in Fig. 12a for a single slide and a single small grid is overall less marked (except along the southern coast of Long Island west of Montauk). For deforming slides, the maximum inundation is reduced to 7 m in Montauk

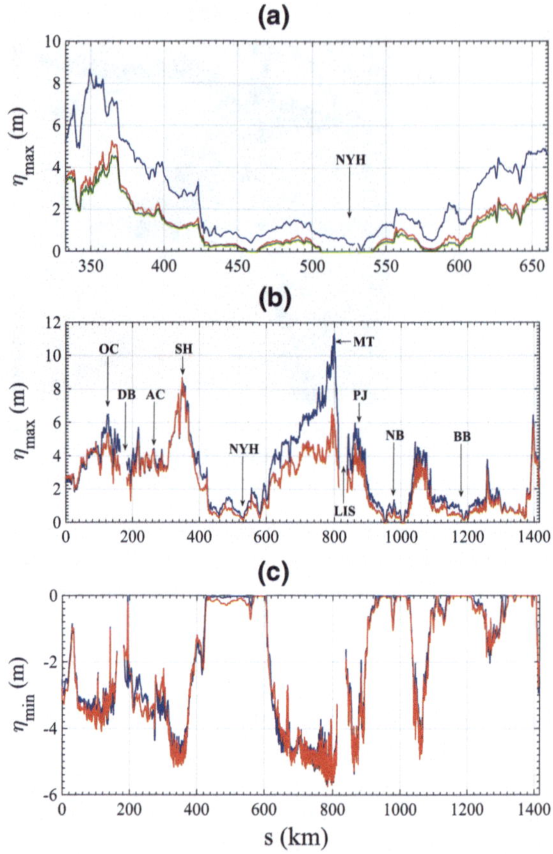

Figure 12
a, **b** Maximum and **c** minimum surface elevation computed in 120 m grids (Fig. 2b), along the 5 m isobath (Fig. 10) for tsunamis generated by: (blue) rigid slump in **a** Area 1, or **b**, **c** the superposition of slumps in Areas 1–4; and (red) deforming slide in **a** Area 1, or **b**, **c** the superposition of all deforming slides in Areas 1–4. All FUNWAVE simulations are performed with $n = 0.025$ s^2/m$^{2/3}$; green line in plot **a** is deforming slide in Area 1 for $n = 0.0375$ s^2/m$^{2/3}$ (Fig. 7). The distance s is the curvilinear distance along the 5 m isobath measured from its southern end. Labels mark entrance to: *OC* Ocean City, MD, *DB* Delaware Bay, *AC* Atlantic City, NJ, *SH* Seaside Heights, NJ, *NYH* New York Harbor, *MT* Montauk, NY, *LIS* Long Island Sound, *PJ* Point Judith, RI, *NB* Narragansett Bay, *BB* Buzzards Bay

and is largest, 4.5–8.5 m, between Atlantic City and Seaside Heights, NJ (Fig. 1). In these combined envelopes, the pattern of wave focusing and defocusing repeats itself, with the lowest impact occurring in or near Bays and estuaries (e.g., Delaware Bay, New York Harbor, Narragansett Bay, Buzzards Bay).

Finally, Fig. 12c shows the envelope of minimum surface elevations computed along the 5 m isobath for the same cases as in Fig. 12b. Such results would be important, for instance, when considering the tsunami impact on the fresh water intakes of a power plant, of which there are a few in this area (e.g., nuclear power plants in Millstone, CT near the mouth of Long Island Sound or Oyster Creek, NJ, just south of Seaside Heights). Here, results appear quite similar for the rigid slump and deforming slide cases; at a few instances, in particular near Atlantic City and Montauk, the minimum elevation would reach down to the local seafloor (i.e., 5 m, with some minor irregularities related to discretization and the moving shoreline algorithm). This similarity in minimum values is expected, since the minimum coastal elevation is typically caused by the depression wave that first arrives on the nearest shores facing Areas 1–4 (Figs. 1, 9), and it is a result of the initial motion of each SMF, i.e., at short time when deformation has not yet played an important role, but instead SMF mass, initial acceleration, and local slope control wave generation (Grilli and Watts 2005).

5. *Importance of Dispersive Effects on the SMF Tsunami Hazard*

In earlier work, frequency dispersion effects were shown to significantly affect tsunami generation by SMFs and subaerial slides when their downslope length b is relatively small with respect to the initial water depth d: as discussed in the introduction, when $b/d < 10$ (e.g., Grilli and Watts 2005; Ma et al. 2012; Abadie et al. 2012). In the far field, dispersion causes changes in surface elevations and lag times that increase with propagation distance (Madsen et al. 2008; Glimsdal et al. 2013; Kirby et al. 2013), and are particularly significant for initially dispersive SMF tsunamis (e.g., Løvholt et al. 2008, 2015; Abadie et al. 2012; Tehranirad et al. 2015). In onshore propagating tsunamis, dispersive effects cause the appearance of undular bores made of shorter waves that may significantly increase nearshore wave elevations and affect the coastal tsunami hazard (Madsen et al. 2008; Glimsdal et al. 2013).

In the present case, $b > 10d$ for the SMFs sited in Areas 1–4 and the waves are not expected to be significantly dispersive initially. However, as tsunamis propagate in the far field over long distances across the Atlantic Ocean, dispersive effects may potentially become significant. On the nearshore side, the N-wave-shaped waves simulated in grid G0 at 20 m depth (Fig. 9) may lead to the the occurrence of undular bores, given a sufficiently fine grid resolution.

In the following, the magnitude of dispersive effects is assessed during tsunami generation, and for both the far- and near-field propagations of the deforming slide sited in Area 1 (Table 1). This will be done by comparing results of additional simulations, with regular dispersive effects or these being turned off in FUNWAVE, in finer nearshore nested grids (NJ30, OC30, NJ7.5; Fig. 2) and in one coarser far-field grid (ATL; Table 2). Without dispersive terms, FUNWAVE becomes a standard Nonlinear Shallow Water Wave equations solver.

5.1. *SMF Tsunami Generation*

The importance of dispersive effects on tsunami generation is first assessed with NHWAVE using the same grid and parameters as discussed above. Figure 13a shows the relative difference (in percent) computed between the non-dispersive and dispersive results at $t = 1200$ s (the time when these waves are passed onto FUNWAVE), and Fig. 13b, c shows surface elevations computed along cross- and along-slope transects marked in (a), at $t = 300$, 600 and 800 s, in the non-dispersive and dispersive results.

In the cross-slope transect of Fig. 13b, surface elevations are almost identical when turning off dispersion for onshore propagating waves, but differences start growing with time/distance for offshore propagating waves. This is consistent with results that could be expected based on long-wave theory. At $t =$ 300 or 600 s, the dominant wavelength in the offshore propagating tsunami is $\lambda_0 \simeq 47.5$ or 60 km (the latter

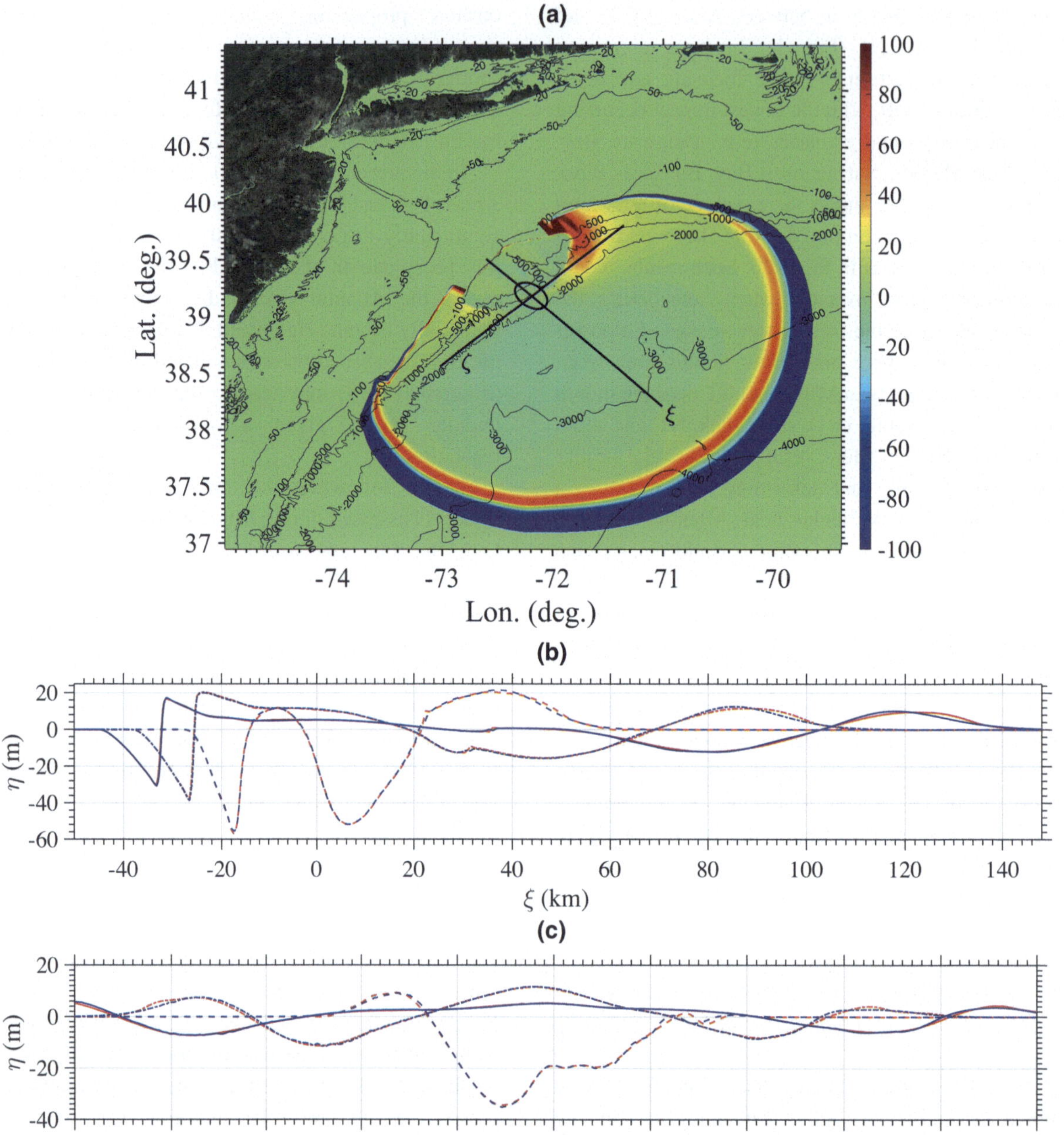

Figure 13
Dispersive effects in simulations with NHWAVE of tsunami generation for the deforming Currituck SMF proxy sited in Area 1 (Fig. 1), propagating in the corresponding 500 m NHWAVE grid (Fig. 2a): **a** relative difference (color scale in %) of maximum surface elevations in non-dispersive versus dispersive computations at $t = 1200$ s; black lines mark the SMF initial elliptical footprint as well as traces of cross- (ξ) and along-(ζ) slope transects. **b**, **c** Instantaneous surface elevation along cross- and along-slope transects marked in **a**, respectively, at $t =$ (dashed) 300, (chained) 600, and (solid) 800 s, for: (blue) dispersive; and (red) non-dispersive simulations

being $\simeq 2b$), in an average depth $d = 1.7$ or 1.9 km, respectively (Fig. 3a); hence, $\mu = 0.22$ or $0.2 < \pi/10$, which is in the long-wave regime. For the onshore propagating waves at $t = 300$ s, $\lambda_0 \simeq 24.5$ km in

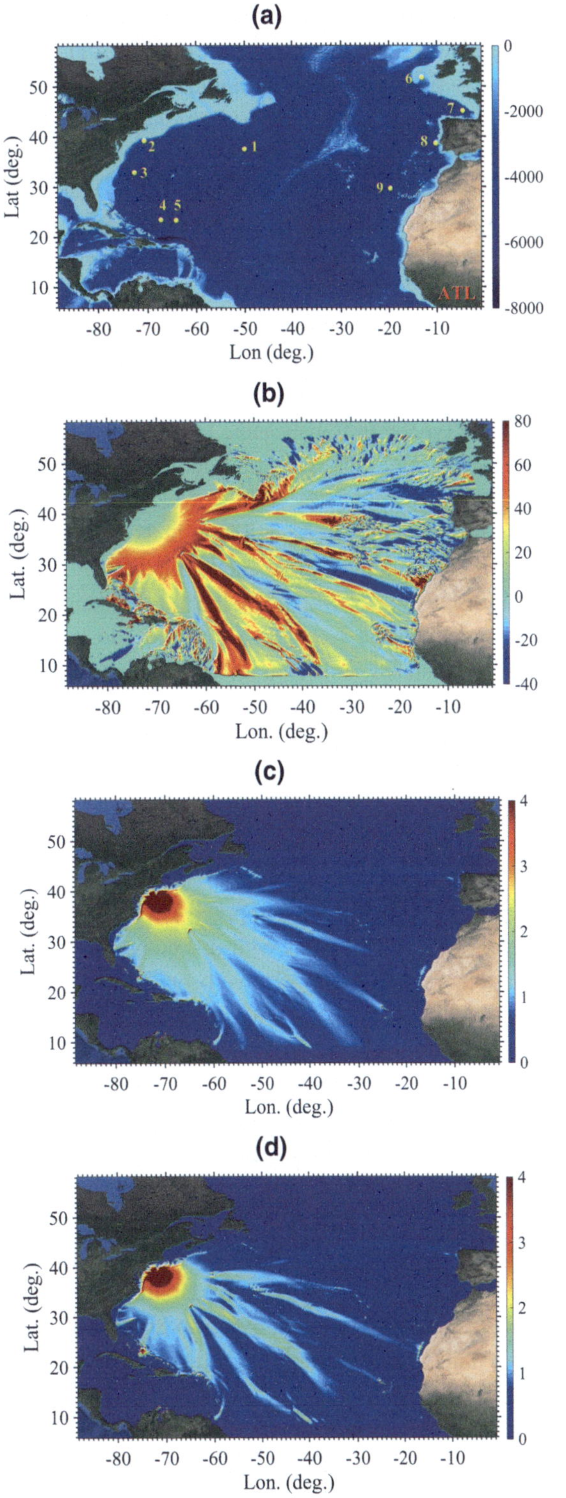

Figure 14
Dispersive effects in FUNWAVE simulations of far-field tsunami propagation and hazard, for deforming Currituck SMF proxy sited in Area 1 (Fig. 1), propagating in 1 arc-min Atlantic grid ATL (Table 2): **a** boundary of ATL grid with locations of wave gage stations 1–9 (Table 3) where time series of surface elevation are compared (Fig. 15); **b** relative difference (color scale in %) of maximum surface elevations in non-dispersive versus dispersive computations (note, areas with dispersive elevations less than 0.05 m have been avoided for clarity); **c**, **d** envelope of maximum surface elevation computed without/with dispersion

depth $d = 0.7$ km, so $\mu = 0.18$, which is also in the long-wave regime. Hence, this is indeed a very large SMF generating a long tsunami, for which dispersive effects are not significant. Consistent with these observations, in the along-slope transects of Fig. 13c, differences between dispersive and non-dispersive results are negligible at all times in the first 30 km on either side of the SMF initial location, but start slightly growing with time/distance, with larger differences occurring on the eastern side, maybe due to the effect of the Hudson River canyon.

According to the dispersion time criterion of Glimsdal et al. (2013), dispersive effects start being significant for $\tau \geq 0.1$, and a propagation distance $L \geq \tau \lambda_0^3/(6h^2)$. In the near field, based on the initial depth and wavelength of the slide in Area 1, this yields $L = 500$ km, but even larger values as waves are propagating over the shallower shelf; for waves propagating offshore, this distance becomes $L > 618$ km. Consistent with these L values, Fig. 13a shows that, at $t = 1200$ s, differences between the non-dispersive and dispersive results only become significant at a large distance from the source. The annular-shaped area with large relative differences ($\pm$ 100%) in instantaneous elevations is a result more of increasing phase shifts between the two wave trains than from large absolute differences in surface elevation.

5.2. Far-Field Propagation in the Atlantic

Figure 14 compares the non-dispersive and dispersive results for the tsunami far-field propagation. In both cases, FUNWAVE is initialized with the same NHWAVE results, computed at $t = 800$ s with dispersion for the deforming slide in Area 1 (Fig. 4b). Results discussed in the previous section indicate that, at this time, differences with a non-dispersive generation are very small (Fig. 13). Tsunami propagation is simulated in the 1 arc-min resolution (about 1800 m) spherical coordinate ATL grid defined over the Northern Atlantic Ocean (Fig. 14a; Table 2). [A video file, with side-by-side animations of simulated surface elevations for the deforming Currituck SMF proxy sited in Area 1 propagating in grid ATL, modeled without and with dispersion, is provided as online supplementary material: EtaSMF-A1-Dispfar.mp4 (color scale in m).]

Figure 14c, d shows envelopes of maximum surface elevations computed without and with dispersive effects. These have fairly identical patterns and, as is typical in transoceanic tsunami propagation (e.g., Kirby et al. 2013; Tehranirad et al. 2015), both envelopes feature narrow stripes of increased surface elevations due to wave-guiding effects by seafloor topographic features. These cause maximum surface elevations to reach 1–2 m in the far field at some locations, such as the north shores of most of the Caribbean islands, the Dominican Republic, Puerto Rico, and the Leeward Islands. One prominent and most distant location with increased impact is a shore-parallel area off of Western Sahara (Morocco; Lat. 23°), south of the Canary Islands, where maximum elevations reach 2 m over 20–30 m-deep shallow water shoals. Figure 14b shows the relative difference in percent computed between these envelopes (note that areas with dispersive maximum elevations less than 0.05 m have been avoided in this figure, for clarity). As tsunami waves propagate away from the source, positive relative changes in elevation first occur, which gradually grow with radial distance, to reach about 50% in all directions. Then, as stripes of increased elevations start appearing in the envelopes, relative changes in elevation become heterogeneous. In far-field areas spanning the S-to-SE directions, where waves propagate towards South America and South Africa, maximum elevations without dispersion increase by up to 80% within the stripes and do not change or slightly decrease in between. In far-field areas spanning the SE-to-NE directions, however, this pattern reverses itself with relative amplitude differences mostly decreasing by 20–40% along the upper West coast of Africa, and the coasts of Portugal and Spain, implying that dispersion increases far-field tsunami elevations in these areas.

Figure 15 ▶
Case of Fig. 14. Time series of surface elevation computed in grid ATL at deep water wave gage stations 1–9 (Table 3; Fig. 14a) in: (red) non-dispersive; and (blue) dispersive simulations

To better assess dispersive effects on tsunami wave trains, Fig. 15 shows time series of surface elevations computed without and with dispersion at deep water wave gage stations 1–9 specified in the ATL grid, the first 5 being collocated with existing DART buoys (deep water tsunami sensors operated by NOAA; Table 3; Fig. 14a). In all cases, differences between these wave trains and, hence, dispersive effects, increase with the propagation distance from the SMF source. Dispersion causes a gradual reduction in leading wave elevation, with a corresponding increase in the second wave elevation, which becomes the largest wave in the train, followed by an increasingly long dispersive wave tail. [The appearance of an increasingly long tail of leading waves in the dispersive simulations is quite apparent in the provided online video.] Note that a shorter tail of smaller waves also appears in the non-dispersive simulations due numerical dispersion in the model. While at the nearest stations (e.g., 3, 4, and 5), the maximum elevations of non-dispersive waves are larger than those of dispersive waves, at the more distant stations (e.g., 1, 6, 7, and 8), these are either about equal or the trend reverses itself, with dispersive waves being larger (e.g., 8). Finally, as noted in earlier work (Kirby et al. 2013; their Fig. 14), at all stations, there is a time lag between the leading non-dispersive wave in the trains, whose phase speed is faster, and the leading dispersive wave, which becomes increasingly large at the most distant gages (e.g., 6, 7.5, and 6 min at stations 6, 8, and 9, respectively). Such time lags further confirm the significance of frequency dispersion effects during far-field tsunami propagation.

5.3. *Near-Field Propagation*

5.3.1 *Directly Onshore of the Source*

Dispersive effects are first assessed on the onshore wave propagation by simulating tsunami propagation without and with dispersion in the 500 m G0 and 120

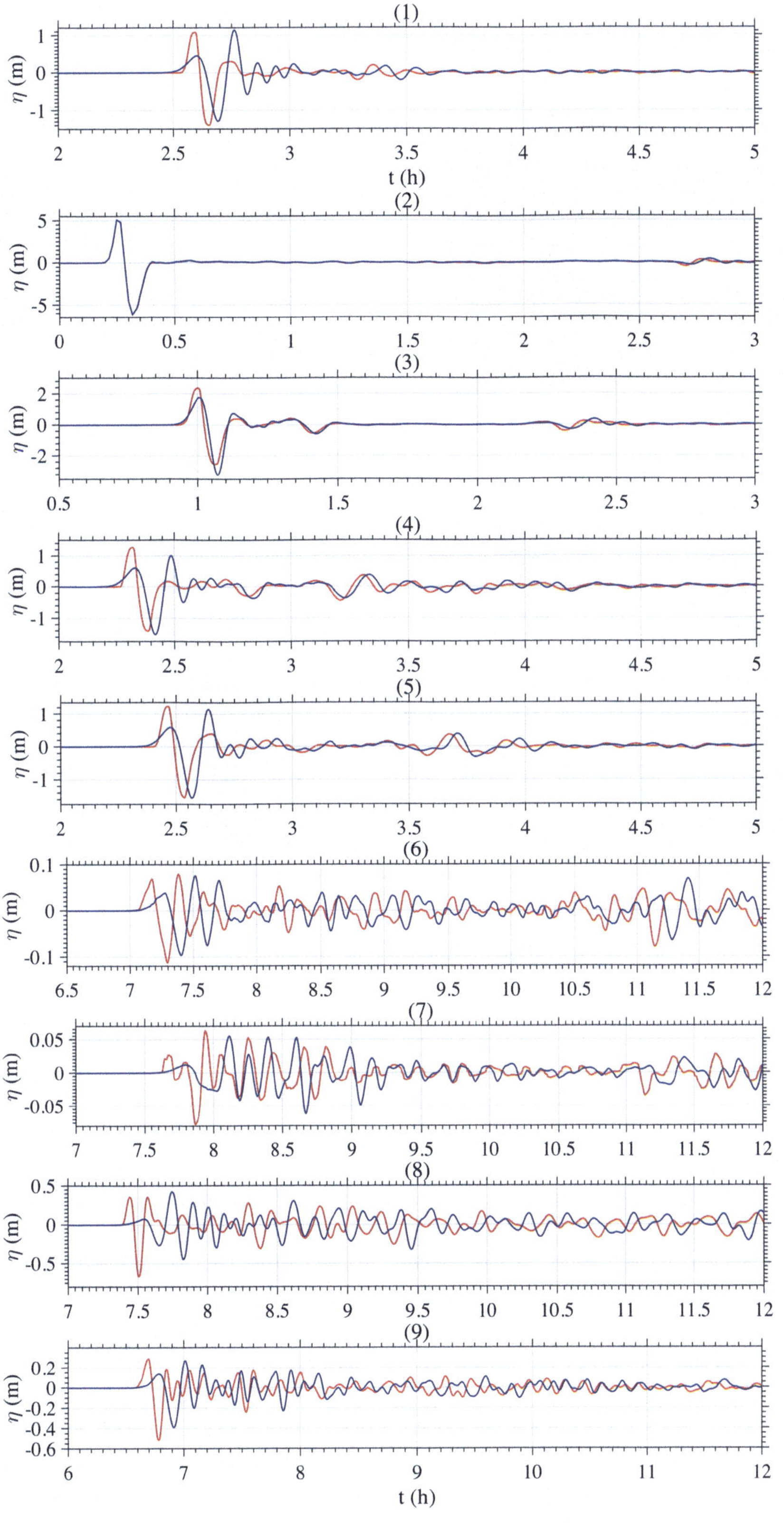
(1)
η (m)
t (h)
(2)
η (m)
(3)
η (m)
(4)
η (m)
(5)
η (m)
(6)
η (m)
(7)
η (m)
(8)
η (m)
(9)
η (m)
t (h)

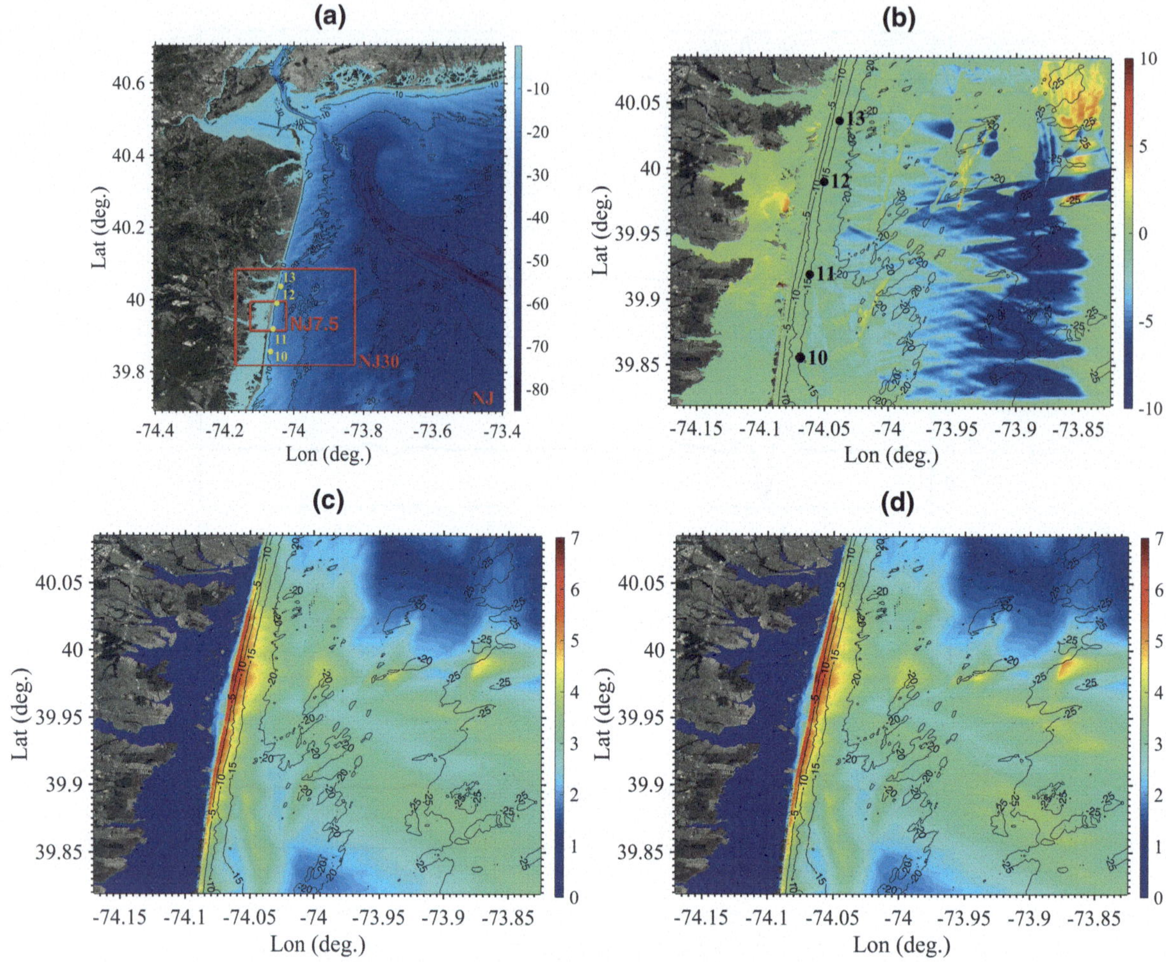

Figure 16

Dispersive effects in FUNWAVE simulations of nearshore tsunami propagation and hazard, for deforming SMF sited in Area 1 (Fig. 1) propagating in 30 and 7.5 m resolution nested grids NJ30 and NJ7.5 (Table 2): **a** boundaries of NJ30 and NJ7.5 grids sited within 120 m NJ grid, with location of wave gage stations 10–13 (Table 3); **b** relative difference (color scale in %) of maximum surface elevations in NJ30 non-dispersive versus dispersive computations; **c**, **d** envelope of maximum surface elevation computed without/with dispersion in NJ30

m resolution NJ grids, which are directly onshore of Area 1 (Table 2; Fig. 2b). Then, to more accurately resolve the shorter waves that dispersive effects might generate and evaluate the effect of grid resolution on tsunami runup, tsunami propagation to shore is simulated without and with dispersion in two additional nested grid levels: a 30 m resolution grid NJ30 and a 7.5 m resolution grid NJ7.5, defined within grid NJ and centered on Seaside Heights (Table 2; Fig. 16a). [A video file, with side-by-side animations of simulated surface elevations for the deforming Currituck SMF proxy sited in Area 1 propagating in grids G0, NJ, NJ30 and NJ7.5, modeled without and with dispersion, is provided as online supplementary material: EtaSMF-A1-Dispnear.mp4 (color scale in m).]

Figures 16c, d and 17b, c show envelopes of maximum surface elevations computed in grids NJ30 and NJ7.5, without and with dispersion, respectively, and Figs. 16b and 17a show their relative difference in percent, respectively. Results in Fig. 16c, d in the 30 m grid NJ30 show fairly identical patterns for both envelopes, with moderate maximum elevation changes when turning off dispersive effects: (1)

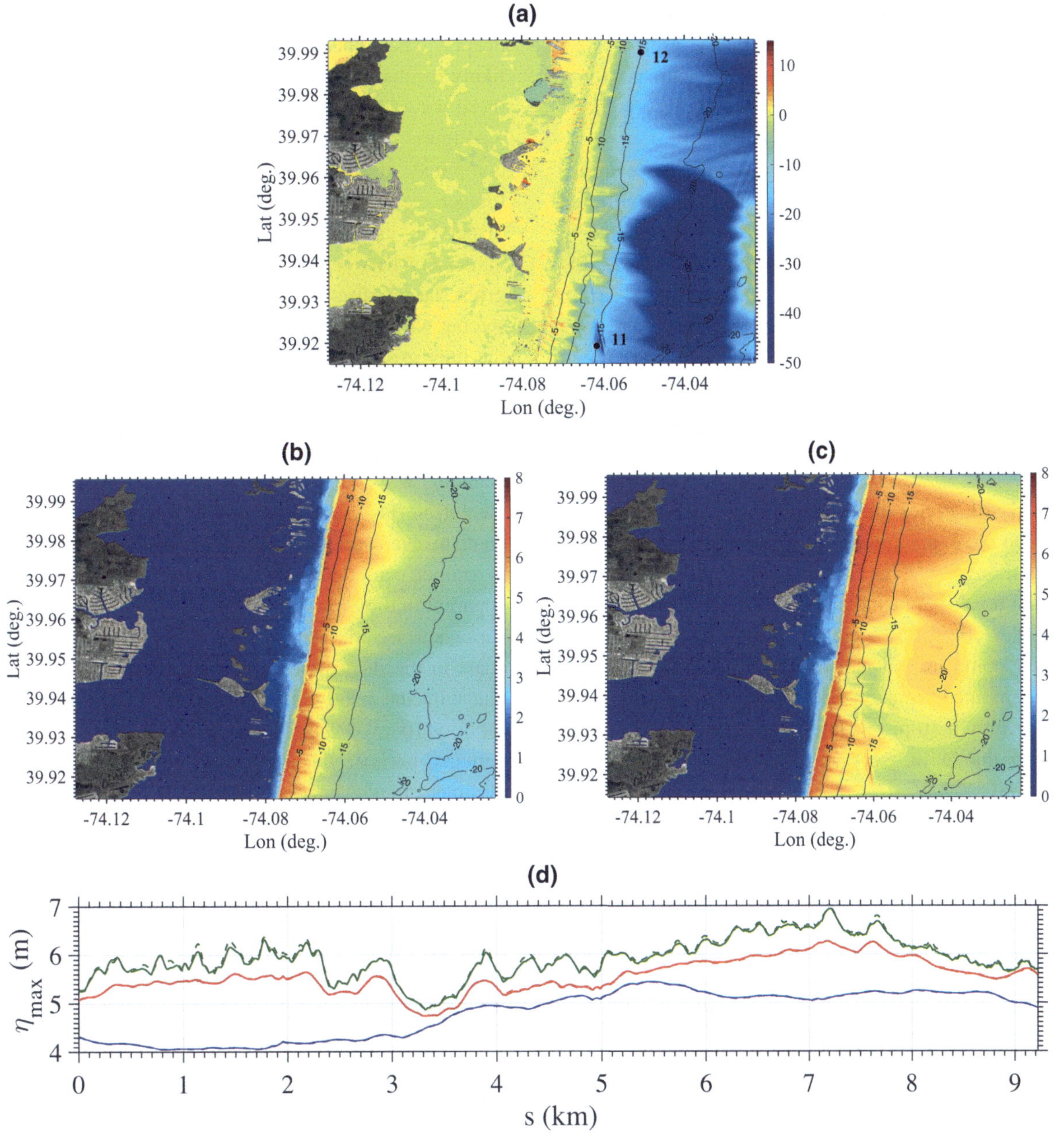

Figure 17
Case of Fig. 16: **a** relative difference (color scale in %) of maximum surface elevations in NJ7.5 non-dispersive versus dispersive computations; **b**, **c** envelope of maximum surface elevation computed without/with dispersion in NJ7.5; **d** maximum surface elevation along the 5 m isobath computed in: (solid) dispersive versus (dashed) non-dispersive simulations in: (blue) 120 m NJ; (red) 30 m NJ30; and (green) 7.5 m NJ7.5, grids (as a function of the distance along the contour, from S to N)

offshore, a decrease by up to 10% in a nearly shore-parallel area, in 20–25 m depth and (2) nearshore, an increase (over and beyond the barrier island) of 2–5%, likely due to shallow water steepening effects (Fig. 16b). In the finer resolution, 7.5 m grid NJ7.5, however, Fig. 17a–c, show larger relative differences and more different patterns for both envelopes of surface elevations when turning off dispersive effects: (1) offshore, a decrease by up to 40% in a nearly shore-parallel area, in depth greater than 10–15 m, and (2) nearshore, a slight increase (over and beyond the barrier island) of 2–5%, likely due to shallow water steepening effects (Fig. 17a).

Figure 17d shows maximum surface elevations computed in grid NJ7.5 at the 5 m isobath with and without dispersion, compared to those computed in the 30 and 120 m resolution grids NJ30 and NJ (the latter results being identical to those of Fig. 12a, b). Consistent with the above observations, there is a small increase in maximum elevation in the 7.5 m grid NJ7.5 when turning off dispersion; however, results are almost identical with or without dispersion in the 30 and 120 m resolution grids, in which dispersive effects are underresolved. The figure also clearly illustrates the important effect of grid resolution on the maximum computed nearshore surface elevations, showing a large increase in elevation when resolution is increased from 120 to 30 m, and a smaller additional increase from 30 to 7.5 m. The latter confirms that tsunami inundation maps should be computed using at least a 30 m resolution grid, but preferably a 10 m resolution [as done by the authors along the USEC as part of NTHMP; see ECMAP (2017)].

These observations are further confirmed and detailed in Fig. 18, which shows time series of surface elevations computed without and with dispersion in the 30 m resolution NJ30 grid, at wave gage stations 10–13, all located in 15 m depth (Fig. 16a; Table 3). For comparison and validation, time series computed at the same locations in the 120 m resolution NJ grid (with dispersion) are also shown. Overall, the latter agree well with the higher resolution results, except that the steeper and higher parts of wave crests are not as sharply captured as in the finer resolution grid. In the NJ30 grid results, in all cases, there are no noticeable differences between dispersive and non-dispersive results at the front of the wave trains (up to $t \simeq 2.6$–2.7 h), where waves are higher and steeper; these are waves that have only propagated a short distance from the onshore part of the SMF tsunami source. Small differences between dispersive and non-dispersive results start appearing at all stations later in the wave train ($t > 2.6$–2.7 h). When dispersion is turned off, waves' crests become slightly higher and steeper, and trough slightly deeper. The appearance of these dispersive effects likely results from the longer propagation distance of these waves, from their generation area.

Figure 19 shows time series of surface elevations computed at stations 11 and 12 in grids NJ30 (dispersive case) and NJ7.5 (dispersive and non-dispersive cases). While there is a good overall agreement between simulations in the 30 and 7.5 m grids, larger discrepancies can be seen around the steeper wave crests and some troughs. Specifically, in the finer grid, well-developed undular bores made of shorter oscillatory waves of $T \simeq 20$ s period appear at both gages near $t = 2.2$ h. These waves are much shorter than the 3–5 min shortest period waves in the tsunami wave train. By contrast, in the 30 m resolution grid, undular bores are absent, which confirms the need for a high-resolution nearshore grid to properly model them. [Undular bores are quite apparent in the provided online video.] Undular bores cause surface elevations to be much larger off the coast in the dispersive simulations, which is consistent with the relative differences shown in Fig. 17a, for $h > 10$ m or so. In shallower water, however, perhaps, because the undular bore waves break and dissipate their energy nearshore, the maximum surface elevations in the dispersive and non-dispersive simulations become nearly identical, consistent with Fig. 17d.

While surface elevations near the coast and at the shoreline do not significantly differ in the non-dispersive and dispersive results of grid NJ7.5 (Figs. 16, 17, 18, 19), Fig. 20 shows that this is not the case for the horizontal velocity module $U = |\mathbf{U}|$ (m/s) and impulse force $F_I = \rho_{\mathrm{w}}(h + \eta)U^2$ (N/m; with $\rho_{\mathrm{w}} = 1025$ kg/m^3, the average sea water density, and η the instantaneous surface elevation). Figure 20a, b shows the maximum envelopes of these parameters computed with dispersion in grid NJ7.5 and Fig. 20c, d shows the relative differences (%) of non-dispersive versus dispersive results. Consistent with Fig. 17a, large decreases in both parameters occur offshore of the 5 m isobath. Onshore of this depth, the pattern becomes more complex but, unlike surface elevations, near the most impacted stretch of coastline around Seaside Heights, turning off dispersion decreases both U and F_I at many locations (see

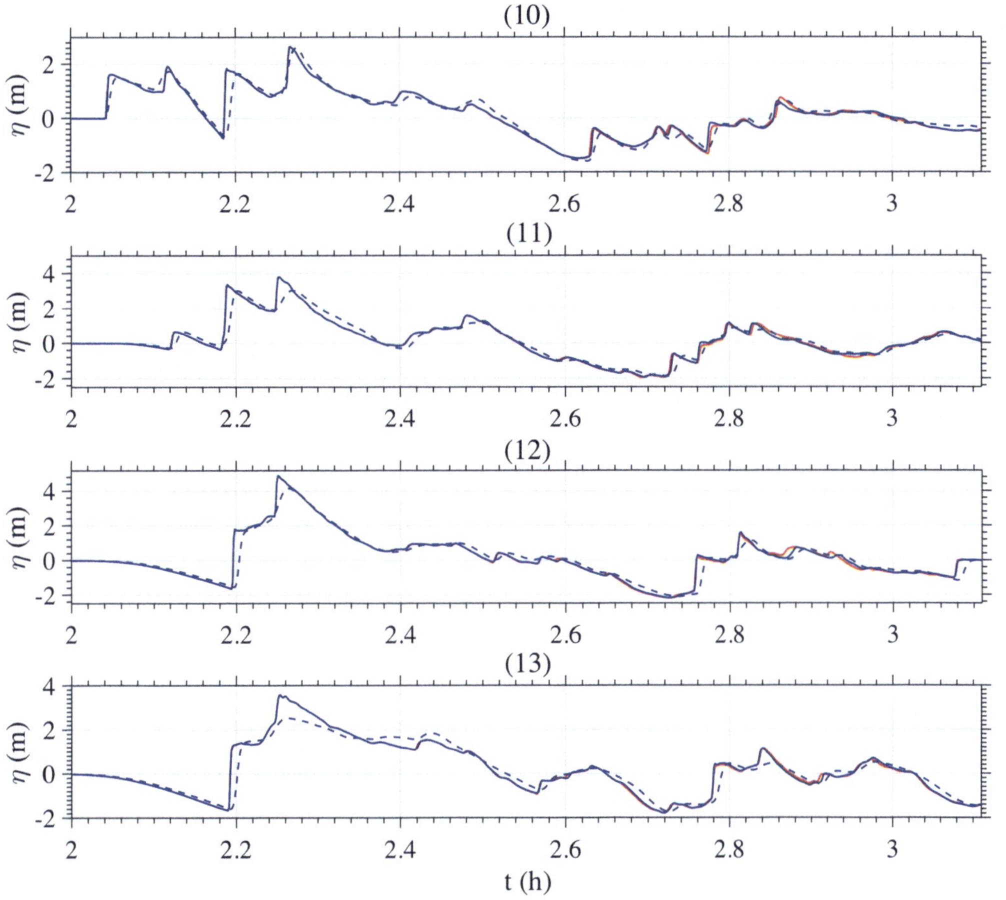

Figure 18
Case of Fig. 16. Time series of surface elevation computed in: (solid) 30 m NJ30; (dashed) 120 m NJ, grid at shallow water wave gage stations 10–13 (Fig. 16a; Table 3), for: (red) non-dispersive; and (blue) dispersive simulations

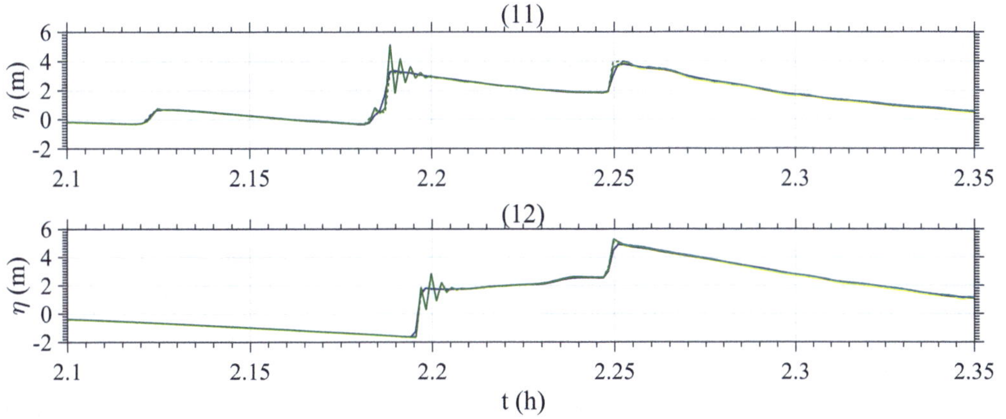

Figure 19
Case of Fig. 16. Time series of surface elevation computed in: (blue) 30 m NJ30; and (green) 7.5 m NJ7.5, grids at shallow water wave gage stations 11–12 (Fig. 16a; Table 3), for: (dashed) non-dispersive; and (solid) dispersive simulations

zoom-ins of Fig. 20e, f). In particular, near the shoreline and in the overland flow slightly onshore of it, maximum velocities and corresponding impulse forces are underpredicted by up to 20 and 40%, respectively. In other areas, turning off dispersion still leads to overpredicting these parameters. It should be pointed out that a significant underprediction of the maximum F_I and U in build-up areas highly impacted by the tsunami-induced overland flow would have important implications for coastal hazard assessment if these results were used to design coastal structures. [A video file, with side-by-side animations of simulated impulse forces for the deforming Currituck SMF proxy sited in Area 1 propagating in grids NJ30 and NJ7.5, modeled without and with dispersion, is provided as online supplementary material: IFSMF-A1-Dispnear.mp4 (color scale in N/m).]

5.3.2 *Towards Coastlines Away from the Source*

As shown in Fig. 8b–g, due to wave-guiding effects by the shelf break and slope, tsunami waves generated by the deforming SMF in Area 1 "open-up" and end up propagating onshore along the entire length of the shelf in grid G0. In particular, some of these waves propagate onshore towards Ocean City, MD, which is located about 325 km west of the SMF source (Fig. 1). Since such propagation takes place over longer distances than for waves propagating directly onshore in grid NJ30, albeit over the shallow shelf, we evaluate whether this may cause increased dispersive effects. To do so, nearshore wave propagation is simulated in a 30 m resolution grid OC30 centered around Ocean City, nested within the 120 m grid OC (Table 2; Figs. 2b, 21a). As before, Fig. 21b–d shows envelopes of maximum surface elevations and their relative difference computed in grid OC30 without and with dispersion. The SMF tsunami impact on Ocean City is almost as severe as in Seaside Heights, with a maximum inundation reaching over 6.5 m and overtopping the barrier beach. Both envelopes, however, are fairly similar with almost no change in surface elevation nearshore and an up to 8% decrease in elevation offshore when turning off dispersion in the model. This is confirmed by time series of surface elevations at 4 gages stations (14–17) located in 10 m depth shown in Fig. 22, for a 1 h time period. At grid OCE30's resolution, similar to results in NJ30, only small changes in instantaneous surface elevation occur when turning off dispersion, and these all correspond to a slight increase and a steepening of the wave crest fronts. It is equally clear here that the first wave generated by this large SMF is quite long, with a 30 min period, and becomes N-wave-shaped in shallow water. For such a long wave propagating from its source in shallow water, one would not have expected significant dispersive effects, as confirmed by the results presented here. However, as for simulations around Seaside Heights and in view of the waves' steep fronts, it is likely that further increasing the grid resolution would cause the appearance of undular bores in numerical results. Such additional simulations are not performed here for the sake of brevity.

6. *Discussion*

In their probabilistic Monte Carlo Simulations (MCS) of the landslide tsunami hazard along the USEC, Grilli et al. (2009) concluded that the 500 year tsunami runup along the upper USEC was largest from Virginia to Cape Cod (about 6–7 m maximum). By performing additional MCS and geotechnical analyses, Krauss (2011) and Eggeling (2012) identified 4 areas (Fig. 1), where large SMFs were both possible due to sufficient sediment thickness and most probable due to low safety factors in slope stability analyses. In light of these conclusions and considering Probable Maximum Tsunamis (PMTs), Grilli et al. (2015) simulated extreme SMFs sited in these 4 areas, which had the lumped characteristics of the historical Currituck slide complex (i.e., "Currituck SMF proxies"; 30 km long by 20 m wide by 0.75 km maximum thickness; with a 158 km^3

Figure 20 ▶
Case of Fig. 16 in 7.5 m grid NJ7.5: **a, b** maximum velocity module (color scale in m/s)/impulse force (color scale in N/m) computed with dispersion; **c, d** relative difference of non-dispersive versus dispersive computations in **a, b** (color scale in %); **e, f** zoom-in of **c, d** in maximum impacted area around Seaside Heights. Black bullets mark locations of wave gage stations 11 (bottom) and 12 (top)

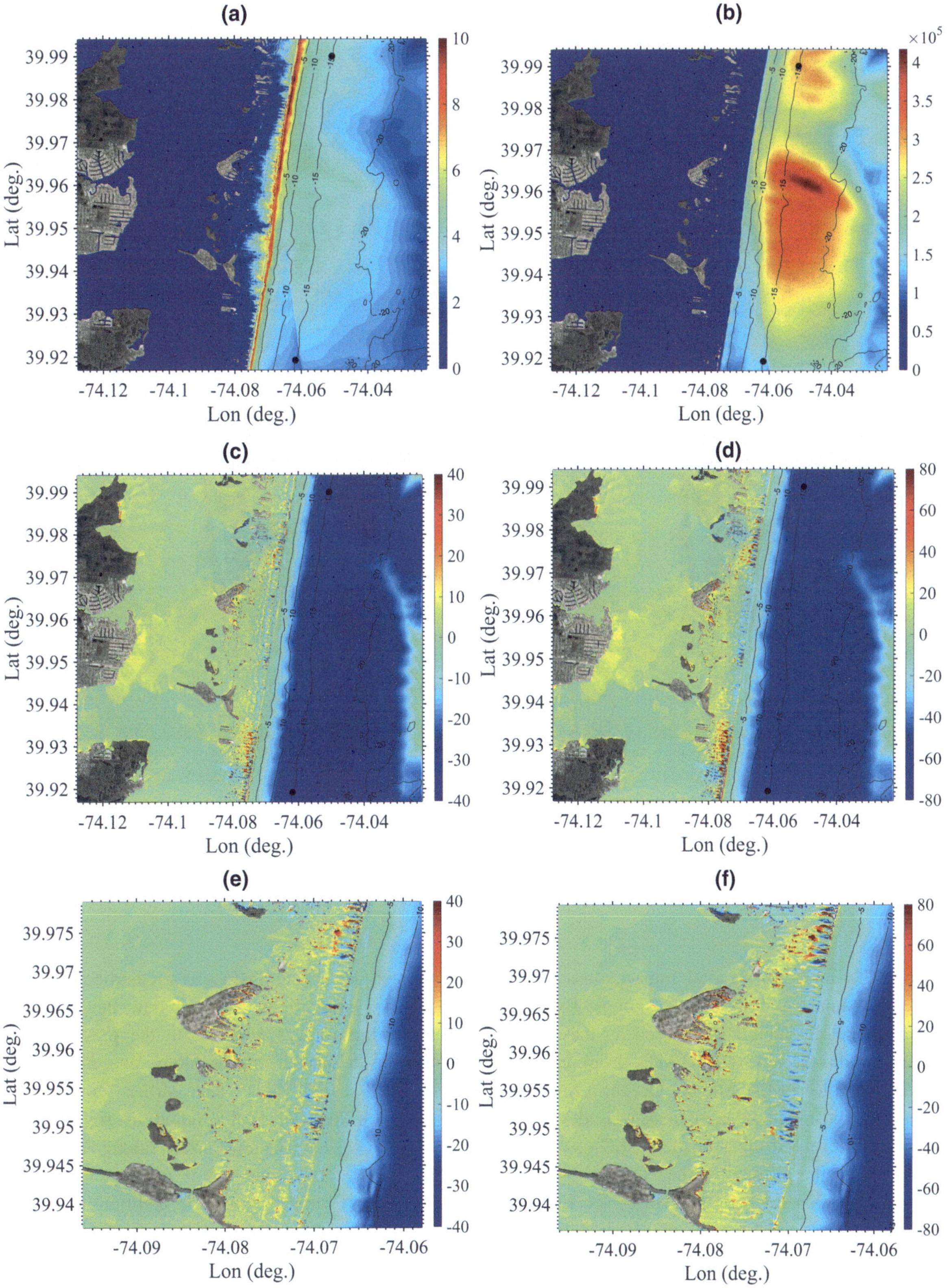
(a)
(b)
(c)
(d)
(e)
(f)
Lat (deg.)
Lon (deg.)
×10^5

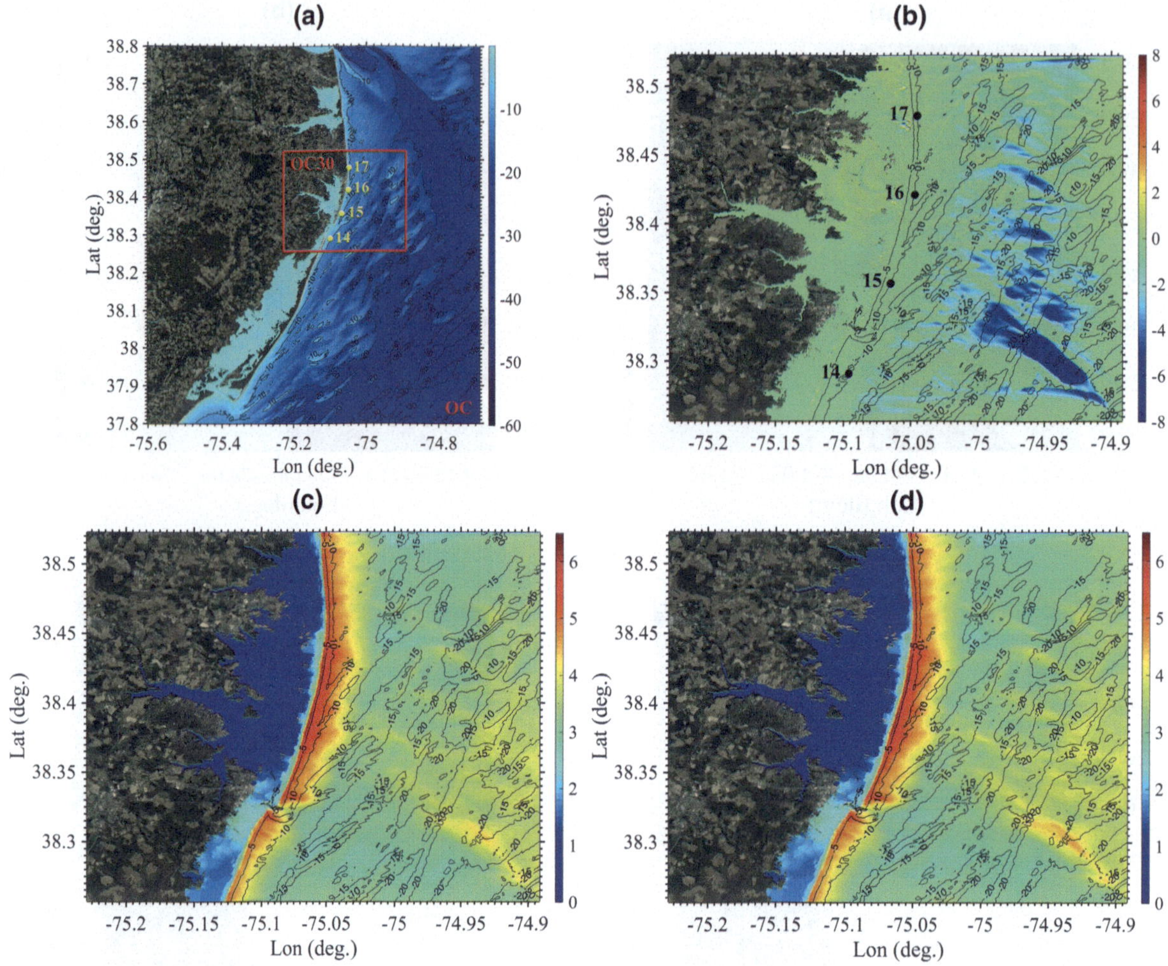

Figure 21
Dispersive effects in FUNWAVE simulations of nearshore tsunami propagation and hazard, for deforming SMF sited in Area 1 (Fig. 1) propagating in nested 30 m grid OC30 (Table 2): **a** boundary of OC30 grid sited within 120 m OC grid, with location of wave gage stations 14–17 (Table 3); **b** relative difference (color scale in %) of maximum surface elevations in non-dispersive versus dispersive computations; **c**, **d** envelope of maximum surface elevation computed without/with dispersion

volume). To maximize coastal impact, these SMFs were all assumed to fail as rigid slumps. By combining the coastal inundation caused by these SMF tsunamis with that of other PMTs from extreme sources identified in the Atlantic Ocean, a first generation of tsunami inundation maps was developed for the most exposed areas of the USEC to tsunami hazard, under the auspice of the National Tsunami Hazard Mitigation Program (NTHMP; see ECMAP 2017). Using a newer model, in which deforming slides are modeled as a dense viscous fluid layer (Kirby et al. 2016), Grilli et al. (2017b) had simulated deforming SMFs of various rheology sited in Area 1 and briefly assessed how the nearshore tsunami hazard was affected. In the present work, following a similar methodology, we more thoroughly assessed how the earlier conservative assumption that all SMFs failed as rigid slumps affected coastal tsunami hazard along the upper USEC, in terms maximum inundation, current velocity, and impulse forces. In addition, effects on these results of including bottom friction and frequency dispersion in the simulations were evaluated.

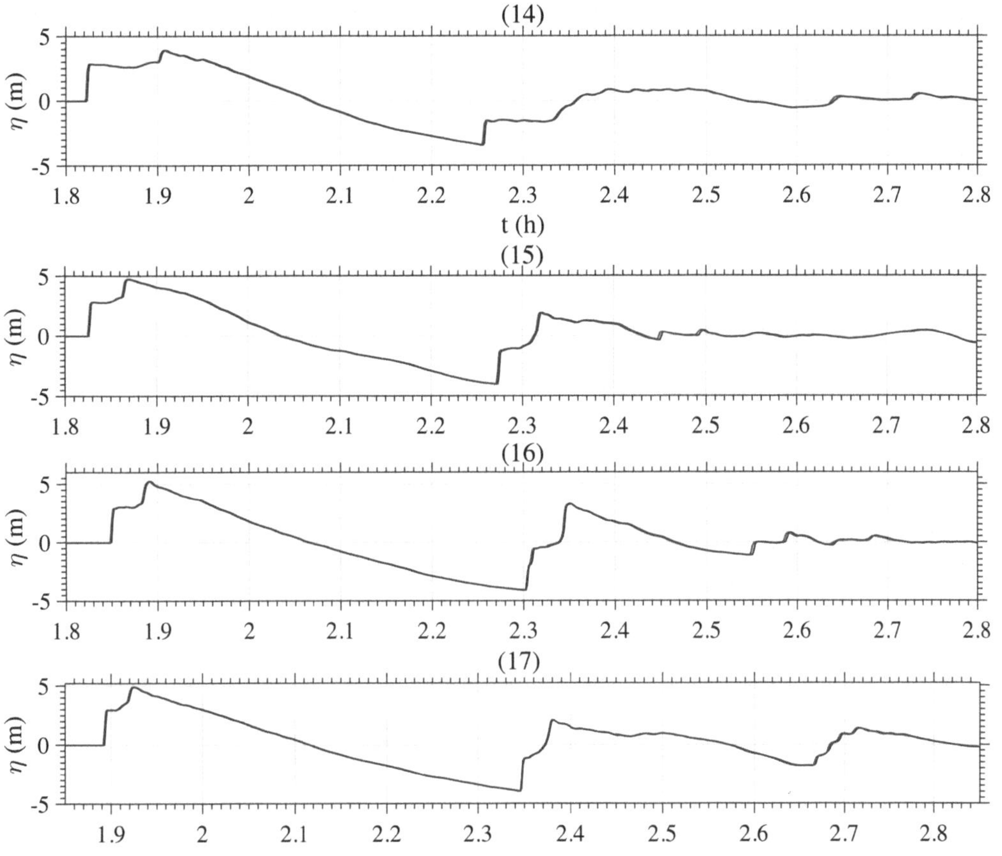

Figure 22
Case of Fig. 21. Time series of surface elevation computed in 30 m OC30 grid at shallow water wave gage stations 14–17 (Fig. 21a, Table 3), for: (red) non-dispersive; and (blue) dispersive simulations

As in earlier work, both slumps and slides were assumed to all have the Currituck SMF proxy characteristics, with a quasi-Gaussian initial geometry and bulk density $\rho_s = 1900$ kg/m^3. Slides were assumed to have a fairly large viscosity, in the upper range of debris flows ($\nu_s = 0.5$ m/s^2), and a substrate-to-slide Manning friction coefficient $n = 0.1$. To more accurately compare nearshore results and evaluate dispersive effects in tsunami wave trains, simulations were performed at much higher resolution than before [in nested grids down to a 7.5 m resolution versus a 120 m resolution in Grilli et al. (2017b)]. Tsunami generation was simulated using the 3D non-hydrostatic model NHWAVE, either with a single layer for slumps (Ma et al. 2012) or with two layers for deforming slides (Kirby et al. 2016). For rigid slumps, both time varying geometry and law of motion were specified as bottom boundary conditions (Grilli and Watts 2005; Watts et al. 2005; Grilli et al. 2015), and for deforming slides, SMF motion and deformation were directly modeled as a depth-integrated bottom layer of dense Newtonian fluid, fully coupled to the overlying fluid motion modeled with the standard σ-layer NHWAVE. Once SMFs were no-longer tsunamigenic (i.e., the slumps had stopped moving or the slides were deep and thin enough), tsunami propagation simulations were continued using the 2D fully nonlinear and dispersive long-wave model FUNWAVE. Simulations of onshore tsunami propagation and inundation along the entire stretch of USEC considered here (from Virginia to Cape Cod) were performed in nested 500 and 120 m resolution grids, with a standard one-way coupling methodology. To assess dispersive effects, selected simulations were similarly performed nearshore in 30 and 7.5 m resolution nested grids (off of Seaside Heights, NJ and Ocean City, MD), and offshore in a 1 arc-min spherical coordinate grid covering the Northern Atlantic Ocean.

Results of tsunami generation simulations showed that deforming slides, while having a slightly larger initial acceleration, generated smaller onshore propagating tsunamis than rigid slumps. Because deforming slide parameters were selected to ensure that the center of mass motion of both types of SMFs was quite similar, at least during the first half of their motion (see Fig. 5), these observed differences in onshore tsunami waves should essentially be associated with slide deformation. This was also the conclusion of Grilli and Watts (2005), who modeled deforming slides by superimposing a rate of deformation on top of the center of mass law of motion of rigid (block) slides. As shown in Figs. 4 and 6, when deforming slides move down the slope, their material spreads and thins out, which gradually makes them less tsunamigenic; by contrast, rigid slumps keep their specified shape during their pendulum-like motion. In addition, since they flow down the slope following the terrain, deforming slides cause more asymmetric tsunami waves alongshore, with respect to their initial direction of motion, than rigid slumps (Fig. 6). The offshore propagating tsunami waves caused by slides were usually as large as or even larger in some cases than for slumps (Fig. 4).

Effects of SMF deformation on the coastal tsunami impact were evaluated by comparing the combined maximum envelope of surface elevations caused nearshore and along the shore, and over the 5 m isobath, by tsunamis generated by the 4 rigid slumps or deforming slides, in seven 120 m resolution grids covering the coast (Fig. 12). Although a 120 m resolution is not sufficient to accurately compute coastal inundation, it is adequate to perform this comparison. Consistent with earlier work (Tehranirad et al. 2015, 2017; Grilli et al. 2017b), the bathymetry of the wide shelf was found to have strong, first-order, control on the magnitude of tsunami coastal inundation, as it induces wave focusing and defocusing effects. In addition, tsunami propagation and refraction over the shelf and shelf break, both north and south of each source area, causes non-trivial variations in surface elevation and coastal inundation. This implies that a given SMF can generate tsunami waves causing a significant coastal inundation far alongshore from their source area (see, e.g., in Fig. 21, the large inundation caused at Ocean City, MD, by the SMF sited in Area 1 in the Hudson River Canyon).

Overall, as expected from earlier work (Grilli and Watts 2005; Grilli et al. 2017b), tsunamis caused by rigid slumps were found to be worst case scenarios, causing the largest maximum inundation at all sites (with a maximum of about 11.5 m around Montauk, NY), and up to 50% more inundation directly onshore than for the slides considered here, which only had a moderate level of deformation (Fig. 12). However, the slightly larger offshore propagating tsunamis generated by deforming slides could affect far-distant nearshore areas, through shelf refraction, thus increasing the relative coastal impact of such tsunamis and reducing the difference in maximum inundation with that of tsunamis generated by rigid slumps. For rigid slumps, the maximum computed inundation of 8–11.5 m at the 5 m isobath is larger than the 500 year runup of 6–7 m estimated by Grilli et al. (2009), also based on rigid slumps (or slides) assumptions. This could be expected, since the Currituck SMF proxies considered here were aimed at causing PMTs for the upper USEC, with potentially thousands of years return periods (the estimated age of the Currituck slide complex is about 16Ka). By contrast, the maximum inundation of 8.5 m computed here for deforming slides (around Atlantic City) is more consistent with the 500 year runup estimated in earlier work based on MCS. This could indicate that the return period of or the maximum runup caused by the largest events were underestimated in the MCS of Grilli et al. (2009). Regarding minimum elevations at the coast, which affect power plant intakes, tsunamis from both types of SMFs were shown to be able to cause water withdrawal to the 5 m isobath or slightly deeper.

The effects of bottom friction on tsunami coastal inundation were assessed by performing simulations for the Area 1 deforming slide, in the 120 m NJ grid, using two different Manning coefficients, one 50% larger than the other (from $n = 0.025$–0.375). Using the increased friction in simulations, i.e., considering a gravel/boulder rather than a sandy substrate, led to a reduction of the largest maximum tsunami inundation at the coast, in some cases, by up to 15%. Smaller inundation levels were less noticeably affected (Fig. 12a).

Dispersive effects were assessed during tsunami generation, for the deforming Currituck SMF proxy sited in Area 1, by comparing NHWAVE simulations with and without non-hydrostatic pressure terms in the model. Similarly, dispersive effects were assessed on both near- and far-field tsunami propagations by comparing FUNWAVE simulations with and without dispersion. For the very large SMFs considered here, which generate tsunamis having a very long initial wavelength compared to the local depth ($\mu < \pi/10$), non-hydrostatic (dispersive) effects in NHWAVE were found to be negligible for tsunami generation (Fig. 13). During tsunami propagation (Figs. 14, 15, 16, 17, 18, 19, 20, 21, 22), overall, turning off dispersion in FUNWAVE only led to moderate changes in maximum inundation at the coast (2–5% changes), but to more significant effects in the far field (− 40 to 80% changes). In addition, in far-field dispersive wave trains, the largest wave, which is initially the leading crest, gradually becomes the second wave in the train and is associated with a significant dispersive wave tail. In non-dispersive wave trains the largest wave is always the leading crest and the number of significant waves in the tail is more limited. Finally, there is an increasing time lag with propagation distance of the dispersive waves, as compared to the non-dispersive case.

The nearshore simulations in finer grids clearly confirmed the need to perform tsunami inundation mapping using at least a 30 m resolution, or preferably a 10 m resolution grid as recommended by NTHMP, to have sufficiently converged maximum inundation results (see Fig. 17; note a 10 m resolution grid was not used here, but the 7.5 m grid is deemed close enough to it to draw this conclusion). However, equally important, the 7.5 m resolution results with dispersion showed the appearance nearshore of undular bores made of shorter ($\sim$ 20 s period) waves, which significantly affect the dynamics of incoming tsunami waves; such shorter waves are absent in the coarser resolution grids, hence further justifying the need for using high-enough nearshore resolution grids (here 7.5 m). While these bores likely break near or at the coast and thus have little effect on maximum inundation, they significantly increase maximum flow velocity and impulse forces, particularly in the overland flow, which both increase the tsunami hazard on civil infrastructures. Without dispersive effects, the undular bores do not occur in model results at any resolution, and maximum flow velocities and corresponding impulse forces can be underestimated at the coast, by up to 20 and 40%, respectively (note, this conclusion is only based on limited results around the highly impacted area of Seaside Heights, NJ; see Fig. 20).

Because onshore surface elevations with or without dispersion are almost identical, the large underprediction of impulse forces mostly results from the underprediction of velocities. When undular bores are present, flow velocities are larger near the long-wave crests, including when the steep individual bores end up breaking in very shallow water [breaking undular bores riding longer tsunami waves can be seen, for instance, in pictures of the Indian Ocean tsunami approaching the coast reported by Madsen et al. (2008); see also the various discussion regarding undular bores in this paper]. Hence, incident undular bores intensify the dynamics and kinematics of the overall flow. This is an important finding for tsunami hazard assessment, particularly in view of the recently released structural design code ASCE 7–16, which recommends using impulse forces and maximum speed of transported debris for designing tsunami resistant structures (Chock 2016).

These results imply that while using a NSW model may be adequate to predict maximum coastal inundations, this may not capture the required physics to model flow velocities and corresponding impulse forces. This conclusion is consistent with those of a recent tsunami velocity benchmarking workshop (Lynett et al. 2017).

7. *Conclusions*

Unless considering critical coastal facilities (such as a nuclear power plant), which require considering the most extreme PMTs, it appears adequate in coastal tsunami hazard assessment to simulate more realistic moderately deforming SMFs, whose center of mass may follow a similar kinematics as that of rigid slumps, but whose deformation leads to a reduced coastal inundation. Using a moderate level of SMF deformation is also consistent with paleo-SMF observations made on the upper East coast continental shelf and margin (Booth et al. 2003; ten Brink et al. 2014). Considering its marked effect on maximum inundation, values of the bottom friction coefficient should be carefully selected as a function of the local substrate, particularly nearshore, and onshore land cover. In the absence of detailed information, assuming that a coarse sand substrate should be conservative. The need to use a dispersive model for landslide tsunami generation depends on the relative size of the SMF and waves it is generating, as compared to the local depth. For the very large SMFs considered here, a non-dispersive model would have been adequate. For tsunami propagation, using a dispersive model such as a long-wave Boussinesq model is important to accurately simulate both far-field and near-field tsunami propagation, together with sufficiently high-resolution model grids. In the near field, even if the initial SMF tsunami is made of non-dispersive long waves (as in the present case), this is necessary to accurately model the undular bores made of shorter dispersive waves that may occur in shallow water near the steep fronts of incoming long waves. Such bores cause rapid changes in surface elevation and intensify wave dynamics in a way that may significantly increase maximum forces on structures. However, using a NSW model is adequate if maximum onshore inundation is the only parameter of interest in the simulations. While in future tsunami inundation mapping work, it would be desirable to perform a PTHA, as is currently being done in other regions of the US (e.g., West Coast), the current lack of sufficient information on return periods of tsunamigenic SMF sources precludes considering this approach for the USEC. However, should a sufficient number of past tsunamigenic SMF sources be dated, this could provide relevant hazard information, making it possible to perform a PTHA.

Acknowledgements

This work was supported by the National Tsunami Hazard Mitigation Program (NTHMP), NOAA, through Grants NA-15-NWS4670029 and NA-16-NWS4670034 to the University of Delaware (with subaward to the University of Rhode Island). Additional support at the University of Rhode Island and the University of Delaware came from Grants CMMI-15-35568 and CMMI-15-37568 from the Engineering for Natural Hazards Program, National Science Foundation, respectively. Numerical simulations reported in this work used HPC resources, as part of the Extreme Science and Engineering Discovery Environment (XSEDE) (project BCS-170006), which is supported by the National Science Foundation (NSF) Grant number ACI-1548562. FUNWAVE-TVD is open source software, available at http://github.com/fengyanshi/FUNWAVE-TVD/. NHWAVE is open source software, available at http://github.com/jimkirby/nhwave/. Finally, the authors acknowledge anonymous reviewers for their thorough and constructive reviews of this work.

Appendix: SMF Geometry and Slump Law of Motion

For rigid slumps, kinematics is specified based on the analytical laws developed by Grilli and Watts (1999, 2005); Watts et al. (2005). Additionally, as in Enet and Grilli (2007), the SMF geometry is idealized as having a "Quasi-Gaussian" shape (below seafloor) of elevation $\zeta(x,y)$, whose steepness is controlled by a shape parameter ε (here $\varepsilon = 0.717$), and elliptical footprint of downslope length b, width w, and maximum thickness T defined as (Fig. 23)

$$\zeta(x,y) = \frac{T}{1-\varepsilon}\max\{0, \mathrm{sech}(\mathrm{k_b}\xi)\,\mathrm{sech}(\mathrm{k_w}\chi) - \varepsilon\}, \tag{1}$$

where (ξ, χ) are the local downslope and spanwise horizontal coordinates, rotated in the direction of SMF motion θ, and $k_b = 2C/b$, $k_w = 2C/w$, with $C = acosh(1/\varepsilon)$. With this geometry and parameters, the SMF volume is given by

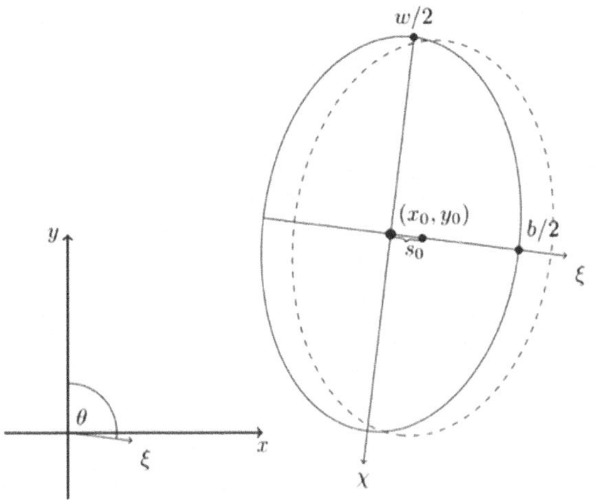

Figure 23
Geometric parameterization of a SMF initially centered at (x_0, y_0) moving in direction ξ, with an azimuth angle θ from North and center of mass motion $S(t)$ measured parallel to the mean local slope of angle α; (x, y) denote the longitudinal and latitudinal horizontal directions, respectively

$$V_s = bwT\frac{I_2}{C^2}\left(\frac{\frac{I_1}{I_2} - \varepsilon}{1 - \varepsilon}\right) \quad \text{with} \qquad I_{1;2} = \int_0^C f(\mu)\,\mathrm{d}\mu; g(\mu)\,\mathrm{d}\mu \tag{2}$$

and

$$f(\mu) = \mathrm{sech}\mu\,\mathrm{a}\tan(\sinh\mathrm{g}(\mu)), \mathrm{g}(\mu) = \mathrm{acosh}\left(\frac{\mathrm{sech}\mu}{\varepsilon}\right). \tag{3}$$

[Note that Eqs. 2 and 3 have been corrected and are different from those reported in earlier papers (e.g., Enet and Grilli 2007; Grilli et al. 2015), which resulted from a mistake in the volume calculation.] For the specified ε, we find, $C = 0.8616$, $I_1 = 0.4804$, $I_2 = 0.5672$, and $V_s = 0.3508\,bwT$.

Earlier modeling work (Locat et al. 2009) indicates that, during its tsunamigenic period of motion, the Currituck SMF achieved a relatively small maximum displacement (runout) $S_f < b$ in its main direction of motion down the slope, over an unknown time of motion t_f. The combination of rigid block SMF and small displacement parallel to the slope supports modeling the SMF kinematics as a rigid slump or a deforming slide with moderate deformation achieving a similar runout over the same time. In either cases, one can assume a constant basal friction (i.e., slide-to-substrate friction) and negligible hydrodynamic drag (Grilli and Watts 2005). This type of rigid-body motion kinematics was investigated in earlier work (see above-listed references), leading for the slump to a pendulum-like center of mass motion $S(t)$ parallel to the local mean slope of angle α. Here, this simple law of motion for rigid slumps is used, which reads

$$S(t) = S_0\left(1 - \cos\frac{t}{t_0}\right) \quad \text{for} \quad 0 \le t \le t_f \tag{4}$$

with $S_0 = S_f/2$ and $t_0 = t_f/\pi$, and $S = S_f$ for $t > t_f$ (assuming SMF triggering occurs at $t = 0$).

At $t = 0$, the SMF elevation is specified below the current seafloor bathymetry $h_0(x, y)$. Given the SMF initial center of mass location (x_0, y_0) in global axes (x, y) (i.e., coordinates of the center of the elliptical footprint) and azimuth angle of SMF motion θ, the coordinate transformation to the local SMF slope-parallel coordinate system (ξ, χ) (Fig. 23) is defined as

$$\begin{aligned} \xi &= (x - x_o)\cos\theta - (y - y_0)\sin\theta - S(t)\cos\alpha \\ \chi &= (x - x_o)\sin\theta + (y - y_0)\cos\theta \end{aligned} \tag{5}$$

with $S(t)$ given by Eq. (4).

Then, assuming $\sin\alpha \simeq 0$ for small slopes, the instantaneous seafloor depth above the SMF is given by

$$\begin{aligned} h(x, y, t) = h_0(x, y) &+ \zeta\{\xi(x, y, t), \chi(x, y, t)\} \\ &- \zeta\{\xi(x, y, 0), \chi(x, y, 0)\} \end{aligned} \tag{6}$$

with $\Delta h = h - h_0$. The seafloor motion described by Eq. (6) is similar to a translation parallel to the average slope of part of the seabed, over the actual bathymetry. The vertical seafloor velocity (used in NHWAVE as a bottom boundary condition) is computed as

$$\frac{\mathrm{dh}}{\mathrm{dt}}(x, y, t) = \frac{\mathrm{d}\zeta}{\mathrm{dt}}\{\xi(x, y, S(t)), \chi(x, y, t)\} \tag{7}$$

which can be easily derived from Eqs. (1) to (6) as

$$\frac{\mathrm{dh}}{\mathrm{dt}}(x, y, t) = k_b\cos\alpha\left(\zeta + \frac{\varepsilon T}{1 - \varepsilon}\right)U\tan h(k_b\xi) \quad \text{with} \qquad U(t) = \frac{\mathrm{dS}}{\mathrm{dt}} = U_{max}\sin\frac{t}{t_0} \tag{8}$$

the slump velocity obtained from Eq. (4), with $U_{\max} = S_0/t_0$ the maximum velocity. Similarly, the slump acceleration is found as

$$A(t) = \frac{d^2S}{dt^2} = A_0 \cos\frac{t}{t_0} \quad \text{with} \quad A_0 = \frac{S_0}{t_0^2} \tag{9}$$

the initial acceleration.

For rigid slumps, hydrodynamic drag can be neglected due to low velocity and small amplitude of motion, and inertia includes both the SMF mass $M_s = \rho_s V_s$, with ρ_s denoting the sediment bulk density, and the specific density being defined as $\gamma = \rho_s/\rho_w$, with ρ_w the water density, and an added mass $\Delta M_s = C_M \rho_w V_s$, defined by way of an added mass coefficient C_M. Assuming a constant basal friction, a nearly circular rupture surface of radius R, and a small angular displacement $\Delta\Phi$, Grilli and Watts (2005) derived the characteristic distance and time of motion for rigid slumps as

$$S_0 = \frac{R\Delta\Phi}{2} \quad \text{and} \quad t_0 = \sqrt{\frac{R}{g}\frac{\gamma + C_M}{\gamma - 1}} \quad \text{with} \quad R \simeq \frac{b^2}{8T} \tag{10}$$

with g denoting the gravitational acceleration. Equation (8), proposed by Watts et al. (2005), is a semi-empirical relationship to estimate the radius of slump motion as a function of slump downslope length and maximum thickness.

References

Abadie, S., Harris, J. C., Grilli, S. T., & Fabre, R. (2012). Numerical modeling of tsunami waves generated by the flank collapse of the Cumbre Vieja Volcano (La Palma, Canary Islands): Tsunami source and near field effects. *Journal of Geophysical Research, 117*, C05030. https://doi.org/10.1029/2011JC007646.

Amante, C, & Eakins, B. W. (2009). ETOPO1 1 arc-minute global relief model: Procedures, data sources and analysis. NOAA Technical Memorandum NESDIS NGDC-24. National Geophysical Data Center, NOAA. https://doi.org/10.7289/V5C8276M(**March 15, 2017**).

Booth, J. S., O'Leary, D. W., Popenoe, P., & Danforth, W. W. (1993). US Atlantic Continental Slope Landslides: Their distribution, general attributes, and implications. In: Schwab, W. C., Lee, H. J., & Twichell, D. C. (Eds.) *Submarine landslides: Selected studies in the US exclusive economic zone*. US Geological Survey Bulletin, 2002, pp. 14–22.

ten Brink, U. S., Barkan, R., Andrews, B. D., & Chaytor, J. D. (2009b). Size distributions and failure initiation of submarine and subaerial landslides. *Earth and Planetary Science Letters, 287*, 31–42.

ten Brink, U. S., Chaytor, J. D., Geist, E. L., Brothers, D. S., & Andrews, B. D. (2014). Assessment of tsunami hazard to the US Atlantic margin. *Marine Geology, 353*, 31–54.

ten Brink, U. S., Lee, H. J., Geist, E. L., & Twichell, D. (2009a). Assessment of tsunami hazard to the US East Coast using relationships between submarine landslides and earthquakes. *Marine Geology, 264*, 65–73.

ten Brink, U., Twichell, D., Geist, E., Chaytor, J., Locat, J., Lee, H., Buczkowski, B., Barkan, R., Solow, A., Andrews, B., Parsons, T., Lynett, P., Lin, J., & Sansoucy, M. (2008). *Evaluation of tsunami sources with the potential to impact the US Atlantic and Gulf coasts*. USGS Administrative report to the US Nuclear Regulatory Commission, p. 300.

Chaytor, J. D., ten Brink, U. S., Solow, A. R., & Andrews, B. D. (2009). Size distribution of submarine landslides along the US Atlantic margin. *Mar Geol, 264*(12), 16–27.

Chock, G. Y. (2016). Design for tsunami loads and effects in the ASCE 7–16 standard. *Journal of Structural Engineering, 142*(11), 04016093. https://doi.org/10.1061/(ASCE)ST.1943-541X.0001565.

ECMAP. (2017). NTHMP tsunami inundation maps for the US East Coast. https://www.udel.edu/kirby/nthmp.html.

Eggeling, T. (2012). Analysis of earthquake triggered submarine landslides at four locations along the US east coast. Masters Thesis. University of Rhode Island.

Enet, F., & Grilli, S. T. (2007). Experimental study of tsunami generation by three-dimensional rigid underwater landslides. *Journal of Waterway Port Coastal and Ocean Engineering, 133*(6), 442–454. https://doi.org/10.1061/(ASCE)0733-950X(2007).

Fine, I. V., Rabinovich, A. B., Bornhold, B. D., Thomson, R., & Kulikov, E. A. (2005). The Grand Banks landslide-generated tsunami of November 18, 1929: Preliminary analysis and numerical modeling. *Marine Geology, 215*, 45–57.

Geist, E. L., & Lynett, P. J. (2014). Source processes for the probabilistic assessment of tsunami hazards. *Oceanography, 27*(2), 86–93.

Geist, E., Lynett, P., & Chaytor, J. (2009). Hydrodynamic modeling of tsunamis from the Currituck landslide. *Marine Geology, 264*, 41–52. https://doi.org/10.1016/j.margeo.2008.09.005.

Glimsdal, S., Pedersen, G. K., Harbitz, C. B., & Løvholt, F. (2013). Dispersion of tsunamis: Does it really matter? *Natural Hazards and Earth System Sciences, 13*, 1507–1526. https://doi.org/10.5194/nhess-13-1507-2013.

González, F. I., Geist, E. L., Jaffe, B., Kânoglu, U., Mofjeld, H., Synolakis, C. E., et al. (2009). Probabilistic tsunami hazard assessment at seaside, Oregon, for near- and far-field seismic sources. *Journal of Geophysical Research Oceans, 114*, C11.

Grezio, A., Babeyko, A., Baptista, M. A., Behrens, J., Costa, A., Davies, G., et al. (2017). Probabilistic tsunami hazard analysis: Multiple sources and global applications. *Reviews of Geophysics, 55*(4), 1158–1198. https://doi.org/10.1002/2017RG000579.

Grilli, S. T., & Watts, P. (1999). Modeling of waves generated by a moving submerged body. Applications to underwater landslides. *Engineering Analysis with Boundary Elements, 23*, 645–656.

Grilli, S. T., & Watts, P. (2005). Tsunami generation by submarine mass failure. I: Modeling, experimental validation, and sensitivity analyses. *Journal of Waterway Port Coastal and Ocean Engineering, 131*(6), 283–297. https://doi.org/10.1061/(ASCE)0733-950X(2005).

Grilli, S. T., Dubosq, S., Pophet, N., Pérignon, Y., Kirby, J. T., & Shi, F. (2010). Numerical simulation and first-order hazard analysis of large co-seismic tsunamis generated in the Puerto Rico trench: Near-field impact on the North shore of Puerto Rico and far-field impact on the US East Coast. *Natural Hazards and Earth System Sciences, 10*, 2109–2125. https://doi.org/10.5194/nhess-2109-2010.

Grilli, S. T., Grilli, A. R., Tehranirad, B., & Kirby, J. T. (2017a). Modeling tsunami sources and their propagation in the Atlantic Ocean for coastal tsunami hazard assessment and inundation mapping along the US East Coast. In *Proceedings of coastal structures and solutions to coastal disasters 2015: Tsunamis (pp. 1–12)*. Boston: American Soc. Civil Eng, 2015.

Grilli, S. T., Harris, J. C., Tajalibakhsh, T., Masterlark, T. L., Kyriakopoulos, C., Kirby, J. T., et al. (2013b). Numerical simulation of the 2011 Tohoku tsunami based on a new transient FEM co-seismic source: Comparison to far- and near-field observations. *Pure and Applied Geophysics, 170*, 1333–1359. https://doi.org/10.1007/s00024-012-0528-y.

Grilli, S. T., Harris, J. C., Tappin, D. R., Masterlark, T., Kirby, J. T., Shi, F., & Ma, G. (2013a). Numerical modeling of coastal tsunami dissipation and impact. In P. Lynett, & J. Mc Kee Smith (Eds.), *Proceedings of 33rd international coastal engineering conference (ICCE12, Santander, Spain, July, 2012)* (p. 12). World Scientific Publishing Co. Pte. Ltd.

Grilli, S. T., O'Reilly, C., Harris, J. C., Tajalli Bakhsh, T., Tehranirad, B., Banihashemi, S., et al. (2015). Modeling of SMF tsunami hazard along the upper US East Coast: Detailed impact around Ocean City, MD. *Natural Hazards, 76*(2), 705–746. https://doi.org/10.1007/s11069-014-1522-8.

Grilli, S. T., Shelby, M., Kimmoun, O., Dupont, G., Nicolsky, D., Ma, G., et al. (2017b). Modeling coastal tsunami hazard from submarine mass failures: Effect of slide rheology, experimental validation, and case studies off the US East Coast. *Natural Hazards, 86*(1), 353–391. https://doi.org/10.1007/s11069-016-2692-3.

Grilli, S. T., Taylor, O.-D. S., Baxter, C. D. P., & Maretzki, S. (2009). Probabilistic approach for determining submarine landslide tsunami hazard along the upper East Coast of the United States. *Marine Geology, 264*(1–2), 74–97.

Grilli, S. T., Vogelmann, S., & Watts, P. (2002). Development of a 3D Numerical Wave Tank for modeling tsunami generation by underwater landslides. *Engineering Analysis with Boundary Elements, 26*(4), 301–313.

Grothe, P. R., Taylor, L. A., Eakins, B. W., Warnken, R. R., Carignan, K. S., Lim, E., Caldwell, R.J., & Friday, D. Z. (2010). *Digital elevation model of Ocean City, Maryland: Procedures, data sources and analysis. NOAA Technical Memorandum NESDIS NGDC-37*. National Geophysical Data Center, NOAA.

Heidarzadeh, M., & Kijko, Andrzej. (2011). A probabilistic tsunami hazard assessment for the Makran subduction zone at the northwestern Indian Ocean. *Natural Hazard, 56*(3), 577–593.

Hill, J. C., Brothers, D. S., Craig, B. K., ten Brink, Uri S., Chaytor, J. D., & Flores, C. H. (2017). Geologic controls on submarine slope failure along the central US Atlantic margin: Insights from the Currituck Slide Complex. *Marine Geology, 385*, 114–130. https://doi.org/10.1016/j.margeo.2016.10.007.

Horspool, N., Pranantyo, I., Griffin, J., Latief, H., Natawidjaja, D. H., Kongko, W., et al. (2014). A probabilistic tsunami hazard assessment for Indonesia. *Natural Hazards and Earth System Sciences, 14*(11), 3105–3122.

Jakeman, J. D., Nielsen, O. M., van Putten, K., Mleczko, R., Burbidge, D., & Horspool, N. (2010). Towards spatially distributed quantitative assessment of tsunami inundation models. *Ocean Dynamics, 60*(5), 11151138. https://doi.org/10.1007/s10236-010-0312-4.

Kaiser, G., Scheele, L., Kortenhaus, A., Løvholt, F., Römer, H., & Leschka, S. (2011). The influence of land cover roughness on the results of high resolution tsunami inundation modeling. *Natural Hazards and Earth System Sciences, 11*, 2521–2540. https://doi.org/10.5194/nhess-11-2521-2011.

Kirby, J. T., Shi, F., Nicolsky, D., & Misra, S. (2016). The 27 April 1975 Kitimat, British Columbia submarine landslide tsunami: A comparison of modeling approaches. *Landslides, 13*(6), 1421–1434. https://doi.org/10.1007/s10346-016-0682-x.

Kirby, J. T., Shi, F., Tehranirad, B., Harris, J. C., & Grilli, S. T. (2013). Dispersive tsunami waves in the ocean: Model equations and sensitivity to dispersion and Coriolis effects. *Ocean Modelling, 62*, 39–55. https://doi.org/10.1016/j.ocemod.2012.11.009.

Krauss, T, (2011). *Probabilistic tsunami hazard assessment for the United States East Coast*. Masters Thesis, University of Rhode Island. http://chinacat.coastal.udel.edu/nthmp/krause-ms-uri11.pdf.

Løvholt, F., Bondevik, S., Laberg, J. S., Kim, J., & Boylan, N. (2017). Some giant submarine landslides do not produce large tsunamis. *Geophysical Research Letters, 44*(16), 8463–8472. https://doi.org/10.1002/2017GL074062.

Løvholt, F., Pedersen, G., & Gisler, G. (2008). Oceanic propagation of a potential tsunami from the La Palma Island. *Journal of Geophysical Research, 113*, C09026.

Løvholt, F., Pedersen, G., Harbitz, C. B., Glimsdal, S., & Kim, J. (2015). On the characteristics of landslide tsunamis. *Philosophical Transactions of the Royal Society A, 373*, 20140376. https://doi.org/10.1098/rsta.2014.0376.

Løvholt, F., Schulten, I., Mosher, D., Harbitz, C., & Kraste, S. (2018). Modelling the 1929 Grand Banks slump and landslide tsunami. In D. G. Lintern, D. C. Mosher, L. G. Moscardelli, P. T. Bobrowsky, C. Campbell, J. D. Chaytor, J. J. Clague, A. Georgiopoulou, P. Lajeunesse, A. Normandeau, D. J. W. Piper, M. Scherwath, C. Stacey, & D. Turmel (Eds.), *Subaqueous mass movements* (p. 477). London: Geological Society London Special Publications. https://doi.org/10.1144/SP477.28.

Locat, J., Lee, H., ten Brink, U. S., Twitchell, D., Geist, E., & Sansoucy, M. (2009). Geomorphology, stability and mobility of the Currituck slide. *Marine Geology, 264*, 28–40.

Lynett, P., et al. (2017). Inter-model analysis of tsunami-induced coastal currents. *Ocean Modeling, 114*, 14–32. https://doi.org/10.1016/j.ocemod.2017.04.003.

Ma, G., Shi, F., & Kirby, J. T. (2012). Shock-capturing non-hydrostatic model for fully dispersive surface wave processes. *Ocean Modellimg, 43–44*, 22–35.

Madsen, P. A., Fuhrman, D. R., & Schaffer, H. A. (2008). On the solitary wave paradigm for tsunamis. *Journal of Geophysical Research, 113*(C12012), 22.

Matsuyama, M., Ikeno, M., Sakakiyama, T., & Takeda, T. (2007). A study of tsunami wave fission in an undistorted experiment. Tsunami and Its Hazards in the Indian and Pacific Oceans. *Pure and Applied Geophysics, 164*, 617–631.

McMaster, R. L. (1960). Sediments of Narragansett Bay system and Rhode Island Sound, Rhode Island. *Journal of Sedimentary Research, 30*(2), 249–274.

National Geophysical Data Center. (1999a). *US coastal relief model—Northeast Atlantic*. National Geophysical Data Center, NOAA. https://doi.org/10.7289/V5MS3QNZ(**March 15, 2017**)

National Geophysical Data Center. (1999b). *US coastal relief model—Southeast Atlantic*. National Geophysical Data Center, NOAA. https://doi.org/10.7289/V53R0QR5(**March 15, 2017**).

National Centers for Environmental Information. (2014). *Tiled coastal elevation models*. NOAA National Centers for Environmental Information. Dataset. https://doi.org/10.7289/V5P55KQC.

Piper, D. J. W., Cochonat, P., & Morrison, M. L. (1999). The sequence of events around the epicentre of the 1929 Grand Banks earthquake: Initiation of the debris flows and turbidity current inferred from side scan sonar. *Sedimentology, 46*, 79–97.

Shelby, M., Grilli, S. T., & Grilli, A. R. (2016). Tsunami hazard assessment in the Hudson River Estuary based on dynamic tsunami tide simulations. *Pure and Applied Geophysics, 173*(12), 3999–4037. https://doi.org/10.1007/s00024-016-1315-y.

Shi, F., Kirby, J. T., Harris, J. C., Geiman, J. D., & Grilli, S. T. (2012). A high-order adaptive time-stepping TVD solver for Boussinesq modeling of breaking waves and coastal inundation. *Ocean Modelling, 43–44*, 36–51. https://doi.org/10.1016/j.ocemod.2011.12.004.

SLIDE. (2017). Slope Stability Analysis Model. RockScience Inc. https://www.rocscience.com/rocscience/products/slide.

Tappin, D. R., Grilli, S. T., Harris, J. C., Geller, R. J., Masterlark, T., Kirby, J. T., et al. (2014). Did a submarine landslide contribute to the 2011 Tohoku tsunami? *Marine Geology, 357*, 344–361. https://doi.org/10.1016/j.margeo.2014.09.043.

Tehranirad, B., Harris, J. C., Grilli, A. R., Grilli, S. T., Abadie, S., Kirby, J. T., et al. (2015). Far-field tsunami hazard in the north Atlantic basin from large scale flank collapses of the Cumbre Vieja volcano, La Palma. *Pure and Applied Geophysics, 172*(12), 3589–3616. https://doi.org/10.1007/s00024-015-1135-5.

Tehranirad, B., Kirby, J. T., Grilli, S. T., Grilli, A. R, & SHi, F. (2017). Continental shelf bathymetry controls the spatial distribution of tsunami hazard for the US East Coast. Geophysical Research Letters **(in preparation)**.

Towns, J., Cockerill, T., Dahan, M., Foster, I., Gaither, K., Grimshaw, A., et al. (2014). XSEDE: Accelerating scientific discovery. *Computing in Science and Engineering, 16*(5), 62–74. https://doi.org/10.1109/MCSE.2014.80.

Twichell, D. C., McClennen, C. E., & Butman, B. (1981). Morphology and processes associated with the accumulation of the fine-grained sediment deposit on the southern New England shelf. *Journal of Sedimentary Research, 51*(1), 269–280.

USGS. (2002). US Geological Survey "2002 interactive deaggregations". https://geohazards.usgs.gov/deaggint/2002/. Accessed 2012.

Watts, P. & Grilli, S. T. (2003). Underwater landslide shape, motion, deformation, and tsunami generation. In: *Proceedings of 13th ISOPE, Honolulu* (pp. 364–371). http://personal.egr.uri.edu/grilli/3055p364.pdf.

Watts, P., Grilli, S. T., Tappin, D. R., & Fryer, G. (2005). Tsunami generation by submarine mass failure, II: Predictive equations and case studies. *Journal of Waterway Port Coastal and Ocean Engineering, 131*(6), 298–310.

Yavari-Ramshe, S., & Ataie-Ashtiani, B. (2016). Numerical simulation of subaerial and submarine landslide generated tsunami waves—recent advances and future challenges. *Landslides, 13*(6), 1325–1368. https://doi.org/10.1007/s10346-016-0734-2.

(Received January 11, 2018, revised May 22, 2018, accepted August 14, 2018, Published online September 3, 2018)

Pure Appl. Geophys. 176 (2019), 3099–3114

https://doi.org/10.1007/s00024-019-02187-3

Pure and Applied Geophysics

Probabilistic Landslide-Generated Tsunamis in the Indus Canyon, NW Indian Ocean, Using Statistical Emulation

Dimitra M. Salmanidou,[1] Mohammad Heidarzadeh,[2] and Serge Guillas[1]

Abstract—The Indus Canyon in the northwestern Indian Ocean has been reported to be the site of numerous submarine mass failures in the past. This study is the first to investigate potential tsunami hazards associated with such mass failures in this region. We employed statistical emulation, i.e. surrogate modelling, to efficiently quantify uncertainties associated with slump-generated tsunamis at the slopes of the canyon. We simulated 60 slump scenarios with thickness of 100–300 m, width of 6–10.5 km, travel distances of 500–2000 m and submergence depth of 250–450 m. These scenarios were then used to train the emulator and predict 500,000 trial scenarios in order to study probabilistically the tsunami hazard over the near field. Due to narrow–deep canyon walls and the shallow continental shelf in the adjacent regions (<100 m water depth), the tsunami propagation has a unique pattern as an ellipse stretched in the NE–SW direction. The results show that the most likely tsunami amplitudes and velocities are approximately 0.2–1.0 m and 2.5–13 m/s, respectively, which can potentially impact vessels and maritime facilities. We demonstrate that the emulator-based approach is an important tool for probabilistic hazard analysis since it can generate thousands of tsunami scenarios in few seconds, compared to days of computations on High Performance Computing facilities for a single run of the dispersive tsunami solver that we use here.

Key words: Indian Ocean, Indus Canyon, landslide-generated tsunami, statistical emulation, uncertainty quantification.

1. Introduction

Tsunamigenic submarine landslides may occur in different geological settings around the world, from the gentle slopes of continental margins to steep slopes along volcano-tectonic islands (Harbitz et al. 2014). Such landslides have been the sources of catastrophic tsunamis worldwide. Some examples are the 1929 Grand Banks (Canada) (Fine et al. 2005; Løvholt et al. 2018) and the 1998 Papua New Guinea (more than 2100 deaths) tsunami events (Synolakis et al. 2002; Tappin et al. 2008; Heidarzadeh and Satake 2015). The Storegga Slide, 8 ka BP, generated tsunami waves with run-up heights reaching elevations of ca. 20 m at the Shetland islands (Bondevik et al. 2005). Landslide-induced tsunamis can be destructive in the near field but they tend to lose energy during propagation due to their relatively shorter wavelengths; their far field catastrophic potential, therefore, tends to be small (Fritz et al. 2003; Glimsdal et al. 2013; Heidarzadeh and Satake 2017). The complexity of the source mechanism and the lack of data render the study of landslide tsunamis cumbersome. As in-situ observations of the submarine sliding process are rare, physical and numerical experiments are often utilised to shed light on these events. Various numerical approaches have been employed up to date for landslide tsunami propagation, such as 3D Navier Stokes equation solvers for the generation process and shallow water equation solvers, often including a dispersive term (Boussinesq approximation) for tsunami propagation (Heidarzadeh et al. 2014; Løvholt et al. 2015; Yavari-Ramshe and Ataie-Ashtiani 2016).

Submarine canyons can be found all over the world; at many cases their morphology results from continuous episodes of sediment collapse (Normark and Carlson 2003). Canyons located in the vicinity of shorelines can pose a risk to populated areas when flank collapse gives way to tsunami generation. For example, submarine collapse in the Cook Strait Canyon could lead to tsunami inundation on the coasts of Wellington, in New Zealand (Lane et al. 2016). The steep slopes and narrow widths of

[1] Department of Statistical Science, University College London, London, UK. E-mail: d.salmanidou.12@ucl.ac.uk
[2] Department of Civil and Environmental Engineering, Brunel University London, London, UK.

submarine canyons are capable of producing destructive tsunamis. Tsunamigenesis by sliding on open continental margins differs from those in submarine canyons (Power et al. 2016). The abrupt depth changes between the canyon and the shelf have an effect on the wave speed; refraction and reflection may also occur when the waves propagate over the canyon walls (Power et al. 2016). Tsunami propagation is sensitive to the canyon size, with the width and length being the most critical parameters; the canyon depth and the distance between the shore and the canyon head play also an important role (Aranguiz and Shibayama 2013). Tsunami run-up and arrival are affected by the geomorphology of the canyon: smaller but faster approaching waves are observed in the locations adjacent to the canyon head, whereas wave amplitude magnification occurs at the sides of the canyon (Aranguiz and Shibayama 2013; Iglesias et al. 2014).

In this work, we assess the potential of submarine collapses in the Indus Canyon, NW Indian Ocean (Fig. 1), to generate tsunami in the near field and quantify the associated uncertainties of future events. The Indus Canyon is approximately 120 km long, 12–17 km wide and its maximum water depth is at the range of 700–1000 m (Fig. 1). Slump deposits and gullies, indicative of submarine mass wasting, have been identified at several locations along the canyon thanks to past bathymetric surveys (von Rad and Tahir 1997; Clift et al. 2014). We use these previously-identified slope failures as landslide tsunami sources to account for a realistic representation of the slump's geomorphological characteristics. Due to the limited knowledge of these events, we do not attempt to reconstruct past collapse episodes but we quantify the uncertainty of potential tsunamigenic scenarios. To do so, we build a statistical emulator that predicts probabilistic maximum tsunami amplitudes and velocities at specific locations.

2. *The Indus Canyon: Geological Setting and Past Slumping Activities*

The Indus Canyon is a submarine canyon cutting the Indus shelf on the passive continental margin of India and Pakistan (Figs. 1, 2a). The head of the canyon is located in front of the delta of the Indus River, one of the longest rivers in Asia. The upper and middle parts of the Indus Canyon are characterised by steep erosional walls which gradually give way to an extensive channel-levee system at the lower part of the canyon (von Rad and Tahir 1997). The channel-levee system feeds with sediment the Indus Fan, the second largest deep-sea fan in the world, possibly being active since the late Eocene (Clift et al. 2001). Several canyon systems on the Indus shelf, that contributed to the sediment supply of the Indus Fan, have progressively become inactive and were filled with sediments (Kolla and Coumes 1987; Bourget et al. 2013). The Indus Canyon forms the most recently active canyon at the shelf. Large volumes of sediment from the Indus River basin are being stored in the form of submarine clinoforms at the continental shelf east and west of the canyon (Clift et al. 2014). Among the scientific expeditions that took place in the region, the cruise with R.V. Sonne in 1993 and the most recent expedition (winter 2008/2009) on board R.V. Pelagia have studied the morphology of the Indus Canyon collecting swath bathymetry and seismic reflection data.

The data from the R. V. Pelagia survey provided with a high resolution bathymetric map of the canyon (Clift et al. 2014). One of the major findings of the R. V. Pelagia cruise was that the head of the canyon acts as a depositional area with rapid sediment accumulation rates (Clift et al. 2014). Discharged sediments from the Indus River are accumulating on the head of the canyon. They are then transported towards the lowest parts of the canyon without, however, evidence of significant sedimentation inside the canyon in the last 7 ka (Clift et al. 2014). Changes in the sedimentation rates are possibly linked with falling sea-levels or heavy monsoon periods (Clift et al. 2014). In-situ observations, seismic profiles and sedimentary cores obtained during the 1995 and 2008/2009 scientific expeditions provide evidence for slumping in different parts of the Indus Canyon (von Rad and Tahir 1997; Clift et al. 2014). von Rad and Tahir (1997) linked mass wasting processes on the continental shelf and the steep walls of the canyon with the low sea levels during the Last Glacial Maximum.

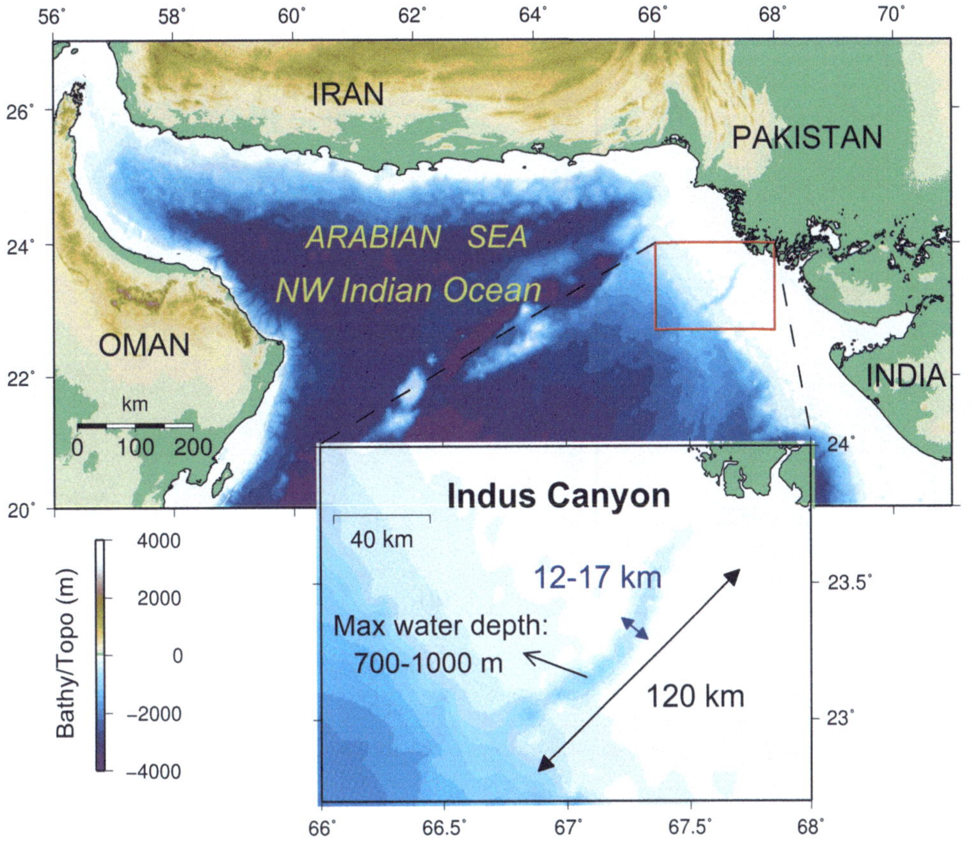

Figure 1
General location map of the study area showing the Indus Canyon in the Arabian Sea, NW Indian Ocean. The inset at the bottom provides a zoomed view of the Indus Canyon and indicates its approximate dimensions (length: 120 km, width: 12–17 km, maximum water depth: 700–1000 m)

Four plausible failure locations discussed in Clift et al. (2014) and von Rad and Tahir (1997) are shown in Fig. 2a by red/grey stars. Slump terraces were observed in the middle part of the canyon (von Rad and Tahir 1997). Scarps at depths of 325 and 620 m, but also as shallow as 200 m, and slump masses have been identified by von Rad and Tahir (1997). At the head of the canyon, erosional features observed at seismic reflection profiles provide evidence of major reworking through slumping (Clift et al. 2014). A seismic profile across the axis of the upper part of the canyon revealed slump deposits on the seabed (Clift et al. 2014). The study area and a bathymetric profile across the axis of the canyon are shown in Fig. 2b–d. The 1D profile A–A′ is similar to the seismic profile collected by R.V. Pelagia (Fig. 2b, d). Some thick sedimentary deposits can be observed at the bottom of the slope, NE of the profile (Fig. 2b), that may constitute the slump deposits observed by Clift et al. (2014).

3. Numerical and Statistical Methods

3.1. Numerical Modelling of Tsunami and Slump Sources

Tsunami generation is strongly affected by the shape parameters and the landslide kinematics (Watts et al. 2005). The rheological properties of the sliding material play an important role during the depositional process (Salmanidou et al. 2018). In this work, we consider the effect of variance of the shape parameters in tsunami generation. We use a slump model based on the empirical formulas developed by

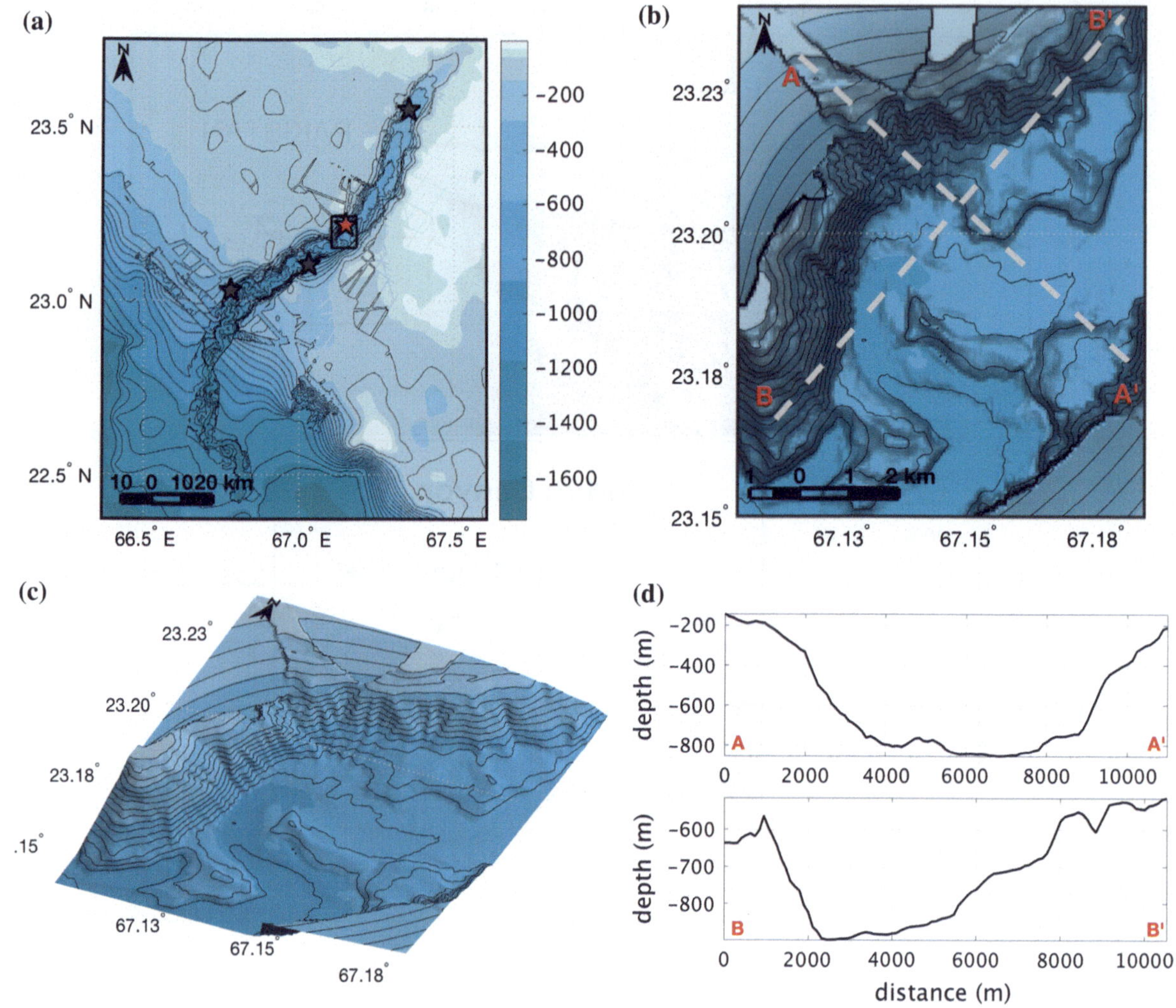

Figure 2

a The Indus Canyon. The red and grey stars denote locations of slumping described in von Rad and Tahir (1997) and Clift et al. (2014). The black rectangle and the red star show the location of the slope region in this study (**b**, **c**). The colorscale shows the depth in meters. **b** 2D view of the slope region. The grey dashed lines indicate the bathymetric profile A–A′ [which follows closely the seismic profile described in Clift et al. (2014)] and the profile B–B′ (**d**). **c** 3D view of the slope region. **d** Bathymetric profiles across the section A–A′ and B–B′ in the Indus Canyon

Grilli and Watts (2005) and Watts et al. (2005) to compute the tsunami waves induced by rotational slumping or translational sliding. The initial slump model is a simultaneous dipole Gaussian (Fig. 3) with varying length, width and height depending on the source scenario characteristics. This model has been used in several studies of landslide tsunamis in the past (Tappin et al. 2001, 2014; Synolakis et al. 2002; Heidarzadeh and Satake 2015, 2017; Løvholt et al. 2018).

For modelling tsunami propagation, the numerical code JAGURS is used (Baba et al. 2015, 2016). The code solves the nonlinear dispersive long-wave equations using a leapfrog (staggered-grid) scheme on Finite Differences (Baba et al. 2015). A domain-decomposition method was implemented for the parallelisation of the code (Baba et al. 2015). Dispersion of landslide tsunami propagation can be accounted by adding a dispersive term in the long wave equations (Boussinesq approach) (Baba et al.

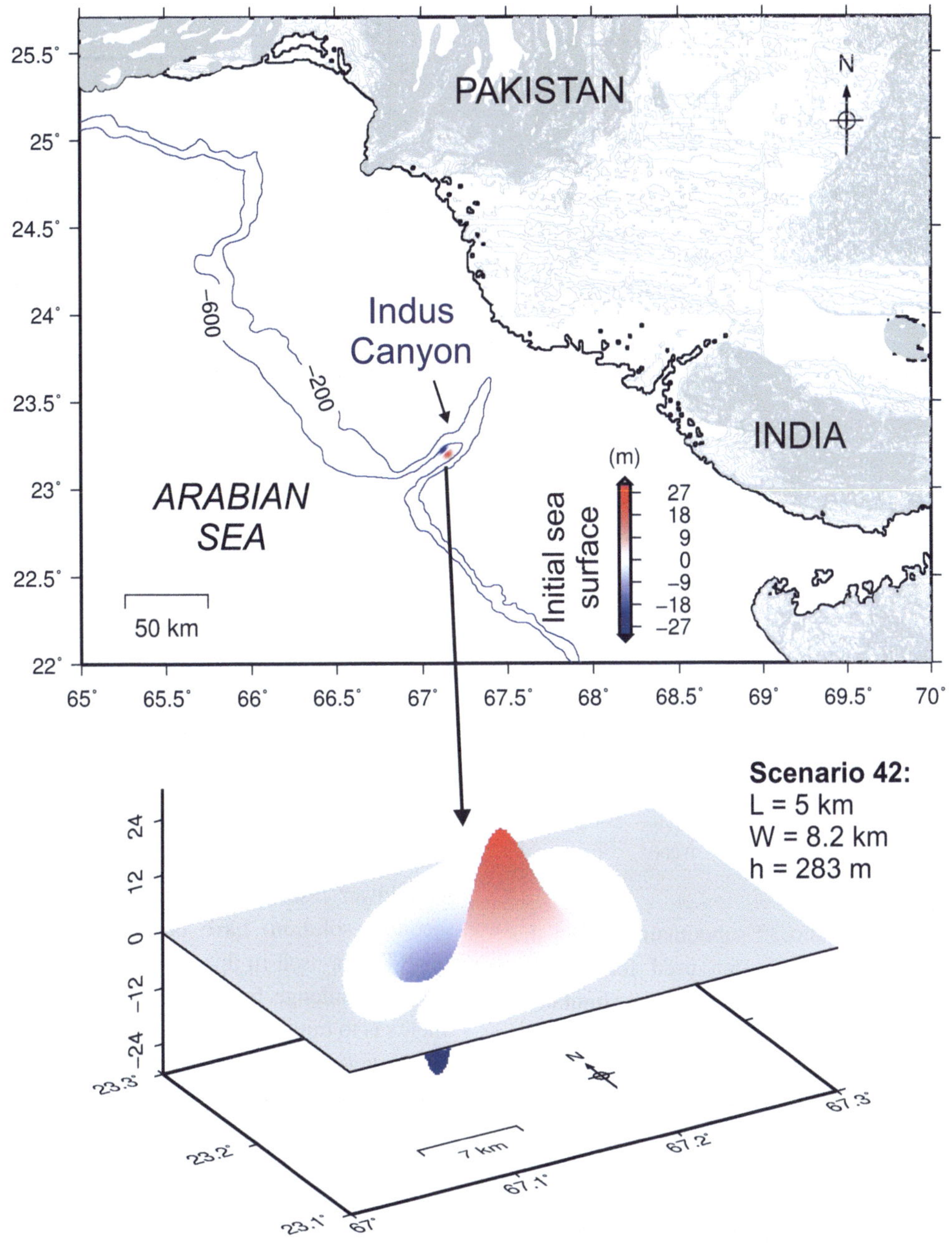

Figure 3
Map showing the location and 3D projection of one of the landslide scenarios (scenario 42 in Table 2) considered in this study

2015). The use of nested grids allows the refinement of the regions of interest in the computational domain (Baba et al. 2015, 2016). Four grids are used in the numerical experiments with spatial resolutions varying from $dx = dy = 0.0045°$ down to $0.0005°$ (Fig. 4). The largest domain (grid A) has the coarsest resolution and grid B has a spatial resolution of $dx = dy = 0.0015°$. Grids C1 and C2 have the finest resolution (Table 1). Grid B acts as the transition between grids A and C1–C2; a grid ratio of 1 : 3 is chosen in accordance with other numerical studies with JAGURS (Baba et al. 2015, 2016).

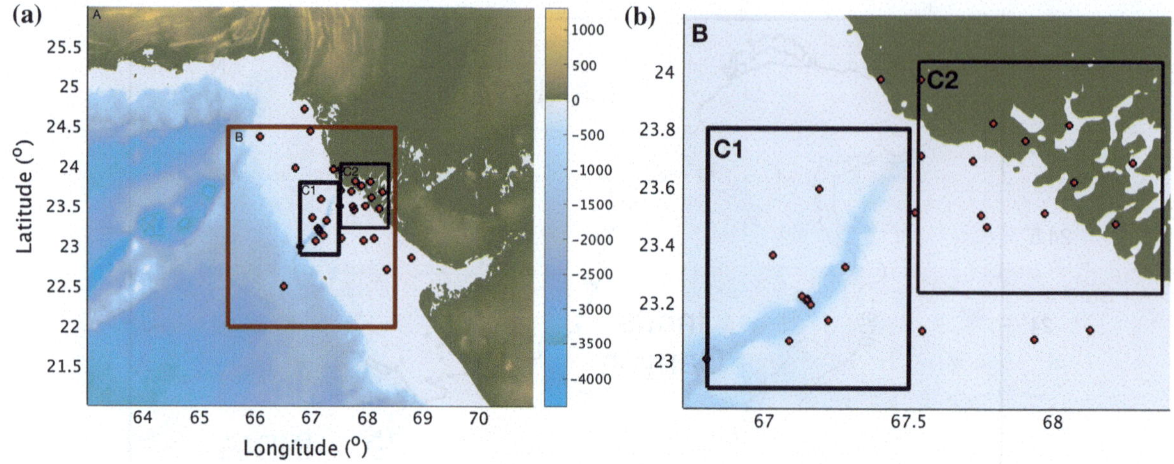

Figure 4
The computational domain used for the simulations. The four grids used in this study (**a**, **b**, C1 & C2) have different spatial resolutions (**a** 0.0045°, **b** 0.0015° and C1–C2: 0.0005°). The red circles indicate the location of the numerical gauges

Table 1

Nested computational grids used in this study

Grid	Lat (°)	Lon (°)	dx, dy (°)	dx, dy (m)
A	21.01–25.996	63.01–70.9885	0.0045	~300
B	22–24.4975	65.503–68.5	0.0015	~100
C1	22.9–23.8	66.8035–67.501	0.0005	~50
C2	23.233–24.034	67.5325–68.3785	0.0005	~50

Each grid is split into 32 subdomains to run the code in parallel. The time step used for tsunami simulation is $dt = 0.1$ seconds with a total simulation time of $t_{fin} = 150$ min. A tsunami propagation scenario of 150 min may take roughly 85 h of KNL (KNights Landing) time on the computer cluster used in this study (32 processors on the Cambridge Service for Data Driven Discovery, CSD3, facility). Thirty-four artificial wave gauges measure the free surface elevation and velocity at various locations within the computational domain (Fig. 4). To study wave propagation toward the shoreline, we show here the records of three gauges located at lon. 68.129E, lat. 23.105N in water depth of – 19.22 m (G24 in Fig. 7); lon. 67.97E, lat. 23.508N in water depth of – 15.27 m (G30 in Fig. 7); and lon. 68.347E, lat. 22.711N in water depth of – 30.28 m (G31 in Fig. 7).

We make use of the Global Multi-Resolution Topography (GMRT) synthesis for the bathymetry and elevation data used in the modelling (Ryan et al. 2009). The GMRT synthesis is a product of different sources at multiple resolutions, multibeam sonar data assembled by scientists and institutions have been used to compile the dataset. Thanks to the scientific survey on board R.V. Pelagia (Clift et al. 2014), the Indus Canyon has been mapped in detail with a resolution of 0.000125°. These data in combination with lower resolution data (GEBCO 2014, 30 arc-second resolution) have been used to compile the GMRT data used in this study.

The challenge for the design of computer experiments is to minimize the number of runs, considering the high computational cost of each run. A Latin Hypercube Sampling (LHS) approach was taken for its suitable coverage and space filling properties (Morris 1991). It is a random design used to build the emulator over the whole input ranges, irrespective of the subsequent chosen distributions of input parameters that will represent the subset of realistic cases within these ranges. The LHS is only based on the sensitivity of the tsunami response to the landslide uncertain inputs. As a rule of thumb, the minimum sampling size n for an experiment can be chosen as $n = 10p$ where p is the number of input parameters (Loeppky et al. 2009). For this study, we run 60 numerical experiments (Table 2) for the four input parameters: the initial maximum thickness of the material, h; the initial slump width, w; the initial

Table 2

Slump scenarios used in this study. Parameters are: h, thickness of the slump; w, width; d, water depth; and s_0, the travel distance of the slump

No	h (m)	w (m)	d (m)	s_0 (m)	No	h (m)	w (m)	d (m)	s_0 (m)
1	126	8038.5	287	959	31	198	6531	399	1275.5
2	253	7836	382	1376	32	161	6072	330	1347.5
3	244	9370.5	371	884	33	104	10486.5	288	1224.5
4	260	9150	418	1019	34	190	8997	376	698
5	222	9460.5	320	1802	35	279	7242	412	1730
6	155	7566	306	1614.5	36	263	10342.5	378	716
7	157	6945	363	1457	37	186	8578.5	309	1401.5
8	233	7377	265	1724	38	192	6679.5	267	978.5
9	168	8673	401	1568	39	292	6616.5	291	923
10	265	9537	311	1524.5	40	226	8362.5	386	1251.5
11	218	8457	424	668	41	108	8254.5	323	774.5
12	270	10027.5	390	1545.5	42	283	8241	273	1641.5
13	151	9375	282	1079	43	214	7044	395	540.5
14	272	7309.5	390	743	44	249	10419	261	1448
15	207	9906	423	857	45	102	6315	415	753.5
16	237	7642.5	316	1322	46	114	7692	364	1484
17	196	9766.5	297	641	47	242	10171.5	438	1353.5
18	140	7921.5	278	1151	48	297	6108	428	1182.5
19	210	7728	331	846.5	49	178	10221	349	1142
20	287	9078	357	1116.5	50	227	6225	321	1761.5
21	148	7120.5	334	927.5	51	299	8533.5	447	602
22	173	8016	435	1029.5	52	135	10068	259	560
23	140	8875.5	346	1854.5	53	119	9622.5	431	1902.5
24	236	6441	341	1053.5	54	110	6508.5	251	1682
25	285	7165.5	339	1586	55	133	6184.5	301	516.5
26	203	8173.5	406	1847	56	276	9064.5	356	1962.5
27	173	7471.5	254	1877	57	127	8808	440	585.5
28	166	6877.5	351	1989.5	58	146	9870	275	1944.5
29	121	9280.5	410	1203.5	59	183	6765	447	1782.5
30	209	9681	370	1650.5	60	254	8745	296	822.5

Table 3

Parameter ranges for slump scenarios considered in this study

h (m)	w (km)	d (m)	s_0 (m)
100–300	6–10.5	250–450	500–2000

Parameters are: h thickness of the slump, w width, d water depth, and s_0 the travel distance of the slump

depth of submergence, d; and the characteristic travel distance of the sliding mass, s_0. The ranges of the parameters are selected in close agreement with the observations and the range of validity of the empirical equations (Grilli and Watts 2005; Watts et al. 2005). The range of the aforesaid four parameters are shown in Table 3.

The thickness of the material has a range of [100–300 m] which suggests that a substantial part of the slope may have failed and is in agreement with a reference threshold of $h<0.2L$ (Watts et al. 2005). From observations of the collapsed slope width geometry, we select a slump width range of [6–10.5 km], where 10.5 km corresponds to the maximum possible width from the one edge of the slope to another (B–B′ profile, Fig. 2d). The higher scarp in the canyon is observed at a depth of 200 m (von Rad and Tahir 1997). The lower boundary of the initial depth range is considered to be relatively shallow with a total range of $[250-450\text{ m}]$. Yielding a minimum factor of $d/L=0.005$, this value comes in close agreement with the reference threshold ($d/L=0.006$) and the studies of real events (Løvholt et al. 2018). The characteristic distance travelled by the slump s_0 is estimated by: $2s_0=Rd\phi$ where R is the radius of curvature and $d\phi$ is the angular displacement (Watts et al. 2005). The distance that the mass travels is often smaller than half of the length of the slump (Watts et al. 2005); here we assume a range of [500–2000 m] for s_0.

The initial length of the slump, the slope angle and the density of the material are kept constant for all the simulation scenarios. As the width of the Indus Canyon is narrow (12–17 km, Fig. 1), a length of $L = 5$ km is considered for the pre-failure slump length. The length is fixed for all the landslide scenarios. The slope angle of the failure plane is roughly 11.3°. Submarine landslides can occur in both steep and gentle slopes, with the largest occurrences in slopes that can be as small as 1° (Masson et al. 2006; Harbitz et al. 2014). The slump orientation was taken in rough agreement with the slope orientation. Considering that the y-axis has a North–South alignment, the direction of the slump movement was 225° counter-clockwise. The density of the slumping materials is considered to be 2000 kg/m^3.

3.2. Statistical Emulation

Probabilistic Hazard Tsunami Assessment (PTHA) quantify the uncertainty in tsunami hazard and thus risks and impacts (Grezio et al. 2017). Novel stochastic tsunami inversion methods rely on a breadth of observational data (Gopinathan et al. 2017). Statistical emulators form a prominent tool in these forward and inverse problems, as they can be used to assess the results of numerous scenarios in a fast and efficient way. Emulators are stochastic approximations of a computer model (also known as simulator) and thus can help compute the probabilistic distributions of tsunami hazards. In the past few years, novel statistical emulation methods have gained ground (Sarri et al. 2012; Beck and Guillas 2016; Liu and Guillas 2017). Statistical emulators have been used to study earthquake and landslide-induced tsunami hazards in the past (Sraj et al. 2014; Salmanidou et al. 2017; Guillas et al. 2018). For instance, Guillas et al. (2018) used an emulator to predict entire time series of wave heights over the Cascadia subduction zone whereas de Baar and Roberts (2017) used a multi-fidelity approach for the Hokkaido Nansei-oki Tsunami.

Here, we build a Gaussian Process emulator to compute probabilistic results of slumping in the Indus Canyon and quantify the uncertainty of such events. Statistical emulation using Gaussian Process regression is a highly flexible and non-linear interpolation approach. It assumes that the uncertainty in the interpolation follows a multivariate Gaussian distribution in between points in the multidimensional design, so the overall uncertainty distribution can be very complex. We employ here the squared exponential covariance function as it is a standard choice, but the choice of covariance structure usually has a very limited effect, particularly here compared to the overall levels of uncertainties in the hazard itself.

We use 60 tsunami simulation scenarios to train the emulator, a statistical approximation of the tsunami model. Then, 5000 scenarios will be generated using the emulator to enable a more complete (and fast) propagation of the uncertainties from the model parameters (submergence depth, width, thickness and travel distance of the slumps) to the outputs. To predict with the emulator, we use Kriging (Lophaven et al. 2002) and follow an approach similar to the one used in Salmanidou et al. (2017). The emulator approximates the deterministic input–output relationship between the response $y(x)$ and the input parameters $x = (x_1, \cdots, x_p)$ through a stochastic relationship written down as:

$$y(x) \sim \sum_{i=1}^{p} \beta_i x_i + z(x). \qquad (1)$$

It means that the predictor is the sum of a linear regression part—the mean of the Gaussian Process (GP)—and a more complex zero-mean GP z; see for instance Beck and Guillas (2016) for more details on the fitting and predictive abilities of GPs for computer experiments. The number of input parameters is denoted by p (here $p = 4$). The β_i's are the coefficients of the linear regression on all the inputs, used as a mean, and z is a Gaussian Process (GP) representing the non-linear part of the variations, unexplained by the mean, which employs another set of coefficients associated with the covariance structure of the GP. Formally, a Gaussian Process is a continuous stochastic process over the input space (hence an uncertain function), for which each

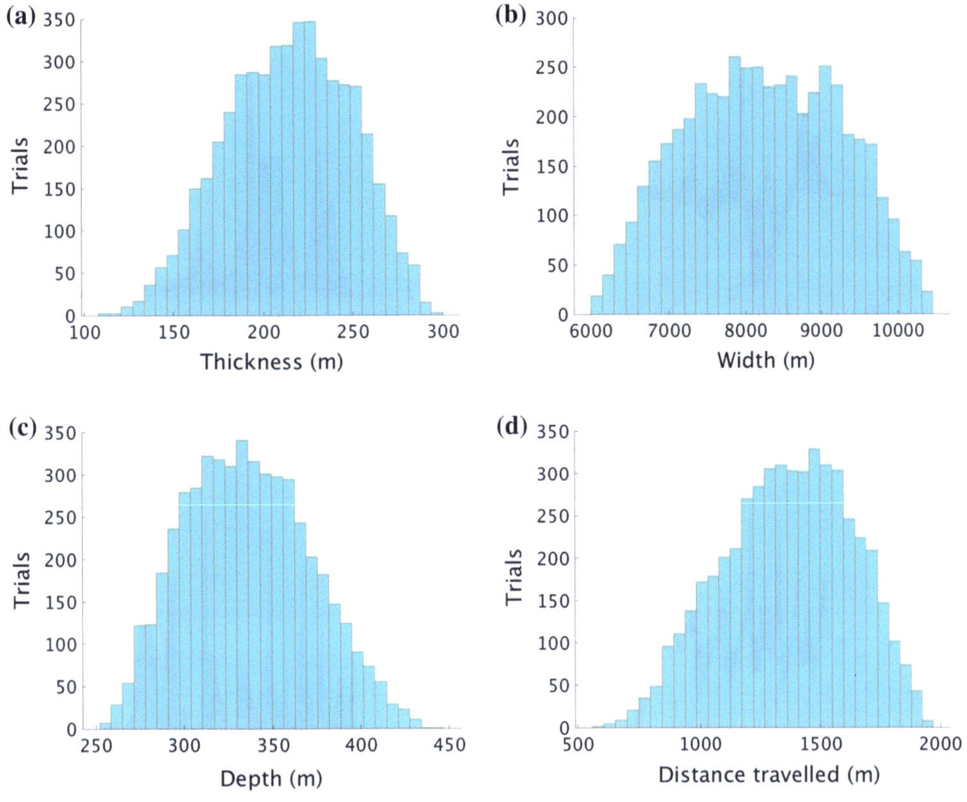

Figure 5
Histograms for 5000 samples of the beta distributions of the input parameters thickness (**a**), width (**b**), submergence depth (**c**) and travelled distance (**d**)

marginal distribution of a finite set of unobserved outputs at selected inputs, constrained by the set of observed inputs–outputs, is multivariate normal. For more general details about GPs, see Rasmussen and Williams (2005).

Since there is a limited amount of information about the potential landslide characteristics in the Indus Canyon, we create probability distributions of these inputs that reflect our uncertain knowledge. Since these ranges are finite, we employ a standard beta distribution $Be(\alpha, c)$, where α and c values refer to the shape (skewness) of the distribution (see also Fig. 5), which is commonly used in the literature for such efforts (Sarri et al. 2012; Guillas et al. 2018). We show samples from these distributions for 5000 scenarios in Fig. 5. We chose distributions that spread over the initially attributed ranges exactly. Hence, we selected $Be(4, 3)$ for the maximum thickness and the distance travelled, $Be(2, 2)$ for the width of the slump and $Be(3, 4)$ for the depth above the slump centre (Fig. 5). The distributions present some skewness that reflects possible values taken by these parameters. For example, the input depths in this case have values between 250 and 450 m (as in Table 3) but due to the shape of the distribution ($Be(3, 4)$, Fig. 5c) the most likely values are in the upper part of the slope between 300–350 m, in agreement also with slump scars observations in the canyon (von Rad and Tahir 1997; Clift et al. 2014). Using the emulator, we propagate the uncertainty from these distributions for the 5000 scenarios.

4. *Probabilistic Landslide Tsunami Analysis*

The results of the tsunami simulations for our initial 60 slump scenarios and statistical emulation are shown in this section. The simulation results

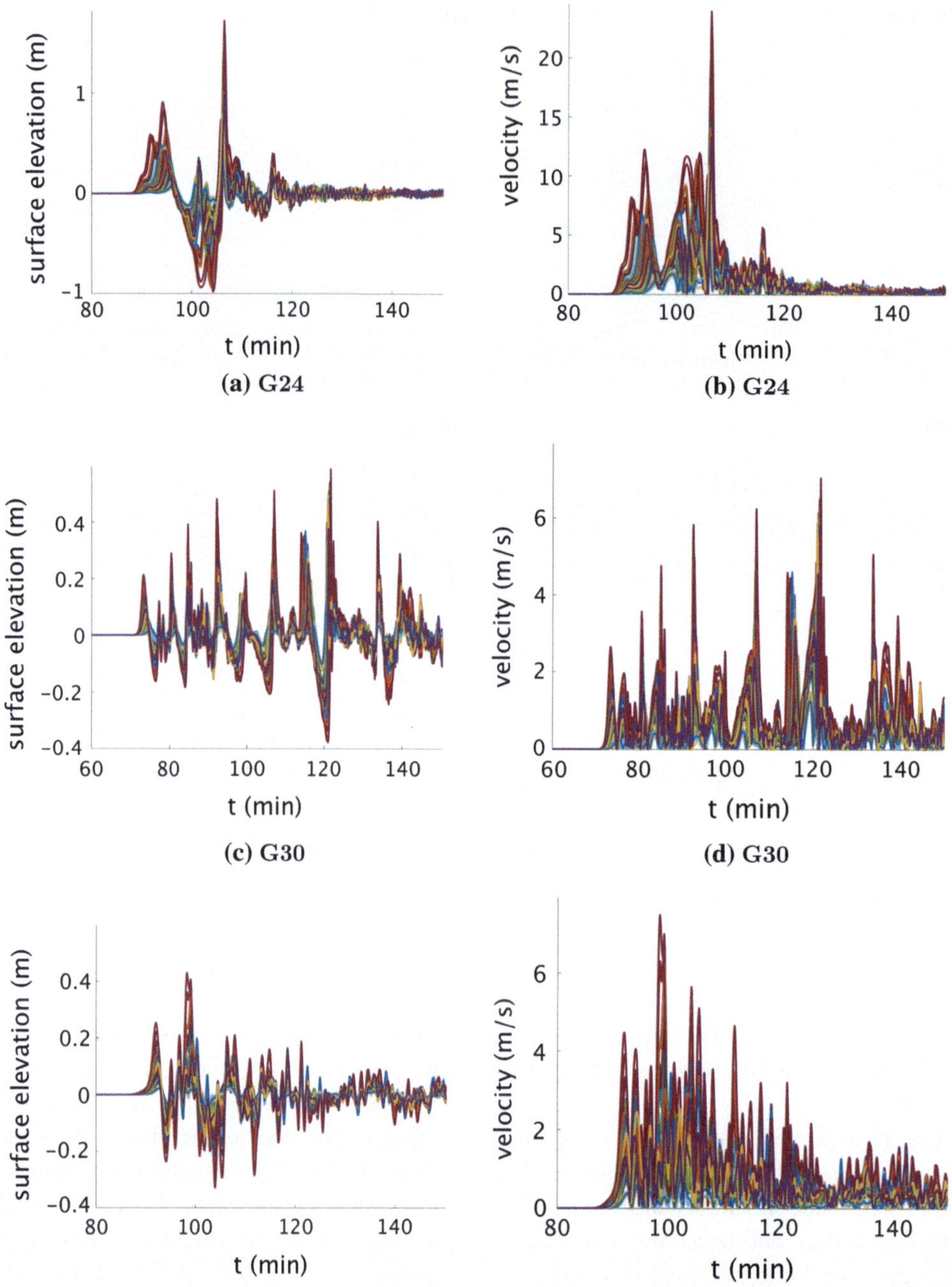

Figure 6
The free surface elevation (**a**, **c** and **e**) and wave velocities (**b**, **d** and **f**) recorded at the three gauges G24, G30 and G31. See also Fig. 7 for gauge locations

demonstrate a large variability in the tsunami amplitudes and velocities (Fig. 6). We present here the tsunami waveforms and velocities at three gauges (see locations of gauges G24, G30 and G31 in Fig. 7). Each panel of Fig. 7 includes a snapshot from one simulation (scenario 42), but a similar pattern appears

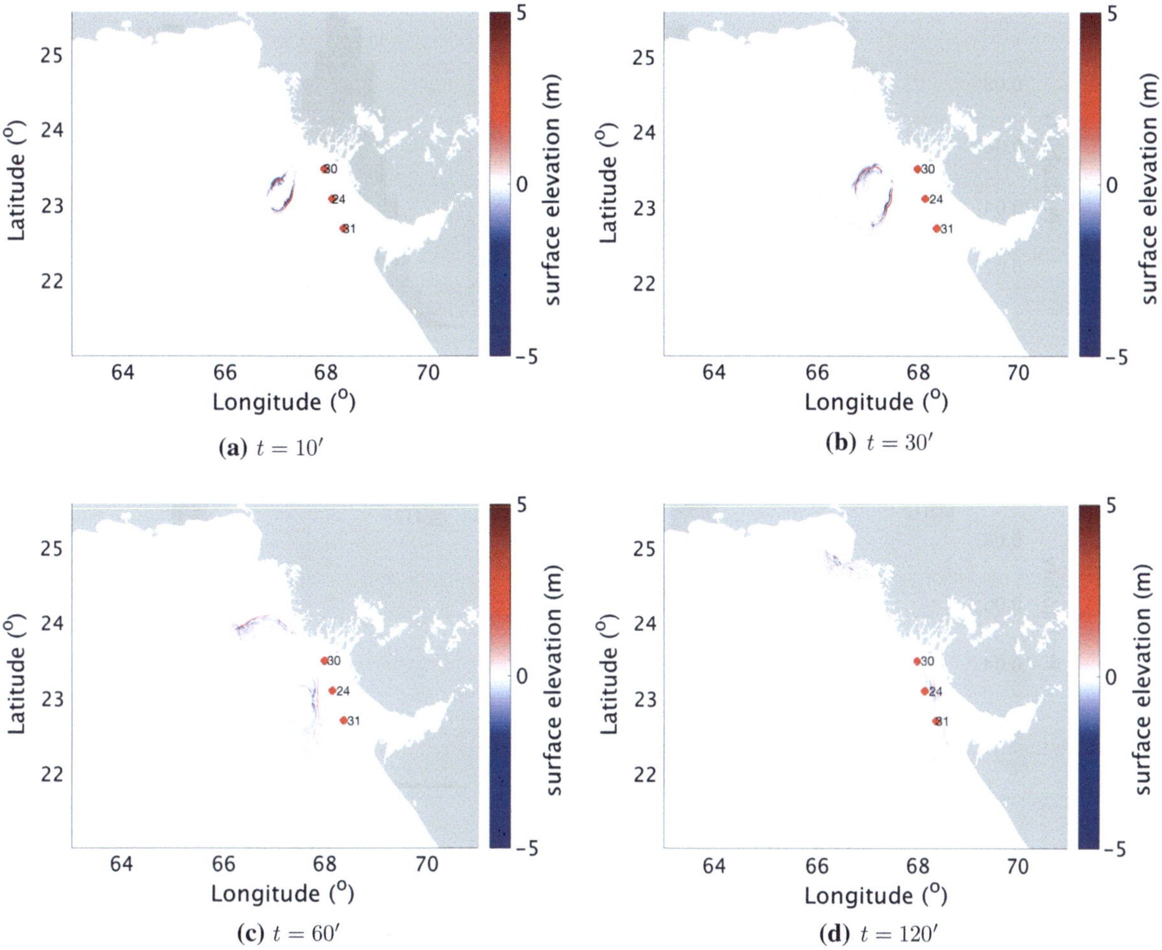

Figure 7
Snapshots of the tsunami propagation at t = 10 min (**a**), 30 min (**b**), 1 h (**c**) and 2 h (**d**) for scenario 42 (see Table 2 for details of this scenario)

across all simulations. As it has been observed in several other tsunamis (Rabinovich et al. 2006), the first wave is not always the largest in all gauges shown in Fig. 6. Unlike tectonic tsunamis whose energetic waves persist for long times (Rabinovich and Thomson 2007), the waves generated by our slump scenarios generally lose in amplitude and velocity rapidly, mostly after a few cycles; except for G30 which could potentially be due to some wave resonance. The waves reach the northern part of the shelf at approximately 2 h from the generation time, such slow tsunami propagation speed can be attributed to the shallow depth of the area (a depth of 50–100 m, Fig. 1) which yields a tsunami phase velocity of approximately 100 km/h. The maximum tsunami amplitudes are in the range of 0.4–1.5 m in these three coastal gauges while the wave velocities vary from 2 to 12 m/s (neglecting a single peak velocity at 22 m/s).

Tsunami snapshots in Fig. 7 reveal a unique pattern for tsunami propagation in the Indus Canyon. These snapshots belong to Scenario 42 (Table 2) which is a worst-case scenario with maximum free surface elevation of ca. 20 m in the generation region (Fig. 7). The snapshots show that tsunami wave fronts approximately form an ellipse with its major axis in the NE–SW direction while the minor axis strikes NW–SE. Such a shape is controlled by the deeper water depth of the canyon's axis (NE–SW direction) which provides faster propagation speed

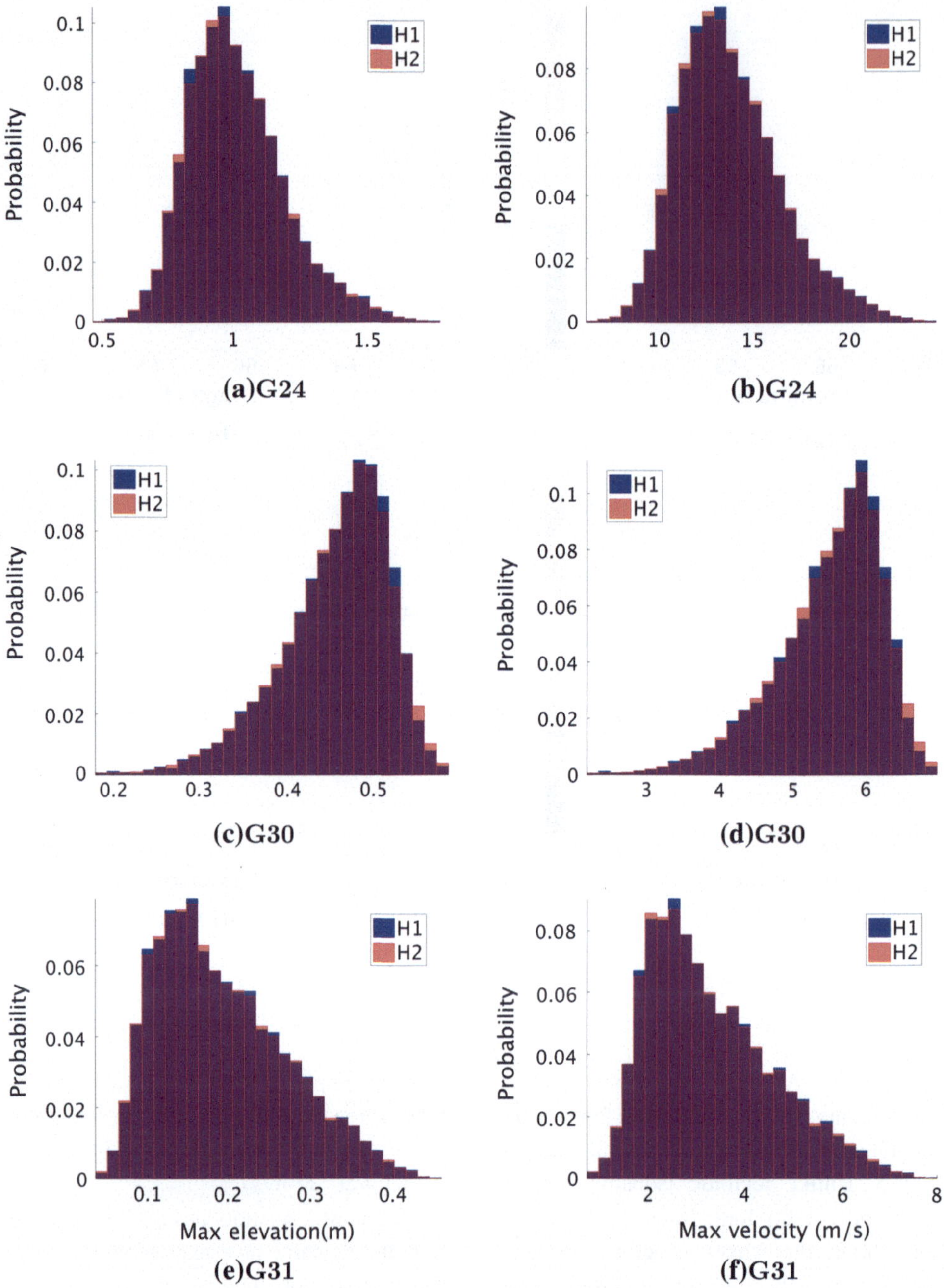

Figure 8
Probabilistic predictions of the maximum free surface elevation (**a**, **c**, **e**) and maximum wave velocities (**b**, **d**, **f**) at three gauges. Histograms H1 & H2 show the predictions with (H2-500,000) and without (H1-5000) incorporating uncertainty of the emulator. The purple colour of the histograms results from the overlapping of H1 and H2

along the canyon's axis. Several secondary and tertiary wave fronts are seen in the snapshots which are the reflected or refracted waves by the canyon's walls.

Trained by the results of 60 tsunami simulations, the emulator can now be built and used for prediction. Histogram H1 in Fig. 8 represents the predictions of the probability of maximum tsunami amplitudes and

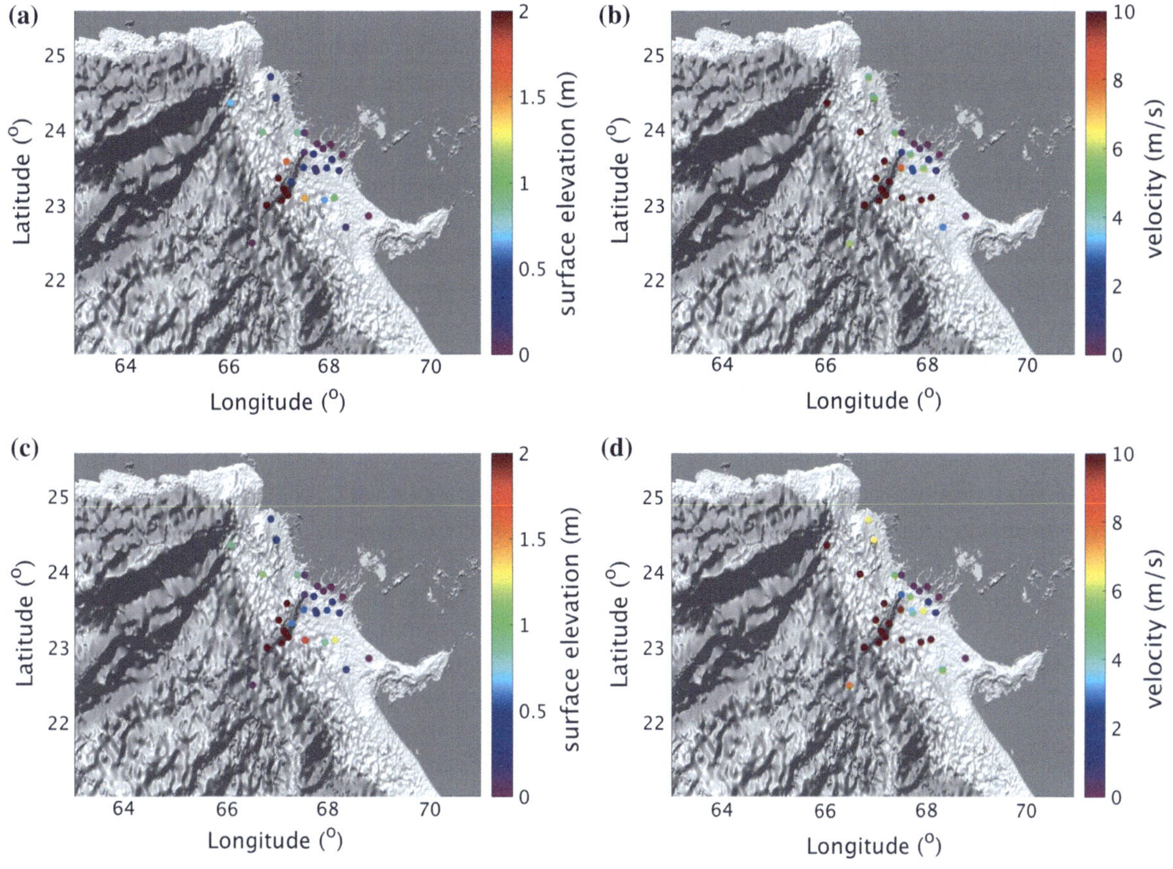

Figure 9
The expected maximum tsunami amplitudes and expected maximum tsunami velocities are given based on the 50th (**a**, **b** respectively) and 90th (**c**, **d** respectively) percentiles of the probabilistic predictions

maximum velocities at gauges 24, 30, and 31 drawn from 5000 slump scenarios (Fig. 5) According to Fig. 8, the most likely amplitude at G30 is approximately 0.5 m with most likely velocity of approximately 6 m/s out of 5000 tsunami trials.

To incorporate uncertainty from the predictions of the emulator itself, we compute 100 additional predictions for each prediction resulting in $5000 \times 100 = 500,000$ trials. The predictions are sampled from the normal distributions that characterise uncertainties in GP predictions, i.e. the part $z(x)$ in (1). These new predictions are shown as H2 in Fig. 8. It can be seen in Fig. 8 that the two histograms H1 and H2 are very close to each other indicating that the results are robust to the uncertainty of the predictive GP distributions: our sample of 60 initial simulations was informative enough to obtain reliable emulations. Naturally, the spread of H2 is slightly larger than the spread of H1 due to this additional level of uncertainty incorporated.

Having as guidance the probabilistic distributions of the maximum free surface elevation and maximum wave velocities, we can produce hazard maps as shown in Fig. 9. The 34 gauges show the distributions of maximum elevation and velocity over the domain. To make these maps, we consider the 50th and 90th percentiles of the probabilistic predictions at the gauges (i.e. partial summaries of the histograms in Fig. 8). The results shown in Fig. 9 are the outcomes of 5000 and 500,000 trials for tsunami amplitude and velocity, respectively. We note that these results are indicative of variations of tsunami amplitude and velocity depending on the variations of input parameters as shown in Fig. 5. The percentiles show that as the waves propagate toward the coastal areas, the maximum tsunami amplitude is up to 1 m and the tsunami velocity achieves a maximum value of approximately 7 m/s (Fig. 9). These values are not the result of a single simulation, but they are the outcomes of 500,000 trials. The most likely amplitude and velocity overall in all three gauges is approximately in the range 0.2–1.0 m and 2.5–13 m/s, respectively. Our results once more demonstrate the applicability of the statistical emulator for probabilistic predictions of landslide tsunamis, as thousands of scenarios were computed in few seconds and the parameter space was explored fully. The emulator-based probabilistic approach in this study computes 500,000 predictions in ca. 2–3 s per gauge as opposed to around 85 h of computation on a state-of-the art cluster of the numerical code when the dispersive simulation is carried out. It can thus be a prominent solution to replace the numerical runs when fully exploring the parameter space.

5. Conclusions

We produced the first realistic landslide-generated tsunami waves in the Indus Canyon, NW Indian Ocean by dispersive modelling of 60 slump scenarios. We also introduced statistical emulation to this context, in order to create 500,000 scenarios for probabilistic analysis. The slump scenarios had thickness of 100–300 m, width of 6–10.5 km, travel distances of 500–2000 m, and submergence depth of 250–450 m. An uncommon propagation pattern, having an ellipse-shaped feature stretched in the NE–SW direction, was observed including several wave fronts resulting from reflections/refractions from the walls of the canyon and propagation over the shallow shelf.

From our probabilistic analysis, the most likely amplitude and velocity at the study locations is approximately 0.2–1.0 m and 2.5–13 m/s, respectively. Hazards maps based on the emulator's predictions compute maximum tsunami amplitudes and velocities of approximately 1 m and 7 m/s, respectively, at close to shore locations. The emulator-based probabilistic approach used in this study for landslide tsunami is a powerful tool for probabilistic hazard studies capable of generating hundreds of thousands of tsunami scenarios in few seconds. A major extension of this work would be to produce a more complete hazard mapping at the coast, given more detailed bathymetry and topography data as well as considering (probabilistically) all possible slump locations along the Indus Canyon through enhanced geophysical surveys.

Acknowledgements

This research was supported by the EPSRC (EP/P016774/1) network M2D (Models-to-Decisions): Decision making under uncertainty, and the EPSRC Impact Acceleration Account Grant (EP/R511638l1). SG and MH acknowledge support from the NERC project (NE/P016367/1) "Tsunami risk for the Western Indian Ocean: steps toward the integration of science into policy and practice" under the Global Challenges Research Fund: Building Resilience programme. SG also acknowledges support from the Alan Turing Institute project "Uncertainty Quantification of multi-scale and multiphysics computer models: applications to hazard and climate models". This work has been performed using resources provided by the Cambridge Tier-2 system operated by the University of Cambridge Research Computing Service (http://www.hpc.cam.ac.uk) funded by EPSRC Tier-2 capital grant EP/P020259/1. Some of the figures are drafted using the open-source code GMT, (Wessel and Smith 1998). The bathymetry and topography data are derived from the Global Multi-Resolution Topography (GMRT) dataset (Ryan et al. 2009). The authors would like to thank Dr Devaraj Gopinathan for the fruitful discussions and Dr Toshitaka Baba for his advice and help with the numerical simulations.

Publisher's Note Springer Nature remains neutral with regard to jurisdictional claims in published maps and institutional affiliations'.

References

Aranguiz, R., & Shibayama, T. (2013). Effect of submarine canyons on tsunami propagation: A case study of the Biobio canyon, Chile. *Coastal Engineering Journal*, *55*, 1350016. https://doi.org/10.1142/S0578563413500162.

Baba, T., Ando, K., Matsuoka, D., Hyodo, M., Hori, T., Takahashi, N., et al. (2016). Large-scale, high-speed tsunami prediction for the Great Nankai Trough Earthquake on the K computer. *The International Journal of High Performance Computing Applications*, *30*(1), 71–84. https://doi.org/10.1177/1094342015584090.

Baba, T., Takahashi, N., Kaneda, Y., Ando, K., Matsuoka, D., & Kato, T. (2015). Parallel implementation of dispersive tsunami wave modeling with a nesting algorithm for the 2011 Tohoku Tsunami. *Pure and Applied Geophysics*, *172*(12), 3455–3472. https://doi.org/10.1007/s00024-015-1049-2.

Beck, J., & Guillas, S. (2016). Sequential design with mutual information for computer experiments (MICE): Emulation of a tsunami model. *SIAM/ASA Journal on Uncertainty Quantification*, *4*(1), 739–766. https://doi.org/10.1137/140989613.

Bondevik, S., Løvholt, F., Harbitz, C., Mangerud, J., Dawson, A., & Svendsen, J. I. (2005). The storegga slide tsunami-comparing field observations with numerical simulations. *Marine and Petroleum Geology*, *22*(1), 195–208. https://doi.org/10.1016/j.marpetgeo.2004.10.003.

Bourget, J., Zaragosi, S., Rodriguez, M., Fournier, M., Garlan, T., & Chamot - Rooke, N. (2013). Late quaternary megaturbidites of the Indus fan: Origin and stratigraphic significance. *Marine Geology*, *336*, 10–23. https://doi.org/10.1016/j.margeo.2012.11.011.

Clift, P. D., Giosan, L., Henstock, T. J., & Tabrez, A. R. (2014). Sediment storage and reworking on the shelf and in the Canyon of the Indus River—Fan system since the last glacial maximum. *Basin Research*, *26*(1), 183–202. https://doi.org/10.1111/bre.12041.

Clift, P., Shimizu, N., Layne, G., Blusztajn, J., Gaedicke, C., Schlüter, H. U., et al. (2001). Development of the Indus Fan and its significance for the erosional history of the Western Himalaya and Karakoram. *GSA Bulletin*, *113*(8), 1039.https://doi.org/10.1130/0016-7606(2001)113<1039:DOTIFA>2.0.CO;2

de Baar, J. H. S., & Roberts, S. G. (2017). Multifidelity sparse-grid-based uncertainty quantification for the Hokkaido Nansei-oki Tsunami. *Pure and Applied Geophysics*, *174*(8), 3107–3121. https://doi.org/10.1007/s00024-017-1606-y.

Fine, I., Rabinovich, A., Bornhold, B., Thomson, R., & Kulikov, E. (2005). The Grand Banks landslide-generated tsunami of November 18, 1929: Preliminary analysis and numerical modeling. *Marine Geology*, *215*(1–2), 45–57.

Fritz, H. M., Hager, W. H., & Minor, H. E. (2003). Landslide generated impulse waves. *Experiments in Fluids*, *35*(6), 505–519.

Glimsdal, S., Pedersen, G. K., Harbitz, C. B., & Løvholt, F. (2013). Dispersion of tsunamis: Does it really matter? *Natural Hazards and Earth System Sciences*, *13*(6), 1507–1526. https://doi.org/10.5194/nhess-13-1507-2013.

Gopinathan, D., Venugopal, M., Roy, D., Rajendran, K., Guillas, S., & Dias, F. (2017). Uncertainties in the 2004 Sumatra-Andaman source through nonlinear stochastic inversion of tsunami waves. *Proceedings of the Royal Society A: Mathematical, Physical and Engineering Sciences*, *473*(2205), 20170353. https://doi.org/10.1098/rspa.2017.0353.

Grezio, A., Babeyko, A., Baptista, M. A., Behrens, J., Costa, A., Davies, G., et al. (2017). Probabilistic tsunami hazard analysis: Multiple sources and global applications. *Reviews of Geophysics*, *55*(4), 1158–1198. https://doi.org/10.1002/2017RG000579.

Grilli, S. T., & Watts, P. (2005). Tsunami generation by submarine mass failure. I: Modeling, experimental validation, and sensitivity analyses. *Journal of Waterway, Port, Coastal, and Ocean Engineering*, *131*(6), 283–297. https://doi.org/10.1061/(ASCE)0733-950X(2005)131:6(283).

Guillas, S., Sarri, A., Day, S., Liu, X., & Frederic, D. (2018). Functional emulation of the high resolution tsunami modelling over Cascadia. *Annals of Applied Statistics*, *12*(4), 2023–2053. https://doi.org/10.1214/18-AOAS1142.

Harbitz, C., Løvholt, F., & Bungum, H. (2014). Submarine landslide tsunamis: How extreme and how likely? *Natural Hazards*, *72*, 1341–1374.

Heidarzadeh M., Krastel S., & Yalciner A. C. (2014) The state-of-the-art numerical tools for modeling landslide Tsunamis: a short review. In: S. Krastel et al. (Eds.), *Submarine mass movements and their consequences*, Chap. 43. Advances in natural and technological hazards research (pp. 483–495), vol. 37. Cham: Springer (**ISBN: 978-3-319-00971-1**).

Heidarzadeh, M., & Satake, K. (2015). Source properties of the 1998 July 17 Papua New Guinea tsunami based on tide gauge records. *Geophysical Journal International*, *202*(1), 361–369.

Heidarzadeh, M., & Satake, K. (2017). A Combined earthquake–landslide source model for the tsunami from the 27 November 1945 Mw 8.1 Makran earthquake. *Bulletin of the Seismological Society of America*, *107*(2), 1033–1040.

Iglesias, O., Lastras, G., Souto, C., Costa, S., & Canals, M. (2014). Effects of coastal submarine canyons on tsunami propagation and impact. *Marine Geology*, *350*, 39–51.

Kolla, V., & Coumes, F. (1987). Morphology, internal structure, seismic stratigraphy, and sedimentation of Indus Fan. *American Association of Petroleum Geologists Bulletin*, *71*, 650–677.

Lane, E., Mountjoy, J., Power, W., & Mueller, C. (2016). Probabilistic hazard of tsunamis generated by submarine landslides in the Cook Strait Canyon (New Zealand). In: *Pure and Applied Geophysics*, pp. 3757–3774

Liu, X., & Guillas, S. (2017). Dimension reduction for Gaussian process emulation: An application to the influence of bathymetry on tsunami heights. *SIAM/ASA Journal on Uncertainty Quantification*, *5*(1), 787–812. https://doi.org/10.1137/16M1090648.

Loeppky, J. L., Sacks, J., & Welch, W. J. (2009). Choosing the sample size of a computer experiment: A practical guide. *Technometrics*, *51*(4), 366–376. https://doi.org/10.1198/TECH.2009.08040.

Lophaven, S., Nielsen, H., & Sondergaard, J. (2002). DACE—A MATLAB kriging toolbox, version 2.0. IMM-REP, Technical Report

Løvholt, F., Pedersen, G., Harbitz, C. B., Glimsdal, S., & Kim, J. (2015). On the characteristics of landslide tsunamis. *Philosophical Transactions of the Royal Society of London A:*

Mathematical, Physical and Engineering Sciences, 373(2053), 20140376. https://doi.org/10.1098/rsta.2014.0376.

Løvholt, F., Schulten, I., Mosher, D., Harbitz, C., & Krastel, S. (2018). Modelling the 1929 Grand Banks slump and landslide tsunami. *Geological Society, London, Special Publications, 477*, S477-28. https://doi.org/10.1144/SP477.28.

Masson, D., Harbitz, C., Wynn, R., Pedersen, G., & Løvholt, F. (2006). Submarine landslides: Processes, triggers and hazard prediction. *Philosophical Transactions of the Royal Society A: Mathematical, Physical and Engineering Sciences, 364*(1845), 2009–2039. https://doi.org/10.1098/rsta.2006.1810.

Morris, M. D. (1991). Factorial sampling plans for preliminary computational experiments. *Technometrics, 33*(2), 161–174.

Normark, W. R., & Carlson, P. R. (2003). Giant submarine canyons: Is size any clue to their importance in the rock record? *Special Paper of the Geological Society of America, 370*, 175–190.

Power, W., Mountjoy, J., Lane, E., Popinet, S., & Wang, X. (2016). Assessing landslide-tsunami hazard in submarine canyons, using the Cook strait canyon system as an example. *Science of Tsunami Hazards, 35*, 145–166.

Rabinovich, A.B., & Thomson, R.E. (2007). The 26 December 2004 Sumatra tsunami: Analysis of tide gauge data from the world ocean Part 1. Indian Ocean and South Africa. In: Tsunami and Its Hazards in the Indian and Pacific Oceans, Springer, pp 261–308

Rabinovich, A. B., Thomson, R. E., & Stephenson, F. E. (2006). The Sumatra tsunami of 26 december 2004 as observed in the North Pacific and North Atlantic oceans. *Surveys in Geophysics, 27*(6), 647–677.

Rasmussen, C. E., & Williams, C. K. I. (2005). *Gaussian processes for machine learning (Adaptive Computation and Machine Learning)*. Cambridge, Massachusetts: The MIT Press (**ISBN: 026218253X**).

Ryan, W., Carbotte, S., Coplan, J., O'Hara, S., Melkonian, A., Arko, R., Weissel, R., Ferrini, V., Goodwillie, A., Nitsche, F., Bonczkowski, J., & Zemsky, R. (2009). Global multi-resolution topography synthesis. *Geochem Geophys Geosyst*, 10(Q03014), https://doi.org/10.1029/2008GC002332

Salmanidou, D. M., Georgiopoulou, A., Guillas, S., & Dias, F. (2018). Rheological considerations for the modelling of submarine sliding at Rockall Bank, NE Atlantic Ocean. *Physics of Fluids, 30*(3), 030705. https://doi.org/10.1063/1.5009552.

Salmanidou, D. M., Guillas, S., Georgiopoulou, A., & Dias, F. (2017). Statistical emulation of landslide-induced tsunamis at the Rockall Bank, NE Atlantic. *Proceedings of the Royal Society of London A: Mathematical, Physical and Engineering Sciences, 473*(2200), 20170026. https://doi.org/10.1098/rspa.2017.0026.

Sarri, A., Guillas, S., & Dias, F. (2012). Statistical emulation of a tsunami model for sensitivity analysis and uncertainty quantification. *Natural Hazards and Earth System Sciences, 12*(6), 2003–2018. https://doi.org/10.5194/nhess-12-2003-2012.

Sraj, I., Mandli, K. T., Knio, O. M., Dawson, C. N., & Hoteit, I. (2014). Uncertainty quantification and inference of Manning's friction coefficients using DART buoy data during the Tōhoku tsunami. *Ocean Modelling, 83*, 82–97. https://doi.org/10.1016/j.ocemod.2014.09.001.

Synolakis, C. E., Bardet, J. P., Borrero, J. C., Davies, H. L., Okal, E. A., Silver, E. A., et al. (2002). The slump origin of the 1998 Papua New Guinea tsunami. *Proceedings of the Royal Society of London. Series A: Mathematical, Physical and Engineering Sciences, 458*(2020), 763–789.

Tappin, D. R., Grilli, S. T., Harris, J. C., Geller, R. J., Masterlark, T., Kirby, J. T., et al. (2014). Did a submarine landslide contribute to the 2011 Tohoku tsunami? *Marine Geology, 357*, 344–361. https://doi.org/10.1016/j.margeo.2014.09.043.

Tappin, D. R., Watts, P., & Grilli, S. T. (2008). The Papua New Guinea tsunami of 17 July 1998: Anatomy of a catastrophic event. *Natural Hazards and Earth System Sciences, 8*(2), 243–266. https://doi.org/10.5194/nhess-8-243-2008.

Tappin, D., Watts, P., McMurtry, G., Lafoy, Y., & Matsumoto, T. (2001). The Sissano, Papua New Guinea tsunami of July 1998—offshore evidence on the source mechanism. *Marine Geology, 175*(1–4), 1–23.

von Rad, U., & Tahir, M. (1997). Late Quaternary sedimentation on the outer Indus shelf and slope (Pakistan): Evidence from high-resolution seismic data and coring. *Marine Geology, 138*, 193–236.

Watts, P., Grilli, S. T., Tappin, D. R., & Fryer, G. J. (2005). Tsunami generation by submarine mass failure. II: Predictive equations and case studies. *Journal of Waterway, Port, Coastal, and Ocean Engineering, 131*(6), 298–310. https://doi.org/10.1061/(ASCE)0733-950X(2005)131:6(298).

Wessel, P., & Smith, W. H. (1998). New, improved version of generic mapping tools released. *Eos, Transactions American Geophysical Union, 79*(47), 579–579.

Yavari-Ramshe, S., & Ataie-Ashtiani, B. (2016). Numerical modeling of subaerial and submarine landslide-generated tsunami waves–recent advances and future challenges. *Landslides, 13*(6), 1325–1368. https://doi.org/10.1007/s10346-016-0734-2.

(Received November 17, 2018, revised April 1, 2019, accepted April 5, 2019, Published online April 16, 2019)

Pure Appl. Geophys. 176 (2019), 3115–3153

https://doi.org/10.1007/s00024-018-1907-9

Pure and Applied Geophysics

Performance Comparison of NAMI DANCE and FLOW-3D® Models in Tsunami Propagation, Inundation and Currents using NTHMP Benchmark Problems

DENIZ VELIOGLU SOGUT[1,2] and AHMET CEVDET YALCINER[1]

Abstract—Field observations provide valuable data regarding nearshore tsunami impact, yet only in inundation areas where tsunami waves have already flooded. Therefore, tsunami modeling is essential to understand tsunami behavior and prepare for tsunami inundation. It is necessary that all numerical models used in tsunami emergency planning be subject to benchmark tests for validation and verification. This study focuses on two numerical codes, NAMI DANCE and FLOW-3D®, for validation and performance comparison. NAMI DANCE is an in-house tsunami numerical model developed by the Ocean Engineering Research Center of Middle East Technical University, Turkey and Laboratory of Special Research Bureau for Automation of Marine Research, Russia. FLOW-3D® is a general purpose computational fluid dynamics software, which was developed by scientists who pioneered in the design of the Volume-of-Fluid technique. The codes are validated and their performances are compared via analytical, experimental and field benchmark problems, which are documented in the "Proceedings and Results of the 2011 National Tsunami Hazard Mitigation Program (NTHMP) Model Benchmarking Workshop" and the "Proceedings and Results of the NTHMP 2015 Tsunami Current Modeling Workshop". The variations between the numerical solutions of these two models are evaluated through statistical error analysis.

Key words: Tsunami, depth-averaged shallow water, Reynolds-averaged Navier–Stokes, benchmarking, NAMI DANCE, FLOW-3D®.

1. Introduction

Coastal regions have always been attractive settling grounds for human populations since they have provided a great deal of marine resources and opportunities for transportation and trade. Today, a large part of the world's population inhabits coastal areas, and the population density on these regions is still growing rapidly. However, coastal environments are dynamic natural systems and they pose a great challenge to human habitation since they are prone to water-related hazards such as floods, storms and tsunamis, tsunamis being one of the most devastating hazards related to these dynamics.

Theoretical approaches taken towards the nearshore tsunami motion are difficult to apply due to strong nonlinearities of equations that govern tsunami behaviors, the three dimensionality of the flow, and the turbulence that develops due to shoaling effects. Numerical modeling has proved to be an accurate and useful method of modeling tsunami inundation on a coastline with the rapid development of computing technology. Over the past several decades, many numerical models have been developed to compute wave transformation from deep water to shallow water (Liu 1994; Wu 2001). The computational cost of three-dimensional (3D) numerical models, which require the numerical solution of the full Navier–Stokes equations, is higher than that of two-dimensional (2D) numerical models, which are based on the shallow water approximation. Determining which model would be suitable for a particular problem requires knowledge and experience.

It is necessary that all numerical models used in tsunami emergency planning be validated and verified. Validation/verification of a numerical code must never cease; even proven models must be tested continuously as new knowledge and data are acquired (Synolakis et al. 2008). Careful and explicit validation/verification of tsunami models through the comparison of their predictions with benchmark analytical solutions, laboratory experiments and field

[1] Middle East Technical University, 06800 Ankara, Turkey. E-mail: deniz.velioglusogut@stonybrook.edu; yalciner@metu.edu.tr

[2] *Present Address*: 1212 Computer Science, Department of Civil Engineering, Stony Brook University, Stony Brook, NY 11794, USA.

measurements has helped them evolve in the last two decades. A numerical code that has performed well in all benchmark tests will not necessarily make realistic inundation predictions in every instance. However, in the results of validated/verified codes, the level of uncertainty is largely reduced to the uncertainty in the geophysical initial conditions.

This study focuses on two numerical tools; NAMI DANCE and FLOW-3D®. The codes are applied to analytical, experimental and field benchmark problems for validation and performance comparison. These problems are capable of identifying the shortcomings of numerical models in terms of tsunami motion and inundation parameters such as water surface elevation, current velocity and run-up. A total of six benchmark problems are considered in this study. The first three benchmark tests were proposed in the ''Proceedings and Results of the 2011 NTHMP Model Benchmarking Workshop'' (NTHMP 2012). The rest of the benchmark problems was documented in the ''Proceedings and Results of the NTHMP 2015 Tsunami Current Modeling Workshop" (NTHMP 2015). The first benchmark problem (BMP 1) covers analytical and experimental investigations of the evolution, propagation and run-up of a single solitary wave climbing up a plane beach. Benchmark problem 2 (BMP 2) is composed of physical model studies which were carried out to investigate the run-up of a solitary wave on a conically shaped island and to provide a better understanding of three-dimensional tsunami run-up. An extreme run-up height of 31.7 m was measured near the village of Monai, in Okushiri Island after the 1993 Okushiri tsunami. Benchmark problem 3 (BMP 3) provides the data of a 1:400 scale laboratory experiment of the Monai run-up. Benchmark problem 4 (BMP 4) describes a series of experiments that analyze the transformation of a single solitary wave as it propagates up a triangular-shaped shelf with an island feature located at the offshore point of the shelf. Benchmark problem 5 (BMP 5) covers a series of experiments having a single long period wave that propagates up a piecewise linear slope and onto a small-scale model of the town of Seaside, Oregon. The last benchmark problem, Benchmark problem 6 (BMP 6), is a case study, consisting of a field data set of the Japan 2011 tsunami recorded in Hilo Harbor, Hawaii. This study is also the first in-depth study to employ such a large number of benchmark problems to validate the commercial numerical model FLOW-3D® in tsunami wave evolution, propagation and inundation.

2. *Recent Studies in the Literature*

This section summarizes the most prominent and recent studies which are conducted using NAMI DANCE and FLOW-3D®. A more detailed literature review can be found in Velioglu (2017).

Yalciner et al. (2005) issued a report about the field survey results concerning run-up, arrival time and damages of the 2004 Indian Ocean tsunami. In addition, they simulated the 2004 Indian Ocean tsunami using NAMI DANCE and compared the numerical results with the observed field data of other tsunami survey teams at Sumatra, Thailand, Sri Lanka, India and Maldives. Zaitsev et al. (2009) presented the data of the instrumental registration of the tsunami on Sakhalin and Hokkaidō islands as well as performing numerical simulations of the tsunami propagation in the Tatar Strait using NAMI DANCE software. They found that the numerical results were in satisfactory agreement with the observed data. Yalciner et al. (2014) conducted a numerical modeling study, whose main focus was the western part of Peloponnese in Greece, in the framework of European Union funded project, SEAHELLARC. In the study, simulations of possible tsunami source scenarios for the Pylos, Kyparissia, Filiatra and Zakynthos regions, located at the western part of Greece, were carried out using NAMI DANCE to estimate extreme but possible tsunami wave effects in these regions. Necmioglu and Ozel (2014) studied the sensitivity of earthquake source parameters in tsunami generation and propagation, focusing on the Eastern Mediterranean, using NAMI DANCE. Patel et al. (2016) modeled the Makran 1945 tsunami using NAMI DANCE for the coasts of Pakistan, Iran, India and Oman. The results were consistent with the available reports and published data.

Choi et al. (2007) employed FLOW-3D® to simulate wave run-up on a conical island, based on the unexpectedly large tsunami run-up elevations which had been observed in the circular Babi Island in the

1992 Flores tsunami and in the pear-shaped Okushiri Island in the 1993 Okushiri tsunami. They maintained that the results obtained using two- and three-dimensional models were in good agreement with each other as well as the laboratory data of the physical model studies conducted earlier by Briggs et al. in 1995. Basu et al. (2009) assessed the impulse waves generated by the subaerial landslide at Lituya Bay, Alaska by conducting numerical simulations. FLOW-3D® software was used to track the free surface and shoreline movements. The simulation results were compared with the experimental ones obtained in a physical model of a real case and they were found to be in good agreement with the measured data. It was concluded that the model captured the basic flow properties which were associated with the time-dependent evolution of the flowfield as the landslide interacts with the bay. Pelinovsky et al. (2013) cited both the highest wave run-up which was recorded at Lhok Nga Twin Peaks after the 2004 Sumatra–Andaman earthquake and the 2011 earthquake, which generated extremely destructive tsunami waves up to 37.9 m in height at the tiny fishery port Koborinai situated north of Miyako City in the Iwate Prefecture. The researchers reproduced these extremely large run-ups of tsunami waves in both cases via FLOW-3D® and discovered that the water flow that climbed the coast featured a strong vertical velocity component. Kim et al. (2015) reproduced the dynamics of extreme run-up height, which was measured as 31.7 m at Monai Valley as a result of the 1993 Hokkaidō Nansei-Oki tsunami. They used FLOW-3D® to investigate the role of diffraction in the extreme run-up. The results revealed that the shape and layout of the coast line as well as the diffraction of tsunami waves by the two islands located in front of the valley had led tsunami waves to converge, leading to the extreme run-up height. Wu et al. (2016) focused on the generation of solitary waves in a wave flume using a piston-type wavemaker. They made experimental observations to evaluate the stability of the generated solitary waves, defining stability as the generated solitary wave travelling a longer distance without an apparent decay. They carried out numerical simulations using FLOW-3D® to support their experimental results.

At this point, other studies on NTHMP tsunami benchmark problems should also be referred to. A number of researchers have performed similar work on NTHMP benchmark problems with their particular models. A case in point is Apotsos et al. (2011), Nicolsky et al. (2011), Tolkova (2014), Horrillo et al. (2015), Lynett et al. (2017) and Macias et al. (2017). Furthermore, the proceedings of NTHMP (2012) and NTHMP (2015) are also good resources that describe the detailed work of other authors.

3. Theoretical Framework of the Numerical Models

NAMI DANCE is an in-house tsunami numerical model developed by the Ocean Engineering Research Center (OERC) of Middle East Technical University (METU), Turkey and Laboratory of Special Research Bureau for Automation of Marine Research, Far Eastern Branch of Russian Academy of Sciences, Russia. NAMI DANCE gained importance and popularity in a short span of time in the area of coastal modeling and was recommended by the United Nations Educational, Scientific and Cultural Organization (UNESCO) to be used in the simulations of possible tsunami scenarios.

NAMI DANCE solves nonlinear forms of non-dispersive two-dimensional depth-averaged shallow water (2D-NSW) equations. Navier–Stokes equations, which are the governing equations in the prediction of fluid motion, form the basis of long wave/shallow water theory (Shuto 1991), in which the vertical motion of water particles is not taken into consideration because of its negligible effect on the pressure distribution. Therefore, the pressure distribution is accepted to be hydrostatic. Given this approximation, necessary dynamic and kinematic conditions are used and the bottom friction terms (for nonlinearity) are included so that the fundamental equations of NAMI DANCE are obtained and are discretized by means of the staggered leapfrog scheme. The equations of continuity (conservation of mass) and momentum are given by the following set of equations (Velioglu et al. 2016):

$$\frac{\partial \eta}{\partial t}+\frac{\partial M}{\partial x}+\frac{\partial N}{\partial y}=0, \tag{1}$$

$$\frac{\partial M}{\partial t}+\frac{\partial}{\partial x}\left(\frac{M^2}{D}\right)+\frac{\partial}{\partial y}\left(\frac{MN}{D}\right)+gD\frac{\partial \eta}{\partial x}+\frac{n^2}{D^{7/3}}M\sqrt{M^2+N^2}=0, \tag{2}$$

$$\frac{\partial N}{\partial t}+\frac{\partial}{\partial x}\left(\frac{MN}{D}\right)+\frac{\partial}{\partial y}\left(\frac{N^2}{D}\right)+gD\frac{\partial \eta}{\partial y}+\frac{n^2}{D^{7/3}}N\sqrt{M^2+N^2}=0, \tag{3}$$

where x and y are the horizontal x and y axes in the Cartesian coordinate system, z is the vertical axis in the Cartesian coordinate system, t is the time, g is the gravitational acceleration, n is the Manning's roughness coefficient, η is the free surface elevation, M and N are the discharge fluxes on horizontal plane along x and y axes, τ_{ij} is the normal or tangential shear stress in i direction on the j normal plane, and D is the total water depth given by $D = \eta + d$.

FLOW-3D®, which was developed and commercialized by Flow Sciences, Inc., is a general purpose CFD software. Flow Sciences, Inc. was founded in Santa Fe, New Mexico U.S. in 1980 by Dr. C. W. (Tony) Hirt. Dr. Hirt was one of the scientists who pioneered in the design of the Volume-of-Fluid (VOF) technique, which was first reported in Nichols and Hirt (1975), and more completely in Hirt and Nichols (1981). FLOW-3D® is a direct descendant of this work. The model solves the fully three-dimensional mass continuity and momentum (Navier–Stokes) equations via the finite difference or finite volume method in the Cartesian coordinate system (x, y, z) to describe fluid motion. The general mass continuity equation is given by (Flow Science 2002):

$$V_{\mathrm{F}}\frac{\partial \rho}{\partial t}+\frac{\partial}{\partial x}(\rho u A_x)+\mathcal{R}\frac{\partial}{\partial y}(\rho v A_y)+\frac{\partial}{\partial z}(\rho w A_z)+\xi\frac{\rho u A_x}{x}=R_{\mathrm{DIF}}+R_{\mathrm{SOR}}, \tag{4}$$

where ρ is the fluid density, u, v and w are the water particle velocities in x, y and z directions, V_{F} is the fractional volume, A_x, A_y, A_z are the fractional areas in x, y, z directions, $\mathcal{R}$ and ξ are the coordinate conversion coefficients, R_{DIF} is a turbulent diffusion term and R_{SOR} is a mass source. The equations of motion are the Navier–Stokes equations with some additional terms (Flow Science 2002):

$$\frac{\partial u}{\partial t}+\frac{1}{V_{\mathrm{F}}}\left\{uA_x\frac{\partial u}{\partial x}+vA_y\mathcal{R}\frac{\partial u}{\partial y}+wA_z\frac{\partial u}{\partial z}\right\}-\xi\frac{A_y v^2}{xV_{\mathrm{F}}}=-\frac{1}{\rho}\frac{\partial P}{\partial x}+G_x+f_x-b_x-\frac{R_{\mathrm{SOR}}}{\rho V_{\mathrm{F}}}(u-u_w-\Delta u_s), \tag{5}$$

$$\frac{\partial v}{\partial t}+\frac{1}{V_{\mathrm{F}}}\left\{uA_x\frac{\partial v}{\partial x}+vA_y\mathcal{R}\frac{\partial v}{\partial y}+wA_z\frac{\partial v}{\partial z}\right\}+\xi\frac{A_y uv}{xV_{\mathrm{F}}}=-\frac{1}{\rho}\left(\mathcal{R}\frac{\partial P}{\partial y}\right)+G_y+\mathrm{f}_y-b_y-\frac{R_{SOR}}{\rho V_{\mathrm{F}}}(v-v_w-\Delta v_s), \tag{6}$$

$$\frac{\partial w}{\partial t}+\frac{1}{V_F}\left\{uA_x\frac{\partial w}{\partial x}+vA_y\mathcal{R}\frac{\partial w}{\partial y}+wA_z\frac{\partial w}{\partial z}\right\}=-\frac{1}{\rho}\frac{\partial P}{\partial z}+G_z+\mathrm{f}_z-b_z-\frac{R_{SOR}}{\rho V_F}(w-w_w-\Delta w_s), \tag{7}$$

where P is the pressure, G_x, G_y and G_z are the body accelerations, f_x, f_y and f_z are the viscous accelerations, b_x, b_y and b_z are the flow losses across porous media, u_{w}, v_{w} and w_{w} represent the velocity of the source component and u_{s}, v_{s} and w_{s} represent the fluid velocities at the surface of the source relative to the source itself.

FLOW-3D® uses the quadratic law for bottom shear stress calculation for turbulent flow (Flow Science 2002):

$$\tau_b=\rho C_D|\vec{u}|\vec{u}, \tag{8}$$

where $\vec{u}$ is the depth-averaged velocity and C_{D} is the drag coefficient.

The default value of C_{D} is 0.0026, which is a typical value in hydraulic and coastal engineering applications. However, it is also possible to define component specific C_{D} or let the code to calculate C_{D} for a component based on its surface roughness. For the latter option, the surface roughness, flow depth, and drag coefficient are related by (Flow Science 2002):

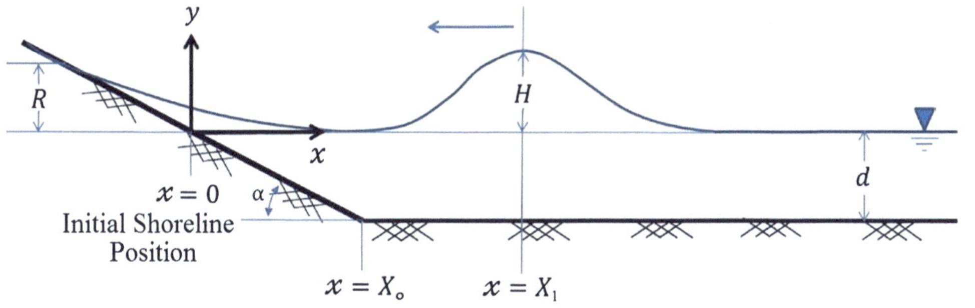

Figure 1
A sketch of the 1:19.85 sloping beach connected to a constant depth region (distorted scale) (Horrillo et al. 2015)

$$C_D = \left[\frac{\kappa}{0.71 + ln\left(\frac{k_s}{30d}\right)}\right]^2, \tag{9}$$

where κ is the von Kármán constant, which is equal to 0.40 and k_s is the Nikuradse equivalent surface roughness, which can be related to Manning's roughness coefficient by (Yen 1991):

$$k_s = \left(\frac{n}{0.0389}\right)^6. \tag{10}$$

4. Benchmarking

Relating the numerical results and benchmark data within a statistical framework is a reasonable way to assess the capability of numerical models. National Oceanic and Atmospheric Administration (NOAA) introduces different types of statistical errors to determine the correlation between the calculated and observed values. The most commonly used statistical errors are: (1) the normalized root mean square error (NRMSE) and (2) the error of the maximum value (MAX). NRMSE is a tool used to assess the overall model performance and it is valid when applied within a space segment or time period to all observed data points.

$$\text{NRMSE} = \frac{1}{f(x_i)_{\max} - f(x_i)_{\min}} \sqrt{\frac{\sum(f(x_i) - y_i)^2}{n}}, \tag{11}$$

where $f(x_i)$ and y_i represent the observed and predicted data, respectively, and the difference between the maximum and minimum values of the observed data is used for normalization.

MAX error, which is a relative error based on the maximum magnitude of the observed values, is employed to quantify the predictive accuracy of a numerical model for the maximum value regardless of time or location:

$$\text{MAX} = \frac{|f(x_i)_{\max} - y_{i\max}|}{f(x_i)_{\max}}. \tag{12}$$

The most recent allowable error limits can be found in the NOAA Technical Memorandum OAR-PMEL-135 standard (Synolakis et al. 2007).

This section contains the NAMI DANCE and FLOW-3D® results for each of the six benchmark problems that are proposed in the scope of NTHMP Tsunami Inundation activities. Detailed descriptions of all benchmarks, as well as topography data when required and laboratory or field data for comparison when applicable, can be found in the NCTR repository, https://nctr.pmel.noaa.gov/benchmark/, and in the repository of University of Southern California (USC), http://coastal.usc.edu/currents_workshop/. For the sake of completeness, a brief description of each benchmark problem is provided. The computational domains, input parameters and applied boundary conditions are identical in both models for each BMP.

4.1. Benchmark Problem 1

In this section, the numerical results for solitary wave shoaling on a plane beach having an angle of $\alpha = $ arccot (19.85) (Fig. 1) are compared with the

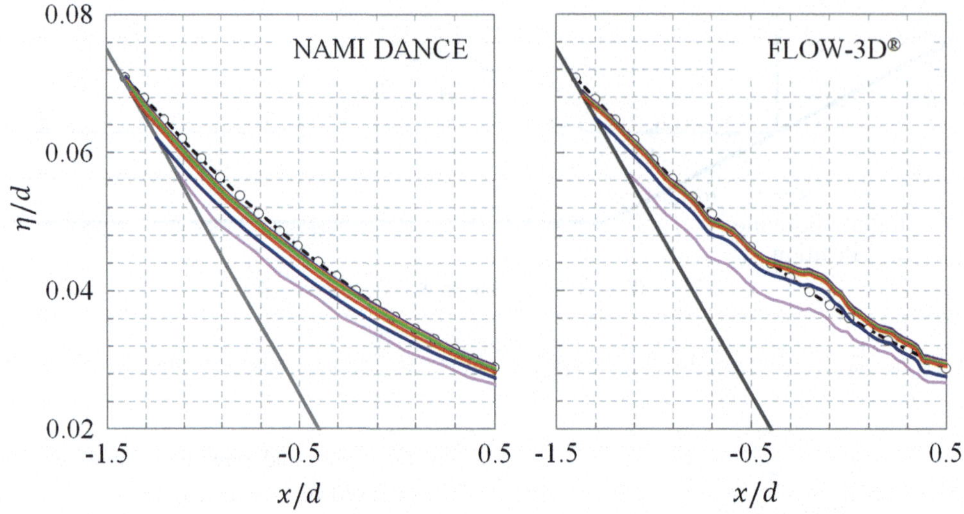

Figure 2
Non-dimensional free surface profile for frictionless bottom and $H/d = 0.0185$ at $t = 60\sqrt{d/g}$—the time of maximum run-up. The circles and dashed black line represent the analytical solution; the purple line: $\Delta x = 0.125$; the orange line: $\Delta x = 0.25$; the green line: $\Delta x = 0.5$; the red line: $\Delta x = 1.0$; the blue line: $\Delta x = 2.0$; the pink line: $\Delta x = 5.0$

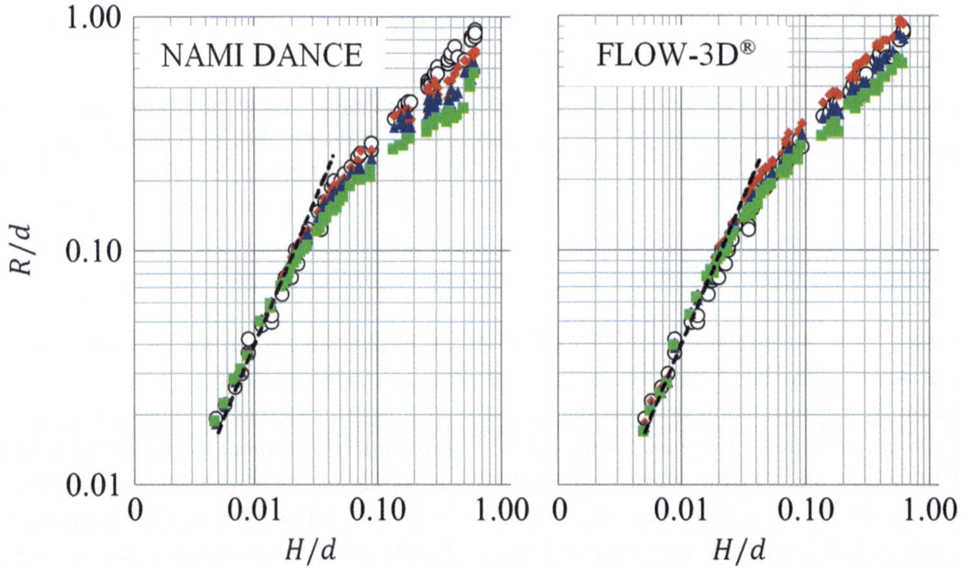

Figure 3
BMP 1 friction analysis. The dashed black line and circles represent the analytical solution and laboratory data (Synolakis 1986), the red markers represent the results when $n = 0$; the blue markers represent the results when $n = 0.01$; the green markers represent the results when $n = 0.03$. $\Delta x = 1.0$

analytic solution for frictionless bottom based on the shallow water equations. Moreover, the maximum run-up of solitary waves, R, on the same plane beach is compared with the laboratory data and the solution of the following expression, derived by Synolakis (1986):

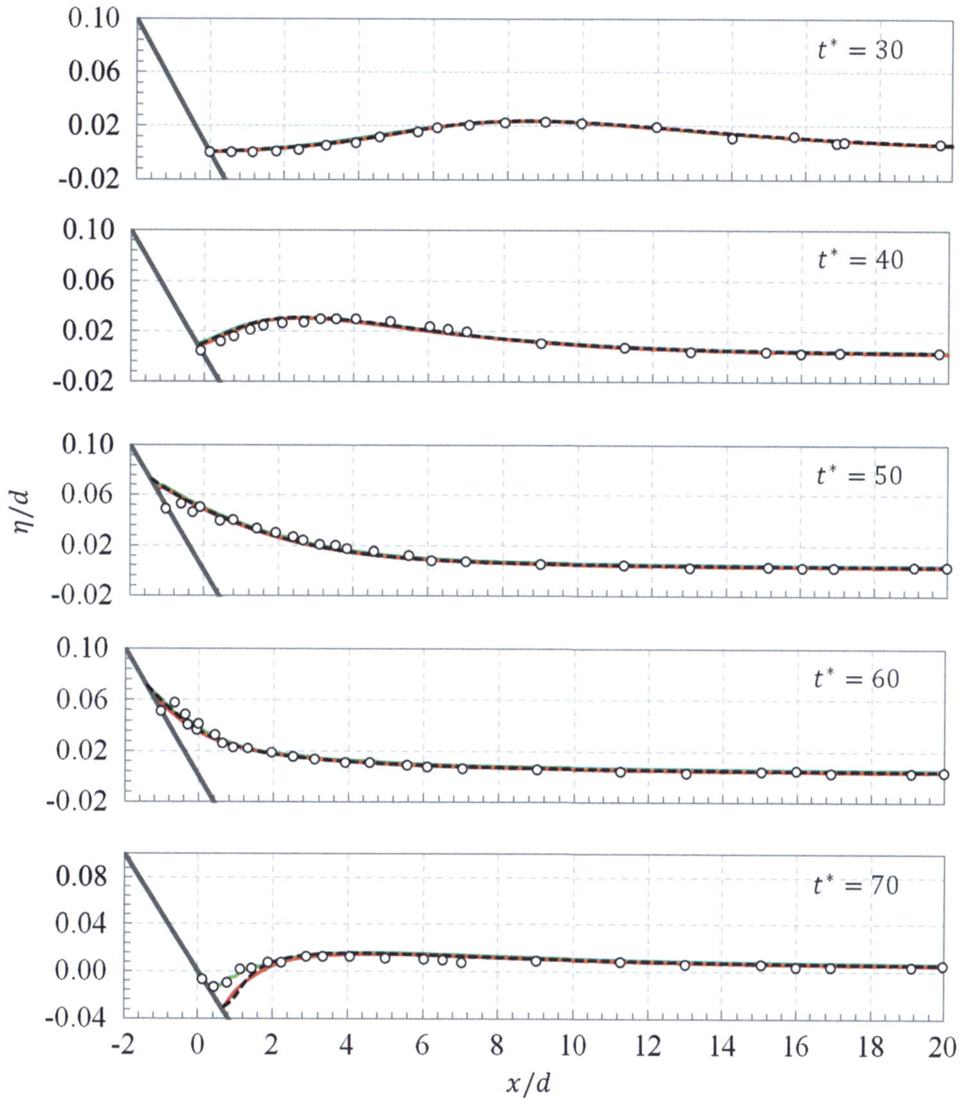

Figure 4
Non-dimensional free surface profiles for frictionless bottom and $H/d = 0.0185$. The circles represent the experimental data; the dashed black line represents the analytical solution; the red line represents the NAMI DANCE results; the green line represents the FLOW-3D® results. $\Delta x = 1.0$

$$\frac{R}{d} = 2.831\sqrt{\cot\alpha}\left(\frac{H}{d}\right)^{5/4}, \qquad (13)$$

where d is the constant offshore depth and H is the solitary wave height.

All variables used in this problem are non-dimensional and the computations are performed accordingly. The computational domain along x is between $x = 10$ and $x = 70$. The slope is located on the left side. Wall boundary condition is applied for both models, where the normal velocity is taken as zero and the tangential velocity is set to any value for a no-slip type of wall. The incident wave at time $t = 0$ is given by (Synolakis 1986)

$$\eta^*(x,0) = \frac{H}{d}\mathrm{sech}^2\left[\sqrt{\frac{3H}{4d}}(x - X_1)\right], \qquad (14)$$

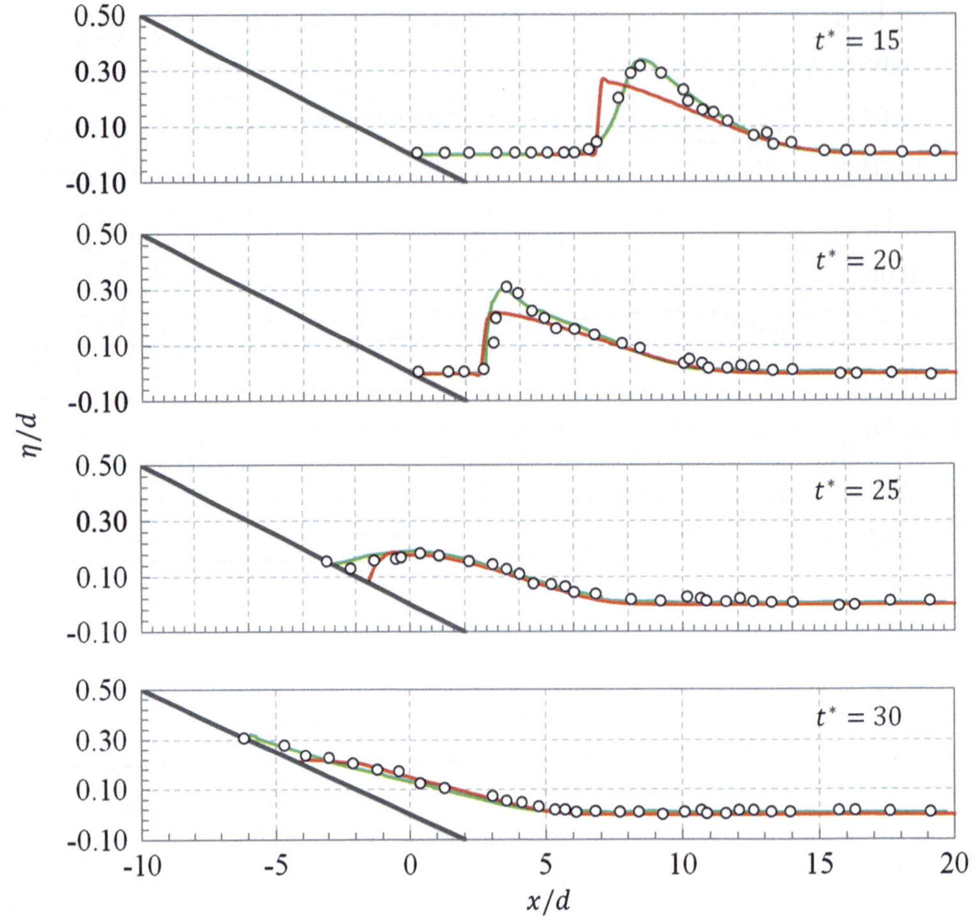

Figure 5
Non-dimensional free surface profiles for $H/d = 0.3$. The circles represent the experimental data; the red line represents the NAMI DANCE results; the green line represents the FLOW-3D® results. $\Delta x = 1.0$ and $n = 0.01$

where x is the dimensionless horizontal coordinate and increases seaward and X_1 is the dimensionless x location where the offshore height of a solitary wave is defined. In addition, $x = 0$ and $x = X_0 = d\cot(\alpha)$ define the positions of the initial shoreline and the toe of the beach, respectively. The non-dimensional effective wavelength, L, is estimated as (Synolakis 1986)

$$L = \frac{2}{\sqrt{\frac{3H}{4d}}}\cosh^{-1}\left(\sqrt{\frac{1}{0.05}}\right). \qquad (15)$$

The benchmark data were obtained from Synolakis (1986). Frictionless bottom was used for comparison of model results with the analytical solution. On the other hand, for the maximum run-up comparison, a friction analysis was made. The water surface elevation time series used at the upstream end of the computational domain (i.e., minimum x boundary) was located $L/2$ distance away from the toe of the slope.

The non-dimensional free surface profile at $t^* = 60$ was predicted using grid resolutions of $\Delta x = 0.125$, $\Delta x = 0.25$, $\Delta x = 0.5$, $\Delta x = 1.0$, $\Delta x = 2.0$ and $\Delta x = 5.0$ (Fig. 2). The results of the grid analysis show that the finer the grid size, the closer are the predicted results to the exact solution. No considerable change was observed in model results for resolutions finer than $\Delta x = 1.0$. The effect of the vertical grid size, Δz, on the FLOW-3D® results was

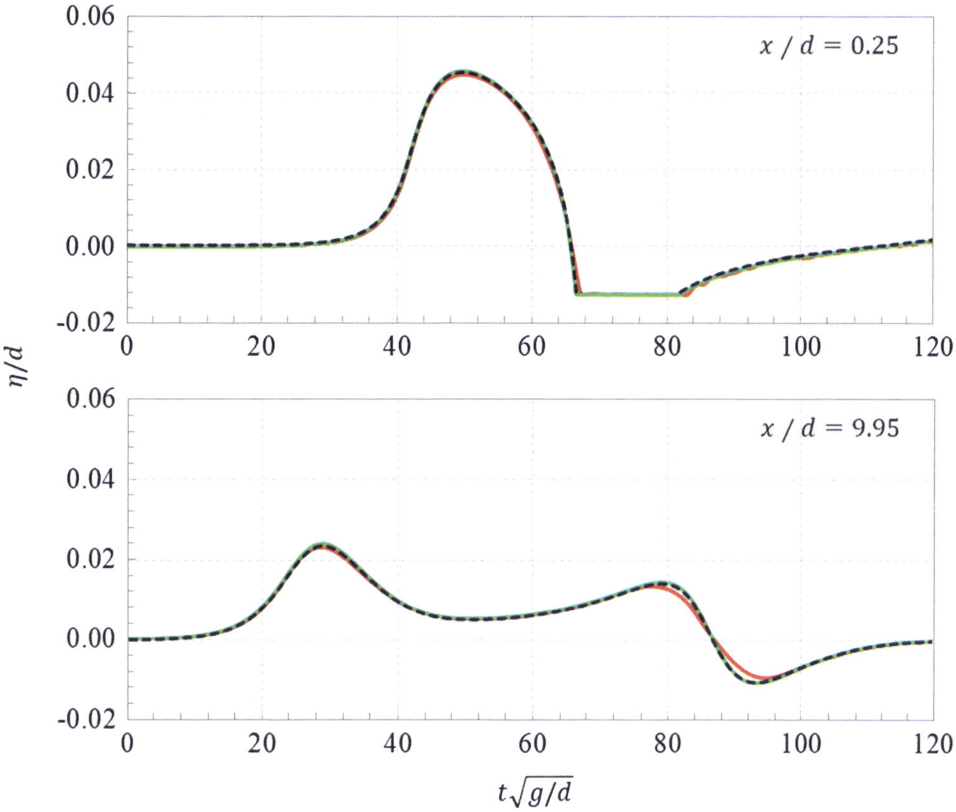

Figure 6
Water level time series at $x/d = 0.25$ (upper panel) and $x/d = 9.95$ (lower panel). $\Delta x = 1.0$ and $n = 0$

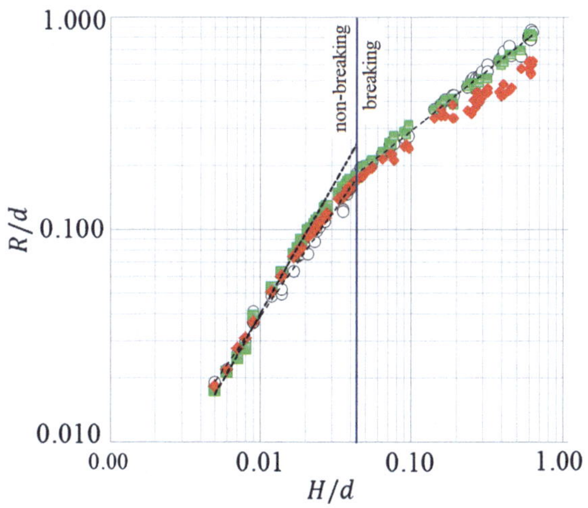

Figure 7
Maximum run-up values on 1:19.85 slope when $n = 0.01$: The dashed black line represents the Synolakis's (1986) run-up law; the circles represent the Synolakis's (1986) experimental data; the red markers represent the NAMI DANCE results; the green markers represent the FLOW-3D® results. $\Delta x = 1.0$ and $n = 0.01$

also analyzed. The predictions got closer to the analytical solution for finer vertical grid sizes with a constant Δx. No considerable change was observed for vertical resolutions finer than $\Delta z = 1.0$. Thus, a uniform grid size of 1.0 unit was used in the rest of the simulations. Variable time stepping based on a CFL condition was used for both models.

The friction analysis of BMP 1 reveals that the predicted maximum run-up values are not dependent on n for small amplitude waves, $H/d \leq 0.01$. The computed run-up heights are slightly affected by the bottom friction when $0.01 < H/d < 0.044$. In the case of breaking solitary waves, $H/d \geq 0.044$, the maximum run-up strongly depends on n (Fig. 3). Similar results were reported by Lynett et al. (2002). For large wave heights, it is obvious that an appropriate choice of n improves the FLOW-3D® results, whereas the NAMI DANCE results, which have already been deteriorated by the excessive dissipation

Table 1

NAMI DANCE and FLOW-3D® model free surface profile errors with respect to the analytical solution for frictionless bottom, where $H/d = 0.0185$

$t\sqrt{g/d}$	NAMI DANCE		FLOW-3D®	
	% NRMSE	% MAX	% NRMSE	% MAX
30	0.3	0.2	2.3	0.3
40	1.2	0.8	1.7	0.5
50	0.8	5.0	1.5	4.0
60	0.8	1.7	1.7	0.1
70	2.0	4.2	6.5	2.3

Table 2

NAMI DANCE and FLOW-3D® model free surface profile errors with respect to laboratory data for $n = 0.01$, where $H/d = 0.3$

$t\sqrt{g/d}$	NAMI DANCE		FLOW-3D®	
	% NRMSE	% MAX	% NRMSE	% MAX
15	11.0	19.9	1.5	5.0
20	9.5	20.0	5.0	3.5
25	6.3	4.5	3.5	1.0
30	4.0	17.0	4.4	4.2

Table 3

NAMI DANCE and FLOW-3D® model water level dynamics errors with respect to the analytical solution for frictionless bottom at $x/d = 0.25$ and $x/d = 9.95$, where $H/d = 0.0185$

x/d	NAMI DANCE		FLOW-3D®	
	% NRMSE	% MAX	% NRMSE	% MAX
0.25	0.4	1.2	0.5	1.1
9.95	3.3	5.4	1.1	0.6

during wave breaking, got worse with the addition of the friction term.

To accomplish BMP 1:

1. Numerically computed water level profiles at $t = 30\sqrt{d/g}$, $t = 40\sqrt{d/g}$, $t = 50\sqrt{d/g}$, $t = 60\sqrt{d/g}$ and $t = 70\sqrt{d/g}$ were compared with both analytical solutions and laboratory data for frictionless bottom, where $H/d = 0.0185$ (Fig. 4).
2. Numerically computed water level profiles at $t = 15\sqrt{d/g}$, $t = 20\sqrt{d/g}$, $t = 25\sqrt{d/g}$ and $t = 30\sqrt{d/g}$ were compared with the laboratory data, where $H/d = 0.3$ and $n = 0.01$ (Fig. 5).
3. Numerically and analytically computed water level dynamics at locations $x/d = 0.25$ and $x/d = 9.95$ were compared for frictionless bottom (Fig. 6).
4. Maximum run-up of the solitary wave was numerically computed and R/d vs. H/d graph was constructed for $n = 0.01$ (Fig. 7).

The existence of strong wave breaking hinders the agreement of the predicted values with the laboratory measurements for the two-dimensional depth-averaged model, NAMI DANCE. On the other hand, the predicted wave profiles obtained via the three-

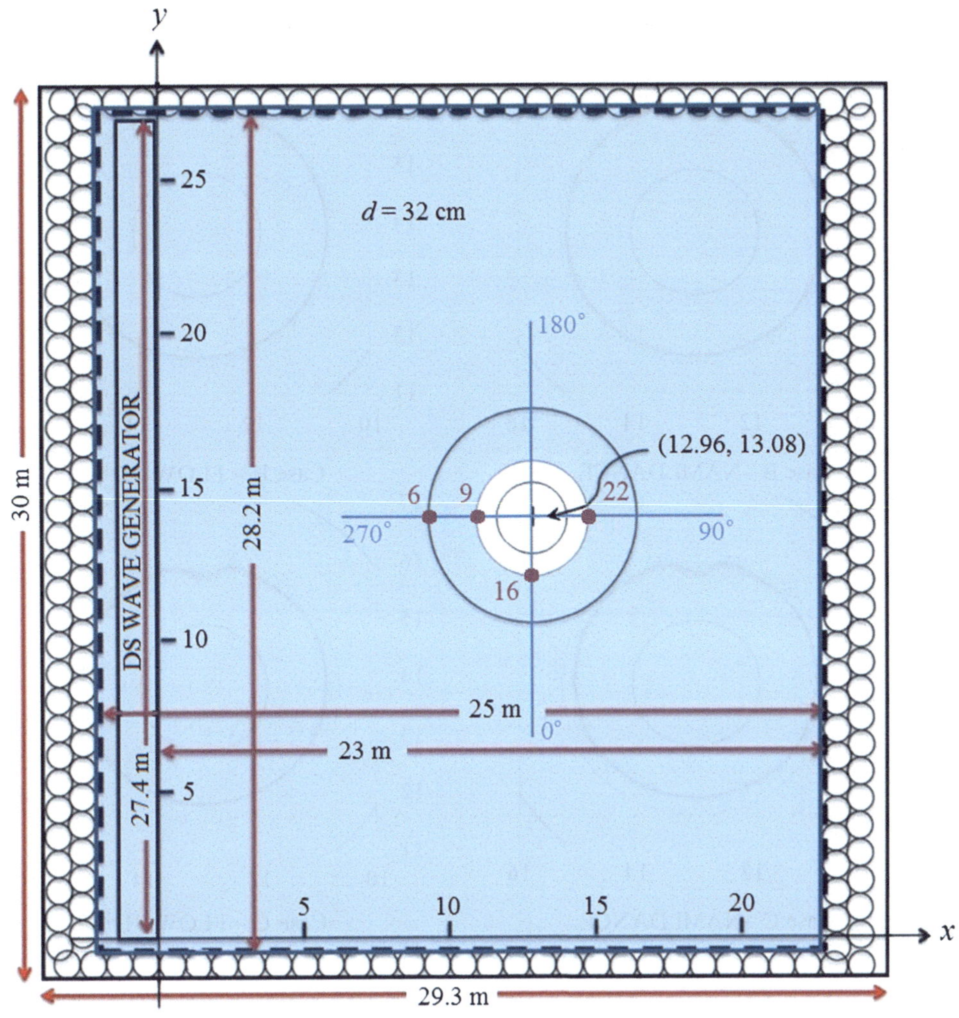

Figure 8
Basin geometry, coordinate system and location of gauges (not to scale) (NTHMP 2012)

Table 4

BMP 2 gauge locations

Gauge ID	x (m)	y (m)	z (m)	Transect
6	9.36	13.80	31.7	270^0
9	10.36	13.80	8.2	270^0
16	12.96	11.22	7.9	0^0
22	15.56	13.80	8.3	90^0

dimensional model, FLOW-3D®, are in good agreement with the laboratory data (Tables 1, 2).

When the analytical and numerical solutions at $x/d = 0.25$ and $x/d = 9.95$ are compared for $H/d = 0.0185$ case, it is found that there is good agreement between the two solutions during the propagation and reflection of the wave (Table 3).

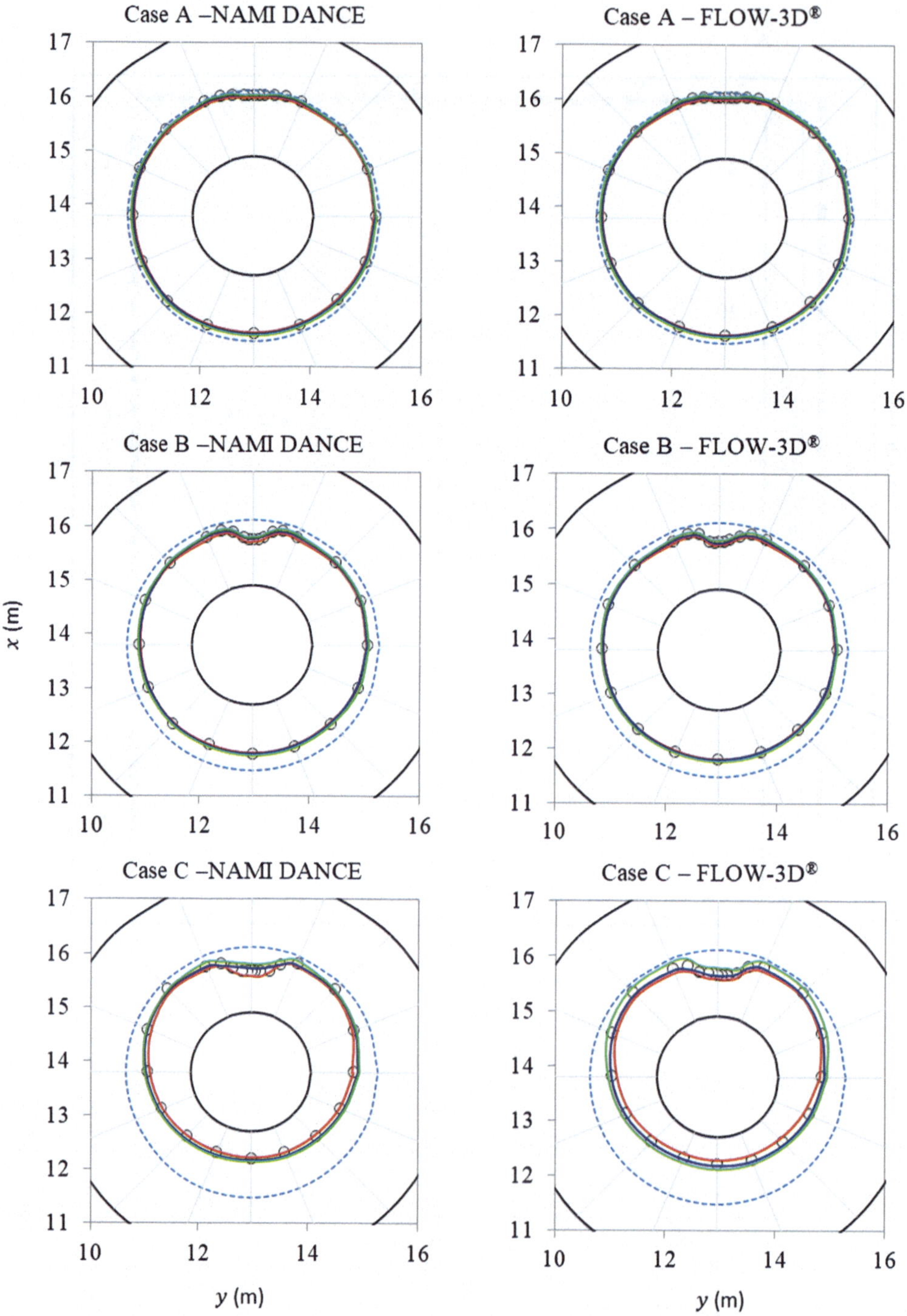

Figure 9
Run-up predictions according to different n values. The dashed blue line represents the initial shoreline; the red line represents the results when $n = 0$; the blue line represents the results when $n = 0.01$; the green line represents the results when $n = 0.03$, $\Delta x = \Delta z = 0.1$ m

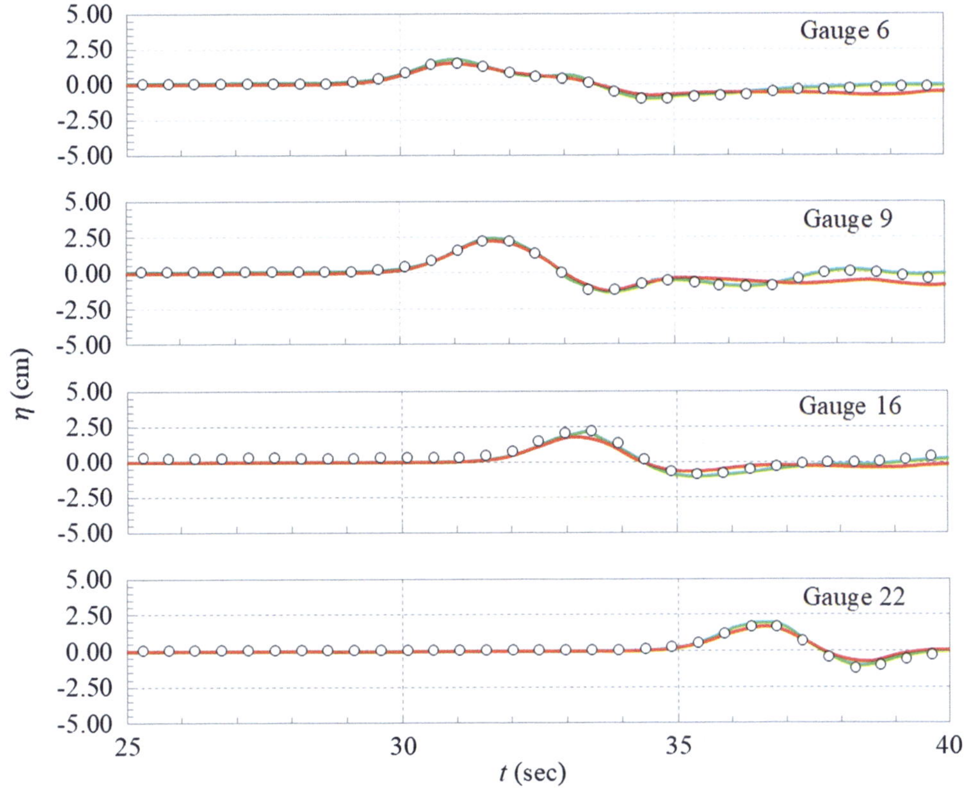

Figure 10
Comparison between the computed and measured water level at Gauges 6, 9, 16, and 22 for Case A ($H/d = 0.045$). The circles represent the laboratory measurements; the red line represents the NAMI DANCE results; the green line represents the FLOW-3D® results

The inundation of the 1:19.85 sloped beach by solitary waves with different H/d ratios is simulated in several series of numerical model runs. In each series, the bottom friction is parameterized by Manning's roughness coefficient, $n = 0.01$ (Fig. 7).

NAMI DANCE predicted the maximum run-up values with an NRMSE of 13%, whereas FLOW-3D® results have an NRMSE of 4%. Due to the fact that the experimental parameters violate the shallow water wave assumption, the predicted wavefront is smaller and steeper than that of the measured wave amplitude. Consequently, NAMI DANCE underestimates the run-up distances of large amplitude solitary waves by approximately 20%.

4.2. *Benchmark Problem 2*

The eastern region of Flores Island was struck by an earthquake with a magnitude of 7.8 M_w on December 12, 1992, which was followed by 25-m-high tsunami waves. Babi Island, which is about 5 km offshore from Flores, was also severely affected by the tsunamis. The fact that an extremely high inundation was observed in the back (lee) side of island suggested that there was a need for a better understanding of the important physical parameters involved in a three-dimensional tsunami run-up. Therefore, a study was initiated in 1992 by the National Science Foundation to analyze the parameters involved in three-dimensional tsunami run-up (Briggs et al. 1995) (Fig. 8) (Table 4).

To perform BMP 2, it is necessary to demonstrate that the modeled wavefronts split in front of the island and collide behind it producing an extreme localized run-up. Moreover, the computed water levels at the four gauges around the island; namely, Gauges 6—in the front of the island at the toe, 9, 16 and 22—closest to the shoreline located at the 0^0, 90^0,

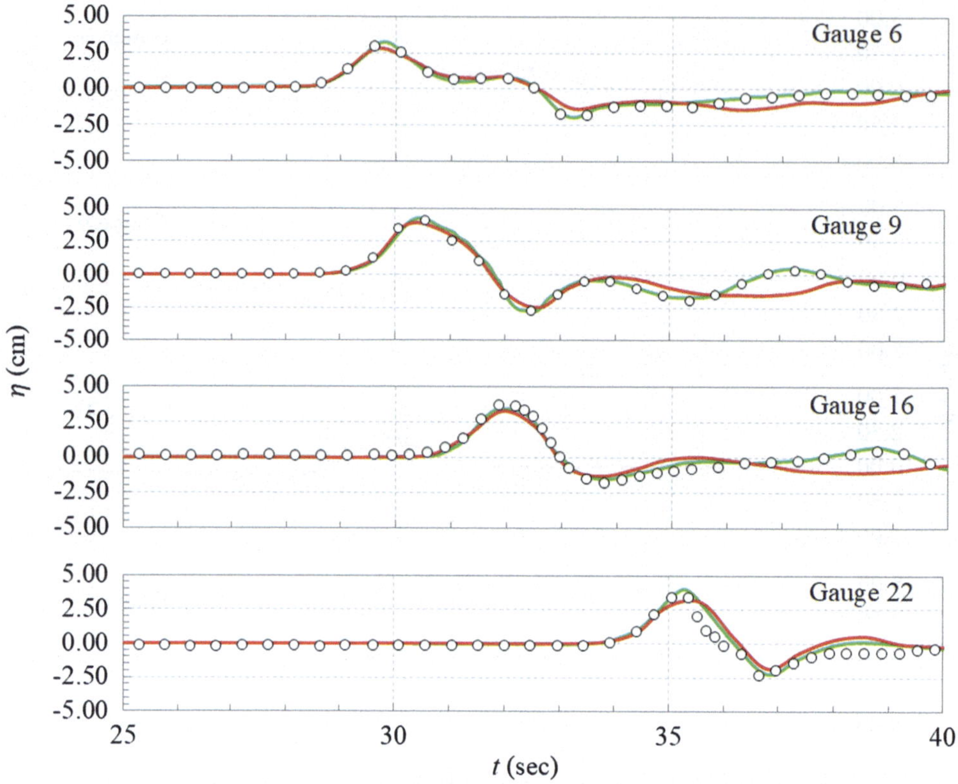

Figure 11
Comparison between the computed and measured water level at Gauges 6, 9, 16, and 22 for Case B ($H/d = 0.091$). The circles represent the laboratory measurements; the red line represents the NAMI DANCE results; the green line represents the FLOW-3D® results

and 180^0 radial lines, are replicated and the angular distribution of run-up is compared with the laboratory data for the various cases of incident waves. Three different cases are selected from the laboratory experiments, namely Case A, Case B and Case C, where the initial H/d ratios are equal to 0.045, 0.091 and 0.181 at 32 cm water depth, respectively.

The computational domain is 23.5 m along x direction and 25 m along y direction. The maximum x boundary has a rigid wall boundary condition, which defines the downstream end of the wave tank. The minimum and maximum y boundaries are also defined as rigid walls to prevent outflow from the computational domain. The initial wave is a solitary wave whose profile and length at time $t = 0$ are given by Eqs. (14) and (15), respectively. It is extended to two dimensions with a constant wave amplitude and zero velocity in y-direction.

The convergence of the numerical models is tested by comparing the predicted run-up values at Gauge 9, for Cases A, B and C, respectively. The match between the predicted data and laboratory measurements increases as the computational grid resolution is decreased from 1.0 to 0.01 m. The most obvious manifestation of this convergence is the improved form of the first wave. No considerable change is observed for resolutions finer than $\Delta x =$ 0.1 m. The predicted run-up values differ by less than 0.5% for $\Delta x \leq 0.1$ m. When different vertical grid sizes are used, no improvement was observed for $\Delta z \leq 0.1$ m. Thus, a uniform grid size of 0.1 m is used in the simulations. Variable time stepping based on a CFL condition is used for both models.

The effect of friction varies spatially over the computational domain since the friction term is a function of water depth. Introduction of the friction coefficient improves the numerical predictions. It is

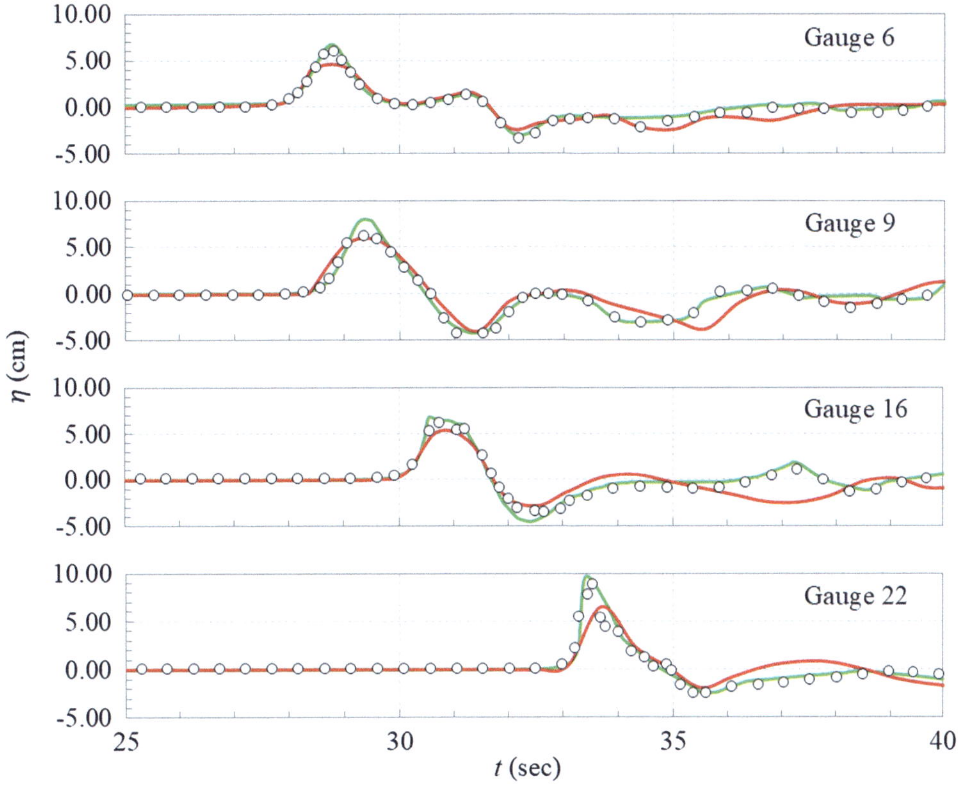

Figure 12
Comparison between the computed and measured water level at Gauges 6, 9, 16, and 22 for Case C (H/d = 0.181). The circles represent the laboratory measurements; the red line represents the NAMI DANCE results; the green line represents the FLOW-3D® results

Table 5

NAMI DANCE and FLOW-3D® model water level time series error with respect to laboratory data for Case A ($H/d = 0.045$)

Gauge	NAMI DANCE		FLOW-3D®	
	% NRMSE	% MAX	% NRMSE	% MAX
6	7.3	7.1	3.4	7.2
9	7.9	7.8	4.1	5.8
16	8.3	10.0	3.5	5.6
22	7.0	7.3	4.7	7.1

seen that, around the perimeter of the conical island, the predicted and measured run-up values match when $n = 0.01$, including the extreme run-up behind the conical island. Accordingly, the predicted inundation level on the frictionless surface is higher than the measured one. These situations are most spotted in Cases B and C, where $H/d = 0.091$ and $H/d = 0.181$, respectively (Fig. 9). In this study, to reflect the real laboratory conditions as much as possible, the run-up predictions are made for a Manning's roughness coefficient $n = 0.01$, which is used for smooth concrete.

The wave transformations around the conical island that are simulated by both models are given in Figs. 10, 11 and 12 along with the statistical errors (Tables 5, 6, 7).

Table 6

NAMI DANCE and FLOW-3D® model water level time series error with respect to laboratory data for Case B ($H/d = 0.091$)

Gauge	NAMI DANCE		FLOW-3D®	
	% NRMSE	% MAX	% NRMSE	% MAX
6	8.2	10.6	3.0	2.2
9	8.7	7.6	3.8	4.2
16	11.5	10.7	5.5	3.3
22	8.7	10.7	6.1	0.9

Table 7

NAMI DANCE and FLOW-3D® model water level time series error with respect to laboratory data for Case C ($H/d = 0.181$)

Gauge	NAMI DANCE		FLOW-3D®	
	% NRMSE	% MAX	% NRMSE	% MAX
6	7.0	9.1	3.9	5.9
9	9.3	7.8	5.3	9. 8
16	12.2	4.3	5.3	8.4
22	10.5	17.9	5.9	6.1

The inundation around the perimeter of the island for the three cases is given in Fig. 13. The predicted inundation levels are fairly uniform around the perimeter of the island for Case A.

Good agreement is seen overall, and in particular, between predicted and measured time series for the first wave. The agreement for the later wave details becomes progressively worse, as multiple reflections and refraction occur at the basin boundaries. In Case C, it is obvious that 3D model shows an appreciable improvement over the 2D one. Case C is an extreme case, in which a distinctive pattern of run-up emerges because of the edge waves that propagate around the island. The run-up on the back side of the island is almost as large as that on the front side. The results reveal that NAMI DANCE predicts the run-up on the back of the island approximately 15% less than the measured value (Table 8).

4.3. Benchmark Problem 3

A 7.7 M_w earthquake occurred west of Hokkaidō, Japan on July 12, 1993 at 13:17 GMT. The damage caused by the earthquake was not very severe; yet, the tsunami destroyed an entire coastal community on Okushiri Island before it hit the west coast of Hokkaidō. What made this tsunami remarkable was the maximum run-up height of 31.7 m measured at the bottom of the Monai Valley, which opens onto a small pocket beach. Matsuyama and Tanaka (2001) carried out a laboratory study at the Central Research Institute for Electric Power Industry (CRIEPI) to investigate the 31.7 m run-up in Monai zone during the 1993 Hokkaidō Nansei-Oki earthquake tsunami. The experimental setup was a 1:400-scale model of the bathymetry and topography around Monai. The computational domain recommended for the simulation of BMP 3 is a 5.488 m by 3.402 m portion of the wave tank near the shore, which is divided into 0.014 m $\times$ 0.014 m grid cells (Fig. 14). The recommended time step size, Δt, is 0.05 s (NOAA Center for Tsunami Research, n.d). It is worth to mention that variable time stepping based on a CFL condition was also used for both models. However, no significant change was observed in the numerical predictions.

The free surface elevation time series of the incident wave at the water depth 13.5 cm is known. The initial wave, which is a LDN wave, has a leading-depression height of 2.5 mm and a crest of

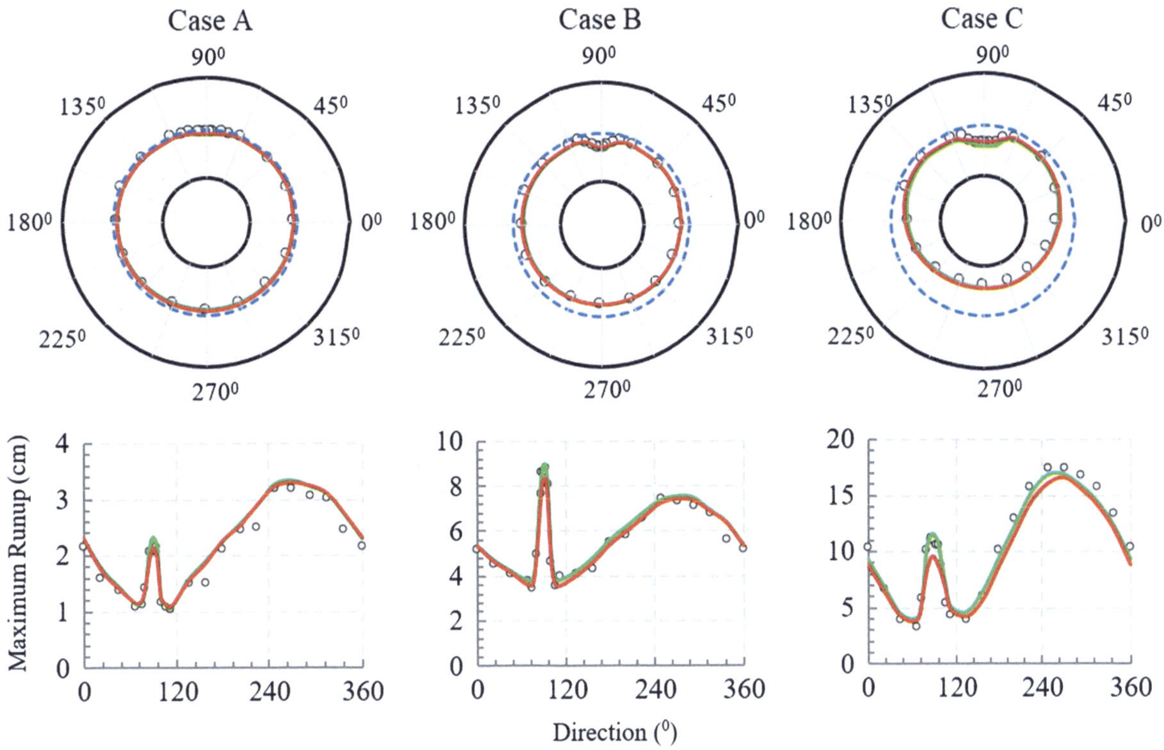

Figure 13
Inundation around the perimeter of the conical island for Cases A, B and C. The circles represent the laboratory measurements; the dashed blue line represents the initial shoreline; the red line represents the NAMI DANCE results; the green line represents the FLOW-3D® results

Table 8

NAMI DANCE and FLOW-3D® model run-up errors with respect to laboratory data

Gauge	NAMI DANCE		FLOW-3D®	
	% NRMSE	% MAX	% NRMSE	% MAX
Case A	7.2	9.3	4.1	1.4
Case B	11.9	6.0	5.0	1.0
Case C	15.3	13.4	8.2	1.1

1.6 cm, which is produced in the model using FFT analysis and the WK TIME SERIES wavemaker option (NTHMP 2012) (Fig. 15). The input data only covers 22.5 s and it is defined as the left boundary condition.

The primary theme of this benchmark problem is to model the propagation of the incident and reflective wave, to compare the numerical and experimental water level dynamics at Gauges 5, 7 and 9 and to show snapshots of the numerically computed water level at time synchronous with those of the video frames.

The convergence of both models are tested by comparing the run-up measurements along the transects at $y = 1.88$ m—where the maximum run-up was recorded, $y = 2.2062$ m and $y = 2.32$ m for $\Delta x = 0.007$ m, $\Delta x = 0.014$ m, $\Delta x = 0.028$ m and $\Delta x = 0.056$ m. The results reveal that the grid size has a negligible effect on the run-up predictions up to $\Delta x = 0.028$ m and that a much coarser grid size (i.e.,

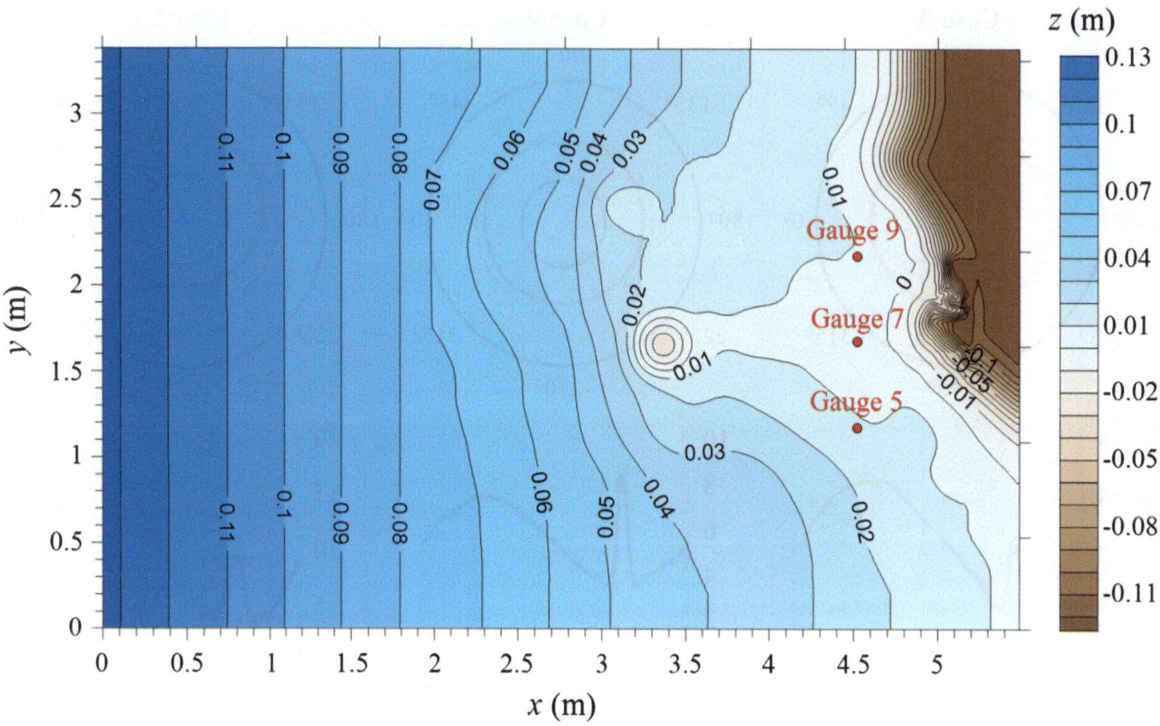

Figure 14
The computational domain—1:400 scale model around Monai

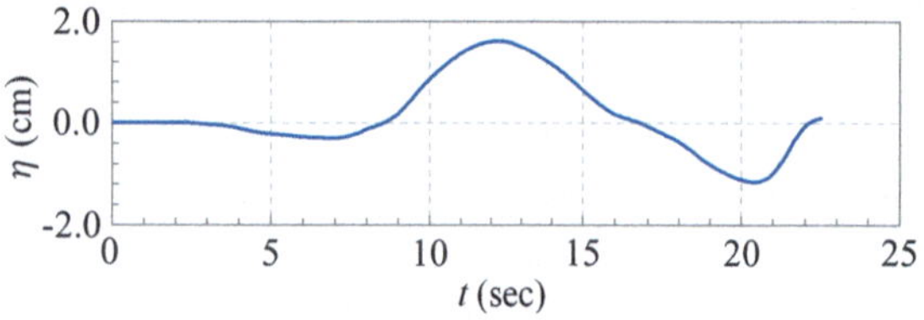

Figure 15
Incident wave profile defined from $t = 0$ to $t = 22.5$ s

$\Delta x \geq 0.028$ m) may lead to unreasonable run-up values. The same values are tested for the vertical grid size and almost identical results are obtained. As a result, the recommended grid size, which is 1.4 cm, is used in the simulations (NTHMP 2012). The time step size is 0.05 s, as recommended.

To assess the effect of friction on the maximum run-up values, different Manning's roughness coefficients are considered: (1) $n = 0$ (i.e., frictionless bottom), (2) $n = 0.012$ (i.e., painted plywood), (3) $n = 0.03$ (i.e., very rough surface). The results are compared along the transects at $y =$ 1.88 m, $y =$ 2.2062 m and $y = 2.32$ m. The analysis results reveal that the effect of varying friction is only felt during the inundation, as expected. The predicted run-up values are slightly larger on a frictionless bottom, whereas when $n = 0.03$, lower run-up values are computed, which are approximately 10–15% less than the laboratory measurements. These results indicate that friction plays an important role in the predicted run-up values and $n = 0.012$ is a good choice for the simulations.

The NAMI DANCE and FLOW-3D® results regarding the wave transformation in the shallow zone, at Gauges 5, 7 and 9, where $(x, y) =$ (4.521, 1.196 m), (4.521, 1.696 m) and (4.521, 2.196 m), are given in Fig. 16 and Table 9.

Figure 17 compares the movie frames that are extracted from the overhead camera used during the laboratory experiments with the results of numerical simulations. The frames focused on the narrow gully where the highest run-up is observed. The actual shoreline location is ambiguous; therefore, the dashed yellow lines on the frames are provided as part of the

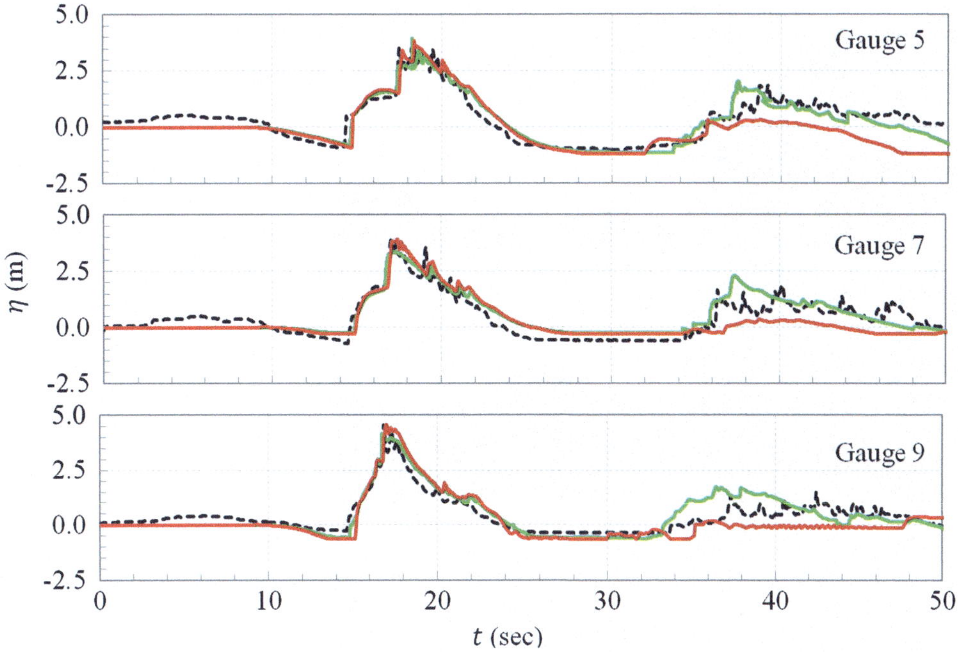

Figure 16
The predicted and measured free surface profiles at: **a** gauge 5; **b** gauge 7; **c** gauge 9. The dashed black line represents the laboratory measurements; the red line represents the NAMI DANCE results; the green line represents the FLOW-3D® results

Table 9

NAMI DANCE and FLOW-3D® model errors for water level time series with respect to laboratory data

Gauge	NAMI DANCE		FLOW-3D®	
	% NRMSE	% MAX	% NRMSE	% MAX
5	7.1	4.2	6.5	2.7
7	7.5	4.3	5.8	1.3
9	7.7	0.7	6.0	0.5

benchmark specification and show the approximate shoreline (NTHMP 2012).

The run-up and inundation comparisons of the NAMI DANCE and FLOW-3D® results are shown in Fig. 18.

Both NAMI DANCE and FLOW-3D® reproduce the overall flow and inundation patterns in fully hydrostatic and non-breaking events considerably well and produce similar results to the previous studies (NTHMP 2012). Both models are able to capture the rapid sequence of run-up and rundown and reproduce the extreme characteristics of the tsunami in the Monai Valley, including the observed extreme run-up value.

4.4. Benchmark Problem 4

BMP 4 is a test case whose experiments were carried out at the O.H. Hinsdale Wave Research Laboratory (HWRL), Oregon State University (OSU) in 2009. This benchmark problem serves to investigate the three-dimensional kinematic properties associated with a breaking solitary wave as it propagates, evolves and breaks over an irregular

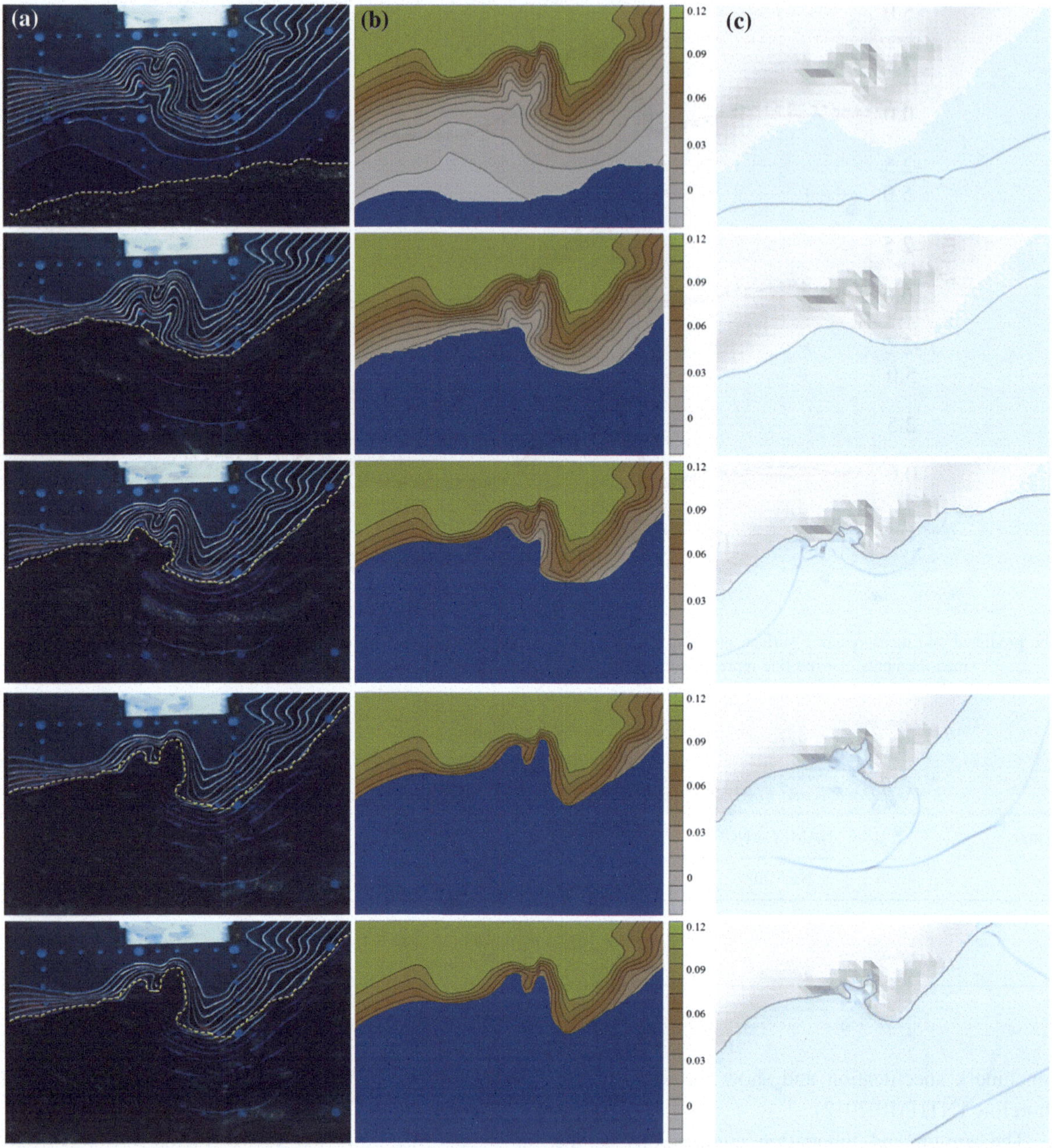

Figure 17
a Frames 10, 25, 40, 55, and 70 from the overhead movie of the laboratory experiment (Liu et al. 2008; NTHMP 2012); **b** snapshots of the NAMI DANCE simulation at the time intervals corresponding to the movie frames; **c** snapshots of the FLOW-3D® simulation at the time intervals corresponding to the movie frames

shallow water bathymetry. It also deals with currents induced by breaking and overtopping of a solitary wave on a circular island and a complex shelf. The NAMI DANCE results for BMP 4 were published previously (Velioglu et al. 2016). The improved results are presented here.

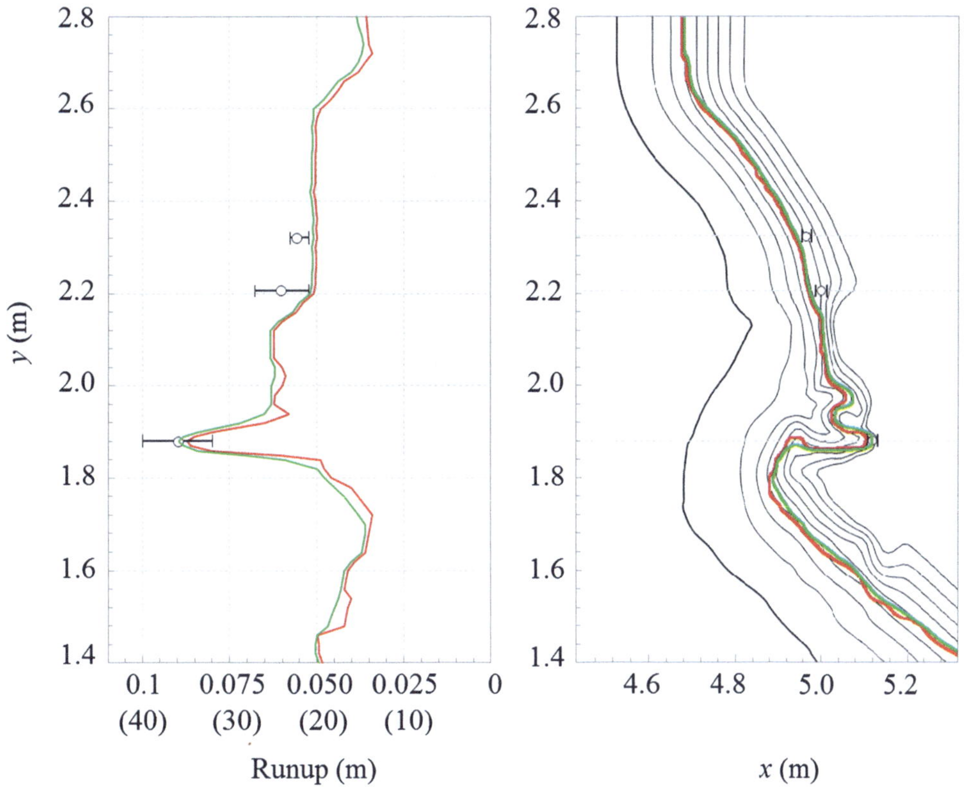

Figure 18
Run-up and inundation comparisons: **a** run-up; **b** inundation. The circles represent the laboratory data of Matsuyama and Tanaka (2001); the red line represents the NAMI DANCE results; the green line represents the FLOW-3D® results

The walls and underlying bathymetry of the basin were made of finished concrete. The bathymetry used in BMP 4 is given in Fig. 19.

The free surface elevation time series recorded at Gauge 1 is defined as the input wave. It propagates over a uniform water depth of 0.78 m before transforming over the shelf system and inundating the backshore area (Fig. 20).

The grid analysis reveals that both the horizontal and vertical numerical resolutions have no evident influence on the free surface profiles, even in the areas with high vorticity concentration (i.e., Gauges 3 and 7) (Fig. 21). A uniform grid size of 0.1 m is used in the rest of the simulations. Variable time stepping based on a CFL condition is used for both models.

The varying friction is mostly effective at the gauges located farther behind the obstacle, where smaller friction produces results closer to the measured data. It is observed that, in general, frictionless bottom or $n < 0.01$ leads to an overestimation of flow speed, and in turn, to a small mismatch in the wave amplitude and arrival times at the wave gauges (Fig. 22).

The NAMI DANCE and FLOW-3D® results regarding the wave transformation in the shallow zone and in the vicinity of the conical island are given in Fig. 23. Moreover, the predicted velocity components are given in Figs. 24, 25 and 26.

The computations are terminated after the first reflection of the wave from the island. FLOW-3D® accurately describes the evolution and transformation of the wave, as well as the complex free surface profiles occurring behind the island. On the other hand, NAMI DANCE underestimates the propagation speed of the wave and somewhat misrepresents the waveform, especially behind the obstacle, on the centerline. The results reveal that the depth-averaged

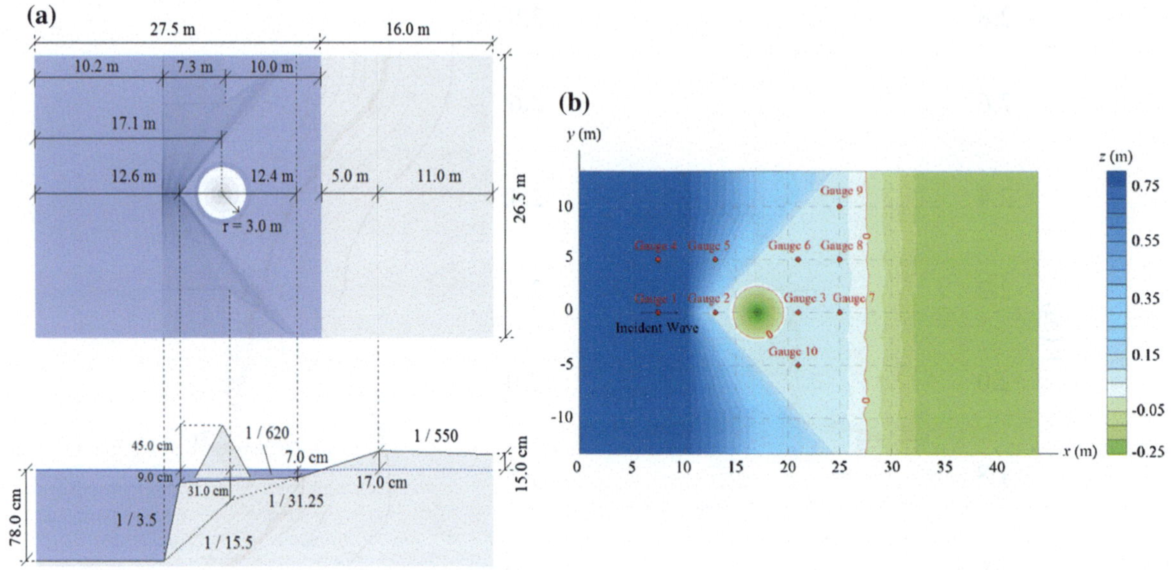

Figure 19
BMP 4: **a** schematic sketch of the experimental setup; **b** gauge locations (Velioglu et al. 2016; Velioglu 2017)

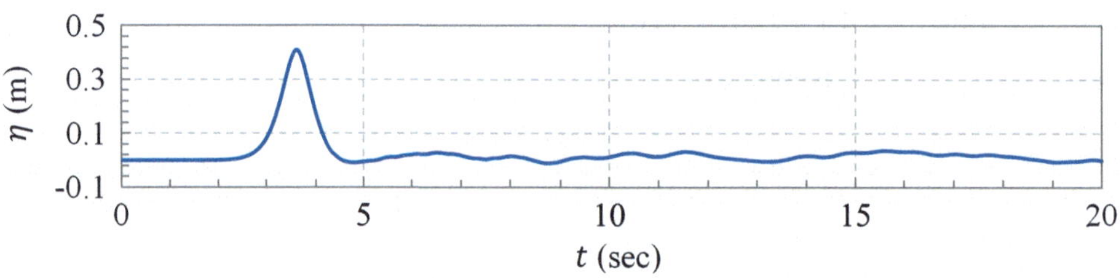

Figure 20
The free surface elevation time series recorded at Gauge 1 (NTHMP 2015)

model predicts the wave amplitude nearly 10% less prior to wave breaking (Tables 10, 11).

NAMI DANCE model results show a noticeable delay in the wave arrival times at Gauges 3 and 7. The wave overtopping the island is properly modeled with FLOW-3D®, hence giving a direct path to the gauges located behind the island. However, NAMI DANCE cannot simulate the wave overtopping properly, leading to a longer path around the island and delaying the arrival time.

The discrepancy between the 3D model result and the measured data regarding the estimate of vertical velocity, *w*, at Gauge 2 may be attributed to an instrumental and/or human error (probably related to the current meter). This can be further investigated further if needed.

4.5. *Benchmark Problem 5*

Several studies have been conducted into macro-roughness and tsunami velocity variation. Cox et al. (2008) carried out physical model tests of Seaside, Oregon, which showed that the macro-roughness reduced the tsunami inundation velocity by 40% (Rueben et al. 2011). BMP 5, which investigates the effect of macro-roughness elements on inundation flow depth, cross-shore velocity and momentum flux, is based on these tests. The model, constructed at

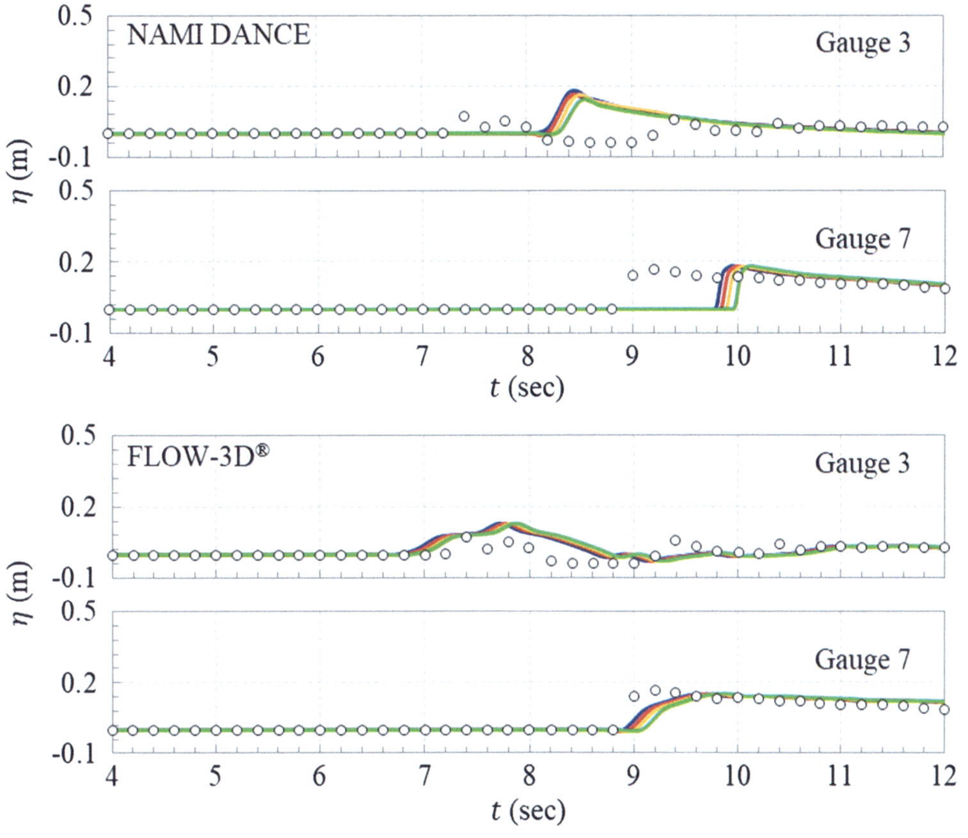

Figure 21
Convergence test at Gauges 3 and 7. The circles represent the laboratory measurements; the blue, red, orange and green lines represent model results for $\Delta x = 0.01$ m, $\Delta x = 0.05$ m, $\Delta x = 0.1$ m and $\Delta x = 0.2$ m; respectively

1:50 scale, was an idealization of the town of Seaside, Oregon and featured several of the characteristic structures, including a 2-m (prototype) high seawall, large hotels along the water front, and smaller commercial and residential structures (Fig. 27). The measurement locations are divided into four lines: A to D. Lines A, B and C are located on a city street parallel to the primary inundation flow direction and numbered from 1 to 9. Line D is located mostly behind the buildings and only has four measurement locations (Park et al. 2013).

A synthetic time series of incident wave elevation at $x = 5$ m is used to force the numerical model at $x = 5$ m (Fig. 28). The overland flow depth, cross-shore velocity and cross-shore specific momentum flux at four locations, B1, B4, B6 and B9 are compared to efficiently tun BMP 5.

To simulate BMP 5 efficiently, it is necessary to model the macro-roughness elements properly. To achieve this, the FAVOR™ module of FLOW-3D® (Flow Science, 2002) is used to decide whether the elements are modeled accurately enough. It is found that the horizontal and vertical grid sizes have to be smaller than 0.1 m, i.e., $\Delta x \leq 0.1$ m and $\Delta z \leq 0.1$, since for larger grid resolutions, macro-roughness elements can not be modeled accurately. A finer grid size of 0.05 m is preferred for the simulations. For the sake of consistency, the same grid size is used in NAMI DANCE simulations. Variable time stepping is preferred in the simulations.

The results reveal that the effect of varying friction is stronger in more inland areas (Fig. 29). When the bottom surface conditions are very rough, in the furthest areas inland, there is a decrease in the overland flow depth by 20%, in the cross-shore

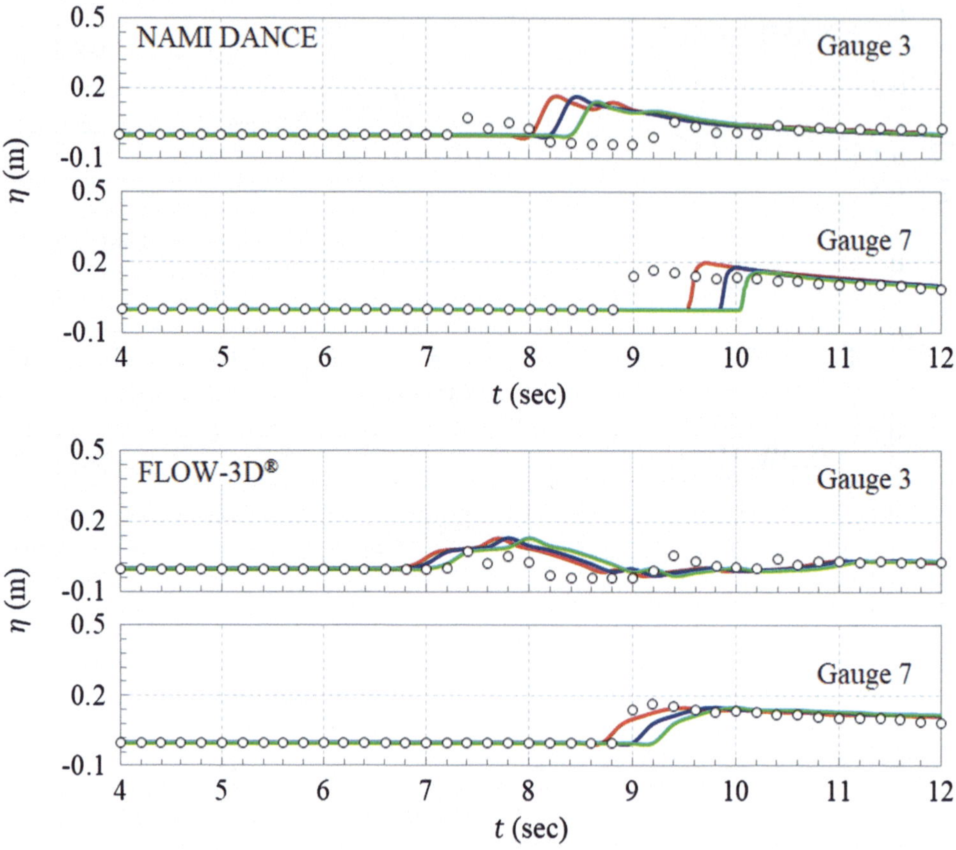

Figure 22
Frictional analysis at Gauges 3 and 7. The circles represent the laboratory measurements; the red line represents the results when $n = 0$; the blue line represents the results when $n = 0.01$; the green line represents the results when $n = 0.03$

velocity by 60%, and in the specific momentum flux by 80%. This fact reveals that the predicted velocity and momentum flux terms are highly sensitive to the bottom friction factor further inland. The arrival time of the inundation wave is earliest when no friction is set in the model.

The NAMI DANCE and FLOW-3D® model predictions regarding the overland flow depth, cross-shore velocity and cross-shore specific momentum flux are compared at B1, B4, B6 and B9 locations (Figs. 30, 31, 32, 33). It is a fact that inundation parameters are very sensitive to bottom friction for this specific case; therefore, it is important to calibrate the models according to the most appropriate roughness coefficient. The numerical simulations are conducted using $n = 0.01$ since it is found that the results most similar to the physical model are produced by this value, which is then used for all subsequent analyses with both models.

The overland flow depth computed by NAMI DANCE fits better to the measurements compared to FLOW 3D results at location B1 (Fig. 30). In terms of cross-shore momentum flux, both models compute lower values. The agreement between measured and computed results is in acceptable limits at locations B4 and B6, for both models (Figs. 31, 32). However, the results for location B9 show delayed arrival of predicted wave by both models compared to the measured data (Fig. 33).

The general tendencies and magnitudes of predicted overland flow depth and cross-shore velocity are well matched with the physical model results for the locations that are close to the shoreline. In both models, it is observed that the maximum overland

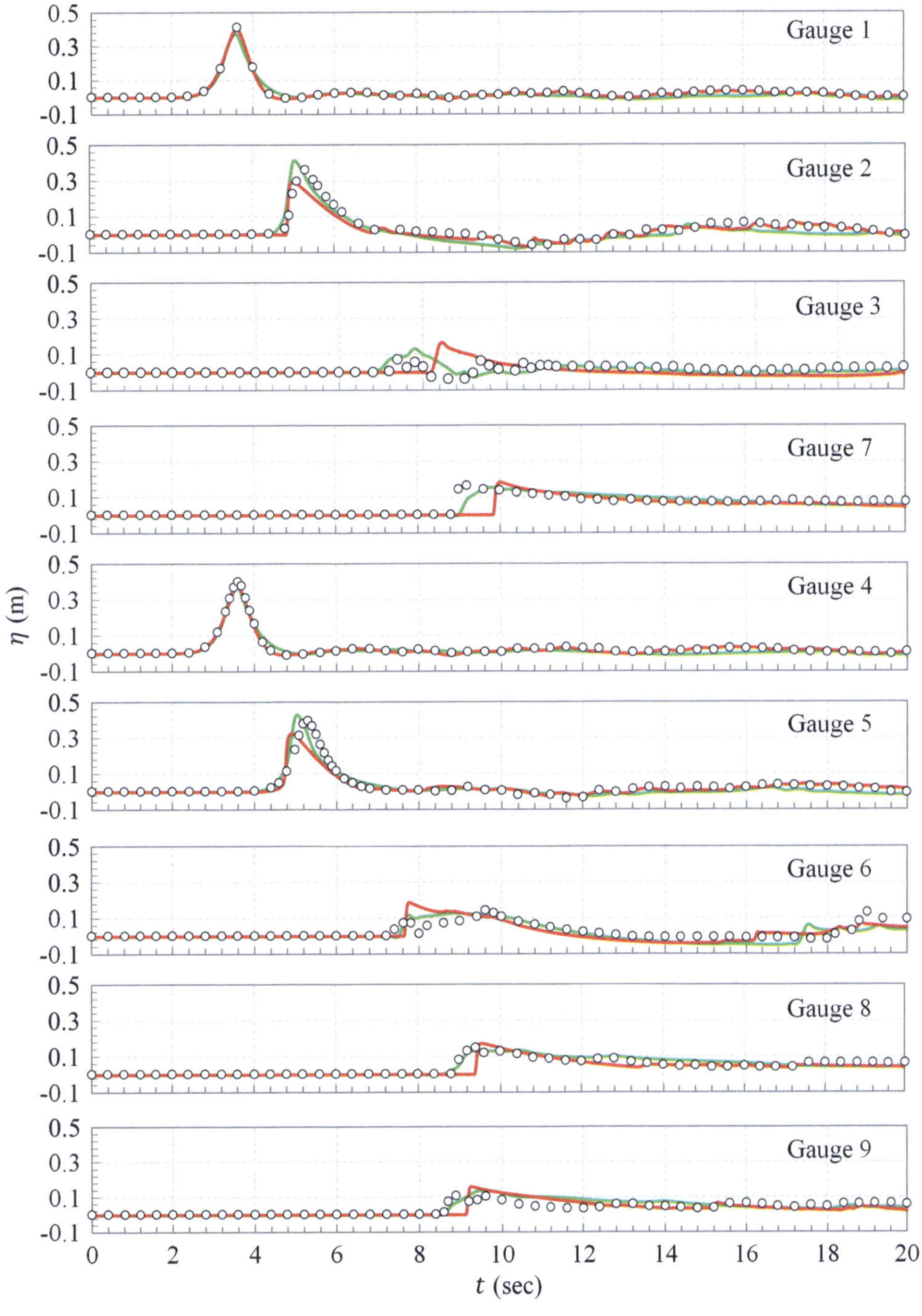

Figure 23

The predicted and measured free surface profiles at Gauges 1–9. The circles represent the laboratory data; the red line represents the NAMI DANCE results; the green line represents the FLOW-3D® results

flow depth and cross-shore momentum flux rapidly decrease as the wave moves landward, while the cross-shore velocity slowly decreases. Specifically, from B1 to B9, the maximum overland flow depth decreases by 77%, the maximum cross-shore velocity decreases by 30% and the maximum momentum flux

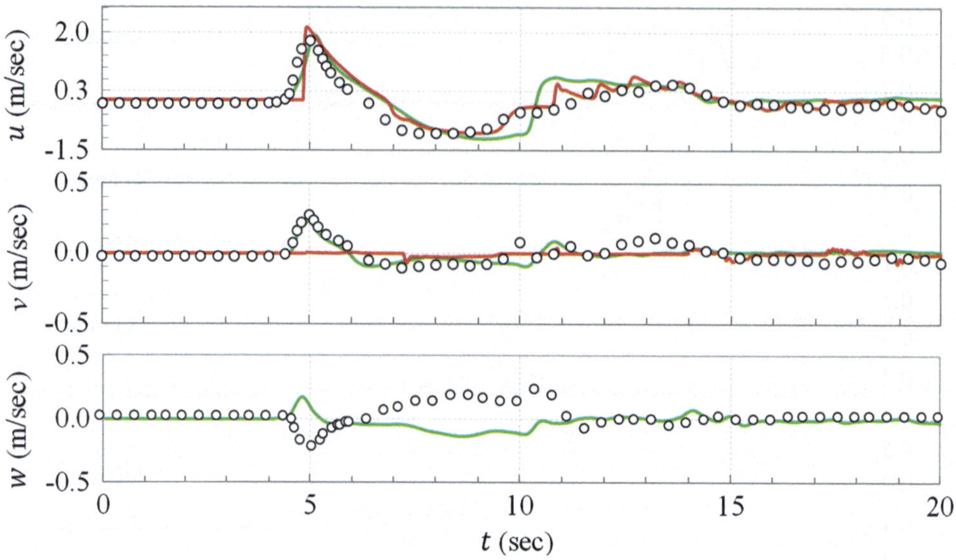

Figure 24
The predicted and measured velocity components at Gauge 2. The circles represent the laboratory measurements; the red line represents the NAMI DANCE results; the green line represents the FLOW-3D® results

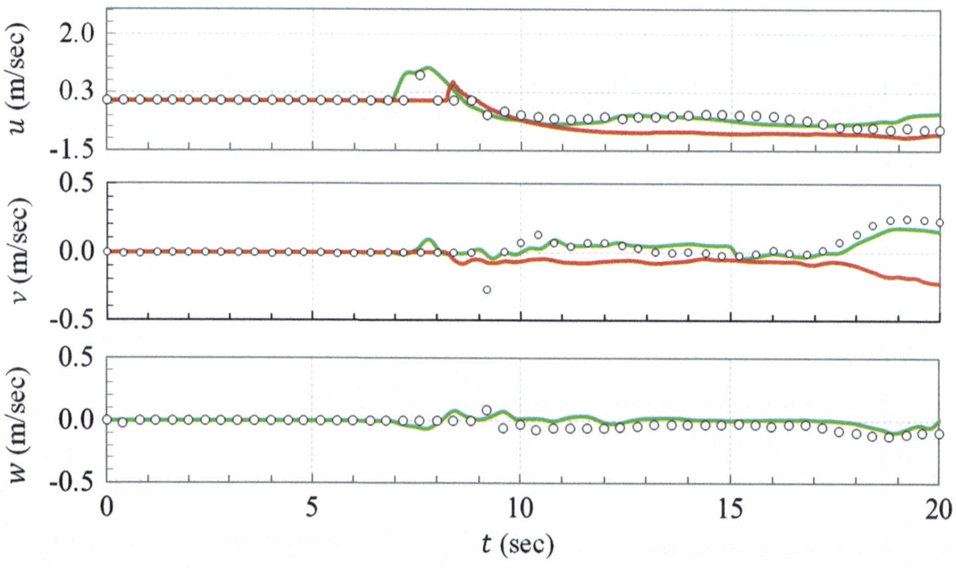

Figure 25
The predicted and measured velocity components at Gauge 3. The circles represent the laboratory measurements; the red line represents the NAMI DANCE results; the green line represents the FLOW-3D® results

decreases by 97% in the physical model. According to the measurements of the cross-shore velocities at B1, B4, B6 and B9, the 2D depth-averaged model NAMI DANCE predicts the cross-shore velocity approximately 20% less than the 3D model FLOW-3D®.

The NRMSE of the NAMI DANCE and FLOW-3D® predictions for the overland flow depth, cross-

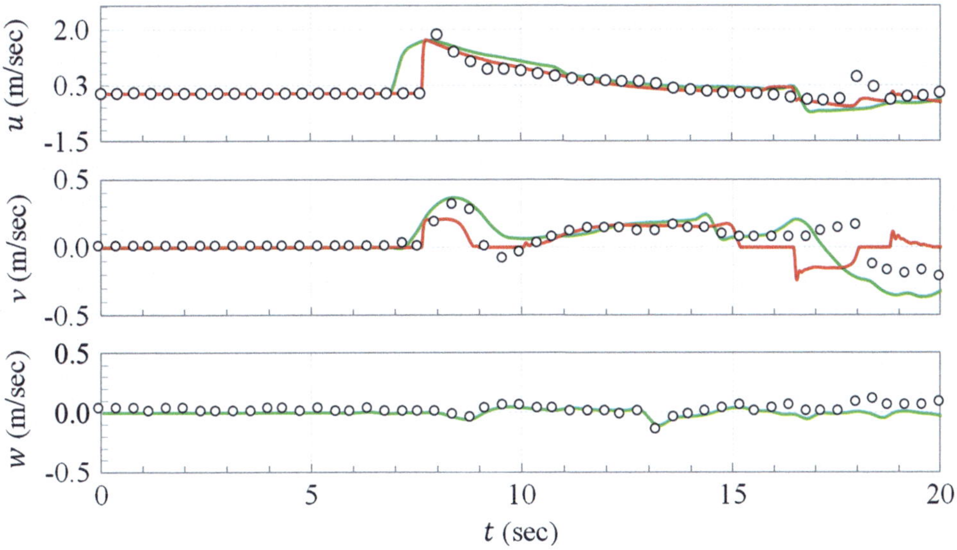

Figure 26
The predicted and measured velocity components at Gauge 10. The circles represent the laboratory measurements; the red line represents the NAMI DANCE results; the green line represents the FLOW-3D® results

Table 10
NAMI DANCE and FLOW-3D® model errors for water level time series with respect to laboratory data

Gauge	NAMI DANCE		FLOW-3D®	
	% NRMSE	% MAX	% NRMSE	% MAX
1	0.6	0.0	0.9	0.0
2	8.5	9.7	6.2	6.2
3	19.2	17.7	7.5	5.0
4	4.4	7.6	2.8	7.4
5	10.7	13.2	9.1	5.3
6	13.8	9.2	9.1	4.7
7	14.2	9.2	8.6	4.9
8	15.2	11.9	5.1	3.6
9	15.9	12.6	9.8	8.4

Table 11
NAMI DANCE and FLOW-3D® model errors for velocity components with respect to laboratory data

Gauge	Velocity component	NAMI DANCE		FLOW-3D®	
		% NRMSE	% MAX	% NRMSE	% MAX
2	*u*	12.0	18.4	9.8	3.3
	v	19.9	34.3	9.7	6.8
	w	*NA*	*NA*	11.2	9.6
3	*u*	19.5	11.9	9.8	7.6
	v	21.3	30.1	10.7	10.2
	w	*NA*	*NA*	8.3	10.1
10	*u*	8.5	16.1	11.0	9.3
	v	14.1	23.5	10.7	11.2
	w	NA	NA	10.5	9.2

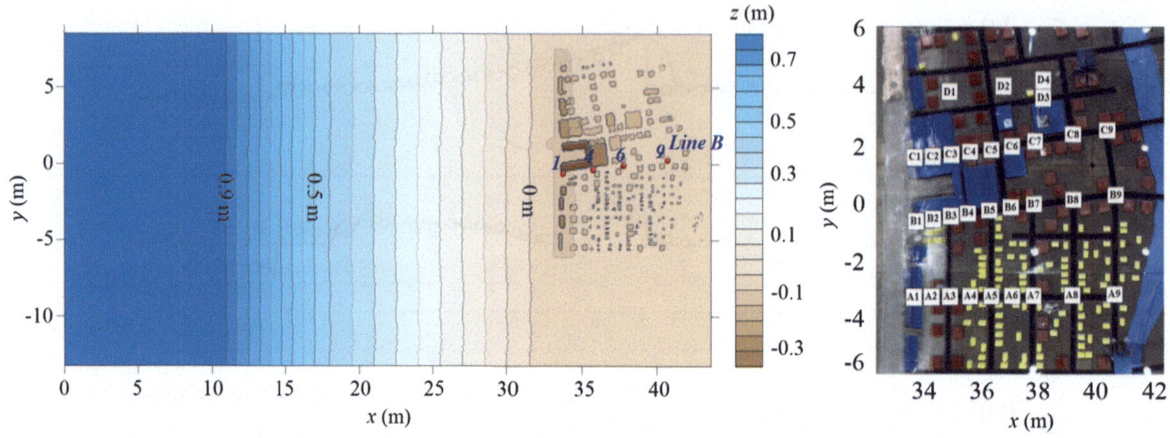

Figure 27
BMP 5 computational domain (left panel) (Velioglu 2017) and detailed plan view of macro-roughness elements with measurement locations (right panel) (Park et al. 2013)

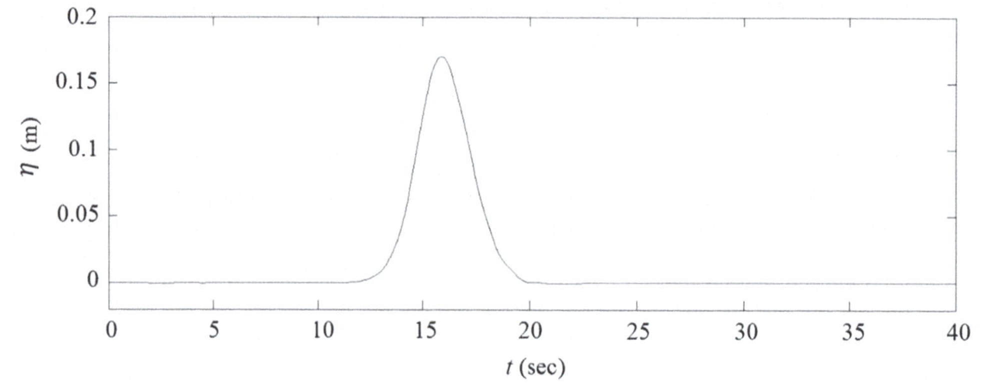

Figure 28
Incident wave time series at $x = 5$ m—simulated data, which can be used to drive a stationary input wave boundary condition at $x = 5$ m (NTHMP 2015)

shore velocity and cross-shore momentum flux at lines A, B, C and D are given in Fig. 34. The NRMSE for line D is relatively large. This anomaly may be attributed to the difference in the measurement locations. In the experimental model, the gauges along lines A, B, and C are located on the road, with no obstructions between the locations and the ocean, while along line D the gauges are situated mostly behind the buildings.

4.6. Benchmark Problem 6

During the 11 March 2011 Tōhoku tsunami, the 328 current meters that had been placed around the islands of Hawaii recorded the time series of the fluid velocity at varying depths of water (Lynett et al. 2014). BMP 6 introduces a part of this data set from the recordings at Hilo Harbor. To accomplish BMP 6, the free surface elevation time series from the Hilo tide station and the velocity information from the two ADCPs, namely HA1125 and HA1126 are to be compared and reliable agreements with the measured data are to be achieved. The first ADCP, HA1125, is located at the Hilo Harbor entrance, at (lat, long) = (19.745, 204.918). The second ADCP, HA1126, is located inside the Hilo Harbor, at (lat, long) = (19.742, 204.930). Here, it is important to note that the data recorded at the ADCPs are averaged over

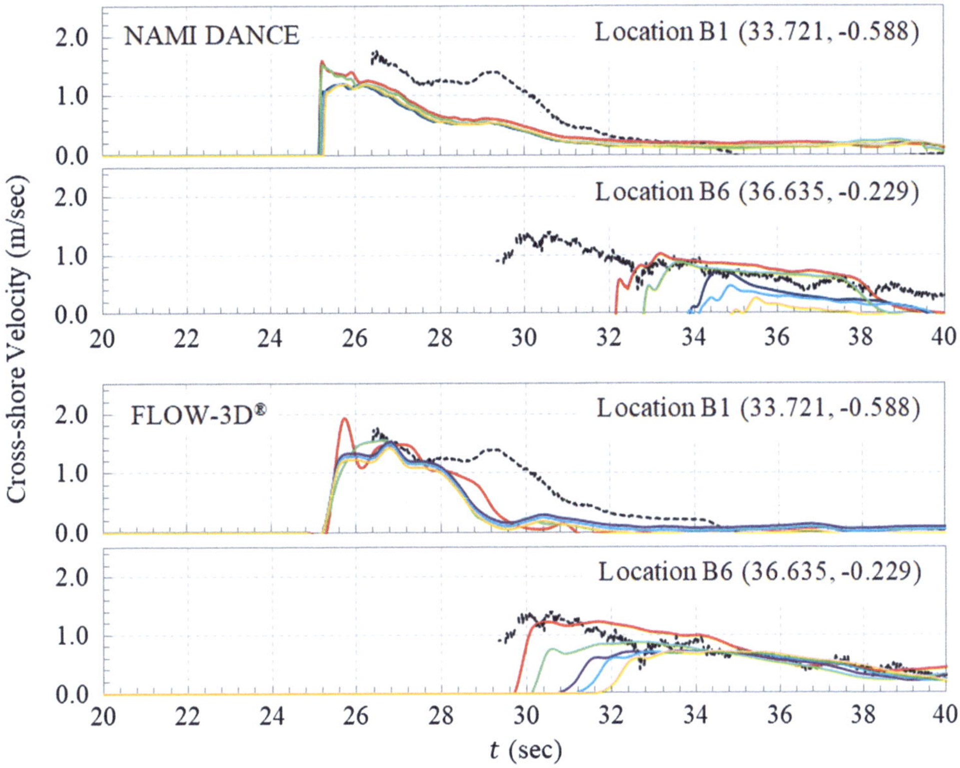

Figure 29
Comparison of predicted and measured cross-shore velocity for locations B1 and B6. The dashed black line represents the measured data; the red, green, purple, blue and orange lines represent the model predictions when $n = 0$, $n = 0.005$, $n = 0.01$, $n = 0.02$ and $n = 0.03$, respectively

depth and filtered to remove the long period tidal components.

The computational domain is the full geographic footprint of the 1/3 arc second benchmark bathymetry at longitude $204.901° \leq x \leq 204.965°$ and latitude $19.710° \leq y \leq 19.774°$. The x axis of the Cartesian coordinate system is parallel to the Hilo breakwater and the y axis is perpendicular to the wave propagation direction (Fig. 35).

In BMP 6, the simulations are driven with an offshore simulated free surface elevation time series which is supposed to be located at (lat, long) = (19.758, 204.930) (i.e., Control Point) (Fig. 36).

The computed and measured time series of the water surface elevation at the Hilo tide station (Hilo_TS) and the average flow speeds recorded at the two ADCP locations, HA1125 and HA1126 are compared using different spatial grid resolutions to identify the grid resolution effect. The numerical modeling of BMP 6 is carried out at 5, 10 and 20 m grid resolutions with a Manning's roughness coefficient of 0.025. The study shows reasonable overall convergence of the results, going down from 20 m resolution to 10 m and then a 5 m (Fig. 37).

The effect of friction is investigated through the application of different roughness coefficients to the BMP 6 numerical model. The results indicate that the effect of varying friction is very limited and only slightly felt in the maximum and minimum values of the time series in the current speed and surface elevation (Fig. 38).

The incident wave propagates southward and enters the harbor, in the middle of which ADCP HA1125 is located. The waves refract and diffract around the breakwater before reaching HA1126 and the Hilo tide station. The reflected waves interact with the subsequent arrivals to produce a complex wave field in Hilo Bay (Fig. 39). The predicted u and v components of current velocities in N–S and E–W directions are favorably compared with their

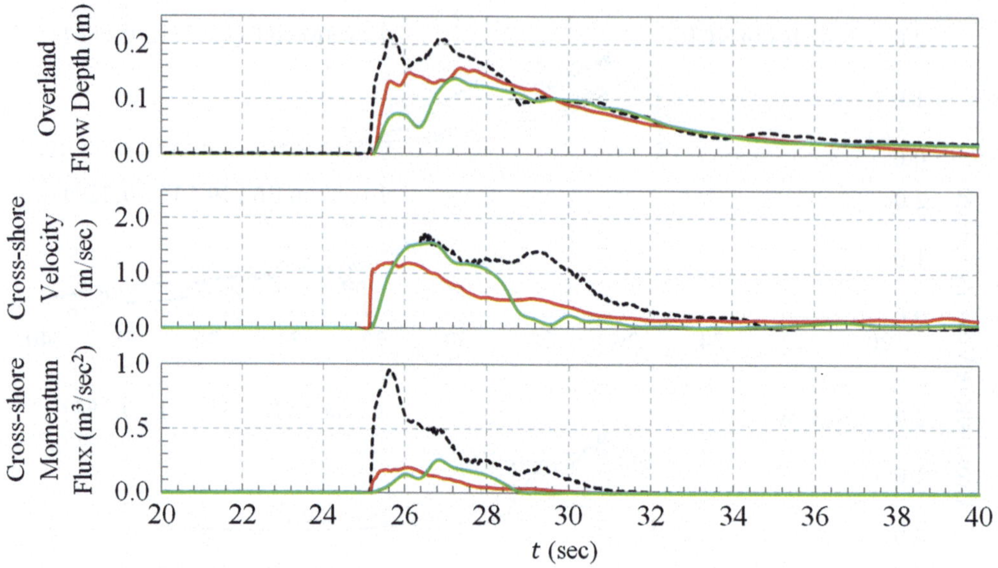

Figure 30
Comparison of predicted and measured overland flow depth, cross-shore velocity and cross-shore momentum flux for location B1 (33.721, − 0.588). The black line represents experimental data; the red and green lines represent the NAMI DANCE and FLOW-3D® results, respectively

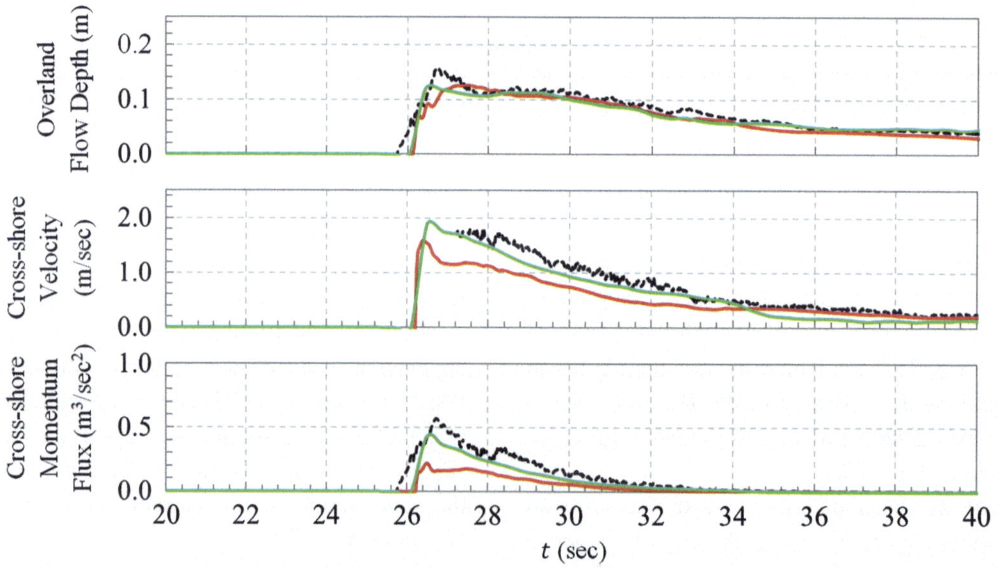

Figure 31
Comparison of predicted and measured overland flow depth, cross-shore velocity and cross-shore momentum flux for location B4 (35.176, − 0.406). The black line represents experimental data; the red and green lines represent the NAMI DANCE and FLOW-3D® results, respectively

observed counterparts. It is seen that the predicted velocity components have lots of fluctuation and thus several maxima/minima after $t = 9$ h (Fig. 40). These values are not observed in the recorded data. Because of the coarse time sampling of the tsunami currents during the Tōhoku 2011 event, some peak

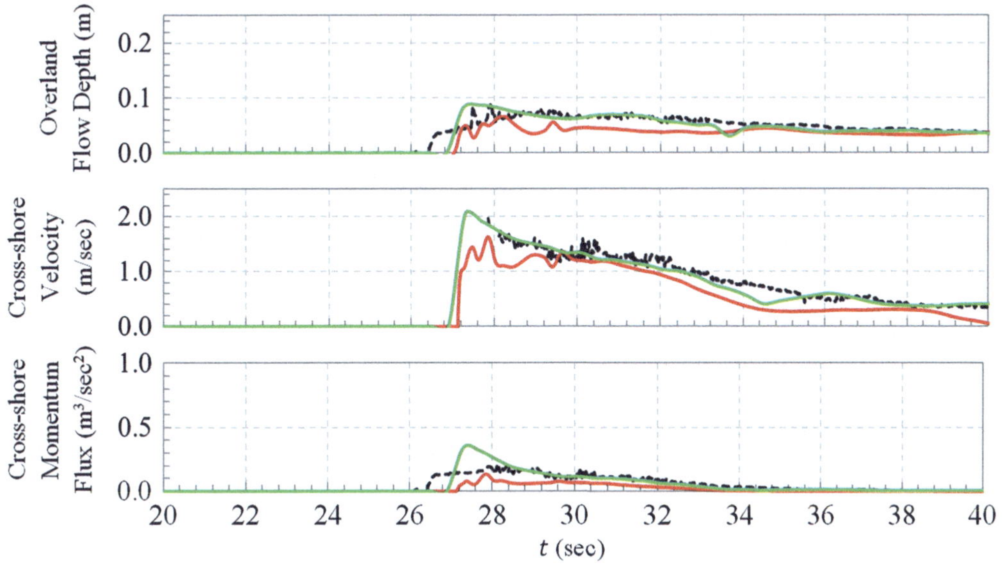

Figure 32
Comparison of predicted and measured overland flow depth, cross-shore velocity and cross-shore momentum flux for location B6 (36.635, − 0.229). The black line represents experimental data; the red and green lines represent the NAMI DANCE and FLOW-3D® results, respectively

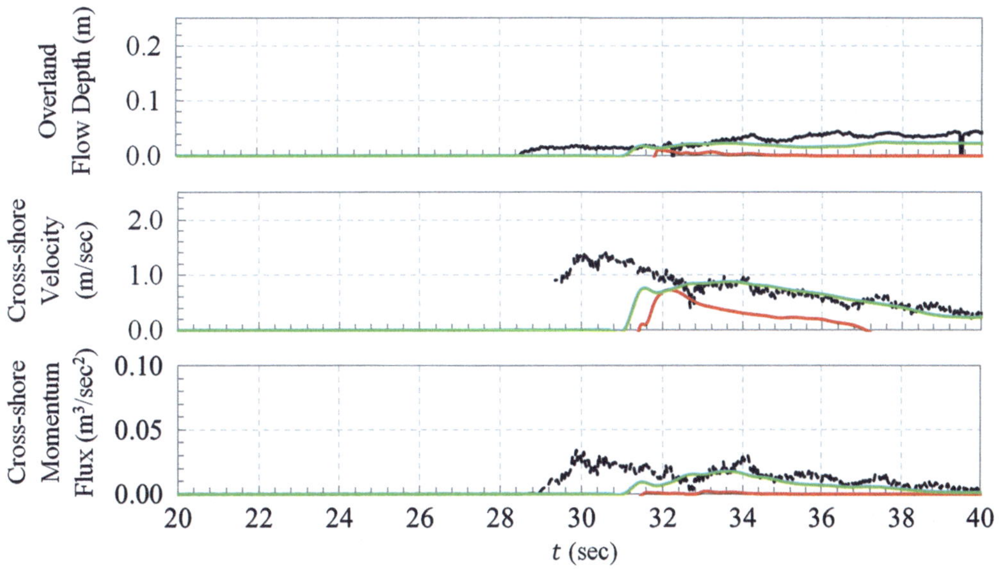

Figure 33
Comparison of predicted and measured overland flow depth, cross-shore velocity and cross-shore momentum flux for location B9 (40.668, − 0.269). The black line represents experimental data; the red and green lines represent the NAMI DANCE and FLOW-3D® results, respectively

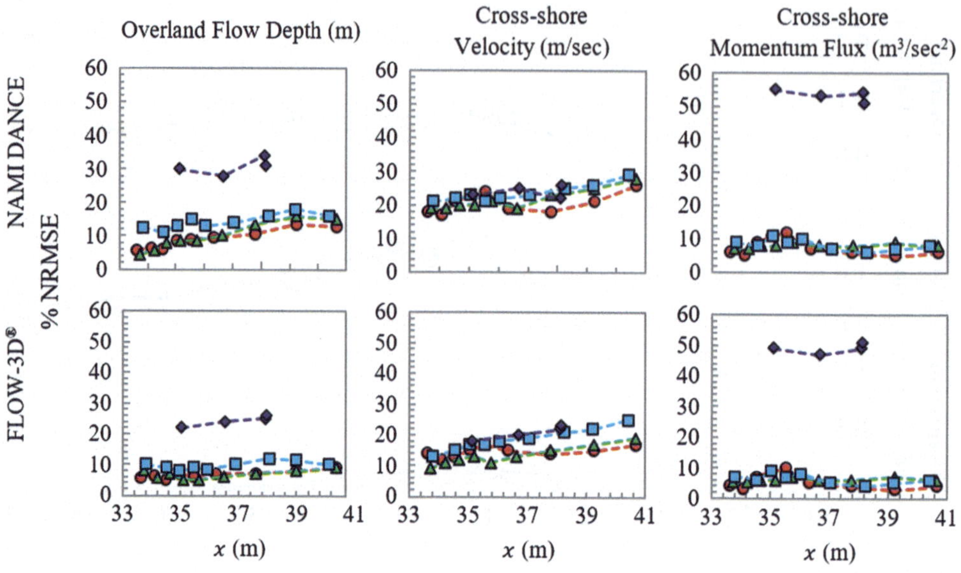

Figure 34
%NRMSE of NAMI DANCE and FLOW-3D® predictions along lines A, B, C, and D. The circles, triangles, squares and diamonds represent the predictions along lines A, B, C and D, respectively

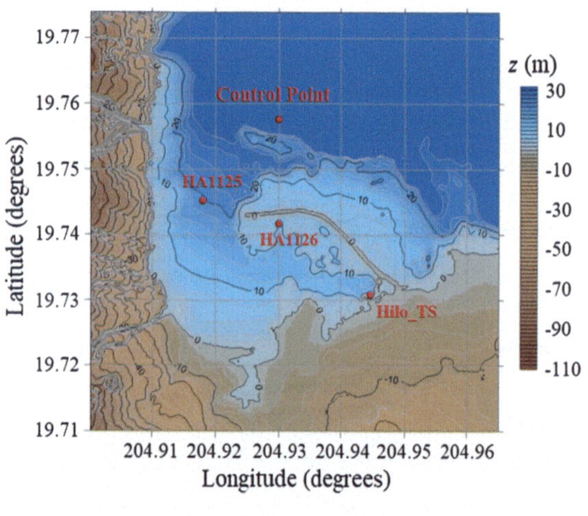

Figure 35
Computational domain for BMP 6 (top view)

velocities might be missing in the measurements. The sampling of the predicted current velocity components is 3 s for both models, while the sampling interval of the recorded velocity data is 6 min. This may be the main reason why there is a mismatch between the predicted and recorded velocity values after 9 h.

The NRMSE of the NAMI DANCE and FLOW-3D® results for the free surface dynamics recorded at the Hilo tide station is calculated as 24 and 21%, respectively, and the MAX errors are 33 and 27%, respectively.

While time series comparisons are useful for understanding how errors and variations evolve temporally, they do not help to understand how quickly flow properties change spatially. To examine spatial variability, maps for the predicted maximum current distribution are constructed for both models (Fig. 41). The greatest speeds are found in the area around breakwater head and along the coast where the depths are shallow. In these areas, the predicted current speeds are in an excess of 4 m/s. These results are consistent with the findings of the previous studies (Allan et al. 2012; Cheung et al. 2013; Barberopoulou et al. 2015; Lynett et al. 2017).

Figure 41 also reveals that the spatial variability results of the two models differ considerably. This may be attributed to the effect of the vertical velocity component, which provides a more detailed spatial distribution of the maximum current speed for the 3D model results. Furthermore, it is seen that the eddies take different paths in the models in accordance with their strength; therefore, in areas that are affected by eddies, the models are expected to show significant

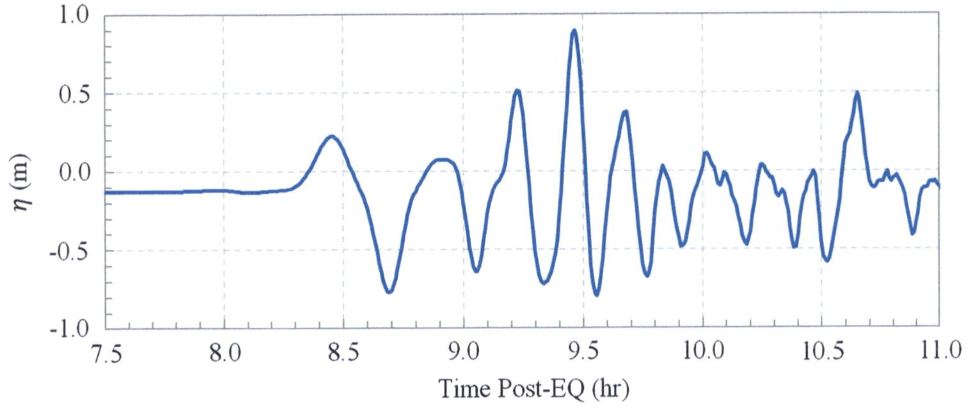

Figure 36
The incident wave of BMP 6: time series of ocean surface elevation at Control Point (NTHMP 2015)

differences. Similar results are reported in Lynett et al. (2017).

5. Results and Discussion

The NAMI DANCE and FLOW-3D® numerical models are validated and verified using some of the NOAA benchmarks for tsunami propagation, inundation and currents. The numerical solutions are tested against analytical solutions, laboratory measurements and field observations.

In general, the numerical results are found to be below the established thresholds given in the OAR PMEL-135 standard for both models. However, the 2D model gave relatively large errors when non-hydrostatic effects were present. BMP 1 analyses reveal that the 2D model cannot accurately describe the wave form of the large amplitude solitary wave in very shallow areas. The wavefront steepens, which is typical of any hydrostatic model, and the predicted wave heights are approximately 30% smaller than the laboratory measurements. This leads to the underestimation of the run-up height of large amplitude solitary waves, by approximately 20%. When the results of BMP 2 are assessed, it is seen that the 3D model shows an appreciable improvement over the 2D one in the case of the large amplitude solitary wave. BMP 3 test case shows that both models are able to reproduce the rapid sequence of run-up and rundown as well as the maximum run-up in the narrow gully near the village of Monai. BMP 4 involves energetic breaking waves with overtopping, vortex formation and run-up. NAMI DANCE underestimates the propagation speed of the wave and somewhat misrepresents the waveform at the gauge located behind the obstacle, on the centerline, where strong turbulence and nonlinearities occur. On the other hand, FLOW-3D® predicts the cross-shore and long-shore velocity components reasonably well with no mismatch in the phase at all gauges. In BMP 5, the time series and maximum values of overland flow depth, cross-shore velocity and cross-shore momentum flux are considered. Compared to the measurements along line B, NAMI DANCE computes the cross-shore velocity $\sim$ 20% and cross-shore momentum flux $\sim$ 50% less than FLOW-3D® at the farthest landward locations. This is a huge deviation in terms of force calculations. BMP 6 is a case study since it provides a comprehensive tsunami current data set recorded after the 2011 Tōhoku tsunami at Hilo Harbor, Hawaii. Particularly, at both ADCPs, there is good agreement between the FLOW-3D® results and the recorded data for the first and second peaks of the current velocities in N–S and E–W directions.

The mesh properties and process time for all BMPs are given in Table 12.

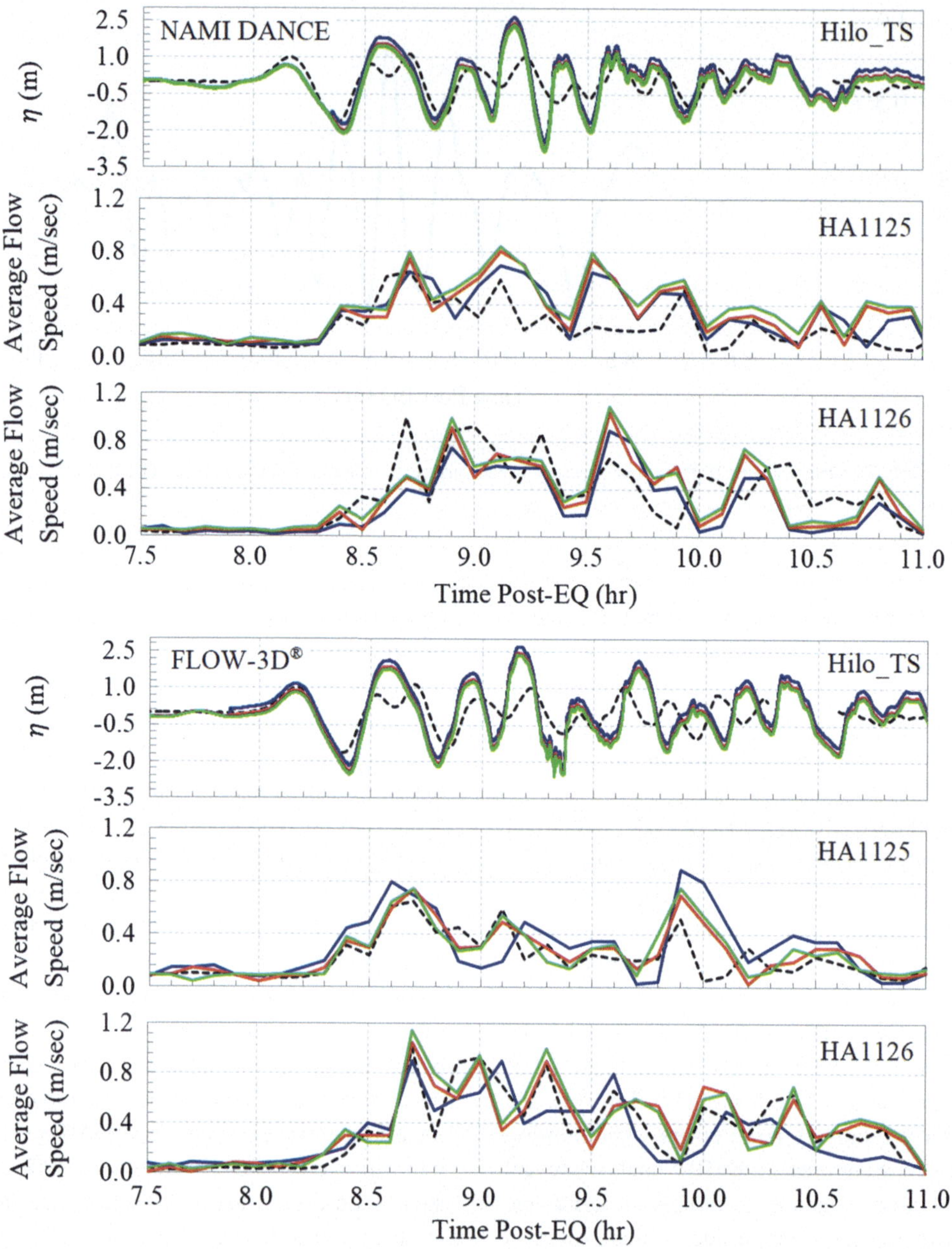

Figure 37
Convergence test for BMP 6 at Hilo tide station, HA1125 and HA1126 for $n = 0.025$. The dashed black line represents the field data; the solid blue, red and green lines represent the numerical predictions for $\Delta x = 20$ m, $\Delta x = 10$ m, and $\Delta x = 5$ m, respectively

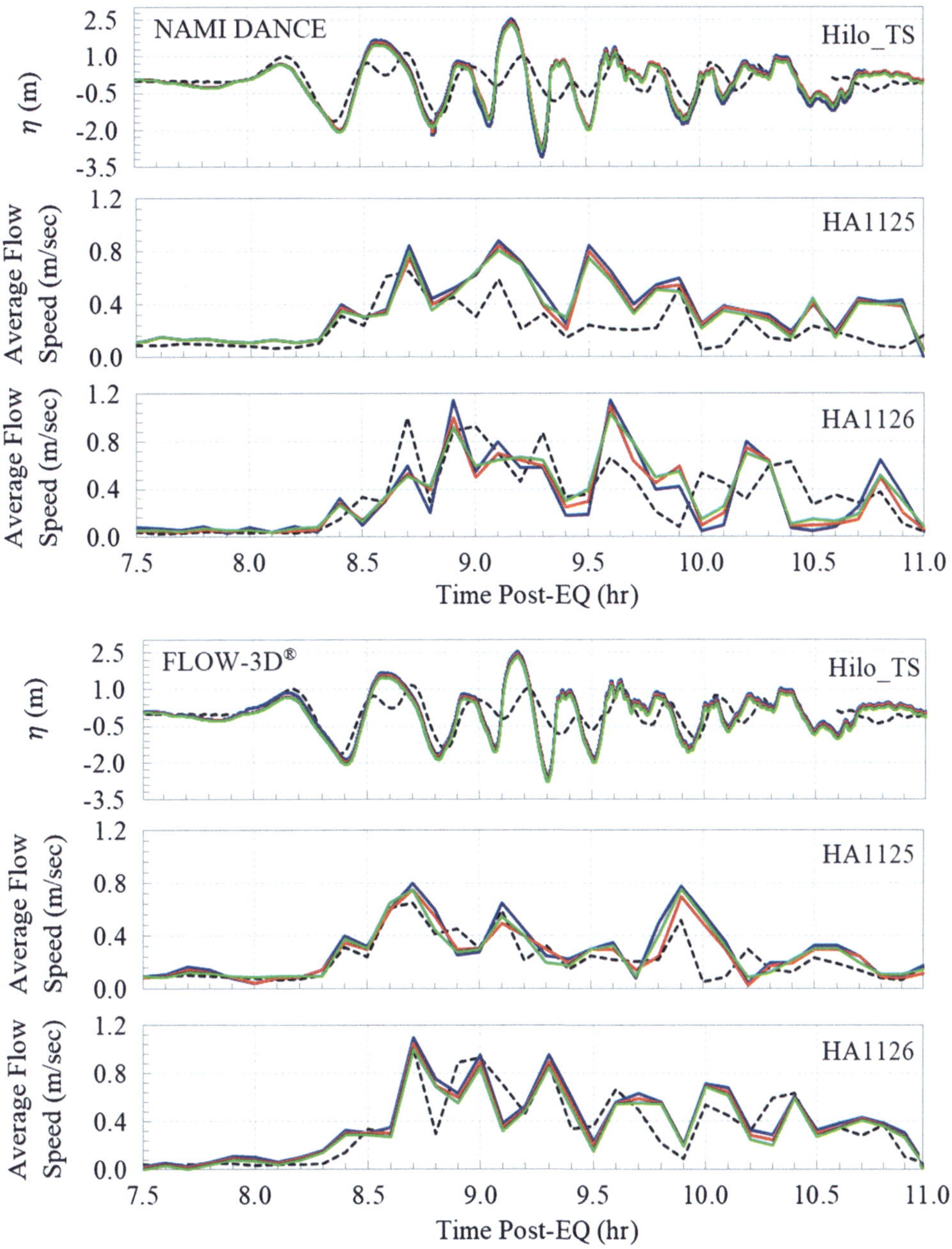

Figure 38
Model results according to different n at Hilo tide station, HA1125 and HA1126, where $\Delta x = 10$ m. The dashed black line represents the field data; the solid blue, red and green lines represent the numerical predictions when $n = 0$, $n = 0.015$ and $n = 0.025$, respectively

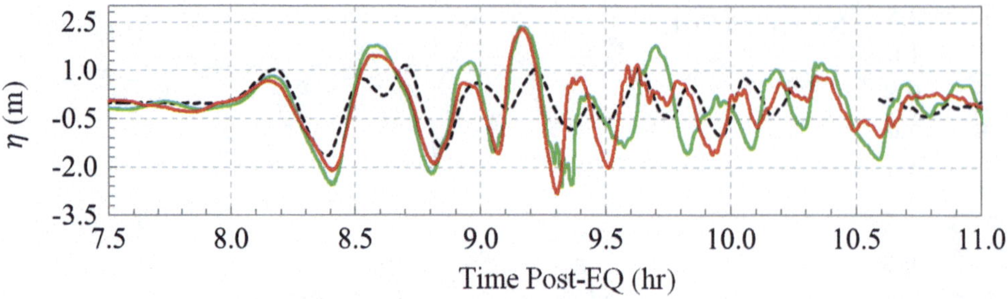

Figure 39
Comparison of the free surface elevation time series at Hilo tide station where $\Delta x = 10$ m and $n = 0.025$. The dashed black line represents the field data; the red line represents the NAMI DANCE results; the green line represents the FLOW-3D® results

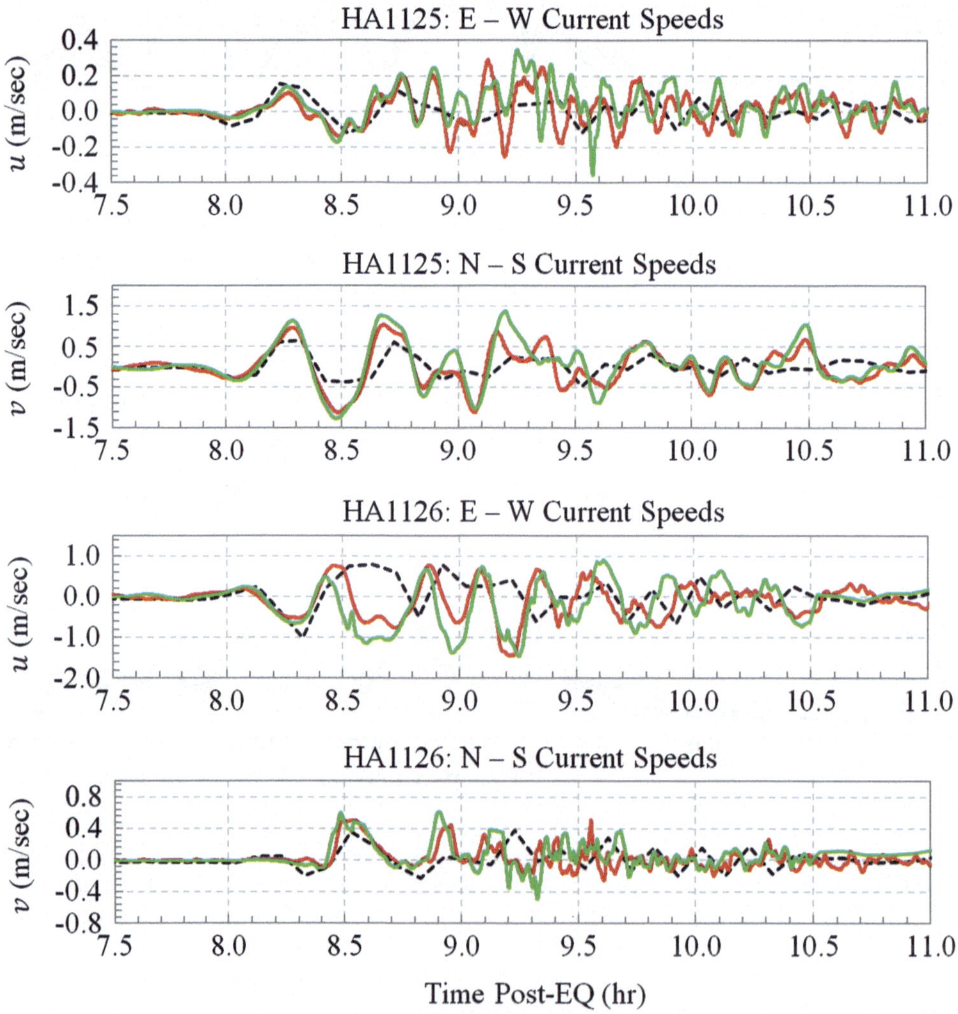

Figure 40
Comparison of the velocity components in E–W and N–S directions at HA1125 and HA1126 where $\Delta x = 10$ m and $n = 0.025$. The dashed black line represents the field data; the red line represents the NAMI DANCE results; the green line represents the FLOW-3D® results

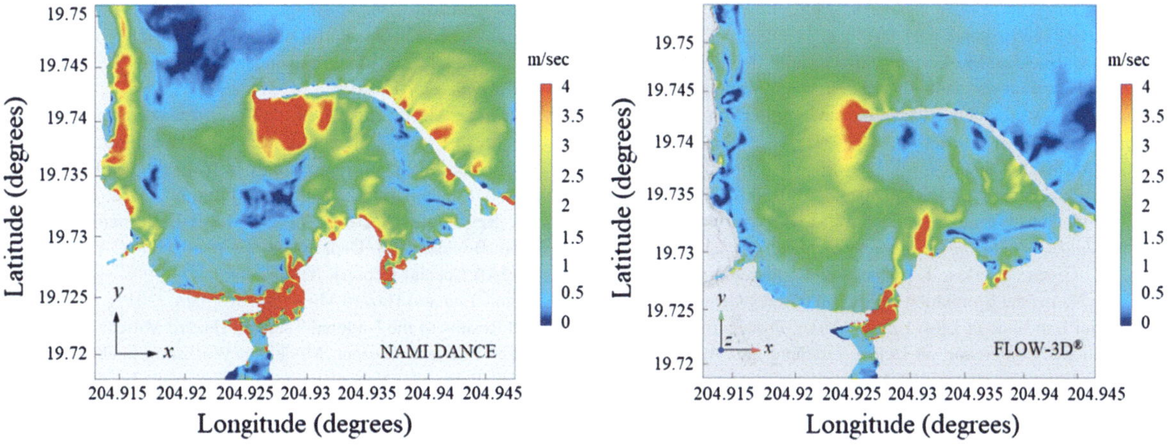

Figure 41
The distribution of the computed maximum current speed during the entire duration of the NAMI DANCE and FLOW-3D® simulations. The resolution of computational domain is 10 m

Table 12

Mesh properties and process time for all BMPs

BM test	Category	Tested quantity	Δx	Simulation time (sec)	NAMI DANCE Process time Intel® Xeon® E5-1650 v3 3.5 GHz processor	FLOW-3D® Process time Intel® Xeon® E5-1650 v3 3.5 GHz processor
BMP1	Analytical solution and laboratory measurements	Run-up/ amplitude	0.01	60	10 min	4 h
BMP2	Laboratory measurements	Run-up/ amplitude	0.05	40	40 min	6 h
BMP3	Laboratory measurements	Amplitude/ velocity	0.05	20	15 h	2 days
BMP4	Laboratory measurements	Run-up/ amplitude	0.014	50	3 h	3 days
BMP5	Laboratory measurements	Run-up/ velocity	0.05	40	21 h	3 days
BMP6	Field measurements	Amplitude/ velocity	10	48 600	2 days	15 days

Acknowledgements

The authors wish to thank Dr. Andrey Zaytsev due to his undeniable contributions to the development of in-house numerical model, NAMI DANCE. The Turkish branch of Flow Science, Inc. is also acknowledged. Finally, the National Tsunami Hazard Mitigation Program (NTHMP), who provided most of the benchmark data, is appreciated. This research did not receive any specific grant from funding agencies in the public, commercial, or not-for-profit sectors.

Conflicts of Interest The authors declare no conflict of interest.

References

Allan, J. C., Komar, P. D., Ruggiero, P., & Witter, R. (2012). The March 2011 Tohoku tsunami and its impacts along the U.S. West

Coast. *Journal of Coastal Research, 28*(5), 1142–1153. https://doi.org/10.2112/jcoastres-d-11-00115.1.

Apotsos, A., Buckley, M., Gelfenbaum, G., Jafe, B., & Vatvani, D. (2011). Nearshore tsunami inundation and sediment transport modeling: towards model validation and application. *Pure and Applied Geophysics, 168*(11), 2097–2119. https://doi.org/10.1007/s00024-011-0291-5.

Barberopoulou, A., Legg, M. R., & Gica, E. (2015). Time evolution of man-made harbor modifications in San Diego: effects on Tsunamis. *Journal of Marine Science and Engineering, 3,* 1382–1403.

Basu, D., Green, S., Das, K., Janetzke, R. and Stamatakos, J. (2009). Numerical Simulation of Surface Waves Generated by a Subaerial Landslide at Lituya Bay, Alaska. Proceedings of 28th International Conference on Ocean, Offshore and Arctic Engineering. Honolulu, Hawaii, USA.

Briggs, M. J., Synolakis, C. E., Harkins, G. S., & Green, D. R. (1995). Laboratory experiments of tsunami run-up on a circular island. *Pure and Applied Geophysics, 144*(3/4), 569–593.

Cheung, K. F., Bai, Y., & Yamazaki, Y. (2013). Surges around the Hawaiian Islands from the 2011 Tohoku Tsunami. *Journal of Geophysical Research: Oceans, 118,* 5703–5719. https://doi.org/10.1002/jgrc.20413.

Choi, B. H., Dong, C. K., Pelinovsky, E., & Woo, S. B. (2007). Three-dimensional Simulation of Tsunami Run-up Around Conical Island. *Coastal Engineering, 54,* 618–629.

Cox, D., T. Tomita, P. Lynett, R.A., Holman. (2008). Tsunami Inundation with Macroroughness in the Constructed Environment. Proceedings of 31st International Conference on Coastal Engineering, ASCE, pp. 1421–1432.

Flow Science. (2002). FLOW-3D User's Manual.

Hirt, C. W., & Nichols, B. D. (1981). Volume of fluid (VOF) method for the dynamics of free boundaries. *Journal of Computational Physics, 39,* 201–225.

Horrillo, J., Grilli, S. T., Nicolsky, D., Roeber, V., & Zang, J. (2015). Performance benchmarking Tsunami models for NTHMP's inundation mapping activities. *Pure and Applied Geophysics, 172,* 869–884.

http://nws.weather.gov/nthmp/documents/nthmpWorkshopProcMerged.pdf.

http://nws.weather.gov/nthmp/documents/NTHMP_Currents_Workshop_Report.pdf.

Kim, K. O., Kim, D. C., Choi, B.-H., Jung, T. K., Yuk, J. H., & Pelinovsky, E. (2015). The role of diffraction effects in extreme run-up inundation at Okushiri Island due to 1993 Tsunami. *Natural Hazards and Earth Systems Sciences, 15,* 747–755.

Liu, P. L.-F. (1994). Model equations for wave propagations from deep to shallow water. (P.-F. Liu, Ed.) Advances in Coastal and Ocean Engineering, 1, 125–158.

Liu, P. L.-F., Yeh, H., & Synolakis, C. E. (2008). Advanced numerical models for simulating Tsunami waves and run-up. *Advances in Coastal and Ocean Engineering, 10,* 344.

Lynett, P. J., Borrero, J., Son, S., Wilson, R., & Miller, K. (2014). Assessment of the tsunami-induced current hazard. *Geophysical Research Letters, 41,* 2048–2055. https://doi.org/10.1002/2013GL058680.

Lynett, P. J., Gately, K., Wilson, R., Montoya, L., Arcas, D., Aytore, B., et al. (2017). Inter-model analysis of Tsunami-induced coastal currents. *Ocean Modelling, 114,* 14–32.

Lynett, P. J., Wu, T.-R., & Liu, P. L.-F. (2002). Modeling wave run-up with depth-integrated equations. *Coastal Engineering, 46*(2), 89–107.

Macias, J., Castro, M. J., Ortega, S., Escalante, C., & Gonzalez-Vida, J. M. (2017). Performance benchmarking of Tsunami-HySEA model for nthmp's inundation mapping activities. *Pure and Applied Geophysics, 174,* 3147–3183.

Matsuyama, M., & Tanaka, H. (2001). An experimental study of the highest run-up height in the 1993 Hokkaidō Nansei-Oki Earthquake Tsunami. *Proceedings of ITS, 2001,* 879–889.

National Tsunami Hazard Mitigation Program. 2012. Proceedings and Results of the 2011 NTHMP Model Benchmarking Workshop. Boulder: U.S. Department of Commerce/NOAA/NTHMP; (NOAA Special Report). p. 436.

National Tsunami Hazard Mitigation Program. (2017). Proceedings and Results of the National Tsunami Hazard Mitigation Program 2015 Tsunami Current Modeling Workshop, February 9-10, 2015, Portland, Oregon: compiled by Patrick Lynett and Rick Wilson, p 194.

Necmioglu, O., & Ozel, N. M. (2014). An earthquake source sensitivity analysis for Tsunami propagation in the Eastern Mediterranean. *Oceanography, 27*(2), 76–85.

Nichols, B.D. and Hirt, C.W. (1975). Methods for Calculating Multi-Dimensional, Transient Free Surface Flows Past Bodies. Proceedings of 1st International Conference Num. Ship Hydrodynamics. Gaithersburg.

Nicolsky, D. J., Suleimani, E. N., & Hansen, R. A. (2011). Validation and verification of a numerical model for Tsunami propagation and run-up. *Pure and Applied Geophysics, 168*(6), 1199–1222.

NOAA Center for Tsunami Research: Tsunami Run-up onto a Complex Three-dimensional Beach; Monai Valley. (n.d). Retrieved from: https://nctr.pmel.noaa.gov/benchmark/Laboratory/Laboratory_MonaiValley/.

Park, H., Cox, D. T., Lynett, P. J., Wiebe, D. M., & Shin, S. (2013). Tsunami inundation modeling in constructed environments: a physical and numerical comparison of free-surface elevation, velocity, and momentum flux. *Coastal Engineering, 79,* 9–21.

Patel, V. M., Dholakia, M. B., & Singh, A. P. (2016). Emergency preparedness in the case of Makran Tsunami: a case study on Tsunami risk visualization for the Western Parts of Gujarat, India. *Geomatics Natural Hazard and Risk, 7*(2), 826–842.

Pelinovsky, E., Kim, D.-C., Kim, K.-O., & Choi, B.-H. (2013). *Three-dimensional simulation of extreme run-up heights during the 2004 Indonesian and 2011 Japanese Tsunamis.* Vienna: EGU General Assembly.

Rueben, M., Holman, R., Cox, D., Shin, S., Killian, J., & Stanley, J. (2011). Optical measurements of Tsunami inundation through an urban waterfront modeled in a large-scale laboratory basin. *Coastal Engineering, 58,* 229–238.

Shuto, N. (1991). Numerical simulation of Tsunamis—its present and near future. *Natural Hazards, 4,* 171–191.

Synolakis, C. E. (1986). The run-up of long waves. Ph.D. Thesis. California Institute of Technology, Pasadena, California.

Synolakis, C. E., Bernard, E. N., Titov, V. V., Kanoglu, U. & Gonzalez, F. (2007). Standards, criteria, and procedures for NOAA evaluation of Tsunami Numerical Models. 55 p. Seattle, Washington: NOAA OAR Special Report, Contribution No 3053, NOAA/OAR/PMEL.

Synolakis, C. E., Bernard, E. N., Titov, V. V., Kanoglu, U., & Gonzalez, F. I. (2008). Validation and verification of Tsunami numerical models. *Pure and Applied Geophysics, 165,* 2197–2228.

Tolkova, E. (2014). Land-water boundary treatment for a tsunami model with dimensional splitting. *Pure and Applied Geophysics, 171*(9), 2289–2314.

Velioglu, D. (2017). Advanced two- and three-dimensional Tsunami models: benchmarking and validation. Ph.D. Thesis. Middle East Technical University, Ankara.

Velioglu, D., Kian, R., Yalciner, A.C. and Zaytsev, A. (2016). Performance assessment of NAMI DANCE in Tsunami evolution and currents using a benchmark problem. (R. Signell, Ed.) J. Mar. Sci. Eng., 4(3), 49.

Wu, T. (2001). A unified theory for modeling water waves. *Advances in Applied Mechanics, 37,* 1–88.

Wu, N.-J., Hsiao, S.-C., Chen, H.-H., & Yang, R.-Y. (2016). The study on solitary waves generated by a piston-type wave maker. *Ocean Engineering, 117,* 114–129.

Yalciner, A. C., Dogan, P. and Sukru. E. (2005). December 26 2004, Indian Ocean Tsunami Field Survey, North of Sumatra Island. UNESCO.

Yalciner, A. C., Gülkan, P., Dilmen, I., Aytore, B., Ayca, A., Insel, I., et al. (2014). Evaluation of Tsunami scenarios For Western Peloponnese, Greece. *Bollettino di Geofisica Teorica ed Applicata, 55,* 485–500.

Yen, B. C. (1991). Hydraulic resistance in open channels. In B. C. Yen (Ed.), *Channel flow resistance: centennial of manning's formula* (pp. 1–135). Highlands Ranch: Water Resource Publications.

Zaitsev, A. I., Kovalev, D. P., Kurkin, A. A., Levin, B. V., Pelinovskii, E. N., Chernov, A. G., et al. (2009). The Tsunami on Sakhalin on August 2, 2007: mareograph evidence and numerical simulation. *Tikhookeanskaya Geologiya, 28,* 30–35.

(Received December 22, 2017, revised May 16, 2018, accepted May 24, 2018, Published online June 7, 2018)

Pure Appl. Geophys. 176 (2019), 3155–3166

https://doi.org/10.1007/s00024-018-2057-9

Pure and Applied Geophysics

A Modelling Study on Tsunami Propagation in the Caspian Sea

Raúl Periáñez[1] and Carmen Cortés[2]

Abstract—A numerical model that simulates the propagation of tsunamis produced by submarine earthquakes was applied to the Caspian Sea. The model was first applied to the June 1990 (Rudbar) earthquake and our results proved to be coherent with previous simulations. It was also applied to the Krasnovodsk earthquake (1895), which had not been simulated before, and it was found that the coastal area was flooded by a tsunami, as reported in literature. Nevertheless, it was a rather local effect since the tsunami was generated in very shallow waters. Some worst-case hypothetical earthquakes were simulated in the most seismically active regions of the Caspian. In general, both the east and west coasts of the central Caspian Sea would be the most affected by these earthquake-induced tsunamis, with potential significant effects in some cases in cities such as Baku.

Key words: Caspian Sea, tsunami, earthquake, numerical simulation.

1. Introduction

The Caspian Sea is an area of relatively high seismic activity, determined by its geotectonics. It can by divided into northern, central and southern parts. The last two parts are separated by the so-called Cheleken–Apsheron Threshold (CAT). This is a narrow strip crossing the sea between these two peninsulas from west to east (see Fig. 1) which is characterized by high seismic activity (Kulikov et al. 2014). This high activity has been attributed to normal fault-focal mechanisms related to subduction of the lithosphere as it sinks into the mantle below the northern region of the Caspian Sea (Kadirov and Gadirov 2014). The northern part of the Caspian basin belongs to the Russian platform, bounded in the south by the Turan platform. The Alpine folded system comprises the central and southern parts of the sea. These regions are separated by deep faults (Kulikov et al. 2014). Details are presented in the cited references and not repeated here.

A large zone of seismic activity is located in the eastern side of the CAT, where high recurrence of strong earthquakes is reported (Dotsenko et al. 2002). Kulikov et al. (2014) and Dotsenko et al. (2002) report other areas of high seismicity where tsunami events have been recorded. They are located in the western side of the Caspian Sea, both south and north of the CAT. Earthquakes produced in the southern Caspian have been studied in detail by Jackson et al. (2002), who provide focal mechanisms for a large number of earthquakes occurred during the second half of the 20th century. Areas where tsunamis have been observed were identified in Dotsenko et al. (2002) and Kulikov et al. (2014), from 17th to 20th centuries.

There is intensive infrastructure development for oil and gas in the Caspian Sea (Kulikov et al. 2014), as well as a nuclear power plant (Aktau, now decommissioned) on the northeast shore. These facilities may constitute a possible source of hazard due to pollution; emissions could be caused by an earthquake and the potential associated tsunami, given the high seismicity of the region, as occurred in the Fukushima Daiichi nuclear power plant (Japan) after the 2011 tsunami (Kaeriyama et al. 2013). In addition to potential pollution events, the damage to infrastructure caused directly by the tsunami waves and consequential high risk to the population living in cities bordering the Caspian Sea should also be considered. Thus, it is relevant to study the characteristics of tsunamis in the Caspian Sea and to

[1] Dpto. Física Aplicada I, ETSIA, Universidad de Sevilla, Ctra. Utrera km 1, 41013 Sevilla, Spain. E-mail: rperianez@us.es
[2] Dpto. Matemática Aplicada I, ETSIA, Universidad de Sevilla, Ctra. Utrera km 1, 41013 Sevilla, Spain. E-mail: ccortes@us.es

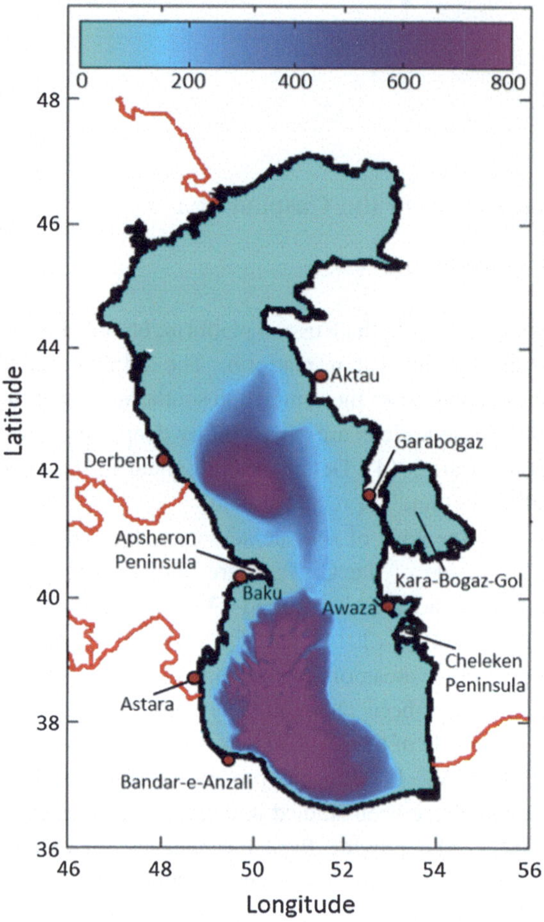

Figure 1
Model domain with water depths (m) and places mentioned in the text. Red lines are political boundaries

estimate the maximum waves which could be expected.

The numerical modelling of tsunami propagation is a relatively well established methodology that has been validated against recorded data from historical events over the world (Choi et al. 2008; Alasset et al. 2006; Ioualalen et al. 2010; Periáñez and Abril 2013, 2014; Titov et al. 2016; Abril and Periáñez 2017) as well as against laboratory experiments. In spite of this, few modelling works on tsunami propagation in the Caspian Sea can be found in current literature. For instance, Dotsenko et al. (2002) and Kulikov et al. (2014) present some simulations, but for very idealized and non-realistic sources: these authors consider the tsunami source as an ellipse with a reducing water elevation as the distance from the ellipse center increases. Simulations with a realistic sea floor deformation, for instance as evaluated from the classical Okada (1985) formulation, can be found only (to the authors knowledge) in Salaree and Okal (2015). These authors conducted one simulation for the earthquake occurring in June 20th, 1990, (known as the Rudbar earthquake) in north Iran. They found that the simulated tsunami resulting from this deformation was too weak and did not agree with inundations reported by eye witnesses during a field survey. Thus, the authors concluded that the tsunami was actually produced by a submarine landslide triggered by the earthquake.

This paper is aimed at studying tsunami propagation in the Caspian Sea by adapting previously tested numerical tools (Periáñez and Abril 2013, 2014; Abril and Periáñez 2017). These numerical exercises are expected to provide some insight into the main characteristics, such as maximum wave amplitude, of the tsunami propagation in this marine system. Initially, the model was applied to the June 20th, 1990 tsunami to compare results with those of Salaree and Okal (2015) and assess its correct customization to the Caspian Sea. Then it was applied to simulate the tsunami triggered by the Krasnovodsk earthquake in July 8th, 1895, magnitude $M = 7.8$, (Ambraseys 1997; Balakina and Moskvina 2007), which had not been done before. Finally, some hypothetical tsunamis (based upon realistic sources) occurring in the most seismically active areas of the Caspian Sea were simulated to estimate their potential impacts along the borders of the sea.

The model is briefly described in Sect. 2. Results are described in Sect. 3. A comparison of this model with a previous simulation of the June 20th, 1990 earthquake and tsunami is initially presented. Next, the simulation of the Krasnovodsk tsunami is described. Finally, hypothetical tsunami scenarios are discussed.

2. *Model Description*

The tsunami propagation model is based on 2D depth-averaged barotropic shallow water equations, which describe the propagation of surface shallow water gravity waves. The numerical tool was adapted from previous works, and proved to be a very robust computational tool. It has been applied to simulate tsunamis in the Atlantic Ocean (Periáñez and Abril 2013), Mediterranean Sea (Periáñez and Abril 2014) and the Red Sea (Abril and Periáñez 2017). It also has been used to simulate the Zanclean Flood of the Mediterranean, an extreme flooding episode (Periáñez and Abril 2015; Abril and Periáñez 2016). A detailed description and involved equations may be seen in these references, and a summary is given in Appendix 1.

The model domain (Fig. 1) extends from 46.0°E to 56.0°E, and from 36.0°N to 47.5°N, with a spatial resolution of 30 seconds of arc. The bathymetry was obtained from the GEBCO08 database, available online. Water depths were corrected to take into account variability of the Caspian Sea level with respect to the ocean (Naderi Beni et al. 2013).

3. *Results and Discussion*

3.1. *The June 20th, 1990, Rudbar Tsunami*

The Rudbar earthquake, with magnitude $M_w = 7.4$, was one of the most destructive earthquakes ever recorded in Iran (Berberian et al. 1992). Several studies were conducted to determine the source mechanism, which are listed in Salaree and Okal (2015). These two authors simulated the tsunami triggered by this earthquake using the standard Harvard CMT solution for the source (Table 1, HRV source). However, there is no consensus about the geometry of the fault system where the earthquake was produced. Thus, scaling laws which empirically relate the fault size and slip with the earthquake magnitude (Geller 1976) were applied by these authors. Salaree and Okal (2015) also carried out a simulation considering the source as a composition of nine different subevents. Differences between both simulations were negligible. Hence, only a single-event source was applied in the present model, and results are compared with those of Salaree and Okal (2015) to evaluate the correct customization of the model to the Caspian Sea. Nevertheless, there are recent empirical relationships between the fault size and slip with the earthquake magnitude, as for instance those proposed by Wells and Coppersmith (1994) among many others,—see Appendix 2. These equations lead to a slightly larger fault length and smaller width (source HRV2 in Table 1). This source was tested with the present model. The purpose of this additional simulation was simply to investigate whether the differences in source geometry could lead to conclusions other than those of Salaree and Okal (2015).

Results of simulations may be seen in Fig. 2, where only the southern part of the Caspian Sea is shown to allow a better comparison with Fig. 7a in Salaree and Okal (2015). Results using the HRV

Table 1

Fault parameters used in simulations

	M_w	Depth (km)	ϕ_E°	λ_N°	Length (km)	Width (km)	Slip (m)	Strike (°)	Dip (°)	Rake (°)
HRV	7.4	15	49.33	36.96	65	32	1.9	300	73	32
HRV2	7.4	15	49.33	36.96	84	23	2.0	300	73	32
KRA	7.8	20	53.7	39.5	110	40	3.8	292	80	− 90
CAT1	8	33	51.55	40.36	190	35	5.24	299	87	− 86
CAT2	8	55	51.53	40.33	190	35	5.24	332	64	− 64
MIDC1	8	27	48.63	42.72	190	35	5.24	124	88	13
MIDC2	8	82	49.40	41.87	190	35	5.24	101	68	124
MIDC3	8	63	48.70	42.04	190	35	5.24	322	86	− 50
BAK1	8	50	49.94	40.24	190	35	5.24	337	85	− 53
BAK2	8	33	49.95	40.16	190	35	5.24	324	59	− 138

ϕ_E° and λ_N° are longitude and latitude, respectively, of the fault center. HRV sources refer to the June 20, 1990 Rudbar earthquake (see text). KRA is the source for the Krasnovodsk earthquake, July 8th, 1895. BAK, CAT and MIDC are variations for sources south the CAT, along the CAT and north the CAT, respectively. Magnitude in the case of KRA should be understood as M, and depth refers directly to the fault centroid depth in this case source are essentially the same as in Salaree and Okal (2015): only a narrow band along the southwest shore is affected by the tsunami, and its amplitude (wave maximum height) is extremely small, below 3 cm. Thus, it seems that the present model works correctly for the Caspian Sea.

HRV2 leads to essentially the same results. The differences in amplitudes between both simulations (HRV–HRV2) are also shown in Fig. 2. These differences are of the order of 1 mm over most of the sea. Maximum values are of the order of 1 cm in the southwestern corner of the sea. Consequently, it seems that conclusions of Salaree and Okal (2015) are confirmed: the tsunami triggered directly by the earthquake is too small to agree with visual observations of inundation. Thus, the observed tsunami could have been produced by a submarine landslide triggered by the earthquake, as stated by these authors.

An additional simulation was carried out with a slight variation of the source, as reported in ISC[1] bulletin (event ID is 366754 in this database). This source was obtained by the USGS[2] and implies a larger rake angle (54° instead of 32°), which could enhance the vertical displacements. Nevertheless, results are essentially the same as before (thus they are not shown) and the conclusions above do not change.

After comparing model results with those obtained by a previous calculation using an identical source, the model was used to evaluate the propagation of tsunamis not simulated before.

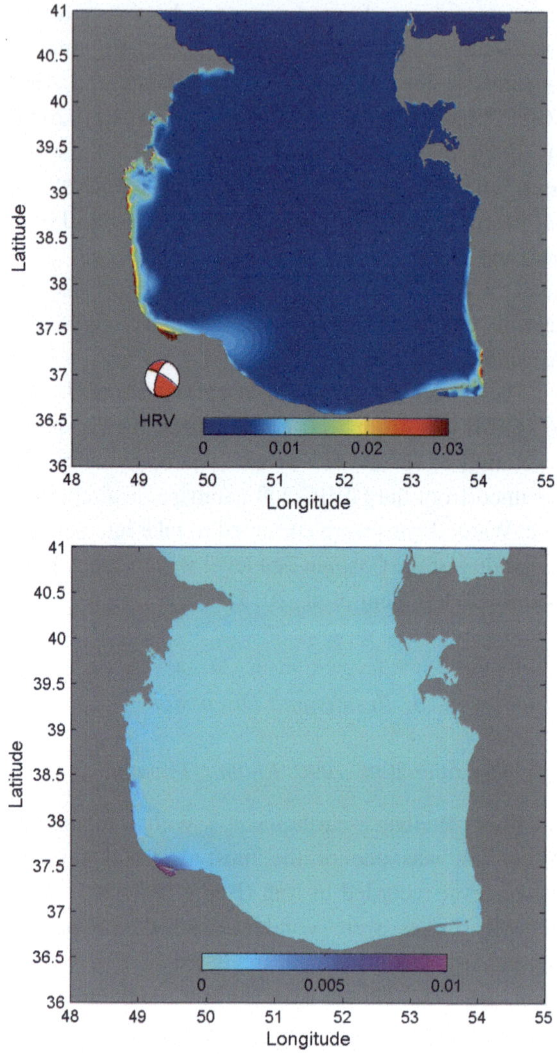

Figure 2
Top: tsunami maximum wave height (m) for the June 20th, 1990 earthquake computed with HRV source. Bottom: differences (m) in amplitudes for this tsunami calculated for HRV and HRV2 sources (HRV–HRV2). A beach ball representing the focal mechanism is drawn at the epicenter location

3.2. The Krasnovodsk Earthquake, July 8th, 1895

This was the strongest earthquake of the Transcaspian region and has been studied in detail by Ambraseys (1997) and Balakina and Moskvina (2007). In this case there is no clear solution for the source due to insufficient amount and quality of seismic data. Thus, plausible values for required parameters have been compiled from information in current literature. According to those authors, the most likely origin of the earthquake is a normal fault in direction WNW and dipping northwards at an angle of 80°. Consequently, strike, dip and rake are fixed as 292°, 80° and $-90°$, respectively (source KRA in Table 1). Balakina and Moskvina (2007) found that 7.8 is a realistic value for the earthquake magnitude. They add that fault length should be between 100 and 120 km and fault width between 30 and 50 km. Rupture takes place from the surface, thus

[1] International Seismological Centre; http://www.isc.ac.uk/iscbulletin/search/fmechanisms/interactive/..

[2] US Geological Survey; https://earthquake.usgs.gov/earthquakes/search/..

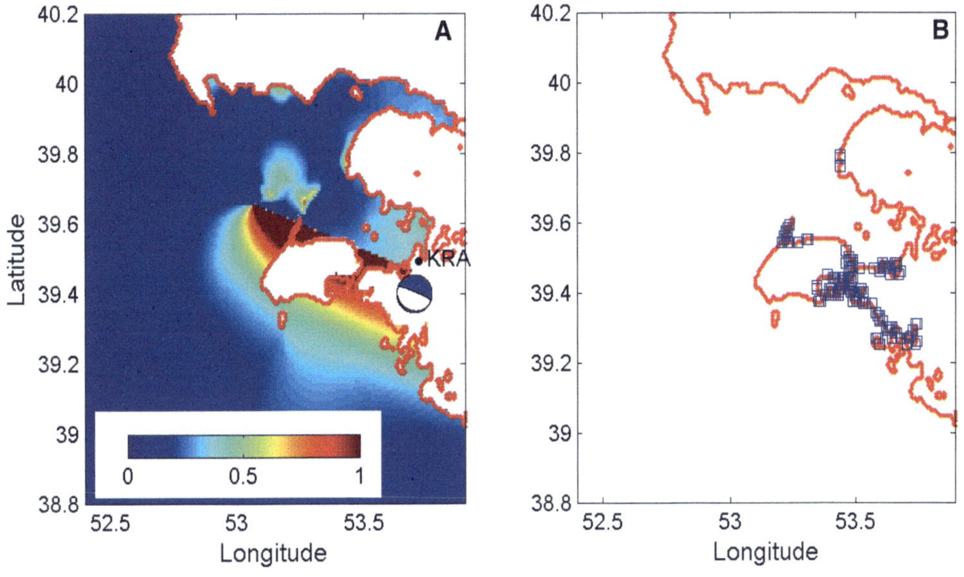

Figure 3
a Tsunami maximum wave height (m) for the Krasnovodsk 1895 earthquake. The red line is the reconstructed 1895 coastline. The beach ball corresponding to the source in Table 1 is drawn; the black dot indicates the epicenter location. **b** Blue squares indicate land grid cells which are flooded after the tsunami

20 km may be a realistic value for the fault centroid depth. There is no information about the slip magnitude. A value equal to 3.8 m is obtained using the equations by Wells and Coppersmith (1994) for a magnitude of 7.8, which agrees with the average displacement of 2.5–3.6 m mentioned by Balakina and Moskvina (2007). Parameters are summarized in Table 1. In addition, the sea level was displaced 2 m upwards according to Ambraseys (1997). Actually, the present-day Cheleken Peninsula was then an island. As mentioned in Appendix 1, topography was also corrected due to subsidence or uplift caused by the fault adding the Okada deformation to the original level. This correction may be relevant in calculating inundation for an earthquake occurring inland.

There are no records of field surveys for this tsunami, but only a few historical notes (Kondorskaya and Shebalin 1982; Balakina and Moskvina 2007): after the earthquake, sea waves flooded Cheleken Island and other coastal areas. Thus, it was suggested that the earthquake might have been accompanied by a tsunami (Balakina and Moskvina 2007). This tsunami was simulated for the first time using the source parameters in Table 1.

Figure 3a shows the amplitude of the computed tsunami, together with the 1895 reconstructed coastline. Blue squares in Fig. 3b indicate the grid cells which are flooded by the tsunami. It may be seen that the eastern coast of the Cheleken Peninsula, which was then an island, was flooded. Significant flooding also occurs over the mainland, in front and south of Cheleken Island. This is consistent with information in Balakina and Moskvina (2007). The tsunami amplitude achieves 1.5 m, but only over a very limited area. Its amplitude quickly decreases as the wave propagates into the open waters of the Caspian Sea [according to Green's law (1838)], leading to a negligible effect far from the source area. This behaviour is attributed to its occurrence in very shallow waters.

3.3. Hypothetical Tsunami Scenarios

Some hypothetical tsunamis were simulated in three other seismically active regions of the Caspian: the CAT, and south and north of the CAT (these two in the western Caspian Sea). The latter two areas are denoted, respectively, as BAK and MIDC (from Baku and Middle Caspian).

Table 2

Earthquake ID in ISC database, magnitude, date and time (UTC) for the real earthquakes which have been used as a basis to simulate the hypothetical tsunamis listed in Table 1

ID	M_w	Date	Time	Source in Table 1
2770732	6.8	25/11/2000	18:09	BAK1
1742450	6.5	25/11/2000	18:10	BAK2
499877	6.5	06/03/1986	00:05	CAT1
392584	6.5	16/09/1989	02:05	CAT2
1860504	5.0	05/06/2001	15:33	MIDC1
217463	5.3	31/08/1993	06:55	MIDC2
1734652	4.6	22/04/2000	08:46	MIDC3

The procedure, to allow a simple estimation of the source parameters, was as follows. The largest magnitude earthquakes which have occurred in the areas to date were identified from databases of the ISC bulletin and USGS. These events are listed in Table 2. Focal mechanisms are provided in these databases and their parameters (epicenter location and depth, fault strike, rake and dip) were used in the simulations, but earthquake magnitudes were increased to 8, considered as representative of the worst expected earthquake in the area according to Ulomov (2008). This author states that the largest magnitude of the seismic shock in the Caspian region is not expected to exceed 8. On the other hand, it has been stated that 6.8 may be the minimum magnitude required to trigger a tsunami in the Caspian Sea (Dotsenko et al. 2002)—actually, there is no evidence that the real earthquakes produced any tsunami. The magnitude of 8 was used with the Wells and Coppersmith (1994) relations to deduce the fault size and slip for these hypothetical studies. All sources are detailed in Table 1.

Thus, it has to be kept in mind that these sources remain hypothetical, although based upon real data in these regions.

3.3.1 Middle Caspian

According to data collected in the USGS bulletin since 1990, the largest earthquakes identified in this region have magnitudes 5, 5.3 and 4.6 (Table 2). However, as commented above, the earthquake magnitude was increased to 8. The corresponding sources are given in Table 1.

Amplitudes of the generated tsunamis are shown in Fig. 4. The smallest amplitudes occur for MIDC1. Although this is the shallowest earthquake in this region (27 km), the rake angle of 13° produces a small vertical displacement of the sea floor, thus small waves are generated. In all cases the southern Caspian remains essentially unaffected. The fault strikes prevent energy propagation towards the southeast. Towards the north, the shallow waters produce significant energy dissipation and the wave amplitude reduces (Green 1838); most of the tsunami energy is directed to the west and east shores of the northern Caspian.

3.3.2 Cheleken–Apsheron Threshold

The two largest earthquakes identified in this region by the USGS both have magnitudes of 6.5 (Table 2). Again, the earthquake magnitude was increased to 8. The corresponding sources are given in Table 1.

Resulting amplitudes are presented in Fig. 5. Waves are larger in CAT1 than in CAT2. This may be attributed to the fact that the CAT1 earthquake is shallower than CAT2 (33 and 55 km respectively). Also, displacement is more vertical in the case of the CAT1 earthquake. Energy is mainly directed to the east and west in both cases. The entrance of the Kara-Bogaz-Gol is hit by waves around 1 m high in the case of CAT1.

It is also interesting to find that amplitudes in the central part of the southern Caspian are small, but waves are amplified when reaching the southwest corner of the sea. Here they reach heights over 0.5 m.

As an example, some snapshots of water elevations for the CAT1 earthquake are presented in Fig. 6. The western shore (Apsheron Peninsula) of the Caspian is hit by waves a few minutes after the tsunami. In contrast, sea level initially recedes in the eastern shore. There is a slow propagation towards

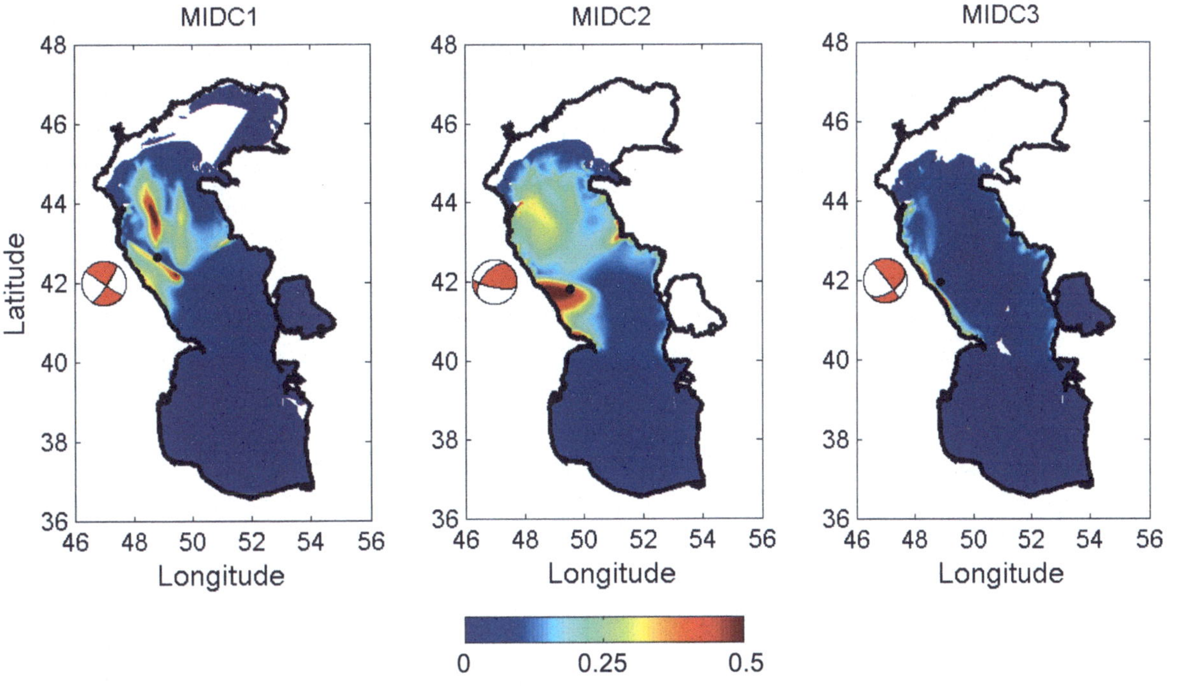

Figure 4
Tsunami maximum wave height (m) for the MIDC1, MIDC2 and MIDC3 sources. The corresponding beach balls to the sources in Table 1 are drawn; the black dots indicate the epicenter location

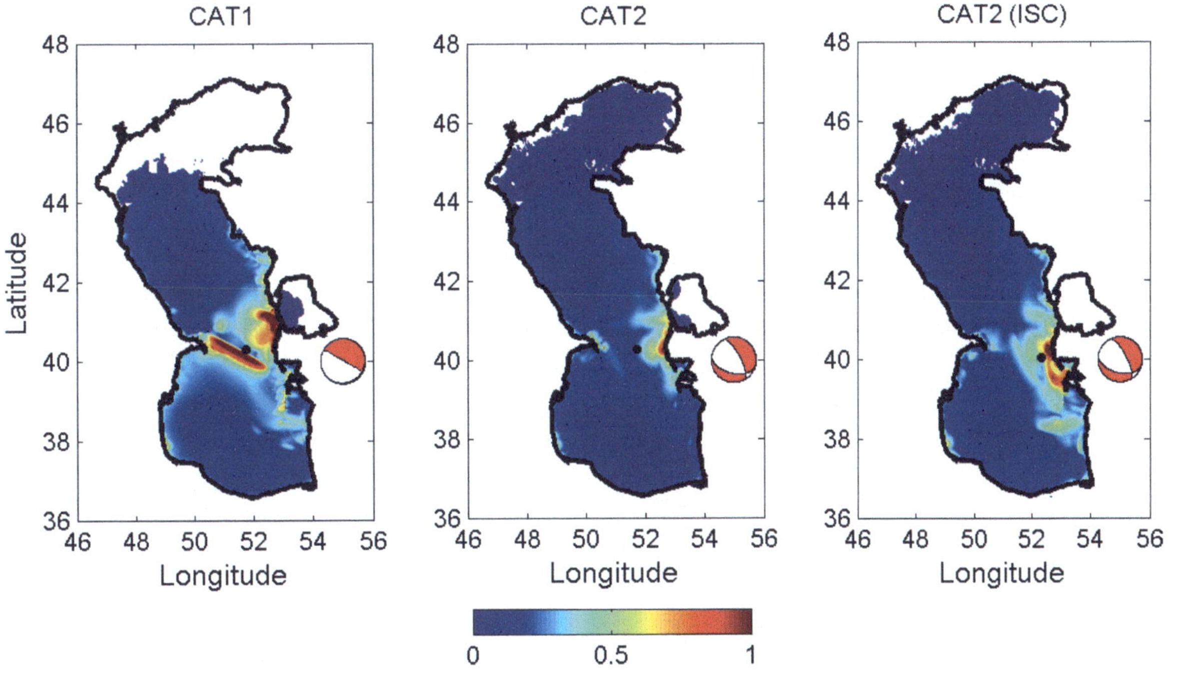

Figure 5
Tsunami maximum wave height (m) for the CAT sources. The ISC map corresponds to CAT2 source but extracted from the ISC bulletin instead of USGS (see text for details). The corresponding beach balls to the sources in Table 1 are drawn; the black dots indicate the epicenter location

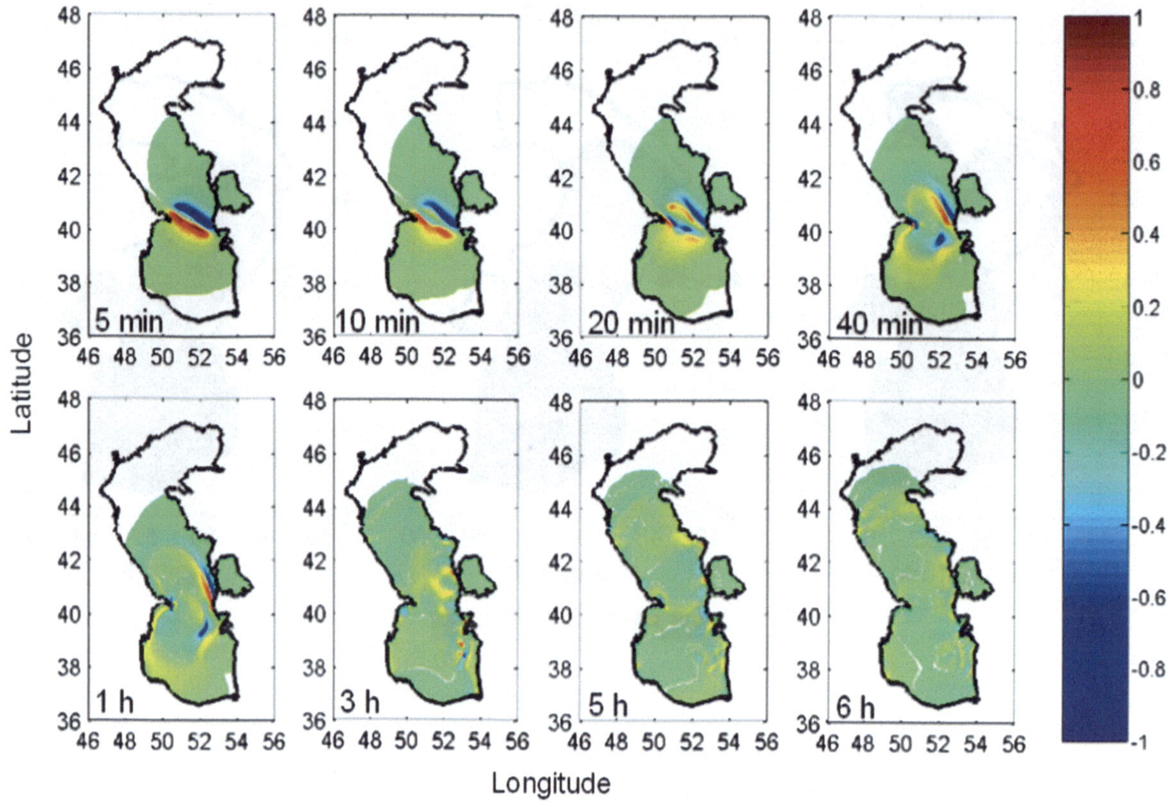

Figure 6
Snapshots of sea surface elevation (m) for CAT1 source

the north due to the shallower waters here: the first wave front (although small) reaches the 45° latitude after some hours. This is in agreement with the calculations in Dotsenko et al. (2002) with a simple idealized circular tsunami. Propagation towards the south is faster due to the deeper waters of the southern Caspian Sea. The tsunami reaches the south coast in about 1 h.

Another example of results, consisting of time series of water surface elevations with respect to the mean sea level for several coastal locations, can be seen in Fig. 7. An interesting feature is that sea level goes down by about 1 m in Garabogaz, approximately 1.5 h after the earthquake. Then a wave about 80 cm high quickly arrives. More waves arrive about 3 and 5 h after the earthquake. Wave maximum height in Baku is about 40 cm. Here, successive waves arrive with an approximate period of 1.5 h. Amplitudes about 50 cm are also apparent in Awaza; smaller waves reach the other locations.

The focal mechanisms of these earthquakes can also be found in the ISC bulletin. All parameters are the same for CAT1. In the case of CAT2 there is a discrepancy in depth: it is 35 km in ISC instead of the 55 km depth which is defined in USGS (Table 1). Also, the location of the epicenter is slightly different, closer to the shore: 52.17°E, 40.10°N. These changes lead to a higher impact south from Cheleken (Fig. 5).

3.3.3 Baku Area

The two largest earthquakes identified in this region in the USGS have magnitudes 6.8 and 6.5 (Table 2). Again, the earthquake magnitude was increased to 8. The corresponding sources are given in Table 1.

Calculated amplitudes are presented in Fig. 8. Waves over 1 m are apparent, and they are slightly larger in BAK2 than in BAK1. As before, this must be attributed to the different depths (50 and 33 km). The largest amplitudes are found in the Apsheron

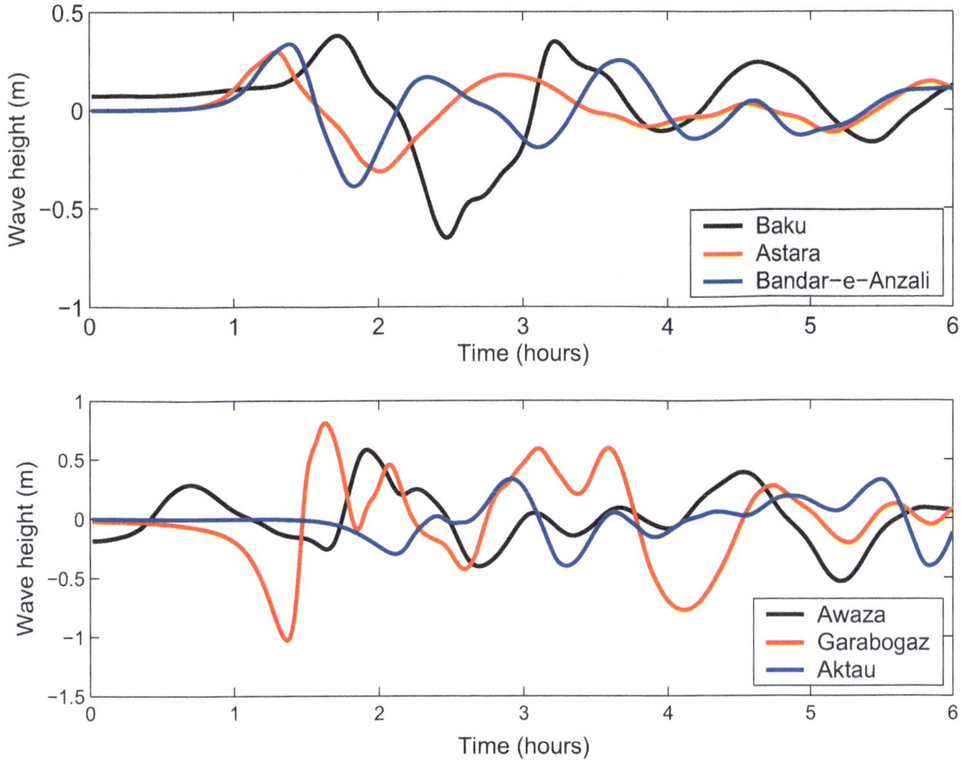

Figure 7
Time series of sea surface elevation (m) for CAT1 source at several coastal locations. These are shown in Fig. 1

Peninsula since the earthquake epicenter is in this area; in particular, the city of Baku would be affected by this potential tsunami.

4. Conclusions

A numerical model which simulates tsunami propagation, previously tested in other marine areas, was applied to simulate the propagation of tsunamis produced by submarine earthquakes in the Caspian Sea. The case of the June 20, 1990, the Rudbar earthquake was used to compare model results with previous calculations. Our results, in agreement with those, indicated that no significant tsunami was produced by that earthquake. Also, this simulation indicated that the model was appropriately set-up for the Caspian Sea.

The 1895 Krasnovodsk earthquake was simulated for the first time. Results show that the Cheleken peninsula, which was an island at that time, suffered flooding along most of its coast. Significant flooding also occurs over the mainland, in front and south of Cheleken Island. This is consistent with information in literature. The tsunami amplitude was significant only over a very limited area, quickly decreasing when entering the Caspian Sea open waters. This has to be attributed to the occurrence of the tsunami in very shallow waters.

Some hypothetical earthquakes occurring in the most seismically active areas of the Caspian Sea were simulated. These areas are the Cheleken–Apsheron Threshold and the west side of the Caspian, both north and south of the Cheleken–Apsheron Threshold. In general, both the east and west coasts of the central Caspian Sea would be the most affected by these earthquake-induced tsunamis. Tsunamis generated by earthquakes in the CAT produce the largest impacts in the shore. Tsunamis generated in the MIDC area generally have a very local effect near the source, not propagating to the southern Caspian Sea. In the case of the BAK area, tsunamis generated here

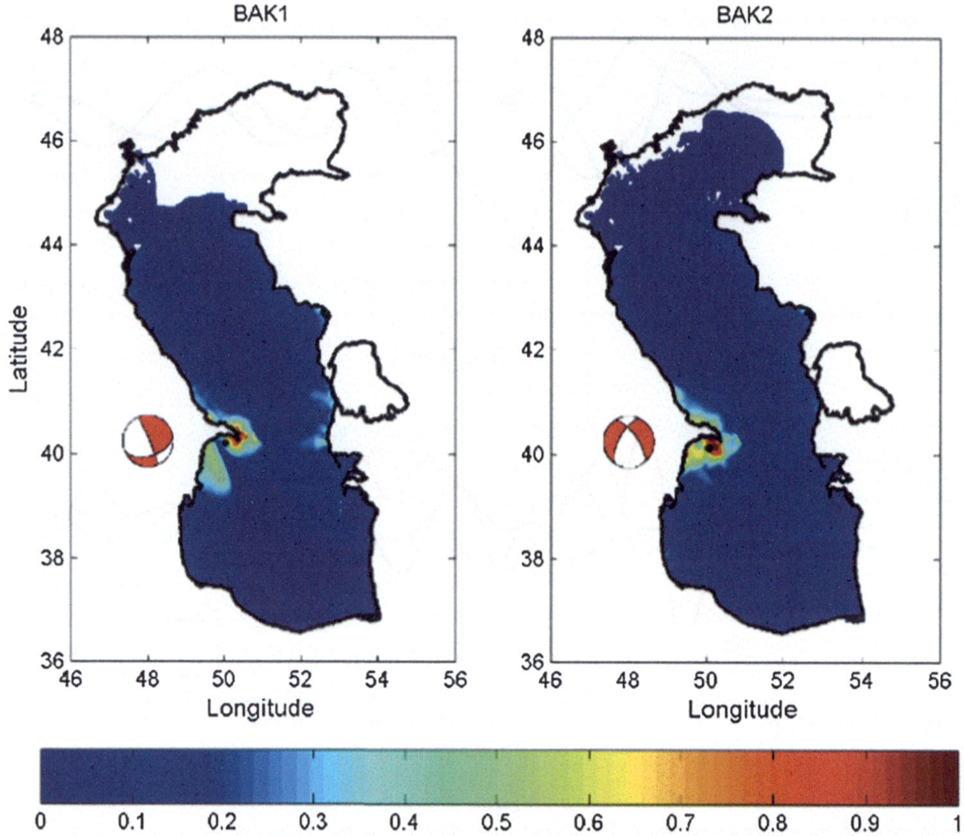

Figure 8
Tsunami maximum wave heights (m) for the BAK sources. The corresponding beach balls to the sources in Table 1 are drawn; the black dots indicate the epicenter location

could produce a significant impact in the close area of the city of Baku.

Acknowledgements

The authors are grateful to Mr Andrew Little, Defence Academy of the UK, for kindly reviewing the English of this paper.

Appendix 1: Model Description

The 2D depth-averaged barotropic hydrodynamic equations, which describe the propagation of surface shallow water gravity waves, are [see for instance (Kowalik and Murty 1993)]:

$$\frac{\partial \zeta}{\partial t} + \frac{\partial}{\partial x}(Hu) + \frac{\partial}{\partial y}(Hv) = 0; \quad (1)$$

$$\frac{\partial u}{\partial t} + u\frac{\partial u}{\partial x} + v\frac{\partial u}{\partial y} + g\frac{\partial \zeta}{\partial x} - \Omega v + \frac{\tau_u}{\rho H} = A\left(\frac{\partial^2 u}{\partial x^2} + \frac{\partial^2 u}{\partial y^2}\right); \quad (2)$$

$$\frac{\partial v}{\partial t} + u\frac{\partial v}{\partial x} + v\frac{\partial v}{\partial y} + g\frac{\partial \zeta}{\partial y} + \Omega u + \frac{\tau_v}{\rho H} = A\left(\frac{\partial^2 v}{\partial x^2} + \frac{\partial^2 v}{\partial y^2}\right), \quad (3)$$

where u and v are the depth averaged water velocities along the x and y axis, h is the depth of water below the mean sea level, ζ is the displacement of the water surface above the mean sea level measured upwards, $H = h + \zeta$ is the total water depth, Ω is the Coriolis parameter ($\Omega = 2\omega \sin \lambda$, where ω is the Earth rotational angular velocity and λ is latitude), g is acceleration due to gravity, ρ is a mean value of water density and A is the horizontal eddy viscosity. τ_u and τ_v are friction stresses which have been written in terms of a quadratic law:

$$\tau_u = k\rho u\sqrt{u^2 + v^2}; \quad \tau_v = k\rho v\sqrt{u^2 + v^2}, \tag{4}$$

where k is the bed friction coefficient. Essentially, these equations express mass and momentum conservation. Parameter values were set to $k = 0.0015$ and $A = 10$ m^2/s (Periáñez and Abril 2013, 2014).

All the equations are solved using explicit finite difference schemes (Kowalik and Murty 1993) with second order accuracy. In particular, the MSOU (Monotonic Second Order Upstream) is used for the advective non-linear terms in the momentum equations. Time step is fixed as 5 s to ensure stability.

Boundary conditions have to be specified. There is no water flow towards a dry grid cell. A flood/dry algorithm is required since when the tsunami reaches the coast new wet or dry grid cells may be generated due to run-up or rundown. The numerical scheme described in Kampf (2009) was adopted. Wet grid cells are defined as those with a total water depth H larger than a threshold value $H_{\min}$ typically set as a few centimeters. Dry cells are defined as cells where $H \leq H_{\min}$. Flooding and drying is implemented in the code via the calculation of the water velocity normal to the interface between wet and dry cells. The calculation is performed when the pressure gradient force is directed towards the dry cell. Otherwise velocity is set to zero at this point. In the case of a non-zero velocity, water level in the dry cell will increase and the cell turns into a wet one once the water depth is larger than $H_{\min}$. In the present study, $H_{\min} = 0.10$ m has been fixed following Kampf (2009) and our previous studies (Periáñez and Abril 2013, 2014; Abril and Periáñez 2017).

Still waters (zero water elevations and velocities over all the domain) are used as initial conditions across the entire domain, except for those areas affected by the deformation in the free surface induced by the earthquake.

The floor deformation produced by the earthquake is computed using the classical Okada formulae (Okada 1985) through a Matlab script.[3] Inputs for this equation are fault plane strike, rake, dip, slip, location, length and width (a rectangular fault is assumed), as well as magnitude of the earthquake and fault centroid depth.

Topography over the domain (both sea and land) is corrected due to subsidence or uplift caused by the fault adding the Okada deformation to the original level. Then, in water cells, the standard procedure assumes that this deformation is instantaneously transferred to the free water surface, and it is imposed as the initial conditions for the numerical simulation, as commented above.

[3] https://es.mathworks.com/matlabcentral/fileexchange/25982-okada–surface-deformation-due-to-a-finite-rectangular-source.

Appendix 2: Empirical Scaling Relationships

Empirical scaling relationships between the fault size and slip with the earthquake magnitude proposed by Wells and Coppersmith (1994) are:

$$\log L = 0.59M_{\mathrm{w}} - 2.44; \tag{5}$$

$$\log W = 0.32M_{\mathrm{w}} - 1.01; \tag{6}$$

$$\log S = 0.69M_{\mathrm{w}} - 4.80, \tag{7}$$

where M_{w} is the earthquake magnitude, L and W are the fault length and width (both in km) and S is the average slip (in m) along the fault.

It should be commented that Stirling et al. (2013) provides scaling laws for the different tectonics of the considered earthquake. However, some of the required laws do not give the values for all parameters which are needed to evaluate deformation according to the Okada (1985) formulae. Consequently, the general laws (which do not differentiate between tectonic regimes) of Wells and Coppersmith (1994) have been used because i) they provide all required parameters (fault length, width and slip) by Okada equations; ii) correlations are better than those of scaling laws, which are specific for given tectonic regimes and iii) the maximum earthquake magnitude used when establishing such specific laws was lower than 8, while the general relations presented above included magnitudes up to 8.1 (note than 8 is the maximum magnitude used in our simulations). In any case, it should be kept in mind that these empirical relations provide only approximations to the real fault geometries.

References

Abril, J. M., & Periáñez, R. (2016). Revisiting the time scale and size of the Zanclean flood of the Mediterranean (5.33 Ma) from CFD simulations. *Marine Geology, 382*, 242–256.

Abril, J. M., & Periáñez, R. (2017). A modelling study on tsunami propagation in the Red Sea: Historical events, potential hazards and spectral analysis. *Ocean Engineering, 134*, 1–12.

Alasset, P. J., Hébert, H., Maouche, S., Galbini, V., & Meghraoui, M. (2006). The tsunami induced by the 2003 Zemmouri earthquake (MW = 6.9, Algeria): Modelling and results. *Geophysical Journal International, 166*, 213–226.

Ambraseys, N. N. (1997). The Krasnovodsk (Turkmenistan) earthquake of 8 July 1895. *Journal of Earthquake Engineering, 1*, 293–317.

Balakina, L. M., & Moskvina, A. G. (2007). Seismogenic zones of the Transcaspian Region: characteristics of sources of the largest earthquakes II. The Krasnovodsk and Kazandzhik earthquakes. *Physics of the Solid Earth, 43*, 378–403.

Berberian, M., Qorashi, M., Jackson, J., Priestley, K., & Wallace, T. (1992). The Rudbar-Tarom earthquake of 20 June 1990 in NW Persia: Preliminary field and seismological observations, and its tectonic significance. *Bulletin of the Seismology Society of America, 82*(4), 1726–1755.

Choi, B. H., Pelinovsky, E., Kim, D Ch., Kim, K. O., & Kim, K. H. (2008). Three-dimensional simulation of the 1983 central East (Japan) Sea earthquake tsunami at the Imwon Port (Korea). *Ocean Engineering, 35*, 1545–1559.

Dotsenko, S.F., Kuzin, I.P., Levin, B.V., & Solovieva, O.N. (2002). Tsunamis in the Caspian Sea: historical events, regional seismicity and numerical modeling. In *Petropavlovks-Kamchatsky Tsunami Workshop*, September 10–15 2002.

Geller, R. J. (1976). Scaling relations for earthquake source parameters and magnitudes. *Bulletin of the Seismological Society of America, 66*, 1501–1523.

Green, G. (1838). On the motion of waves in a variable canal of small depth and width. *Transactions of the Cambridge Philosophical Society, 6*, 457–462.

Ioualalen, M., Arreaga-Vargas, P., Pophet, N., et al. (2010). Numerical modelling of the 26th December 2004 Indian Ocean tsunami for the southeastern coast of India. *Pure and Applied Geophysics, 167*, 1205–1214.

Jackson, J., Priestley, K., Allen, M., & Berberian, M. (2002). Active tectonics of the South Caspian Basin. *Geophysical Journal International, 148*, 214–245.

Kadirov, F. A., & Gadirov, A. H. (2014). A gravity model of the deep structure of South Caspian basin along submeridional profile Alborz-Absheron Sill. *Global and Planetary Change, 114*, 66–74.

Kaeriyama, H., Ambe, D., Shimizu, Y., et al. (2013). Direct observation of ^{134}Cs and ^{137}Cs in surface seawater in the western and central North Pacific after the Fukushima Dai-ichi nuclear power plant accident. *Biogeosciences, 10*, 4287–4295.

Kampf, J. (2009). *Ocean Modelling for Beginners*. Heidelberg: Springer.

Kondorskaya, N.V., & Shebalin, N.V. (1982). New Catalog of Strong Earthquakes in the USRR from ancient times through 1977. Report SE-31. Translated and Published by World Data Center for Solid Earth Geophysics. NOAA, US.

Kowalik, Z., & Murty, T. S. (1993). *Numerical modelling of ocean dynamics*. Singapore: World Scientific.

Kulikov, E. A., Kuzin, I. P., & Yakovenko, O. I. (2014). Tsunamis in the central part of the Caspian Sea. *Oceanology, 54*, 435–444.

Naderi Beni, A., Lahijani, H., & Harami, R. M. (2013). Caspian sea-level changes during the last millennium: Historical and geological evidence from the south Caspian Sea. *Climate of the Past, 9*, 1645–1665.

Okada, Y. (1985). Surface deformation due to shear and tensile faults in a half-space. *Bulletin of the Seismological Society of America, 75*, 1135–1154.

Periáñez, R., & Abril, J. M. (2013). Modelling tsunami propagation in the Iberia-Africa plate boundary: Historical events, regional exposure and the case-study of the former Gulf of Tartessos. *Journal of Marine Systems, 111–112*, 223–234.

Periáñez, R., & Abril, J. M. (2014). Modelling tsunamis in the Eastern Mediterranean Sea. Application to the Minoan Santorini tsunami sequence as a potential scenario for the biblical Exodus. *Journal of Marine Systems, 139*, 91–102.

Periáñez, R., & Abril, J. M. (2015). Computational fluid dynamics simulations of the Zanclean catastrophic flood of the Mediterranean (5.33 Ma). *Palaeogeography, Palaeoclimatology, Palaeoecology, 424*, 49–60.

Salaree, A., & Okal, E. A. (2015). Field survey and modelling of the Caspian Sea tsunami of 1990 June 20. *Geophysical Journal International, 201*, 621–639.

Stirling, M., Goded, T., Berryman, K., & Litchfield, N. (2013). Selection of Earthquake scaling relationships for seismic-hazard analysis. *Bulletin of the Seismological Society of America, 103*(6), 2993–3011.

Titov, V., Kânoğlu, U., & Synolakis, C. (2016). Development of MOST for real-time tsunami forecasting. *Journal of Waterway, Port, Coastal, and Ocean Engineering, 142*(6), 03116004. https://doi.org/10.1061/(ASCE)WW.1943-5460.0000357.

Ulomov, V. I. (2008). Identification of potential spots and long term forecast of strong earthquakes in Northern Caucasus. In A. O. Gliko (Ed.), *Variability of environment and climate. Natural and related technogenic disasters. seismic processes and disasters* (Vol. 1). Moscow: Inst. Fiz., Ross. Akad.

Wells, D. L., & Coppersmith, K. L. (1994). New empirical relationships among magnitude, rupture width, rupture area and surface displacement. *Bullettin of the Seismological Society of America, 84*, 974–1002.

(Received May 22, 2018, revised November 19, 2018, accepted November 20, 2018, Published online November 30, 2018)

Pure Appl. Geophys. 176 (2019), 3167–3183

https://doi.org/10.1007/s00024-018-2047-y

Pure and Applied Geophysics

Magnetic Signals at Easter Island During the 2010 and 2015 Chilean Tsunamis Compared with Numerical Models

CARLOS E. TORRES,[1] IGNACIA CALISTO,[1] and DANTE FIGUEROA[1]

Abstract—Tsunami flow across the geomagnetic field induces electric currents in both the ocean and the ionosphere. These currents generate secondary magnetic fields, observable at the seafloor and on land. Magnetic field disturbances during two Chilean tsunamis (2010 and 2015) were recorded at Easter Island Observatory. These tsunamis were produced after the earthquakes in Chile on February 27, 2010, at 06:34 UTC, and September 16, 2015, at 22:55 UTC, respectively. Differences between existing studies regarding the origin of the tsunamigenic disturbances, and the lack of studies using a numerical model for both the water velocity and the induced magnetic field, motivated us to compare magnetic records with computed magnetic fields for the two events. Modeled disturbances were calculated by applying the Biot–Savart law to tsunami simulations. Our model can calculate tsunami-induced perturbations in the vertical component of the geomagnetic field, obtaining a good agreement with the observations. We hypothesize that the magnetic measurements may be produced by a combination of oceanic and ionospheric electric currents. Our analysis further reveals that even a relatively small tsunami, such as the one produced by the 2015 *Mw* 8.2 earthquake, induces an observable magnetic perturbation due to the water movement. Future improvements to our model may allow us to identify situations in which magnetic variations precede the tsunami arrival, thereby allowing early tsunami detection.

Key words: Tsunami, magnetic field, Biot–Savart, Easter Island, COMCOT, INTERMAGNET.

1. Introduction

The movement of electrically-conductive seawater across the Earth's steady magnetic field induces electric currents in the sea. These currents generate secondary magnetic fields, in the range of a few nanotesla (nT), observable at stations on land, on the seafloor, and sometimes at satellite altitudes (Larsen 1968; Tyler et al. 2003; Zhang et al. 2014). This physical mechanism is called "motional induction". Various ocean processes, such as internal waves (Pukhtyar and Kukushkin 1996), tides (Larsen 1968; Tyler et al. 2003) and circulation (Manoj et al. 2006), among others, produce detectable magnetic signals. Since part of the ocean-induced magnetic field is proportional to conductivity-weighted and depth-integrated water transports, measurements of oceanic motional induction enable the possibility of indirectly observing global oceanic transports of water, heat, and salinity (Tyler and Mysak 1995; Irrgang et al. 2018). In the case of a tsunami flow, ocean-generated magnetic signals could even precede the tsunami arrival, allowing for early detection at land-based observatories (Klausner et al. 2014, 2016; Tatehata et al. 2015).

Tsunamis have historically been a major cause of destruction for several coastal countries, and recent large events have increased the level of concern in coastal communities worldwide. The tsunami in Indonesia on December 26, 2004, causing over 200,000 fatalities, is the largest of all known catastrophes in the recorded history of tsunamis (Grilli et al. 2007). Other twenty-first century tsunami events include the Chilean tsunami of February 27, 2010 (> 500 victims) (Fritz et al. 2011); October 25, 2010 in Sumatra (> 400 victims) (Levin and Nosov 2016); the Tohoku-Oki tsunami in Japan on March 11, 2011 (>18,000 victims) (Mori et al. 2011); and again in Chile on September 16, 2015 (seven victims) (Aránguiz et al. 2017). This has prompted the search for methods able to detect tsunamis before they strike the coastal zones.

In a tsunami event, magnetic variations recorded at a ground observatory may be produced by electric

[1] Departamento de Geofísica, Casilla 160-C, Universidad de Concepción, Concepción, Chile. E-mail: catorrese@udec.cl; ctorresec@gmail.com; icalisto@dgeo.udec.cl; dantefigueroa@udec.cl

currents not only within the ocean but also in the ionosphere. The latter can be caused by atmospheric acoustic/gravity waves excited by the tsunami wavefield. This mechanism is denominated tsunami–atmosphere–ionosphere (TAI) coupling (Kherani et al. 2012). Moreover, at each instant of time, a combination of ocean, ionospheric, and even magnetospheric effects can contribute to a magnetic record (Campbell 2003). Recognizing the origin of a tsunami magnetic signal would represent an advance in incorporating this physical mechanism into a warning system.

The problem of claiming which physical mechanism is responsible for a tsunami magnetic signal is manifested in the discrepancies between existing studies. For the 2010 Mw 8.8 Chile tsunami, a periodic signal of 1 nT in the vertical component of the geomagnetic field was captured by the Easter Island Observatory (code IPM) (Manoj et al. 2011). This perturbation started at 11:35 UTC, when the tsunami wave passed around Easter Island. The oceanic origin of the electric currents generating the magnetic anomaly was reported by Wang et al. (2015). They compared the measurement at IPM with a modeled disturbance, the latter calculated using a formula (derived by Tyler (2005)) which directly relates the expected magnetic field perturbation to the simulated sea surface displacement. However, the model used by Wang et al. (2015) only reproduced the perturbation produced by the leading wave and did not predict the magnetic field induced by the trailing part of the tsunami wave train.

The interpretation of Wang et al. (2015) seems to substantively differ from the interpretation of Klausner et al. (2014). The latter analysed measurements at IPM during the 2010 event using a wavelet transform data-processing technique. Based on the work of Kherani et al. (2012) for the 2011 Mw 9.0 Japan tsunami, Klausner et al. (2014) associated the dominant periods in the wavelet power spectra to the possible influence of the TAI coupling mechanism. However, this interpretation could be better supported by a numerical simulation of electric currents in the sea and/or the ionosphere.

In previous studies (e.g., Manoj et al. 2011; Zhang et al. 2014; Tatehata et al. 2015), tsunami-generated magnetic anomalies have been detected at ground-based stations for events of Mw 8.8 or larger. Utada et al. (2011) concluded that only such large magnitude, tsunami-generating, earthquakes produce detectable ionospheric disturbances. Similarly, Schnepf et al. (2016) argued that magnetic signals generated by small tsunamis are limited to just the ocean and cannot be detected using land-based magnetometers. However, IPM clearly showed a magnetic disturbance of about 1 nT in the vertical component during the 2015 Mw 8.2 Chile tsunami, coincident with the tsunami arrival time. Moreover, based on the work of Kherani et al. (2016) for the 2011 Japan event, Klausner et al. (2016) suggested that the 2015 signal may also arise from TAI coupling, in a similar way to the 2010 event.

Discrepancies between existing studies, and the lack of an oceanic numerical model for the 2015 event in previous reports, produce an incomplete understanding of the physical mechanisms generating the tsunamigenic magnetic anomalies. Furthermore, these factors contribute to a lack of sufficient knowledge for the development of an early magnetic-based warning system in the near future. A practical consequence would be, for example, the triggering of false alarms.

Therefore, the main objective of this study is to confirm that tsunami-forced electric currents in the sea that caused magnetic field perturbations at Easter Island during the 2010 and 2015 events. In order to accomplish this aim, we first calculate seawater volume fluxes from tsunami numerical simulations. Second, we compute the generated electric currents within the ocean. Third, we calculate induced magnetic fields at IPM using the Biot–Savart law. Finally, we perform a comparison between the computed and observed magnetic fields and discuss the results. Our model can calculate tsunami-induced perturbations in the vertical component of the geomagnetic field, obtaining a good agreement with the observations.

2. Methodology

2.1. Tsunami Modeling

The sea level and currents associated with the tsunami were computed using the software COrnell

Multi-grid COupled Tsunami model (COMCOT), version 1.7 (Wang 2009). COMCOT v1.7 uses a nested grid system, which means that in a region of one grid size, one or more regions with smaller grid sizes and higher resolution are imbedded. This system allows the simultaneous calculation of tsunami propagation in the deep ocean and inundation in coastal zones using different grid resolutions, reducing computational costs. COMCOT v1.7 assumes that the uplift motion is much faster than the wave propagation, so the initial vertical water displacement is assumed to be the same as the sea floor vertical deformation, calculated here using the improved elastic finite fault plane theory of Okada (1985).

COMCOT v1.7 adopts staggered leap-frog finite difference schemes to solve linear shallow water (LSW) and nonlinear shallow water (NSW) equations in spherical and Cartesian coordinates (Liu et al. 1998). LSW equations are preferred for propagation in the deep ocean. As a tsunami reaches a coastal region, NSW equations, including bottom friction effects, are required due to the relatively short tsunami wavelengths on the lengthscale of the spatial variation in the bathymetry (Calisto et al. 2016). According to the leap-frog method, the evaluation of water surface elevation and depth-integrated water fluxes is staggered in time and space. For each time step, these aforementioned parameters are output in data files at all the grid points.

2.1.1 Fault Models and Grid Generation for the Chilean Tsunami

Using seafloor deformation data as the initial condition, COMCOT v1.7 is capable of simulating the entire life-span of a tsunami. To quickly estimate tsunami height and the first wave arrival time, homogeneous fault models, in which a constant slip occurs simultaneously, can be used. However, to have a better understanding of the tsunami source, heterogeneous models are necessary. In this case, the fault zone is divided into smaller sub-fault elements, each one with parameters of longitude, latitude, depth, dip, slip, rake, strike and area (Calisto et al. 2015, 2016).

On February 27, 2010 at 06:34 UTC, an earthquake with a moment magnitude of 8.8 occurred offshore Chile's Maule region. The epicenter was located at (36.1S, 72.6W). In another event, on September 16, 2015 at 22:55 UTC, an earthquake with a moment magnitude of 8.2 occurred 46 km offshore Illapel. The epicenter was located at (31.6S, 71.7W). Both earthquakes triggered tsunamis that reached Chilean coastal locations. To simulate tsunami waveforms, heterogeneous fault models were constructed and validated in Koper et al. (2012) for the 2010 tsunami, and constructed in Benavente et al. (2016) and validated in Calisto et al. (2016) for the 2015 event. We used these source models to perform our tsunami numerical simulations.

The first-level grid, with a resolution of 2.16 arc min, is common for all locations (Fig. 1, left). Second-(black rectangle in Fig. 1, left) and third-(Fig. 1, right) level grids, with respective resolutions of 0.27 and 0.054 arc min, were constructed around Easter Island. In order to validate the tsunami source models, accurate simulations near to the rupture zones are necessary. Therefore, second-, third- and fourth-level grids with 0.54, 0.108 and 0.018 arc min, respectively, were constructed for three locations along the Chilean coast that have high resolution bathymetric and tide gauge data available.

A linear numerical tsunami model was used for westward propagation across the Pacific Ocean. For the third- and fourth-level grids at Easter Island and the Chilean coast, respectively, NSW equations, with a bottom friction coefficient of 0.025, were used. This value is widely used in tsunami simulation and is equivalent to a coarse sand with a diameter of 2 cm (Masamura et al. 2000).

2.1.2 Validation

For comparison between the modeled and measured data, three tide gauge locations along the Chilean coast and four offshore DART buoy locations were used (Fig. 1, left). Table 1 summarizes the locations of tide gauges and DART buoys. Virtual wave gauges, as close as possible to the real equivalents, were defined for the validation. Tide gauge data were acquired from coastal stations close to Coquimbo, Valparaíso and Talcahuano. These stations are

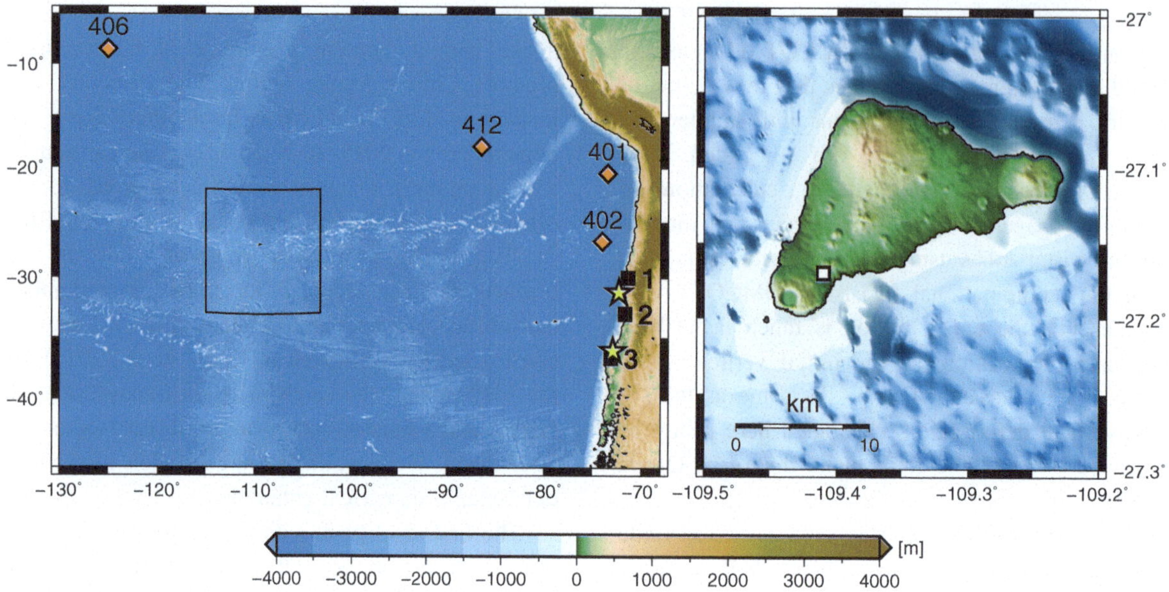

Figure 1
Left: the first-level bathymetric grid, showing 2015 and 2010 epicenters (north and south yellow stars, respectively); the positions of DART buoys 32401, 32402, 32406 and 32412 (orange diamonds); and the three coastal locations from north to south, Coquimbo, Valparaíso and Talcahuano (numbered squares). The black rectangle indicates the position of the second-level grid around Easter Island. Right: the third-level grid around Easter Island, showing the location of the Easter Island Observatory (white square). Abscissa and ordinate are longitude and latitude, respectively

Table 1
Summary of tide gauge and DART buoy locations

	Latitude	Longitude	Code
Coquimbo	29.95S	71.34W	1
Valparaíso	33.03S	71.63W	2
Talcahuano	36.70S	73.11W	3
DART 32412	17.98S	86.37W	412
DART 32401	20.47S	73.42W	401
DART 32402	26.74S	73.98W	402
DART 51406	8.48S	125.03W	406

operated by the Hydrographic and Oceanographic Service of the Chilean Navy (Spanish acronym: SHOA). Data are available in the UNESCO Intergovernmental Oceanographic Commission (IOC) site (http://www.ioc-sealevelmonitoring.org/). DART data were acquired from NOAA's National Data Buoy Center (http://www.ndbc.noaa.gov/). The oceanic tides, simulated using a classical harmonic analysis, were subtracted from the observed data to obtain the tsunami signals.

The observed tide gauge signals, compared to those generated by the numerical models, are shown in Fig. 2. During the 2010 tsunami, Talcahuano station ceased to operate approximately 57 min after the earthquake and the measured signal has a cut-off value due to the low water depth at the installation point.

The simulations can model the arrival times and waveforms of both tsunamis, particularly for the first couple of oscillations. In spite of the different behaviour of the tsunami at the three coastal locations, and independent of their distance to the tsunami source, the sea level evolution is well-modeled. Refraction effects of the bathymetry, such as trapping of long waves on the continental shelf and slope or small scale resonance close to coastlines, have a large influence on the tsunami wave trains and may produce the differences between modeled and observed tsunamis after the first sea level oscillations (Yamazaki and Cheung 2011; Bellotti et al. 2012). These differences are likely to get progressively larger, since they propagate forward through the

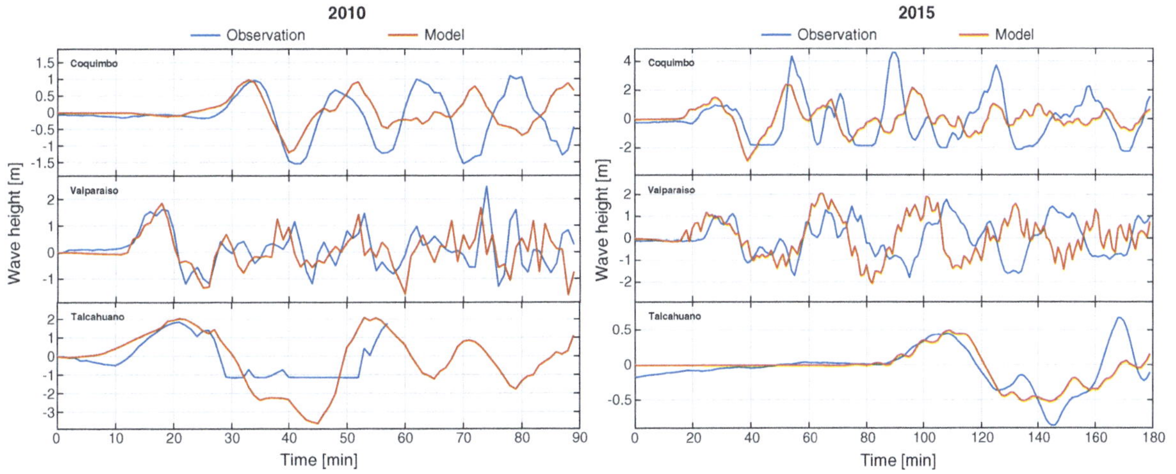

Figure 2
Comparison between observed (blue line) and modeled (red line) signals for three stations along the Chilean Coast. From top to bottom, Coquimbo, Valparaíso and Talcahuano gauges. Left: 2010 Chile tsunami. Right: 2015 Chile tsunami. Time is measured relative to the earthquake origin time. Note the different vertical scales in these figures

successive time steps in a simulation (Calisto et al. 2015). Therefore, our analysis is focused mainly on the first oscillations, with the purpose of validating the tsunami sources. Our comparison between DART data and virtual buoys is shown in Figs. 3 and 4. We obtained a very good agreement for both tsunamis, although our model arrives a couple of minutes before the measurements. The model accurately reproduces not only the leading oscillation but also the trailing part of the tsunami wave train.

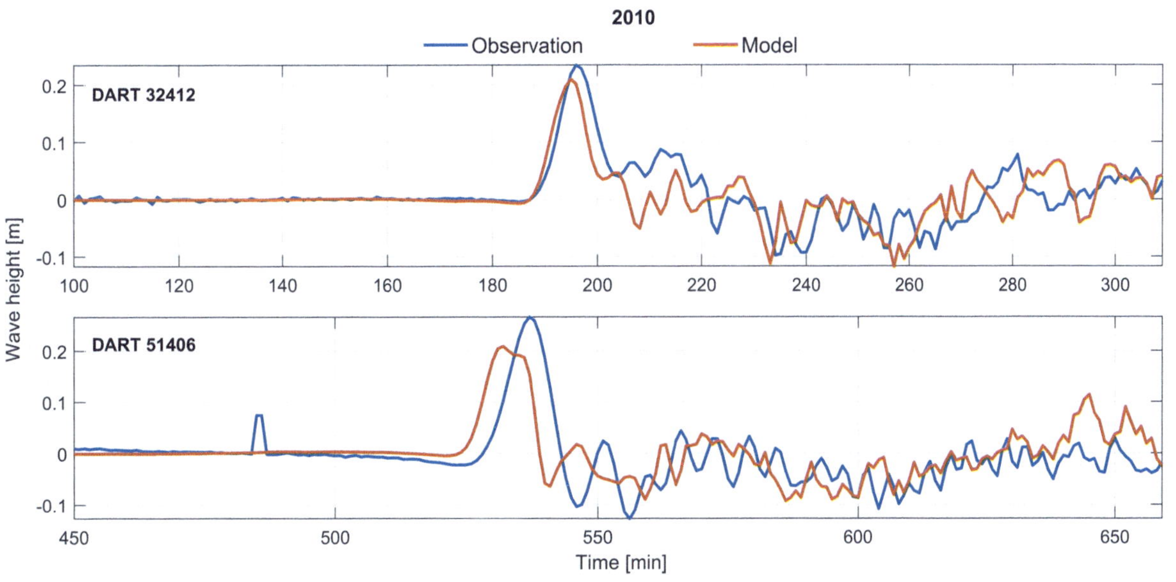

Figure 3
Comparison between the observed (blue line) and modeled (red line) 2010 tsunami signals at the DART buoy locations given in Table 1. Time is measured relative to the earthquake origin time

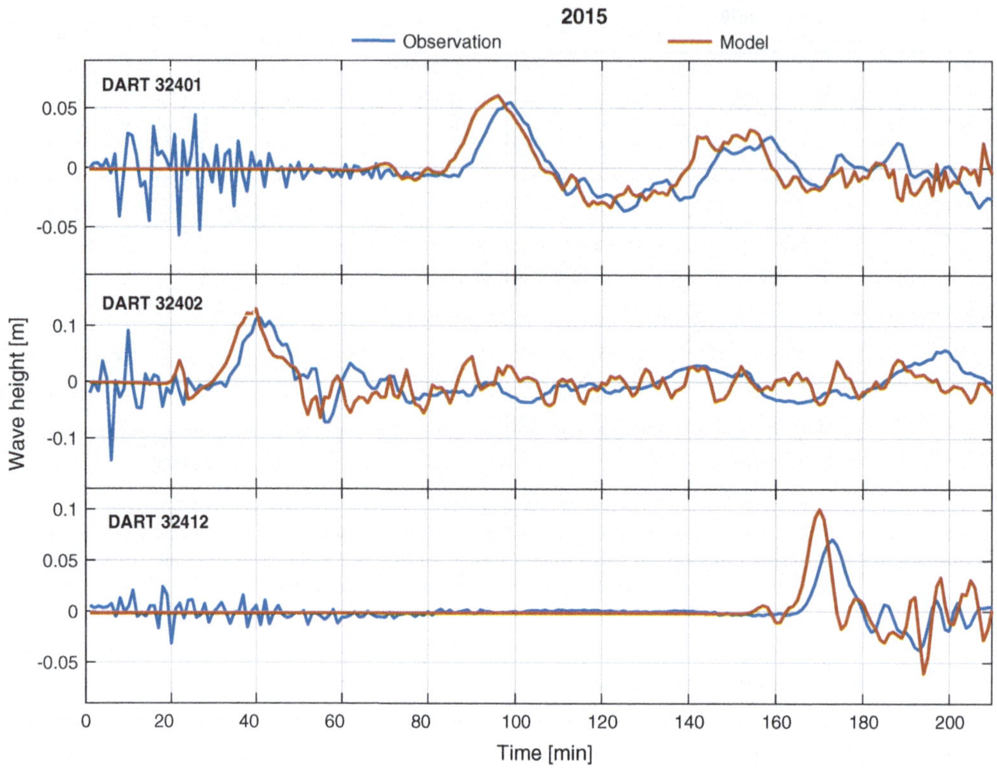

Figure 4
Comparison between the observed (blue line) and modeled (red line) 2015 tsunami signals at the DART buoy locations given in Table 1. Time is measured relative to the earthquake origin time

2.2. Magnetic Measurements

Ocean-induced magnetic fields can be decomposed into their poloidal and toroidal components using the Helmholtz decomposition theorem (Chave 1983). The poloidal magnetic mode results from electric currents flowing in horizontal planes (Minami et al. 2017). This mode is observable outside the oceans at land-based observatories. For a land observation case, induced horizontal components are much weaker than the vertical one. In fact, only the variations in the vertical component of the magnetic field have been reported for 2010 and 2015 events in previous works (Manoj et al. 2011; Klausner et al. 2014; Wang et al. 2015; Klausner et al. 2016). The vertical component is less affected than the horizontal components by solar-magnetospheric interactions, providing a greater opportunity to identify tsunamigenic variations during geomagnetic quiet days (Klausner et al. 2014). No geomagnetic storms were reported during both events. Therefore, the vertical component was used to study the ocean-generated magnetic fields.

For both tsunami events, magnetic measurements were recorded at the IPM station, located at (27.17S, 109.41W) in Easter Island, situated about 3500 km west of the Chilean coast. IPM is operated by Dirección Meteorológica de Chile and Bureau Central de Magnetisme Terrestre, France. The observatory is equipped with a fluxgate magnetometer with a resolution of 0.1 nT. The observatory provides time series of three magnetic components, quoted with respect to a local right-handed geographical tangent plane: X (northward), Y (eastward), and Z (downward), at an interval of 1 min (http://www.intermagnet.org).

When data were corrupted or unavailable, interpolation, using a cubic spline, was carried out. Then, the signals were high-pass filtered in order to remove long periods not related to the tsunami process.

Filtering was done using the wavelet transform, following the procedure described by Torrence and Compo (1998). A Morlet wavelet (with $\omega_0 = 6$) and a cutoff period of 24.77 min were chosen for the filtering process. Reconstructed, wavelet-filtered time series were obtained by adding up the real part of the wavelet transform of magnetic records over the corresponding subset of scales.

2.3. Induced Magnetic Fields

To study motional induction in the sea, simplified governing equations can be obtained by scaling the Maxwell equations for a moving conducting medium. Advection of charge, displacement currents and polarization effects can be ignored, since they are small compared with the true electric density term. By decomposing the magnetic field into a local, steady, irrotational main field part **F** and a secondary ocean-generated perturbation **b**, the differential form of the Biot–Savart law is obtained (Sanford 1971; Larsen 1973):

$$\nabla \times \mathbf{b} = \mu \mathbf{J},, \tag{1}$$

where μ is the magnetic permeability (assumed to be the free-space value, μ_0, in this analysis), **J** is the source current density given by $\mathbf{J} = \sigma(\mathbf{E} + \mathbf{u} \times \mathbf{F})$, **E** is the electric field developed in the seawater, σ corresponds to the electrical conductivity of the seawater and **u** is the seawater velocity field. By solving the Maxwell equations, the ocean-generated electromagnetic fields, **b** and **E**, can be obtained.

Using 2-D time-domain numerical simulations with solitary waves propagating over a flat seafloor, Minami et al. (2015) found that the importance of the self-induction term, $\partial_t \mathbf{b}$, in Faraday's Law of induction, increases with ocean depth. Self-induction generates a counter electric field **E** opposed to $\mathbf{u} \times \mathbf{F}$, attenuating the amplitude of the induced vertical component of magnetic field. Computing Maxwell equations using a realistic bathymetry would involve a great computational cost, therefore we decided to not consider the self-induction term in this work. As pointed out by Minami et al. (2015), a magnetostatic form is obtained by neglecting this term, and the observable magnetic fields are almost the same as those generated by the electric current density $\sigma(\mathbf{u} \times \mathbf{F})$.

According to the coordinate system used by COMCOT v1.7, *X*-, *Y*-, and *Z*-components in eastward, northward and upward directions, respectively, were used in this study. For each grid cell, the magnitude and direction of electric currents were obtained by integrating the local electric current density over both the cell width and the water column. The *X*- and *Y*-components of the electric currents, as a function of location $\mathbf{r}'$, are:

$$\begin{aligned} dI_x(\mathbf{r}') &= \sigma n(\mathbf{r}') F_z \Delta y', \\ dI_y(\mathbf{r}') &= -\sigma m(\mathbf{r}') F_z \Delta x', \end{aligned}, \tag{2}$$

where F_z is the vertical component of geomagnetic field. $\Delta x'$, $\Delta y'$, and m, n are cell widths and volume fluxes in the *X*- and *Y*-directions, respectively, computed from the numerical tsunami model mentioned in Sect. 2.1 Values of σ range from 2 to 6 S/m, depending on the salinity. Following Wang et al. (2015), conductivity was assumed to be uniformly 4 S/m in this work. No appreciable differences were obtained between simulation results by considering other values of conductivity (e.g., 3.5 S/m). Values of F_z were calculated using a mean of 31 days of measurements at IPM, centered on the day of each event. For 2010 and 2015, $F_z =$ 19,610.84 nT and $F_z =$ 19,438.77 nT were obtained, respectively.

Each electric current element induces a small magnetic field at the observation point. The ocean-generated magnetic field at IPM was computed by adding up all these contributions over a rectangular area around Easter Island, using the second and third-level grids described in Sect. 2.1.1. The integration of these magnetic field elements was carried out using the Biot–Savart law for a line current, whose expression in this case is:

$$b_z(\mathbf{r}) = \frac{\mu_0}{4\pi} \int_C \frac{dI_x(\mathbf{r}')(y-y')\Delta x' - dI_y(\mathbf{r}')(x-x')\Delta y'}{((x-x')^2 + (y-y')^2 + (z-z')^2)^{3/2}}. \tag{3}$$

The time series of the induced magnetic fields were obtained by evaluating Eq. 3 for each time step of the entire simulation duration. In order to compare the model with the measurements, the same filtering process and cutoff period described in Sect. 2.2 were applied.

Summarizing, our methodology involved four main steps:

1. For the 2010 and 2015 Chile events, obtaining the seawater volume fluxes from numerical tsunami simulations.
2. Computation of generated electric currents using Eqs. 2.
3. Computation of induced magnetic field time series using Eq. 3 for each time step.
4. Comparison between computed time series and magnetic records observed at IPM.

3. Results

3.1. 2010 Chile Tsunami

The observed and computed magnetic field variations at IPM for the 2010 event are shown in Fig. 5. A clear signal displaying seven peaks between 320 and 390 min after the earthquake was recorded at the observatory, coincident with the passing of the tsunami wave train around Easter Island. The numerical model can predict tsunami-induced perturbations, reproducing the shape and timing of observed peaks between 325 and 360 min, despite some disagreements in amplitude. Between 250 and 300 min, the model shows an initial oscillation and a trough, which appear from the filtering process. After 360 min, the amplitudes and timing only partially agree in both waveforms. Similarity in the dominant period of the observed and modeled waveforms suggests that their frequency content should be analysed.

To extract the frequency components of the data and their time dependency, a wavelet analysis was performed on time series displayed in Fig. 5. Cones of influence are not displayed, since the spectra were obtained for time series much longer than the observed and the computed induced magnetic field. A Morlet wavelet (with $\omega_0 = 6$) was chosen for feature extraction purposes, since it provides a good balance between time and frequency localization (Grinsted et al. 2004).

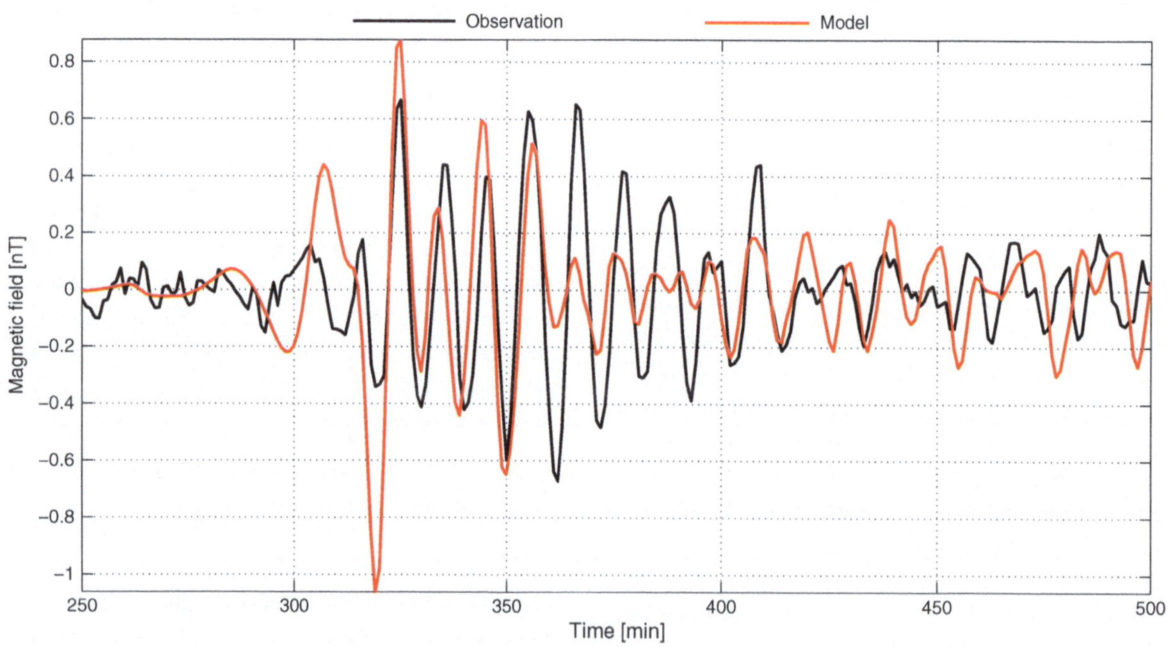

Figure 5
Comparison between the magnetic field measurement (black line) and simulated magnetic field (red line) at IPM for the 2010 Chile tsunami. Time is measured relative to the earthquake origin time

Periodograms for the observed and the computed induced magnetic field at IPM are shown in Fig. 6. Dominant range periods in model (bottom panel) and measurement (top panel) are around 12 min, although some energy around 16 min appears in the model between 309 and 325 min. For the measured data, the power is mainly distributed on the 12-min scale between 320 and 390 min after the earthquake, according to the observed signal described in Fig. 5. On the other hand, the power of the modeled signal is located mainly at the time interval 310–360 min. The model and measurement display similar power between 320 and 360 min on the 12-min scale, coincident with the comparison between time series (Fig. 5) during this time interval. Between 360 and 500 min, a reasonably similar power spectrum exists between the model and measurement, especially at around 12 min periods.

3.2. 2015 Chile Tsunami

The temporal variations in the observed and computed magnetic field at IPM for the 2015 event are shown in Fig 7. A signal displaying several distinctive peaks, against a background oscillation, was recorded at IPM between 325 and 390 min, coincident with the passing of the tsunami around Easter Island. Similar to the 2010 event, the model can reproduce the tsunami-induced perturbations. From 320 to 340 min there is a phase shift of about 5 min between simulated and observed signals, which vanishes at 350 min. With a hypothetical correction of this time lag, the resulting model and measurement would be similar, despite some differences in amplitude values, especially at the leading part of the tsunami-induced signal. From 375 min, amplitudes are similar in both waveforms, but the model predicts shapes not present in the measurement.

Periodograms for the observed induced magnetic field (top panel) and computed waveform (bottom panel) at IPM are shown in Fig. 8. As expected from Fig. 7, dominant periods around 12 min, arising likely from the passing of the tsunami wave around Easter Island, appear between 320 and 370 min for both the model and the measurement.

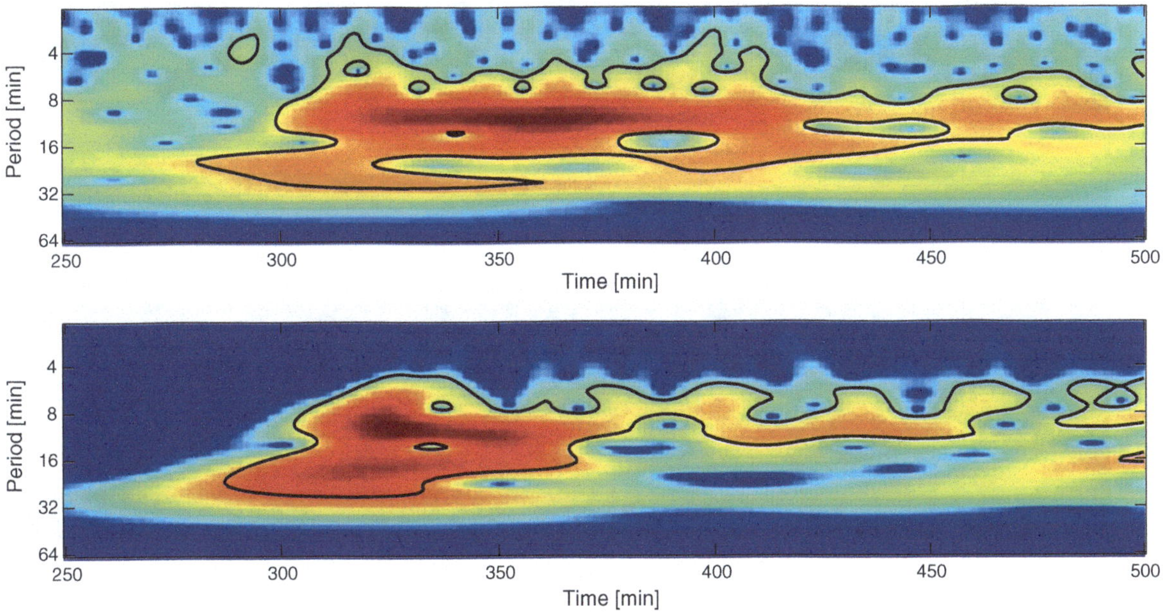

Figure 6
Wavelet power spectra of the magnetic field measurement at IPM (top) and simulated magnetic field (bottom) for the 2010 Chile tsunami. Areas of stronger power are shown in dark red color on a plot of time (horizontal) and period (vertical). The thick black contours enclose areas of greater than 95 percent of confidence for red-noise processes with lag-1 coefficients (Torrence and Compo 1998). Time is measured relative to the earthquake origin time

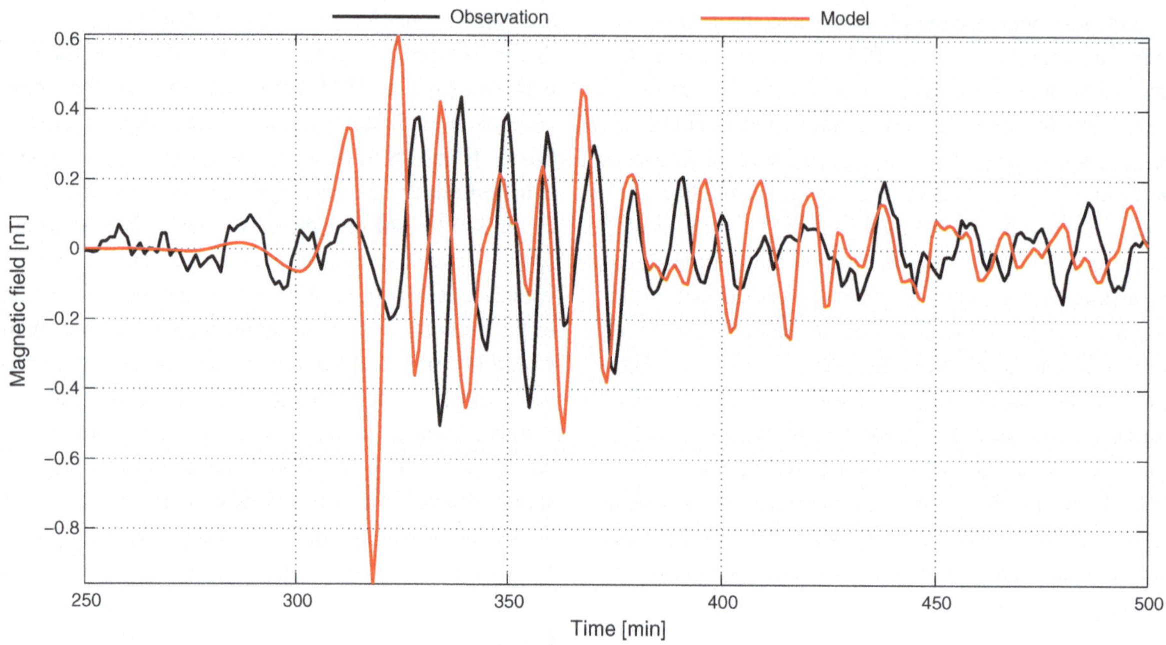

Figure 7
Comparison between the magnetic field measurement (black line) and simulated magnetic field (red line) at IPM for the 2015 Chile tsunami. Time is measured relative to the earthquake origin time

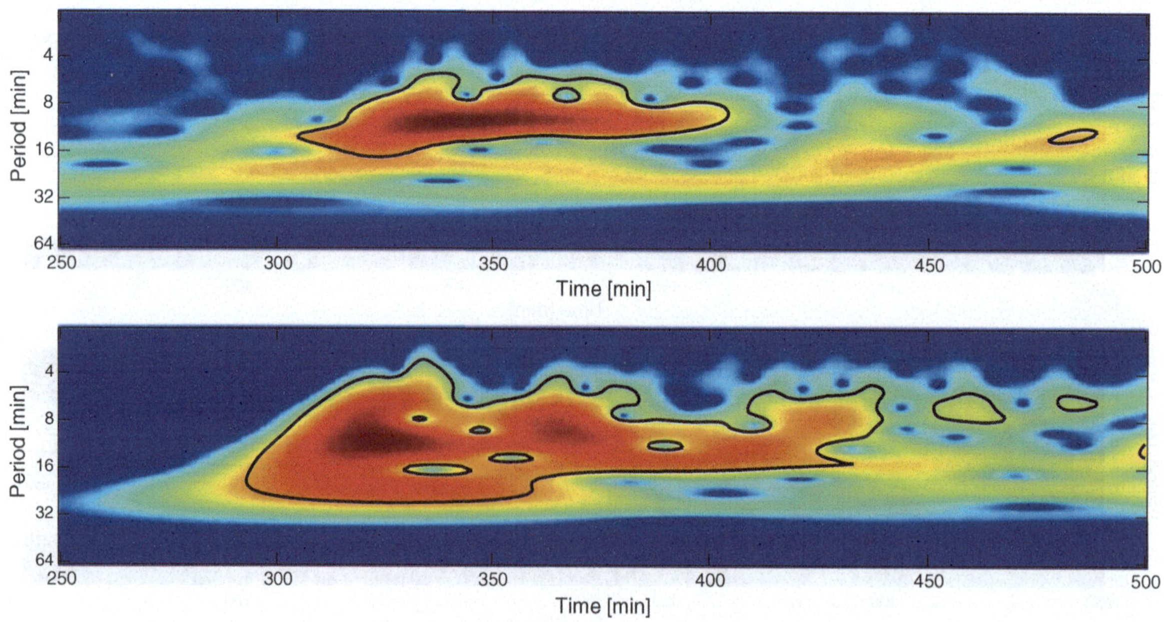

Figure 8
Wavelet power spectra of magnetic field measurement at IPM (top) and simulated magnetic field (bottom) for the 2015 Chile tsunami. The colour scale and contour displayed are the same as for Fig. 6. Time is measured relative to the earthquake origin time

4. Discussion

Electrically-conductive seawater moving across the Earth's magnetic field induces electric currents in the sea that generate secondary magnetic fields. In order to assess the potential for remote sensing of a tsunami event, numerical modeling may provide possibilities for understanding the space–time behaviour of tsunamigenic magnetic disturbances.

Two tsunami events (2010 and 2015) were considered in this study. We compare the magnetic records observed at Easter Island with the computed variations, obtained using a tsunami source model and the Biot–Savart law. Our model can reproduce the magnetic disturbances, coincident with the passing of the tsunami wave train around Easter Island, in both events. Computed waveforms agree well with the measurements, suggesting that magnetic records may be caused in part by oceanic electric currents.

The calculated and observed magnetic signals between 11:00 and 14:00 UTC during the 2010 event were presented by Wang et al. (2015) (see their Fig. 4). This time interval corresponds to 265–445 min in Figs. 5 and 6. When compared with the measured anomaly from 325 to 360 min, our computed waveform represents an improvement to their simulation. To estimate the magnetic field at IPM, Wang et al. (2015) focused the analysis in the leading wave and considered sea surface elevations from virtual mareographs located two degrees away from Easter Island. On the other hand, our model considers information from each grid cell, thereby probably explaining the better fit to the measurement for the trailing part of the tsunami.

In contrast to our model, the formula of Tyler (2005), used by Wang et al. (2015) to estimate the magnetic anomaly, is valid only in the deep ocean, far from coastlines, where the tsunami wavelength can be much smaller than the lengthscale of bathymetric variations. This restriction appears from a simplification of the continuity equation for the tsunami flow. In spite of this limitation, Tatehata et al. (2015) used a modified version of Tyler's formula to calculate electric current densities induced by the 2011 Tohoku-Oki tsunami flow in each grid cell around Chichijima Island station. The magnetic perturbation at this station was then computed using the Biot–Savart law, as in this work. They obtained a good agreement between the model and the measurement for the first four peaks. We also followed the approach of Tatehata et al. (2015) to calculate the magnetic perturbations at IPM for the 2010 and 2015 events, but -due to the computational cost required to apply the steps specified in their work- using a coarser resolution of 1.08 arc min for the second level grid around Easter Island. As we observed that these results did not show a significant improvement of the fit between observations and models (Figs. 9 and 10), we decided not to further pursue this approach in the present study.

A source of the discrepancy between our models and the measurements may be neglecting the self-induction effect. Since a magnetostatic approximation was used in this work, an amplitude attenuation of the induced vertical component of magnetic field (mentioned in Sect. 2.3) is not expected to be described by our results. To illustrate how the self-induction effect could influence our models, in Figs. 11 and 12 we show calculated tsunami wave heights and induced electric currents around Easter Island for two different times after the earthquake origin time. The left and right panels in each figure correspond to the times where our model (red lines in Figs. 5 and 7) predicts the first peaks and troughs. The generation of a counter electric field opposite to the electric current flow (black arrows in Figs. 11 and 12) could result in a reduction of these induced currents, thereby attenuating the induced magnetic field at IPM and resulting in a better fit to measurements. Although no improvements were observed by including the self-induction effect under the approach of Tatehata et al. (2015) (Figs. 9 and 10), possible improvements to our results could be achieved following the method of Minami et al. (2017). This means a refinement of our model, involving the incorporation of additional computational resources, which we intend to apply later in the next steps of our work in this project.

Another factor that might influence our model is the effect of frequency dispersion in the tsunami wave train during oceanic propagation. According to Wang et al. (2015), dispersive properties of the 2010 tsunami would explain the poor agreement between the measurement at IPM and the trailing part of the

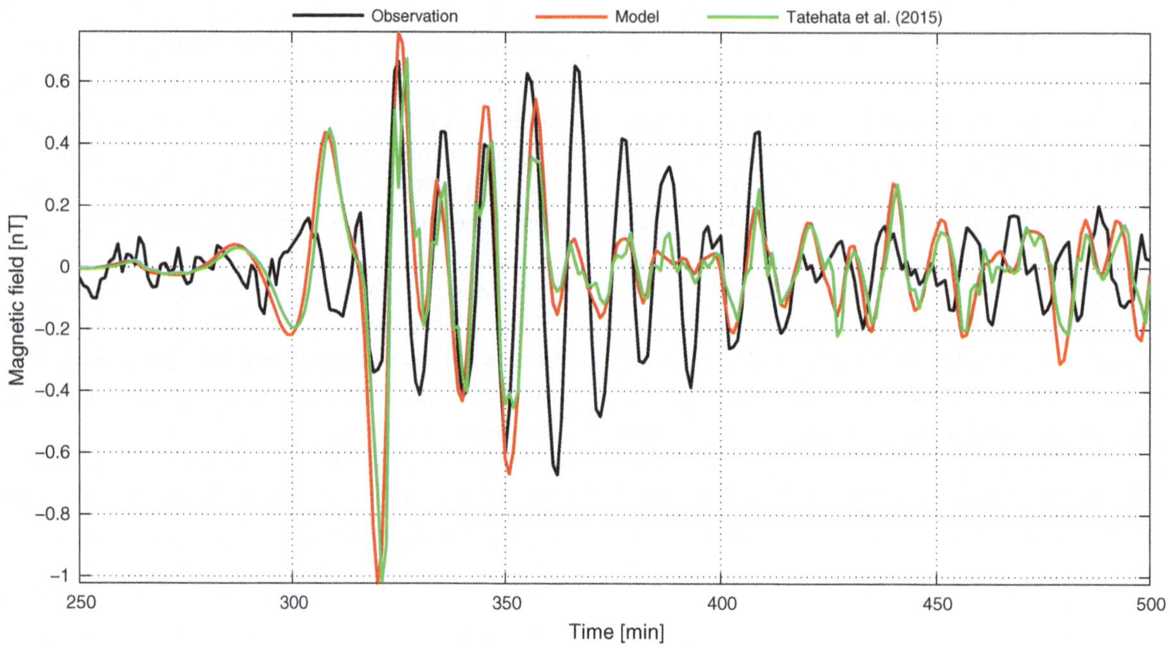

Figure 9
Comparison between the magnetic field measurement (black line), simulated magnetic field using our model (red line), and simulated magnetic field using the model of Tatehata et al. (2015) (green line) at IPM, for the 2010 Chile tsunami, considering a resolution of 1.08 arc min for the second level grid around Easter Island. Time is measured relative to the earthquake origin time

tsunami wave in their model. However, as mentioned in this section, Fig. 5 shows that our model reproduces the observed anomaly, only using shallow water equations, from at least 325 min up to 360 min. On the other hand, the phase shift between the model and the measurement in DART 51406 (right panel in Fig. 3) is also attributed by Wang et al. (2015) to dispersive effects. The 5 min time lag between the simulated and observed signals in Fig. 7 then might also be attributed to the dispersion of the tsunami wave. We simulated the induced magnetic fields at IPM for the two tsunami events using the dispersion-improved numerical scheme provided by COMCOT v1.7, but no significant differences to our original model were found. A more detailed study of the dispersion is necessary, since dispersive effects depend on the relation between the wavelength of the tsunami and the water depth, the travel time of the tsunami wave, the propagation angle from the fault strike, among other factors (Glimsdal et al. 2013; Okal and Synolakis 2016).

Following Sanford (1971), the ocean's electrical conductivity may affect the perturbed Earth's magnetic field, as the poloidal part of the ocean-induced magnetic field is sensitive to the conductivity-weighted average velocity over the vertical coordinate. In our case, we considered a uniform value for the conductivity and that the velocity field within a tsunami is coherent in the water column, as long as we use shallow water equations. Alternatively, a more realistic approximation would be to consider an island-ocean conductivity contrast (Larsen 1968). The toroidal magnetic component couples with this contrast and generates a secondary poloidal component, which contributes to the poloidal magnetic measurements on land (Dostal et al. 2012). Such an effect was found to be negligible by Zhang et al. (2014) for the 2011 Tohoku-Oki tsunami. The influence of conversion from toroidal to poloidal magnetic modes at Easter Island for the 2010 and 2015 events will be investigated in future studies.

The findings in this report are based on a numerical model, and we consider them to represent an improvement to the wavelet transform data-processing technique used by Klausner et al. (2014) and Klausner et al. (2016), who proposed TAI coupling

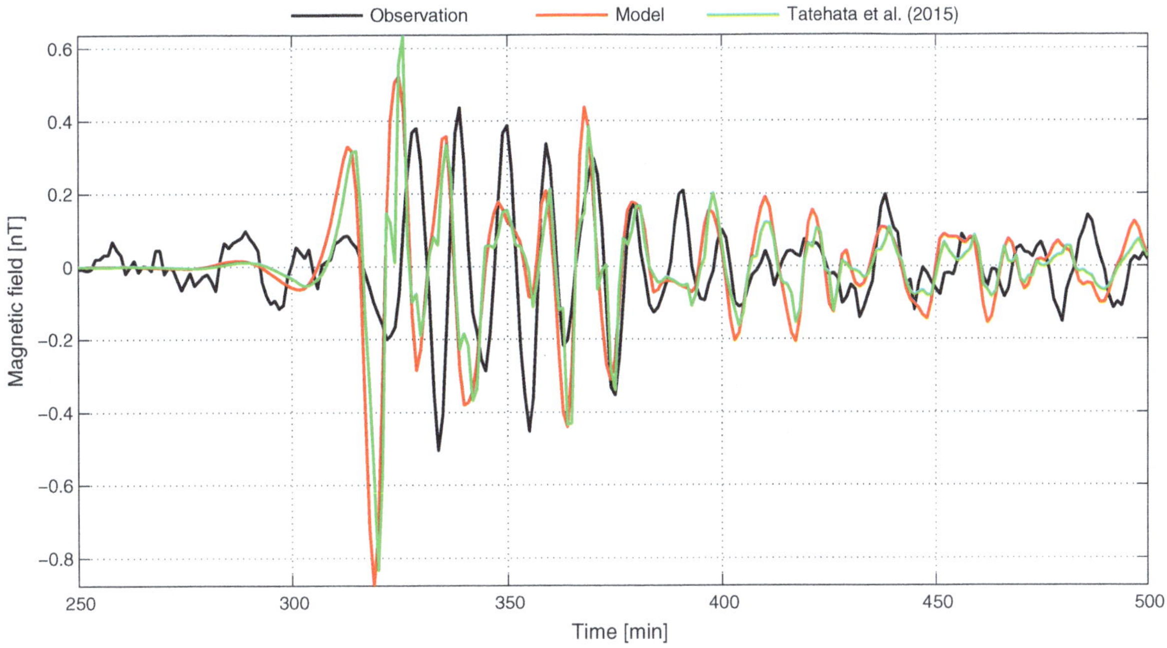

Figure 10
Comparison between the magnetic field measurement (black line), simulated magnetic field using our model (red line), and simulated magnetic field using the model of Tatehata et al. (2015) (green line) at IPM, for the 2015 Chile tsunami, considering a resolution of 1.08 arc min for the second level grid around Easter Island. Time is measured relative to the earthquake origin time

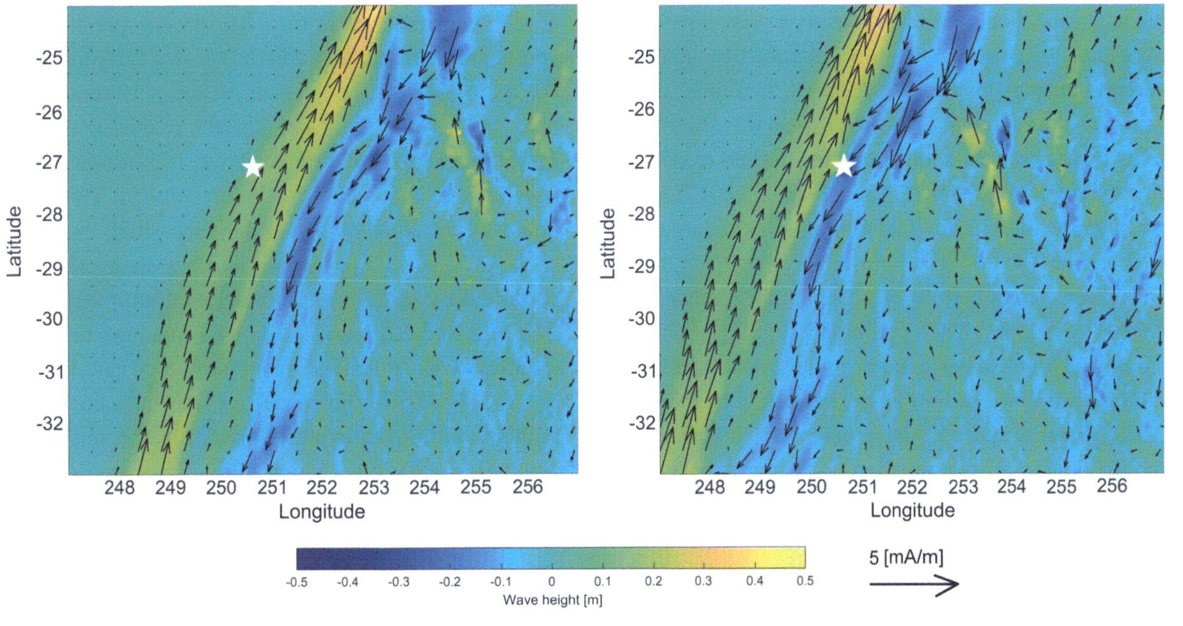

Figure 11
Simulated tsunami wave height (color images) and induced electric currents (black arrows) around Easter Island (white star) for the 2010 Chile tsunami at 308 min (left) and 320 min (right). Time is measured relative to the earthquake origin time

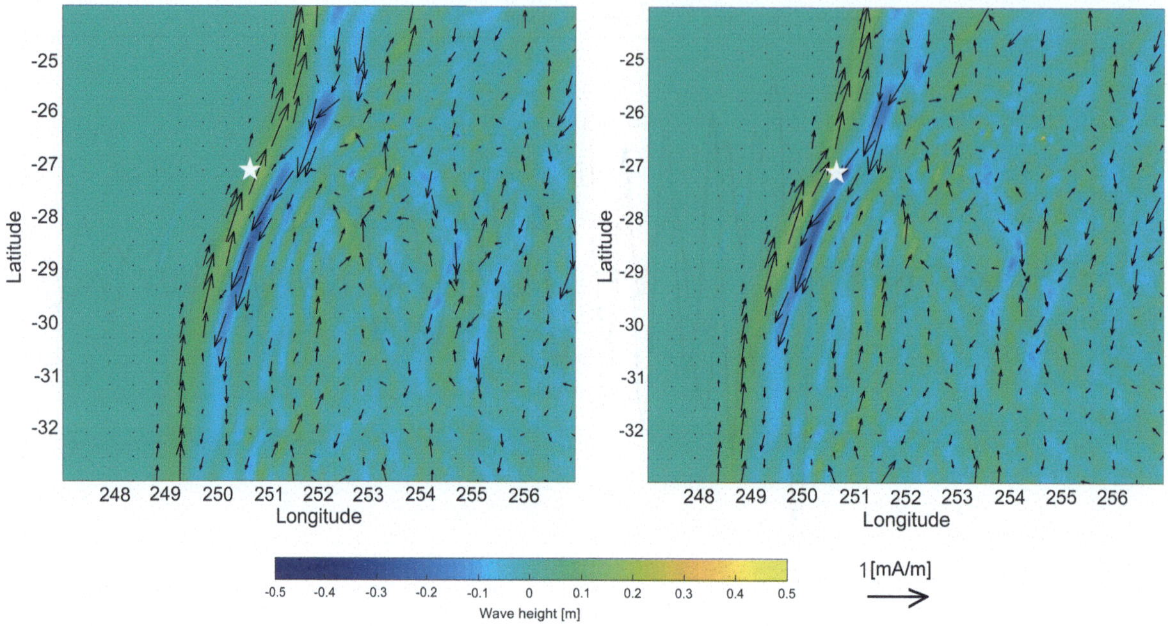

Figure 12
Simulated tsunami wave height (color images) and induced electric currents (black arrows) around Easter Island (white star) for the 2015 Chile tsunami at 313 min (left) and 318 min (right). Time is measured relative to the earthquake origin time. Note the different electric current values compared to those displayed in Fig. 11

as responsible for measurements in 2010 and 2015 events, respectively. In spite of their wavelet technique that can reduce the time required to identify tsunamigenic magnetic disturbances, a source separation of these signals, by subtracting the simulated fields from observed data, could be possible using our time domain approach (Minami et al. 2017). We hypothesize that the magnetic anomalies recorded at IPM station for the 2010 and 2015 events might be caused by a combination of oceanic and ionospheric electric currents. The combined contribution of oceanic and ionospheric electric currents could even validate the first oscillations in the calculated waveforms (Figs. 5 and 7), since magnetic disturbances due to ionospheric contributions can appear both before and after the tsunami arrival (Klausner et al. 2016).

5. Conclusion

In this study, we model the magnetic disturbances at Easter Island Observatory during the 2010 and 2015 Chilean tsunamis. Using a simple magnetostatic model, perturbations in the vertical component of the geomagnetic field are predicted. As we can reproduce the main features, but not all, of the perturbed magnetic field, we hypothesize that the magnetic records for both events are produced by a combination of oceanic and ionospheric electric currents. Our findings suggest that even a relatively small tsunami, such as the one produced by the 2015 Mw 8.2 earthquake (which is small compared with 2010 Mw 8.8 Chile and 2011 Mw 9.0 Japan tsunamis), might be monitored by magnetic records. We also conclude that the success of our model was partially due to the high resolution bathymetry used for the numerical tsunami simulations, which allowed for an accurate reproduction of the tsunami-induced currents.

A numerical solution of the full Maxwell equations for the 2010 and 2015 tsunami events may represent a great improvement to our model. Based on the theoretical work of Minami et al. (2015), this could allow for identifying oceanic zones where magnetic variations could be induced ahead of the tsunami wave. Moreover, for purposes of tsunami warning, not only the tsunami wave front but also the trailing waves are important, since they can trigger unexpected shelf and bay/harbor resonance phenomena in coastal areas (Horrillo et al. 2006; Yamazaki and Cheung 2011; Bellotti et al. 2012).

Especially important for us is monitoring the magnetic activity in the complex coastal zone off Concepción (Chile). Encouraged by the results obtained in this and in previous studies, our research group has decided to install a land-based magnetometer in San Vicente Bay, close to Concepción, whose results will be presented in a future work.

Acknowledgements

C. Torres would like to thank Programa Formación de Capital Humano Avanzado (PFCHA) of the Comisión Nacional de Investigación Científica y Tecnológica (CONICYT), Chile, for the financial support of his PhD (CONICYT PFCHA/DOCTORADO NACIONAL/2015—21150632). The authors would like to thank the International Real-time Magnetic Observatory Network (INTERMAGNET) for the dataset used in this work, Chambon-La-Forêt Observatory and Institut de Physique du Globe de Paris (IPGP) as creators of the Easter Island Observatory (with funding from National Aeronautics and Space Administration, NASA), the Hydrographic and Oceanographic Service of the Chilean Navy (SHOA) for the bathymetry data, Departamento de Ingeniería Matemática of Universidad de Concepción for computational support regarding data processing, and M. Miller for his assistance with the production of figures and the improvement of the manuscript. The DART records are available from NOAA's National Data Buoy Center at http://www.ndbc.noaa.gov/dart.shtml. Wavelet software was provided by C. Torrence and G. Compo, and is available at URL: http://atoc.colorado.edu/research/wavelets/. Figures in this study are produced using Matlab R2017b and the Generic Mapping Toolkit of Wessel et al. (2013). Finally, the authors thank the Editor and the two anonymous reviewers for their constructive comments and suggestions.

References

Aránguiz, R., González, G., González, J., Catalán, P. A., Cienfuegos, R., Yagi, Y., et al. (2017). The 16 September 2015 Chile Tsunami from the post-tsunami survey and numerical modeling perspectives, Pageoph Topical Volumes. In C. Braitenberg & A. B. Rabinovich (Eds.), *The Chile-2015 (Illapel) earthquake and tsunami* (pp. 219–234). Cham: Birkhüuser.

Bellotti, G., Briganti, R., & Beltrami, G. M. (2012). The combined role of bay and shelf modes in tsunami amplification along the coast. *Journal of Geophysical Research: Oceans*. https://doi.org/10.1029/2012JC008061.

Benavente, R., Cummins, P. R., & Dettmer, J. (2016). Rapid automated W-phase slip inversion for the Illapel great earthquake (2015, Mw = 8.3). *Geophysical Research Letters, 43*, 1910–1917. https://doi.org/10.1002/2015GL067418.

Calisto, I., Miller, M., & Constanzo, I. (2016). Comparison between tsunami signals generated by different source models and the observed data of the Illapel 2015 Earthquake. *Pure Applied Geophysics, 173*(4), 1051–1061. https://doi.org/10.1007/s00024-016-1253-8.

Calisto, I., Ortega, M., & Miller, M. (2015). Observed and modeled tsunami signals compared by using different rupture models of the April 1, 2014, Iquique earthquake. *National Hazards, 79*(1), 397–408. https://doi.org/10.1007/s11069-015-1848-x.

Campbell, W. H. (2003). *Introduction to geomagnetic fields*. Cambridge: Cambridge University Press.

Chave, A. D. (1983). On the theory of electromagnetic induction in the Earth by ocean currents. *Journal of Geophysical Research: Solid Earth, 88*(B4), 3531–3542. https://doi.org/10.1029/JB088iB04p03531.

Dostal, J., Martinec, Z., & Thomas, M. (2012). The modelling of the toroidal magnetic field induced by tidal ocean circulation. *Geophysical Journal International, 189*(2), 782–798. https://doi.org/10.1111/j.1365-246X.2012.05407.x.

Fritz, H. M., Petroff, C. M., Catalán, P. A., Cienfuegos, R., Winckler, P., Kalligeris, N., et al. (2011). Field survey of the 27 February 2010 Chile Tsunami. *Pure and Applied Geophysics, 168*(11), 1989–2010. https://doi.org/10.1007/s00024-011-0283-5.

Glimsdal, S., Pedersen, G. K., Harbitz, C. B., & Løvholt, F. (2013). Dispersion of tsunamis: Does it really matter? *National Hazards Earth System Science, 13*(6), 1507–1526. https://doi.org/10.5194/nhess-13-1507-2013.

Grilli, S. T., Ioualalen, M., Asavanant, J., Shi, F., Kirby, J. T., & Watts, P. (2007). Source constraints and model simulation of the December 26, 2004, Indian Ocean Tsunami. *Journal of Waterway, Port, Coastal, and Ocean Engineering, 133*(6), 414–428. https://doi.org/10.1061/(ASCE)0733-950X(2007)133:6(414).

Grinsted, A., Moore, J. C., & Jevrejeva, S. (2004). Application of the cross wavelet transform and wavelet coherence to geophysical time series. *Nonlinear Processes in Geophysics, 11*(5/6), 561–566.

Horrillo, J., Kowalik, Z., & Shigihara, Y. (2006). Wave Dispersion Study in the Indian Ocean-Tsunami of December 26, 2004. *Marine Geodesy, 29*(3), 149–166. https://doi.org/10.1080/01490410600939140.

Irrgang, C., Saynisch-Wagner, J., & Thomas, M. (2018). Depth of origin of ocean-circulation-induced magnetic signals. *Annals of*

Geophysics, *36*(1), 167–180. https://doi.org/10.5194/angeo-36-167-2018.

Kherani, E. A., Lognonné, P., Hébert, H., Rolland, L., Astafyeva, E., Occhipinti, G., et al. (2012). Modelling of the total electronic content and magnetic field anomalies generated by the 2011 Tohoku-Oki tsunami and associated acoustic-gravity waves. *Geophysics Journal International*, *191*(3), 1049–1066. https://doi.org/10.1111/j.1365-246X.2012.05617.x.

Kherani, E. A., Rolland, L., Lognonné, P., Sladen, A., Klausner, V., & de Paula, E. R. (2016). Traveling ionospheric disturbances propagating ahead of the Tohoku-Oki tsunami: A case study. *Geophysics Journal International*, *204*(2), 1148–1158. https://doi.org/10.1093/gji/ggv500.

Klausner, V., Almeida, T., de Meneses, F. C., Kherani, E. A., Pillat, V. G., & Muella, M. T. A. H. (2016). Chile 2015: Induced magnetic fields on the Z component by tsunami wave propagation. *Pure Applied Geophysics*, *173*(5), 1463–1478. https://doi.org/10.1007/s00024-016-1279-y.

Klausner, V., Mendes, O., Domingues, M. O., Papa, A. R. R., Tyler, R. H., Frick, P., et al. (2014). Advantage of wavelet technique to highlight the observed geomagnetic perturbations linked to the Chilean tsunami (2010). *Journal of Geophysical Research: Space Physics*, *119*(4), 3077–3093. https://doi.org/10.1002/2013JA019398.

Koper, K. D., Hutko, A. R., Lay, T., & Sufri, O. (2012). Imaging short-period seismic radiation from the 27 February 2010 Chile (Mw 8.8) earthquake by back-projection of P, PP, and PKIKP waves. *Journal of Geophysical Research: Solid Earth*, *117*(B2), 2156–2202. https://doi.org/10.1029/2011JB008576.

Larsen, J. C. (1968). Electric and magnetic fields induced by deep sea tides. *Geophysical Journal of the Royal Astronomical Society*, *16*(1), 47. https://doi.org/10.1111/j.1365-246X.1968.tb07135.x.

Larsen, J. C. (1973). An introduction to electromagnetic induction in the ocean. *Physics of the Earth and Planetary Interiors*, *7*(3), 389–398. https://doi.org/10.1016/0031-9201(73)90063-0.

Levin, B. W., & Nosov, M. A. (2016). *Physics of tsunamis*. Dordrecht: Springer.

Liu, P. L.-F., Woo, S.-B., & Cho, Y.-S. (1998). *Computer programs for tsunami propagation and inundation*. Ithaca: Cornell University.

Manoj, C., Kuvshinov, A., Maus, S., & Lühr, H. (2006). Ocean circulation generated magnetic signals. *Earth Planets Space*, *58*(4), 429–437. https://doi.org/10.1186/BF03351939.

Manoj, C., Maus, S., & Chulliat, A. (2011). Observation of magnetic fields generated by tsunamis. *Eos, Transactions American Geophysical Union*, *92*(2), 13–14. https://doi.org/10.1029/2011EO020002.

Masamura, K., Fujima, K., Goto, C., Iida, K., & Shigemura, T. (2000). Theoretical solution of long wave considering the structure of bottom boundary layer and examinations on wave decay due to sea bottom friction. *Journal of Hydraulic, Coastal and Environmental Engineering, JSCE, No663/II-53*, 69–78.

Minami, T., Toh, H., Ichihara, H., & Kawashima, I. (2017). Three-dimensional time domain simulation of tsunami-generated electromagnetic fields: Application to the 2011 Tohoku earthquake tsunami. *Journal of Geophysical Research: Solid Earth*, *122*(12), 9559–9579. https://doi.org/10.1002/2017JB014839.

Minami, T., Toh, H., & Tyler, R. H. (2015). Properties of electromagnetic fields generated by tsunami first arrivals: Classification based on the ocean depth. *Geophysical Research Letters*, *42*(7), 2171–2178. https://doi.org/10.1002/2015GL063055.

Mori, N., Takahashi, T., Yasuda, T., & Yanagisawa, H. (2011). Survey of 2011 Tohoku earthquake tsunami inundation and run-up. *Geophysical Research Letters*. https://doi.org/10.1029/2011GL049210.

Okada, Y. (1985). Surface deformation due to shear and tensile faults in a half-space. *Bulletin of the Seismological Society of America*, *75*, 1135–1154.

Okal, E. A., & Synolakis, C. E. (2016). Sequencing of tsunami waves: Why the first wave is not always the largest. *Geophysical Journal International*, *204*(2), 719–735. https://doi.org/10.1093/gji/ggv457.

Pukhtyar, L. D., & Kukushkin, A. S. (1996). Investigation of the electromagnetic fields induced by sea wave motion. *Physical Oceanography*, *7*(1), 33–41. https://doi.org/10.1007/BF02509823.

Sanford, T. (1971). Motionally induced electric and magnetic fields in the sea. *Journal of Geophysical Research*, *76*(15), 3476–3492. https://doi.org/10.1029/JC076i015p03476.

Schnepf, N. R., Manoj, C., An, C., Sugioka, H., & Toh, H. (2016). Time-frequency characteristics of tsunami magnetic signals from four pacific ocean events. *Pure Applied Geophysics*, *173*(12), 3935–3953. https://doi.org/10.1007/s00024-016-1345-5.

Tatehata, H., Ichihara, H., & Hamano, Y. (2015). Tsunami-induced magnetic fields detected at Chichijima Island before the arrival of the 2011 Tohoku earthquake tsunami. *Earth Planets Space*, *67*(1), 185. https://doi.org/10.1186/s40623-015-0347-3.

Torrence, C., & Compo, G. P. (1998). A practical guide to wavelet analysis. *Bulletin of the American Meteorological Society*, *79*(1), 61–78. https://doi.org/10.1175/1520-0477(1998)079<0061:APGTWA>2.0.CO;2.

Tyler, R. H. (2005). A simple formula for estimating the magnetic fields generated by tsunami flow. *Geophysical Research Letters*. https://doi.org/10.1029/2005GL022429.

Tyler, R. H., Maus, S., & Lühr, H. (2003). Satellite observations of magnetic fields due to ocean tidal flow. *Science*, *299*(5604), 239–241. https://doi.org/10.1126/science.1078074.

Tyler, R. H., & Mysak, L. A. (1995). Motionally-induced electromagnetic fields generated by idealized ocean currents. *Geophysical and Astrophysical Fluid Dynamics*, *80*(3–4), 167–204. https://doi.org/10.1080/03091929508228954.

Utada, H., Shimizu, H., Ogawa, T., Maeda, T., Furumura, T., Yamamoto, T., et al. (2011). Geomagnetic field changes in response to the 2011 off the Pacific Coast of Tohoku Earthquake and Tsunami. *Earth and Planetary Science Letters*, *311*, 11–27. https://doi.org/10.1016/j.epsl.2011.09.036.

Wang, X. (2009). *User manual for COMCOT version 1.7 (first draft)*. Ithaca: Cornell University.

Wang, B., Guo, X., Liu, H., & Gong, C. (2015). On the magnetic anomaly at Easter Island during the 2010 Chile tsunami. *Theoretical and Applied Mechanics Letters*, *5*(5), 187–190. https://doi.org/10.1016/j.taml.2015.08.001.

Wessel, P., Smith, W. H. F., Scharroo, R., Luis, J., & Wobbe, F. (2013). Generic mapping tools: Improved version released. *Eos, Transactions American Geophysical Union*, *94*(45), 239–241. https://doi.org/10.1002/2013EO450001.

Yamazaki, Y., & Cheung, K. F. (2011). Shelf resonance and impact of near-field tsunami generated by the 2010 Chile earthquake.

Geophysical Research Letters. https://doi.org/10.1029/2011GL047508.

Zhang, L., Utada, H., Shimizu, H., Baba, K., & Maeda, T. (2014). Three-dimensional simulation of the electromagnetic fields induced by the 2011 Tohoku tsunami. *Journal of Geophysical Research: Solid Earth*, *119*(1), 150–168. https://doi.org/10.1002/2013JB010264.

(Received August 2, 2018, revised October 31, 2018, accepted November 9, 2018, Published online November 20, 2018)

Pure Appl. Geophys. 176 (2019), 3185–3205

https://doi.org/10.1007/s00024-019-02180-w

Pure and Applied Geophysics

Developing an Approximate Tsunami Hazard Zone for Areas with Poor Topographic Coverage in Alaska

D. J. Nicolsky,[1] E. N. Suleimani,[1] R. D. Koehler,[2] and J. B. Salisbury[3]

Abstract—Tsunamigenic earthquakes along the Alaska–Aleutian megathrust pose a major threat to many rural coastal communities in Alaska. To mitigate the impact of tsunamis and minimize loss of life, maximum potential runup and tsunami hazard zones need to be estimated despite a lack of the high-resolution digital elevation models. In this manuscript we present a methodology to approximate tsunami hazard zones in areas with poor topographic and bathymetric coverage. To manage a trade-off between the computational complexity and lack of quality data, we develop three scenario earthquakes with the intention that one of them characterizes impacts of a credible worst-case event. We then apply a safety factor to account for the limited number of considered scenarios and a coarse resolution elevation model used to simulate water dynamics near the community. The developed methodology is illustrated using a case study focused on the community of Adak, in the Andreanof Islands, Alaska. To validate the proposed method, we explore sensitivity of tsunami height with respect to the coseismic slip distribution and develop a comprehensive set of credible worst-case scenarios for the community. The tsunami hazard zone, developed with the use of the safety factor, is validated against high-resolution modeling of potential inundation according to the set of worst-case scenarios in a number of other coastal communities.

Key words: Tsunami modeling, runup, tsunami hazard zone, approximation.

1. Introduction

The Alaska–Aleutian megathrust, the fault formed by the Pacific–North American plate interface, is the most seismically active tsunamigenic fault zone in the United States. Subduction of the Pacific plate under the North American plate along the megathrust has resulted in numerous great earthquakes and has the highest potential to generate tsunamis in Alaska (Dunbar and Weaver 2015). The latest sequence of great earthquakes along the Alaska–Aleutian megathrust began in 1938; rupture extents are shown in Fig. 1. Tsunamis generated by these great earthquakes reached Alaska coastal communities within minutes after the earthquakes and resulted in widespread damage and loss of life (National Centers for Environmental Information/World Data Service [NCEI/WDS] 2019).

In recent years, we have published tsunami inundation maps and reports for more than 20 communities in coastal Alaska. In these studies, published at the Alaska Division of Geological & Geophysical Surveys, see the web-page http://dggs.alaska.gov/pubs/tsunami, we performed numerical modeling of tsunamis using high-resolution digital elevation models (DEMs) of combined bathymetry and topography with a typical spatial resolution of 15 m. However, despite the recent progress in data acquisition and processing, some coastal Alaska communities still do not have high-resolution seamless topographic–bathymetric DEMs. Nevertheless, estimation of the potential flooding in the event of a local or distant tsunami is an integral part of community preparedness—the first step to reduce the loss of life and property damage. Thus, estimation of tsunami inundation for communities that lack high-resolution topographic and bathymetric datasets is essential for tsunami preparedness and loss reduction.

Because our goal is to estimate the maximum credible (i.e., worst-case) tsunami inundation for the community, we think of this problem in terms of a

[1] Geophysical Institute, University of Alaska Fairbanks, 2156 Koyukuk Drive, Fairbanks, AK 99775-7320, USA. E-mail: djnicolsky@alaska.edu

[2] Nevada Bureau of Mines and Geology, Mackay School of Earth Science and Engineering, University of Nevada, Reno, 1664 North Virginia Street, MS 178, Reno, NV 89557, USA.

[3] Alaska Division of Geological and Geophysical Surveys, 3354 College Road, Fairbanks, AK 99709-3707, USA.

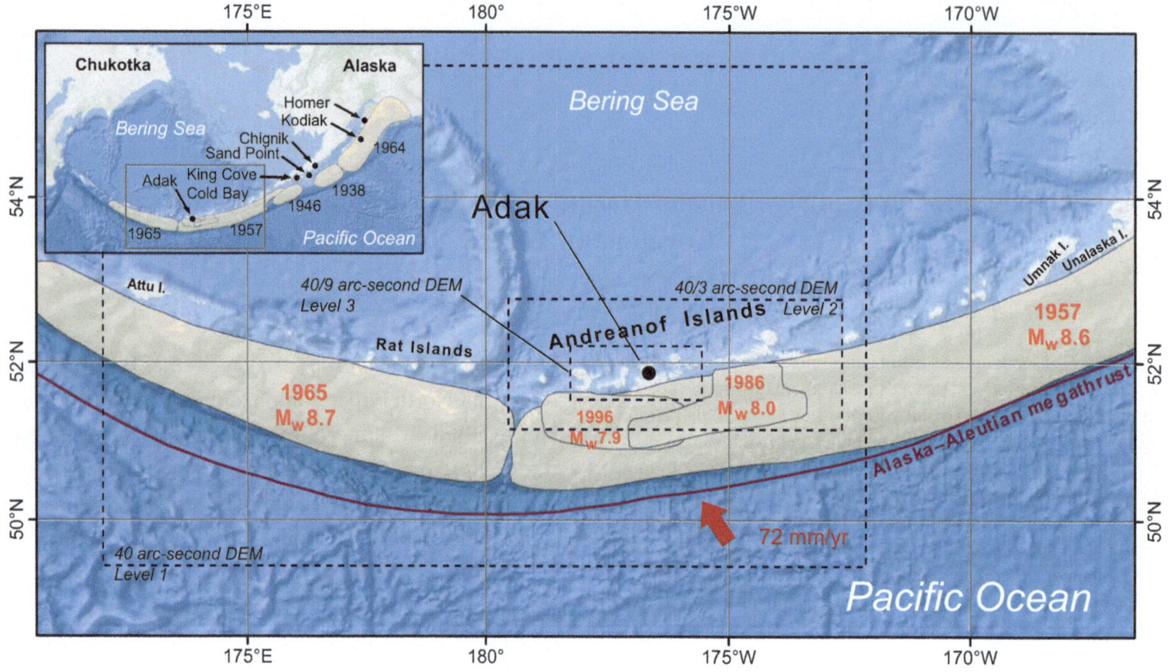

Figure 1
Map of the central Aleutian Islands and the Andreanof Islands group. Rupture areas of the 1938 M_w 8.3 earthquake west of Kodiak Island (Estabrook et al. 1994) and four subsequent events, the 1946 M_w 8.6 Aleutian (Lopez and Okal 2006), the 1957 M_w 8.6 Andreanof Islands (Johnson et al. 1994), the 1964 M_w 9.2 Great Alaska (Kanamori 1970), and the 1965 M_w 8.7 Rat Island (Wu and Kanamori 1973) earthquakes are shown by shaded polygons. Extents of the bathymetry/topography DEMs used for the numerical modeling of tsunami propagation and runup are marked by dashed lines. Location of the Alaska–Aleutian megathrust is shown by the red line. Inset map shows geographic localities referred to in the text and rupture boundaries of historical earthquakes

mathematical optimization: maximize a modeled runup in the community given certain geological constrains on the coseismic slip and plate interface location. The optimization problem can be solved by considering a series of scenarios for coseismic slip distributions on the Alaska–Aleutian plate interface. For each scenario, it is necessary to compute the vertical coseismic deformation and then model propagation and runup of the hypothetical tsunami. Unfortunately, numerical calculations of tsunami runup are a time consuming procedure despite recent advances in parallel computing. Furthermore, an exact solution to the optimization problem, even if computed with great accuracy, could be of limited value because the bathymetry and topography are only known in some communities at a coarse resolution. Therefore, we propose to balance a level of computational complexity, cost effectiveness, and potential trade-offs as follows.

To find an approximate solution to the optimization problem, we consider a set of sample megathrust rupture scenarios. We then assume that one of these scenarios will be sufficiently similar to the optimal solution (a plausible coseismic slip distribution causing the worst-case inundation). In order to account for the limited number of sample scenarios, and because the maximum wave height for a community could be associated with alternative, unknown slip distributions, we introduce a safety factor. The safety factor primarily accounts for tsunami source uncertainties inherent to this method of optimization. To a lesser degree, the safety factor also accounts for an uncertainty in the computed runup, since the bathymetry and topography are only known at a coarse resolution. Although effects of grid resolution and elevation accuracy on the inundation extent are not a primary focus of this study, as a part of the verification study we compare maximum tsunami height modeled on the coarse- and fine-resolution DEMs.

To demonstrate the proposed methodology, we develop an approximate tsunami hazard zone (ATHZ) in the community of Adak, population 247, one of the largest communities on the Andreanof Islands (Fig. 1). In the last 60 years, three major earthquakes occurred near the community of Adak (51.53′0″N, 176.38′42″W). The 1957 (M_w 8.6) earthquake ruptured eastward from the Andreanof Islands to the Unalaska region (Johnson et al. 1994). The 1986 Andreanof Islands earthquake (M_w 8.0) and the 1996 event (M_w 7.9) re-ruptured the western one-third of the 1957 rupture area. These earthquakes, respectively, generated 2.7, 0.88, and 0.51 m tsunami waves in Adak (Lander 1996; NCEI/WDS 2019). We list significant tsunamis that have occurred in Adak since 1946 in Table 1. In addition to these events the community has been affected by numerous smaller tsunamis for a total of about 50 occurrences in the recorded history (Lander 1996; NCEI/WDS 2019). We note that when we developed the ATHZ for Adak, only a coarse-resolution DEM was available. After obtaining a high-resolution DEM (a year later) we verified the accuracy of the ATHZ in Adak using high-resolution modeling of tsunami inundation from a set of fully-developed worst-case scenarios. Similar validation studies were conducted for other communities along the Alaska–Aleutian megathrust. We emphasize that the community of Adak was one of the first Alaska communities that lacked high-resolution DEMs and for which the ATHZ approach was applied. There are other Alaska communities without high-resolution DEMs, but in need of tsunami hazard mitigation products. This work constitutes a step towards ensuring that within the acceptable error of the proposed methodology, the ATHZ is a reasonable approximation and can be used as a model for studies in other communities.

2. Development of Sample Scenarios

Alaska tsunami hazard maps are produced on the basis of maximum credible or worst-case tsunami scenarios for a given segment of the Alaska–Aleutian subduction zone (AASZ) (Nicolsky et al. 2011a, 2013, 2014a, 2015, 2016a; Suleimani et al. 2010, 2013, 2015, 2016). The community of Adak is in the central Aleutian arc, in the Andreanof Islands group, where the Pacific and North American plates converge along the megathrust at a rate of approximately 72 mm/year (DeMets et al. 1990; Argus et al. 2010). Because of friction between the two converging tectonic plates, shear stress builds up on the megathrust and stored energy is released in earthquakes that cause ground surface/ocean bottom displacement and a subsequent tsunami. It is

Table 1

Tsunami effects at Adak. Data from the National Centers for Environmental Information Global Historical Tsunami Database (http://www.ngdc.noaa.gov/hazard/tsu_db.shtml) and comments from Lander (1996)

Date	Magnitude (M_w)	Origin	Maximum water height (m)	Comments
04/01/1946	8.6	Unimak Island, Alaska	0.23	
11/04/1952	9.0	Kamchatka Peninsula	1.1	The waves slightly overflowed the banks of the harbor
03/09/1957	8.6	Aleutian Islands	2.7	Tsunami caused severe damage. All structures at the fuel and oil dock washed away up to 4 m elevation; oil pipelines damaged
05/22/1960	9.5	Chile	1.1	
03/28/1964	9.2	Gulf of Alaska	0.27	
02/04/1965	8.7	Rat Islands	0.46	Sea level fell 1.2 ft. in 15 min
11/22/1969	7.7	Kamchatka Peninsula	0.3	
05/07/1986	8.0	Andreanof Islands	0.88	
06/10/1996	7.9	Andreanof Islands	0.51	
11/15/2006	8.3	Kuril Islands	0.2	
02/27/2010	8.8	Chile	0.38	
03/11/2011	9.0	Honshu, Japan	1.1	

theorized that the stress is primarily acquired in the seismically coupled regions of the megathrust where friction is the greatest (Dixon and Moore 2012), thus making determination of these regions crucial for the development of tsunami scenarios.

Delineating the seismically coupled regions in various subduction zones has been the subject of extensive research (Uyeda 1982; Pacheco et al. 1993; Tichelaar and Ruff 1993; Uchida et al. 2009; Scholz and Campos 2012). Most of this research has focused on determining the spatial extent of the coupled region and its updip and downdip limits. For example, Tichelaar and Ruff (1993, Fig. 8) analyzed previous seismicity along the AASZ and hypothesized that the seismic coupling extends down to 35–41 km near Adak. In another study, Cross and Freymueller (2007) used GPS measurements to study subduction zone coupling in the Andreanof region and found strong along-strike variations in coupling that correlated with the rupture zones of the 1986 and 1996 earthquakes. They showed that the combined rupture area of these events was a region of high slip deficit. The same area represents the western one-third of the 1957 rupture zone, which had high slip during the earthquake (Fig. 15, Johnson et al. 1994). Cross and Freymueller (2007) found that the plate interface is nearly 100% coupled south of Adak (west of the 1986 and 1996 earthquakes) suggesting that this area is susceptible to future earthquake ruptures. The updip limit of the seismic coupling in AASZ is largely uncertain due to the lack of geodetic measurements in the region. In this work and other tsunami modeling studies (Nicolsky et al. 2015; Suleimani et al. 2016), we assume that the updip limit can extend to shallow depths, i.e. the 5–10 km depth.

Once the updip and downdip limits of seismic coupling are established it is possible to consider various tsunami scenarios parameterized by the coseismic slip distribution. Following lessons learned from the 11 March 2011 M_w 9.0 Tohoku earthquake and tsunami (Ito et al. 2011), and similarities between the Alaska and Tohoku margins (Ryan et al. 2012), we assume that coseismic slip close to the trench could reach 50–60 m. The maximum slip in the deeper parts of the megathrust is limited to 35 m, which is consistent with previous tsunami hazard assessments (e.g. Nicolsky et al. 2015, 2016a, b; Suleimani et al. 2016) and also with tsunami scenarios proposed by Butler (2014).

Because local tsunamis propagate predominantly normal to the axis of coseismic ground deformation, the runup of local tsunamis (which could produce damaging tsunami in Adak) primarily depends on the coseismic slip near the community. This warrants a careful consideration of the coseismic slip distribution near the community, whereas the tapering of slip and total length of rupture (as long as it includes areas near Adak) are rather inconsequential to the maximum runup in the community. Hence, we assume a uniform along-strike distribution and bell-shaped parameterization along the down-dip direction. The bell-shaped curve is described by Freund and Barnett's (1976) formulae. The rigidity modulus is assumed to be constant (3.6×10^{10} N/m^2) and independent of depth. Despite these apparent simplifications, there are still many variants for the slip distribution near Adak.

Our experience with tsunami modeling for communities along the AASZ shows that the worst-case scenario (causing a maximum runup) is either a scenario with the maximum slip close to the trench (Nicolsky et al. 2014a, b), a scenario with a rupture at significant depth (Nicolsky et al. 2015, 2017), or rupture at some intermediate depth (Nicolsky et al. 2016b; Suleimani et al. 2017). Therefore, we propose to consider three sample scenarios with the depth range, at which the maximum slip occurs, varying from the shallow region, close to the trench, to the deeper parts of the plate interface.

Below we describe three sample scenarios for Adak, the main rupture parameters are listed in Table 2, while the slip distributions are shown in the left panel of Fig. 2.

Sample Scenario 1 M_w 9.0 earthquake: SAFRR-type event, slip near the trench

This event is a hypothetical M_w 9.0 (M_0 4×10^{21} N m) earthquake rupturing the Alaska–Aleutian megathrust. During the Tohoku earthquake a large amount of slip occurred between the subducting and overriding plates near the Japan trench (Fujii et al. 2011; Shao et al. 2011). The USGS Science Application for Risk Reduction (SAFRR) project, in collaboration with the National

Table 2

All hypothetical tectonic scenarios used to model tsunami runup in Adak

Scenarios	Depth range (km)	Maximum slip depth range (km)	Maximum slip (m)	Maximum subsidence (m)	Maximum uplift (m)	Vertical displacement in Adak (m)
Sample scenarios						
1 M_w 9.0 earthquake: SAFRR-type event, slip near the trench	8–54	11–14	55–65	− 2.7	13.8	− 1.1
2 M_w 9.0 earthquake: maximum slip at 15–25 km depth	5–35	15–25	34–35	− 6.4	13.6	− 2.1
3 M_w 9.0 earthquake: maximum slip at 25–35 km depth	14–45	25–35	34–35	− 5.6	17.6	− 4.7
Validation scenarios						
1 M_w 8.8 earthquake in the Andreanof Islands region: maximum slip at 17–27 km depth	12–42	17–27	34–35	− 3.9	12.1	− 1.9
2 M_w 8.8 earthquake in the Andreanof Islands region: maximum slip at 25–35 km depth	18–48	25–35	34–35	− 3.0	10.4	− 2.7
3 M_w 8.8 earthquake in the Andreanof Islands region: maximum slip at 35–42 km depth	18–49	35–42	34–35	− 2.9	9.5	− 2.7
4 M_w 8.9 earthquake in the Andreanof Islands region: maximum slip at 15–35 km depth	11–42	15–35	34–35	− 4.5	12.8	− 2.2
5 M_w 8.9 earthquake in the Andreanof Islands region: maximum slip at 25–45 km depth	18–49	25–45	34–35	− 3.3	11.1	− 3.0
6 M_w 9.2 earthquake: Uniform 35-m slip	7–50	7–50	34–35	− 3.9	16.2	− 3.7
7 M_w 9.1 earthquake: Uniform 50/20-m slip	6–38	10–34	37–38	− 5.4	22.8	− 0.9

Oceanic and Atmospheric Administration (NOAA) and State of California agencies, developed a plausible hypothetical tsunami scenario (Kirby et al. 2013) to describe the impacts of a tsunami generated by an earthquake in the Alaska Peninsula region (Ross et al. 2013). For Sample Scenario 1 in this paper we assume that the slip distribution in the downdip direction is the same as that in the SAFRR source, where greatest slip occurs close to the trench. This hypothetical rupture is positioned across the shelf area from the community of Adak. The slip is distributed uniformly along strike. The maximum slip of 46 m is at a depth of 5–15 km.

Sample Scenario 2 M_w 9.0 earthquake: Maximum slip at 15–25 km depth

This event is a hypothetical M_w 9.0 (M_0 4×10^{21} N m) earthquake rupturing the Alaska–Aleutian megathrust. The slip is distributed uniformly along strike. The maximum slip of 35 m is at a depth of 15–25 km in order to model a rupture in the middle of the seismically coupled region along the AASZ.

Sample Scenario 3 M_w 9.0 earthquake: Maximum slip at 25–35 km depth

Similar to Scenario 2, this event has the same moment magnitude and along-strike distribution of the slip. However, the maximum slip of 35 m is now located at a depth of 25–35 km. This scenario models a rupture close to the downdip limit of the seismically coupled region.

To compute the co-seismic ground/ocean bottom deformation we use a discretized model (Nicolsky et al. 2015) of the Alaska–Aleutian plate interface (Hayes et al. 2012) between the subducting and overriding plates. The interface is discretized into a set of 10–15 km long rectangles with an upper and lower edge of each rectangle being coincident with a 1-km depth contour of the interface. The value of slip is prescribed to each rectangle (subfault) and its contribution to the overall coseismic deformation is consequently computed by Okada (1985) formulae. The value of rake—the direction a hanging wall block moves during rupture is assumed to be equal to 90° in order to maximize coseismic deformation for the given slip. All contributions from subfaults are consequently added together to derive the co-seismic deformation associated with each considered

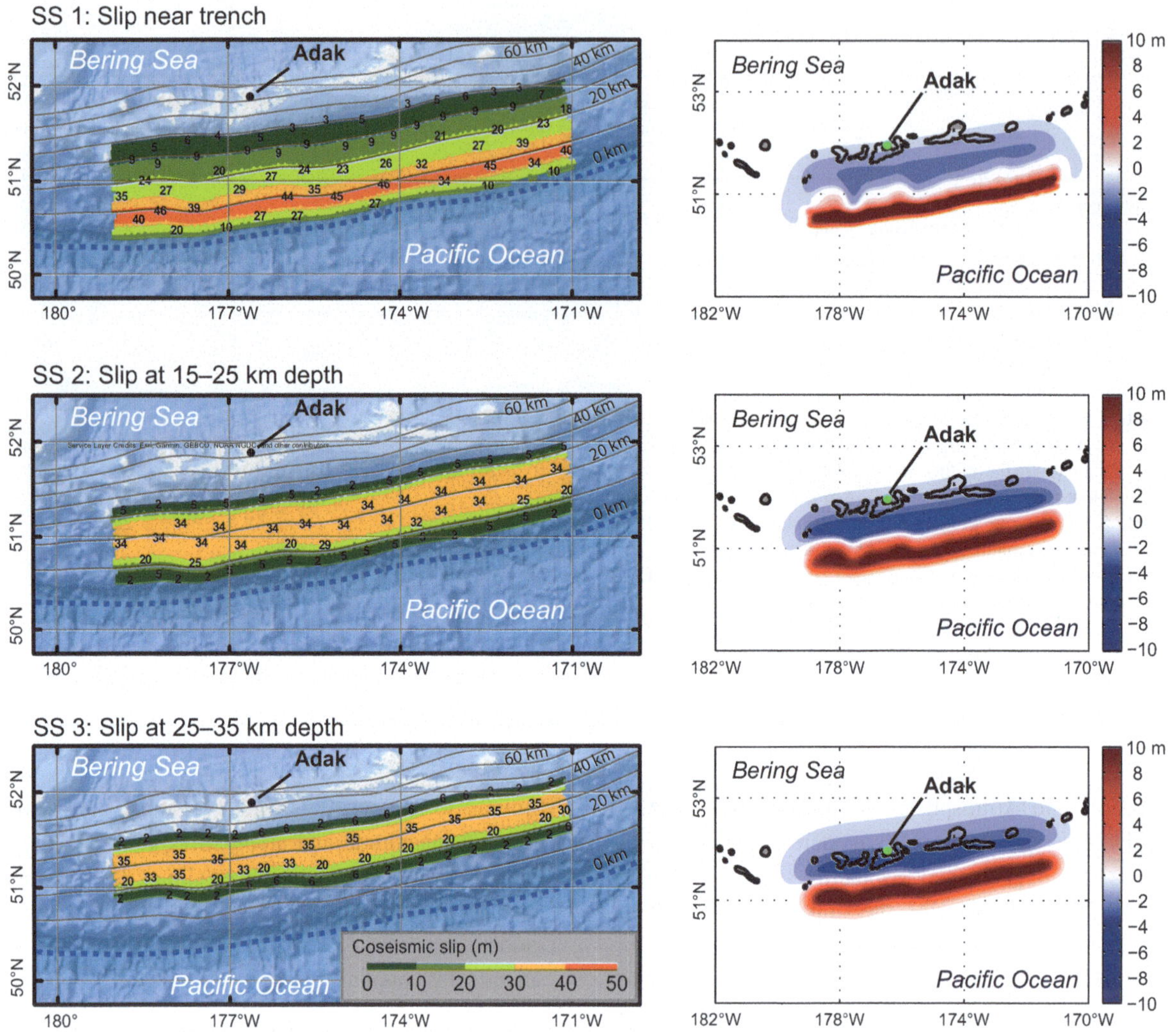

Figure 2

Left: proposed slip distributions along the plate interface for Sample Scenarios (SS) 1–3. The plate interface is discretized into a set of rectangles (subfaults) used to compute the coseismic vertical displacement by Okada's (1985) formulae. Deeper subfaults have a higher dip angles as the plate interface bends downward. Slip values in meters are marked for selected subfaults by small black labels. The depth contours of the Alaska–Aleutian subduction interface are shown by gray lines. Right: vertical coseismic deformations corresponding to the scenarios on the left side. Blue areas are associated with coseismic ground subsidence; areas of uplift are shown in red

scenario. The computed vertical coseismic deformations are shown in the right panel of Fig. 2.

In order to account for the limited number of sample scenarios and for the fact that the maximum wave height could be associated with an alternative slip distribution (potentially unknown), we introduce a safety factor. For each sample scenario, we multiply the computed maximum wave height near the community by the safety factor and then find the maximum across all scenarios. The resulting value is called the maximum estimated runup for the community and is considered the best approximate solution to the overall optimization problem—maximizing runup in the community given geophysical constrains. Because the tsunami dynamics for the sample scenarios are modeled at a coarse resolution, we extrapolate the estimated maximum runup along the coastline by selecting an elevation contour related to this value and thus develop the ATHZ.

Note that the value of safety factor is a priori unknown. According to Synolakis et al. (2007), the benchmarking of a tsunami model with field data requires that numerical predictions for the maximum runup should not differ by more than 20% from the measurements. Therefore, in order to incorporate an uncertainty due to a numerical model, we suggest that the safety factor needs to be at least 1.2. Similarly, in the Federal Emergency Management Agency guidelines (Applied Technology Council 2012; Technical Report FEMA P-646), it is recommended that "the design runup elevation be taken as 1.3 times the predicted maximum runup elevation to envelop the potential variability in the estimates of modeling".

In this work, we employ the same safety factor, 1.3. The safety factor depends on a number of considered sample scenarios, the larger the number of scenarios, the lesser the safety factor needs to be. For example, for optimization problems with "sharp" local maximum the safety factor needs to be large, since a function near the local maximum can be significantly smaller than the maximum. For a rather "flat" maximum, the safety factor can be assumed to be small and a finite sampling of the function near the maximum can be exploited. The safety factor and a number of sample scenarios likely depends on the geometry of plate interface; the safety factor for studies focusing on other subduction zones can differ from that assumed in this work and needs to be independently validated. In Sect. 4, we validate the chosen value of safety factor through comparison with high-resolution modeling results for a number of communities on the AASZ, where we have developed a large suite of maximum credible, scientifically-plausible scenarios.

3. Development of Approximate Tsunami Hazard Zone

To model tsunami dynamics in Adak, we utilize a nonlinear, flux-formulated, shallow-water model by Nicolsky et al. (2011b) that has been used to simulate tsunami runup in a number of coastal communities (e.g., Suleimani et al. 2018) and was validated by NTHMP (2012) through a set of analytical benchmarks and tested against laboratory and field data (Synolakis et al. 2007).

To develop an ATHZ for Adak, we use a series of nested computational grids. A nested grid allows for higher resolution in areas where it is needed without expending computer resources in areas where it is not. The bathymetric and topographic relief in each nested grid is based on digital elevation models (DEMs) developed at the National Centers for Environmental Information (NCEI) of the National Oceanic & Atmospheric Administration (NOAA) in Boulder, Colorado. The extent of each grid is shown in Fig. 1 by dashed rectangles. The coarsest grid, with 2-arc-minute (approximately 3.7 × 2.3 km near Adak) resolution, spans the central and northern Pacific Ocean. Spatial extents of all employed DEMs are listed in Table 3. The bathymetric data for the 2-arc-minute-resolution grid is extracted from the ETOPO2 dataset (NCEI). The data sources and methodology used to develop 40-, 40/3-, and 40/9-arc-second DEMs are described in detail by Lim et al. (2011) and Carignan et al. (2009). The 40/9-arc-second (85 × 140 m) grid covers the western Andreanof Islands including Adak Island; no DEM

Table 3

Nested grids used to compute propagation of tsunami waves generated in the Pacific Ocean to the community of Adak

Grid name	Resolution		West–East boundaries	South–North boundaries
	Arc-seconds	m (near Adak)		
Level 0, Northern Pacific	120 × 120	≈ 2280 × 3700	120°00′E–100°00′W	10°00′N–65°00′N
Level 1, Western Aleutians	40 × 40	≈ 760 × 1230	171°57′28″E–172°10′40″W	48°43′12″N–55°30′58″N
Level 2, coarse resolution Andreanof Islands	40/3 × 40/3	≈ 250 × 410	179°34′57″E–172°39′40″W	51°08′17″N–52°48′58″N
Level 3, fine-resolution	40/9 × 40/9	≈ 85 × 140	178°18′29″W–175°34′21″W	51°32′34″N–52°12′27″N
Level 4, high-resolution	40/27 × 40/45	≈ 28 × 28	176°41′19″W–176°29′34″W	51°49′42″N–51°58′4″N

The fine-resolution grid, Level 3, is used to compute maximum estimated runup and consequently develop the ATHZ. The high-resolution grid, Level 4, was used to validate the ATHZ. Note that the grid resolution in meters is not uniform and is used to illustrate grid fineness in the Andreanof Islands region. The first dimension is the longitudinal grid resolution, while the second is the latitudinal resolution verification efforts were conducted to reduce uncertainties in any grids. In this study, we conduct all model runs using bathymetric data that correspond to the Mean Higher High Water (MHHW) tide level in Adak. The MHHW level is chosen to model the worst-case scenarios—the highest wave arrives on the high tide.

For each considered sample scenario, we model the water dynamics in each computational grid and calculate the maximum tsunami wave heights in the 40/9-arc-second grid over the course of the entire model run. In particular, each model run was performed for 6 h of tsunami propagation to account for all waves in the wave train, as well as for secondary (reflected) waves. Figure 3 shows the maximum tsunami wave heights above the sea level offshore and inland for Sample Scenarios 1–3. The highest tsunami waves are present along the Pacific shore of the Andreanof Islands, while the Bering Sea coast experiences smaller waves due to the natural barrier of the island chain. Consequently, we find the maximum wave height for all sample scenarios; see the first line in Table 4. The absolute maximum value of the tsunami height near the community is 17.6 m. The maximum composite wave height (the maximum value for each grid point among each scenario) in the

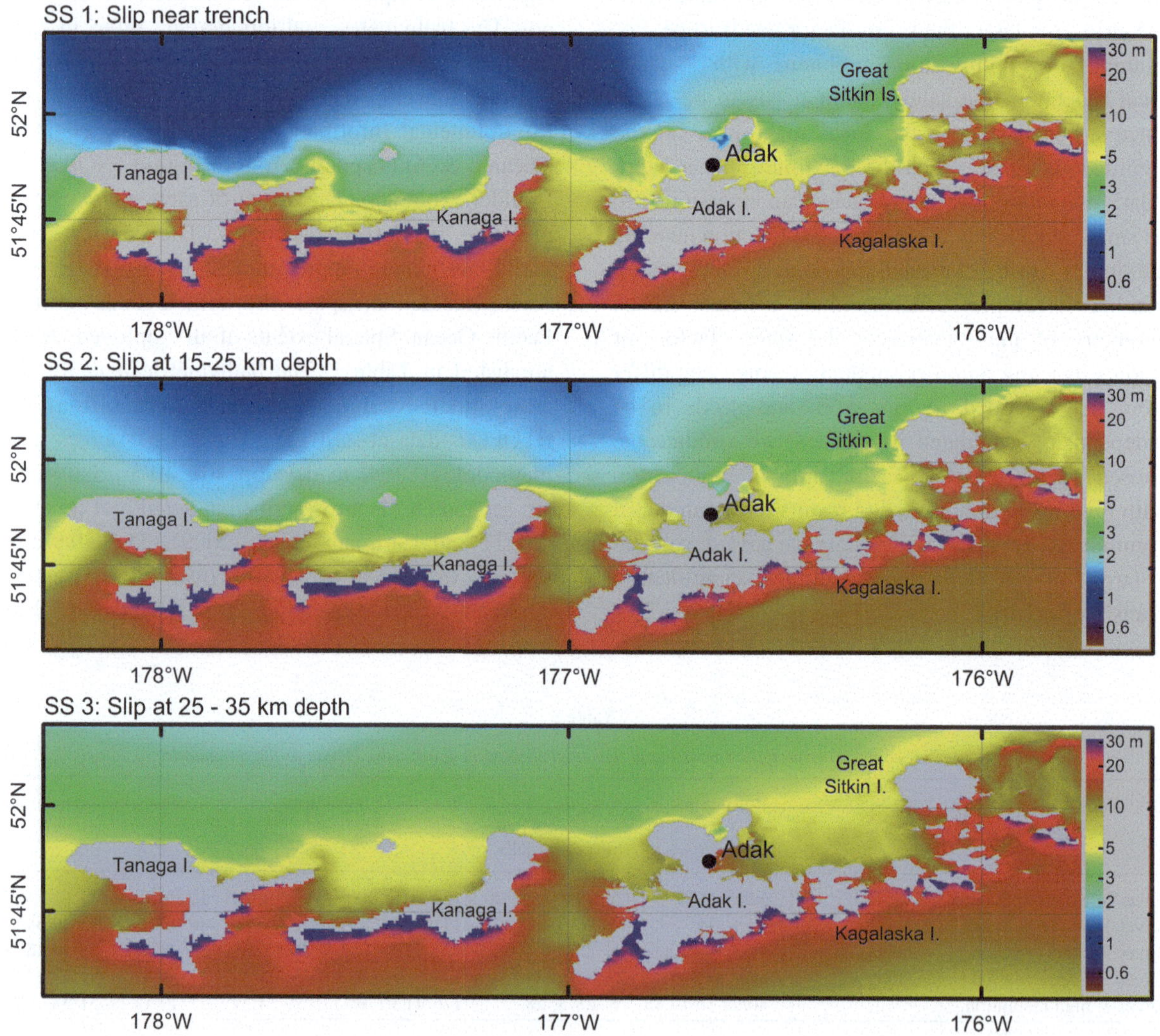

Figure 3
Maximum tsunami wave height for sample scenarios 1–3 within the extent of the 40/9 arc-second DEM

Table 4

Maximum modeled wave height near the community according to the sample scenario and estimated maximum runup after incorporation of the safety factor

Community	Computed maximum wave height (m)			Estimated maximum runup (m)
	Sample Scenario 1: M_w 9.0 earthquake: SAFRR-type event, slip near the trench	Sample Scenario 2: M_w 9.0 earthquake: maximum slip at 15–25 km depth	Sample Scenario 3: M_w 9.0 earthquake: maximum slip at 25–35 km depth	
Adak (this study)	9.2–9.7	11.8–12.3	17.1–17.6	22.8
Sand Point (Nicolsky et al. 2017)	13.5–14	11–11.5	13.5–14	18.2
Cold Bay (Suleimani et al. 2016)	3.8–4.3	3.6–4.2	6.5–7	9.1
King Cove (Suleimani et al. 2016)	8–8.5	9.5–10	9–9.5	13.0
Chignik Bay (Nicolsky et al. 2016b)	15.3–15.8	14–14.5	15–15.5	20.5
Chignik Lagoon (Nicolsky et al. 2016b)	6–6.5	7.7–8.2	12–12.5	16.3
Homer (Suleimani et al. 2018)	4.5–5	5.1–5.6	8.3–8.8	11.5
Kodiak (Suleimani et al. 2017)	10.5–11	15.0–15.5	7.0–7.5	20.1

vicinity of Adak and surrounding waters is shown in Fig. 4.

Multiplying the maximum tsunami height in the vicinity of Adak by the safety factor of 1.3, we estimate 23 m for the maximum estimated runup for the community. This elevation should be used by emergency planners as a guideline for various tsunami mitigation activities, e.g. plotting the 23-m elevation contour above the MHHW level and thus drafting the ATHZ for Adak.

4. *High-Resolution Modeling of Tsunami Runup*

The sample earthquake sources were developed with the intent that one of them is a good approximation to the true worst-case scenario and that the safety factor will account for DEM uncertainties and simplified earthquake scenarios. In this section, we demonstrate that the developed ATHZ is a liberal estimate—one that is cautious to avoid underestimating inundation in the community. Therefore, we conduct a sensitivity study, consider additional scenarios, and model tsunami inundation in the community with the high-resolution DEM, Level 4 grid, in Table 3. This DEM was verified with the RTK GPS survey data collected in the community (Carignan et al. 2017). The vertical datum of the Level 4 DEM is MHHW in order to maximize inundation in the community as previously discussed.

In the sensitivity study, we analyze waves arriving to Adak from various portions of the seismically coupled plate interface. In particular, we consider four idealized hypothetical ruptures (Cases A–D, Fig. 5), located south of Adak—the area where co-seismic slip distribution exerts the most control over wave heights in the community. These four hypothetical rupture cases are used later to develop maximum credible validation scenarios. The relative slip distribution for all four cases is identical: uniform in the along-strike direction with tapering at the ends of the rupture, and a bell-type slip curve in the

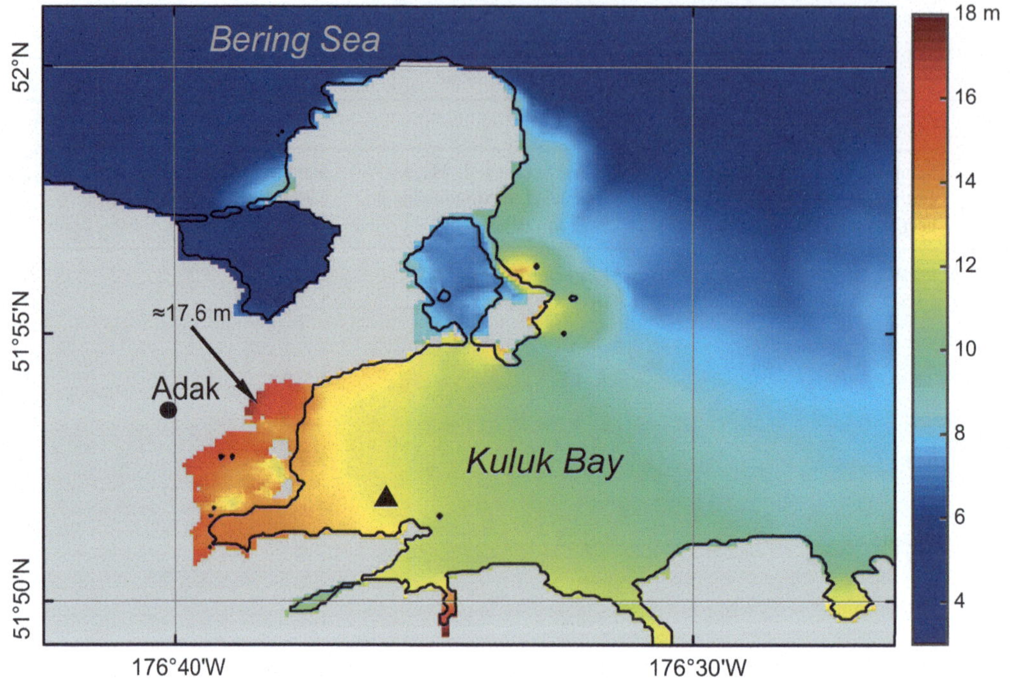

Figure 4
Maximum composite tsunami wave height at Adak. The black triangle indicates the location of the time series point, and the black contour is the MHHW shoreline. The arrow marks a location of the maximum wave height (above MHHW elevation) at 17.6 m

downdip direction. Between any two consecutive cases the rupture is offset by about 10 km in the downdip direction. The left panel in Fig. 5 displays the hypothetical slip distribution for each case; slip at the center of each subfault is in meters and is color-coded. The maximum slip in each case is 1 m and the depth contours of the plate interface models are in kilometers and shown by dashed brown lines. Note that the absolute value of slip is not crucial because we will analyze whether waves corresponding to each rupture can constructively interfere with each other. To compute vertical deformation for each idealized rupture, we employ the plate interface model by Hayes et al. (2012). The latter was discretized into a set of rectangular subfaults (Nicolsky et al. 2015). The vertical deformation associated with each case is displayed in the right panel of Fig. 5. Blue shading indicates ground subsidence while red shading marks areas of uplift.

After simulating water dynamics near Adak, in Fig. 6 we plot the modeled water level for each hypothetical rupture case (Cases A–D) at the location marked by a black triangle in Fig. 4. The water level is displayed from the perspective of an observer watching a tide gauge that subsides together with the land in Adak. The modeling results indicate that Cases C and D result in the most significant waves and the corresponding wave crests arrive in sync with each other. Wave crests for Case B arrive almost in sync with the ones for Cases C and D. The water level time series related to Case A is more erratic; however it is not significantly out of phase with the other time series. This implies that if a hypothetical rupture occurs within the areas related to Cases B–D then waves arriving to Adak from different areas of the hypothetical rupture may superimpose and constructively interfere with each other. Thus, in order to develop the worst-case scenario, it is necessary to place a significant amount of slip in regions related to Cases B–D in order for the originating waves to constructively interfere with each other.

On the basis of the above results, we develop hypothetical ruptures with maximum slip of 35 m in the 20–40 km depth range (related to cases B–D). In Scenarios 1–3, the down-dip width of the maximum slip area is about 10 km, centered at the depth

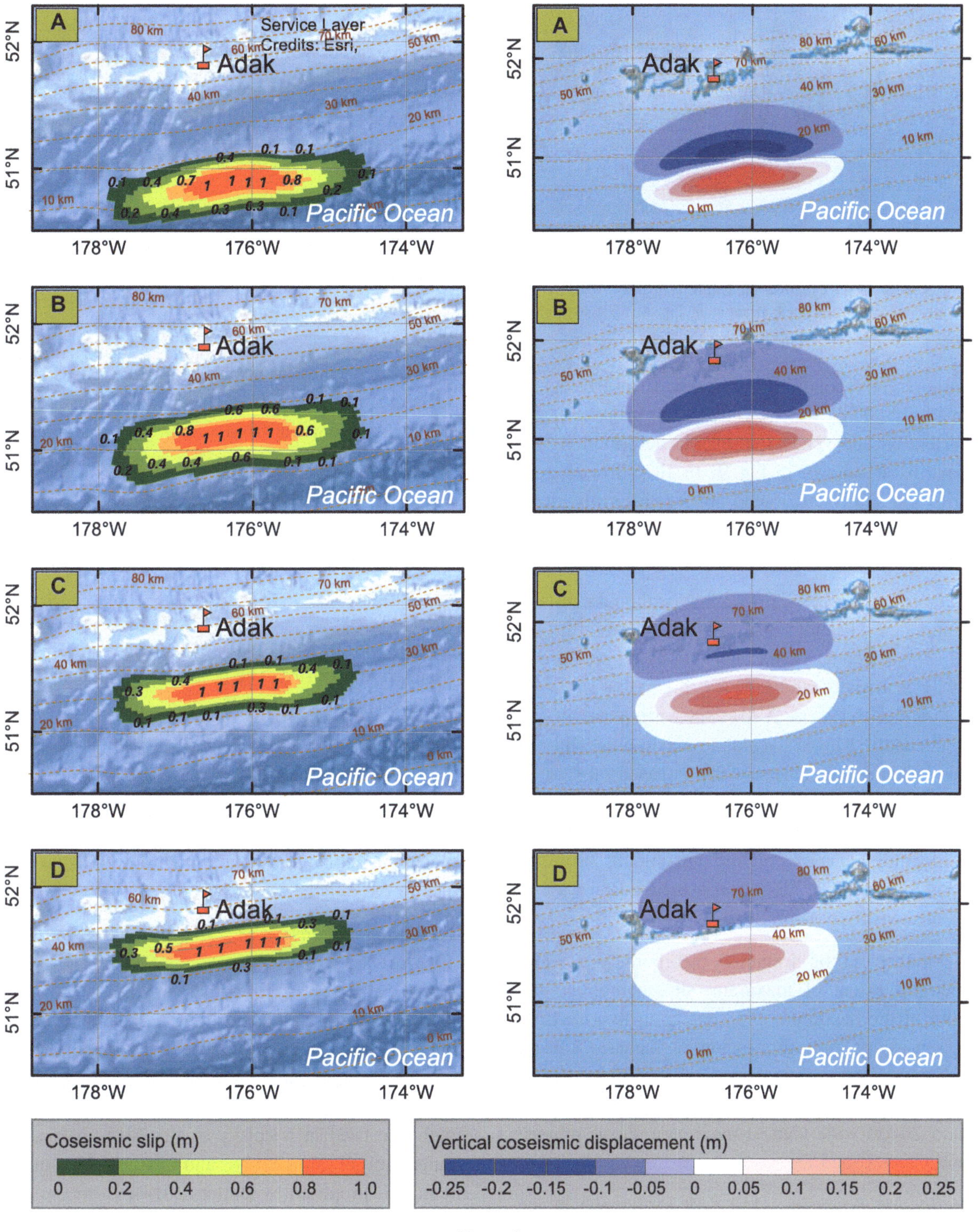

Figure 5

Left: assumed slip distribution along the plate interface for cases A through D. The slip location varies in the downdip direction of the plate interface while preserving the same patch configuration. Dashed brown lines are depth contours in kilometers. Right: computed vertical ground-surface deformation related to cases shown on the left side. Blue areas are associated with coseismic ground subsidence; areas of uplift are shown in red

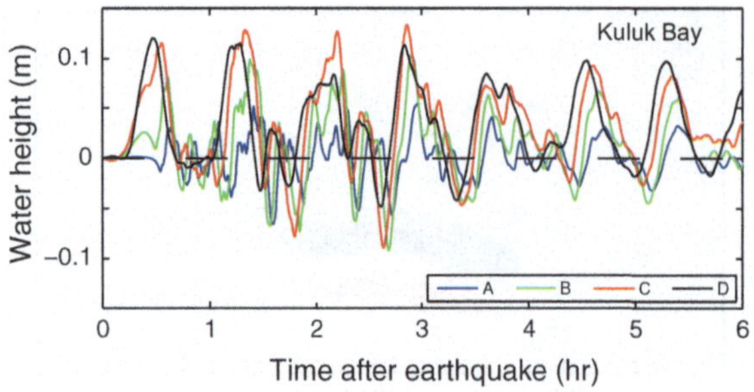

Figure 6
Modeled water-level dynamics in Kuluk Bay near Adak, at the locations marked by a black triangle in Fig. 4, for the ground-surface deformations shown in Fig. 5

contours corresponding to cases B, C and D of the sensitivity study, respectively. The proposed slip distribution and vertical coseismic deformations for scenarios are shown in Fig. 7. In all following scenarios the slip is distributed almost uniformly along strike except for the edges of the rupture where it tapers.

Validation Scenario 1 M_w 8.8 earthquake in the Andreanof Islands region: maximum slip at 17–27 km depth. This scenario is based on Case B of the sensitivity study.
Validation Scenario 2 M_w 8.8 earthquake in the Andreanof Islands region: maximum slip at 25–35 km depth. This scenario is based on Case C of the sensitivity study.
Validation Scenario 3 M_w 8.8 earthquake in the Andreanof Islands region: maximum slip at 35–42 km depth. This scenario is based on Case D of the sensitivity study.

Another set of scenarios incorporates a wider down-dip extent of maximum slip, combining the width of the maximum slip zone from two cases of the sensitivity study. For example, Scenario 4 has a 20-km wide area of maximum slip combined from Cases B and C of the sensitivity study, and Scenario 5 has a 20-km wide area of maximum slip combined from Cases C and D.

Validation Scenario 4 M_w 8. 9 earthquake in the Andreanof Islands region: maximum slip at 15–35 km depth. This scenario is based on the combination of Cases B and C of the sensitivity study.

Figure 7 ▶
Left: proposed slip distribution along the plate interface for Validation Scenarios (VS) 1–7. Slip values in meters are marked by small black labels. The depth contours of the Aleutian interface are shown by gray lines. Right: vertical coseismic deformations corresponding to the scenarios shown on the left side. Blue areas are associated with coseismic ground subsidence; areas of uplift are shown in red

Validation Scenario 5 M_w 8.9 earthquake in the Andreanof Islands region: maximum slip at 25–45 km depth. This scenario is based on the combination of Cases C and D of the sensitivity study.

Finally, because waves arriving from Cases B, C, and D are in phase with each other, we consider an earthquake emanating from a combined source areas of these three cases. Namely, we consider a scenario with the 35 m coseismic slip uniformly distributed south of Adak Island.

Validation Scenario 6 M_w 9.2 earthquake: Uniform 35-m slip
This event is a hypothetical M_w 9.2 (M_0 8×10^{21} N m) earthquake rupturing the Alaska–Aleutian megathrust. The proposed slip is consistent with that used in Nicolsky et al. (2015, 2016a, b) and also with tsunami scenarios proposed by Butler (2014).

Additionally, similar to Nicolsky et al. (2015, 2016a) and Butler (2014), we consider another great hypothetical earthquake with a significant slip close to the trench.

Validation Scenario 7 M_w 9.1 earthquake: Uniform 50/20-m slip
This event is a hypothetical M_w 9.1 (M_0 6×10^{21} N m) earthquake with a 20 m slip on the plate interface between the 18 and 31 km depth, and up to a 50 m slip near the trench, that is, between 5 and 18 km depth. The proposed slip is consistent with that used in Nicolsky et al. (2015, 2016a, b), Suleimani et al. (2017); and also with tsunami scenarios proposed by Butler (2014).

We modeled tsunami dynamics in Adak using the high-resolution DEM (Level 4) for the three sample scenarios and seven validation scenarios. First, in order to understand how the grid resolution affects

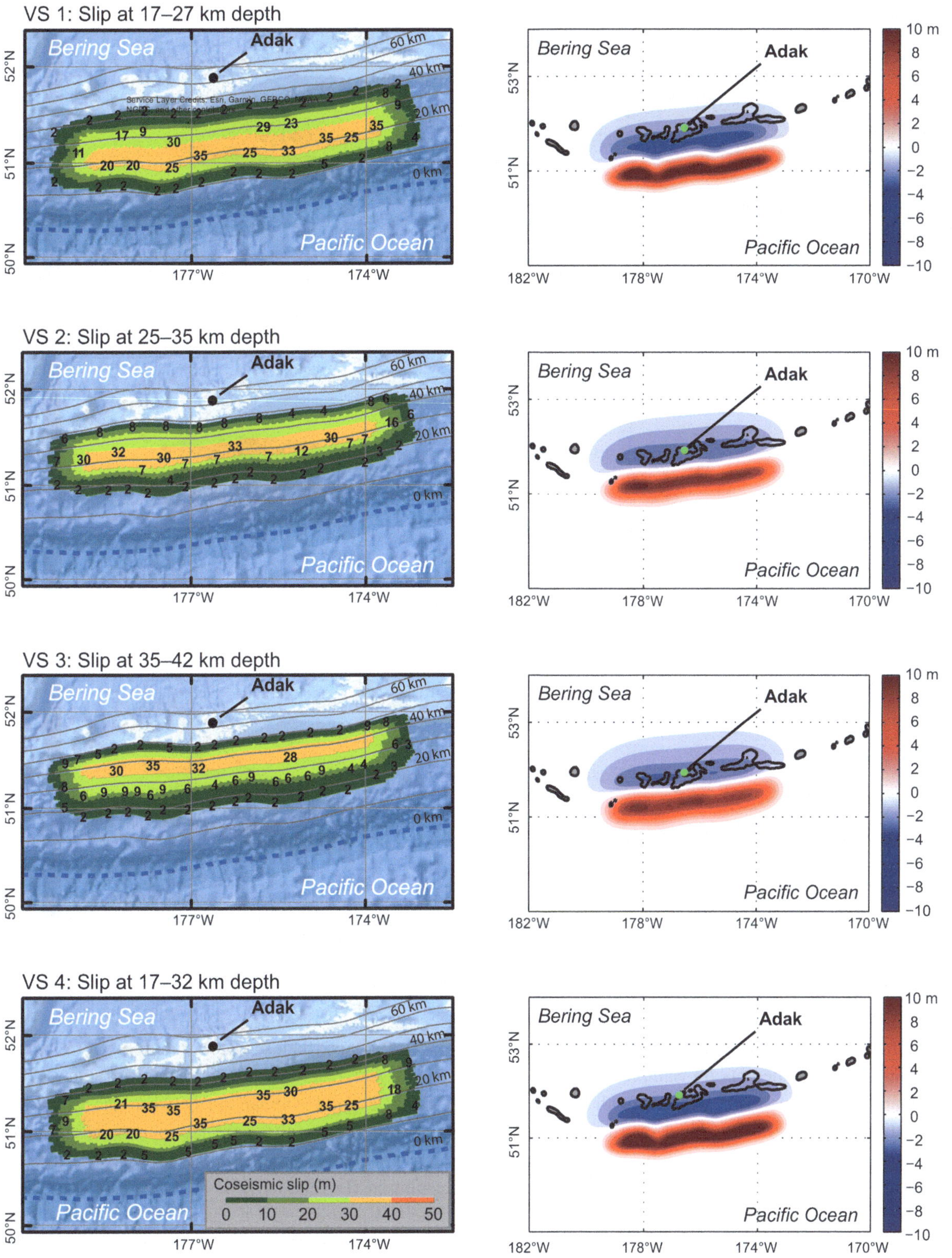
VS 1: Slip at 17–27 km depth
VS 2: Slip at 25–35 km depth
VS 3: Slip at 35–42 km depth
VS 4: Slip at 17–32 km depth
Bering Sea
Adak
Pacific Ocean
60 km
40 km
20 km
0 km
52°N
51°N
50°N
177°W
174°W
53°N
182°W
178°W
170°W
10 m
8
6
4
2
0
−2
−4
−6
−8
−10
Coseismic slip (m)
0 10 20 30 40 50

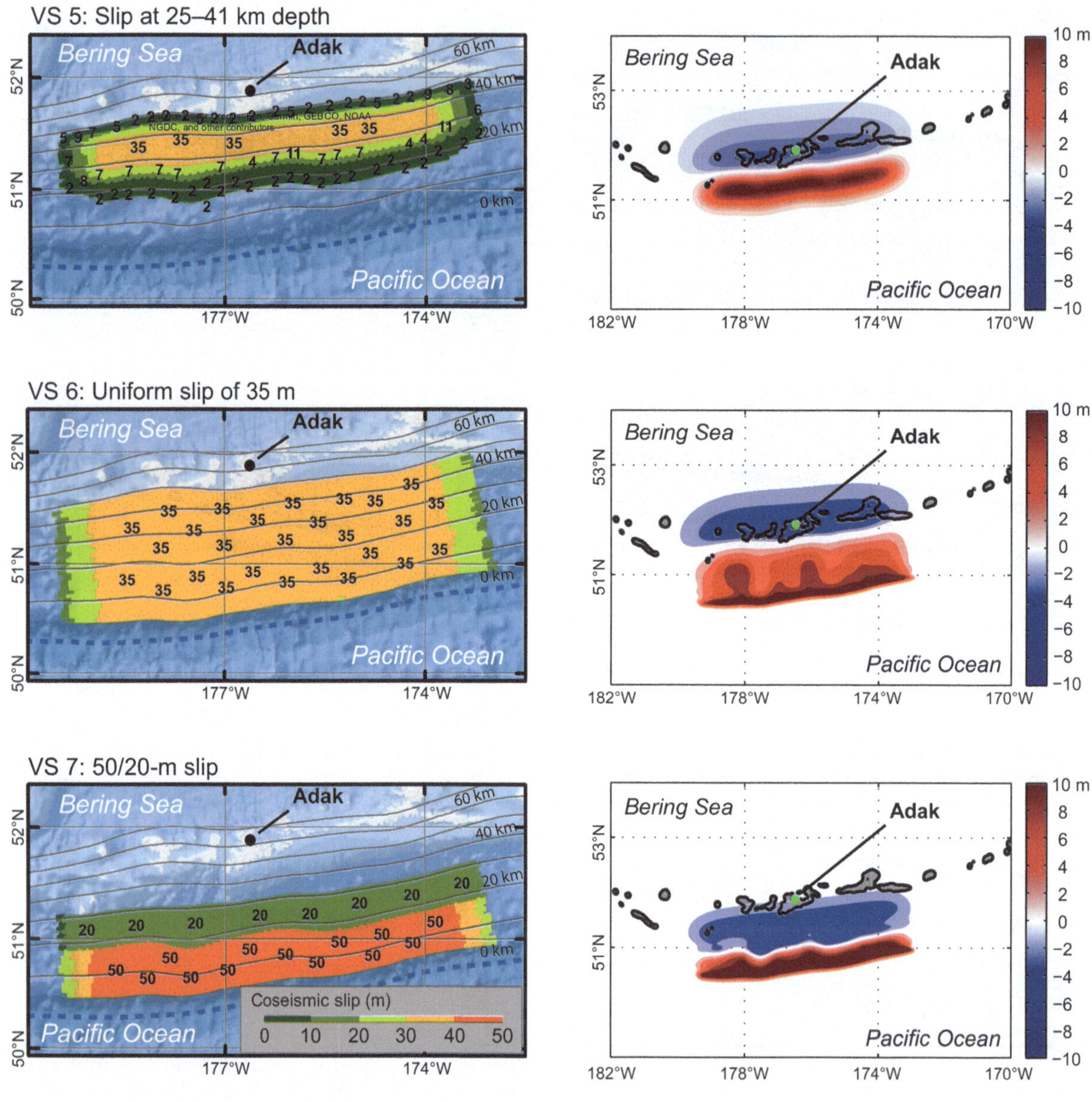

Figure 7
continued

tsunami modeling results, we compare the maximum composite tsunami heights for Sample Scenarios 1-3 simulated on the fine-resolution grid with their high-resolution counterparts. Regional distribution of the maximum composite tsunami heights was previously shown in Fig. 4. Figure 8 now shows comparison of the tsunami heights on the Level 3 and 4 grids near the community. The tsunami heights in both cases agree with each other quite well, i.e. the pattern of inundation and maximum values match. The high-resolution nevertheless allows runup in more narrow gullies and results in farther landward inundation. Despite a good comparison between the high- and fine-resolution results for Adak, our modeling experience shows that computation using the high-resolution (Level 3) DEMs may result in underestimate of tsunami height if narrow channels and canals are present on the way of tsunami towards the shoreline. The community of Adak is located on the

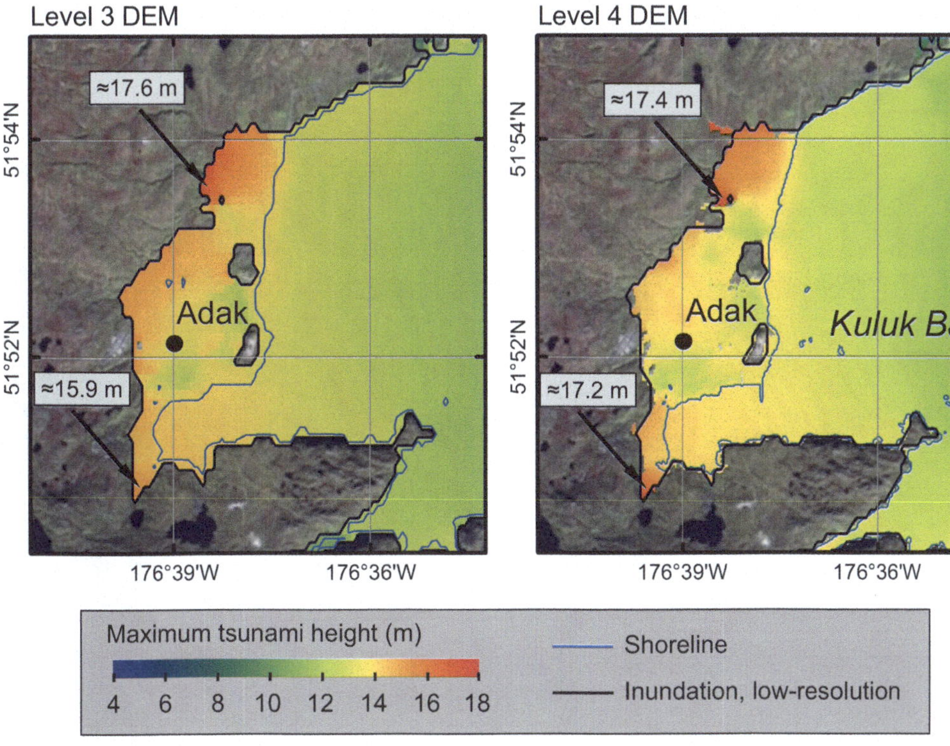

Figure 8
Comparison of the maximum composite tsunami wave height at Adak for Sample Scenarios 1-3 using the fine-resolution Level 3 DEM (left) and high-resolution Level 4 DEM (right). The black line is the maximum extent of inundation computed using the Level 3 DEM

back side of the island and the Kuluk Bay is connected to the Bering Sea by wide water bodies.

The simulated extents of inundation in Adak for all considered scenarios are shown in Fig. 9. The ATHZ, related to the 23-m elevation contour above the MHHW level is shown by the bold dashed red line. Results of the individual scenarios are shown by lines with various colors. Sample Scenarios 1 and 2 do not produce the highest inundation, while Sample Scenario 3 results in inundation that slightly underpredicts an extent of inundation due to Validation Scenario 6. Sample Scenario 1 and Validation Scenario 7, which assume a large slip close to the trench, do not result in the worst-case scenario for the community of Adak, located on the Bering Sea sides of the Aleutian Islands, because the waves are primarily reflected from the Pacific coast. Although we have not considered an outer rise event scenario, we infer that such an event would not result in significant waves near Adak. The outer rise event should be considered for coastal communities located on the Pacific side of the islands (e.g. Suleimani et al. 2016). The comparison reveals that potential inundation according to all considered scenarios is within the ATHZ line and the latter provides a liberal estimate (one that is cautious to avoid underestimating the inundation), if no other information is provided. The safety factor in the ATHZ line construction thus largely plays a role of the safeguard because of the limited number of sample scenarios (recall that Validation Scenario 6 resulted in an inundation greater than that according to sample scenarios).

To further validate the technique, we develop the ATHZs for seven communities (see an inset in Fig. 1 for their locations), where the potential maximum-credible tsunami inundation has been previously estimated on high-resolution DEMs using an exhaustive suite of the plausible scenarios (Suleimani et al. 2016, 2017, 2018; Nicolsky et al. 2016b, 2017). The computed maximum wave heights for each sample scenario are listed in Table 4. Numerical calculations of the maximum wave height in each

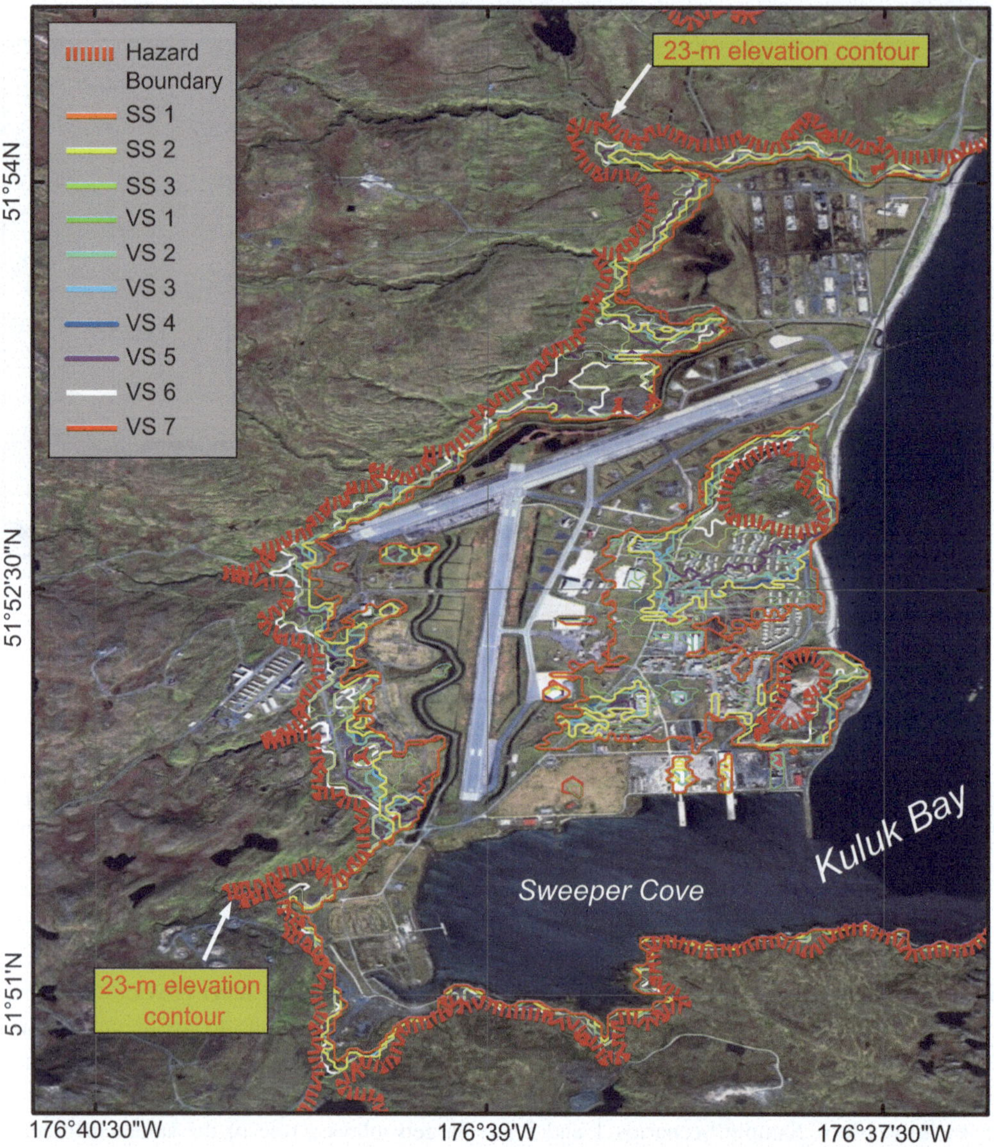

Figure 9
Comparison of the approximate tsunami hazard boundary, related to the 23-m elevation contour, with results of the sample and validation scenarios, denoted by SS and VS, respectively. Note that the 23-m elevation contour encompasses all other lines and provides a liberal estimate of the potential inundation

community were completed in the appropriate Level 3 grid (the 8/3-arcsecond resolution). After incorporating the safety factor, we obtain the estimated maximum runup, consequently construct an elevation contour related to the runup, and thus obtain the ATHZ for the given community. Comparison of the ATHZs and the maximum potential inundation line for the seven communities is shown in Fig. 10. The

Figure 10 ▶
Comparison of the ATHZs (green line) with maximum potential inundation zones (red line) computed with the high-resolution modeling approach in Kodiak (Suleimani et al. 2017), Homer (Suleimani et al. 2018), Sand Point (Nicolsky et al. 2017), Cold Bay and King Cove (Suleimani et al. 2016), Chignik Bay and Chignik Lagoon (Nicolsky et al. 2016b)

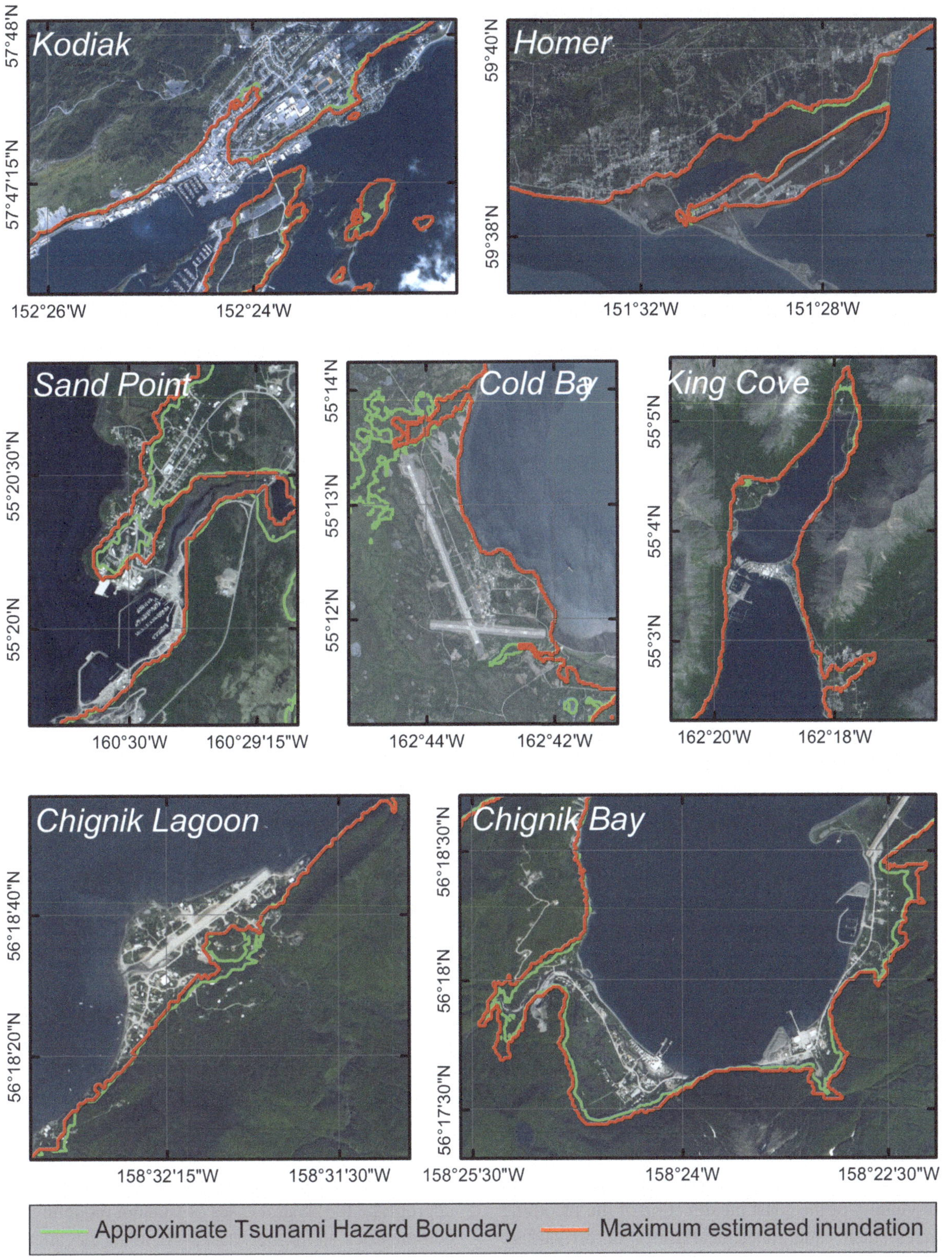
Kodiak
57°48'N
57°47'15"N
57°47'15"N
152°26'W
152°24'W
Homer
59°40'N
59°38'N
151°32'W
151°28'W
Sand Point
55°20'30"N
55°20'N
160°30'W
160°29'15"W
Cold Bay
55°14'N
55°13'N
55°12'N
162°44'W
162°42'W
King Cove
55°5'N
55°4'N
55°3'N
162°20'W
162°18'W
Chignik Lagoon
56°18'40"N
56°18'20"N
158°32'15"W
158°31'30"W
Chignik Bay
56°18'30"N
56°18'N
56°17'30"N
158°25'30"W
158°24'W
158°22'30"W
Approximate Tsunami Hazard Boundary
Maximum estimated inundation

results indicate that the ATHZ line closely approximates or is slightly higher than the maximum potential inundation line for all communities except for Chignik Bay, located at the head of the bay. Similarly the ATHZ underestimates the potential maximum inundation at the head of King Cove Lagoon. Therefore, we conclude that the safety factor for communities located at the heads of bays may need to be increased to accommodate for potential wave focusing; subject to further research. We also hypothesize that in the case of Chignik Bay, a number of sample scenarios need to be increased because of a rather large distance between the trench and community. However, despite some disparities, the ATHZs generally avoid underestimation and provide liberal guidelines for the maximum potential inundation in the communities located along the Alaska–Aleutian subduction zone.

5. *Sources of Errors and Uncertainties*

To simulate tsunami dynamics caused by seafloor deformation from an earthquake we assume some simplifications. First, the initial displacement of the ocean surface is equal to the vertical displacement of the ocean floor induced by the earthquake rupture process. The contribution of horizontal displacements is not considered. Second, the finite speed of rupture propagation along the fault is not taken into account. We consider ocean bottom displacement to be instantaneous. Third, the initial topography is modified to account for coseismic deformation of land due to the earthquake, which can be either uplift or subsidence depending on the community location relative to the earthquake's slip patch. Soil compaction in areas of unconsolidated deposits in the coastal area might occur and the extent of the tsunami inundation could be farther landward.

Because the initial condition for the tsunami modeling is determined by the displacement of the ocean bottom, the largest source of error is the earthquake model. When a tsunami is generated in the vicinity of the coast the direction of the incoming waves, their amplitudes, and times of arrival are determined by the initial displacements of the ocean floor in the source area because the distance to the shore is too small for the waves to dissipate. Furthermore the near-field inundation modeling results is sensitive to the fine structure of the tsunami source. Therefore the ATHZ can be used only as a guideline for predicting an actual inundation event.

Although the ATHZ line provides a credible estimation of tsunami inundation, potential inundation from tsunamis originating outside of Alaska will likely be less. Alaska also has a long recorded history of tsunamis generated by submarine and subaerial landslides, avalanches, and rockfalls (Kulikov et al. 1998). A primary cause of submarine slumps or landslides is the accumulation of sediments on underwater slopes and the consequent oversteepening of unconsolidated deposits. These conditions typically occur at the mouths of glacier-fed rivers, creeks, and streams. Other potential locations for submarine slump failures include the continental shelf. Unfortunately, little is known about the extent, volume, and locations of these potential landslides and even less about the slope stability in these areas. The proposed ATHZ line does not take into account potential inundation due to the mass failures, because of insufficient data on the locations and volumes of these potential hazards.

6. *Summary*

We present the results of modeling earthquake-generated tsunamis in the Andreanof Islands region, Alaska, and their effects on the community of Adak. We estimate the potential tsunami inundation zone using three sample tsunami scenarios. Although the modeled tsunami inundation cannot be considered exhaustive, the results are thought to provide a reliable approximation to the potential tsunami inundation zone. The developed ATHZ was validated using a suite of worst case scenarios constructed on the basis of a sensitivity study of maximum runup with respect to the rupture location and high-resolution modeling of the all considered scenarios.

The proposed safety factor of 1.3 seems to be appropriate for general cases of tsunami modeling for communities along the Alaska–Aleutian megathrust. Based on application of the ATHZ method to other communities where tsunami hazard maps have been developed using high-resolution DEMs, the safety factor may need to be increased for locations at the heads of the bays, where the wave amplification might occur due to the topography/bathymetry focusing effects.

Acknowledgements

This project, which is part of the National Tsunami Hazard Mitigation Program (NTHMP), received support from the National Oceanic and Atmospheric Administration (NOAA) under Reimbursable Services Agreement ADN 952011 with the State of Alaska's Division of Homeland Security & Emergency Management (a division of the Department of Military and Veterans Affairs). Numerical calculations for this work were supported in part by the high-performance computing and data storage resources operated by the Research Computing Systems Group at the University of Alaska Fairbanks, Geophysical Institute. We are grateful to Peter Haeussler (USGS), Ahmet Yalciner (METU) and Jeffrey Freymueller (UAF) for their helpful suggestions and review of potential tsunami sources that helped to improve the manuscript.

REFERENCES

Applied Technology Council. (2012). Guidelines for design of structures for vertical evacuation from tsunamis, second edition: Redwood City, CA, Applied Technology Council Technical Report FEMA P-646, prepared for Federal Emergency Management Agency (FEMA) National Earthquake Hazard Reduction Program (NEHRP) and National Oceanic and Atmospheric Administration (NOAA) National Tsunami Hazard Mitigation Program (NTHMP). https://www.fema.gov/media-library/assets/documents/14708. Accessed 29 Mar 2019.

Argus, D. F., Gordon, R. G., Heflin, M. B., Ma, Chopo, Eanes, R. J., Willis, P., et al. (2010). The angular velocities of the plates and the velocity of the Earth's centre from space geodesy. *Geophysics Journal International, 180*(3), 913–960. https://doi.org/10.1111/j.1365-246x.2009.04463.x.

Butler, R. (2014). Great Aleutian tsunamis. Honolulu, HI, University of Hawai'i at Manoa, Hawai'i Institute of Geophysics & Planetology, Peer-Reviewed Report HIGP-2014-1. http://www.higp.hawaii.edu/reports/2014. Accessed 29 Mar 2019.

Carignan, K. S., Taylor, L. A., Eakins, B. W., Warnken, R. R., Medley, P. R., & Lim, E. (2009). Digital elevation model of Adak, Alaska—Procedures, data sources and analysis. National Oceanic and Atmospheric Administration Technical Memorandum NESDIS NGDC-31, U.S. Dept. of Commerce, Boulder, CO. http://www.ngdc.noaa.gov/dem/squareCellGrid/download/258. Accessed 29 Mar 2019.

Carignan, K. S., Eakins, B. W., Lancaster, M., Loveand M. R., & Sutherland, M. (2017). Digital elevation models of Adak and Atka, Alaska: Procedures, data sources, and analysis. National Oceanic and Atmospheric Administration Technical Report, U.S. Dept. of Commerce, Boulder, CO. http://ciresgroups.colorado.edu/coastalDEM/pubs. Accessed 29 Mar 2019.

Cross, R. S., & Freymueller, J. T. (2007). Plate coupling variation and block translation in the Andreanof segment of the Aleutian arc determined by subduction zone modeling using GPS data. *Geophysical Research Letters, Solid Earth, 34*(6), L06304. https://doi.org/10.1029/2006GL028970.

DeMets, C., Gordon, R. C., Argus, D. F., & Stein, S. (1990). Current plate motions. *Geophysical Journal International, 101*(2), 425–478.

Dixon, T. H., & Moore, C. (2012). *The seismogenic zone of subduction thrust faults*. Columbia: Columbia University Press.

Dunbar, P. K., & Weaver, C. S. (2015). U.S. states and territories national tsunami hazard assessment: Historical record and sources for waves—Update. Report, National Oceanic and Atmospheric Administration, Boulder, CO.

Estabrook, C. H., Jacob, K. H., & Sykes, L. R. (1994). Body wave and surface wave analysis of large and great earthquakes along the eastern Aleutian arc, 1923–1993—Implications for future events. *Journal of Geophysical Research, 99*(B6), 11643–11662. https://doi.org/10.1029/93JB03124.

Freund, L. B., & Barnett, D. M. (1976). A two-dimensional analysis of surface deformation due to dip-slip faulting. *Bulletin of the Seismological Society of America, 66*(3), 667–675.

Fujii, Y., Satake, K., Sakai, S., Shinohara, M., & Kanazawa, T. (2011). Tsunami source of the 2011 off the Pacific coast of Tohoku earthquake. *Earth Planets Space, 63,* 815–820. https://doi.org/10.5047/eps.2011.06.010.

Hayes, G. P., Wald, D. J., & Johnson, R. L. (2012). Slab1.0—A three-dimensional model of global subduction zone geometries. *Journal of Geophysical Research*. https://doi.org/10.1029/2011JB008524.

Ito, Y., Tsuji, T., Osada, Y., Kido, M., Inazu, D., Hayashi, Y., et al. (2011). Frontal wedge deformation near the source region of the 2011 Tohoku-Oki earthquake. *Geophysical Research Letters, 38,* L00G05. https://doi.org/10.1029/2011gl048355.

Johnson, J. M., Tanioka, Yuichiro, Ruff, L. J., Satake, Kenji, Kanamori, Hiroo, & Sykes, L. R. (1994). The 1957 great Aleutian earthquake. *Pure and Applied Geophysics, 142*(1), 3–28. https://doi.org/10.1007/BF00875966.

Kanamori, H. (1970). The Alaska earthquake of 1964—Radiation of long-period surface waves and source mechanism. *Journal of Geophysical Research, 75*(26), 5029–5040. https://doi.org/10.1029/JB075i026p05029.

Kirby, S., Scholl, D., von Huene, R., & Wells, R. (2013). Alaska earthquake source for the SAFRR tsunami scenario, chapter B. In S. L. Ross, & L. M Jones (Eds.), *The SAFRR (Science Application for Risk Reduction) Tsunami Scenario: U.S. Geological*

Survey Open-File Report 2013–1170. http://pubs.usgs.gov/of/2013/1170/b/. Accessed 29 Mar 2019.

Kulikov, E. A., Rabinovich, A. B., Fine, I. V., Bornhold, B. D., & Thomson, R. E. (1998). Tsunami generation by landslides at the Pacific coast of North America and the role of tides. *Oceanology, 38*(3), 323–328.

Lander, J. F. (1996). Tsunamis affecting Alaska, 1737–1996. Boulder, CO, National Oceanic and Atmospheric Administration (NOAA), National Geophysical Data Center (NGDC), Key to Geophysical Research Documentation, vol. 31. ftp://ftp.ngdc.noaa.gov/hazards/publications/Kgrd-31.pdf. Accessed 29 Mar 2019.

Lim, E., Eakins, B. W., & Wigley, R. (2011). Coastal relief model of southern Alaska—Procedures, data sources and analysis. National Oceanic and Atmospheric Administration (NOAA) Technical Memorandum NESDIS NGDC-43. https://www.ngdc.noaa.gov/mgg/coastal/s_alaska.html. Accessed Mar 2019.

Lopez, A. M., & Okal, E. A. (2006). A seismological reassessment of the source of the 1946 Aleutian 'tsunami' earthquake. *Geophysical Journal International, 165*(3), 835–849. https://doi.org/10.1111/j.1365-246X.2006.02899.x.

National Centers for Environmental Information (NCEI)/World Data Service (WDS). (2019). Global historical tsunami database at NGDC, 2100 BC to present (interactive map): National Geophysical Data Center, NOAA. http://doi.org/10.7289/V5PN93H7 **(in progress)**.

National Tsunami Hazard Mapping Program (NTHMP). (2012). Proceedings and results of the 2011 NTHMP model benchmarking workshop. Boulder, CO, U.S. Department of Commerce/NOAA/NTHMP, NOAA Special Report. http://nthmp.tsunami.gov. Accessed 29 Mar 2019.

Nicolsky, D. J., Suleimani, E. N., Combellick, R. A., & Hansen, R. A. (2011a). Tsunami inundation maps of Whittier and western Passage Canal, Alaska. Alaska Division of Geological & Geophysical Surveys Report of Investigation 2011-7. http://doi.org/10.14509/23244.

Nicolsky, D. J., Suleimani, E. N., & Hansen, R. A. (2011b). Validation and verification of a numerical model for tsunami propagation and runup. *Pure and Applied Geophysics, 168*, 1199–1222. https://doi.org/10.1007/s00024-010-0231-9.

Nicolsky, D. J., Suleimani, E. N., Freymueller, J. T., Koehler, R. D. (2015). Tsunami inundation maps of Fox Islands communities, including Dutch Harbor and Akutan, Alaska. Alaska Division of Geological & Geophysical Surveys Report of Investigation 2015-5, 2 sheets, scale 1:12,500. http://doi.org/10.14509/29414.

Nicolsky, D. J., Suleimani, E. N., Haeussler, P. J., Ryan, H. F., Koehler, R. D., Combellick, R. A., et al. (2013). Tsunami inundation maps of Port Valdez, Alaska. Alaska Division of Geological & Geophysical Surveys Report of Investigation 2013-1, 1 sheet, scale 1:12,500. http://doi.org/10.14509/25055.

Nicolsky, D. J., Suleimani, E. N., & Koehler, R. D. (2014a). Tsunami inundation maps of Cordova and Tatitlek, Alaska: Alaska Division of Geological & Geophysical Surveys Report of Investigation 2014-1, 49 p. http://doi.org/10.14509/27241.

Nicolsky, D. J., Suleimani, E. N., & Koehler, R. D. (2014b). Tsunami inundation maps of the villages of Chenega Bay and northern Sawmill Bay, Alaska. Alaska Division of Geological & Geophysical Surveys Report of Investigation 2014-3, 7 sheets, scale. http://doi.org/10.14509/29126.

Nicolsky, D. J., Suleimani, E. N., & Koehler, R. D. (2016a). Tsunami inundation map for the village of Nikolski, Alaska. Alaska Division of Geological & Geophysical Surveys Report of Investigation 2016-7, 1 sheet, scale 1:25,000. http://doi.org/10.14509/29674.

Nicolsky, D. J., Suleimani, E. N., & Koehler, R. D. (2016b). Tsunami inundation maps for the communities of Chignik and Chignik Lagoon, Alaska. Alaska Division of Geological & Geophysical Surveys Report of Investigation 2016-8, 2 sheets, scale 1:12,500. http://doi.org/10.14509/29675.

Nicolsky, D. J., Suleimani, E. N., & Koehler, R. D. (2017). Tsunami inundation maps for the city of Sand Point, Alaska. Alaska Division of Geological & Geophysical Surveys Report of Investigation 2017-3, 4 sheets, scale 1:15,000. http://doi.org/10.14509/29706.

Okada, Y. (1985). Surface deformation due to shear and tensile faults in a half-space. *Bulletin of the Seismological Society of America, 75*(4), 1135–1154.

Pacheco, J. F., Sykes, L. R., & Scholz, C. H. (1993). Nature of seismic coupling along simple plate boundaries of the subduction type. *Journal of Geophysical Research, 98*(B8), 14133–14159.

Ross, S. L., Jones, L. M., Miller, K., Porter, K. A., Wein, A., & Wilson, R. I. et al. (2013). SAFRR (science application for risk reduction) tsunami scenario—Chapter A, executive summary and introduction. In: S. L. Ross, & L. M. Jones (Eds.), *The SAFRR Tsunami Scenario: U.S. Geological Survey Open-File Report 2013–1170* (pp. 1–17). http://pubs.usgs.gov/of/2013/1170/.

Ryan, H., von Huene, R., Scholl, D., & Kirby, S. (2012). Tsunami hazards to US coasts from giant earthquakes in Alaska. *Eos, Transactions American Geophysical Union, 93*(19), 185. https://doi.org/10.1029/2012EO190001.

Scholz, C. H., & Campos, J. (2012). The seismic coupling of subduction zones revisited. *Journal of Geophysical Research—Solid Earth, 117*(B5), 5310. https://doi.org/10.1029/2011jb009003.

Shao, G., Li, X., Ji, C., & Maeda, T. (2011). Focal mechanism and slip history of 2011 M_w 9.1 off the Pacific coast of Tohoku earthquake, constrained with teleseismic body and surface waves. *Earth Planets Space, 63*(7), 559–564. https://doi.org/10.5047/eps.2011.06.028.

Suleimani, E. N., Nicolsky, D. J., & Koehler, R. D. (2013). Tsunami inundation maps of Sitka, Alaska. Alaska Division of Geological & Geophysical Surveys Report of Investigation 2013-3, 1 sheet, scale 1:250,000. http://doi.org/10.14509/26671.

Suleimani, E. N., Nicolsky, D. J., & Koehler, R. D. (2015). Tsunami inundation maps of Elfin Cove, Gustavus, and Hoonah, Alaska. Alaska Division of Geological & Geophysical Surveys Report of Investigation 2015-1. http://doi.org/10.14509/29404.

Suleimani, E. N., Nicolsky, D. J., & Koehler, R. D. (2017). Updated tsunami inundation maps of the Kodiak area, Alaska. Alaska Division of Geological & Geophysical Surveys Report of Investigation 2017-8, 10 sheets. http://doi.org/10.14509/29740.

Suleimani, E. N., Nicolsky, D. J., Koehler, R. D., Freymueller, J. T., & Macpherson, A. E. (2016) Tsunami inundation maps for King Cove and Cold Bay communities, Alaska. Alaska Division of Geological & Geophysical Surveys Report of Investigation 2016-1, 2 sheets, scale 1:12,500. http://doi.org/10.14509/29565.

Suleimani, E. N., Nicolsky, D. J., & Salisbury, J. B. (2018). Updated tsunami inundation maps for Homer and Seldovia, Alaska. Alaska Division of Geological & Geophysical Surveys Report of Investigation 2018-5, 4 sheets. http://doi.org/10.14509/30095.

Suleimani, E. N., Nicolsky, D. J., West, D. A., Combellick, R. A., & Hansen, R. A. (2010). Tsunami inundation maps of Seward and Northern Resurrection Bay, Alaska. Alaska Division of Geological & Geophysical Surveys Report of Investigation 2010-1, 3 sheets, scale 1:12,500. http://doi.org/10.14509/21001.

Synolakis, C. E., Bernard, E. N., Titov, V. V., Kânoğlu, U., & González, F. I. (2007). Standards, criteria, and procedures for NOAA evaluation of tsunami numerical models: Seattle, WA, NOAA/Pacific Marine Environmental Laboratory, Technical Memorandum OAR PMEL-135. http://www.pmel.noaa.gov/pubs/PDF/syno3053/syno3053.pdf. Accessed 29 Mar 2019.

Tichelaar, B. W., & Ruff, L. J. (1993). Depth of seismic coupling along subduction zones. *Journal of Geophysical Research, 98*(B2), 2017–2037. https://doi.org/10.1029/92JB02045.

Uchida, N., Nakajima, J., Hasegawa, A., & Matsuzawa, T. (2009). What controls interplate coupling?—Evidence for abrupt change in coupling across a border between two overlaying plates in the NE Japan subduction zone. *Earth and Planetary Science Letters, 283*(1–4), 111–121. https://doi.org/10.1016/j.epsl.2009.04.003.

Uyeda, S. (1982). Subduction zones—An introduction to comparative subductology. *Tectonophysics, 81*(3–4), 133–159. https://doi.org/10.1016/0040-1951(82)90126-3.

Wu, F. T., & Kanamori, Hiroo. (1973). Source mechanism of February 4, 1965, Rat Island earthquake. *Journal of Geophysical Research, 78*(26), 6082–6092. https://doi.org/10.1029/JB078i026p06082.

(Received February 1, 2019, revised March 15, 2019, accepted March 25, 2019, Published online April 1, 2019)

Pure Appl. Geophys. 176 (2019), 3207–3225

https://doi.org/10.1007/s00024-018-1984-9

Pure and Applied Geophysics

Quantitative Tsunami Risk Assessment in Terms of Building Replacement Cost Based on Tsunami Modelling and GIS Methods: The Case of Crete Isl., Hellenic Arc

I. Triantafyllou,[1] T. Novikova,[1] M. Charalampakis,[1] A. Fokaefs,[1] and G. A. Papadopoulos[1]

Abstract—Tsunami risk assessment is an important component of the planning for risk reduction. We considered tsunami risk as a convolution of tsunami hazard, vulnerability of the assets at risk (e.g. buildings) and the economic value exposed. For testing the model, a coastal segment at the west of Heraklion, capital city of Crete Isl., Greece, was selected. Heraklion was hit in the past by strong tectonic and volcanic tsunamis, generated along the Hellenic Arc in the Mediterranean region. The Minoan tsunami produced by pyroclastic flows during the LBA (17th century BC) great eruption of Thera (Santorini) was selected as an extreme tsunami scenario for the hazard (inundation) zone determination through numerical simulation based on Boussinesq equations for fully non-linear waves. It was found that the wave penetrates inland up to $\sim$ 1.2 km, while the maximum water depth is $\sim$ 14 m. The building stock was obtained from the 2011 national census data and validated with the use of orthophotomaps, field inspection and Google Maps. Building vulnerability was determined with the use of the empirical GIS tool DAMASCHE which is based on 2004 Indian Ocean tsunami building damage data and produces damage level by combining water depth and building construction types. The damage level was translated to absolute monetary loss on the basis of cost flat rates determined officially for building replacement, i.e., either reparation or reconstruction, after the destructive earthquakes in Greece during 2014. The method is applicable in other parts of the Mediterranean and beyond provided that appropriate data are available.

Key words: Tsunami risk, Minoan tsunami, numerical simulation, GIS mapping, building vulnerability, monetary loss.

1. Introduction

Large tsunamis are low-probability but high-impact natural events causing physical and economic destruction as well as social disruption in coastal communities. After the devastating mega tsunamis of 26th December 2004 in the Indian Ocean and of 11th March 2011 in NE Japan, the issue of preparedness against, and management of, the tsunami threat attracted great interest around the globe. The interest is multifold and includes hazard, vulnerability and risk assessment, building-up and operation of tsunami warning systems as well as the development of a series of risk mitigation actions including risk management plans, training and education, exercises and public awareness. Critical elements in the destructiveness of tsunamis are, among others, the impact on buildings, critical facilities and infrastructures as well as on population, such as loss of life, injury, or other health impacts. Therefore, the estimation of expected impacts from future tsunamis is of particular value for risk assessment and emergency planning.

1.1. Definitions

Definitions of the term risk related to earthquakes, as originally defined by UNESCO in 1978 (Algermissen et al. 1979; Fournier d' Albe 1982), considers it as a multidisciplinary issue mathematically expressed by the convolution of hazard, vulnerability, and value exposed to the hazard:

$$\text{Risk} = \text{hazard} \bigstar \text{vulnerability} \bigstar \text{value} \qquad (1)$$

This definition was generalised to other kinds of risks (e.g. Smith 1992) and adopted for tsunami risk as well (e.g. Papadopoulos and Dermentzopoulos 1998; Curtis and Pelinovski 1999), which is consistent with definitions of relevant terms and glossaries (UNISDR 2009; EC-Working Paper 2010; IOC 2013). We adopted such a discrimination between hazard, which is a potentially damaging or destructive natural process or phenomenon, and risk as an expression of the natural event's impact. On the other

[1] Institute of Geodynamics, National Observatory of Athens, Athens, Greece. E-mail: ioannatriantafyllou@yahoo.gr

hand, value may be considered as capital value (e.g. monetary losses) for buildings and other engineered structures, services and properties.

1.2. Tsunami Hazard and Risk Assessment

Several authors were based on a variety of deterministic or probabilistic models to assess tsunami hazard (e.g. Rikitake and Aida 1988; Hébert et al. 2005; Tinti et al. 2005; Geist and Parsons 2006; Burbidge et al. 2008; Lorito et al. 2007, 2009; Power et al. 2012; Sørensen et al. 2012; Løvholt et al. 2015; Li et al. 2016; Davies et al. 2017). The deterministic approach is usually based on the largest sources (extreme tsunami scenario) for the study area or on the so-called worst-case credible tsunami scenario. On the other hand, various models or approaches have been tested to assess vulnerability, exposure and risk (e.g. Pararas-Carayannis 1988; Qinghai and Adams 1988; Karim and Yamazaki 2003; Papathoma and Dominey-Howes 2003; Papathoma et al. 2003; Sato et al. 2003; Kulikov et al. 2005; Dias et al. 2009; Omira et al. 2010; Dall' Osso et al. 2010, 2016; Gardi et al. 2011; Strunz et al. 2011; Tonini et al. 2011; Valencia et al. 2011; Wegscheider et al. 2011; Jelínek et al. 2012; Benchekroun et al. 2013; Suppasri et al. 2013a, b; Yeh et al. 2013; Charvet et al. 2014; González-Riancho et al. 2014; Leelawat et al. 2014; Tarbotton et al. 2015; Pagnoni et al. 2015, Pagnoni and Tinti 2016).

Each one of the several components involved in the tsunami risk assessment is susceptible to a variety of epistemic and aleatory uncertainties. Therefore, risk assessment is a highly complex procedure which also suffers from uncertainties of various origins. As a consequence, the several qualitative, semi-quantitative or fully quantitative approaches, which have been tested so far, are quite variable and closely dependent on the data availability as well as on the reliability and quality of the data. For this reason, no standard methodologies have been concluded so far, although some general recommendations for tsunami risk assessment have been organised (e.g. Jelínek and Krausmann 2008; IOC 2009, 2013; Hettiarachchi et al. 2011). In this context, no abundant studies are available on the quantitative risk assessment in terms of expected economic losses due to tsunami building replacement, which includes both repairable damage and building reconstruction if unrepairable destruction occurs. Such losses have been expressed using as metrics either the probable maximum loss for a given return period of the extreme tsunami event (e.g. Dominey-Howes et al. 2010) or the probabilistic average annual loss and the loss exceedance curve (e.g. Goda and Song 2015; Jaimes et al. 2016). However, the various assets exposed to hazard, e.g. buildings, constitute parameters which change in space and time because of the development of the built environment. Therefore, such approaches projecting the expected tsunami impact in the future become unrealistic due to the future changes of the assets exposed to hazard.

1.3. Objectives of the Study

For the reasons explained before, we preferred to assess tsunami risk on the basis of the worst-case or extreme scenario and calculate the expected economic losses at the moment of the study performance in terms of absolute monetary cost for building replacement. In the Mediterranean and the adjacent seas only a few relevant studies were performed so far. In Istanbul, Marmara Sea, in a study by Hancilar (2012) only the exposure of buildings to tsunami inundation was approached but no estimation of physical tsunami damages to buildings was estimated. Therefore, the total monetary value of the exposed buildings was calculated only as an indicative figure. Pagnoni et al. (2015) performed scenario-based assessment of building damage and population exposure due to earthquake-induced tsunamis for the town of Alexandria, Egypt, and estimated the amount of buildings that may suffer damage of various levels. Tsunami vulnerability and damage models for the town of Syracuse, Sicily (Italy), were tested by Pagnoni and Tinti (2016).

The coastal segment to the west of Heraklion, the capital city of Crete Isl., Greece (Figs. 1, 2) was selected as one of the test sites for the European tsunami research project ASTARTE (EU-FP7 2013–2017) for the reason that the coastal zone of Heraklion and its surroundings are highly exposed to

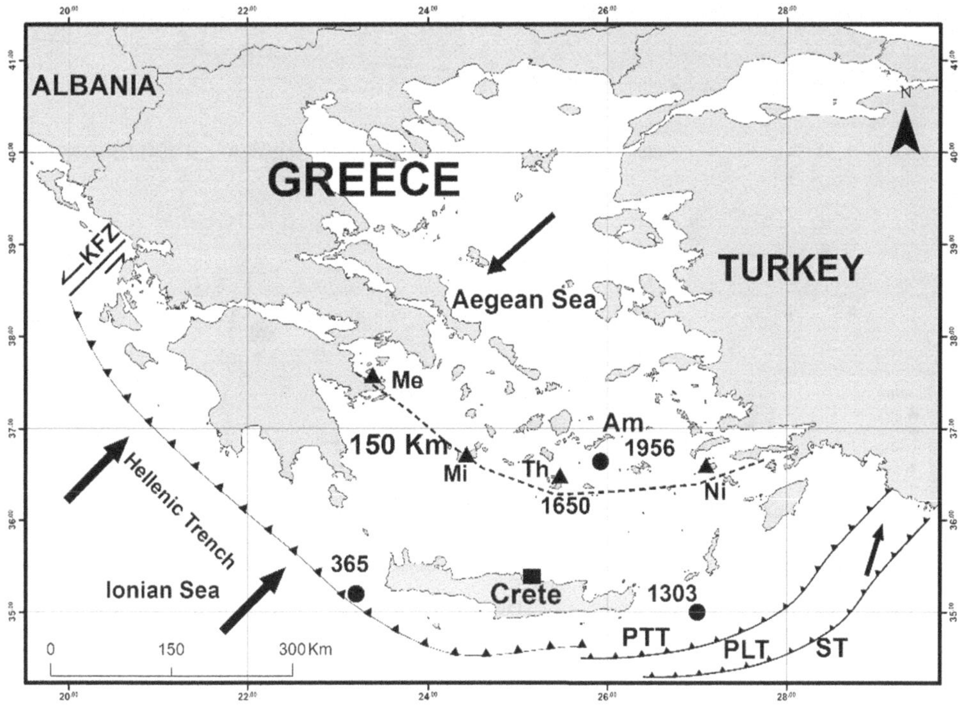

Figure 1
Geodynamic features in the area of the Hellenic Arc (modified from Papadopoulos 2017). Symbol key: arrow, direction of lithospheric plate motion; KFZ, Kephalonia Fault Zone; PTT, Ptolemeus Trench; PLT, Plinius Trench; ST, Strabo Trench; dashed line = 150-km isodepth of the subducting lithospheric slab; solid triangle, volcanic centre, Me, Methana, Mi, Milos, Th, Thera, Ni, Nissyros, Am, Amorgos, solid square, Heraklion. The epicentres of the large tsunamigenic earthquakes of 21 July 365, 8 August 1303 and 9 July 1956 are illustrated by solid circles. Geographic latitude and longitude are north and east, respectively. Projection system is WGS84

tsunamis. Studies towards assessing tsunami risk and vulnerability in Heraklion were performed several years ago (e.g. Papadopoulos and Dermentzopoulos 1998; Papathoma et al. 2003). However, neither building vulnerability nor risk assessments were quantitatively included in those early studies. In this paper, we developed a methodology for the ex-ante tsunami risk estimation in terms of absolute monetary values. The new contribution relies on that we propose to translate the direct building physical damage due to tsunami attack to building replacement cost. Building replacement includes either repair or reconstruction. The method was tested in Heraklion based on the extreme tsunami scenario concept, the determination of the inundation zone through tsunami numerical modelling and the use of GIS tools for mapping and estimation of building damage levels and the resulting replacement costs.

2. *Methodology*

A scenario-based tsunami risk assessment approach is suitable where the available historical tsunami data are not sufficient to organise reliable tsunami statistics, e.g. magnitude–frequency relationships (e.g. Nadim and Glade 2006; Pagnoni and Tinti 2016; Rehman and Cho 2016). Although this approach does not provide probabilistic information it has the advantage that historical data combined with results of wave numerical simulation yield tsunami physical features needed for the determination of the hazard (inundation) zone expected from the selected extreme event. Several assets of the anthropogenic system, such as buildings, infrastructures and life lines, vessels and other property, cultivated land, population, insured and non-insured values, are exposed to tsunami hazard. In this report, we selected to focus on the quantitative risk assessment in terms

Figure 2
The study area of the western segment of Heraklion city (Gazi), Crete Isl., in Google Earth map. The area is characterised by the Yofiros river and the nearby stadium in the eastern side, by the Xeropotamos river and a free of buildings area in the central side and by the motorway running from east to west at the south side. The arrow pinpoints part of the built environment of the study area shown in Fig. 4

of economic losses due to the need for building replacement.

The methodology proposed in this paper for the building risk assessment follows a work flow (Fig. 3) which is adjusted to the adopted concept, i.e. risk is a convolution of three main factors according to formula (1). To this aim, we estimated quantitatively each one of the three main parameters involved in the adopted risk concept: tsunami hazard, building vulnerability, building values exposed to tsunami hazard. Finally, the risk was assessed in terms of expected absolute economic cost for building replacement. The steps followed according to this flow chart are as follows: (a) selection of a testing coastal segment and of the worst-case scenario; (b) tsunami numerical simulation; (c) determination and GIS mapping of the inundation (hazard) zone through tsunami numerical simulation; (d) assessment and GIS mapping of the building exposure based on types of building material and use; (e) calculation with the use of the GIS DAMASCHE tool of the expected damage level as a function of the building type and the water depth, i.e. the depth of the sea water column in the inundation zone; (f) calculation of absolute monetary cost from the number of buildings that would need replacement, i.e. either only reparation or complete reconstruction due to damage beyond repair. Steps (a)–(c) represent the tsunami hazard determination, while building vulnerability assessment is represented by steps (d) and (e). The final step (f) represents the building value exposed to hazard. Then, it is possible to perform an ex-ante calculation of the expected economic losses in terms of absolute monetary cost for building replacement. In the next lines, we explain in more details the procedure followed.

2.1. *Hazard Determination*

The procedure starts with the selection of the testing coastal segment in the study area. This is done by taking into consideration that the testing area should be characterised by high building exposure with the aim to allow for testing the method. In order to determine the worst-case scenario of particular importance is the examination of the tsunami history of the area on the basis of all the available data, such as instrumental records, documentary sources and geological tsunami signatures. However, it is not necessary the extreme scenario selected to be a representation of a particular historical event. A

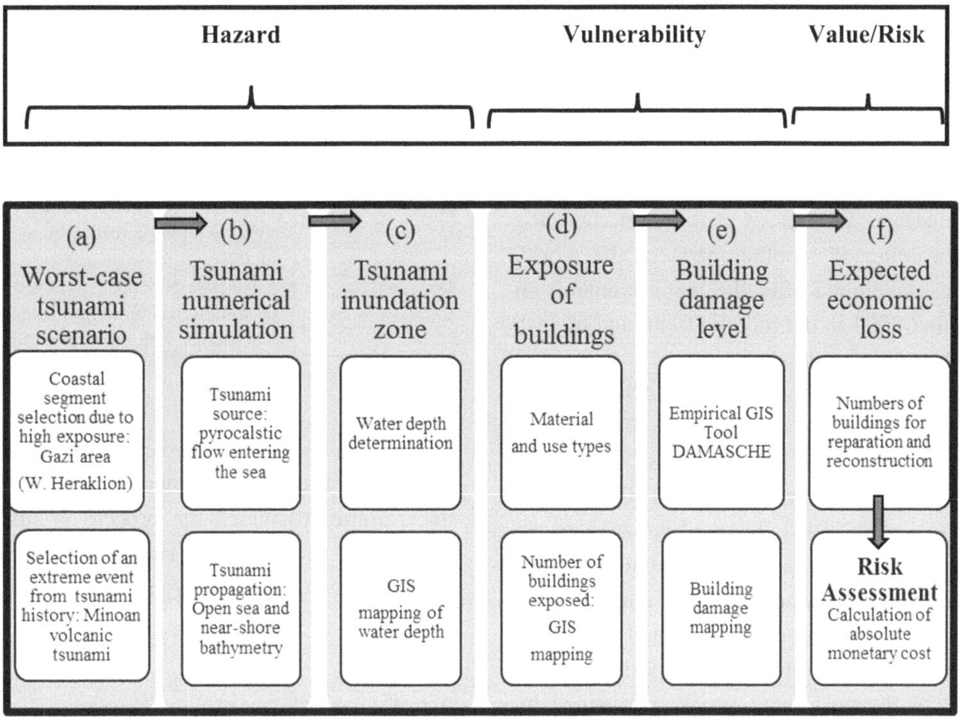

Figure 3
Flow chart of the procedures followed to calculate expected losses due to tsunami building damage in terms of absolute monetary cost

hypothetical scenario inferred from the potential tsunamigenic sources of the area, regardless the type of source, seismic, volcanic or landslide, could be also adopted. The numerical simulation of the generation, propagation and inundation of the tsunami selected for the extreme scenario concludes with the determination of the inundation or hazard zone in the coastal area of interest. Then, metrics of the tsunami hydrodynamics would be available, e.g. water depth, wave velocity, etc. This step of the procedure provides all the necessary information that should be known as regards the hazard component, which is the first step involved in the risk assessment procedure (Fig. 3). For the needs of the present study, the wave simulation was performed based on the Fully Non-linear Wave model (FUNWAVE, Wei and Kirby 1995; Wei et al. 1995; Kennedy et al. 2000; Kirby et al. 1998, 2003) which is a fully non-linear Boussinesq theory using a predictor–corrector scheme. For this model, the initial conditions (Grilli and Watts 1999), i.e. surface elevation and water velocities, were provided by approximate simulation tool known as TOPICS (Tsunami Open and Progressive Initial Conditions System). The combination of TOPICS and FUNWAVE is the software package widely used under the name GEOWAVE (e.g. Watts et al. 2003).

2.2. *Building Vulnerability Assessment*

As the water proceeds inland may cause damage or even complete destruction to buildings and other engineered structures. Therefore, in this step there is need for a model or an empirical tool to translate the water impact into physical building damage. A model widely used is the so-called PTVA (Papathoma Tsunami Vulnerability Assessment) model, e.g. Papathoma et al. (2003) and its later modifications such as the one published by Dall' Osso et al. (2010, 2016). On the other hand, building fragility or damage functions or curves calibrated against data collected after recent tsunamis comprise empirical tools which are also in use for the estimation of the damage level expected to be caused in buildings in the inundation zone. However, no such functions or curves are available in the Mediterranean region for the reason that the last destructive tsunamis occurred many years ago, namely the Messina Straits (South Italy) tsunami in 1908 and the South Aegean (Greece) tsunami in 1956. Therefore, alternative tools are fragility functions or curves based on building damage data collected in other parts of the world, like the ones developed after the mega tsunamis of 26th December 2004 in the Indian Ocean and of 11th March 2011 in NE Japan.

A relevant GIS tool, called DAMASCHE, was developed in the frame of the EU-FP6 project SCHEMA and was tested in various test sites of the European and the Mediterranean region (Gardi et al. 2009; Valencia et al. 2011). The tool is based on the data collected in the aftermath of the Indian Ocean 26 December 2004 tsunami and relates the damage level with the building structural material and the water depth (Table 1). Such a relation is supported also by the findings of the post-event survey performed by Leelawat et al. (2014) after the 11 March 2011 Japanese mega tsunami. The results of that study indicated that the water depth, the structural material, and the function of building were significantly associated with the damage level. DAMASCHE tool is applicable for various types (classes) of buildings as explained in Table 2. From this point of view, the application of the tool is not restricted by the different building types characterising the European-Mediterranean, SE Asia or other regions. Further discussion on tsunami-induced building damage through fragility functions can be found in Charvet et al. (2017).

Once the tsunami hydrodynamics has been translated into building damage levels, the vulnerability becomes available. However, such an assessment should be carefully evaluated in view of the various uncertainties to which the process is susceptible to (e.g. Gardi et al. 2011). The building classification adopted according to the types of buildings is listed in Table 2.

Table 2

Building vulnerability classification (Leone et al. 2010)

Building class	Type of buildings
A	Light constructions on wood or timber without any design
B	Brick not reinforced masonry
C	Brick with reinforced column and masonry filling
D	Collective buildings, no concrete reinforced
E	Well-designed buildings, made of reinforced concrete with columns and infill walls

2.3. *Ex-ante Calculation of Economic Losses*

Economic losses expected from the tsunami impact on buildings can be estimated on the basis of needs for building replacement which is a function of the damage level. According to Table 1, the damage level in a building due to tsunami impact may scale from “no damage” to “total collapse”. No damage implies no losses, while economic losses implied by the “light damage” level could be considered as negligible. Under these assumptions, we considered that building replacement is needed either for building reparation, when repairable damage has occurred, which include “important damage” and “heavy damage”, or for complete reconstruction

Table 1

Building damage level as a function of the water depth (in m) in the inundation zone (Gardi et al. 2009; Valencia et al. 2011)

Damage level		Class A (m)	Class B (m)	Class C (m)	Class D (m)	Class E (m)
No damage	D_0	0	0	0	0	0
Light damage	D_1	0–1.8	0–2	0–2.5	0–2	0–3
Important damage	D_2	1.8–2.2	2–3	2.5–4	2–4.5	3–6
Heavy damage	D_3	2.2–2.6	3–4	4–6	4.5–6.5	6–9.5
Partial collapse	D_4	2.6–3.8	4–5	6–8	6.5–9	9.5–12.5
Total collapse	D_5	> 3.8	> 5	> 8	> 9	> 12.5

The building classification is according to the construction material types listed in Table 2 of buildings that suffer damage beyond repair. This includes the cases of "partial collapse" and "total collapse". It should be noticed that the terminology used for the damage classification is standardised for the DAMASCHE tool as summarised in Table 1. However, one may correspond the terms of light, important and heavy damage to alternative terms such as minor, moderate and major damage, respectively.

For the calculation of the total cost, C, for building replacement, one may take into account that combined damage effects may occur after the tsunami attack. This includes physical damage to the structural system itself mainly due to the water flow as well as nonstructural damage and damage to the building contents (property) due to flood. Here, we considered only the physical damage to the structural system of buildings. Then, the total cost, C, for building replacement is estimated as:

$$C = \sum\nolimits_{n=1}^{i} C_1 + \sum\nolimits_{n=1}^{k} C_2, \qquad (2)$$

where C_1 and C_2 are the costs for reparation and reconstruction of buildings, respectively, and i, k are the numbers of buildings that need reparation or reconstruction. Metrics of replacement cost may include (i) the actual cost needed for full replacement, (ii) the official cost recognised by the state, i.e. the local or central governmental organisations supporting replacement, (iii) insured losses. The use of one or more of these options in the estimation of the building replacement cost depends on local practices and on data availability.

3. Application of the Method in Heraklion City

3.1. The Coastal Zone Tested

Heraklion, the capital city of Crete Isl., is situated at the north coast of the island (Figs. 1, 2). It has a permanent population of about 170,000 residents but during tourist peak season, the population at least doubles. The main infrastructures along or close to the coastal zone of Heraklion and its adjacent municipalities are the new harbour and the international airport, both situated at the eastern side of the city. The central coastal segment is characterised by historical built environment dominated by the presence of the old port and the Venetian protective walls being as high as ∼ 4 m. Along the western coastal segment, which belongs to the municipality of Gazi, the main feature is a long beach 5–6-km long. At the western end of this beach, the public power plant is situated, which is a critical infrastructure for the entire island.

The eastern segment of Gazi coast (Figs. 2, 4), which is adjacent to Heraklion municipality, was selected for testing the method. The study area is characterised by high exposure to tsunamis not only because several seismic and volcanic tsunami sources are situated around the island but also because the relatively flat geomorphology of the coastal zone allows for an easy tsunami penetration inland. On the other hand, the study area is a modern urban entity developed along the coastal zone. The urban planning of the area favours the development of both residential buildings and touristic services, such as hotels, restaurants and other relevant facilities, especially, during the summer season (Fig. 4).

3.2. Extreme Scenario Selection

Crete Isl. occupies the central segment of the Hellenic Arc, which is a highly active geodynamic structure since it marks the plate boundary between the SSW-moving Eurasian lithospheric plate and the NNE-moving African plate subducting underneath the South Aegean region (Fig. 1). The most frequent generation of large tsunamis in the entire North Eastern Atlantic and Mediterranean (NEAM) region has been reported from large magnitude earthquakes and volcanic eruptions occurring in the Hellenic Arc. For the selection of the worst-case scenario, we initially considered five candidate historical events, three of tectonic and two of volcanic origin. These five tsunamis (Table 3) are the most important ones that affected Crete Isl.

3.2.1 Seismic Tsunami Sources

Very large subduction earthquakes, such as the ones of AD 21 July 365 and of 8 August 1303, having estimated magnitudes around 8.0 or more, ruptured to the west and east of Crete, respectively. Of interest is

Figure 4
A view of the study area. Hotels and private houses are the prevailing building types (photo taken on 14th Dec. 2015). The above area is shown by an arrow in Fig. 2

also the 9 July 1956 large earthquake of magnitude 7.5 that occurred in the back-arc seismotectonic environment of the South Aegean Sea. From the historical documentation, no doubt remains that the large AD 365 tsunami affected the east Mediterranean basin with damage and/or loss of life reported from various coastal areas, such as south Peloponnese and western Crete (Greece), Alexandria (Egypt), east Sicily (Italy) and possibly in Dalmatia, Adriatic Sea (see reviews in Guidoboni et al. 1994; Guidoboni and Comastri 1997, 2005; Papazachos and Papazachou 2003; Ambraseys 2009; Papadopoulos 2011, 2015; Papadopoulos et al. 2014a).

However, the impact of this tsunami along the north Cretan coast, particularly in the area of Heraklion, has not been documented very likely due to that the area was sparsely populated at the time. Besides, although there is general consensus that the source of the causative earthquake was placed in the western segment of the Hellenic Arc, different fault

Table 3

Parameters of the most important seismic and volcanic historical tsunamis that affected the study area of Heraklion

Year	Month	Day	φ°N	λ°E	Source	Size	Height (m)	Intensity (K)
~ 1613 BC			35.00	27.00	VO	VEI 8	~ 20 (NorthCrete)	10 (North Crete)
AD 365	07	21	35.20	23.20	EQ	M 8.3	> 7 (West Crete) > 4 (Alexandria)	10 (Alexandria)
1303	08	08	35.00	27.00	EQ	M 8.0	> 4 (Heraklion)	10 (Heraklion)
1650	10	10	36.50	25.50	VO	VEI 7	~ 20 (Cyclades) > 4 (Heraklion)	8 (East Thera Isl., Santorini) 5–6 (Heraklion)
1956	07	09	36.64	25.91	EQ	M 7.5	~ 15 (Astypalea Isl.) 2–3 (North Crete)	8 (Astypalea) 4 (North Crete)

φ°N and λ°E, geographic latitude and longitude of the source, VO, volcanic source, EQ, seismic source, VEI, Volcanic Explosivity Index (scale by Newhall and Self, 1982), M earthquake magnitude, K, tsunami intensity (scale by Papadopoulos and Imamura, 2001). Parameters are taken from Papadopoulos (2011) positions, geometries and mechanisms were suggested (see review in Papadopoulos et al. 2014a), thus introducing a variety of uncertainties.

The documentary sources available support that the 1303 tsunami affected also a large part of the east Mediterranean basin. Heavy building damage and loss of life were reported in Heraklion (Guidoboni and Comastri 1997). However, although the source of the large causative earthquake is traditionally considered that ruptured the eastern segment of the Hellenic Arc (e.g. Papazachos and Papazachou 1997; Yolsal-Çevikbilen and Taymaz 2012; Papadopoulos et al. 2014a; Necmioglu and Özel 2015), the cooperative European earthquake catalogue of SHARE project (http://www.share-eu.org/; https://www.emidius.eu/SHEEC/; last access 28 Oct. 2017) placed the earthquake epicentre further to the north in the Cretan Sea, while other authors argued that several moderate events occurred rather than only one single large event (El-Sayed et al. 2000). Therefore, the epicentre, fault geometry and mechanism of the tsunami source remain quite uncertain.

The 1956 event was a damaging tsunami of $\sim$ 10 to 15 m in height and observed in a long number of coastal sites and recorded by two tide-gauges in the South Aegean Sea (Galanopoulos 1957). The tsunami wave was also recorded with small amplitude by a remote tide-gauge at Yafo, Israel (Beisel et al. 2009). The causative earthquake was of magnitude measuring M_s 7.4 (Makropoulos et al. 2012) or seismic moment 3.9×10^{27} dyn cm (Okal et al. 2009). The earthquake source should be placed within the $\sim$ 80-km-long aftershock zone striking NE–SW in a graben-like structure in the submarine tectonic trough to the east of Santorini (Thera) Isl. (Papadopoulos and Pavlides 1992). However, it is still controversial if the tsunami was produced by co-seismic fault displacement or landslide or a combination of the two mechanisms (see discussions in Galanopoulos 1957; Ambraseys 1960; Perissoratis and Papadopoulos 1999; Beisel et al. 2009; Okal et al. 2009; Papadopoulos et al. 2014a, Papadopoulos 2015). During the 1956 tsunami, the sea level rise in Heraklion was of $\sim$ 2 m causing only little damage (Galanopoulos 1957). A preliminary tsunami simulation performed by us produced only a limited inundation zone in the selected coastal segment of Gazi, which implies that by taking also into account the source ambiguities, the 1956 event does not make an appropriate extreme scenario case for testing the method in the study area.

3.2.2 Volcanic Tsunami Sources

The South Aegean volcanic island arc, which is nearly parallel to the Hellenic Arc at an average distance of about 130 km (Fig. 1), has also been the source of large volcanic tsunamis (see review in Papadopoulos et al. 2014a). The so-called Minoan tsunami, generated by the giant Late Bronge Age (17th century BC) caldera eruption of the volcano of Thera (Santorini, Fig. 1), has been initially supported from archaeological findings (Marinatos 1939). Onshore tsunami sediment deposits attributed to that event were found in several coastal spots in the eastern Mediterranean basin including SW Turkey and north Crete (Minoura et al. 2000; McCoy and Papadopoulos 2001; Bruins et al. 2008), Thera (McCoy and Heiken 2000) and eastern Sicily (De Martini et al. 2010). Offshore coring in the upper shelf (< 20 m depth) sediments in Caesarea, Israel, found by Goodman-Tchernov et al. (2009), as well as in Augusta Harbor, east Sicily (Smedile et al. 2011) revealed sediment structures which were also attributed to the Minoan tsunami. It has been supported that thick layers of megaturbidites found in the Ionian Sea were triggered by the Minoan tsunami (e.g. Kastens and Cita 1981; Hieke 1984; Cita et al. 1984, 1996; Cita and Aloisi 2000). However, Polonia et al. (2013) attributed these characteristic layers to the passage of the AD 365 large tsunami.

In AD 1650, 29th September (Old Style), the extra-caldera eruption at the Kolumbo submarine volcanic edifice produced a very strong tsunami which inundated several coastal zones in the South Aegean Sea as documented by several historical accounts (Dominey-Howes et al. 2000; Nomikou et al. 2012) and tsunami sediment deposits (Papadopoulos 2015, pp. 81–83; Ulrova et al. 2016). The AD 1650 volcanic tsunami was powerful enough in the Heraklion coast where the wave height was in the order of $\sim$ 4 m (Dominey-Howes et al. 2000). However, the wave penetration at considerable distance inland is not documented. This was verified from a preliminary simulation which implies that the inundation zone is not wide enough to allow for considering this event as the extreme scenario for the area.

On the contrary, the big Minoan tsunami produced either by pyroclastic flows or by caldera collapse during the Santorini colossal eruption ($\sim$ 1613 BC) caused high wave amplitudes, exceeding in places $\sim$ 20 m along the north coast of Crete (e.g. Minoura et al. 2000; Bruins et al. 2008; Novikova et al. 2011, Nomikou et al. 2016). Archaeological evidence implies that very likely the Minoan tsunami caused extensive destruction along the north coast of Crete (Marinatos 1939). For these reasons, in the present study, the Minoan tsunami produced by pyroclastic flows was taken as the worst-case scenario.

3.3. Inundation Zone Determination

The inundation zone was determined by following the numerical simulation techniques performed by Novikova et al. (2011) for tsunami waves produced by massive pyroclastic flows during the Late Bronze Age or Minoan (17th century BC) eruption of Thera (Santorini) and penetrating the sea from the south side of the Santorini volcanic cone facing north Crete coast. The software package used is GEOWAVE (Watts et al. 2003), which consists of two components: TOPICS for the tsunami generation and FUNWAVE for the wave propagation. TOPICS is an approximate simulation tool, based on various fitting techniques, providing surface elevations and water velocities as initial conditions (Grilli and Watts 1999). On the other hand, the numerical model FUNWAVE (Wei and Kirby 1995; Wei et al. 1995; Kennedy et al. 2000; Kirby et al. 1998, 2003) performs wave propagation simulation, based on the fully non-linear Boussinesq theory using a predictor–corrector scheme. GEOWAVE allows simulating a wide range of wavelengths and is capable also to model the physics of breaking waves and resulting run-up (Ioualalen et al. 2006, 2007; Grilli et al. 2007). Also, GEOWAVE accurately captures a wide range of wave parameters and characteristics, e.g. non-linear shoaling, refraction, diffraction, and wave–wave interactions.

Important parameters of pyroclastic flows that may control the interaction between flows and water body are the bulk density, velocity, discharge rate of the flow and the angle of incidence between the flow and the water surface (Cas and Wright 1991; De Lange et al. 2001). Details on how such issues were treated in our simulation can be found in the paper by Novikova et al. (2011).

For modelling of the tsunami propagation, the bathymetry file inserted, which is of $\sim$ 300 m resolution, was taken from EMODnet (European Marine Observation and Data Network: http://www.emodnet-bathymetry.eu/). From local nautical maps, the bathymetry was improved in the near-shore domain to $\sim$ 80-m resolution. The Digital Elevation Model (DEM, Fig. 5), being of resolution of 5 m, was provided by the National Cadastre and Mapping Agency S.A., Greece. For the estimation of the inundation zone, the files were integrated in a final grid of resolution of $\sim$ 30 m.

The numerical simulation results showed that the maximum water depth obtained is about 14 m (Fig. 5). Water depth of 0.5 m has been considered as the lowest value for the determination of the inundation zone. This consideration is based on that the 0.5-m value is the lowest limit for a tsunami to become damaging according to the decision matrix decided to be used for tsunami warning purposes in the frame of the North-Eastern and Mediterranean Tsunami Warning System/UNESCO (e.g. Tinti et al. 2012). In the coastal zone examined, the width of the inundation zone with water depth of more than $\sim$ 0.5 m ranges from $\sim$ 350 to $\sim$ 1000 m with the exception of the eastern side of the area where the inundation zone width reached $\sim$ 1.2 km due to the existence of the local Yofiros river mouth.

3.4. Building Vulnerability Assessment

Data of the building stock in the study area were collected from the Hellenic Statistical Survey (ELSTAT, last national census of 2011). The data are organised in polygon features providing the total number of buildings per building block as well as the number of buildings per type of construction material. In the data set obtained, the buildings are

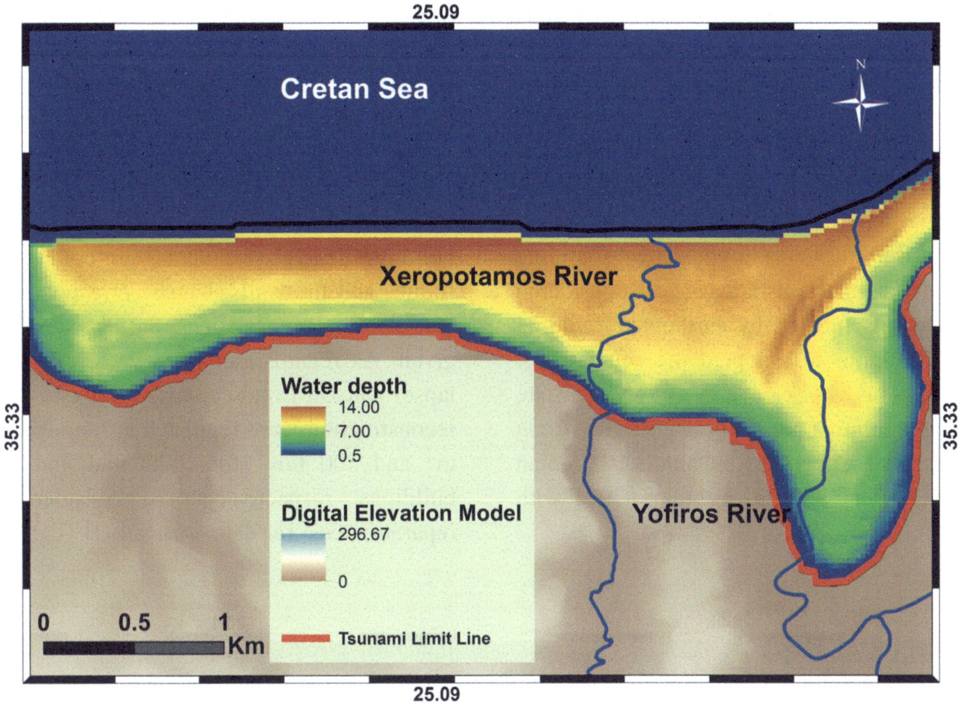

Figure 5
Tsunami inundation zone and the DEM beyond this zone in the study area

classified into six types of construction material: wood, brick, stone, reinforced concrete, metal and other. These types only partly correspond to the five types listed in Table 2 which, however, are the ones that are in use by the DAMASCHE tool (Table 1). To make building types consistent with the ones in Table 2, we assumed that the wood, brick, stone and other types correspond to the classes A, B, C and D classes in Table 2, respectively. We also assumed that reinforced concrete and metallic types correspond to class E. However, no data for individual buildings were provided by ELSTAT due to the law for sensitive personal data. This implies that the vulnerability of individual buildings is not possible to be assessed only from the ELSTAT data. Therefore, the missing building data were completed primarily with the use of a set of orthophotomaps provided by the Urban Planning Division, Heraklion Municipality, and secondarily with the help of Google Maps. In addition, building data in the entire area were validated in situ by performing field inspection during December 2015 and February 2018. Following this procedure, it became possible to organise digital files for mapping individual buildings classified according to their construction material types. It was found that the majority of buildings, at percentage of nearly 90%, are reinforced concrete ones, while only very few belong to other types of construction material.

4. Calculations and Results

After applying the DAMASCHE tool (Table 1), we were able to determine the damage level in buildings affected by the selected tsunami of the worst-case scenario. It was found that the numbers of buildings suffering damage of level D1 (light damage), D2 (important damage), D3 (heavy damage), D4 (partial collapse) and D5 (total collapse) are 81, 281, 1037, 41 and 142, respectively. Only very few buildings suffering of important damage (level D2) or light damage (D1) can be found within the inundation zone close to the inland water depth limit of 0.5 m.

Outside this limit, no damage is expected. The results are illustrated in Fig. 6.

To calculate the cost expected from the tsunami impact to buildings, we took into account that no strong, damaging tsunamis have affected the Mediterranean region in the last 62 years or so and consequently no building damage data are available. Consequently, for the cost implied by building replacement due to tsunami damage, we considered as a proxy the cost implied by earthquake building damage. To this aim, we adopted the official cost rates declared by the Greek government for the State Housing Aid programme put in force after the strong damaging earthquakes that hit Cephalonia Isl., Ionian Sea, on 24th January and 3rd February 2014 with magnitudes M_w 6.0 and M_w 5.9, respectively (Papadopoulos et al. 2014b).

The State Housing Aid regulations predicted flat rates of 250 €/m^2 for reparation of buildings suffering damage only in the walls, which corresponds to damage level D2 in Table 1, regardless the residential or commercial use of buildings. For reparation of buildings that suffered damage in both the reinforced frames and the walls, which corresponds to damage level D3 in Table 1, a cost flat rate of 450 €/m^2 was given. As regards damage level of D4 (partial collapse) and D5 (total collapse), flat rates for building reconstruction were regulated at the level of 1000 €/m^2 and 500 €/m^2 for residential and commercial buildings, respectively. We considered that the reparation cost for damage level D1 (light damage) is

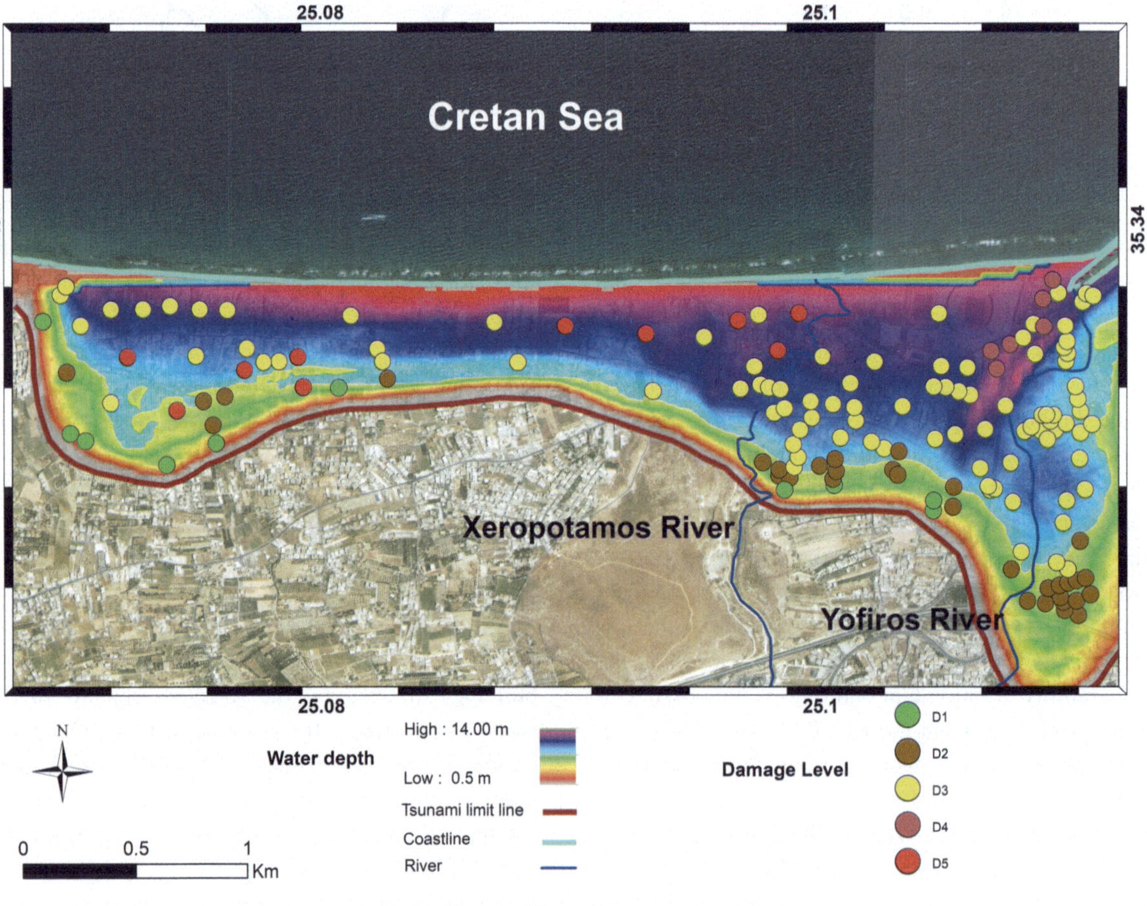

Figure 6
Mapping of the tsunami inundation zone with water depth (in m) and building damage level per building block as determined from the application of the DAMASCHE tool

negligible. In this study, the impact from floating debris has not been considered in our calculations.

For the application of the cost flat rates predicted by the State Housing Aid program of 2014, we assumed that the average building surface is 100 m^2 and 80 m^2 for residential and commercial buildings, respectively. This assumption is consistent with our field validation survey as well as with the earlier one performed in the area by Papathoma (2003). Since the ELSTAT building data set does not include information about the use of each single building, from the field survey we estimated that about 80% are residential buildings while the rest 20% are commercial ones. This estimation was verified with the local authorities. An additional assumption made for reasons of simplicity is that in each individual building, damage occurred only in the ground floor without any involvement of the upper building floors in the calculations. This perhaps underestimates the calculated cost since in the area of high water depth, it is reasonable to assume that upper floors may also suffer some damage. Another source of uncertainty is that single or apartment residential houses may have different replacement cost. However, the State Housing Aid program does not predict different rates for the replacement cost for these two types of residential houses.

The results obtained for building reparation cost are summarised in Table 4, which shows that for the suggested extreme tsunami scenario, the estimated cost to the state is in the order of 52 million Euros. For the calculation of the cost for building reconstruction we considered damage levels of D4 (partial collapse) and D5 (total collapse). In this case the calculated cost for reconstruction (Table 5) is about 5.5 times less than the one for reparation. The total cost for building replacement, both reparation and reconstruction, is in the order of 61 million Euros.

5. Discussion

The quantitative tsunami risk assessment is susceptible to a variety of epistemic and aleatory uncertainties involved in the several methodological steps that one may select to follow (e.g. Didenkulova et al. 2012; Goda and Song 2015; Selva et al. 2016; Sepúlveda et al. 2017). A holistic approach of the risk may include various attributes for the reason that strong tsunamis affect the entire social and economic life, the infrastructures and the life lines as well as the natural environment. In this paper, we selected to approach the tsunami risk assessment from the point of view of building damage based on the concept of the worst-case scenario.

The risk assessment for buildings exposed to tsunami threat could be seen from various points of view depending on the objectives of the study and the data available. Quantitative tsunami risk to buildings may include several metrics such as numbers of buildings damaged or destroyed at various levels, numbers of buildings that need removal or replacement after the tsunami attack, cost needed for removal, reparation or complete reconstruction, total or only insured costs and so on. We selected to calculate the numbers of buildings that after the attack of the worst-case tsunami would need replacement, i.e. either reparation, if only repairable damage is predicted by the model, or reconstruction in case of partial or even total building collapse. Then, the cost implied for building replacement was calculated.

Table 4

Calculated cost for reparation of buildings suffering damage level D2 and D3 due to strong tsunami attack in the inundation area of Gazi, west of Heraklion, Crete Isl

Damage level according to Table 1	Number of buildings	Average building surface (m^2)	Cost/m^2 (€)	Cumulative cost (€)
*D*2	281	96	250	6,744,000
*D*3	1037	96	450	44,798,400
Total	1318			51,542,400

We adopted that the ratio of residential/commercial buildings equals 4 and that the average surface of residential and commercial buildings is 100 m^2 and 80 m^2, respectively. Then, the average building surface is 96 m^2

Table 5

Calculated cost for reconstruction of buildings suffering partial or total collapse (damage levels D4 and D5 combined together) due to strong tsunami attack in the inundation area of Gazi, west of Heraklion, Crete Isl

Building use	Number of buildings	Average building surface (m^2)	Cost/m^2 (€)	Cumulative cost (€)
Residential	37	100	1000	3,700,000
Commercial	146	80	500	5,840,000
Total	183			9,540,000

Lessons learned by post-disaster surveys are much helpful in evaluating the methodology that we followed. However, since the Mediterranean area did not experience strong tsunamis recently, findings of post-disaster surveys in other parts of the world were utilised. The most important example is the 11 March 2011 Great East Japan tsunami given that a large amount of data regarding building damage have been collected and elaborated. Suppasri et al. (2013a,b) found that reinforced concrete and steel buildings presented better resistant performance over wood or masonry buildings. On the other hand, it was found that the damage level of buildings impacted by the 2011 Japan tsunami can be separated into six levels, i.e. from minor damage to wash away (Leelawat et al. 2014). These findings are quite close to the assumptions for building resistant performance and damage level scaling adopted in our study through the DAMASCHE GIS tool which is based on post-disaster data collected after the Indian Ocean 26 December 2004 tsunami. Moreover, our assumption that water depth, structural material and function of buildings are factors that may significantly drive the damage is consistent with the findings of Leelawat et al. (2014). The dependence of the degree of damage primarily on water depth and the building construction type was postulated also by a study in relation to the tsunami of 17 July 2006 that affected the South Java area (Reese et al. 2007).

Building fragility curves developed for Japan by Suppasri et al. (2013a,b) improves fragility curves obtained on the basis of damage data collected after the Indian Ocean 2004 big tsunami. Risk assessment of buildings exposed to tsunamis in the Mediterranean Sea area could be advanced with the use of such a tool for the reason that tsunami building damage data are lacking in this area. Another improvement that could be inserted in the tools currently available for the assessment of building damage levels due to tsunamis is associated with the finding of Suppasri et al. (2013a,b) that some differences in damage states can be explained by the faster flow velocities in one coastal spot with respect to that in another spot at the same water depth. However, the tsunami flow impact on buildings is much more complicated than it was thought before the occurrence of the 2011 tsunami in Japan. For example, in an advanced study, Yeh et al. (2013) supported that engineered reinforced concrete buildings failed by rotation due to the high-velocity and deep tsunami inundation, and that entrapped air in the buildings and soil liquefaction by ground shaking could have contributed to the failure. In addition, the spatial distribution pattern of destroyed and survived buildings indicates that the strength of tsunami was affected significantly by the locations of well-engineered sturdy buildings (Yeh et al. 2013). Namely, weaker buildings situated in the shadow zone tended to survive while it seems that jet and wake formations behind the sturdy buildings enhanced the tsunami forces. Such results should not be ignored in future risk assessments for building damage due to tsunamis.

The calculation of economic losses due to tsunami damage in buildings has been the subject of only few studies. This is explained by the difficulty faced when we try to translate the estimated damage to financial losses. Some efforts have expressed probabilistically the expected losses in different terms such as the average annual loss and the loss exceedance curve (e.g. Goda and Song 2015; Jaimes et al. 2016). To our knowledge, no methods have been tested so far for the ex-ante calculation of the expected absolute monetary losses due to an extreme tsunami scenario. Our contribution towards such an aim has been based on the assumption that the typology of building damage due to tsunamis is similar to that of building damage due to earthquakes. From technical point of view, damage details have similarities and differences in these two cases but they have not studied systematically so far. Our approach smooths out any differences since we focused not on the damage details but on the gross picture which considers three main cases: buildings that did not suffer damage or suffered negligible damage, buildings that need reparation due to serious damage, and buildings that need reconstruction due to partial or total collapse. This concept allows to perform at least gross estimations of the expected absolute cost for building replacement after the tsunami attack. For the translation of building damage into losses in terms of absolute monetary cost, we adopted flat rate costs officially declared after recent earthquakes in Greece.

6. Conclusions

From the review of tsunami history in the area of the Hellenic Arc, it comes out that the coast of north Crete is highly exposed to seismic and volcanic tsunamis. The numerical simulation of the large tsunami produced by the Minoan (17th century BC) eruption of Thera (Santorini) volcano, which was considered as the worst-case scenario for this study, showed that the inundation (hazard) zone in the test coastal area of Gazi, to the west of Heraklion capital city, north Crete Isl. (Greece), is characterised by maximum water depth of $\sim$ 14 m. This is consistent with the results obtained for north Crete and for the same tsunami by other authors (Minoura et al. 2000; Bruins et al. 2008; Novikova et al. 2011). The width of the inundation zone with water depth of no less than $\sim$ 1 m ranges from $\sim$ 350 to $\sim$ 1200 m depending on the coastal geomorphology.

The building stock in the area is highly homogeneous in the sense that the predominant building type at percentage of nearly 90% is the reinforced concrete one, while only very few buildings belong to other types of construction material. With the use of the empirical DAMASCHE GIS tool, it was concluded that most buildings, namely 1318, suffered repairable damage, while partial or total collapse suffered a number of 183 buildings.

Under the assumption that the costs for building replacement due to either earthquake or tsunami damage are equivalent, we adopted the official cost rates declared by the Greek State Housing Aid program put in force after the 2014 strong damaging earthquakes in Cephalonia Isl. Building replacement include either reparation, if only important or heavy but repairable damage occurs, or reconstruction due to partial or total collapse. Then, one may expect that the absolute monetary cost for building reparation would be much higher than the cost for building reconstruction. In fact, we calculated $\sim$ 52 million euros for building reconstruction and only $\sim$ 10 million euros for building reparation.

Our method could be applied to other areas of the Earth which are threatened by tsunamis since it does not depend on the characteristics of the test site examined, such as the type of tsunami sources and building types. Also, the method allows for different building fragility functions or curves to be used in different regions. The only limitation is that appropriate data should be available, e.g. tsunami history, bathymetry, digital elevation model, building data, building fragility functions. Furthermore, additional quantitative risk metrics, such as insured building losses, could be introduced if suitable data are provided.

Acknowledgements

This research is a contribution to the EU-FP7 tsunami research project ASTARTE (Assessment, Strategy And Risk Reduction for Tsunamis in Europe), Grant agreement no.: 603839, 2013-10-30. We are thankful to two anonymous reviewers for their constructive comments that improved substantially the initial manuscript.

References

Algermissen, S.T., Perkins, D.M., Isherwood, W., Gordon, D., Reagor, G., & Howard, C. (1979). Seismic risk evaluation of the Balkan region. In *Proceedings of the Seminar on seismic zoning maps, UNESCO, Skopje 1976*, vol. 2, pp. 172–240.

Ambraseys, N. N. (1960). The seismic sea-wave of July 1956 in the Greek Archipelago. *Journal of Geophysical Research, 65*(4), 1257–1265.

Ambraseys, N. N. (2009). *Earthquakes in the mediterranean and middle east, a multidisciplinary study of seismicity up to 1900* (p. 947). Cambridge: Cambridge University Press.

Beisel, S., Chubarov, L., Didenkulova, I., Kit, E., Levin, A., Pelinovsky, E., et al. (2009). The 1956 Greek tsunami recorded at Yafo, Israel, and its numerical modelling. *Journal of Geophysical Research, 114,* C09002. https://doi.org/10.1029/2008JC005262.

Benchekroun, S., Omira, R., Baptista, M. A., El Mouraouah, A., Iben Brahim, A., & Toto, E. A. (2013). Tsunami impact and vulnerability in the harbor area of Tangier. *Geomatics Natural Hazards and Risk*. https://doi.org/10.1080/19475705.2013.858373.

Bruins, H. J., McGillivray, J. A., Synolakis, C. E., Benjamini, Ch., Keller, J., Kisch, H. J., et al. (2008). Geoarchaeological tsunami deposits at Palaikastro (Crete) and the Late Minoan IA eruption of Santorini. *Journal of Archaeological Science, 35,* 191–212.

Burbidge, D., Cummins, P. R., Mleczko, R., & Thio, H. A. (2008). Probabilistic tsunami hazard assessment for Western Australia. *Pure and Applied Geophysics, 165,* 2059–2088.

Cas, R. A. F., & Wright, J. V. (1991). Subaqueous pyroclastic flows ignimbrites: An assessment. *Bulletin of Volcanology, 53,* 357–380.

Charvet, I., Ioannou, I., Rossetto, T., Suppasri, A., & Imamuara, F. (2014). Empirical fragility assessment of buildings affected by the 2011 Great East Japan tsunami using improved statistical models. *Natural Hazards, 73,* 951–973.

Charvet, I., Macabuag, J., & Rossetto, T. (2017). Estimating tsunami-induced building damage through fragility functions: Critical review and research needs. *Frontiers in Built Environment, 3,* 36. https://doi.org/10.3389/fbuil.2017.00036.

Cita, M. B., & Aloisi, G. (2000). Deep-sea tsunami deposits triggered by the explosion of Santorini (3500 years BP), Eastern Mediterranean. *Sedimentary Geology, 135,* 181–302.

Cita, M. B., Camerlenghi, A., Kastens, K. A., & McCoy, F. W. (1984). New findings of bronze age homogenites in the Ionian sea: Geodynamic implications for the Mediterranean. *Marine Geology, 55,* 47–62.

Cita, M. B., Camerlenghi, A., & Rimoldi, B. (1996). Deep-sea tsunami deposits in the eastern Mediterranean: New evidence and depositional models. *Sedimentary Geology, 104,* 155–173.

Curtis, G. D., & Pelinovski, E. N. (1999). Evaluation of tsunami risk for mitigation and warning. *Science of Tsunami Hazards, 17*(3), 187–192.

Dall'Osso, F., Dominey-Howes, D., Tarbotton, C., et al. (2016). Revision and improvement of the PTVA-3 model for assessing tsunami building vulnerability using "international expert judgment": Introducing the PTVA-4 model. *Natural Hazards, 83,* 1229.

Dall'Osso, F., Maramai, A., Graziani, L., Brizuela, B., Cavalletti, A., Gonella, M., et al. (2010). Applying and validating the PTVA-3 Model at the Aeolian Islands, Italy: Assessment of the vulnerability of buildings to tsunamis. *Natural Hazards and Earth Systems Sciences, 10,* 1547–1562.

Davies, G., Griffin, J., Løvholt, F., Glimsdal, S., Harbitz, C., Thio, H. K., et al. (2017). A global probabilistic tsunami hazard assessment from earthquake sources. In E. M. Scourse, N. A. Chapman, D. R. Tappin, & S. R. Wallis (Eds.), *Tsunamis: Geology, hazards and risks* (p. 456). London: Geological Society, London Special Publications. https://doi.org/10.1144/sp456.5.

De Lange, W., Prasetya, G., & Healy, G. T. (2001). Modelling of tsunami generated by pyroclastic flows (ignimbrites). *Natural Hazards, 24,* 251–266.

De Martini, P. M., Barbano, M. S., Smedile, A., Gerardi, F., Pantosti, D., Del Carlo, P., et al. (2010). A unique 4000 year long geological record of multiple tsunami inundations in the Augusta Bay (eastern Sicily, Italy). *Marine Geology, 276,* 42–57.

Dias, W. P. S., Yapa, H. D., & Peiris, L. M. N. (2009). Tsunami vulnerability functions from field surveys and Monte Carlo simulation. *Civil Engineering and Environmental Systems, 26,* 181–194.

Didenkulova, I., Monserrat, S., & Tinti, S. (2012). New developments in tsunami science: From hazard to risk—a preface. *Natural Hazards and Earth System Sciences, 12,* 2507–2514. www.nat-hazards-earth-syst-sci.net/12/2507/2012/. https://doi.org/10.5194/nhess-12-2507-2012.

Dominey-Howes, D., Dunbar, P., Varner, J., & Papathoma-Kohle, M. (2010). Estimating a probable maximum loss from a Cascadia tsunami. *Natural Hazards, 53*(1), 43–61.

Dominey-Howes, D., Papadopoulos, G. A., & Dawson, A. G. (2000). Geological and historical investigation of the 1650 Mt. Columbo (Thera Island) eruption and tsunami, Aegean Sea, Greece. *Natural Hazards, 21,* 83–96.

EC-Working Paper. (2010). Commission Staff Working Paper-Risk Assessment and Mapping, Guidelines for Disaster Management. SEC (2010), 1626 final, Brussels, p. 42.

El-Sayed, A., Romanelli, F., & Panza, G. (2000). Recent seismicity and realistic waveforms modelling to reduce the ambiguities about the 1303 seismic activity in Egypt. *Tectonophysics, 328,* 341–357.

Fournier d' Albe, E. M. (1982). An approach to earthquake risk management. *Engineering Structures, 4,* 147–152.

Galanopoulos, A. G. (1957). The seismic sea-wave of 9 Iouliou 1956. *Praktika Academy Athens, 32,* 90–101. **(in Greek with Engl. abstr.)**.

Gardi, A., Valencia, N., Guillande, R., & André, C. (2011). Inventory of uncertainties associated with the process of tsunami damage assessment on buildings (SCHEMA FP6 EC co-funded project). *Natural Hazards and Earth Systems Sciences, 11,* 883–893.

Gardi, A., Valencia, N., & Sheer, S. (2009). WP3.3–G.I.S. conceptual model and processing for production of elements required for test sites scenarios, Deliverable 3.1 part II, SCHEMA project.

Geist, E. L., & Parsons, T. (2006). Probabilistic analysis of tsunami hazards. *Natural Hazards, 37,* 277–314.

Goda, K., & Song, J. (2015). *Uncertainty modelling and visualization for tsunami hazard and risk mapping: A case study for the 2011 Tohoku earthquake*. Stoch: Environmental Research and Risk Assessment. https://doi.org/10.1007/s00477-015-1146-x.

González-Riancho, P., Aguirre-Ayerbe, I., García-Aguilar, O., Medina, R., González, M., Aniel-Quiroga, I., et al. (2014). Integrated tsunami vulnerability and risk assessment: Application to the coastal area of El Salvador. *Natural Hazards and Earth Systems Sciences, 14,* 1223–1244. https://doi.org/10.5194/nhess-14-1223-2014.

Goodman-Tchernov, B. N., Dey, H. W., Reinhardt, E. G., McCoy, F., & Mart, Y. (2009). Tsunami waves generated by the Santorini eruption reached Eastern Mediterranean shores. *Geology, 37,* 943–946.

Grilli, S. T., Ioualalen, M., Asavanant, J., Shi, F., Kirby, J. T., & Watts, Ph. (2007). Source constariants and model simulation of the December 26, 2004 Indian Ocean tsunami. *Journal of Waterway Port Coastal and Ocean Engineering, 133*(6), 414–428.

Grilli, S. T., & Watts, Ph. (1999). Modelling of waves generated by a moving submerged body: Applications to underwater landslides. *Engineering Analysis with Boundary Elements, 23*(8), 645–656.

Guidoboni, E., & Comastri, A. (1997). The large earthquake of 8 August 1303 in Crete: Seismic scenario and tsunami in the Mediterranean area. *Journal of Seismology, 1,* 55–72.

Guidoboni, E., & Comastri, A. (2005). *Catalogue of Earthquakes and Tsunamis in the Mediterranean Area from the 11th to the 15th century* (p. 1037). Rome: Istituto Nazionale di Geofisica e Vulcanologia.

Guidoboni, E., Comastri, A., & Traina, G. (1994). *Catalogue of ancient earthquakes in the Mediterranean area up to the 10th Century* (p. 504). Rome-Bologna: ING-SGA.

Hancilar, U. (2012). Identification of elements at risk for a credible tsunami event for Istanbul. *Natural Hazards and Earth Systems Sciences, 12,* 107–119. https://doi.org/10.5194/nhess-12-107-2012.

Hébert, H., Schindele, F., Altinok, Y., Alpar, B., & Gazioglu, C. (2005). Tsunami hazard in the Marmara Sea (Turkey): A numerical approach to discuss active faulting and impact on the Istanbul coastal areas. *Marine Geology, 215,* 23–43.

Hettiarachchi, S. S. L., Samarawickrama, S. P., & Wijeratne, N. (2011). *Risk assessment and management for tsunami hazard-case study of the port city of Galle* (p. 30). New York: UNDP.

Hieke, W. (1984). A thick Holocene homogenite from the Ionian Abyssal Plain (Eastern Mediterranean). *Marine Geology, 55,* 63–78.

IOC Intergovernmental Oceanographic Commission. (2013). *Tsunami Glossary, Technical Series 85* (pp. 1–42). Paris: UNESCO.

IOC-Intergovernmental Oceanographic Commission. (2009). *Tsunami risk assessment and mitigation in the Indian Ocean; Knowing your tsunami risk—and what do about it. IOC Manuals and Guides, 52.* Paris: UNESCO.

Ioualalen, M., Asavanant, J., Shi, F., Kirby, J. T., & Watts, Ph. (2007). Modelling the 26 December 2004 Indian Ocean tsunami: Case study of impact in Thailand. *Journal of Geophysical Research, 112,* C07024. https://doi.org/10.1029/2006JC003850.

Ioualalen, M., Pelletier, B., Regnier, M., & Watts, Ph. (2006). Numerical modelling of the 26th November 1999 Vanuatu tsunami. *Journal of Geophysical Research, 1111,* C06030. https://doi.org/10.1029/2005JC003249.

Jaimes, M. A., Reinoso, E., Ordaz, M., Huerta, B., Silva, R., Mendoza, E., et al. (2016). A new approach to probabilistic earthquake-induced tsunami risk assessment. *Ocean and Coastal Management, 119,* 68–75.

Jelínek, R., & Krausmann, E. (2008). Approaches to tsunami risk assessment. *European Commission Joint Research Centre, Scientific and Technical Reports, EUR 23573 EN,* Ispra, Italy.

Jelínek, R., Krausmann, E., González, M., Álvarez-Gómez, J. A., Birkmann, J., & Welle, T. (2012). Approaches for tsunami risk assessment and application to the city of Cádiz, Spain. *Natural Hazards, 60,* 273–293. https://doi.org/10.1007/s11069-011-0009-0.

Karim, K. R., & Yamazaki, F. (2003). A simplified method of constructing fragility curves for highway bridges. *Earthquake Engineering and Structural Dynamics, 32,* 1603–1626.

Kastens, K. A., & Cita, M. B. (1981). Tsunami-induced transport in the abyssal Mediterranean Sea. *Geological Society of America Bulletin, 92,* 845–857.

Kennedy, A. B., Chen, Q., Kirby, J. T., & Dalrymple, R. A. (2000). Boussinesq modelling of wave transformation, breaking, and run-up I: 1D. *Journal of Waterway Port Coastal and Ocean Engineering ASCE, 126*(1), 39–47.

Kirby, J.T., Chen, Q., Noyes, T.J., Elgar, S., & Guza, R. T. (2003). Evaluationg of low frequency prediction of a Boussinesq wave model: Field cases. In *Proceeding, ISOPE-2003, international society of offshore and polar engineers, Honolulu,* May 25–30, pp. 396–403.

Kirby, J.T., Wei, G., Chen, Q., Kennedy, A. B., & Dalrymple, R. A. (1998). FUNWAVE 1.0. Fully nonlinear Boussinesq wave model, documentation and user's manual, Rep. CACR-98-06, Cent. For Appl. Coastal Res., Department of Civil and Environmental Engineering, University of Delaware, Newark.

Kulikov, E. A., Rabinovich, A. B., & Thomson, R. E. (2005). Estimation of tsunami risk for the coasts of Peru and northern Chile. *Natural Hazards, 35,* 185–209.

Leelawat, N., Suppasri, A., Charvet, I., & Imamura, F. (2014). Building damage from the 2011 Great East Japan tsunami: Quantitative assessment of influential factors. A new perspective on building damage analysis. *Natural Hazards, 73,* 449–471. https://doi.org/10.1007/s11069-014-1081-z.

Leone, F., Lavigne, F., Paris, R., Denain, J. C., & Vinet, F. (2010). A spatial analysis of the December 26th, 2004 tsunami-induced damages: Lessons learned for a better risk assessment integrating buildings vulnerability. *Applied Geography.* https://doi.org/10.1016/j.apgeog.2010.07.009.

Li, L., Switzer, A. D., Chan, Ch-H, Wang, Y., Weiss, R., & Qiu, Q. (2016). How heterogeneous coseismic slip affects regional probabilistic tsunami hazard assessment: A case study in the South China Sea. *Journal of Geophysical Research Solid Earth.* https://doi.org/10.1002/2016JB013111.

Lorito, S., Tiberti, M. M., Basili, R., Piatanesi, A., & Valensise, G. (2007). Earthquake-generated tsunamis in the Mediterranean Sea: Scenarios of potential threats to southern Italy. *Journal of Geophysical Research, 113,* B01301. https://doi.org/10.1029/2007JB004943.

Løvholt, F., Glimsdal, S., Lynett, P., & Pedersen, G. (2015). Simulating tsunami propagation in fjords with long-wave models. *Natural Hazards and Earth Systems Sciences, 15,* 657–669. https://doi.org/10.5194/nhess-15-657-2015.

Makropoulos, K., Kaviris, G., & Kouskouna, V. (2012). An updated and extended earthquake catalogue for Greece and adjacent areas since 1900. *Natural Hazards and Earth Systems Sciences, 12*(1425–1430), 2012. https://doi.org/10.5194/nhess-12-1425-2012.

Marinatos, S. (1939). The volcanic destruction of Minoan Crete. *Antiquity, 13,* 425–439.

McCoy, F., & Heiken, G. (2000). Tsunami generated by the Late Bronze Age eruption of Thera (Santorini), Greece. *Pure and Applied Geophysics, 157,* 1227–1256.

McCoy, F. W., & Papadopoulos, G. A. (2001). Tsunami generation during the Late Bronze Age Eruption of Thera: Evidence from tsunami deposits on Thera, Crete, West Turkey and the Deep Sea. *American Journal of Archives, 105,* 258–259.

Minoura, K., Imamura, F., Kuran, U., Nakamura, T., Papadopoulos, G. A., Takahashi, T., et al. (2000). Discovery of Minoan tsunami deposits. *Geology, 28,* 59–62.

Nadim, F., & Glade, T. (2006). On tsunami risk assessment for the west coast of Thailand. In Nadim, F., Pöttler, R., Einstein, H., Klapperich, H., & Kramer, S. (Eds.), *ECI symposium series*, p. 7.

Necmioglu, O., & Özel, N. M. (2015). Earthquake scenario-based tsunami wave heights in the eastern mediterranean and connected seas. *Pure and Applied Geophysics, 172,* 3617–3638. https://doi.org/10.1007/s00024-015-1069-y.

Newhall, C. G., & Self, S. (1982). The volcanic explosivity index (VEI): An estimate of explosive magnitude for historical volcanism. *Journal of Geophysical Research, 87,* 1231–1238.

Nomikou, P., Carey, S., Papanikolaou, D., Croff Bell, K., Sakellariou, D., Alexandri, M., et al. (2012). Submarine volcanoes of the Kolumbo volcanic zone NE of Santorini Caldera, Greece. *Global Planetary Change, 90–91,* 135–151.

Nomikou, P., Druitt, T. H., Hübscher, H., et al. (2016). Post-eruptive flooding of Santorini caldera and implications for tsunami generation. *Nature Communications, 7,* 13332. https://doi.org/10.1038/ncomms13332.

Novikova, T., Papadopoulos, G. A., & McCoy, F. W. (2011). Modelling of tsunami generated by the Giant Late Bronze Age Eruption of Thera, South Aegean Sea, Greece. *Geophysical Journal International.* https://doi.org/10.1111/j.1365-246x.2011.05062.x.

Okal, E. A., Synolakis, C. E., Uslu, B., Kalligeris, N., & Voukouvalas, E. (2009). The 1956 earthquake and tsunami in Amorgos, Greece. *Geophysical Journal International, 178,* 1533–1554.

Omira, R., Baptista, M. A., Miranda, J. M., Toto, E., Catita, C., & Catalão, J. (2010). Tsunami vulnerability assessment of Casablanca Morocco using numerical modelling and GIS tools. *Natural Hazards and Earth Systems Sciences, 54,* 75–95.

Pagnoni, G., Armigliato, A., & Tinti, S. (2015). Scenario-based assessment of buildings' damage and population exposure due to earthquake-induced tsunamis for the town of Alexandria, Egypt. *Natural Hazards and Earth System Sciences, 15,* 2669–2695. https://doi.org/10.5194/nhess-15-2669-2015.

Pagnoni, G., & Tinti, S. (2016). Application and comparison of tsunami vulnerability and damage models for the Town of Siracusa, Sicily, Italy. *Pure and Applied Geophysics, 173,* 3795–3822. https://doi.org/10.1007/s00024-016-1261-8.

Papadopoulos, G. A. (2011). *A seismic history of Crete: Earthquakes and tsunamis, 2000 B.C.–A.D. 2010* (p. 415). Athens: Ocelotos Publishing.

Papadopoulos, G. A. (2015). *Tsunamis in the European-Mediterranean Region: From historical record to risk mitigation* (p. 271). Amsterdam: Elsevier.

Papadopoulos, G. A. (2017). Earthquake sources and seismotectonics in the area of Crete. In S. Jusseret & M. Sintubin (Eds.), *Minoan earthquakes-breaking the myth through interdisciplinarity. Studies in archaeological sciences* (pp. 165–190). Leuven: Leuven University Press.

Papadopoulos, G. A., & Dermentzopoulos, Th. (1998). A tsunami risk management pilot study in Heraklion, Crete. *Natural Hazards, 18,* 91–118.

Papadopoulos, G. A., Gràcia, E., Urgeles, R., Sallares, V., De Martini, P. M., Pantosti, D., et al. (2014a). Historical and pre-historical tsunamis in the Mediterranean and its connected seas: Geological signatures, generation mechanisms and coastal impacts. *Marine Geology.* https://doi.org/10.1016/j.margeo.2014.04.014.

Papadopoulos, G. A., & Imamura, F. (2001). A proposal for a new tsunami intensity scale. In *Proceedings of International Tsunami Symposium , Aug 7–10, 2001, Seattle*, pp. 569–577.

Papadopoulos, G. A., Karastathis, V. K., Koukouvelas, I., Sachpazi, M., Baskoutas, I., Chouliaras, G., et al. (2014b). The Cephalonia, Ionian Sea (Greece), sequence of strong earthquakes of January-February 2014: a first report. *Research in Geophysics.* https://doi.org/10.4081/rg.2014.5441.

Papadopoulos, G. A., & Pavlides, S. (1992). The 1956 large shock in the South Aegean: Macroseismic field, faulting and neotectonics of the Amorgos Island. *Earth and Planetary Sciences Letters, 113,* 383–396.

Papathoma M. (2003). Tsunami vulnerability assessment using a Geographical Information System with special reference to Greece. PhD Thesis, Coventry University, March 2003, p. 289.

Papathoma, M., & Dominey-Howes, D. (2003). Tsunami vulnerability assessment and its implications for coastal hazard analysis and disaster management planning, Gulf of Corinth, Greece. *Natural Hazards and Earth System Sciences, 3,* 733–747.

Papathoma, M., Dominey-Howes, D., Zong, Y., & Smith, D. (2003). Assessing tsunami vulnerability, an example from Herakleio, Crete. *Natural Hazards and Earth System Sciences, 3,* 377–389.

Papazachos, B. C., & Papazachou, C. B. (1997). *Earthquakes of Greece* (p. 304). Editions Ziti: Thessaloniki.

Papazachos, B. C., & Papazachou, C. B. (2003). *Earthquakes of Greece* (p. 286). Thessaloniki: Editions Ziti. **(in Greek)**.

Pararas-Carayannis, G. (1988). Risk assessment of tsunami hazard. In M. I. El-Sabah & T. S. Murty (Eds.), *Natural and man-made hazards* (pp. 183–191). USA: D. Riedel.

Perissoratis, C., & Papadopoulos, G. A. (1999). Sediment instability and slumping in the southern Aegean Sea and the case history of the 1956 tsunami. *Marine Geology, 161,* 287–305.

Polonia, A., Bonatti, E., Carmelenghi, A., Lucchi, R. G., Panieri, G., & Gasperini, L. (2013). Mediterranean megaturbidite triggered by the AD 365 Crete earthquake and tsunami. *Scientific Reports, 3,* 1285. https://doi.org/10.1038/srep01285.

Power, W., Wallace, L., Wang, X., & Reyners, M. (2012). Tsunami hazard posed to New Zealand by the Kermadec and Southern New Hebrides Subduction Margins: An assessment based on plate boundary kinematics, interseismic coupling, and historical seismicity. *Pure and Applied Geophysics, 169,* 1. https://doi.org/10.1007/s00024-011-0299-x.

Qinghai, Z., & Adams, W. M. (1988). Tsunami risk analysis for China. *Natural Hazards, 1,* 181–185.

Reese, S., Cousins, W. J., Power, W. L., Palmer, N. G., Tejakusuma, I. G., & Nugrahadi, S. (2007). Tsunami vulnerability of buildings and people in South Java? Field observations after the July 2006 Java tsunami. *Natural Hazards and Earth Systems Sciences, 7,* 573–589.

Rehman, K., & Cho, Y. S. (2016). Building damage assessment using scenario based tsunami numerical analysis and fragility curves. *Water, 8*(109), 1–17. https://doi.org/10.3390/w8030109.

Rikitake, T., & Aida, I. (1988). Tsunami hazard probability in Japan. *Bulletin of the Seismological Society of America, 78,* 1268–1278.

Sato, H., Murakami, H., Kozuki, Y., & Yamamoto, N. (2003). Study on a simplified method of tsunami risk assessment. *Natural Hazards, 29,* 325–340.

Selva, J., Tonini, R., Molinari, I., Tiberti, M. M., Romano, F., Grezio, A., et al. (2016). Quantification of source uncertainties in Seismic Probabilistic Tsunami Hazard Analysis (SPTHA). *Geophysical Journal International, 205,* 1780–1803. https://doi.org/10.1093/gji/ggw107.

Sepúlveda, I., Liu, P. L.-F., Grigoriu, M., & Pritchard, M. (2017). Tsunami hazard assessments with consideration of uncertain earthquake slip distribution and location. *Journal of Geophysical Research Solid Earth*. https://doi.org/10.1002/2017JB014430.

Smedile, A., De Martini, P. M., Pantosti, D., Bellucci, L., Del Carlo, P., Gasperini, L., et al. (2011). Possible tsunamis signatures from an integrated study in the Augusta Bay offshore (Eastern Sicily-Italy). *Marine Geology Marine Geology, 281,* 1–13. https://doi.org/10.1016/j.margeo.2011.01.002.

Smith, K. (1992). *Environmental hazards-assessing risk and reducing disaster* (1st ed., p. 324). Abingdon: Routledge.

Sørensen, M. B., Spada, M., Babeyko, A., Wiemer, S., & Grűnthal, G. (2012). Probabilistic tsunami hazard in the Mediterranean Sea. *Journal of Geophysical Research, 117,* B01305. https://doi.org/10.1029/2010JB008169.

Strunz, G., Post, J., Zosseder, K., Wegscheider, S., Mück, M., Riedlinger, T., et al. (2011). Tsunami risk assessment in Indonesia. *Natural Hazards and Earth Systems Sciences, 11,* 67–82. https://doi.org/10.5194/nhess-11-67-2011.

Suppasri, A., Koshimura, S., Imamura, F., Ruangrassamee, A., & Foytong, P. (2013a). A review of tsunami damage assessment methods and building performance in Thailand. *Journal of Earthquake and Tsunami, 7*(5), 1350036.

Suppasri, A., Mas, E., Charvet, I., Gunasekera, R., Imai, K., Fukutani, Y., et al. (2013b). Building damage characteristics based on surveyed data and fragility curves of the 2011 Great East Japan tsunami. *Natural Hazards, 66,* 319–341.

Tarbotton, C., Dall'Osso, F., Dominey-Howes, D., & Goff, J. (2015). The use of empirical vulnerability functions to assess the response of buildings to tsunami impact: Comparative review and summary of best practice. *Earth Science Reviews, 142,* 120–134.

Tinti, S., Armigliato, A., Tonini, R., Maramai, A., & Graziani, L. (2005). Assessing the hazard related to tsunamis of tectonic origin: A hybrid statistical-deterministic method applied to southern Italy coasts. *ISET Journal of Earthquake Technology, 42*(4), 189–201.

Tinti, S., Graziani, L., Brizuela, B., Maramai, A., & Gallazzi, S. (2012). Applicability of the Decision Matrix of North Eastern Atlantic, Mediterranean and connected seas Tsunami Warning System to the Italian tsunamis. *Natural Hazards and Earth Systems Sciences, 12,* 843–857. https://doi.org/10.5194/nhess-12-843-2012.

Tonini, R., Armigliato, A., Pagnoni, G., Zaniboni, F., & Tinti, S. (2011). Tsunami hazard for the city of Catania, eastern Sicily, Italy, assessed by means of Worst-case Credible Tsunami Scenario Analysis (WCTSA). *Natural Hazards and Earth Systems Sciences, 11,* 1217–1232. https://doi.org/10.5194/nhess-11-1217-2011.

Ulrova, M., Paris, R., Nomikou, P., Kelfoun, K., Leibrandt, S., Tappin, D. R., et al. (2016). Source of the tsunami generated by the 1650 AD eruption of Kolumbo submarine volcano (Aegean Sea, Greece). *Journal of Volcanology and Geothermal Research, 321,* 125–139. https://doi.org/10.1016/j.jvolgeores.2016.04.034.

UNISDR. (2009). Terminology on Disaster, Risk Reduction. International Strategy for Disaster Reduction, p. 35. http://www.unisdr.org/eng/library/lib-terminology-eng.htm. Accessed 12 Nov 2017.

Valencia, N., Gardi, A., Gauraz, A., Leone, F., & Guillande, R. (2011). New tsunami damage functions developed in the framework of SCHEMA project: Application to European-Mediterranean coasts. *Natural Hazards and Earth Systems Sciences, 11,* 2835–2846.

Watts, P., Grilli, S. T., Kirby, J. T., Fryer, G. J., & Tappin, D. R. (2003). Landslide tsunami case studies using a Boussinesq model and a fully nonlinear tsunami generation model. *Natural Hazards and Earth Systems Sciences, 3*(5), 391–402.

Wegscheider, S., Post, J., Zosseder, K., Mück, M., Strunz, G., Riedlinger, T., et al. (2011). Generating tsunami risk knowledge at community level as a base for planning and implementation of risk reduction strategies. *Natural Hazards and Earth Systems Sciences, 11,* 249–258. https://doi.org/10.5194/nhess-11-249-2011.

Wei, G., & Kirby, J. T. (1995). Time-depended numerical code for extended Boussinesq equations. *Journal of Waterway Port Coastal and Ocean Engineering ASCE, 121*(5), 251–261.

Wei, G., Kirby, J. T., Grilli, S. T., & Subramanya, R. (1995). A fully nonlinear Boussinesq model for free surface waves. Part 1: Highly nonlinear unsteady waves. *Journal of Fluid Mechanics, 294,* 71–92.

Yeh, H., Sato, S., & Tajima, Y. (2013). The 11 March 2011 East Japan Earthquake and tsunami: Tsunami effects on coastal infrastructure and buildings. *Pure and Applied Geophysics, 170,* 1019–1031.

Yolsal-Çevikbilen, S., & Taymaz, T. (2012). Earthquake source parameters along the Hellenic subduction zone and numerical simulations of historical tsunamis in the Eastern Mediterranean. *Tectonophysics, 536–537,* 61–100.

(Received November 13, 2017, revised August 20, 2018, accepted August 24, 2018, Published online September 13, 2018)

Pure Appl. Geophys. 176 (2019), 3227–3246
© 2019 The Author(s)
https://doi.org/10.1007/s00024-019-02091-w

Pure and Applied Geophysics

A New Approximate Method for Quantifying Tsunami Maximum Inundation Height Probability

S. Glimsdal,[1] F. Løvholt,[1] C. B. Harbitz,[1] F. Romano,[2] S. Lorito,[2] S. Orefice,[2] B. Brizuela,[2] J. Selva,[2] A. Hoechner,[3] M. Volpe,[2] A. Babeyko,[3] R. Tonini,[2] M. Wronna,[4] and R. Omira[4]

Abstract—Regional and global tsunami hazard analysis requires simplified and efficient methods for estimating the tsunami inundation height and its related uncertainty. One such approach is the amplification factor (AF) method. Amplification factors describe the relation between offshore wave height and the maximum inundation height, as predicted by linearized plane wave models employed for incident waves with different wave characteristics. In this study, a new amplification factor method is developed that takes into account the offshore bathymetry proximal to the coastal site. The present AFs cover the North-Eastern Atlantic and Mediterranean (NEAM) region. The model is the first general approximate model that quantifies inundation height uncertainty. Uncertainty quantification is carried out by analyzing the inundation height variability in more than 500 high-resolution inundation simulations at six different coastal sites. The inundation simulations are undertaken with different earthquake sources in order to produce different wave period and polarity. We show that the probability density of the maximum inundation height can be modeled with a log-normal distribution, whose median is quite well predicted by the AF. It is further demonstrated that the associated maximum inundation height uncertainties are significant and must be accounted for in tsunami hazard analysis. The application to the recently developed TSUMAPS-NEAM probabilistic tsunami hazard analysis (PTHA) is presented as a use case.

Key words: Tsunamis, inundation, probabilistic tsunami hazard analysis, amplification factors, uncertainty quantification.

1. Introduction

The standard way to estimate tsunami inundation maps is to apply numerical nonlinear shallow water (NLSW) models that include drying-wetting schemes (Titov and Gonzalez 1997; LeVeque and George 2008; Løvholt et al. 2010; Dutykh et al. 2011; de la Asunción et al. 2013; Wronna et al. 2015; Macías et al. 2017). However, if the inundation needs to be quantified over large coastal stretches (e.g. country scale or larger), NLSW inundation simulations are most often not feasible. This is due either to the large computational cost or to the lack of required high-resolution topo-bathymetric models. This issue is particularly relevant for probabilistic tsunami hazard analysis (PTHA, Geist and Parsons 2006; Grezio et al. 2017), in which a large number of scenario simulations must be carried out to take into account the natural potential variability of the tsunami sources (e.g., Selva et al. 2016). Intermediate methods can limit the number of scenario simulations for local PTHA (Gonzalez et al. 2009; Lorito et al. 2015), but they still require high-resolution coastal bathymetric and topographical data that are typically not available over large geographical scales. We still need simpler methods for estimating tsunami inundation heights in PTHA, particularly over large regions.

A faster, yet more approximate, method than NLSW is to relate the nearshore surface elevation to the water elevation at the shoreline, which then acts as an approximation for the maximum inundation height (MIH). We refer to this method as the amplification factor (AF) method. Combined with results from offshore tsunami simulations, AF methods can be used to estimate the tsunami maximum inundation height at a coastal location.

The first version of the AF method (Løvholt et al. 2012, 2015) considered a set of only a few amplification factors based on idealized and overly

[1] Norwegian Geotechnical Institute, Oslo, Norway. E-mail: finn.lovholt@ngi.no
[2] Istituto Nazionale di Geofisica e Vulcanologia, Catania, Italy.
[3] Deutsches GeoForschungsZentrum, Potsdam, Germany.
[4] Instituto Portugues do Mar e da Atmosfera, Lisbon, Portugal.

simplified bathymetric profiles. The idea was that, in principle, these could be fitted to a coastline at any given place. A related AF model that also enables a fast calculation of tsunami heights for tsunami early warning was recently developed by Gailler et al. (2018). In contrast to the method of Løvholt et al. (2012), which is the basis for the model in this paper, the method developed by Gailler et al. (2018) needs to be calibrated with detailed bathymetric/topographic information and with high-resolution NLSW simulations for each site where it is applied.

For practical purposes, and due to a lack of alternatives, the application of Green's law amplification to a given reference water depth (often 0.5 m) has previously been employed to estimate tsunami inundation heights from offshore points as an alternative to the AF method (Sørensen et al. 2012; Brizuela et al. 2014; Horspool et al. 2014; Hébert and Schindelé 2015). However, as pointed out by Hébert and Schindelé (2015) and Davies et al. (2018), the choice of the reference water depth in Green's law makes the method subjective. In this paper, it will be demonstrated that the AF method represents a more accurate method for estimating the coastal amplification than Green's law, because local factors influencing the run-up, such as local bathymetry, wave polarity, and wave length, enable a more accurate approach. In addition, a series of new facilities, which will be described in a moment, makes the new AF method more suitable for use in regional PTHA than other previous approximate methods.

The AF method developed here has several new characteristics. Rather than using a predefined set of idealized bathymetric profiles, our method employs cross-sections from actual local bathymetry. The amplified height (AH) at the shoreline (see Fig. 1) is then quantified with a linear shallow water (LSW) model along a 1D profile close to orthogonal to the shoreline, much like the previous AF method. For a specific point on the coastline, the AH acts as a proxy for the MIH. As discussed by Løvholt et al. (2012), the AF method gives exact estimates for the MIH even under the special condition of plane non-breaking waves. Consequently, as will be demonstrated in this paper, the AH is rather a good estimator of the median of the whole MIH distribution along a set of onshore transects, over a coastline stretch of, say, a few kilometers. On the other hand, the AF method is designed to operate over large stretches of coastline, and the spatial resolution is much coarser than in numerical tsunami inundation models. Moreover, local phenomena such as wave focusing and refraction usually give rise to significant local variability in the MIH that cannot be captured by the AF method, which assumes plane wave amplification. As shown in the lower panel in Fig. 1, there is typically significant alongshore coastal variability in MIH. It was therefore necessary to upgrade the AF method to capture this inundation height variability as a model uncertainty.

Previous attempts to quantify AF-related tsunami inundation height uncertainty and bias are limited to a few studies (Davies et al. 2018, and references therein). Davies et al. (2018) estimated the AF uncertainty and bias by comparing offshore tsunami simulations combined with AF analysis against observed run-up for four tsunamigenic earthquakes (Chile, 1960; Alaska, 1964; Indian Ocean, 2004; Tohoku, 2011) using the previous AF version based on idealized and much simplified bathymetric transects. The authors found that tsunami simulations combined with the AF gave a relatively small bias compared to observed run-up heights, whereas a large log-normal root mean square ($\sigma \sim 0.9$) was found. However, their analysis merged uncertainty from the source with the one associated with the AF method. Therefore, their uncertainties combined the variability due to tsunami source complexity (e.g., heterogeneous coseismic slip or other simplifications in seismic source and tsunami generation) and propagation modeling, and possibly even other sources such as landslides, in addition to the inherent variability in the inundation process we try to quantify here. Thus, their variability due to the inundation process was not separated from the variability due to the source and tsunami propagation. In this study, we quantify the uncertainty of the inundation process separately by employing detailed high-resolution inundation simulations, tying the observed uncertainties from the inundation simulations directly to the AF method.

This paper is organized as follows: Section 2 describes the computation of the local amplification

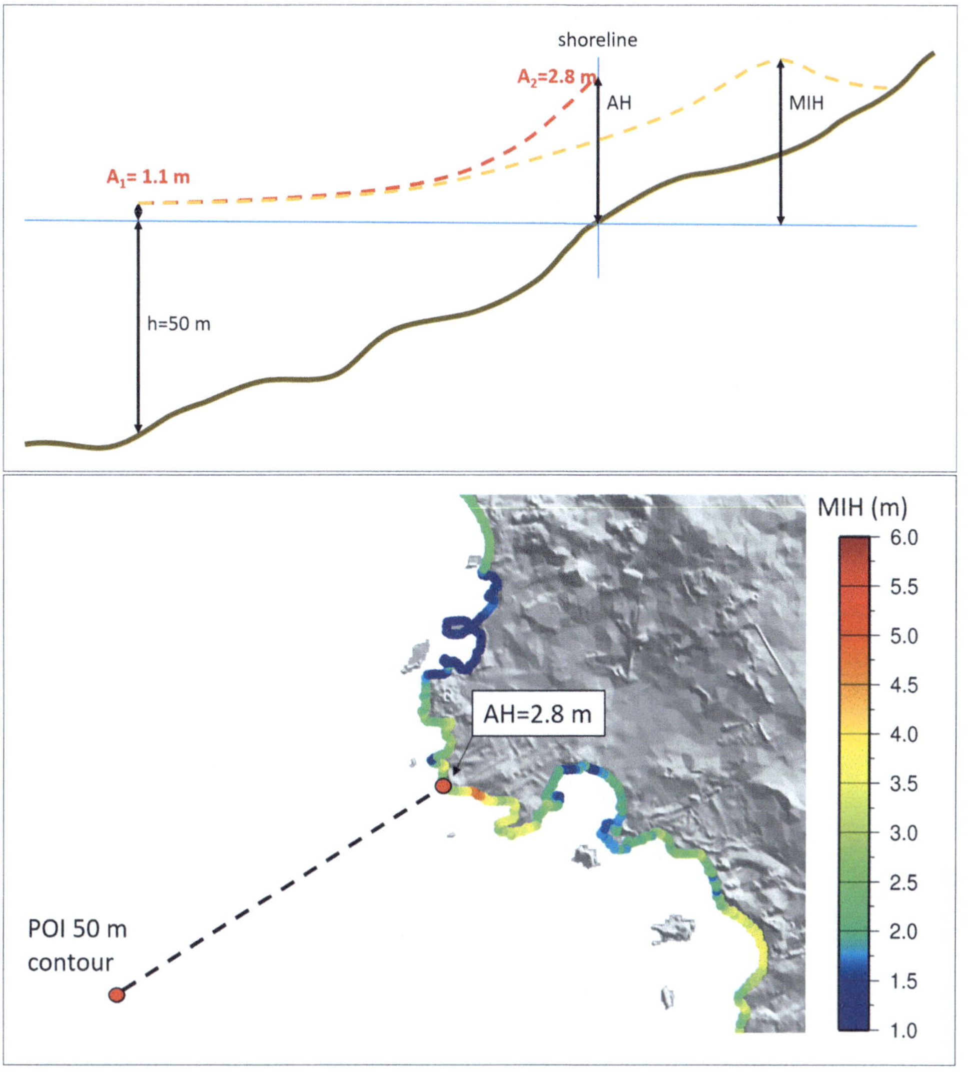

Figure 1
Upper panel: Sketch exemplifying the amplified height (AH) obtained from the amplification factor method as well as the maximum inundation height (MIH) obtained from a reference model over a local transect. The MIH represents the reference value, whereas the AH is a proxy for the MIH. The dashed red curve is the envelope of the amplified wave signal using the AF method (we denote the height at the shoreline AH). The yellow curve shows the corresponding maximum envelope of the coastal amplitude (offshore) and the inundation height (onshore) obtained from an NLSW reference model, and we denote the maximum in this curve MIH. The vertical value where the inundation height intersects the topography is the run-up height, and the corresponding horizontal distance is the inundation distance. Lower panel: example of a coastal distribution of the MIH derived from the 2HD (depth-averaged 2D model) reference model, drawn together with the single AH value derived for a single transect available for this coastal stretch

factors. In Sect. 3, we describe the inundation simulations for six different test sites in southern Europe. In Sect. 4, we estimate and analyze the bias and uncertainty of the method by comparing the AF performance at the test sites with the NLSW inundation models. In Sect. 5, we demonstrate with an example how the results can be used in a PTHA developed for production of tsunami hazard maps in the North-Eastern Atlantic, Mediterranean and connected seas (NEAM) region.

2. *Amplification Factors Based on Local Bathymetric Transects*

2.1. *New Amplification Factors Derived from Transect Simulations*

Løvholt et al. (2012) used a very limited set of idealized bilinear profiles as a basis for the AF computations. In the present study, instead, transects of the coastal water depth were extracted directly from the local bathymetry in the Mediterranean Sea, the Black Sea, and the North-Eastern Atlantic (see Fig. 2). The procedure for extracting these local transects is explained in Appendix 1. Following Løvholt et al. (2012), a 1D LSW model, solved over a staggered Arakawa C-grid with variable spacing, is used for the AF production runs. The resolution of the grid depends on the wave period and the water depth and is tuned to give a constant Courant number $Cr = 0.9$ over the entire domain. The Courant number is here defined as

$$Cr = \frac{\sqrt{gh}\Delta t}{\Delta x} \quad (1)$$

where h is the local water depth, Δx is horizontal grid resolution, and Δt the time step.

For waves with a period larger than 600 s, the grid size ranges from about 300 m in the deepest part of the domain (water depth of 2000 m) down to 7 m at 1 m depth; for a wave period shorter than 300 s, the grid resolution varies from 80 m to about 2 m, respectively. The grid spacing is not further refined in water depths less than 1 m. The resolution applied in

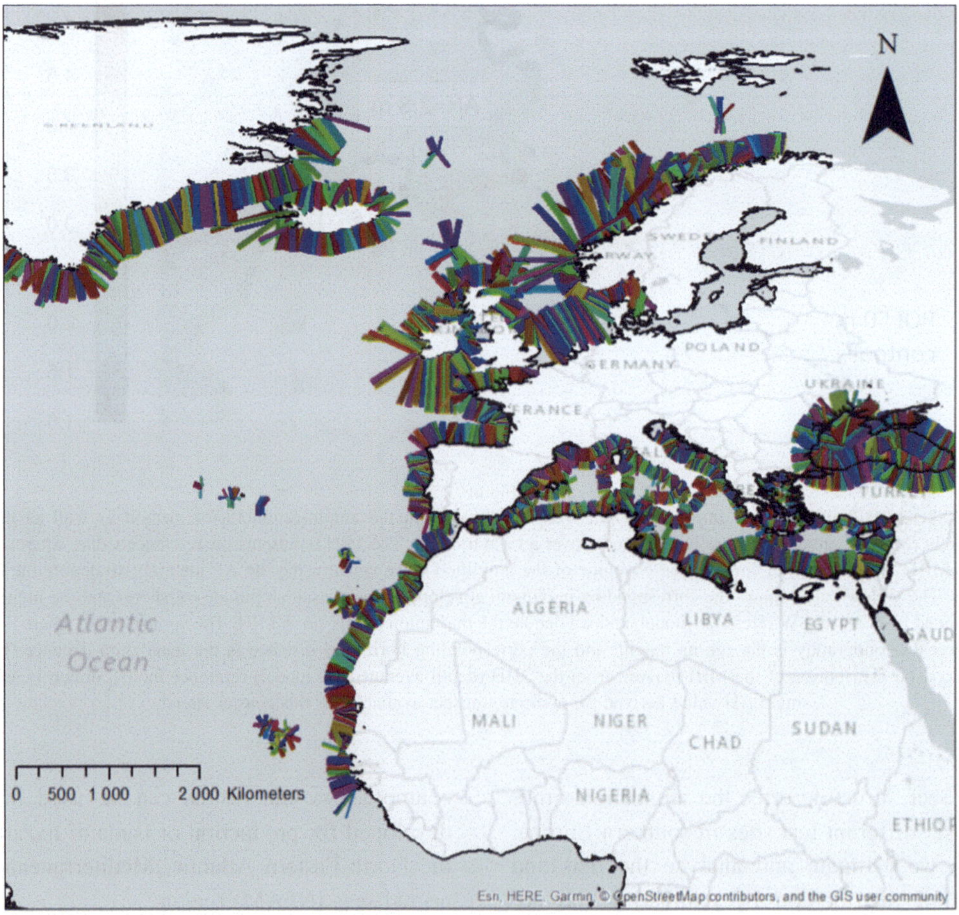

Figure 2
Map showing all profiles applied in the study

the simulations is justified by convergence tests. At the shoreline we have applied a no-flux boundary condition, and the waves are fed into the model from the offshore boundary.

The new method for computing the amplification factors starts with defining the AF points of interest (POIs). The POIs are located at longitude-latitude coordinates inside the computational domain employed for the 2HD tsunami propagation simulations. In this study, amplification factors are computed for almost evenly spaced POIs located every 20 km along the shoreline. For each POI along the 50 m isobath, about 40 depth profiles (exemplified in the upper panels of Figs. 3, 4, 5) are extracted approximately normal to the shoreline, each with a distance of 1 km apart, as described in Appendix 1. However, in the case of complex coastal geometries, deviation from the normal incidence may take place.

The amplification factors are computed along seven subjectively selected profiles out of the 40 profiles. The subjective selection was made to enable computational feasibility. An initial wave of 1 m height in deep water, shaped as a single-period sinusoidal wave pulse (N-shaped wave), is fed over the deep water boundary of the model (see Fig. 1). Linear shallow water (LSW) simulations were carried out for all profiles, both for leading trough or leading peak and for a set of wave periods (120, 300, 600, 1000, 1800, and 3600 s).

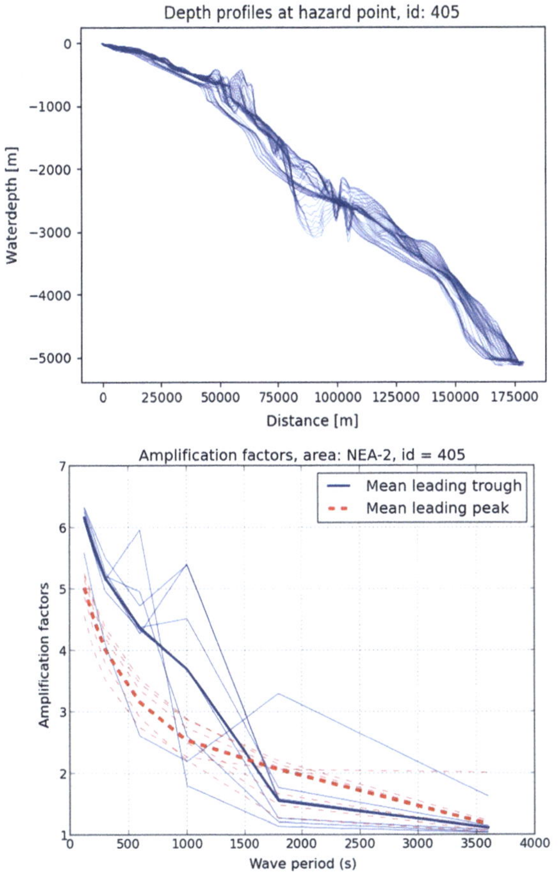

Figure 3
Example of depth profiles (top) and amplification factors (bottom) representing the hazard point 405 (at Sines, Portugal). The amplification factors are shown for the seven depth profiles as explained in the text, versus the wave period (seconds). The red and blue curves are factors for leading peak and leading trough, respectively. The thin curves are the factors for the seven local profiles, while the thick lines are the median values for leading peak and leading trough at this location

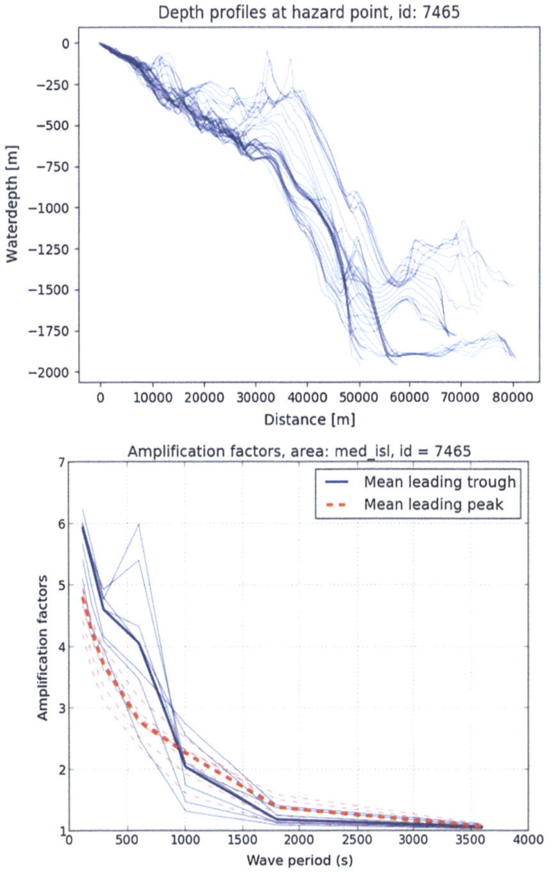

Figure 4
Example of depth profiles and amplification factors representing the hazard point 7465 (at Heraklion, Greece). See also caption in Fig. 3

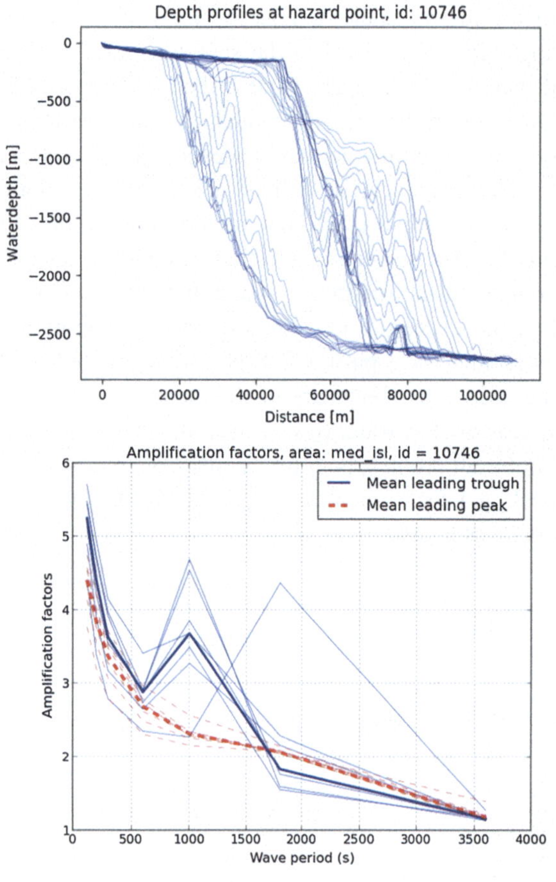

Figure 5
Example of depth profiles and amplification factors representing the hazard point 10746 (at Sant Jordi, Mallorca). See also caption in Fig. 3

Maximum surface elevations from the LSW simulation at 50 m depth and at the shoreline (0 m depth) are extracted. For each profile, we compute the amplification factor, which is defined as the ratio of the height between the latter and the former (i.e. A_1 and A_2 in the upper panel of Fig. 1). We use the median value of the seven amplification factors to avoid unrealistic alongshore fluctuations. Examples of amplification factors and related median values as a function of the incident wave periods are shown in the lower panels of Figs. 3, 4, 5. AF values for all combinations of wave polarities and wave periods are stored in lookup tables. As an example, amplification factors for a leading trough polarity and wave period of 600 s for the Mediterranean and Black Sea are depicted in Fig. 6.

Two versions of the factors were produced, one set using the raw amplification factor values and one set of factors smoothed along the shoreline with a median filter (using a filter length of five points). The median filtering was performed as an additional measure to avoid artificially short amplification fluctuations along the shoreline (see example for the Black Sea in Fig. 7).

As expected, we do observe that the shorter waves and waves with leading trough are amplified more than the longer waves and waves with a leading peak (Figs. 3, 4, 5, lower panels). We also observe that the amplification factors decrease as the period increases; in some cases the behavior appears different and the amplification factors are locally higher with respect to lower periods such as in Fig. 5 (see around period of 1000 s). This stems from interference between incident and reflected waves in the linear shallow water simulations: when the travel time from the POI to the shoreline and back to the POI for the leading trough matches the arrival of the trailing peak here, the measured wave height at the POI is low (trough is canceling the peak) and the amplification factor is high.

3. *Local Inundation Simulations for Bias and Uncertainty Estimation*

In order to evaluate uncertainties of the method of amplification factors, we systematically compare its results against direct NLSW inundation simulations. The employed NLSW model (Tsunami-HySEA, de la Asunción et al. 2013; Macías et al. 2017) is properly benchmarked, and is hence a suitable reference model towards the less accurate AF method. Still, a limitation with this approach is that the inundation simulations are not compared directly to field observations.

Comparisons between the AF and the NLSW models are undertaken at the six test sites for which a suitable and sufficiently detailed DEM existed. The test sites include one in the Atlantic Ocean, namely Sines, Portugal, with the remainder in the Mediterranean: Colonia Sant Jordi (Mallorca) and SE Iberia in Spain, Siracusa and the Catania plain in Italy, and Heraklion on the island of Crete in Greece (Fig. 8).

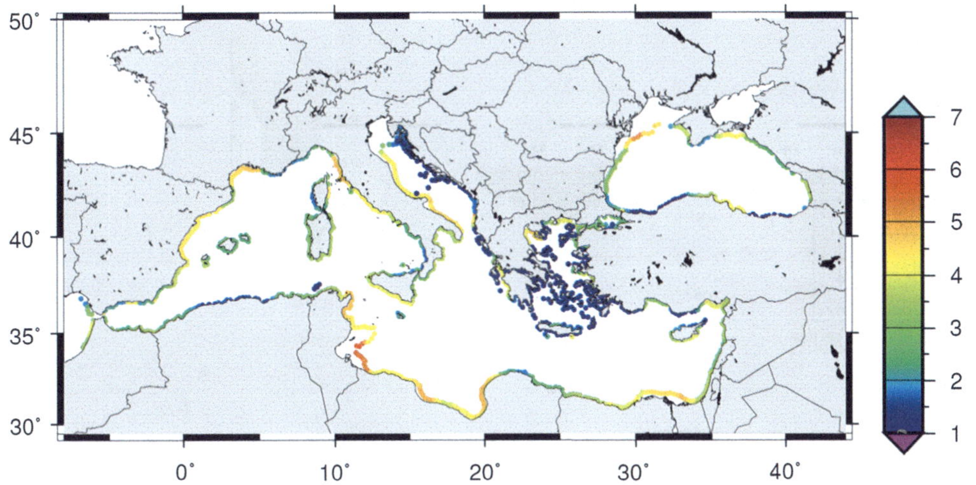

Figure 6
The amplification factors for the Mediterranean and Black Seas for the case with a leading trough and a wave period of 600 s. The factors are filtered along the shoreline with a median filter as explained in the text

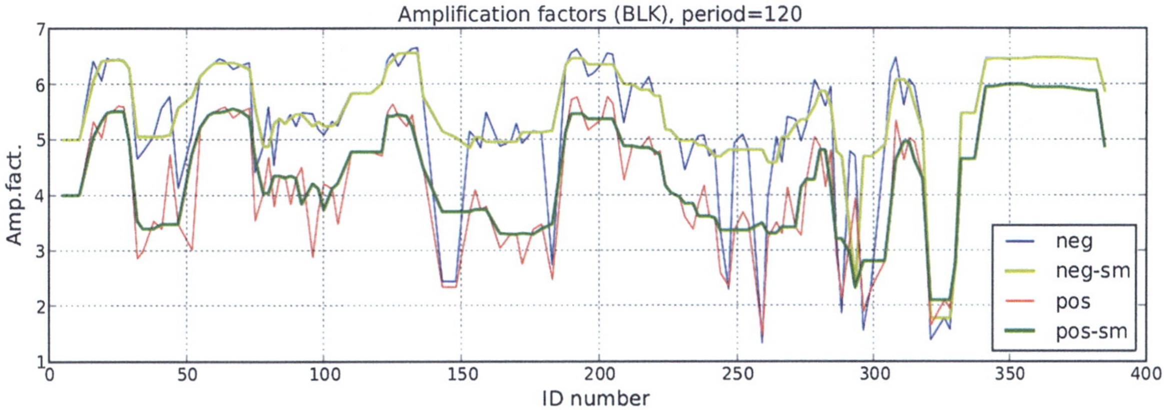

Figure 7
The amplification factors for Black Sea as function of the hazard points lying along the shoreline (ID numbers). The figure shows the effect of filtering the factors along the shoreline. The labels "neg" and "pos" relate to leading trough and leading peak, respectively. The tag "-sm" indicates filtered values (median filter), as explained in the text

For each test site, we use 96 earthquake sources with varying magnitude, strike, dip, and rake in order to explore a variety of situations as far as the source mechanism is concerned and, as a consequence, the wave period and wave polarity variability. In particular, we use six different moment magnitudes (7.1, 7.5, 8.1, 8.5, 8.8, 9.0), four strike angles (22.5, 112.5, 202.5, 292.5), and four pairs of dip/rake angles (10/90, 30/90, 50/270, 70/270). Empirical earthquake scaling laws (Strasser et al. 2010) are then employed to define the fault size (length and width) for each earthquake source. The average slip (D) for each scenario is established by considering the classical relationship $M = \mu AD$, where M is the seismic moment, μ is the shear modulus (30 GPa), and A is the rupture area of the seismic source.

We use the NLSW Tsunami-HySEA numerical code (see, e.g., de la Asunción et al. 2013; Macías et al. 2017) to simulate the distribution of the tsunami inundation height over the extent of the different

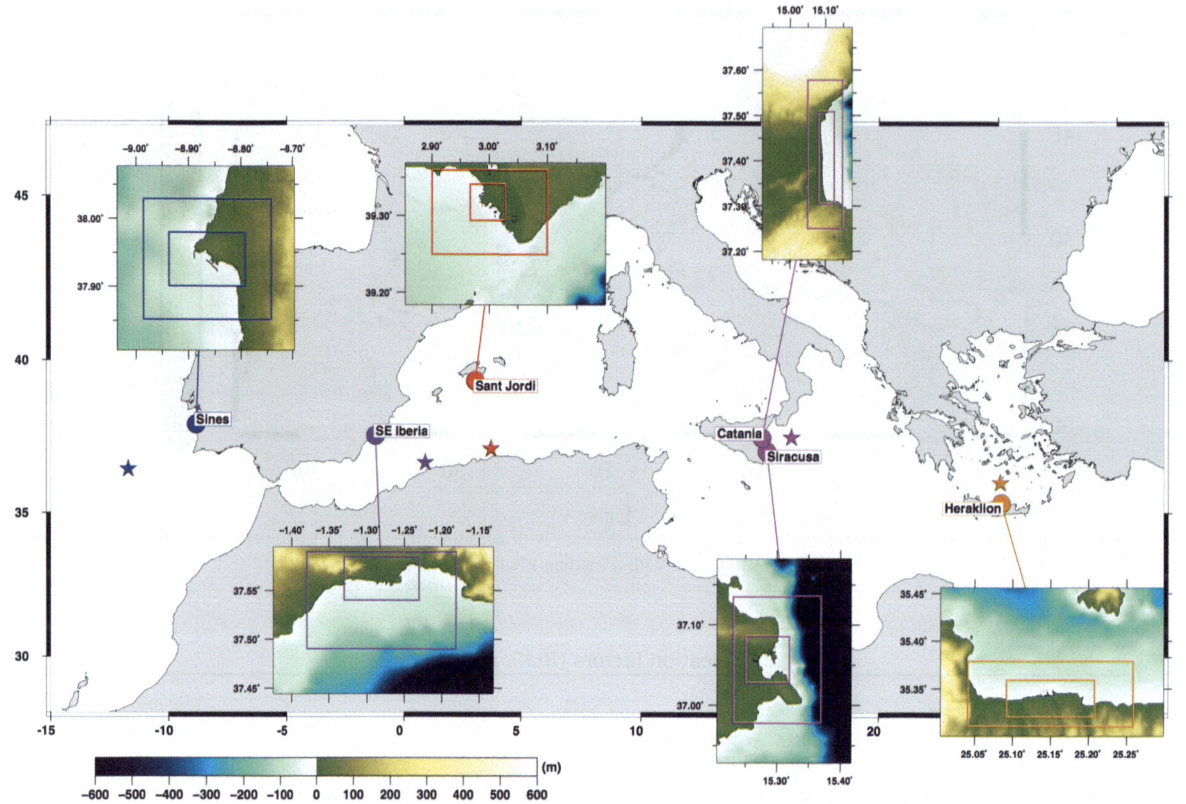

Figure 8
Location map showing the six test sites (dots) used for local inundation simulations and the epicenters of the earthquake scenarios (stars). The insets show the topo-bathymetric nested grids used for each test site in the simulations

sites. All offshore tsunami simulations are conducted on regular grids with a spatial resolution of 30 arc-sec. The NLSW models additionally use nested grids to simulate detailed local inundation at test sites. The resolution of the finest grid is about 10 m at all locations. The Manning-friction (n) is set to $n = 0.03$ in all simulations. We note that while $n = 0.03$ is typically a friction value for overland flow (e.g. Kaiser et al. 2011), this value will be site-dependent and vary spatially as well, and hence using a constant value represents a simplification.

For each test site and source scenario we apply the new AF method to quantify MIHs. As prescribed by the method, MIH is quantified by multiplying the maximum offshore surface elevation retrieved at the POI, taking wave characteristics retrieved from time series at the POI, with specific AF retrieved from the relevant lookup table. The method for selecting the maximum offshore surface elevations and related wave periods and polarities at the POI are described in Appendix 2.

4. *Maximum Inundation Height Uncertainty and Bias*

As explained in Sect. 3, the comparison between the AF method and corresponding NLSW simulations is made from six locations, each site subject to about 96 earthquake-induced tsunami scenario simulations. The new AFs depend on the coastal slope configuration, the tsunami wave period, and the tsunami wave polarity, as simulated in a single POI located offshore along the 50 m isobath. As the AFs are based on a linear method, they are amplitude-independent. For each scenario, we extracted the largest MIHs from the NLSW simulations for a given coastline location. MIH maxima were extracted in either the north–south or east–west direction, line by line, for each grid cell along the preferred coordinate axis. That is, if the shoreline is oriented mostly east–west for a given site, we search along lines oriented in the north–south direction, while if the shoreline is predominantly oriented more north–south for another site, we search along lines oriented in the east–west direction. In some special cases, the height of the terrain landside of the shoreline is too high and too steep to be inundated. In these situations, we also include values of maximum surface elevation for a small distance seaside from the shoreline (at least one cell away).

From counting all MIHs over the computational domain (NLSW) at each test site, we obtain an MIH empirical probability density function (ePDF) for each simulation. Each MIH ePDF is then fitted to a log-normal probability density function (PDF), using standard fitting tools in Matlab. We note that the log-normal PDF implies a normal distribution of the natural logarithm of the random variable with median e^{μ} and variance σ^2, in this case representing the MIH. The log-normal distribution of the MIH reads:

$$p(\mathrm{MIH}) = \frac{1}{\sqrt{2\pi}\mathrm{MIH}\sigma}\exp - \left(\frac{(\ln \mathrm{MIH} - \mu)^2}{2\sigma^2}\right) \quad (2)$$

We stress that the log-normal median value of the MIH over the computational domain for the NLSW simulation is e^{μ}. These fitted medians, arising from each NLSW simulation, are compared with amplified heights ($A \times \eta$, where A is the amplification factor and η is the height offshore) resulting from the AF method, in order to quantify biases for the individual simulations (and later on bias distribution). To this end, we define the normalized bias

$$\epsilon = \frac{(A \times \eta) - e^{\mu}}{e^{\mu}}, \quad (3)$$

to quantify the MIH bias due to the AF method towards the more accurate NLSW model for each model run and site. Examples of PDF for three individual model runs for the Catania, Heraklion, and SE Iberia sites are shown in Fig. 9. As demonstrated, the MIH distributions can be quite heterogeneous in terms of both their shape and/or their widths. To this end, we performed a Kolmogorov–Smirnov test

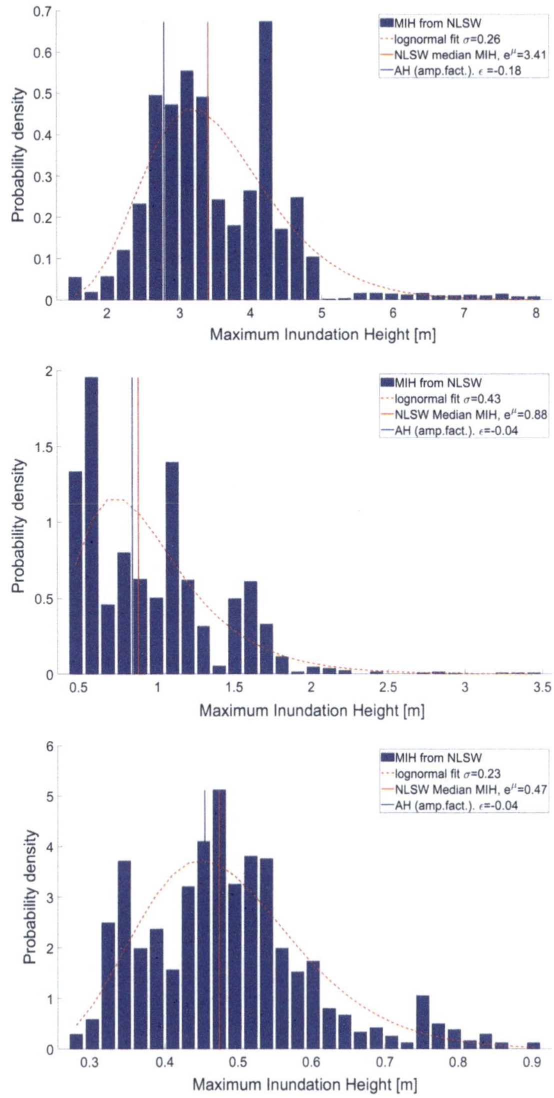

Figure 9
Distribution of MIHs from three individual NLSW simulations, including median values from each distribution and the related AH. Upper panel, MIH distribution in Catania due to a M_w 8.5 earthquake. Middle panel, MIH distribution in Heraklion due to a M_w 7.5 earthquake. Lower panel, MIH distribution in SE Iberia due to a M_w 7.1 earthquake

(Massey 1951) (using the logarithm of the MIH as the density variable) to investigate the validity of the fit of each empirical distribution towards a log-normal distribution. Below, we refer to this as the KS test. Using the KS test, we found that just 7% of the distributions from individual model runs had p values higher than 1% (i.e. more than 1% probability of exhibiting a log-normal distribution). Some sites, in particular Colonia Sant Jordi, had higher representation of statistically significant log-normal fits than other sites. We speculate that heterogeneous spatial characteristics in some of the other sites have given rise to different mean MIH in different parts of the domain, implying that these sites are not easily fitted into one single distribution. This deserves to be better investigated. Nonetheless, the median MIHs from the NLSW simulations have values close to the simplified AHs in the present examples.

4.1. Inundation Height Statistics for the Full Parameter Set

A first impression of the overall results can be drawn by inspecting Fig. 10, showing a histogram of all the median MIHs derived by the NLSW simulations compared with respective AHs. This figure shows that while overall distributions are comparable for the two models, there seems to be a weak tendency for median MIH values from NLSW simulations to exceed the AHs.

To gain a deeper and more quantitative insight, we carried out a statistical analysis based on the MIH PDFs, biases, and log-normal uncertainties. In the statistical analysis, both the merged results and statistics isolating the effects of different factors were analyzed.

The distribution of the normalized AF bias for all model simulations at all sites is shown in the upper panel of Fig. 11. The individual biases are close to being normally distributed, and the KS test returned a p value of 41%. The mean bias is $\bar{\epsilon}_{AF} = -5\%$, which

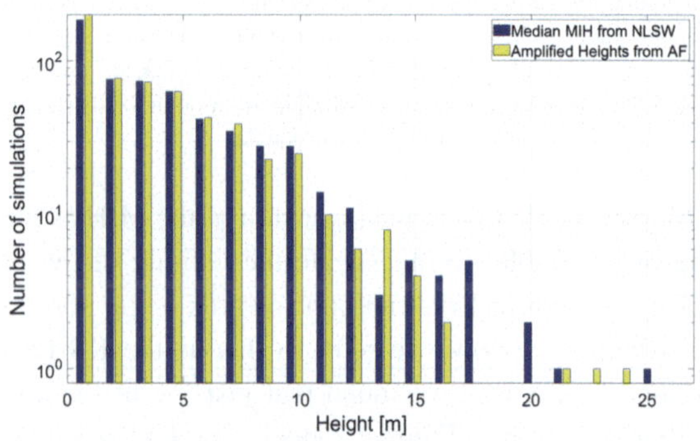

Figure 10
Distribution of median MIHs from the NLSW simulations compared to all the corresponding AHs

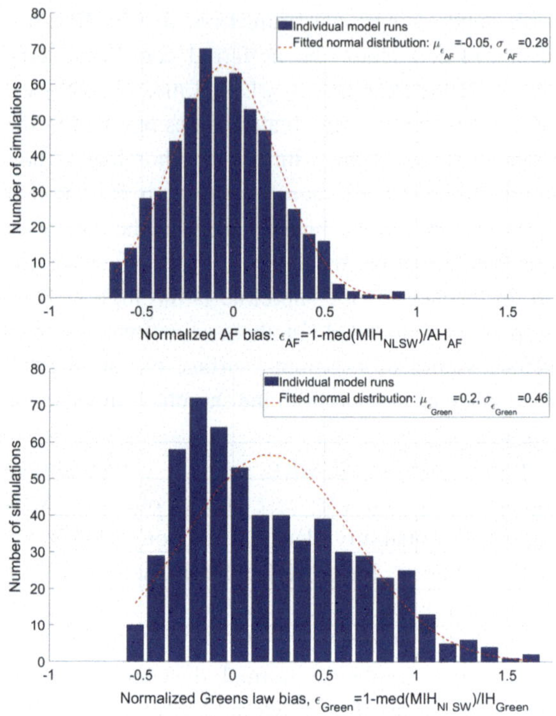

Figure 11
Normalized bias: upper panel, using amplification factors; lower panel, using Green's law

indicates that the AH is slightly lower than the median MIH predicted by the NLSW model. This is in contrast to the first version of the AF method (Løvholt et al. 2012), for which a positive bias was quantified, though against observations rather than simulations (Davies et al. 2018), as already noted. Figure 11 further shows that the standard deviation of the normalized AF bias is $\sigma_{AF} = 0.28$. Similarly, we computed the normalized bias using Green's law amplification to a reference depth of 0.5 m. As shown in the lower panel of Fig. 11, Green's law applied to this reference depth provides a strong positive bias of $\bar{\epsilon}_{Green} = 18\%$, meaning that Green's law would strongly overestimate the inundation height predicted by the NLSW reference model. Furthermore, the Green's law biases are associated with a significantly larger uncertainty ($\sigma_{Green} = 0.46$). The Green's law biases are moreover more strongly skewed than AF biases and definitely not normally distributed according to the KS test. This difference may reflect, among other things, the reference depth subjectivity that is not present in the AF method.

The bias from various model runs represents just one component of the variability. In addition to the variability of the bias, as shown in Fig. 9, the tsunami inundation height does vary spatially across each site. This variability is related only to the properties of the NLSW simulations at the site, and not to the AF method as such. However, when applying the AF method, we must take this uncertainty into account. For the examples shown in Fig. 9, we obtain log-normal standard deviations σ of 0.26, 0.43, and 0.23, respectively. The distribution of all the σ values is shown in Fig. 12, displaying a mean value of $\bar{\sigma} = 0.25$ and standard error of $\sigma_{\sigma} = 0.41$. By visual inspection, the distribution of local standard deviations seems to follow a log-normal distribution relatively closely, despite the data displaying a more peaky behavior than the fitted log-normal PDF. It is worth noting that of the three sources of uncertainty contributing to the AF (bias uncertainty σ_{ϵ}, mean local uncertainty $\bar{\sigma}$, and standard error σ_{σ}), the standard error σ_{σ} is the largest contributor. In this study, σ_{σ} reflects the variability owing to different sites, tsunami sources, wave polarity, wave period, etc. In the next subsection, we will demonstrate with examples how bias and variability differ.

Taking the root mean sum of the different σ's ($\sigma_{\epsilon} = 0.28$, $\bar{\sigma} = 0.25$, and $\sigma_{\sigma} = 0.41$) for the full parameter set simulations, we arrive at a root mean square sum of 0.55 due to the inundation process. It is noted that Davies et al. (2018) found an overall uncertainty of 0.92 when comparing PTHA results with observations from past events. If the uncertainty found in this paper is representative for the inundation process, the

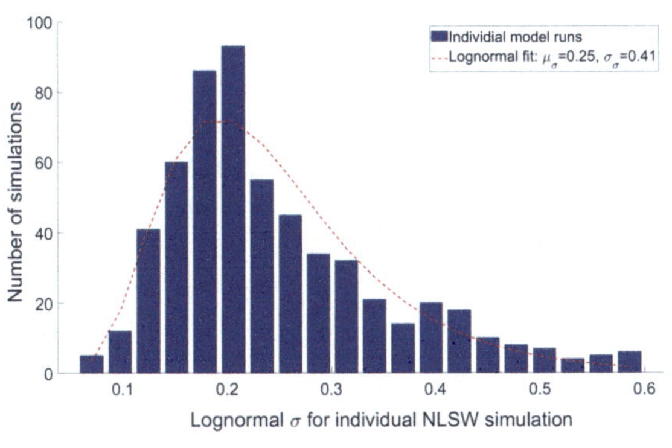

Figure 12
Distribution of the variability from all model runs

overall uncertainty of 0.92 in Davies et al. (2018) would be the sum of the inundation uncertainty and a residual uncertainty due to other sources. By removing the inundation uncertainty from the uncertainty obtained by Davies et al. (2018), we obtain a residual uncertainty of 0.73, from which we may speculate that the variability due to the inundation would then contribute to just less than half of the total uncertainty found by Davies et al. (2018). In this context, we note that Geist (2002) reports coefficients of variation in the range of 0.25–0.35 due to variable slip, which is thought to be one of the governing factors in inundation height uncertainty. These aspects deserves to be better investigated. On the other hand, Davies et al. (2018) used idealized AF profiles, which are most likely have a larger overall bias due to a more crude bathymetric representation. In any case, the present analysis shows that the variability in inundation is a strong overall component in the inundation height uncertainty, which needs to be accounted for in a regional PTHA method.

4.2. Inundation Height Statistics for Parameter Subsets

The bias and variability are further examined by carrying out the statistics for certain sets of parameters (e.g. only for positive wave polarity, only for short-period waves, for an individual site).

First, we examine the bias and local MIH variability separately for the two wave polarities (leading trough and leading peak). The statistical distributions for the two different polarities are quite similar to those shown in Figs. 11 and 12, and hence we choose not to display them. We also note that the p values of the bias distributions are 91 % and 14 %, respectively. However, the small to moderate average biases (3% for leading trough and 8% for leading peak) suggest that the distinction between leading trough and leading peak was necessary. Such a distinction cannot be made with simpler methods such as Green's law amplification.

Second, we examine the bias by making statistics for different ranges of incident wave periods T from the offshore tsunami simulations, that is $T \leq 600$ s, 600 s $< T \leq 1800$ s, and 1800 s $< T$. The bias distributions are all normally distributed, with p

values from KS tests ranging from 53 % to 84 %. The bias distributions in Fig. 13 show that the shortest wave periods display a positive bias (AH overestimation), and the intermediate wave periods a small negative bias (AH underestimation), whereas for the longest wave periods a strong negative bias is observed (AH underestimation). The strong bias for the longest wave periods may be due to several factors. Determining the causes for these offsets should be subject to further investigation. However,

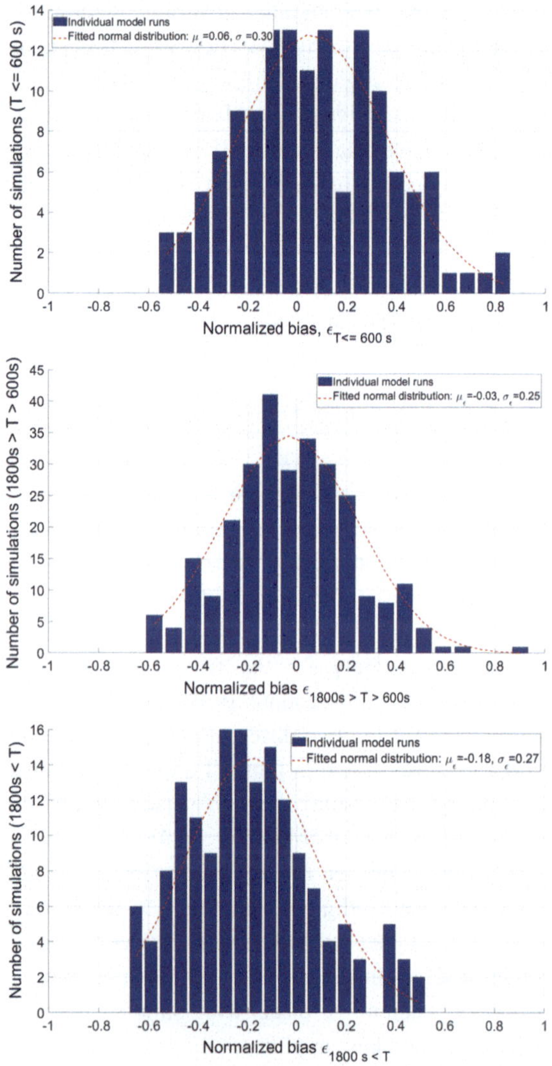

Figure 13
Normalized bias distributions for different ranges of the wave period obtained from offshore tsunami simulations. Upper panel, short wave periods ($T \leq 600$ s); Middle panel, intermediate wave periods (600 s $< T \leq 1800$ s); Lower panel, long wave periods ($T > 1800$ s)

we assume that the limited number of wave periods covered by the AF look-up tables can provide biases for instances when an input wave period from a tsunami simulation is pronouncedly different from the ones in the look-up table. Further, the filtering and zero-crossing techniques sometimes introduce wave lengths that are too large, which reduces inundation heights in the AF method. We also see that the largest bias variability is observed for the shortest and longest wave periods. However, the largest fraction of simulations are located in the intermediate regime.

Third, we examine the bias and local variability for three sites (Catania, Heraklion, and SE Iberia) in Fig. 14. Strong negative average biases (AH underestimation) are shown for Catania and Heraklion, whereas a slight positive bias (AH overestimation) is shown for SE Iberia. We further see that the three sites have pronouncedly different spreads with respect to both the bias and local variability, with the Heraklion site being clearly more heterogeneous, and SE Iberia the site with the smallest variability. This is also reflected in the p values from the KS tests using the bias distributions as input, giving lowest values for Heraklion (1%) and highest values for SE Iberia (61%). It is noted that from a practical point of view, separating the variability and bias from different locations is not feasible, as we cannot perform NLSW simulations at all relevant locations. Consequently, merging uncertainties from different sites would increase the overall MIH uncertainty in a PTHA analysis.

5. Uncertainty Treatment in PTHA

Summarizing from previous sections, the AF depends on both the characteristics of local bathymetry (coastal slope configuration) and the characteristics of the tsunami scenario (dominant period and polarity), as simulated in a single POI located offshore. Multiplying the AF to the tsunami simulated in this POI, we obtain an approximate median MIH for the nearby coast due to the considered individual tsunami scenario.

For the i-th tsunami scenario in the j-th POI, we indicate the maximum offshore elevation at the POI and the corresponding MIH value with η_{ij} and MIH_{ij},

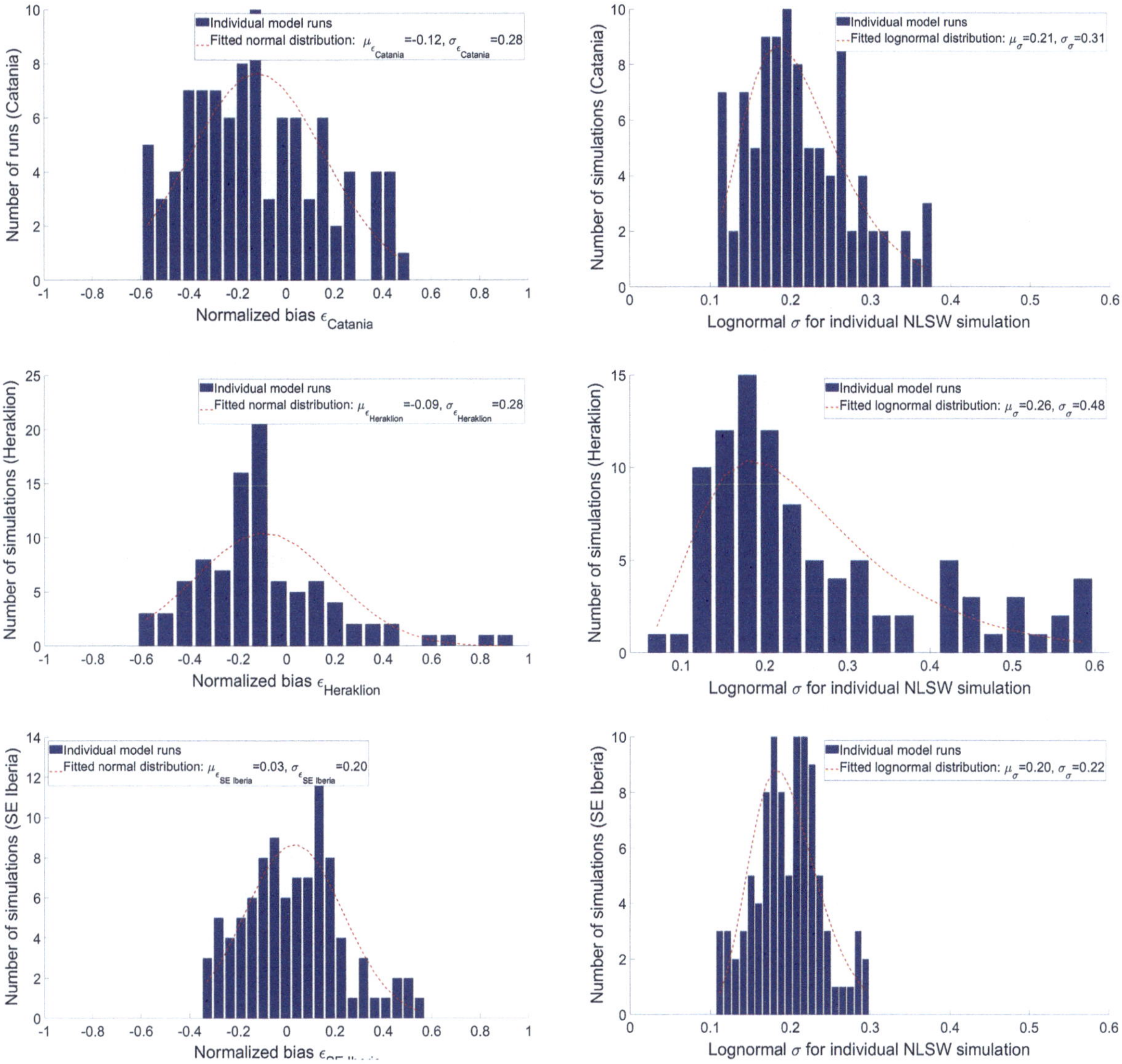

Figure 14

Bias and variability for three different sites. Left panels, model bias. Right panels, model variability. Upper panels, Catania site. Middle panels, Heraklion site. Lower panels, SE Iberia site

respectively. MIH_{ij} corresponds, after correction for a generally quite small bias, to the median value of the distribution of MIH values that the *i*-th tsunami scenario causes in stripes of land perpendicular to the coast and located around the projection to the coast and inland of the *j*-th POI.

In fact, as demonstrated by the NLSW simulations, this distribution can be approximated by a log-normal distribution for MIH, with a median that is well approximated by MIH_{ij} and a reasonably small σ (the standard deviation of the natural logarithm of MIH). These parameters are different for different scenarios and POIs. In particular, for the *i*-th tsunami scenario η_{ij} at the *j*-th POI, we indicated these parameters with μ_{ij} and σ_{ij}, and, considering the AF A_{ij} and the normalized bias ϵ_{ij} (compare Eq. (2)), we have $\mu_{ij} = \ln(A_{ij}\ \eta_{ij}/(1+\ \epsilon_{ij}))$. Among these parameters, the AF A_{ij} and η_{ij} depend on both the coastal slope around the POI and the tsunami scenario. Instead, the parameters σ_{ij} and ϵ_{ij} depend primarily on the local coastal configuration around the j-th POI and can be estimated only with an NLSW simulation of the i-th tsunami scenario in the j-th location.

For the purpose of PTHA, a large number of POIs are defined, each POI uniquely representing an entire stretch of the nearby coastline. Assuming that this coast is reasonably close to the POI, we may assume that, for each tsunami scenario, the MIH distribution on this coast is again well represented by a log-normal distribution. Given one randomly selected coastal point represented by the j-th POI, the probability that a given MIH value is exceeded due to the i-th tsunami scenario in the stripe to which this point belongs, can be computed from this log-normal distribution, that is:

$$Pr_{ij}(>\mathrm{MIH}|\mathrm{MIH_{ij}}) = 1 - \Phi\left(\frac{\ln \mathrm{MIH} - \ln \mu_{\mathrm{ij}}}{\sigma_{\mathrm{ij}}}\right) = 1 - \Phi\left(\frac{\ln \mathrm{MIH} - \ln\left(A_{ij}\eta_{ij}/(1+\epsilon_{ij})\right)}{\sigma_{ij}}\right), \quad (4)$$

where $\Phi(.)$ indicates the cumulative distribution function of a normal distribution.

This curve can be interpreted as the hazard curve for the hazard intensity MIH of one randomly selected point within the stretch of coastline near the j-th POI, conditional to the occurrence of the i-th tsunami scenario with best guess $\mathrm{MIH} = \mathrm{MIH_{ij}} = \mathrm{A_{ij}}\eta_{\mathrm{ij}}/(1+\epsilon_{\mathrm{ij}})$.

There is, however, a certain degree of epistemic uncertainty regarding the parameters of this equation. Indeed, all these parameters are reasonably well known only for the locations in which we have the NLSW simulations. In fact, the computational cost for achieving this for all POIs of a regional hazard quantification is typically unaffordable, since they are a very large number, and moreover precise local topographic/bathymetric data are usually not available. The parameters A_{ij}, μ_{ij}, ϵ_{ij}, and σ_{ij} must then be estimated from the available general information. More specifically:

- A_{ij} for a set of different tsunami scenarios has been extracted as shown in previous sections.
- The time history of the tsunami and the local η_{ij} may be estimated from linear simulations of the propagation in deep sea of each individual tsunami scenario.
- In regional PTHA, these simulations are barely affordable; therefore, both tsunami time history and η_{ij} are here approximated as a linear combination of unit sources (Molinari et al. 2016). Here we introduce z_{ij} as a further source of epistemic uncertainty, which is introduced by combining unit sources rather than simulating all of them as a single source. The treatment of this uncertainty is elaborated below.
- For the estimation of the parameters ϵ_{ij} and σ_{ij}, we assume that combining all the information from selected test sites reasonably approximates the variability of these parameters for all the POIs.

To propagate this epistemic uncertainty for the i-th tsunami scenario at the j-th POI, we adopt a Monte Carlo simulation scheme. We take as input the applicable A_{ij} and z_{ij} obtained from the simulation of the i-th tsunami scenario based on unit source combination for the j-th POI. Then, for each POI and each tsunami scenario, we sample:

- The correction δ^* from the empirical distribution $\delta = (z-\eta)/\eta$. This is the distribution of the relative error of the approximated η values due to the use of unit sources (as defined in Molinari et al. (2016), see their Figure 4d). This distribution is obtained by aggregation of a very large number of NLSW simulations with variable source and site, over a quite large range of tsunami intensities (up to $> 10\,m$), and we may assume that they reasonably approximate this uncertainty source for PTHA purposes.
- The parameters ϵ^* and σ^* from the empirical distributions of b and σ of Fig. 11, upper panel, and Fig. 12.

To obtain the consequent hazard curve, we can substitute each sampled $\{\delta^*,\epsilon^*,\sigma^*\}$ to the corresponding quantities into Eq. (3) as:

$$Pr_{ij}(>\mathrm{MIH}|\mathrm{MIH_{ij}}) = 1 - \Phi\left(\frac{\ln \mathrm{MIH} - \ln\left(A_{ij}z_{ij}/((1+\delta^*)(1+\epsilon^*))\right)}{\sigma^*}\right); \quad (5)$$

in particular, we set $\mathrm{MIH_{ij}} = A_{ij}\, z_{ij}$ as the best-guess reference MIH from the modeled i-th scenario in the j-th POI.

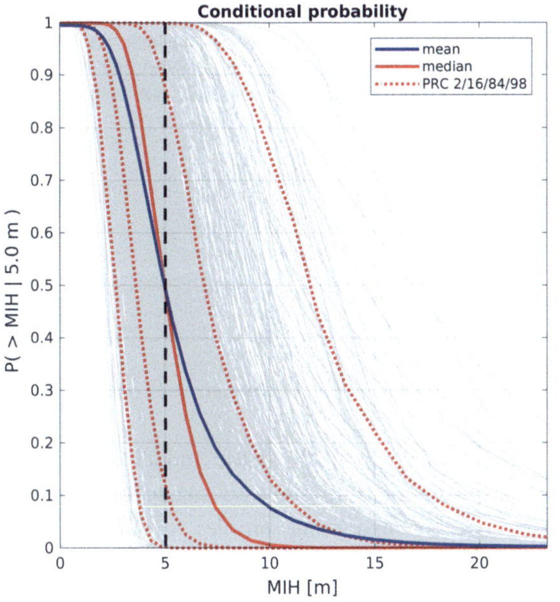

Figure 15
Conditional hazard curve for a MIH_{ij} = 5 m and relative epistemic uncertainty. In gray, we report the N = 1000 sampled alternative curves, while in blue and red we report different statistics of the epistemic uncertainty (blue, mean; red median; dotted red 2nd, 16th, 84th, and 98th percentiles)

Repeating this procedure N (e.g. 1000) times, we obtain N alternative conditional hazard curves, representing the sampled epistemic uncertainty in the conditional hazard. To increase the computational efficiency of this estimation, since $\{\delta^*, \epsilon^*, \sigma^*\}$ do not depend on the selected POI or tsunami scenario, they can be pre-computed for a discrete number of possible MIH values, to be subsequently applied to the different scenarios and POIs.

In Fig. 15, we report one example of the obtained epistemic uncertainty for a MIH_{ij} value of $5m$. In this example, we see that for the mean of the epistemic uncertainty of the model, the frequency of MIH exceeding $5m$ in one randomly selected point over the coastline is 50%, with resulting wide confidence ranges of $\sim 15\%$ considering the 16th and 84th percentiles.

6. Conclusion

In this paper, a new version of the amplification factor (AF) method for tsunami inundation height estimation has been presented. The new version represents a major upgrade compared to previous AF models, by taking into account the local bathymetry and the uncertainty introduced by using an approximated approach to the inundation process estimation through NLSW models. Because of the new AF capabilities, we can for the first time properly estimate the probability distribution of the maximum inundation height for PTHA. Another step forward is that the resulting MIHs emerging from the model are anchored towards a comprehensive set of more accurate reference models. The new method is implemented for the whole NEAM region, for the purpose of probabilistic tsunami hazard analysis for the TSUMAPS-NEAM project.

The AF method is set up to estimate the median value of the maximum inundation height at an area along the coast, when combined with tsunami propagation models. By linking the AF results with local inundation NLSW simulation results, we have also been able to add the inundation uncertainty into the AF model. Statistical analysis of the different sources of uncertainty is performed through 576 local inundation simulations for different earthquake sources and coastal sites. For example, the quantified bias indicates that the AF method may yield a slight underestimation (5%) of the median inundation height at a given site. However, the AF method yields much more accurate results than the Green's law method with respect to both a smaller overall bias and reduced uncertainty.

By comparing results from the inundation simulations with the AF results, we find mainly three sources of uncertainty related to employing the AF method, namely variability in positive and negative biases, local variability in the maximum inundation height (owing to the fact that AF is non-local), and the standard error, i.e. the uncertainty of the variability (because different sites and sources produce different inundation variability). Overall, the results show that the inundation variability is strong and needs to be accounted for in a regional PTHA method.

When the inundation statistics is conducted for smaller subsets of the simulation data, for instance for a single site, statistical distributions tend to be rather heterogeneous. When the results are aggregated, more well-defined normal distributions (for the bias) and log-normal distributions (for the local variability) are obtained. Taking into account that the AF method is designed for regional applications, at present state, merging uncertainties from different sites and sources into a single uncertainty seems to be the most robust option, although this increases the overall uncertainty in a hazard estimate. However, by including many more NLSW simulations in the future, for example covering more sites and sources, we would expect to improve run-up estimation capabilities and to reduce related uncertainties, for instance by better taking into account individual bias factors.

Hence, in its present form, the AF method is primarily set up to assist regional hazard quantification, and cannot be used for estimation of maximum run-up heights or as a replacement for local detailed inundation maps. For correct use, both the uncertainty and the bias associated with the method and with the specific hazard metric need to be taken into account.

Acknowledgements

Our warmest thanks go to Roberto Basili, the TSUMAPS-NEAM coordinator, who created a pleasant collaborative environment. Test sites used in the present work were part of the European FP7 project ASTARTE (www.astarte-project.eu). For this purpose, we thank the ASTARTE participants, Galderic Lastras, Miquel Canals, and Gerassimos Papadopoulos, for providing input data. Topo-bathymetric profile extraction and analysis was performed using ArcGIS (www.esri.com). PDF analysis was performed using MATLAB fitting tools (www.mathworks.com/). Figures in the manuscript were produced using GMT (gmt.soest.hawaii.edu), Matplotlib (https://matplotlib.org) and MATLAB. The new amplification factors (i.e. the data emerging from the work leading to this paper) can be found at https://github.com/sylfest/Ampfactors.git. This work received financial support through the EU ASTARTE project within the FP7-ENV2013 6.4-3 Grant 603839, and through the TSUMAPS-NEAM project, co-financed by the European Union Civil Protection Mechanism, Agreement Number ECHO/SUB/2015/718568/PREV26.

Appendix

1 Extraction of Bathymetric Profiles

The SRTM15+ bathymetric model (15 arc-sec spatial resolution, http://topex.ucsd.edu) was used as the basis for extracting profiles within both the Mediterranean Sea and the Black Sea, while for the North-Eastern Atlantic the SRTM30+ (30 arc-sec spatial resolution), improved with local data in Portugal (Zitellini et al. 2009), was used. The procedure for acquiring the bathymetric profiles began with extracting offshore points along a 50 m depth isobath with an initial separation distance of 2 km; then the POIs were down-sampled to about 20 km for the sake of computational feasibility and because they provide a non-local, averaged representation of the coastal hazard points suitable for the TSUMAPS-NEAM model resolution.

A nearest-neighbor algorithm was used to identify corresponding coastline points, i.e. connecting the offshore to the onshore points. These identified shoreline points were then applied to define a piecewise linear shoreline contour. A set of 40 transects spaced at about 1 km and perpendicular to this contour line were then created (i.e. 20 km to each side of the offshore hazard points, see Figs. 2 and 16). All profiles that intersected islands were deleted in order to avoid positive values (i.e., land). Profiles with anomalous orientation with respect to the shoreline were then identified and manually corrected.

However, we could not straightforwardly apply the above procedure in areas characterized by complex non-planar shorelines with many islands, such as the Aegean islands, deep and narrow bays (e.g., fjords), and the Croatian islands. Therefore, in these areas, some POIs were left out without transects in the automatic procedure and for most areas are

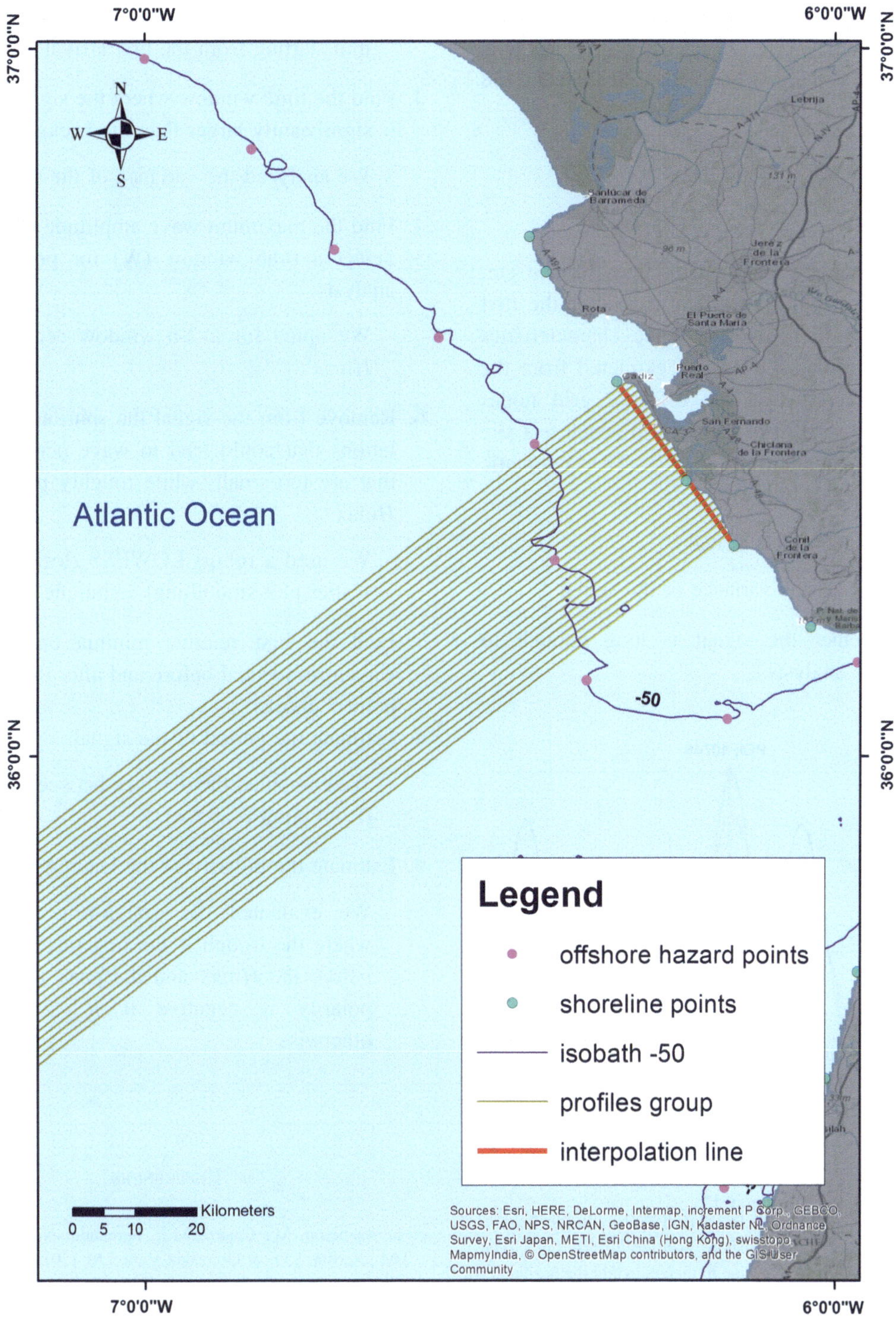

Figure 16
Map depicting the main features used to trace the transects. The offshore points at 50 m depth are colored magenta, and the corresponding shoreline points are colored green. The piecewise linear shoreline contour passes through three points on the shoreline, and a group of profiles perpendicular to the line and to the shoreline is displayed

transect positions drawn manually. For our study, a total of 149 profiles were extracted for the Black Sea, 935 profiles for the Mediterranean Sea, and 1158 profiles for the North-Eastern Atlantic region (Fig. 2).

2 Extraction of Wave Periods

In a search for a maximal offshore surface elevation, waves arriving later than 2 h after the first wave arrival were neglected. Wave characteristics needed to determine the AF are extracted from the mareograms recorded at the offshore grid nodes placed along the 50 m isobath outside each test site location (an example in Fig. 17). The automatic procedure works as follows:

1. Find the tsunami first arrival
   - We analyzed the variance of the signal
2. Check whether the signal is long enough to perform the analysis
   - We set the minimum length of the signal to 90 min starting from the first arrival
3. Find the time window where the signal amplitude is significantly larger than the background level
   - We analyzed the variance of the signal
4. Find the maximum wave amplitude (*H*max)
5. Select a time window (W) for performing the analysis
   - We opted for a 2-h window centered around *H*max
6. Remove from the signal the spurious short oscillations that could lead to wave period estimates that are too small, while roughly preserving the *H*max
   - We used a robust LOWESS (locally weighted scatter plot smoothing) technique within W
7. Find the first relative minima on the filtered waveform located before and after the *H*max (see point 3 above)
8. Estimate the period of the signal
   - We computed the distance between the minima (from point 6 above)
9. Estimate the polarity of the leading wave
   - We evaluated the trough-to-peak ratio (*R*), where the trough is the first relative minimum before the *H*max and the peak is just *H*max; polarity is negative if $R > 25\%$, positive otherwise

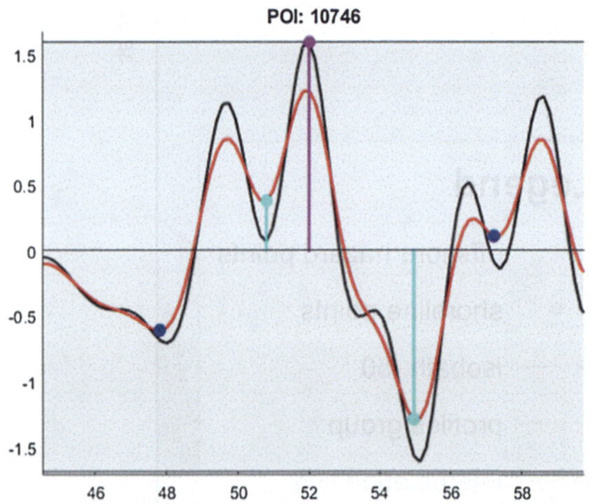

Figure 17
Example of extracting the wave period and polarity from a synthetic mareogram at an offshore position (black line). The vertical axis is the surface elevation in meters, while the horizontal axis is the time in minutes. The maximum surface elevation value (*H*max) is taken at the unfiltered maximum location (marked with a magenta bullet). The troughs of the filtered time series are marked with (blue and cyan) bullets lying on the red line. The wave period is the duration between the troughs ahead of and behind the leading peak measured on the filtered time series (red line), and the wave period interval is indicated by cyan-colored bars. We stress that the filtering algorithm is only used to derive wave periods, and not the simulated offshore wave amplitudes

References

de la Asunción, M., Castro, M.J., Fernández-Nieto, E., Mantas, J.M., Acosta, S.O., & González-Vida, J.M. (2013). Efficient GPU implementation of a two waves TVD-WAF method for the two-dimensional one layer shallow water system on structured meshes. Computers & Fluids 80(Supplement C):441– 452, https://doi.org/10.1016/j.compfluid.2012.01.012. http://www.sciencedirect.com/science/article/pii/S0045793012000217 (**selected contributions of the 23rd International Conference on Parallel Fluid Dynamics ParCFD2011**).

Brizuela, B., Armigliato, A., & Tinti, S. (2014). Assessment of tsunami hazards for the Central American Pacific coast from southern Mexico to northern Peru. *Natural Hazards and Earth System Sciences*, *14*(7), 1889–1903. https://doi.org/10.5194/nhess-14-1889-2014. https://www.nat-hazards-earth-syst-sci.net/14/1889/2014/.

Davies, G., Griffin, J., Løvholt, F., Glimsdal, S., Harbitz, C.B., Thio, H.K., Lorito, S., Basili, R., Selva, J., Geist, E., & Baptista, M.A. (2018). A global probabilistic tsunami hazard assessment from earthquake sources. Geological Society, London, Special Publications 456, https://doi.org/10.1144/SP456.5. http://sp.lyellcollection.org/content/early/2017/03/01/SP456.5.full.pdf.

Dutykh, D., Poncet, R., & Dias, F. (2011). The VOLNA code for the numerical modeling of tsunami waves: Generation, propagation and inundation. *European Journal of Mechanics - B/Fluids*, *30*(6), 598–615. https://doi.org/10.1016/j.euromechflu.2011.05.005. http://www.sciencedirect.com/science/article/pii/S0997754611000574, special Issue: Nearshore Hydrodynamics.

Gailler, A., Hébert, H., Schindelé, F., & Reymond, D. (2018). Coastal amplification laws for the french tsunami warning center: Numerical modeling and fast estimate of tsunami wave heights along the French riviera. *Pure and Applied Geophysics*, *175*(4), 1429–1444. https://doi.org/10.1007/s00024-017-1713-9.

Geist, E. (2002). Complex earthquake rupture and local tsunamis. Journal of Geophysical Research: Solid Earth 107(B5):ESE 2–1–ESE 2–15, https://doi.org/10.1029/2000JB000139. https://agupubs.onlinelibrary.wiley.com/doi/abs/10.1029/2000JB000139, https://agupubs.onlinelibrary.wiley.com/doi/pdf/10.1029/2000JB000139.

Geist, E. L., & Parsons, T. (2006). Probabilistic analysis of tsunami hazards. *Natural Hazards*, *37*(3), 277–314. https://doi.org/10.1007/s11069-005-4646-z.

Gonzalez, F.I., Geist, E.L., Jaffe, B., Kanoglu, U., Mofjeld, H., Synolakis, C.E., Titov, V.V., Arcas, D., Bellomo, D., Carlton, D., Horning, T., Johnson, J., Newman, J., Parsons, T., Peters, R., Peterson, C., Priest, G., Venturato, A., Weber, J., Wong, F., & Yalciner, A. (2009). Probabilistic tsunami hazard assessment at Seaside, Oregon, for near- and far-field seismic sources. Journal of Geophysical Research: Oceans 114(C11):n/a–n/a, https://doi.org/10.1029/2008JC005132, c11023.

Grezio, A., Babeyko, A., Baptista, M. A., Behrens, J., Costa, A., Davies, G., et al. (2017). Probabilistic tsunami hazard analysis: Multiple sources and global applications. *Reviews of Geophysics*, *55*(4), 1158–1198. https://doi.org/10.1002/2017RG000579. https://agupubs.onlinelibrary.wiley.com/doi/abs/10.1002/2017RG000579, https://agupubs.onlinelibrary.wiley.com/doi/pdf/10.1002/2017RG000579

Hébert, H., & Schindelé, F. (2015). Tsunami impact computed from offshore modeling and coastal amplification laws: insights from the 2004 Indian Ocean Tsunami. *Pure and Applied Geophysics*, *172*(12), 3385–3407.

Horspool, N., Pranantyo, I., Griffin, J., Latief, H., Natawidjaja, D., Kongko, W., et al. (2014). A probabilistic tsunami hazard assessment for Indonesia. *Natural Hazards and Earth System Sciences*, *14*(11), 3105–3122.

Kaiser, G., Scheele, L., Kortenhaus, A., Løvholt, F., Römer, H., & Leschka, S. (2011). The influence of land cover roughness on the results of high resolution tsunami inundation modeling. *Natural Hazards and Earth System Sciences*, *11*(9), 2521–2540.

LeVeque, R. J., & George, D. L. (2008). High-resolution Finite Volume methods for the shallow water equations with bathymetry and dry states. *Advanced numerical models for simulating tsunami waves and runup* (pp. 43–73). Singapore: World Scientific.

Lorito, S., Selva, J., Basili, R., Romano, F., Tiberti, M., & Piatanesi, A. (2015). Probabilistic hazard for seismically induced tsunamis: accuracy and feasibility of inundation maps. *Geophysical Journal International*, *200*(1), 574–588. https://doi.org/10.1093/gji/ggu408, http://dx.doi.org/10.1093/gji/ggu408, /oup/backfile/content_public/journal/gji/200/1/10.1093_gji_ggu408/2/ggu408.pdf

Løvholt, F., Pedersen, G., & Glimsdal, S. (2010). Coupling of dispersive tsunami propagation and shallow water coastal response. *The Open Oceanography Journal*, *4*, 71–82. https://doi.org/10.2174/1874252101004020071.

Løvholt, F., Glimsdal, S., Harbitz, C. B., Zamora, N., Nadim, F., Peduzzi, P., et al. (2012). Tsunami hazard and exposure on the global scale. *Earth-Science Reviews*, *110*(1), 58–73. https://doi.org/10.1016/j.earscirev.2011.10.002.

Løvholt, F., Griffin, J., & Salgado-Gálvez, M. (2015). *Tsunami Hazard and Risk Assessment on the Global Scale* (pp. 1–34). Berlin, Heidelberg: Springer Berlin Heidelberg. https://doi.org/10.1007/978-3-642-27737-5_642-1.

Macías, J., Castro, M. J., Ortega, S., Escalante, C., & González-Vida, J. M. (2017). Performance benchmarking of tsunami-hysea model for NTHMP's inundation mapping activities. *Pure and Applied Geophysics*, *174*(8), 3147–3183. https://doi.org/10.1007/s00024-017-1583-1.

Massey, F. J. (1951). The kolmogorov-smirnov test for goodness of fit. *Journal of the American Statistical Association*, *46*(253), 68–78.

Molinari, I., Tonini, R., Piatanesi, A., Lorito, S., Romano, F., Melini, D., et al. (2016). Fast evaluation of tsunami scenarios: uncertainty assessment for a Mediterranean Sea database. *Natural Hazards and Earth System Sciences*, *16*, 2593–2602. https://doi.org/10.5194/nhess-2016-145.

Selva, J., Tonini, R., Molinari, I., Tiberti, M. M., Romano, F., Grezio, A., et al. (2016). Quantification of source uncertainties in Seismic Probabilistic Tsunami Hazard Analysis (SPTHA). *Geophysical Journal International*, *205*, 1780–1803. https://doi.org/10.1093/gji/ggw107.

Sørensen, M., Spada, M., Babeyko, A., Wiemer, S., & Grunthal, G. (2012). Probabilistic tsunami hazard in the Mediterranean Sea. Journal of Geophysical Research: Solid Earth 117(B1), https://doi.org/10.1029/2010JB008169. https://agupubs.onlinelibrary.wiley.com/doi/abs/10.1029/2010JB008169, https://agupubs.onlinelibrary.wiley.com/doi/pdf/10.1029/2010JB008169.

Strasser, F. O., Arango, M. C., & Bommer, J. J. (2010). Scaling of the source dimensions of interface and intraslab subduction-zone earthquakes with moment magnitude. *Seismological Research Letters*, *81*(6), 941–950. https://doi.org/10.1785/gssrl.81.941.

Titov, V.V., & Gonzalez, F.I. (1997). Implementation and testing of the method of splitting tsunami (most) model. Tech. rep., NOAA Technical Memorandum ERL PMEL-112, 11 pp UNIDATA

Wronna, M., Omira, R., & Baptista, M. A. (2015). Deterministic approach for multiple-source tsunami hazard assessment for

Sines, Portugal. *Natural Hazards and Earth System Sciences*, *15*(11), 2557–2568. https://doi.org/10.5194/nhess-15-2557-2015. https://www.nat-hazards-earth-syst-sci.net/15/2557/2015/.

Zitellini, N., Gracia, E., Matias, L., Terrinha, P., Abreu, M., DeAlteriis, G., et al. (2009). The quest for the Africa-Eurasia plate boundary west of the Strait of Gibraltar. *Earth and Planetary Science Letters*, *280*(1), 13–50. https://doi.org/10.1016/j.epsl.2008.12.005. http://www.sciencedirect.com/science/article/pii/S0012821X0800753X.

(Received August 22, 2018, revised November 15, 2018, accepted January 4, 2019, Published online January 25, 2019)

Pure Appl. Geophys. 176 (2019), 3247–3262

https://doi.org/10.1007/s00024-018-1898-6

Pure and Applied Geophysics

Suitability of Open-Ocean Instrumentation for Use in Near-Field Tsunami Early Warning Along Seismically Active Subduction Zones

Amy L. Williamson[1] and Andrew V. Newman[1]

Abstract—Over the past decade, the number of open-ocean gauges capable of parsing information about a passing tsunami has steadily increased, particularly through national cable networks and international buoyed efforts such as the Deep-ocean Assessment and Reporting of Tsunami (DART). This information is analyzed to disseminate tsunami warnings to affected regions. However, most current warnings that incorporate tsunami are directed at mid- and far-field localities. In this study, we analyze the region surrounding four seismically active subduction zones, Cascadia, Japan, Chile, and Java, for their potential to facilitate local tsunami early warning using such systems. We assess which locations currently have instrumentation in the right locations for direct tsunami observations with enough time to provide useful warning to the nearest affected coastline—and which are poorly suited for such systems. Our primary findings are that while some regions are ill-suited for this type of early warning, such as the coastlines of Chile, other localities, like Java, Indonesia, could incorporate direct tsunami observations into their hazard forecasts with enough lead time to be effective for coastal community emergency response. We take into account the effect of tsunami propagation with regard to shallow bathymetry on the fore-arc as well as the effect of earthquake source placement. While it is impossible to account for every type of off-shore tsunamigenic event in these locales, this study aims to characterize a typical large tsunamigenic event occurring in the shallow part of the megathrust as a guide in what is feasible with early tsunami warning.

Key words: Tsunami early warning, near-field, DART gauge, earthquakes.

1. Introduction

Many tsunami warnings stem form an evaluation of the seismic energy released from tsunami-generating earthquakes. Then, if available, additional data about the tsunami are incorporated to update mid and far-field tsunami warnings. While a seismic dataset is valuable for understanding the earthquake source dynamics, the relationship between the seismic energy determined earthquake magnitude and tsunami energy used for event warnings and forecasts is non-linear and poorly understood (Tang et al. 2012). Even with a well constrained seismically determined magnitude, the application to tsunami hazard is not always absolute. As highlighted in Titov et al. (2016a), the tsunami warnings issued for many recent earthquakes based on seismic evaluations did not accurately describe the oncoming wave. In the cases of the moment magnitude, M_w, 8.6 2005 Nias Island earthquake and the 2012 M_w 8.6 Sumatra earthquake, the tsunami threats were overestimated, in some cases leading to unnecessary and costly evacuations with little to no eventual damage related to the tsunami (Song 2007; Titov et al. 2016a). On the other hand, in the cases of the 2006 West Java earthquake (M_w 7.7), no official tsunami warning was issued and an eventual run-up of up to 21 m was recorded in Pangandaran (Okal 2015; Fritz et al. 2007). This earthquake, like the 2010 Mentawai Islands earthquake (M_w 7.8), was a 'tsunami earthquake' (Newman et al. 2011). Tsunami earthquakes oftentimes produce tsunami that are much larger than seismic models rapidly predict (Kanamori 1972; Kanamori and Kikuchi 1993; Newman and Okal 1998) with slip concentrated in the shallow megathrust environment (Polet and Kanamori 2000). While seismic evaluations of the tsunami are valuable—particularly in the near-field where tsunami landfall can be within tens of minutes, a direct measurement of the tsunami for forecasting purposes provides a

Electronic supplementary material The online version of this article (https://doi.org/10.1007/s00024-018-1898-6) contains supplementary material, which is available to authorized users.

[1] School of Earth and Atmospheric Sciences, Georgia Institute of Technology, 311 Ferst Drive, Atlanta, GA 30332, USA. E-mail: amy.williamson@gatech.edu

useful dataset for emergency management. Direct observations lessen the chance of a misestimate of the tsunami, which can be devastating for the near-field environments where the majority of tsunami victims are located (Titov et al. 2016b).

Unlike far-field warnings where source to arrival is on the scale of hours, near-field warnings need to be computed on a severely abbreviated timeframe. One potential aid is the travel time differential between the open-ocean and shallow coastal environments. A tsunami will propagate with a higher velocity in deep water than it will over the shallower continental shelf. This discrepancy was identified for use in far-field tsunami forecasts in Titov et al. (2005) where the placement of sensors in the central Pacific allowed for the computation of lead times between measurement and inundation for the Hawaiian coastal city of Hilo. The efficacy of real-time open-ocean measurements of tsunami generated from far-field sources for forecasts in Hilo was demonstrated through idealized placements of then test-phase tsunameters. Idealized placements allowed for measurement and processing of the oncoming wave prior to inundation at the city—but focused solely on distantly generated events. This study operates under a similar premise as the Titov et al. (2005) study, but with a focus on near-field, rather than far-field generated events over multiple regions where tsunameters could be placed.

After considerable attention on far-field inundation forecasts, near-field early warning is the latest frontier for tsunami warning systems (Song 2007; Titov et al. 2016b). Prior work has assessed the usage of coastal GPS and assimilation of Deep-Ocean Assessment and Reporting of Tsunami (DART) gauges for improved early warning by rapidly approximating tsunami energy (Titov et al. 2016a). With these prior works in mind, this paper analyzes the suitability for localized near-field tsunami warning systems incorporating direct open-ocean tsunami observations, like those measured by DART gauges. We focus on four seismically-active regions, highlighting unique characteristics to take into account for early warning systems including coastal morphology, subducting plate geometry, tsunami source-to-landfall travel times, and currently deployed instrumentation. The goal of this study is not to reformat current warning system operations, but to investigate the feasibility of open-ocean data incorporation.

The study is divided into the following sections. First we outline currently deployed open-ocean instrumentation and its use in tsunami warning systems. Second we provide our methodology and analysis of the feasibility for localized near-field early warning for our four study regions—Cascadia, Japan, Indonesia, and Chile. Third, we discuss the variability of certain model parameters used in tsunami early warning and how they affect our results.

2. *Data*

Open-ocean pressure sensors provide a clean dataset that directly records the passage of tsunami waves. Within the past decade, DART gauge data have been incorporated in real time for tsunami warnings in the United States (Titov et al. 2016b). Each gauge measures the pressure exerted on a seafloor sensor from the overlying water column, translating it to a water column height. Data from each gauge are transmitted to warning centers via satellite and are openly available in semi-real time, offering a direct observation of a passing tsunami that is incorporated into tsunami warning forecasts. The gauges monitor for not only seismically induced tsunamis (Wei et al. 2014), but also from non-seismic events such as landslide tsunamis and meteotsunamis. Currently, there is a globally distributed array of 60 DART gauges owned by a consortium of countries. Most currently deployed gauges sit far from potential tsunami sources. Newer DART 4G instruments, which are in test operations offshore Oregon and Chile, are capable of up to a 1 Hz sampling frequency (Rabinovich and Eblé 2015; Tang et al. 2016) and are deployed closer to the trench and potential tsunami sources.

It is important to note that DART gauges are not deployed with the goal of providing tsunami forecasts for the immediate near-field. Current station capabilities are tailored for mid- to far-field forecasting. Furthermore, DART gauges are not the only instruments available in the open-ocean for tsunami modeling. Cabled networks that include similar pressure gauge technology exist in a few locations. We highlight the deployment and capabilities of these observatories in subsequent sections of this paper. While this paper investigates near-field warnings using any open-ocean tsunami data, we use the current distribution of DART as a reference point because of their ubiquity in tsunami modeling and global presence.

3. *Methodology*

In order to examine the tsunami lead time (between data acquisition and earliest tsunami landfall), we generate a series of trench-parallel synthetic seafloor displacements. For each event, we track the travel time of the crest of the leading wave to determine the difference in travel time towards the open-ocean versus the shoreline (shown in Fig. 1). The earliest arrival time of the tsunami at the shoreline is recorded and used to determine the open-ocean boundary of which a station must be placed in order to capture the first one-quarter of the leading wave prior to its coastal inundation. We then highlight zones inside this boundary where a station must be placed in order to record the tsunami with variable lead times. By combining the results of multiple sources along the strike of the subduction zone, we can generate a composite view for each of our four study regions, to analyze which parts of the subduction zone are ideal for early warning, and which regions inhibit such early forecasts.

The sea-surface displacement generated from each synthetic tsunami source is used as an initial condition for tsunami propagation. Each seafloor displacement is described as a two-dimensional concave surface defined by:

$$H = \frac{1}{4}\left(1 + \cos\left(\frac{\pi x}{L}\right)\right)\left(1 + \cos\left(\frac{\pi y}{L}\right)\right)$$

where L is the diameter of the source and H is the two-dimensional water column profile over a spatial domain governed by x and y (Hossen et al. 2015). We place each of our synthetic sources above the region where we expect the subducting slab to be at a depth of 15 km (Hayes et al. 2012). The reasons for this placement are (1) this depth typically considered the upper extent of the seismogenic zone that nucleates large earthquakes (Lay 2015); and (2) the shallow environment is close to what can be expected for dangerous tsunami earthquakes.

We model the propagating tsunami from each synthetic event over a regional spatial domain with a 1 arc minute bathymetry. We run each simulation using the Method of Splitting Tsunami (MOST) model, which solves the non-linear shallow water wave equations (Titov and González 1997) and is currently an operational tool at the NOAA Tsunami Warning Centers and at the NOAA Center for Tsunami Research.

An example of our analysis using a single synthetic source is shown in Fig. 1. The earliest landfall of the crest of the leading wave, 18 min after nucleation, is recorded for determining instrument warning zones. The simultaneous location of the tsunami's leading wave in open-ocean identifies the furthest extent of the near-field early warning capabilities, acting as a threshold for instrument placement. This threshold contour acts as a 0-min warning zone. The 0-min warning region means any station placed in this shaded region will be measured at about the same time as landfall for local events. Anything inside of this contour will register the tsunami before landfall; however, some areas will record the tsunami with a greater lead time than other areas. To calculate the lead times inside of this threshold for this synthetic event is fairly simple. The lead time for any location is equal to the coastal arrival time subtracted from the tsunami travel time. In Fig. 1, we highlight zones that have 5 and 10 min of lead time before coastal inundation. For each of the subsequent composite figures, we draw shaded regions of expected lead times from 0 min up to 20 min at 5-min increments, where possible.

It is extremely important to note that the location of our synthetic source has a role in the timing of the earliest inundation on land and this threshold of open-ocean detection. We pick our source location, as mentioned earlier, for its consideration as the 'edge' of the typical seismogenic zone. However, any event rupturing primarily up-dip of this synthetic event will have a longer time until inundation, and events rupturing primarily down-dip will have a quicker inundation. We stress this variability and uncertainty

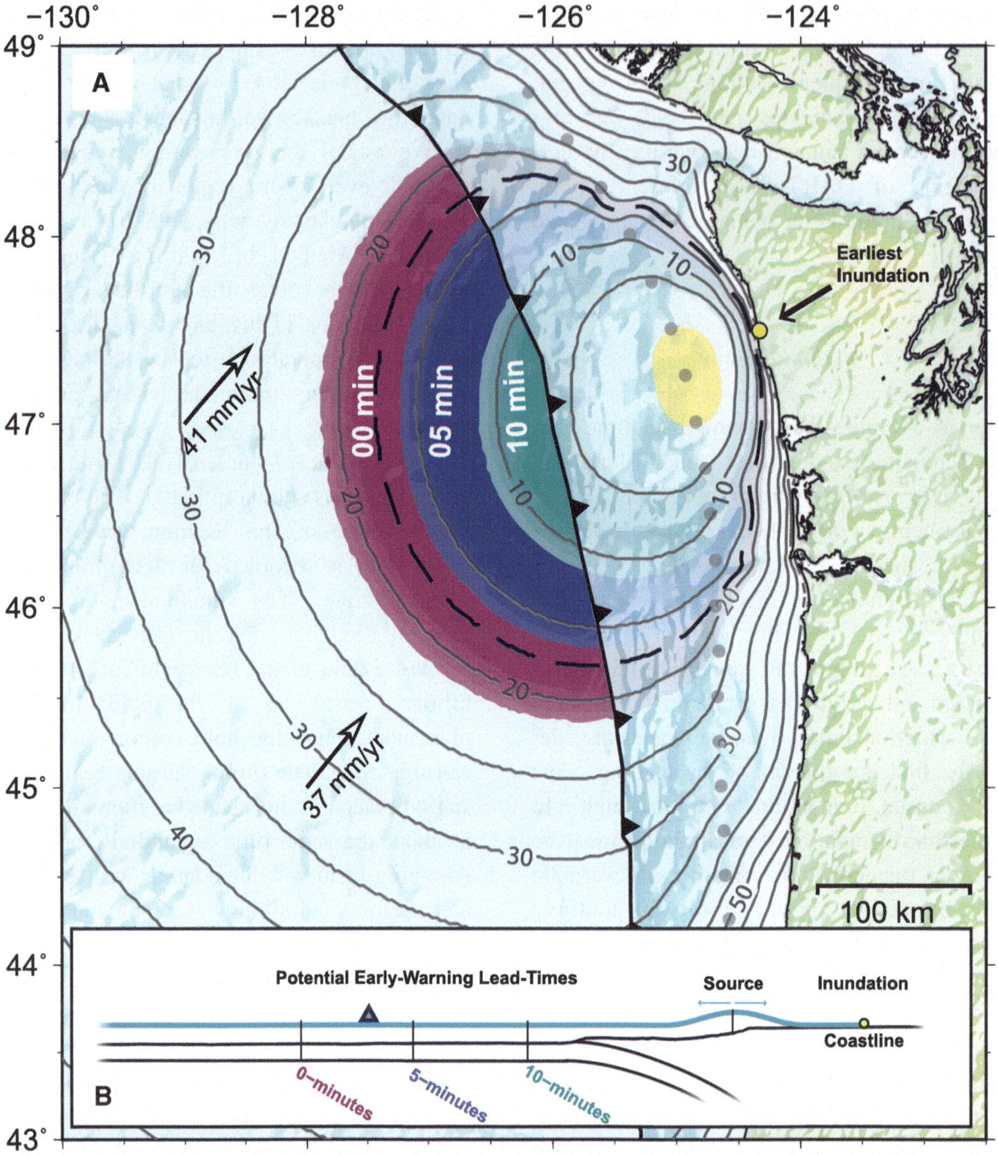

Figure 1

a Propagation of a single synthetic event to determine instrument zones for local early tsunami warning. The synthetic event's source is shown as the light-yellow polygon. The travel time, in minutes, of the crest of the leading wave from this source is plotted by the gray contours. The location of the earliest inundation (at 18 min) is marked with a yellow dot on the coastline. The location of the tsunami's leading wave at this time is outlined by the black dashed line. The magenta, navy, and aqua shaded regions on the subducting plate indicate 0-, 5-, and 10-min zones for possible instrument placement. The gray dots along strike of the trench shows the rest of the synthetic locations that will be combined in the composite image (Fig. 2). Arrows indicate the plate motion between the Juan de Fuca and North American plates using MORVEL-2010 motions (DeMets et al. 2010). **b** Cross-section schematic of the source, inundation, and warning zones for near-field tsunami early warning, generalizing the map shown in **a**

in source location by creating a 2.5-min window on either side of our lead time estimate, as a way to take into account slightly shallower and slightly deeper events. Therefore, each lead time estimate, as shown in Fig. 1, is displayed as a zone rather than a single line.

Figure 1 shows only a single-source solution to demonstrate the methodology, but each final composite image is composed of multiple synthetic events. The spacing of each of these synthetic events is 15′ latitude and longitude in the trench-normal direction. At an equatorial site this amounts to one synthetic source about every 30 km along strike. This spacing allows for the characterization of each section of the subduction zone for its near-field warning capabilities from local offshore data. Regional early-warning images in subsequent figures combine many synthetic events into one composite. Each event's output is weighted equally, where the center of each lead time zone (treated as independent polygons) is assigned the value '1' and areas not in the warning zone are assigned a value of '0'. After all of the same lead time zones are combined via a simple summation into the composite and normalized; areas in each warning zone with a value lower than a threshold 0.1 are discarded. This acts as a filter to avoid the inclusion of spurious areas that do not overlap, or overlap little with neighboring events, due to either poorly resolved bathymetry or extremely fine scaled features, and to smooth small irregularities in lead times. After the composite for each warning zone is created, a final image merges each warning zone together, as shown in Figs. 2, 3, 4, and 5. We want to reiterate that the use of lead times is applicable for a local and trench-normal relationship between the offshore instrumentation, tsunami nucleation point, and nearby coast. A lead time of 10 min in Cascadia, for example, is drawn for an event's immediate surroundings and combined with neighboring events to show the general extent of that level of warning for any area, but should not be extended to assume that an instrument at the southernmost point of Cascadia can provide a 10-min warning for an event originating at the northernmost point for coastal communities also at the northernmost extent of the model-space.

4. Results

The results of five composite images for four near-field tsunami warning cases are explored. Each region is picked due to either unique megathrust geometries, coastal morphology, the threat of a future large event, or the preponderance of large tsunamigenic events in the past. We focus entirely on subduction zone megathrust environments because of their tsunamigenic potential due to megathrust earthquakes and current open-ocean tsunami instrumentation.

4.1. Cascadia

Attention has been focused on the Cascadia subduction zone in recent years in preparation for the possibility of a great tsunamigenic earthquake rupturing along the trench (Satake et al. 1996; Satake et al. 2003). The most recent large tsunami generated from the region occurred in 1700 A.D., supported from Japanese tsunami deposit records and tree ring analysis along the Pacific Northwest (Satake et al. 1996; Atwater 1992). Estimated sources of the 1700 tsunami have been modeled to compare with Japanese coastal records, supporting a claim that the earthquake which generated the tsunami ruptured over 1100 km of the megathrust with a magnitude close to 9.0 (Satake et al. 2003). While the megathrust remains seismically quiet, the region is known for frequent slow-slip events near the deeper portion of the subduction interface (Rogers and Dragert 2003). Currently, with land-based tools, it is unclear if the shallower portion of this interface is locked, or freely slipping, giving little guidance to the seismic hazard. There have been many recent studies on this matter, producing many models of a possible large megathrust event. These models aid in our understanding of possible hazards associated from the tsunami, such as inundation extent, as well as seismic characteristics, such as peak-ground-acceleration that could be expected on land (Witter et al. 2013; Melgar et al. 2016).

Building on the methodology example from Figs. 1, 2 shows the regional composite view of early warning for the Cascadia subduction zone in the Pacific Northwest. The young Juan de Fuca plate subducts underneath the North American plate at about 40 mm/year (DeMets et al. 2010). The offshore coastal region between the trench and coastline is fairly broad with distances that can exceed 100 km. The regional bathymetry includes a wide and broad continental shelf that extends far off the Oregon and Washington coastlines before tapering towards Vancouver Island. This composite solution merges the

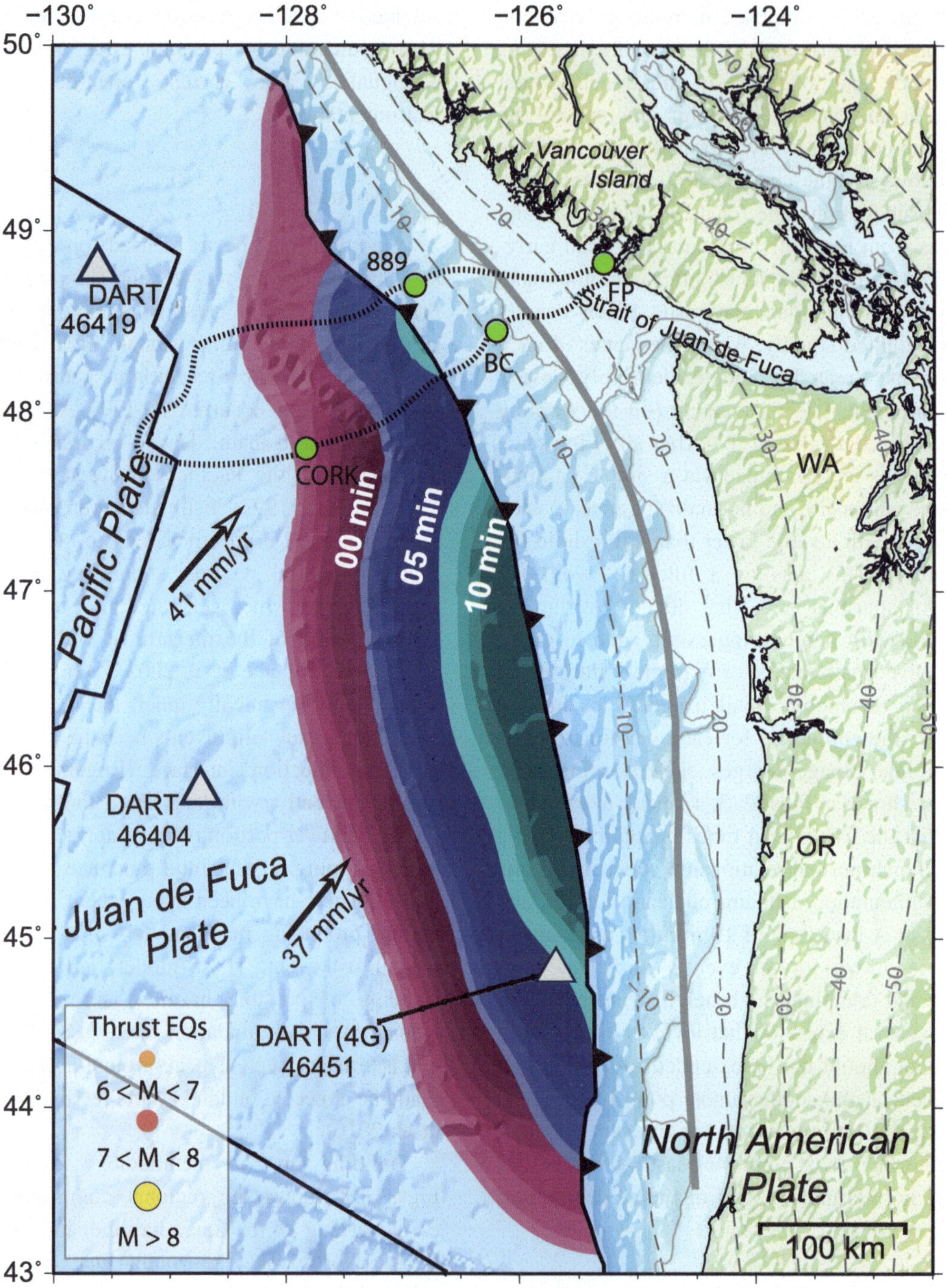

Figure 2

Composite view of Cascadia early warning zones. Magenta, navy, and aqua shading refers to 0 min, 5-min, and 10-min lead times, respectively. Arrows indicate the plate motion between the Juan de Fuca and North American plates using MORVEL-2010 motions (DeMets et al. 2010), gray triangles show approximate locations of DART gauges, and green dots indicate the location of NEPTUNE bottom pressure sensors used for tsunami studies. The thick, gray line that runs trench parallel shows the location of synthetic sources used in the composite. Unlike subsequent figures, there are no recorded earthquakes greater than M 6 with a thrust focal mechanism to display to illustrate seismicity around the megathrust

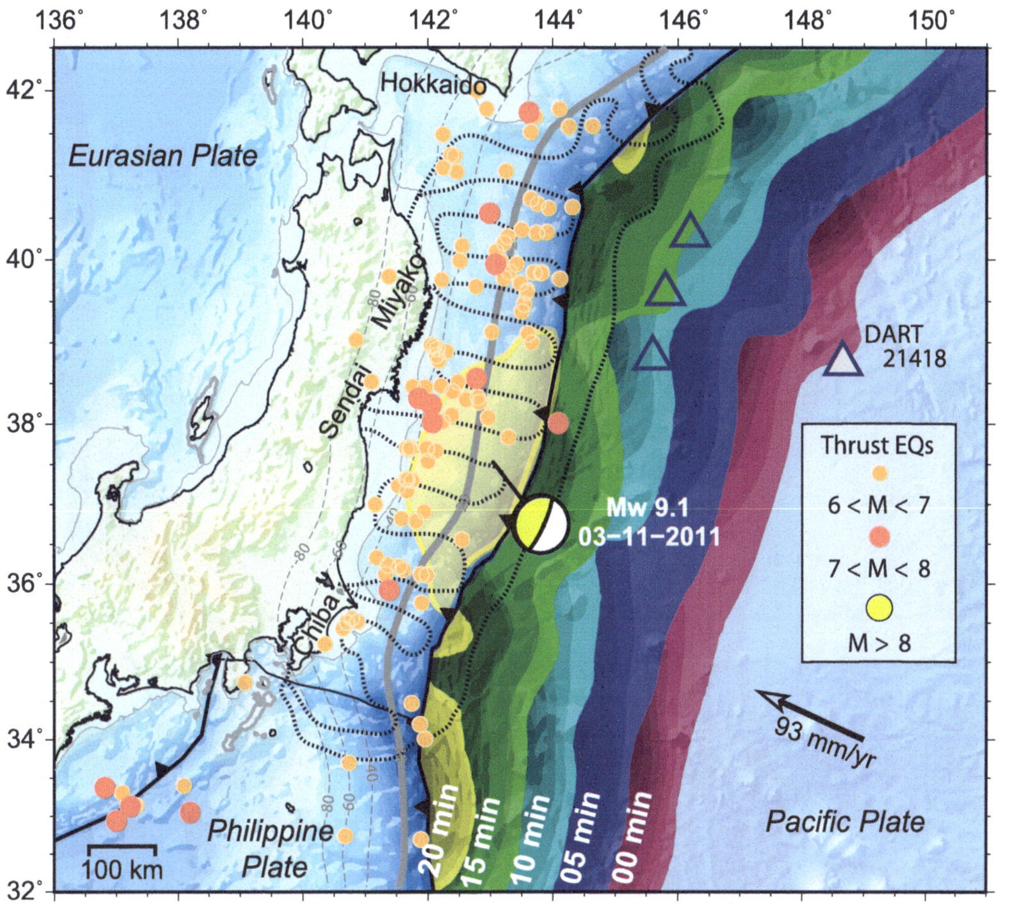

Figure 3

Composite view of northern Japan early warning zones. Magenta, navy, aqua, green, and yellow shading refers to 0 min, 5-min, 10-min, 15-min, and 20-min lead times, respectively. Arrows indicate the plate motion between the Pacific and Eurasian plates using MORVEL-2010 motions (DeMets et al. 2010) and gray triangles show approximate locations of DART gauges. The thick, gray line that runs trench parallel shows the location of synthetic sources used in the composite. Small dots indicate seismicity recorded in the Harvard CMT catalog over the past 25 years. Yellow shaded region is the approximate rupture extent of the 11 March 2011 Tohoku earthquake (Wei et al. 2014). Hollow triangles indicate DART gauges deployed following the 2011 Tohoku tsunami but have since been decommissioned. Black dashed line outlines the cabled S-NET array

warning zones from 22 synthetic trench-parallel events along the megathrust interface. The megathrust interface has a fairly shallow inferred dip, placing the synthetic event boundary close to the continental shelf. Each lead time zone (0-, 5- and 10-mins) highlighted in Fig. 2 is composed by first determining the warning zone on each individual model, as shown in Fig. 1 and then combining each zone together for a regional view.

The best zone for near-field deep-ocean instrumentation extends from 44.5° to 47.5°N where a 10-min lead time is possible for local events. To the south of this zone, land extends further along the interface, precluding shallow earthquakes from displacing enough of the water column be tsunamigenic. Additionally, the continental shelf's width tapers, locally affecting the wave velocity. To the north, the mouth of the Strait of Juan de Fuca inhibits the shoaling and slowing of the tsunami, leading to a quicker inundation at the opening of the strait. Further north along Vancouver Island, the ability to include near-field open ocean datasets declines as the distance to the coast diminishes. Outside of the 10-min warning, the 5-min zone extends further

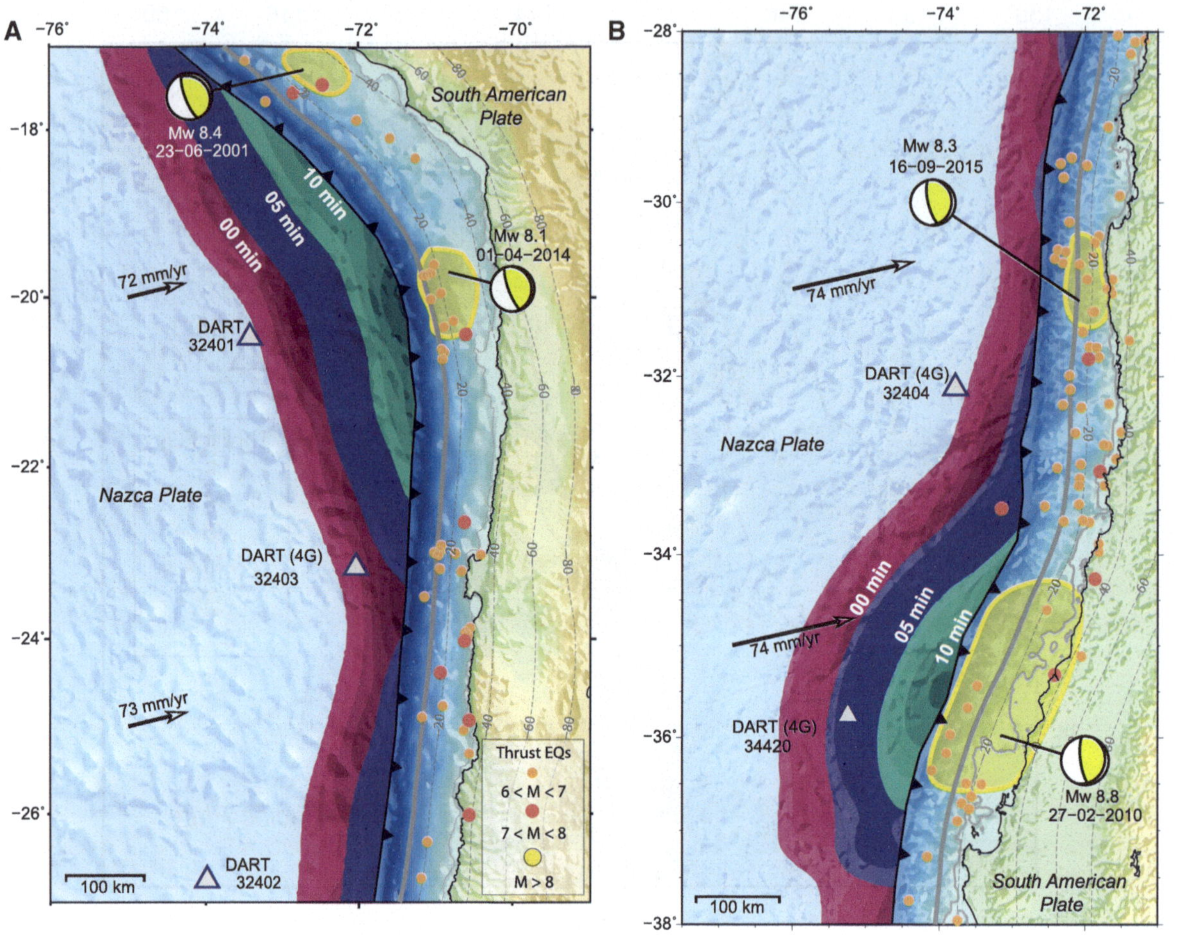

Figure 4

Composite view of northern and central Chile early warning zones. Magenta, navy, and aqua shading refers to 0 min, 5-min, and 10-min lead times, respectively. Arrows indicate the plate motion between the Nazca and South American plates using MORVEL-2010 motions (DeMets et al. 2010) and gray triangles show approximate locations of DART gauges. The thick, gray line that runs trench parallel shows the location of synthetic sources used in the composite. Small dots indicate seismicity recorded in the Harvard CMT catalog over the past 25 years. **a** Northern Peru–Chile trench: yellow polygons show rupture zones from the 23 June 2001 Peru earthquake (Bilek and Ruff 2002) and the 1 April 2014 Iquique earthquake (Lay et al. 2014). **b** Central Peru–Chile trench: yellow polygons show rupture zones from the 27 February 2010 Maule earthquake (Moreno et al. 2010) and 16 September 2015 Illapel earthquake (Williamson et al. 2017)

towards the open ocean and can accommodate a wider region. Localized, near-field tsunami early warning that uses direct tsunami observations in Cascadia are likely limited to this region highlighted in green on Fig. 2. Areas to the south where the shelf tapers do not have the ability for such localized warnings if a rupture were to occur in the immediate vicinity. However, for regional events—such as a Cascadia rupture that is dominantly to the north on the megathrust, the zoning that allows for a feasible tsunami warning system grows substantially.

Three regional DART gauges are included in the composite image to illustrate current operational gauges (DART 46149 and DART 46404) and the DART 4G sensor (DART 46451) which is currently in a testing phase. The 4G gauge's currently deployed location has the ability to offer a larger lead time in warnings along the coast from Washington to Vancouver Island, than conventional DART, aiding in assessment and evacuation efforts. Instrumentation placed further north of DART 46451 in this same trench-proximal locality would be able to not only

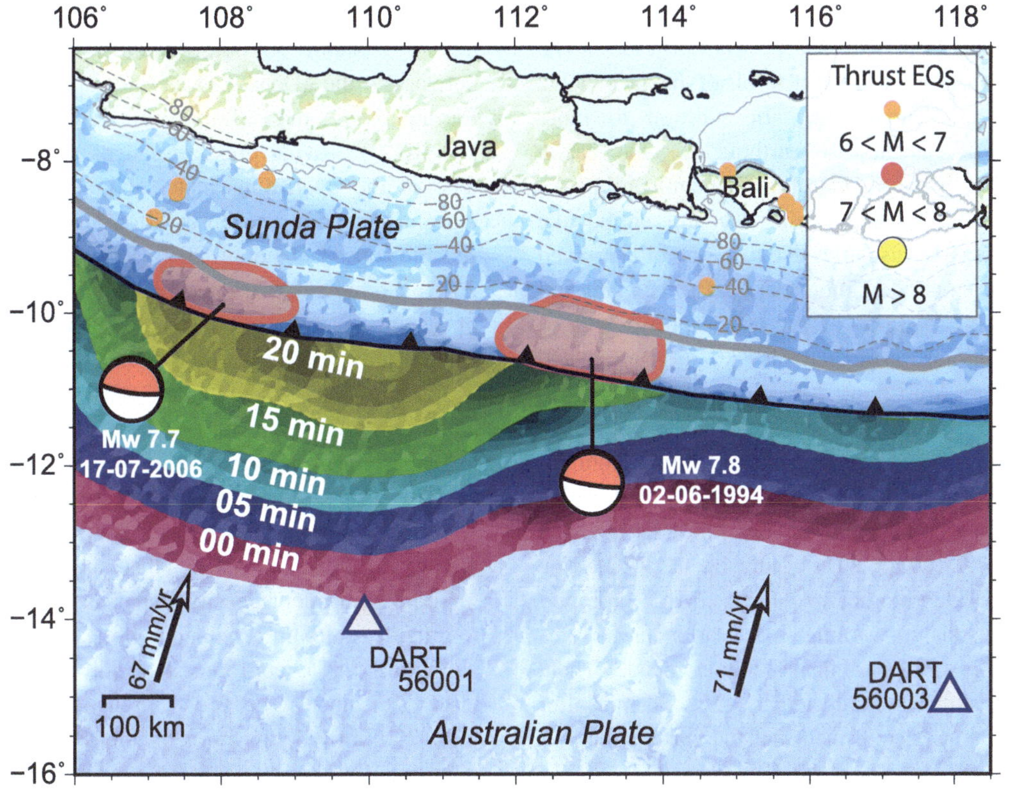

Figure 5

Composite view of Java early warning zones. Magenta, navy, aqua, green, and yellow shading refers to 0 min, 5-min, 10-min, 15-min, and 20-min lead times, respectively. Arrows indicate the plate motion between the Australian and Sunda plates using MORVEL-2010 motions (DeMets et al. 2010) and gray triangles show approximate locations of DART gauges. The thick, gray line that runs trench parallel shows the location of synthetic sources used in the composite. Small dots indicate seismicity recorded in the Harvard CMT catalog over the past 25 years. Pink polygons show the rupture extents from the 2 June 1994 Java tsunami earthquake (Bilek and Engdahl 2007) and the 17 July 2006 Pangandaran tsunami earthquake (Ammon et al. 2006)

provide advanced warning to coastal regions in the Pacific Northwest, but also provide the same mid-and far-field assessments as the older model DART gauges.

In the past decade, the Canadian North-East Pacific Underwater Networked Experiments (NEPTUNE) deployed a cabled network that among other geophysical pursuits, transmits pressure data in real time and can be used for tsunami forecasting (Barnes et al. 2013; Rabinovich and Eblé 2015). The network sits on the accretionary prism offshore from Vancouver Island and provides an intermediate dataset between open-ocean sensors and coastal tide gauges. This network was tested as it recorded the passing of the tsunami generated from the 30 September 2009 Samoa earthquake (M_w 8.1) as it propagated towards the coast of the Pacific Northwest (Thomson et al. 2011). This additional cabled dataset provides useful near-field information that could be integrated into a local tsunami early warning program outside of the DART framework particularly for events affecting the northern portion of the subduction zone.

4.2. Japan

Japan has a strong tradition of tsunami research and hazard resilience, built through centuries of observing local and distantly generated tsunami. The linkage between strong shaking from an earthquake and coastal evacuation from an expected tsunami led to the creation of Japan's first tsunami warning center in 1941 (Bernard and Titov 2015). This first center in

Sendai detected earthquakes to determine if evacuation was necessary in nearby coastal cities. The program was later expanded to all of Japan in 1952 (Bernard and Titov 2015). In addition to more characteristic megathrust earthquakes, the Japan trench has also hosted at least one tsunami earthquake in its past. The 1896 Sanriku tsunami earthquake caused extensive casualties due to a lack of perceived shaking on land (Kanamori 1972). The tsunami had estimates of 25–38 m of run-up despite a surface wave-magnitude estimated at 7.0 (Tanioka and Sataka 1996; Mori et al. 2011).

Japan's tsunami warning program was put to test on March 11, 2011 during the devastating Tohoku-Oki earthquake and tsunami. The M_w 9.0 earthquake, rupturing at 14:46:24 local time was initially assessed as a M_w 7.9 (Ozaki 2011; Tang et al. 2012). The Japanese Meteorological Agency (JMA) promptly issued a tsunami warning for nearby prefectures 3 min after the earthquake commenced. Then, W-phase estimates conducted by JMA 20 min after the event's onset confirmed that the earthquake was much larger-closer to a 9.0 (Ritsema et al. 2012). However, the first tsunami models of the event were not available during the initial tsunami warning, but rather were created an hour after the fact, once the tsunami reached the nearest open-ocean gauge (Wei et al. 2014). By this point, the first waves had already reached the coast and additional tsunami information was then put forth for tsunami forecasts for Hawaii and the far-field. It is also important to note this was the first tsunami to be modeled in real time for its far-field inundation hazard (Titov et al. 2016b) through the Pacific Tsunami Warning System. The possible lead times achieved with the data present in 2011 as well as more recent datasets are shown in Fig. S1.

After the 2011 tsunami, three DART gauges were temporarily deployed off the coast of Japan. The more recent, but smaller 2012 M_w 7.3 Japan earthquake and tsunami benefited from the proximal DART gauges sitting closer to the trench, allowing for a shorter time between generation and observation (Bernard et al. 2014). While those three DART sensors have now been removed, a cabled network of 154 seafloor pressure gauges, called S-net, spanning the overriding plate has been completed (Baba et al. 2014; Rabinovich and Eblé 2015). This network's aim is to provide real-time pressure measurements to aid in tsunami warning. Another set of cabled networks, the Dense Ocean-floor Networks for Earthquakes and Tsunamis (DONET-1 and DONET-2) provide pressure gauges and seismometers on the overriding plate, and aid in real-time tsunami detection and observation along the Nankai seismic zone (Baba et al. 2014; Rabinovich and Eblé 2015). Combined, these networks have the possibility of capturing deformation and tsunami generation from locally sourced earthquakes.

Taking into account the density of pressure gauges on the overriding plate as well as nearby DART gauges, the Japan trench provides one of the best locations for the implementation of an early warning system, as shown in Fig. 3. Nowhere else in the world has a network that approaches their offshore instrument density. Japan is already poised to provide rapid direct tsunami observations, possibly aiding in future event forecasts. Additionally, Japan's large distance between trench and coast (100–200 km) along most of the Pacific-Eurasian plate interface facilitates a greater area of potential site locations on the subducting plate than other regions. Even along some parts of the subduction zone, where the trench to shoreline distance is narrower, the presence of a shallow and expansive continental shelf can also effectively slow the oncoming wave, allowing for effective open-ocean instrumentation.

One of the best examples of this is near the Chiba peninsula, east of Tokyo, shown in Fig. 3. The yellow shaded region offshore indicates the instrumentation zone for up to 20 min of lead time before inundation. The strong shelf south of Sendai also acts to curtail the tsunami wave. As a contrast, the tapering and disappearance of the shelf north of Sendai near Miyako leads to slightly less lead time, about 15 min. Along the entire strike of the trench from Chiba peninsula to Hokkaido, a 10- to 15-min lead time is possible. The presence of a dense cabled network also allows for an earlier recording of the tsunami than can be achieved through instruments on the subducting plate, east of the trench. We highlight the warning times achieved for cabled networks on the overriding plate in Fig. S2.

4.3. Peru–Chile

Of the regions analyzed in this study, the Peru–Chile trench is the least feasible for near-field early warning of this type. The region has a relatively fast plate convergence and a history of great ($M_w > 8$) tsunamigenic earthquakes. The 1960 Valdivia earthquake and tsunami originating off the coast of Southern Chile is the largest instrumentally recorded earthquake (M_w 9.5) generating a trans-oceanic tsunami with local maximum run-up heights ranging from 10 to 25 m (NGDC 2017). Within the 1960 rupture zone was the 2016 M_w 7.7 Chiloé earthquake (Ruiz et al. 2017; Moreno et al. 2018). Additional contemporary earthquakes include the 2010 M_w 8.8 Maule earthquake and the 2014 M_w 8.1 Iquique earthquake, both of which generated tsunami. The most recent addition to the Peru–Chile trench sequence is the 2015 M_w 8.3 Illapel earthquake and tsunami. All three of these recent events have included open-ocean DART derived waveforms in conjunction with seismic and/or geodetic observations for detailed analysis of the fault slip post-event, either through time series inversion (Yue et al. 2014; Gusman et al. 2015; Williamson et al. 2017) or forward tsunami modeling for model validation (Heidarzadeh et al. 2016; Li et al. 2016). However, the capability of real-time near-field modeling has not yet been explored in great detail.

Near-field instrumentation feasibility for this region is displayed in Fig. 4. Unlike Cascadia and Japan, the average distance between trench and coastline is small at under 75 km. This shortens the response time before inundation as it places the shallow seismic interface close to shore. The linear coastal morphology with limited and narrow continental shelves does not slow the coastal tsunami propagation as it does elsewhere. Most of the trench only has 5 min or less in lead time, which exceptions near the bend in the subduction zone around the rupture area near the Iquique tsunami, and further south in the former Maule rupture area. This makes most of the region prohibitive of open-ocean early-warning. However, the narrow submarine zone means this area is more advantageous for the inclusion of geodetic datasets to aid in early warning.

The Chilean agency that oversees tsunami data and warnings, the Hydrographic and Oceanic Service of the Chilean Navy (SHOA), has deployed three DART 4G gauges in the near-trench environment. The southernmost gauge, DART 34420, is within a 5-min warning range; however, the gauge immediately to the north (DART 32404), placed just south of the Illapel rupture zone is in the 0-min warning zone. The third gauge, located just south of the Iquique rupture zone, while close to the trench, is also just out of range. It is important to note that these three new sensors, even if not currently suited for direct near-field warning, will provide more open ocean tsunami observations along the strike of the Peru–Chile trench. Before their emplacement, the entire region was served by three older DART sensors (DARTs 32402, 32401, and 32412) located in northern Chile and more distal sensors near Ecuador and Central America. The inclusion of this new dataset, while not ideal for near-field community warnings, could be instrumental in warnings along Chile's extensive coastline outside of the near-field range. The array of DART gauges also provides a useful dataset for far-field warnings affecting Australia and New Zealand.

4.4. Java-Bali, Indonesia

Acting as a catalyst for a global approach to tsunami awareness and forecasting, the 2004 Sumatra earthquake illuminated the possibility of large megathrust earthquakes and tsunami generated in the Indian Ocean basin. Additional $M > 8$ earthquakes have continued to rupture along the southern extent of Sumatra in the years following the 2004 Boxing Day event. Despite their large magnitude, not every large earthquake in the region has generated a hazardous tsunami, making warnings more difficult. Examples include the 2005 Nias earthquake (M_w 8.6). This event, while generating a tsunami that did damage some coastal communities, did not bring the impact that would have been expected of an earthquake rupturing in this region—given the recent 2004 Boxing day earthquake (Song 2007; Borrero et al. 2011). The distinction between hazardous and non-hazardous activity can be difficult to quickly determine remotely without a direct observation. In addition to being a tectonically active and complex region, half of Indonesia's population resides on the island of Java. The region surrounding Java has not had a great megathrust rupture in recent history but it has had smaller, damaging tsunami earthquakes.

The 1994 Java tsunami earthquake (M_w 7.6) killed over 250 people and was the first thrust mechanism earthquake listed along that region of the Java subduction zone (Abercrombie et al. 2001). It occurred between 16 and 20 km depth (highlighted in Fig. 5) at a far distance from the shoreline. The more recent 2006 Pangandaran tsunami earthquake (M_w 7.8) killed over 600 people. It was particularly deadly because the earthquake was not well felt on land, limiting the reaction time for communities to evacuate from the coastline (Fritz et al. 2007; Ammon et al. 2006). Both events are rarities—the region produces a high number of normal faulting earthquakes which were less productive in generating tsunami (however, tsunami can be generated from normal faulting events). However, the uncommon occurrence of large thrust events—particularly tsunamigenic events, may give communities a false sense of security up until the next earthquake hits.

The region spanning westward from Java towards Bali poses a difficult challenge for some early warning methodologies. The distance from possible shallow source to land-based receivers is high (> 200 km) and the dip of the subducting plate is steep enough that the entirety of the expected seismogenic zone sits offshore, with megathrust interface depths under land reaching upwards of 80 km. This distance poses a far more difficult challenge when using static geodetic (GPS, InSAR) datasets to resolve deformation and tsunami threat than when compared to other regions, such as the Peru–Chile trench. However, this large distance also increases the travel time between source and shoreline for shallow events making it an excellent candidate for tsunami early warning. The region offshore of Pangandaran, just east of the 1996 tsunami earthquake has the opportunity for a near-field tsunami warning system that could be fortuitous for central Java. Synthetic unit sources in this region line-up well with past tsunami earthquakes (Fig. 5). Within central Java, open-ocean datasets could allow for up to 15–20 min of lead time. The main factor that allows for tsunami early warning to be more feasible along central Java and less feasible further east towards Bali is the increase in continental shelf width near Pangandaran. The absence of shallow coastal structures offshore Bali and the rest of the Lesser Sunda Islands increases tsunami arrival time in comparison. Current instrumentation is limited to two Australian owned DART stations (DART 56001 and 56003). These gauges are located too far to be effective for Indonesian early warning.

Unlike other regions examined in this study where the eventual open-ocean propagation of the tsunami remains unhindered for thousands of kilometers, a near-field tsunami warning system for this region adds additional information that can be used to issue warnings for the nearby northwestern shores of Australia. In the event of a larger tsunami stemming from a large Sumatra-style megathrust earthquake, additional infrastructure beyond Australia's two local DART gauges could provide a quicker initial assessment for the country's northern coastline.

5. Discussion

Because the location and structure of the next big earthquake is unknown, global models used for assessing tsunami hazard always need to take into account what model assumptions are made. In this paper, two of the biggest assumptions are in the synthetic event source location and the amount of data that can be efficiently incorporated. We also track the crest of the unit source's tsunami wave, both towards the open-ocean and the coast, which at what is expected as peak inundation (excluding reflection), instead of initial inundation. Of these assumptions, the tsunami source location has the largest effect on tsunami travel time and thus the results presented. One end member scenario for tsunami early warning is an event rupturing proximal to the coastline with an immediate or close to immediate inundation. The other end member (for megathrust events) is a near-trench rupture typical of tsunami earthquakes. Rather than make a judgement call on the potential future tsunamigenic sources for each of our study zones, we adopted a streamlined approach of placing all of our sources at the 15 km depth contour of the subducting slab's assumed geometry. While some may argue that a particular region usually ruptures in a certain location, the earthquake catalog of large megathrust events is not so well defined in any given area that we can definitively point to a local characteristic earthquake that can be confidently used for modeling future hazard. Our general assumed events allow us to focus on the early-warning potential for the region of the megathrust that has a high tsunami hazard. The near-trench environment is also less likely to be resolved form coastal geodetic datasets due to the large distance, lending to the threat of misestimates of tsunami hazard in special tsunami earthquake cases.

We recognize that not all tsunamigenic events occur on the megathrust. Our general source location used for all our regional models also is placed in a way that we can extrapolate the effect of splay faulting on the fore-arc by focusing less on the features any particular faulting pattern and more on the tsunami travel time for a simple disturbed water column. However, steeply dipping splay faults, if activated, have the potential to contribute to large amounts of vertical deformation of the seafloor as has been extrapolated for the Nankai trough (Moore et al. 2007). Splay faults of this nature, located in the accretionary wedge, have also been identified along the Cascadia subduction zone although their future activity is difficult to constrain (Geist and Yoshioka 1996). Tsunami generated on the outer-rise, while also less common, poses a threat for tsunami warning. Recent outer-rise events include 2007 Kuril M_w 8.1 event (Fujii and Satake 2008; Rabinovich et al. 2008). The greater distance between outer rise sources and most coastlines when compared to megathrust events means an extended time before inundation is likely. This should facilitate some of the greatest lead times possible for early warning. Other sources not considered in this work include non-seismically generated tsunamis—the most common of which are landslide-generated tsunamis. Without a strong seismic signature or magnitude to relate to, landslide-generated tsunamis require a direct tsunami observation for an effective tsunami warning.

A second important factor to consider for early-warning feasibility is the timely assimilation of observed data into forecast models. This is dependent on how much data are used. In this study we use the first one-quarter of a tsunami waveform as our threshold for data incorporation. While this is a smaller amount of data than what is currently used to forecast tsunami warnings, the extra error associated with less data is at the trade-off of a quicker warning to affected communities when time is a restricting factor. The use of one-quarter wave periods was assessed for the 2015 Illapel, Chile tsunami with the goal of determining the minimum quantity of open-ocean data required for the earliest possible accurate tsunami forecast (Tang et al. 2016). In this study, the use of the first one-quarter of the recorded waveform was effectively inverted to determine a source model that could be applied for near-field early warning forecasts. This is an abbreviation of the half-wave model (Titov et al. 2016b), which can also be used and yields a slight improvement to fit. The full-wave inversion provides additional unique information, but only a marginally better forecast at the expense of a longer wait time between source and final recording.

The amount of data used in a rapid inversion also affects the processing required. When handling data derived from the open-ocean, tidal signals need to be removed to provide a meaningful dataset. The uncertainty involved in de-tiding datasets should be low enough that a reasonable forecast can still be made. In Percival et al. (2015), the de-tiding of tsunami signals is assessed with synthetic data to view the degree of error from using different filtering methods as well as differing quantities of data. An important takeaway from the study is that while the best results are attained by using over one full wavelength of data, methodologies that incorporated only one-quarter of the wavelength did as sufficient of a job at resolving the unknown slip parameters as half- and three-quarter wavelength solutions. While, like Tang et al. (2016) showed, the best possible solutions are attained with the most data, the increased time needed did not substantially change the final results. In terms of early warning, a slight change in the forecast solution at the cost of tens of minutes of warning time may not always be substantiated.

An additional factor that has been omitted from the work thus far is commentary on the cost of deploying new instrumentation in the open-ocean. The deployment and maintenance on any instrument, including ship-time for servicing can be substantial.

We do not focus on the deployment logistics of adding new stations further as it is outside the scope of this paper. As mentioned in Bernard and Titov (2015) approximated costs of new DART stations can reach $0.5 million per station. Cabled observatories, like those found in Japan have a starting cost of $500 million for the entire network. We do note that by focusing on which regions can facilitate early-warnings, we can reduce the total number of stations that would need to be deployed to be as cost-effective as possible.

6. Conclusion

Through this study, we explore four regions to examine where a localized near-field tsunami system would be effective. Ideally this system would incorporate near-trench, open-ocean, tsunami observations, through pressure gauges such as the currently deployed DART system or through cabled pressure stations. The advantages of direct tsunami measurements over seismic and geodetically focused approaches are listed below. First, tsunami datasets do not rely on the accuracy of an earthquake magnitude. This can be particularly important for tsunami earthquakes and large earthquakes with long rupture durations. Second, tsunami-based models do not need to understand the relationship between the transfer of earthquake energy to tsunami energy. This is crucial, because estimates in earthquake energy can vary greatly in the first hour after rupture and so little of that energy (0.1%) is transferred to the ocean (Titov et al. 2016a). Third, tsunami-based models can directly observe the oncoming threat for forecasting purposes without making assumptions on the fault geometry. Fourth, tsunami-based models can incorporate non-seismic contributions to the tsunami wave such as submarine mass failures. As Bernard and Titov (2015) have noted—20% of tsunamis are generated from non-seismic sources.

The main limitation to this direct observation approach is having enough instrumentation to cover high-risk areas near the trench. This is both a resource and financial limitation. If newer DART systems or comparable open-ocean datasets prove robust and yield fruitful tsunami information despite strong shaking from the earthquake, regions like the Cascadia subduction zone are viable for near-field tsunami warning. Other regions, such as proximal to Java, Indonesia, do not have the near-field infrastructure but could benefit from an early warning of this variety—particularly due to their contemporary history of tsunami earthquakes. Chile, while having near-trench open-ocean tsunami data, is poorly suited for local early warning, given its subducting fault geometry and prohibitive continental shelf geometry. However, those observations are still useful for mid-field warnings within the Peru–Chile trench. Based on this study, our primary findings are that while some regions are ill-suited for this type of early warning, other localities could incorporate direct tsunami observations into their hazard forecasts with enough lead time to be effective for coastal community emergency response. Particular coastal areas with associated high potential tsunami lead times could have a localized system which may require only one to two open-ocean stations.

We do not advocate to dismantle or delay current tsunami warnings and advisories already in place and rooted in public knowledge for coastal communities. However, the vast majority of potential tsunami victims are located in the near-field coastal areas. Improvements on near-field tsunami warnings can save lives (Titov et al. 2016a). The inclusion of additional tsunami data can create an added push for further evacuation if needed in the case of underestimations of the perceived hazard. It can also add confidence to accurate warnings already in place and act as an additional dataset for mid- and far-field forecasts.

Acknowledgements

This research was supported through State Funds through Georgia Tech to AVN. Figures were generated using Generic Mapping Tools from Wessel et al. (2013). Training on usage and implementation of the MOST tsunami model was made possible through the assistance of V. Titov, D. Arcas, and Y. Wei at the Pacific Marine Environmental Laboratory.

References

Abercrombie, R. E., Antolik, M., Felzer, K., & Ekström, G. (2001). The 1994 Java tsunami earthquake: Slip over a subducting seamount. *Journal of Geophysical Research: Solid Earth, 106*(B4), 6595–6607. https://doi.org/10.1029/2000JB900403.

Ammon, C. J., Kanamori, H., Lay, T., & Velasco, A. A. (2006). The 17 July 2006 Java tsunami earthquake. *Geophysical Research Letters*. https://doi.org/10.1029/2006GL028005.

Atwater, B. F. (1992). Geologic evidence for earthquakes during the past 2000 years along the Copalis River, southern coastal Washington. *Journal of Geophysical Research: Solid Earth, 97*(B2), 1901–1919. https://doi.org/10.1029/91JB02346.

Baba, T., Takahashi, N., & Kaneda, Y. (2014). Near-field tsunami amplification factors in the Kii Peninsula, Japan for Dense Oceanfloor Network for Earthquakes and Tsunamis (DONET). *Marine Geophysical Research, 35*(3), 319–325. https://doi.org/10.1007/s11001-013-9189-1.

Barnes, C. R., Best, M. M., Johnson, F. R., Pautet, L., & Pirenne, B. (2013). Challenges, benefits, and opportunities in installing and operating cabled ocean observatories: Perspectives from NEPTUNE Canada. *IEEE Journal of Oceanic Engineering, 38*(1), 144–157.

Bernard, E. N., & Titov, V. V. (2015). Evolution of tsunami warning systems and products. *Philosophical Transactions of the Royal Society A, 373*(2053), 20140371.

Bernard, E. N., Wei, Y., Tang, L., & Titov, V. V. (2014). Impact of near-field, deep-ocean tsunami observations on forecasting the 7 December 2012 Japanese tsunami. *Pure and Applied Geophysics, 171*(12), 3483–3491. https://doi.org/10.1007/s00024-013-0720-8.

Bilek, S. L., & Engdahl, E. R. (2007). Rupture characterization and aftershock relocations for the 1994 and 2006 tsunami earthquakes in the Java subduction zone. *Geophysical Research Letters, 34,* L20311. https://doi.org/10.1029/2007GL031357.

Bilek, S. L., & Ruff, L. J. (2002). Analysis of the 23 June 2001 $M_w = 8.4$ Peru underthrusting earthquake and its aftershocks. *Geophysical Research Letters, 29*(20), 1960. https://doi.org/10.1029/2002GL015543.

Borrero, J. C., McAdoo, B., Jaffe, B., Dengler, L., Gelfenbaum, G., Higman, B., et al. (2011). Field survey of the March 28, 2005 Nias-Simeulue earthquake and tsunami. *Pure and Applied Geophysics, 168*(6–7), 1075–1088.

DeMets, C., Gordon, R. G., & Argus, D. F. (2010). Geologically current plate motions. *Geophysical Journal International, 181*(1), 1–80. https://doi.org/10.1111/j.1365-246X.2009.04491.x.

Fritz, H. M., Kongko, W., Moore, A., McAdoo, B., Goff, J., Harbitz, C., et al. (2007). Extreme runup from the 17 July 2006 Java tsunami. *Geophysical Research Letters*. https://doi.org/10.1029/2007GL029404.

Fujii, Y., & Satake, K. (2008). Tsunami sources of the November 2006 and January 2007 great Kuril earthquakes. *Bulletin of the Seismological Society of America, 98*(3), 1559–1571.

Geist, E., & Yoshioka, S. (1996). Source parameters controlling the generation and propagation of potential local tsunamis along the Cascadia margin. *Natural Hazards, 13*(2), 151–177.

Gusman, A. R., Murotani, S., Satake, K., Heidarzadeh, M., Gunawan, E., Watada, S., et al. (2015). Fault slip distribution of the 2014 Iquique. *Chile, earthquake estimated from ocean-wide tsunami waveforms and GPS data, Geophysical Research Letters, 42*(4), 1053–1060. https://doi.org/10.1002/2014gl062604.

Gusman, A. R., Tanioka, Y., Sakai, S., & Tsushima, H. (2012). Source model of the great 2011 Tohoku earthquake estimated from tsunami waveforms and crustal deformation data. *Earth and Planetary Science Letters, 341,* 234–242.

Hayes, G. P., Wald, D. J., & Johnson, R. L. (2012). Slab1.0: A three-dimensional model of global subduction zone geometries. *Journal of Geophysical Research, 117,* B01302. https://doi.org/10.1029/2011JB008524.

Heidarzadeh, M., Murotani, S., Satake, K., Ishibe, T., & Gusman, A. R. (2016). Source model of the 16 September 2015 Illapel, Chile, M_w 8.4 earthquake based on teleseismic and tsunami data. *Geophysical Research Letters, 43,* 643–650. https://doi.org/10.1002/2015GL067297.

Hossen, M. J., Cummins, P. R., Dettmer, J., & Baba, T. (2015). Time reverse imaging for far-field tsunami forecasting: 2011 Tohoku earthquake case study. *Geophysical Research Letters, 42,* 9906–9915. https://doi.org/10.1002/2015GL065868.

Kanamori, H. (1972). Mechanism of tsunami earthquakes. *Physics of the Earth and Planetary Interiors, 6*(5), 346–359. https://doi.org/10.1016/0031-9201(72)90058-1.

Kanamori, H., & Kikuchi, M. (1993). The 1992 Nicaragua earthquake: A slow tsunami earthquake associated with subducted sediments. *Nature, 361*(6414), 714–716. https://doi.org/10.1038/361714a0.

Lay, T. (2015). The surge of great earthquakes from 2004 to 2014. *Earth and Planetary Science Letters, 409,* 133–146. https://doi.org/10.1016/j.epsl.2014.10.047.

Lay, T., Yue, H., Brodsky, E. E., & An, C. (2014). The 1 April 2014 Iquique, Chile, M_w 8.1 earthquake rupture sequence. *Geophysical Research Letters, 41,* 3818–3825. https://doi.org/10.1002/2014GL060238.

Li, L., Lay, T., Cheung, K. F., & Ye, L. (2016). Joint modeling of teleseismic and tsunami wave observations to constrain the 16 September 2015 Illapel, Chile, Mw 8.3 earthquake rupture process. *Geophysical Research Letters, 43,* 4303–4312. https://doi.org/10.1002/2016GL068674.

Melgar, D., LeVeque, R. J., Dreger, D. S., & Allen, R. M. (2016). Kinematic rupture scenarios and synthetic displacement data: An example application to the Cascadia subduction zone. *Journal of Geophysical Research: Solid Earth, 121*(9), 6658–6674.

Moore, G. F., Bangs, N. L., Taira, A., Kuramoto, S., Pangborn, E., & Tobin, H. J. (2007). Three-dimensional splay fault geometry and implications for tsunami generation. *Science, 318*(5853), 1128–1131.

Moreno, M., Li, S., Melnick, D., Bedford, J. R., Baez, J. C., Motagh, M., et al. (2018). Chilean megathrust earthquake recurrence linked to frictional contrast at depth. *Nature Geoscience, 11,* 285.

Moreno, M., Rosenau, M., & Oncken, O. (2010). 2010 Maule earthquake slip correlates with pre-seismic locking of Andean subduction zone. *Nature, 467*(7312), 198. https://doi.org/10.1038/nature09349.

Mori, N., Takahashi, T., Yasuda, T., & Yanagisawa, H. (2011). Survey of 2011 Tohoku earthquake tsunami inundation and runup. *Geophysical Research Letters*. https://doi.org/10.1029/2011GL049210.

National Geophysical Data Center database (NGDC). (2017). https://www.ngdc.noaa.gov. Accessed 19 Sept 17.

Newman, A. V., Hayes, G., Wei, Y., & Convers, J. (2011). The 25 October 2010 Mentawai tsunami earthquake, from real-time discriminants, finite-fault rupture, and tsunami excitation.

Geophysical Research Letters. https://doi.org/10.1029/2010GL046498.

Newman, A. V., & Okal, E. A. (1998). Teleseismic estimates of radiated seismic energy: The E/M 0 discriminant for tsunami earthquakes. *Journal of Geophysical Research: Solid Earth, 103*(B11), 26885–26898. https://doi.org/10.1029/98jb02236.

Okal, E. A. (2015). The quest for wisdom: Lessons from 17 tsunamis, 2004–2014. *Philosophical Transactions of the Royal Society A, 373*(2053), 20140370. https://doi.org/10.1098/rsta.2014.0370.

Ozaki, T. (2011). Outline of the 2011 off the Pacific coast of Tohoku Earthquake (M_w 9.0). *Earth, Planets and Space, 63*(7), 57.

Percival, D. B., Denbo, D. W., Eblé, M. C., Gica, E., Huang, P. Y., Mofjeld, H. O., et al. (2015). Detiding DART® buoy data for real-time extraction of source coefficients for operational tsunami forecasting. *Pure and Applied Geophysics, 172*(6), 1653–1678.

Polet, J., & Kanamori, H. (2000). Shallow subduction zone earthquakes and their tsunamigenic potential. *Geophysical Journal International, 142*(3), 684–702. https://doi.org/10.1046/j.1365-246x.2000.00205.x.

Rabinovich, A. B., & Eblé, M. C. (2015). Deep-ocean measurements of tsunami waves. *Pure and Applied Geophysics, 172*(12), 3281–3312.

Rabinovich, A. B., Lobkovsky, L. I., Fine, I. V., Thomson, R. E., Ivelskaya, T. N., & Kulikov, E. A. (2008). Near-source observations and modeling of the Kuril Islands tsunamis of 15 November 2006 and 13 January 2007. *Advances in Geosciences, 14,* 105–116.

Ritsema, J., Lay, T., & Kanamori, H. (2012). The 2011 Tohoku Earthquake. *Elements, 8*(3), 183–188.

Rogers, G., & Dragert, H. (2003). Episodic tremor and slip on the Cascadia subduction zone: The chatter of silent slip. *Science, 300*(5627), 1942–1943. https://doi.org/10.1126/science.1084783.

Ruiz, S., Moreno, M., Melnick, D., Campo, F., Poli, P., Baez, J. C., Leyton F., & Madariaga, R. (2017). Reawakening of large earthquakes in South-Central Chile: The 2016 M_w 7.6 Chiloé event. *Geophysical Research Letters. 44*(13), 6633–6640.

Satake, K., Shimazaki, K., Tsuji, Y., & Ueda, K. (1996). Time and size of a giant earthquake in Cascadia inferred from Japanese tsunami records of January 1700. *Nature, 379*(6562), 246–249. https://doi.org/10.1038/379246a0.

Satake, K., Wang, K., & Atwater, B. F. (2003). Fault slip and seismic moment of the 1700 Cascadia earthquake inferred from Japanese tsunami descriptions. *Journal of Geophysical Research: Solid Earth.* https://doi.org/10.1029/2003JB002521.

Song, Y. T. (2007). Detecting tsunami genesis and scales directly from coastal GPS stations. *Geophysical Research Letters*. https://doi.org/10.1029/2007GL031681.

Tang, L., Titov, V. V., Bernard, E. N., Wei, Y., Chamberlin, C. D., & Newman, J. C., et al. (2012). Direct energy estimation of the 2011 Japan tsunami using deep-ocean pressure measurements. *Journal of Geophysical Research: Oceans, 117*(C8), C08008.

Tang, L., Titov, V. V., Moore, C., & Wei, Y. (2016). Real-time assessment of the 16 September 2015 Chile tsunami and implications for near-field forecast. *Pure and Applied Geophysics, 173*(2), 369–387.

Tanioka, Y., & Sataka, K. (1996). Fault parameters of the 1896 Sanriku tsunami earthquake estimated from tsunami numerical modeling. *Geophysical Research Letters, 23*(13), 1549–1552. https://doi.org/10.1029/96gl01479.

Thomson, R., Fine, I., Rabinovich, A., Mihály, S., Davis, E., Heesemann, M., et al. (2011). Observation of the 2009 Samoa tsunami by the NEPTUNE-Canada cabled observatory: Test data for an operational regional tsunami forecast model. *Geophysical Research Letters, 38,* L11701. https://doi.org/10.1029/2011GL046728.

Titov, V.V., & González F.I. (1997). Implementation and testing of the method of splitting tsunami (MOST) model. US Department of Commerce, National Oceanic and Atmospheric Administration, Environmental Research Laboratories, Pacific Marine Environmental Laboratory.

Titov, V.V., Kânoğlu U., & Synolakis C. (2016b). *Development of MOST for real-time tsunami forecasting*. Diss. American Society of Civil Engineers. https://doi.org/10.1061/(asce)ww.1943-5460.0000357031 16004.

Titov, V., Rabinovich, A. B., Mofjeld, H. O., Thomson, R. E., & González, F. I. (2005). The global reach of the 26 December 2004 Sumatra tsunami. *Science, 309*(5743), 2045–2048.

Titov, V. V., Song, Y. T., Tang, L., Bernard, E. N., Bar-Sever, Y., & Wei, Y. (2016). Consistent estimates of tsunami energy show promise for improved early warning. *Pure and Applied Geophysics, 173*(12), 3863–3880. https://doi.org/10.1007/s00024-016-1312-1.

Wei, Y., Newman, A. V., Hayes, G. P., Titov, V. V., & Tang, L. (2014). Tsunami forecast by joint inversion of real-time tsunami waveforms and seismic or GPS data: Application to the Tohoku 2011 tsunami. *Pure and Applied Geophysics, 171*(12), 3281–3305. https://doi.org/10.1007/s00024-014-0777-z.

Wessel, P., Smith, W. H. F., Scharroo, R., Luis, J. F., & Wobbe, F. (2013). Generic Mapping Tools: Improved version released. *EOS Trans AGU, 94,* 409–410.

Williamson, A., Newman, A. V., & Cummins, P. R. (2017). Reconstruction of coseismic slip from the 2015 Illapel earthquake using combined geodetic and tsunami waveform data. *Journal of Geophysical Research: Solid Earth, 122*(3), 2119–2130. https://doi.org/10.1002/2016jb013883.

Witter, R. C., Zhang, Y. J., Wang, K., Priest, G. R., Goldfinger, C., Stimely, L., et al. (2013). Simulated tsunami inundation for a range of Cascadia megathrust earthquake scenarios at Bandon, Oregon, USA. *Geosphere, 9*(6), 1783–1803.

Yue, H., Lay, T., Rivera, L., An, C., Vigny, C., Tong, X., et al. (2014). Localized fault slip to the trench in the 2010 Maule, Chile $M_w = 8.8$ earthquake from joint inversion of high-rate GPS, teleseismic body waves, InSAR, campaign GPS, and tsunami observations. *Journal of Geophysical Research: Solid Earth, 119*(10), 7786–7804. https://doi.org/10.1002/2014jb011340.

(Received October 25, 2017, revised May 2, 2018, accepted May 4, 2018, Published online May 15, 2018)

Pure Appl. Geophys. 176 (2019), 3263–3275

https://doi.org/10.1007/s00024-018-2004-9

Pure and Applied Geophysics

An Evaluation of Modelled Tsunami Arrival Times

Diana J. M. Greenslade,[1] Robert I. Greenwood,[2] and Stewart C. R. Allen[1]

Abstract—The Joint Australian Tsunami Warning Centre provides forecasts of tsunami arrival times based on the Tsunami Travel Times (TTT) software. In this work, travel times from a pre-computed scenario database and TTT are evaluated through comparison with observed arrival times at tide gauges and tsunameters. It is found that travel times from the scenario database are at least as accurate as those from TTT and in most cases, more accurate. The 'best' scenario-based method is Tp20—the time of arrival of the first peak that is at least 20% of the maximum amplitude at that point.

Key words: Tsunami, tsunami warning, arrival times.

1. Introduction

The Joint Australian Tsunami Warning Centre (JATWC) is jointly operated by the Australian Bureau of Meteorology and Geoscience Australia. It provides a comprehensive, independent tsunami warning service to advise the media, public and emergency authorities of any tsunami threat to Australia and its offshore territories. JATWC tsunami warnings are issued according to a technique described in Allen and Greenslade (2010). Determination of the threat level is based on wave amplitudes near the coast from a database of pre-computed tsunami scenarios (T2; Greenslade et al. 2009, 2010).

In addition to the threat level, a key component of the JATWC tsunami warning service is the prediction of tsunami arrival times. These are based on the Tsunami Travel Times (TTT) software (Geoware 2007) which provides estimates of tsunami arrival time for locations of interest.

While TTT has been shown to perform very well, there are some benefits to be gained if the travel times could be derived from the T2 scenario database. If the scenarios can be shown to perform as well as, or better than TTT, then travel times from T2 or from a future model-based forecast system may be used to complement or replace TTT, leading to improved warning guidance. An advantage of using model-based travel times is that it is possible to extract a much wider range of travel time information from model-based forecasts, such as time of arrival of maximum wave and/or first time over a relevant warning threshold. In addition, model-based travel times would provide consistency between the underlying data used for travel times and for warning thresholds.

The aim of this work is to evaluate travel times from the T2 scenario database through comparison with observed arrival times at tide gauges and tsunameters. Travel times from TTT are also evaluated in order to provide a comparison with current operational practice at the JATWC. Section 2 presents some previous related work, and Sect. 3 describes the data set of observed travel times. Section 4 provides a description of the modelled travel times examined, both from the T2 scenario database and from TTT. Section 5 presents the results, including an analysis of the results as a function of earthquake magnitude, and Sect. 6 includes a discussion of various issues arising from this work, including some limitations to the use of model-based travel times. Finally, the conclusions are presented in Sect. 7.

2. Previous Work

The existing published literature relating to the quantitative evaluation of predicted tsunami arrival

[1] Australian Bureau of Meteorology, Science to Services, Melbourne, Australia. E-mail: diana.greenslade@bom.gov.au
[2] Australian Bureau of Meteorology, National Operations Centre, Melbourne, Australia.

times is surprisingly limited. One key piece of work is by Wessel (2009) who evaluated the TTT software against observations from the NGDC/WDS database. He found a mean travel time difference of 19 min with a standard deviation of 131 min. He concluded that the main causes of discrepancies were: (1) the definition of the epicentre as a point source, which may not represent the long ruptures seen in larger earthquakes; (2) the coarse spatial resolution of the propagation grid required for operational use; and (3) uncertainties in bathymetry data in shallow regions. In relation to the first of these points, it is worth noting that TTT can be initialised with multiple epicentres to represent a longer rupture, although this is not currently done operationally at the JATWC.

Syamsidik et al. (2015) produced 'worst-case' arrival times for a number of tsunami-prone cities in Aceh, Indonesia as an input into the development of improved evacuation procedures. Arrival times were computed from a dynamical model and defined as the first time the tsunami reaches an elevation of 0.5 m, in water of 10 m depth, as defined by the Indonesia Geophysics Climate, and Meteorological Agency (BMKG). The forecast sea level was verified against observations, but arrival times were not specifically evaluated.

In a study of the 2011 Tohoku tsunami, Cholifah and Prastowo (2017) examined the difference between observed and modelled arrival times, where the modelled arrival times were defined as the arrival time of the first peak derived through visual inspection. They found that the modelled arrival times were persistently earlier than the observed arrival times, and they attributed this to the omission of the effects of elastic loading of the seafloor by the tsunami, as proposed by, for example, Allgeyer and Cummins (2014). This issue is discussed in Sect. 6.

The definition of tsunami arrival time itself is not straightforward, particularly when dealing with observations. In an analysis of sea-level observations from the 2004 Indian Ocean event, Merrifield et al. (2005) define arrival time as the first measured increase or decrease before the first crest or trough. Rabinovich and Thomson (2007) use a similar approach but take the arrival time to be midway between the last time of 'no tsunami' and the first time of a measurable increase or decrease. Bressan and Tinti (2011) developed an automatic algorithm for the detection of tsunamis in real-time in sea-level observations. This algorithm essentially incorporates two definitions of arrival time—one based on a dynamic threshold that varies according to the level of the background signal, and one based on a filtered sea-level amplitude threshold. Both definitions are calibrated for individual locations. Within the Indian Ocean Tsunami Warning and Mitigation System (IOTWMS) there are a number of arrival times that a Tsunami Service Provider (TSP) is required to provide (IOC 2015), including time of arrival of maximum wave. Of most relevance here is 'first arrival' which is defined to be the first time the wave reaches an elevation of 2 cm, either positive or negative.

Notwithstanding the requirements for TSPs within the IOTWMS, travel times from dynamical model-based systems are not presently used extensively in operational warning centres. This is partially because TTT already provides good results, but, in addition, there are some challenges in deriving travel times from dynamical model output. For example, model output is often contaminated by 'numerical noise' which makes the detection of an appropriate signal difficult for small events. There has been some work on the use of modelled tsunami travel times for non-warning purposes, in particular for the assessment of observing networks. For example, Spillane et al. (2008) define the arrival time to be the time of the first positive peak that is greater than 20% of the maximum amplitude, and Greenslade and Warne (2012) use the arrival of the first positive peak that is larger than a given (small) threshold.

3. *Observed Travel Times*

The observed arrival times used in this study were extracted from the the NGDC/WDS Global Historical Tsunami Database (NGDC/WDC). This database includes 2606 total events (at the time of accessing, November 2017). For this study, we selected only events that occurred from 26 December 2004 onwards (179 total events), under the assumption that this is during the era when there were many more observations available per event. In addition, tide gauge data were predominantly digital rather than analogue.

The events were filtered in order to select only those that were due to earthquake sources, and we removed any that were flagged 'doubtful'. The T2 scenario database is limited to the Pacific and Indian Oceans (see next section) so any earthquake sources that were outside this domain were excluded.

For the observed arrival times, we selected only tide gauge or tsunameter observations, and we removed data for which there was no detail on the location of the observation or the time of arrival. We also eliminated locations that were outside the T2 domain.

A check was undertaken to determine the closest T2 model grid point to each observation location, and we removed those locations for which the closest T2 grid point was more than four grid points away. This could occur in very shallow regions, since the T2 model scenarios are limited to 20 m depth. Given the limited forecast range of the T2 scenarios (24 h), we also excluded any observed travel times that were longer than 24 h.

The dataset also underwent an extensive manual check. The observed arrival times were checked against distance from the source. Any significant outliers were inspected to ascertain if it was a physical effect or possible error in the dataset. The NGDC database includes references to the original documentation where possible so in this way, we were able to determine that, for example, the arrival times for one event had been inadvertently included as local time rather than UTC, and we could correct for this. We also ensured that individual locations had consistent latitude and longitude details for all events and made adjustments where necessary. All the obvious errors that we identified were provided back to NGDC.

Finally, some locations had arrival time but no travel time so we recalculated travel times for all locations. This left a total of 1298 individual observations of travel time from 50 events. A histogram of the number of observations according to event magnitude is shown in Fig. 1. It can be seen that there is a relatively wide range of magnitudes represented in the data set, ranging from $M_w = 6.5$ up to $M_w = 9.1$.

In the following analysis, we make the assumption that the observed travel times are 'the truth'. It should be noted that the observed travel times in the NGDC/WDS database are collated from a variety of sources, and there will be a variety of different ways of defining 'arrival time' from tide gauge or tsunameter data within this dataset. We do not attempt here to identify or record which technique has been used for each observation, but we do acknowledge that the observed arrival time dataset will have a certain amount of noise. We deem this level of noise to be tolerable for the purposes of this work, however, it needs to be borne in mind throughout the following analysis.

4. *Modelled Travel Times*

4.1. *Scenario Database Travel Times*

Modelled travel times in this study are derived from the T2 scenario database, which is used for the determination of threat level by the JATWC. T2 has been developed using the MOST model (Titov et al. 2016). Some relevant details of the scenario database are described here, and further details can be found in Greenslade et al. (2009) and Greenslade et al. (2010).

The main spatial domain of the scenario database covers only the Pacific and Indian Oceans and extends from 69.4°S to 66°N and 30°E to 300°E, i.e., it is not a global domain. The forecast range for each scenario is 24 h from the time of the generating earthquake. The spatial resolution is 4 arc min, which ranges from $\sim$ 7 km at the equator to $\sim$ 2.6 km at the southern boundary. Model grid points are limited to water depths greater than 20 m. The scenario time series are archived at all model grid points at 2 min intervals.

The T2 scenario database includes five earthquake magnitudes of $M_w = 7.0, 7.5, 8.0, 8.5$ and 9.0 located at 100 km intervals along subduction zones, although the $M_w = 7.0$ earthquakes are included only at locations closer to Australia. Intermediate magnitude events are derived from the existing scenarios through linear scaling of the wave amplitudes. The earthquake ruptures that initialise each scenario are each made up of a number of rupture elements of

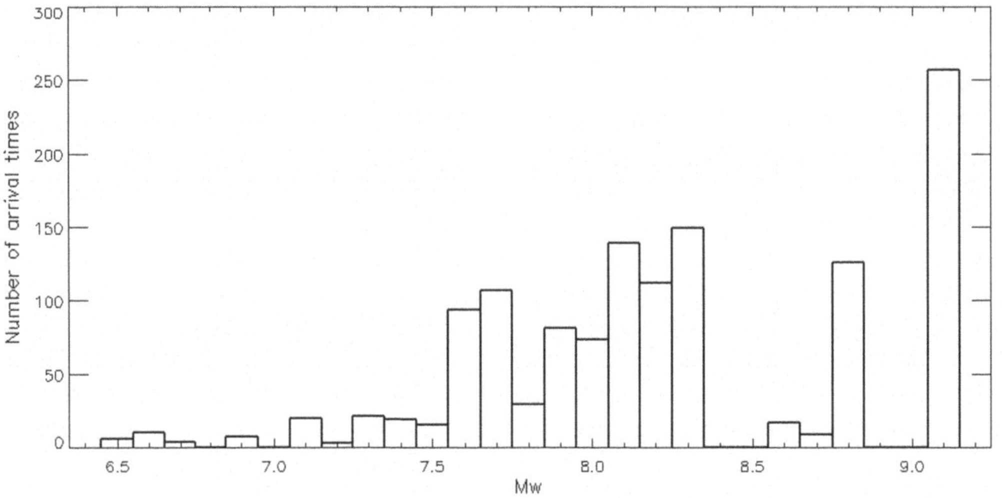

Figure 1
Number of observed arrival times as a function of event magnitude

100 km length, with larger magnitude earthquakes consisting of a larger number of rupture elements, as they typically have longer ruptures. When an earthquake occurs, the closest rupture element to the earthquake epicentre is found. Then, in order to account for the fact that rupture direction is generally not known, all scenarios that include that rupture element are used in the determination of the threat level. This means that for $M_w \le 7.7$, one scenario is used; two scenarios for $7.8 \le M_w \le 8.2$; four scenarios for $8.3 \le M_w \le 8.7$ and ten scenarios for $M_w \ge 8.8$.

We have attempted to replicate current JATWC operational usage insofar as possible, so for each of the 50 events in the observed arrival times dataset, we assess the set of scenarios that would be used operationally, rather than the single scenario that, in hindsight, was the best match for the event. For example, Allen and Greenslade (2016) found that the T2 scenario that best matched the deep-water observations for the $M_w = 9.1$ 2011 Honshu event was a $M_w = 8.5$ event scaled up, but here, we use the set of 10 $M_w = 9.0$ scenarios (scaled to $M_w = 9.1$) which include the epicentre in their ruptures.

For each event, we found the 'envelope' of scenarios that would be used operationally according to the epicentre of the event. Then the closest model grid point to the observed location was found (as long as it is within four grid points) and a range of travel times were calculated for each of the scenarios in the envelope. These different travel times were as follows:

1. *Tp* The first peak in the time series. For some cases, the first peak as defined here has a comparable amplitude to the maximum amplitude, while for other cases, there are small peaks prior to the arrival of this peak and so Tp occurs at the 'foot' of the first major peak. Some examples of this are shown below.
2. *T2cm* The first time the wave amplitude reaches 2 cm. This arrival time was examined as it is the definition of arrival time used within the IOTWMS, as noted in Sect. 2. Note that there will be fewer instances of this arrival time from the scenario database than for Tp because the model may not reach 2 cm for small magnitude events, and/or large propagation distances.
3. *T1cm* This is the same as T2cm, but with a threshold of 1 cm, in order to increase the number of possible data pairs.
4. *Tp10* This is defined as the first peak with the constraint that the amplitude needs to be > 10% of the maximum amplitude at that point. This is to eliminate numerical noise, and also to avoid instances when the first peak occurs at the 'foot' of the first major peak.

5. *Tp20* Same as Tp10 but with 20% instead of 10%.

In each case, we considered the absolute value of the wave amplitude, i.e., the wave amplitude was allowed to be either positive or negative.

For each of the five different arrival times, we also consider two options in relation to the scenario 'envelope' as described above. We consider (1) the minimum (i.e., earliest) arrival time of the envelope, and (2) the mean arrival time of the envelope. The earliest arrival will be the most conservative option, which is often preferable in operational environments. This results in ten different ways to calculate travel time from the T2 scenario database.

Figure 2 shows two examples of the derived travel times for individual scenarios. In the top panel of Fig. 2, the earliest arrival time is T1cm, with T2cm occurring immediately after it. Both of these occur prior to the first significant peak, which has an amplitude of almost 10 cm. Tp, Tp10 and Tp20 all occur at the same time, at the maximum of this first significant peak, as it is relatively large compared to the maximum amplitude in the time series. In the lower panel of Fig. 2, Tp occurs first, as it has picked up a small 'wobble' at the base of the first significant peak. This first significant peak is considerably smaller than the maximum amplitude in the time series so only Tp10 occurs at this time. Tp20 occurs at the maximum of a larger peak, several hours later. Similarly, T1cm and T2cm are also quite distinct, with T1cm occurring just prior to the first peak, and T2cm occurring just prior to the second peak. These examples serve to highlight how variable the different arrival times can be, and they are strongly dependent on the characteristics of the time series at any particular point.

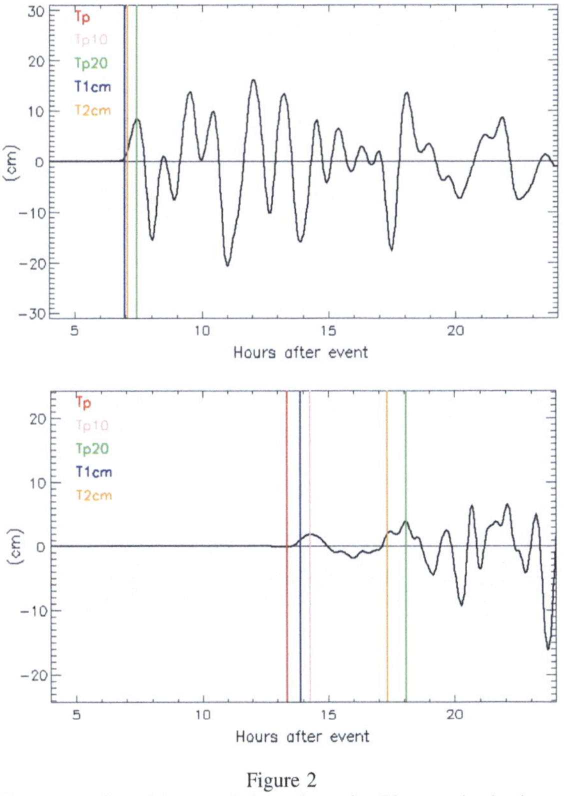

Figure 2
Two examples of the travel times from the T2 scenario database. Top panel is at Pointe La Rue, Seychelles for the $M_w = 8.7$ event on the 28th March 2005. Bottom panel is at Manakau Harbour, New Zealand for the $M_w = 9.1$ event on the 26th December 2004

4.2. *TTT Travel Times*

Within the JATWC, version 3.1 of TTT is currently implemented (Bureau National Operations Centre 2014) with a spatial resolution of 10 arc min. Whilst this is a relatively coarse resolution, it is very rapid to calculate, taking approximately 25 s for a global run. In this work, we replicate this operational implementation and also assess any differences with TTT implemented with a resolution of 5 arc min, as this is close to the spatial resolution of the T2 scenarios.

In addition, we consider two different options for the set of observation locations. Firstly, the actual lat/lons as specified in the NGDC/WDS database, and secondly, the lat/lons of the closest T2 model grid point. This will allow us to eliminate any differences between the T2 and TTT results that might occur as a result of slightly different propagation paths near the point of interest. So in summary, there are four different TTT travel times:

1. *TTT_10_exact* This is 10 arc min resolution with exact lat/lon locations. This replicates operational JATWC usage.
2. *TTT_10_mod* 10 arc min resolution with closest T2 grid point locations.
3. *TTT_5_exact* 5 arc min resolution with exact lat/lon locations.

4. *TTT_5_mod* 5 arc min resolution with closest T2 grid point locations. This is the closest implementation to the T2 scenarios.

5. *Results*

Travel times were computed from the T2 scenarios (according to Sect. 4.1) and from TTT (according to Sect. 4.2) and compared to observed travel times (Sect. 3). An example of the results for Tp (mean of envelope) is shown in Fig. 3 as a scatter plot and also as a box-and-whisker plot. From the scatter plot, it can be seen that, in general, the modelled travel times correlate very well with the observed travel times. However, there are quite a few occasions when the model is considerably earlier than the observation. These are discussed below. In the right panel, the travel time error is calculated as (model − obs) so a negative value represents the model being early. From both plots it can be seen that on average, the model is somewhat earlier than the observed travel time, with only a few instances of the model being late. In the context of tsunami warnings, it is preferable to provide an arrival time that is too early, rather than one that is too late; however, the accuracy overall for this method is generally poor compared to the alternate methods of defining travel time from the model defined earlier, as shall be seen below.

Results for each of the 14 methods to estimate travel times are shown in Figs. 4, 5 and 6 and verification statistics are provided in Table 1. The top row in Fig. 4 shows the 'first peak' methods, with the left panel showing the mean of the envelope (Tp_mean—replicating the right hand panel of Fig. 3) and the right panel showing the minimum of the envelope (Tp_min). It can clearly be seen that there are many instances of Tp_min providing travel times that are significantly early, even by more than 15 h in a few cases. Inspection of the relevant individual time series and arrival times for these cases shows that these are related to numerical noise, or very small peaks that occur early on in the time series, well before the tsunami has arrived. This numerical noise does not occur for all scenarios in an envelope, so taking the mean of the envelope tends to smooth these out somewhat compared to the minimum of the envelope.

These early arrivals can be eliminated by considering the tsunami to have arrived only if the amplitude is above a certain absolute threshold, or if

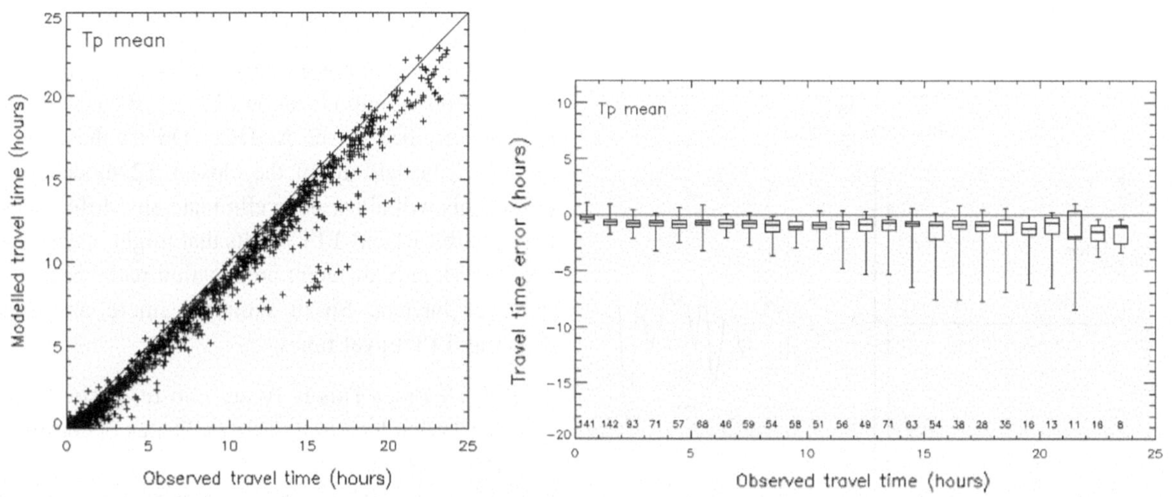

Figure 3
Comparison between modelled and observed travel times presented as a scatter plot (left panel) and a box-and-whisker plot (right panel) of the differences for the 'Tp mean' method of determining travel times (see Sect. 4.1). In the box-and-whisker plot, the 'whiskers' show the maximum and minimum values, the red line shows the median value and the top and bottom of the box show the 25th and 75th percentile values. Numbers along the bottom of the plot show the number of data pairs in each bin. Travel time error is calculated as (model − obs) so a negative value represents the model being early

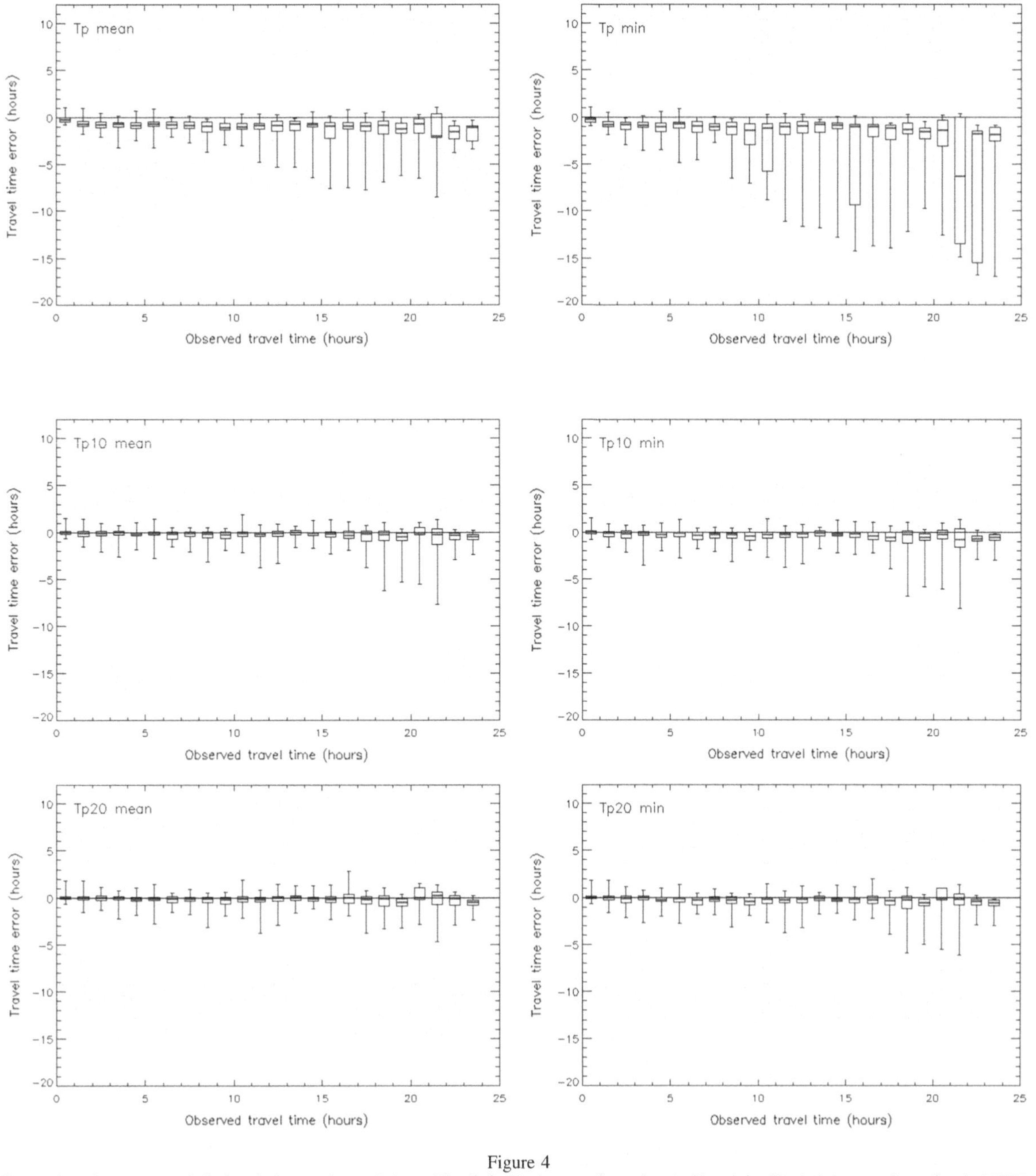

Figure 4
Comparison between modelled and observed travel times. Top left: Tp (mean of envelope); Top right: Tp (minimum of envelope); Middle left: Tp10 (mean); Middle right: Tp10 (minimum); Bottom left: Tp20 (mean); Bottom right: Tp20 (minimum)

the first crest or trough is above a certain percentage of the maximum amplitude at that location. The lower four panels in Fig. 4 show the results for the 'percentage-threshold' techniques, i.e., those requiring the first peak amplitude to be a certain percentage of the maximum amplitude in the time series. It can be seen here that the large early outliers do not appear, and results for the minimum of the envelope are more similar to results for the mean of the envelope, suggesting that there are no outliers within each envelope.

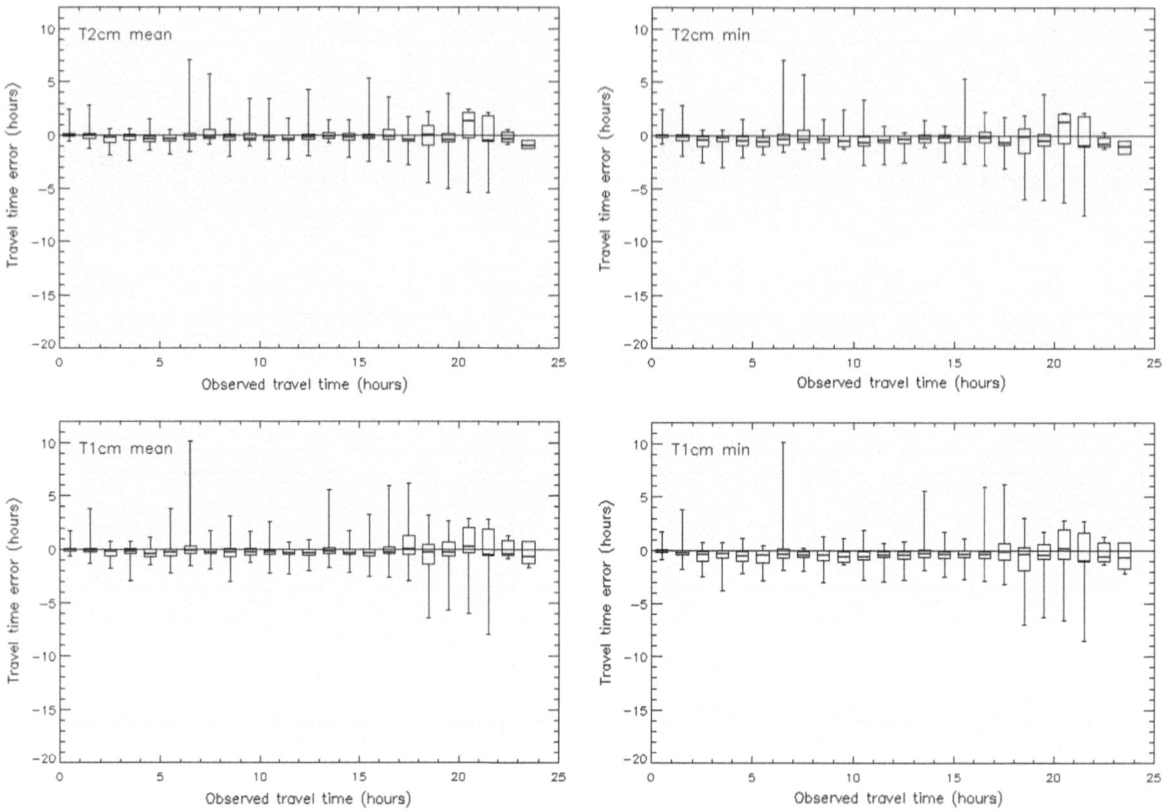

Figure 5
Comparison between modelled and observed travel times. Top left: T2cm (mean of envelope); Top right: T2cm (minimum of envelope); Bottom left: T1cm (mean); Bottom right: T1cm (minimum)

The 'absolute threshold' techniques are shown in Fig. 5. It can be seen here that there are more cases of the model being late compared to the observed arrival time. This will occur when the peaks in the model do not reach the designated thresholds until later in the time series. Finally, the results for the various implementations of TTT are shown in Fig. 6.

Standard verification statistics [bias, root-mean-square error (Rms), standard deviation (SD), linear correlation coefficient (R)] are shown in Table 1. Two key performance indicators (KPI) are also presented here. These are intended to represent possible KPIs for operational use. KPI1 is defined to be the percentage of predicted travel times that are either within half an hour too early, or 15 min too late of the observed travel time, and KPI2 is defined to be the percentage of travel times that are no later than 15 min. Higher KPIs are desired.

It can be seen that overall, the scenario-based 'percentage threshold' technique Tp20 with the mean of the envelope performs best, being in the top 2 for almost all the verification statistics. In terms of comparison to TTT, we see that the scenario–based technique provides a significant improvement over TTT.

It can also be seen that when considering an envelope of scenarios, the mean value over the arrival times from each scenario within the envelope is generally a better estimate of observed arrival time than the earliest arrival time of the scenarios. This is true even for the cases when the minimum is not related to numerical noise as discussed above with respect to Tp.

The bias can be seen to be negative in all cases, showing that all model techniques, including TTT, provide travel times that are early compared to observations. This may be due to physical factors that are not incorporated in the shallow-water equations, and is discussed further in Sect. 6.

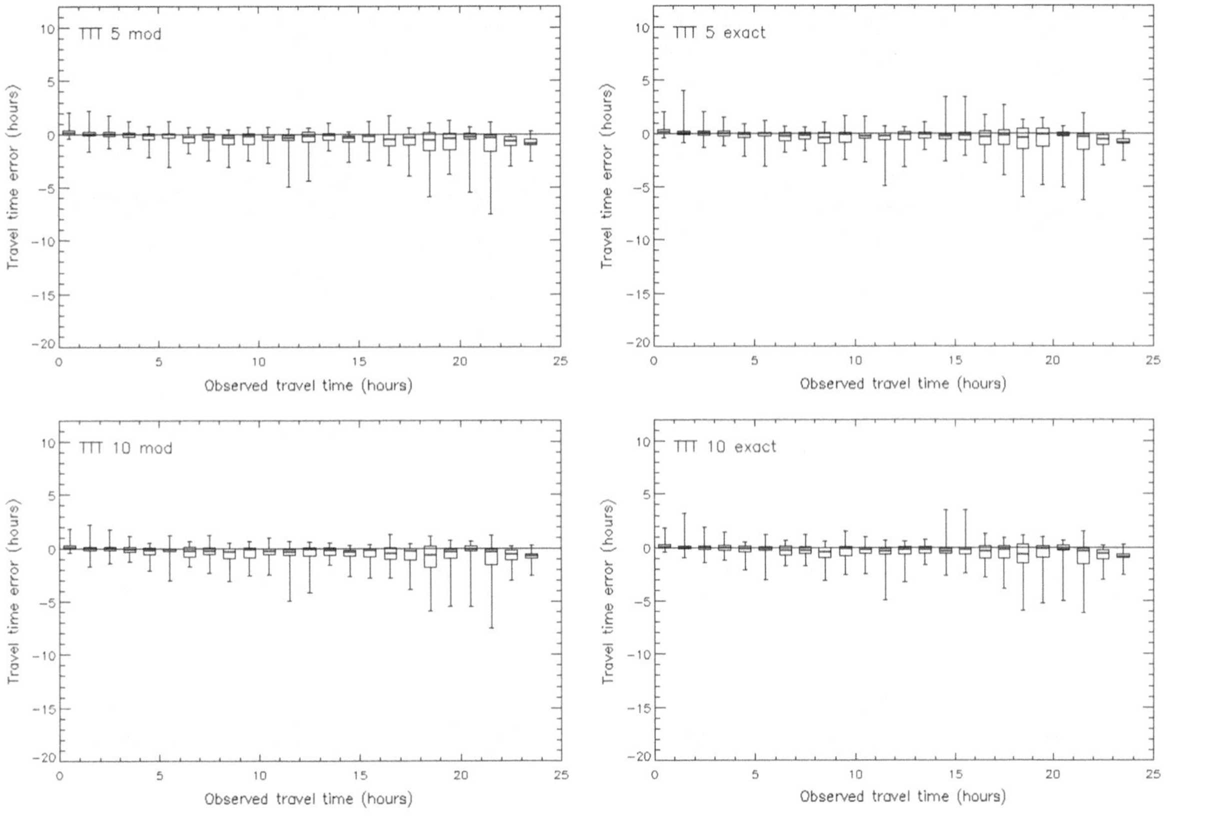

Figure 6
Comparison between TTT and observed travel times. Top left: TTT_5_mod; Top right: TTT_5_exact; Bottom left: TTT_10_mod; Bottom right: TTT_10_exact

For the operational implementation of TTT (TTT_10_exact), it can be seen that the bias is − 0.25 h (15 min) and the standard deviation is 0.79 h (47.4 min). This compares favourably with the results of Wessel (2009) noted in Sect. 2. It is likely that this is related to the observation dataset used. In the present work, we use only recent observations (events since 2004) and many of the outliers identified in Wessel (2009) were from earlier events.

Interestingly, in relation to the spatial resolution of TTT, when we compare TTT_10_exact and TTT_5_exact, or TTT_10_mod and TTT_5_mod, there is no clear indication that a higher spatial resolution improves the results. Also, examining the impact of the observation locations (comparing TTT_10_exact to TTT_10_mod, or TTT_5_exact to TTT_5_mod), again, we can not conclude that moving the lat/lon of the observation locations has a major effect on TTT. It may be that this is because, as mentioned earlier, the observed dataset is rather noisy, and it is not possible to make detailed conclusions due to the lack of precision in the observed travel times.

5.1. The Effect of Earthquake Magnitude

As noted by Wessel (2009), one of the key limitations related to a point-based travel time estimation is that it should perform poorly for travel times from earthquakes with longer ruptures, generally larger magnitude earthquakes. To investigate this, we examined errors in travel times as a function of earthquake magnitude.

We compare here the operational implementation of TTT, i.e., TTT_10_exact with the best performing scenario-based method, Tp20_mean. Travel time differences as a function of earthquake magnitude are shown in the top panel of Fig. 7. It can be seen

Table 1

Statistics of the comparisons between observed and modelled travel times

Travel time method	N	Bias (h)	Rms (h)	SD (h)	R	KPI1 (%)	KPI2 (%)
Tp min	1298	− 1.71	3.20	2.71	0.900	22.6	**98.8**
Tp mean	1298	− 0.95	1.40	1.03	0.987	28.1	**97.9**
T2cm min	770	− 0.33	1.05	0.99	0.987	54.9	89.6
T2cm mean	770	**− 0.14**	0.92	0.91	0.990	65.7	86.6
T1cm min	955	− 0.38	1.13	1.06	0.986	56.0	90.7
T1cm mean	955	− 0.21	1.03	1.01	0.987	64.1	88.6
Tp10 min	1298	− 0.36	0.82	0.73	0.993	64.6	92.6
Tp10 mean	1298	− 0.24	**0.73**	**0.69**	**0.994**	**70.1**	89.6
Tp20 min	1298	− 0.29	0.75	**0.69**	**0.994**	66.6	90.8
Tp20 mean	1298	**− 0.16**	**0.63**	**0.61**	**0.995**	**69.0**	86.3
TTT_10_exact	1298	− 0.25	0.83	0.79	0.992	62.6	85.9
TTT_10_mod	1298	− 0.31	0.85	0.79	0.992	63.5	88.4
TTT_5_exact	1298	− 0.22	0.84	0.82	0.992	61.8	83.5
TTT_5_mod	1298	− 0.30	0.85	0.80	0.992	63.0	87.2

The best two values in each column are in bold

that the errors do get larger for larger earthquakes, but it is not clear from this full dataset that the scenario method is better than TTT. In order to highlight the possible effect of a point source versus a longer rupture, we consider just observed travel times that are less than 3 h, as any differences should be more evident closer to the source. These results are shown in the lower panel of Fig. 7. In this case, we consider just the mean value of the travel time errors. It can be seen that there is a large positive bias for TTT for the larger earthquake magnitudes. A positive bias will result from TTT being late compared to the observation, which is as expected for the point source. However, it is worth noting that the only events contributing to the $M_w = 9.1$ bin are the 2004 Indian Ocean event (which had a very long rupture) and the 2011 Honshu event (which had a relatively short rupture).

6. Discussion

We have shown in this work that a technique to predict tsunami arrival times based on a dynamical model is able to perform better than TTT. However, there are a number of limitations to the JATWC's current model-based system, T2, that preclude an immediate change to operational practice. For example, the T2 domain covers only the Indian and Pacific Oceans, and has a forecast range of only 24 h. This means that arrival times cannot be produced for all events and all locations, so the JATWC will need to continue to produce arrival times based predominantly on TTT in the medium term. This may change in the future if and when new and extended model-based systems are implemented within the JATWC.

There are a number of differences in the two techniques for estimating travel time that we have evaluated here. Firstly, and most obviously, T2 is based on a dynamical model, incorporating the non-linear depth-averaged shallow-water equations, while TTT is a simple propagation of a wave front using the shallow-water approximation propagation speed $c = \sqrt{gh}$, where g is the acceleration of gravity and h is the water depth. There are also differences in how any particular earthquake is initialised. TTT, as implemented here, uses a point source epicentre, while T2 is initialised with a rupture that has length and width dimensions that depend on earthquake magnitude. Furthermore, T2 incorporates a technique to take account of the uncertainty in the rupture direction in real-time, and uses an 'envelope' of scenarios, while TTT does not. And finally, the two models have different underlying bathymetric datasets.

It would be possible to ascertain which of these factors contributes the most to the differences seen here. This could be done through implementing TTT with multiple epicentres to represent a longer rupture, and/or by running multiple versions of TTT to

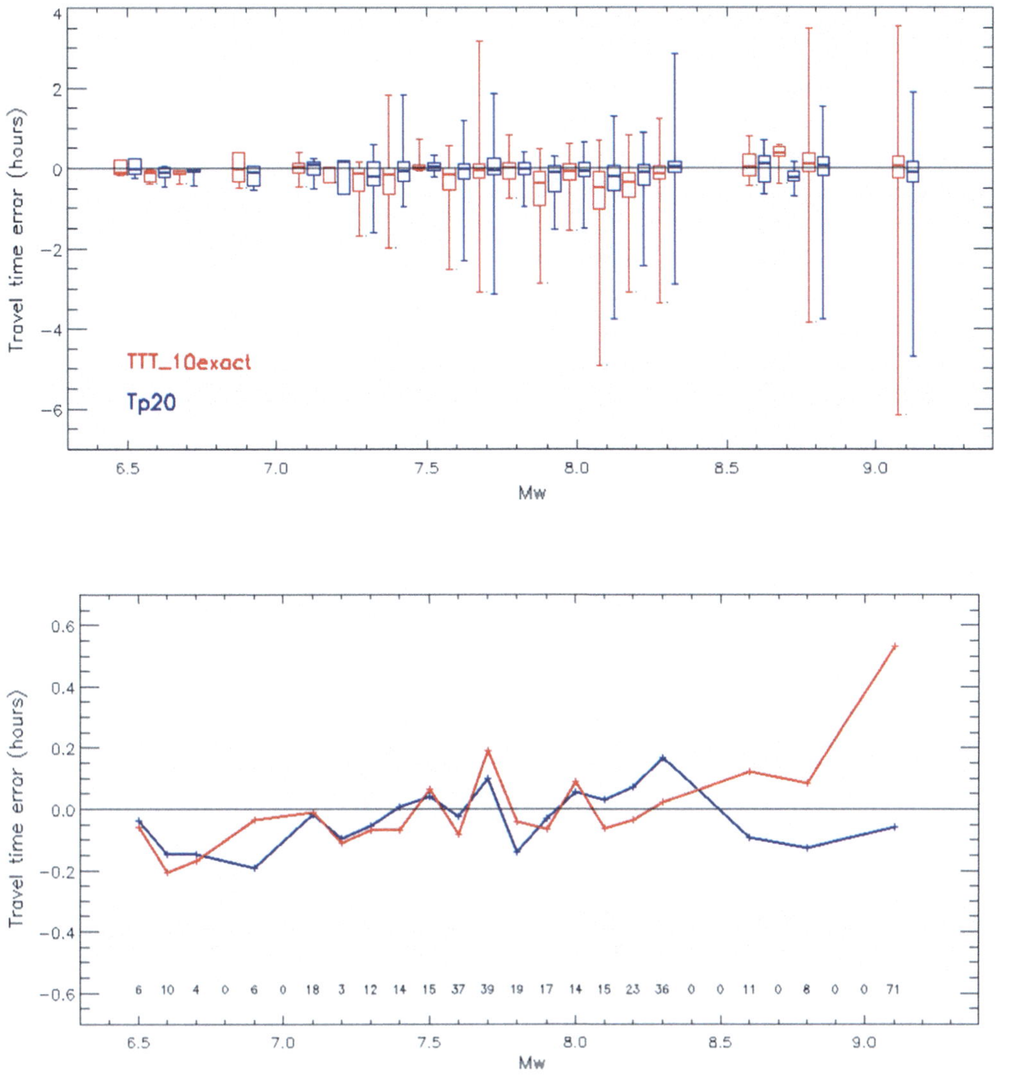

Figure 7
Top panel: travel time errors as a function of earthquake magnitude for TTT_10_exact (red) and Tp20_mean (blue) for all 1298 observed travel times. Bottom panel: mean of the travel time errors limited to observed travel times less than 3 h. Numbers along the bottom of the plot are the number of data pairs in each bin

replicate the 'envelope' of T2 scenarios. It would also be possible to incorporate the T2 bathymetric dataset in TTT. The aim of the present work has been to evaluate T2 in order to assess its suitability for operational use, so at this stage, we do not attempt to investigate any further the differences with TTT. However, this would be an interesting avenue for further work.

It is worth re-iterating that the observed travel times from the NGDC/WDS database are assumed here to be the 'truth'. Given the variety of sources of that data, these observed travel times are unlikely to correspond directly to the various model-based travel times. However, it is the best dataset available at present, and so this issue must simply be noted, and the potential differences acknowledged as a contribution to the noise.

In terms of improving the model-based estimates of travel time, recent work has shown that tsunami propagation speeds can be affected by the effects of elastic loading of the seafloor by the tsunami, and also by sea-water compressibility, neither of which are included in the non-linear shallow water equations on which T2 is based (Tsai et al. 2013; Watada 2013; Allgeyer and Cummins 2014). Allgeyer and Cummins (2014) show that incorporating these effects can delay the tsunami arrival time by 10–30 min for trans-oceanic distances, which could explain the negative bias seen in the results here. Incorporation of these effects requires significant computational resources, and it is unlikely to be feasible for real-time forecasting in the near-term, but could potentially be considered if further pre-computed scenario databases were to be developed.

This work has focussed on evaluating tsunami arrival times from the T2 scenario database. The results are promising and suggest that further work could involve the assessment and evaluation of other arrival times, such as arrival time of maximum wave, or first time over a particular warning-related threshold, as required for IOTWMS TSPs. One of the issues that would need to be addressed with this further work is the availability of an appropriate data set to use for verification.

7. Conclusion

Tsunami travel times from the T2 scenario database have been evaluated through comparison with observed travel times at tide gauges and tsunameters from the NGDC/WDS database. Travel times from TTT are also evaluated in order to provide a comparison with current operational practice at the JATWC. It was found that travel times from the T2 scenario database are at least as accurate as those from TTT and, in most cases, more accurate. The 'best' scenario-based method is Tp20—the time of arrival of the first peak that is at least 20% of the maximum amplitude at that point.

Acknowledgements

We would like to acknowledge NOAA/NCEP for making their global tsunami historical database freely available. We would also like to thank Simon Allen, Eric Schulz and two anonymous reviewers for useful comments on the text.

References

Allen, S. C. R., & Greenslade, D. J. M. (2010). Model-based tsunami warnings derived from observed impacts. *Natural Hazards & Earth System Sciences, 10,* 2631–2642. https://doi.org/10.5194/nhess-10-2631-2010.

Allen, S. C. R., & Greenslade, D. J. M. (2016). A pilot tsunami inundation forecast system for Australia. *Pure and Applied Geophysics*. https://doi.org/10.1007/s00024-016-1392-y.

Allgeyer, S., & Cummins, P. (2014). Numerical tsunami simulation including elastic loading and seawater density stratification. *Geophysical Research Letters, 41,* 2368–2375. https://doi.org/10.1002/2014gl059348.

Bressan, L., & Tinti, S. (2011). Structure and performance of a real-time algorithm to detect tsunami or tsunami-like alert conditions based on sea-level records analysis. *Natural Hazards Earth System Sciences, 11,* 1499–1521. https://doi.org/10.5194/nhess-11-1499-2011.

Bureau National Operations Centre. (2014). Operational upgrade to the ATWS tsunami travel time sub-system (TTT), *Operations Bulletin no 100* (p. 22). Australia: Bur. Met.

Cholifah, L., & Prastowo, T. (2017). Travel time difference between estimated and observed values of the 2011 Trans-Oceanic Tohoku Tsunami. In *Proc. 7th annual basic science int. conf*, Malang, Indonesia, 7–8 March 2017.

GEOWARE. (2007). TTT—A tsunami travel-time calculator. http://www.geoware-online.com. Accessed 24 May 2018.

Greenslade, D. J. M., Allen, S. C. R., & Simanjuntak, M. A. (2010). An evaluation of tsunami forecasts from the T2 scenario database. *Pure and Applied Geophysics, 168*(6–7), 1137–1151. https://doi.org/10.1007/s00024-010-0229-3.

Greenslade, D. J. M., Simanjuntak, M. A., & Allen, S. C. R. (2009). An enhanced tsunami scenario database: T2, *CAWCR Technical Teport no 14* (p. 30). Australia: Bur. Met.

Greenslade, D. J. M, & Warne, J. O. (2012). Assessment of the effectiveness of a sea-level observing network for tsunami warning. *Journal of Waterway Port Coastal and Ocean Engineering, 138*(3), 246–255.

Intergovernmental Oceanographic Commission. (2015). IOTWS-TSP Service Definition Document. In *Tenth session of the intergovernmental coordination group for the Indian Ocean tsunami warning and mitigation system (ICG/IOTWS-X),* Muscat, Oman, 24–26 March 2015, p. 85.

Merrifield, M. A., et al. (2005). Tide gauge observations of the Indian Ocean Tsunami, December 26, 2004. *Geophysical Research Letters, 32,* L09603. https://doi.org/10.1029/2005gl022610.

National Geophysical Data Center/World Data Service (NGDC/WDS): Global Historical Tsunami Database. National Geophysical Data Center, NOAA. https://doi.org/10.7289/v5pn93h7. https://www.ngdc.noaa.gov/hazard/tsu_db.shtml. Accessed 9 Nov 2017.

Rabinovich, A. B., & Thomson, R. E. (2007). The 26 December 2004 Sumatra tsunami: Analysis of tide gauge data from the world ocean part 1. Indian Ocean and South Africa. *Pure and Applied Geophysics, 164,* 261–308.

Spillane, M. C., Gica, E., Titov, V. V., & Mofjeld, H. O. (2008). Tsunameter network design for the U.S. DART® arrays in the Pacific and Atlantic oceans. *National Oceanic and Atmospheric*

Administration (NOAA) Technical Memorandum OAR PMEL-143, National Ocean Service (NOS), Washington, DC.

Syamsidik, Teuku, T. M., & Kato, S. (2015). Development of accurate tsunami estimated times of arrival for tsunami-prone cities in Aceh, Indonesia. *International Journal of Disaster Risk Reduction, 14,* 403–410. https://doi.org/10.1016/j.ijdrr.2015.09.006

Titov, V., Kanoglu, U., & Synolakis, C. (2016). Development of MOST for real-time tsunami forecasting. *Journal of Waterway Port Coastal and Ocean Engineering*. https://doi.org/10.1061/(asce)ww.1943-5460.000035703116004.

Tsai, V. C., Ampuero, J.-P., Kanamori, H., & Stevenson, D. J. (2013). Estimating the effect of Earth elasticity and variable water density on tsunami speeds. *Geophysical Research Letters, 40,* 492–496. https://doi.org/10.1002/grl.50147.

Watada, S. (2013). Tsunami speed variations in density-stratified compressible global oceans. *Geophysical Research Letters, 40,* 4001–4006. https://doi.org/10.1002/grl.50785.

Wessel, P. (2009). Analysis of observed and predicted tsunami travel times for the Pacific and Indian Oceans. *Pure and Applied Geophysics, 166,* 301–324.

(Received June 11, 2018, revised September 6, 2018, accepted September 21, 2018, Published online September 28, 2018)

The manufacturer's authorised representative in the EU is Springer
Nature Customer Service Centre GmbH, Europaplatz 3, 69115 Heidelberg,
Germany. If you have any concerns regarding our products, please
contact ProductSafety@springernature.com

Printed and bound by CPI Group (UK) Ltd, Croydon, CR0 4YY
30/09/2026
02225434-0020